Gas Phase Ion Chemistry

Volume 1

CONTRIBUTORS

N. G. ADAMS
D. L. ALBRITTON
TOMAS BAER
MICHAEL T. BOWERS
WALTER J. CHESNAVICH
F. C. FEHSENFELD
E. E. FERGUSON
J. L. FRANKLIN
I. K. LARIN
MICHAEL MEOT-NER (MAUTNER)
D. SMITH
TIMOTHY SU
V. L. TALROSE
P. S. VINOGRADOV

Gas Phase Ion Chemistry

Edited by MICHAEL T. BOWERS

DEPARTMENT OF CHEMISTRY
UNIVERSITY OF CALIFORNIA
SANTA BARBARA, CALIFORNIA

Volume 1

ACADEMIC PRESS New York San Francisco London 1979
A Subsidiary of Harcourt Brace Jovanovich, Publishers

ACADEMIC PRESS, INC.
111 Fifth Avenue, New York, New York 10003

United Kingdom Edition published by
ACADEMIC PRESS, INC. (LONDON) LTD.
24/28 Oval Road, London NW1 7DX

Library of Congress Cataloging in Publication Data
Main entry under title:

Gas phase ion chemistry.

Includes bibliographies.
1. Chemical reaction, Conditions and laws of -- Addresses, essays, lectures. 2. Gases Ionized--Addresses, essays, lectures. I. Bowers, Michael T.
QD501.G324 541'.39 78-19666
ISBN 0-12-120801-X

PRINTED IN THE UNITED STATES OF AMERICA

79 80 81 82 9 8 7 6 5 4 3 2 1

CONTENTS

Chapter 3 Classical Ion–Molecule Collision Theory

TIMOTHY SU AND MICHAEL T. BOWERS

Chapter 4 Statistical Methods in Reaction Dynamics

WALTER J. CHENAVICH AND MICHAEL T. BOWERS

Chapter 5 State Selection by Photoion–Photoelectron Coincidence

TOMAS BAER

Chapter 6 Temperature and Pressure Effects in the Kinetics of Ion–Molecule Reactions

MICHAEL MEOT-NER (MAUTNER)

Chapter 7 **Energy Distribution in the Unimolecular Decomposition of Ions**

J. L. FRANKLIN

Chapter 8 **On the Rapidity of Ion–Molecule Reactions**

V. L. TALROSE, P. S. VINOGRADOV, AND I. K. LARIN

LIST OF CONTRIBUTORS

Numbers in parentheses indicate the pages on which the authors' contributions begin.

N. G. ADAMS (1), Department of Space Research, University of Birmingham, England

D. L. ALBRITTON (45), Aeronomy Laboratory, Environmental Research Laboratories, National Oceanic and Atmospheric Administration, Boulder, Colorado 80303

TOMAS BAER (153), Department of Chemistry, University of North Carolina, Chapel Hill, North Carolina 27514

MICHAEL T. BOWERS (83, 119), Department of Chemistry, University of California, Santa Barbara, California 93106

WALTER J. CHESNAVICH (119), Department of Chemistry, University of California, Santa Barbara, California 93106

F. C. FEHSENFELD (45), Aeronomy Laboratory, Environmental Research Laboratories, National Oceanic and Atmospheric Administration, Boulder, Colorado 80303

E. E. FERGUSON (45), Aeronomy Laboratory, Environmental Research Laboratories, National Oceanic and Atmospheric Administration, Boulder, Colorado 80303

J. L. FRANKLIN (273), Chemistry Department, Rice University, Houston, Texas 77001

I. K. LARIN (305), Institute of Chemical Physics of the Academy of Sciences of the USSR, 117977 GSP-1 Moscow V-334, U.S.S.R.

MICHAEL MEOT-NER (MAUTNER) (197), The Rockefeller University, New York, New York 10021

D. SMITH (1), Department of Space Research, University of Birmingham, England

TIMOTHY SU (83), Department of Chemistry, Southeastern Massachusetts University, North Dartmouth, Massachusetts 02747

V. L. TALROSE (305), Institute of Chemical Physics of the Academy of Sciences of the USSR, 117977 GSP-1 Moscow V-334, U.S.S.R.

P. S. VINOGRADOV (305), Institute of Chemical Physics of the Academy of Sciences of the USSR, 117977 GSP-1 Moscow V-334, U.S.S.R.

PREFACE

Fifteen years ago there was a handful of research groups involved in research on the chemistry of ions in the gas phase. Today there are hundreds of groups spanning a discipline that stretches from biochemistry to theoretical chemical physics. The growth has been dramatic, almost explosive, in the past decade, and the field is currently at the point where no one can even superficially keep up with all of it. There are three journals totally devoted to ion chemistry of one sort or another, and ion chemistry papers are making major contributions to most of the main line chemical journals.

The roots of gas phase ion chemistry, perhaps more than those of some disciplines, are found in technique development. The rapid technological advances in the post-Sputnik era provided the materials and vacuum and electronic capabilities to develop the sophisticated instrumentation we take for granted today. The chapter by Gentry on molecular beams in Volume 2 is illustrative of the inventive methods used to probe microscopic reaction mechanisms. Previous editors recognized this relationship and built their books around chapters that discussed the various techniques in use at that time.

The question arose, in attempting to put the present volumes together, of what topics should be emphasized and how they should be organized. In the end, the decision was to emphasize the topics I found most interesting and relevant, and to assign the organization of these topics to the various authors who graciously agreed to write the chapters. Each of the authors is a world leader in his respective area, and each area presents different requirements of both length and style presentation. The areas I chose to emphasize are theory (3), kinetics and mechanism (6), thermochemistry (3), and technique and applications (4). The numbers in parentheses give the number of chapters in each section. The dominant theme of the treatise is the chemical physics aspect of gas phase ion chemistry; but a broad range of chemists should find most of the chapters of interest, particularly those on thermochemistry where comprehensive compendiums of gas phase basicities, acidities, and electron affinities are given.

Finally, I would like to thank the many people who made these volumes possible. To the authors go the primary thanks for preparing high quality chapters and forwarding them promptly. Members of my research group all deserve thanks for carrying on during the editing process. The helpful, competent, and professional attitude of everyone at Academic Press minimized all of the difficulties encountered, and my thanks go to them. Last, my family, and especially my wife Diane, gave support and encouragement throughout, and I am deeply grateful and thankful for their presence in my life.

CONTENTS OF VOLUME 2

Chapter 1

Recent advances in flow tubes: Measurement of ion–molecule rate coefficients and product distributions

D. SMITH AND N. G. ADAMS

DEPARTMENT OF SPACE RESEARCH
UNIVERSITY OF BIRMINGHAM
BIRMINGHAM, ENGLAND

ISBN-0-120801-X

I. INTRODUCTION

The use of flow tubes to obtain quantitative data on ion–neutral reactions dates back to 1963 with the development of the flowing afterglow plasma technique by Ferguson, Fehsenfeld and Schmeltekopf in the U.S. Department of Commerce Research Laboratories, formerly ESSA now NOAA, at Boulder, Colorado (Fehsenfeld *et al.*, 1966a; Ferguson *et al.*, 1969). The technique was developed in response to the need for critical data relating to ionospheric chemistry which demands that the ion–neutral interaction energies be very low ($\lesssim 0.1$ eV). At that time very little relevant data were available, although a start had been made in the time-resolved afterglow (often termed the stationary afterglow) by Sayers and colleagues at Birmingham University, England (Dickinson and Sayers, 1960; Sayers and Smith, 1964). In principle, the collision dominated afterglow plasma is an ideal medium in which to study both binary and ternary ion–neutral reactions at truly thermal energies since the reacting species possess a Maxwellian distribution of kinetic energies $f(E)$ characterized by the system temperature, and the internal energy states of the ions and neutrals will have the corresponding Boltzmann distribution. Thus, the rate coefficient k for the reaction can be determined under controlled temperature conditions. This contrasts with measurements made using beam techniques in which reaction cross sections $\sigma(E)$ over a much wider range of energies are determined but for which, in practice, accurate measurements are limited to interaction energies above an electron volt or so (McDaniel *et al.*, 1970). In some cases such beam data can be utilized to determine thermal energy rate coefficients by energy extrapolation and application of the appropriate mathematics ($k = \int \sigma(E) f(E)\, dE$), an approach which has been discussed in detail by Henchman (1972). However, ion beams are also rather prone to contain significant fractions of excited ions, which can have greatly different cross sections for reaction than the ground state ion species, and thus the uncertainty in the composition of the ion beam has led researchers requiring thermal energy reaction rate data to favor those obtained from collision dominated experiments.

The stationary afterglow can be criticized for the measurement of ion–molecule reaction rate coefficients. The technique involves the ionization of a gas or a mixture of gases by a short time duration pulse of an rf or dc electric field and then the time history of the ions in the afterglow of the electrical discharge is studied by mass spectrometric sampling of the diffusive wall currents of the positive ions. The loss rate of the ions is related to the partial pressure of the reactant neutral gas. It is in the nature of the technique that the reactant molecular gas is exposed to the gas discharge which produces the afterglow plasma, and hence the possibility of significant population of internal energy states of the molecular gas arises which can result in misleading rate coefficients. Problems of this kind can be minimized by monitoring the ion loss rates at times late into the afterglow, thus allowing the excited states to relax by collision with plasma electrons and neutral gas particles before measurements are made. This phenomenon was graphically illustrated by the stationary afterglow data relating to the reaction of O^+ ions with N_2 molecules. The "early" afterglow data (Batey *et al.*, 1965) provided a rate coefficient which was more than an order of magnitude greater than the "late" afterglow data (Copsey *et al.*, 1966), presumably due to the presence of significant concentrations of vibrationally excited N_2 molecules in the early afterglow as was suggested to us at the time by Ferguson. Subsequent flowing afterglow studies by Schmeltekopf *et al.* (1967) strongly supported this argument. Nevertheless, the stationary afterglow can provide accurate reaction rate coefficients for both binary and ternary reactions if used with care and has been exploited to provide a great deal of accurate data concerning positive-ion–molecule reactions, notably by the Birmingham University group (Smith and Fouracre, 1968; Adams *et al.*, 1972; Smith *et al.*, 1972) and the Ballistics Research Laboratory group (Lineberger and Puckett, 1969a,b). However, it is not so useful for the study of negative-ion–molecule reactions. This is because the simultaneous presence of electrons as well as negative ions in the afterglow plasma results in an appreciable ambipolar field which prevents the flow of negative ions to the plasma boundary and so the essential property that the wall current of ions should be proportional to the number density of ions in the plasma breaks down (Smith and Plumb, 1973; Smith *et al.*, 1974). However, in the presence of an appreciable concentration of an electron-attaching gas, the electron/positive-ion afterglow plasma will convert to a negative-ion/positive-ion plasma and the ambipolar field will collapse to an insignificant intensity allowing the negative ions to flow toward the plasma boundary. Under such conditions, negative-ion–molecule reactions can be studied, as has been demonstrated in the experiments of Puckett and Lineberger (1970). Similar phenomena have been observed by Kebarle (1975) in the pulsed high pressure mass spectrometer. Nevertheless, the high pressure mass spectrometer has been used very successfully in the study of ion equilibrium in positive- and negative-ion–neutral reactions (Kebarle, 1975).

Several of the disadvantages of the stationary afterglow have been overcome by the development of the flowing afterglow and with its development came a great upsurge in the amount of data relating to low energy ion–neutral reactions. A brief discussion of the principles of the technique and some of its many successes will be given in Section II and so they will only be mentioned here. The essence of the technique is that ionization is created upstream in a fast flowing carrier gas and an afterglow plasma is distributed along a flow tube remote from the source of ionization. The ions, either positive or negative, are sampled at a point downstream in the flow and the decrease in the primary ion signal is observed as a function of the rate of flow of a reactant gas, which is introduced into the afterglow at a position remote from the ionization source. The essential point is that the neutral reactant gas is never exposed to the extreme conditions of the ionization source but only to the thermalized afterglow plasma, and so significant internal excitation is prevented. Several other important advantages are apparent. By the sequential addition of gases at different positions along the length of the afterglow column, a wide variety of ion types can be generated and their reactions studied. The method of ion sampling ensures that negative-ion reactions can also readily be studied (Section II.B).

Following its development and exploitation to study a wide range of reactions at the Boulder laboratories, flowing afterglow apparatuses have been built and are being used in several laboratories throughout the world, notably by Schiff and Bohme at York University, Toronto (Burt *et al.*, 1970; Bohme, 1975), Fite and Kaufman at Pittsburgh University (Farragher *et al.*, 1969; Howard *et al.*, 1971), Twiddy at the University College of Wales at Aberystwyth (Bolden and Twiddy, 1972), Bierbaum and DePuy at Colorado University, Boulder (Bierbaum *et al.*, 1976), and Freeman and McEwan at the University of Canterbury, Christchurch, New Zealand (Liddy *et al.*, 1977) for the study of ion–neutral reactions, and by Smith and colleagues at the University of Birmingham, England for the study of ion–ion mutual neutralization (Smith and Church, 1976; Smith *et al.*, 1976) and electron temperature relaxation (Smith and Dean, 1975).

Some energy variation can be obtained in the flowing afterglow by varying the flow tube wall temperature and preheating or precooling the carrier gas (Dunkin *et al.*, 1968; Lindinger *et al.*, 1974) but a more successful and quite recent development is the inclusion of a drift-tube section downstream in a flowing afterglow tube. This so-called flow–drift tube experiment, again developed at Boulder, allows the study of ion–neutral reactions over an interaction energy range from thermal to a few electron volts (McFarland *et al.*, 1973a,b,c). This new development is discussed in more detail in Section III, and so it is sufficient to note here that it combines all the flexibility of the flowing afterglow with the essential features of drift tubes and as such extends the study of ion–neutral reactions into an important energy regime.

A restrictive feature of the flowing afterglow as used conventionally is that the ions to be studied have to be generated in the flow tube (see Section II). Although it has been possible to study many hundreds of different ionic species by thoughtful choice of suitable systems, many other species for which data are urgently required, for example radical-ion species such as CH^+ and $CH_2{}^+$, have defied the attempts of the best flowing afterglow researchers to generate them without simultaneously generating other ions, which renders the mass spectrometer data intractible. The presence in the flow tube of the gas from which the required ionic species is generated also complicates the interpretation of the mass spectrometer data and these factors render the technique unsatisfactory for the study of many species. In response to the desire to study ion–neutral reactions thought to be of importance in the synthesis of the molecules observed in interstellar clouds (e.g., Dalgarno and Black, 1976), the authors conceived of, and have recently constructed and exploited at Birmingham University, the selected ion flow tube (SIFT) (Adams and Smith, 1976a,b). This most recent development in flow tube instrumentation, which is discussed in some detail in Section IV, essentially involves the injection of a mass selected ion beam at low energy into a flowing gas where the method of studying the ionic reactions is very similar to that of the flowing afterglow technique. Thus, in principle with the SIFT, the reactions at thermal energies of any ion which can be generated in an ion source and injected into a relatively high pressure gas without undergoing collisional fragmentation can be studied. This has greatly extended the versatility of the flow tube technique. At the time of writing, the rate coefficients and product ion distributions of several hundred positive-ion reactions have been determined using the SIFT (Section IV.C). We wish to stress at the onset that the SIFT reactive medium is *not* an afterglow plasma, no electrons are present, only a low concentration of ions convected along the flow tube by the carrier gas.

In summary, having placed the topic of flow tube studies of ion–neutral reactions in perspective, the following sections involve a brief description of the well-established flowing afterglow technique (Section II), mentioning some of its more noteworthy successes, in order to prepare the ground for a discussion of the flow–drift technique (Section III) and of the SIFT technique (Section IV). Finally in Section V, we allude to our most recent development, the temperature variable SIFT–DRIFT apparatus, which combines the essential features of the SIFT and the flow–drift tube and which promises to provide a great deal of critical data concerning ion–neutral reactions at thermal energies.

II. THE FLOWING AFTERGLOW

The more recent developments in flow tube technology rely heavily on the now standard flowing afterglow technique which has been reviewed in detail by the originators (Ferguson *et al.*, 1969) and has also been the subject of several more recent reviews (McDaniel *et al.*, 1970; Twiddy, 1974; Fehsenfeld, 1975a; Meisels, 1975). In this short discourse only those aspects of the technique which are relevant to the more recent developments are discussed in any detail.

A. General Description of the Apparatus and Technique

A schematic diagram of a typical flowing afterglow apparatus is shown in Fig. 1. A carrier gas, usually helium (flow rate about 200 torr liter s^{-1}), is made to flow along a glass or stainless steel tube (about 8 cm internal diameter, 1–2 m long) at a velocity of about 10^4 cm s^{-1} by the action of a large Roots pump. Typical flow tube pressures range from ~0.3 to several torr limited by the requirements for an adequate ion number density, an established flow pattern, and a tolerable pressure in the ion detection system. Ionization is produced either in the pure carrier gas alone or in the carrier gas containing a small admixture of an ion source gas, usually by a hot cathode electron emitter, although microwave (Farragher, 1970) and cold cathode (Bolden *et al.*, 1970a) discharges have also been used. The resulting plasma of electrons and positive ions (and often negative ions and metastable neutrals) is then constrained to flow with the carrier gas and the charged particles thermalize as they move away from the ion source and along the flow tube. The ion source gas is often added downstream of the ionization region, and thus into the afterglow of the discharge, creating the desired primary ion species by reaction with ions or metastable atoms or electrons rather than by direct electron impact in the upstream ion source. By these various reactions, a great variety of primary ions can be produced. More than one primary ion is often present and although this does not generally influence the determination of reaction rate coefficients, it does complicate and, in some cases, prevent the identification of the ionized reaction products. Also, the presence of energetic metastable species such as 2^3S helium atoms

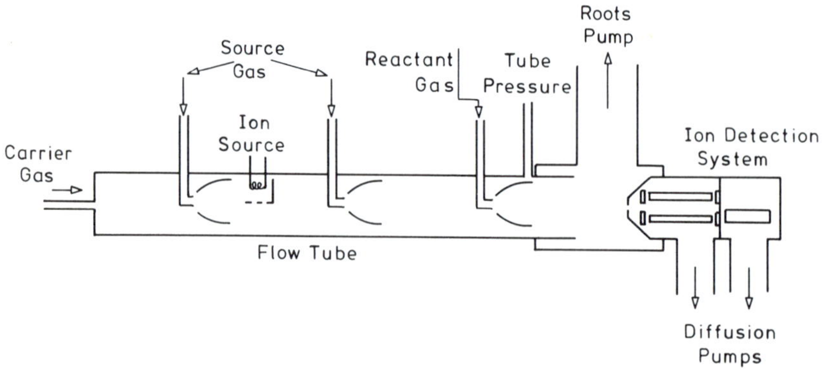

FIG. 1 Schematic diagram of a typical flowing afterglow apparatus.

can greatly add to the "product" ion spectrum as a result of Penning ionization and also complicate product identification. However, the presence of metastable atoms in the flowing afterglow has been very successfully exploited to determine many Penning ionization rate coefficients (Bolden *et al.*, 1970b; Schmeltekopf and Fehsenfeld, 1970).

On their passage along the flow tube, the ions pass an axially positioned entry port through which the reactant gas is added (Fig. 1). This method of reactant gas injection contrasts with that employed in the flow drift and SIFT (Sections III and IV, respectively). The flow rate of the reactant gas is usually determined using pressure transducers in conjunction with calibrated capillaries (Ferguson *et al.*, 1969), although in a few cases the rate of pressure drop in a known volume of reactant gas has been used (Bolden and Twiddy, 1972). The former method is also usually used to determine the much larger flow rates of the carrier gas. At the end of the reaction zone (typically 25–75 cm long), the ions are sampled through an orifice (typically 0.5–1.5 mm diameter) in a disk located at the tip of a truncated nose cone, passed into a differentially pumped quadrupole mass filter, and are detected by a particle-multiplier/ion-counting system. The orifice disk is usually constructed of molybdenum since only with this metal is reliable sampling consistently obtained, especially when reactive neutrals like O and H are present (Ferguson *et al.*, 1969). The particle multiplier is located off axis to reduce the background count rate due to photons from the ionization source, due to metastable atoms and—in the case of negative-ion studies—due to electrons which are not efficiently filtered out by the quadrupole spectrometer. Rate coefficients are determined and ionized products identified from observations of the primary and product ion count rates as a function of reactant gas concentration by the procedure outlined below (Section II.C). In one apparatus, the position of the entry port is varied at constant reactant gas flow rate in order to obtain the rate coefficient (Farragher, 1970; Howard *et al.*, 1970). Using either method, binary rate coefficients of the order of 10^{-9}–10^{-13} $cm^3\ s^{-1}$ can be accurately determined for reactant gas flow rates varying from typically 10^{-4} to several torr liter s^{-1}.

A schematic diagram illustrating many of the features often obtained in flowing afterglow data is shown in Fig. 2 for the reaction in which two primary ion products result, viz,

$$\begin{aligned} A^+ + B &\rightarrow C^+ + D \\ &\rightarrow E^+ + F \end{aligned} \tag{I}$$

The ion count rates are plotted logarithmically as a function of the flow rate of reactant gas B. Numerous examples of actual flowing afterglow data are available in the literature. The decay of the primary ion A^+, from an initial count rate of typically 10^5–10^6 counts s^{-1}, is usually linear over one

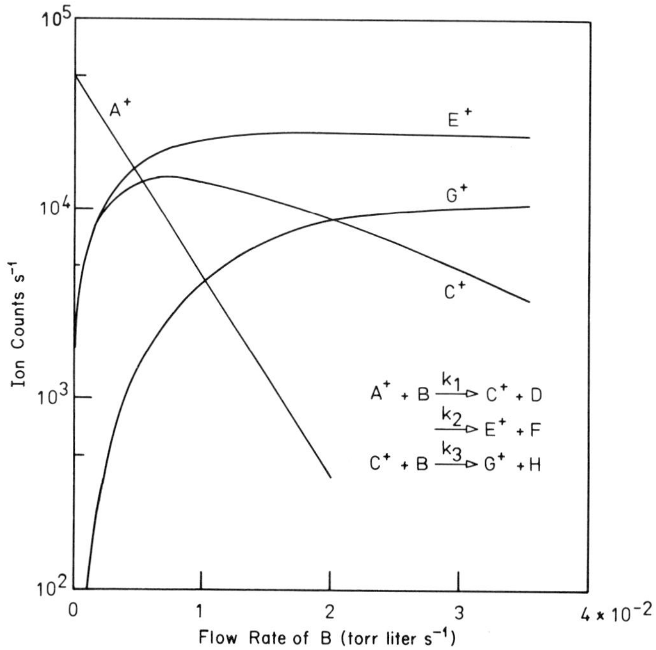

FIG. 2 Diagram illustrating features commonly observed in flowing afterglow mass spectrometer data. Here the A^+ ion reacts with molecule B and the ions C^+ and E^+ are primary products. C^+ then reacts with B to produce G^+. The rate coefficient for the primary reaction is obtained from the exponential loss of A^+. Mass discrimination in the detection system is evident since the total count rate ($E^+ + G^+ + C^+$) for large flow rates of B is less than the initial A^+ count rate.

to several orders of magnitude. The count rates of C^+ and E^+ are shown to increase with that of C^+ reaching a maximum and the declining due to further reaction with B producing a secondary ion G^+. The total primary ion count rate at zero flow and the total product ion count rate at maximum flow are deliberately made unequal to illustrate the inevitable mass discrimination in the ion sampling system. It is difficult (Govers *et al.*, 1974) to accurately account for mass discrimination in the flowing afterglow. Such is however readily possible when a SIFT apparatus is used, as is described in Section IV.A.3, and thus enables the relative proportions of the product ions to be accurately determined.

B. Plasma Properties and Ion Sampling

In determining rate coefficients from flowing plasma experiments, it is important to have an appreciation of other ion loss processes and of the method of ion sampling from a plasma. In a flowing afterglow plasma containing positive ions and electrons, the overall ion number density decreases with distance along the flow tube mainly as a result of ambipolar diffusion to the walls, but also at the higher charge densities ($\gtrsim 10^8$ cm^{-3}) due to electron–ion recombination (Smith *et al.*, 1975). Thus in rate coefficient measurements, it is necessary to operate with small charge densities in the reaction zone to minimize ion loss by recombination to simplify the interpretation of the mass spectrometer data. Under such conditions, the ion density is typically 10^7 cm^{-3} at the sampling orifice. In this plasma situation, the space charge sheath at the orifice is such as to accelerate positive ions through the orifice. A detailed discussion of positive-ion sampling with orifice probes has been given by Smith and Plumb (1973).

When negative ions are also present the situation is very different. For small relative concentrations of negative ions to electrons, the potentials within the plasma are still controlled by the electrons and so the radial ambipolar field is retarding to negative ions, reducing their diffusive loss much below that of the positive ions and effectively preventing them from reaching the flow tube walls. This type of plasma situation has been studied experimentally in the stationary afterglow using electrostatic probes to map out the distributions of the charged particle densities and the potential (Smith *et al.*, 1974). However, the problem which this presents to negative-ion sampling from stationary afterglows is overcome in the flowing afterglow by sampling on axis in the region where the negative-ion number density and the gas velocity are greatest and by using large sampling orifices. For electron densities $\geq 10^7$ cm^{-3} and electron temperatures of about 300 K, the Debye screening length is $\lesssim 0.5$ mm, resulting in a sheath thickness of the same order as the orifice diameter. Under these conditions, some distortion of the sheath must occur allowing at least the high energy tail of the negative-ion energy distribution to penetrate the sheath and pass through the orifice with the convective assistance of a small fraction of the carrier gas. The details of this sampling process are still not completely understood.

However, if the electron density n_e to the negative-ion density, n_- decreases below a critical ratio ($n_e/n_- \lesssim 10^{-2}$), the electron–positive-ion ambipolar field collapses and the remaining electrons are rapidly lost to the walls, resulting in a positive-ion/negative-ion plasma completely devoid of electrons. Such a transition has been observed both in stationary afterglows (Lineberger and Puckett, 1969a; Smith *et al.*, 1974) and flowing afterglows (Smith and Church, 1976). Under these conditions, the positive and negative ions can both be readily sampled by the orifice since the sheath potential difference is much less than the average ion energy (equivalent). The orifice sampling problems discussed above do not arise in SIFT ion sampling (Section IV.A.1) since there are no electrons in the reaction zone. In the study of ion–neutral reactions in flowing afterglows containing negative ions it is necessary to ensure that a transition from electron–ion to ion–ion plasma does not occur in the reaction zone otherwise anomalous rate coefficients may be obtained because of the resultant variation in the diffusion loss rate of the ions. In some negative–ion reaction studies, this has been prevented by deliberately creating an ion–ion plasma upstream of the reaction region by the addition of a rapidly attaching gas such as SF_6 (Howard *et al.*, 1974). Clearly, this approach is only useful if the electron scavenging gas does not interfere with the reactions under study.

C. *Plasma Dynamics and Determination of Rate Coefficients*

In order to obtain a rate coefficient from the experimental decrease in the reactant ion count rate (Fig. 2), it is necessary to appreciate the fluid dynamics of the flowing carrier gas and afterglow plasma. For typical flow tube dimensions and carrier gas throughputs Q and pressures p, the developed carrier gas flow will be almost laminar with only a few percent slip at the walls. Under these conditions, the radial velocity profile $v_g(r)$ of carrier gas is given by

$$v_g(r) = 2v_0 \frac{(1 - (r^2/a^2) + (2s/a))}{(1 + (4s/a))}, \tag{1}$$

where a is the radius of the tube, v_0 the carrier gas bulk velocity which equals $Q/\pi a^2 p$ and $s = s'/p$, where s' is the pressure independent slip coefficient. Pitot tube measurements (Ferguson *et al.*, 1969) have shown that the form of the velocity profile is as predicted by Eq. (1). Also using this equation, Q can be expressed as

$$Q = \frac{\pi a^4}{8\eta}\left(\bar{p} + \frac{4s'}{a}\right)\Delta p, \tag{2}$$

where η is the carrier gas viscosity, Δp the pressure drop per unit length of flow tube, and $\bar{p}$ the average pressure. From experimental measurements of $\bar{p}$ and Δp as a function of Q, this relationship has been shown to be valid down to helium pressures of ~ 0.1 torr where the effects of molecular flow begin to dominate (Adams *et al.*, 1975). For predominantly viscous conditions and making the reasonable assumption that the instantaneous ion-velocity radial profile is identical to that of the carrier gas, the continuity equation for a given ion type of number density $[A^+]$, which is lost only by ambipolar diffusion and binary reaction with a neutral of concentration $[B]$, is given (see, e.g., Bolden *et al.*, 1970a) by

$$v_g(r)\frac{\partial[A^+]}{\partial z} = \underbrace{\frac{D_a}{r}\frac{\partial}{\partial r}\left(r\frac{\partial[A^+]}{\partial r}\right)}_{A} + \underbrace{D_a\frac{\partial^2[A^+]}{\partial z^2}}_{B} - \underbrace{k[B][A^+]}_{C}, \tag{3}$$

where the terms *A*, *B*, and *C* are the radial diffusive, axial diffusive, and reactive contributions, respectively, D_a is the ambipolar diffusion coefficient, *k* the reaction rate coefficient, and *r* and *z* are the radial and axial coordinates, respectively. This equation can readily be solved analytically if term *B* is neglected (this results in only a small error, e.g., 2% at a carrier gas pressure of 0.3 torr), giving the solution (Bolden *et al.*, 1970a; Adams *et al.*, 1975)

$$[A^+]_z = [A^+]_0 \exp - \left(\frac{\Delta D_a}{a^2} + \Gamma k[B] \right) \frac{z}{v_0}, \quad (4)$$

where $[A^+]_0$ and $[A^+]_z$ are the ion densities at the reactant gas entry point and at the sampling orifice, respectively, and Δ and Γ are pressure dependent factors also related to the slip coefficient s'. That the fluid dynamics and diffusive ion loss have been correctly treated, has been confirmed experimentally by Smith *et al.* (1975) who monitored the ion density as a function of axial position for a pure helium flowing afterglow plasma in which ambipolar diffusion was the only ion loss process and deduced a value of D_a within a few percent of the accepted value. Measurements of plasma flow velocities (Ferguson *et al.*, 1969; Adams *et al.*, 1975) are also consistent with the above analysis.

In the above analysis, it has been assumed that the reactant gas is introduced uniformly into the flow tube at a plane $z = 0$, an obvious approximation since the gas is usually introduced at a point. However, this can largely be corrected for by using an effective reaction length $(z + \epsilon)$ which is greater than the physical reaction length. The value of ϵ, the "end correction", has been obtained experimentally by measuring a given ion–neutral reaction for several different reaction lengths (Bohme *et al.*, 1973). ϵ has also been determined from optical absorption measurements of the concentration of helium 2^3S metastable atoms as a function of distance along the flow tube for a given reactant gas addition (Bolden *et al.*, 1970b). These experiments yield a pressure-dependent end correction which typically varies from 4 to 11 cm over the pressure range 0.2–1.5 torr (Bohme *et al.*, 1973). The corrections necessary to account for the addition of the reactant gas at a point (and other smaller corrections) have been treated numerically in detail by Ferguson *et al.* (1969). Thus from the experimental decay in the primary ion count rate with reactant neutral addition, the reaction rate coefficient can be calculated. Various workers associate differing error figures with their measured rate coefficients ranging from ± 15 to $\pm 30\%$ for a straightforward measurement. Comparison of the values obtained by flow tube techniques and by other well-established techniques such as stationary afterglows, high pressure mass spectrometry, and ion cyclotron resonance (ICR) indicates that these error figures are generally conservative and that the flow tube data analysis for reaction rate coefficients has been correctly applied.

This analysis for determining reaction rate coefficients from flowing afterglow data can equally well be applied to the analysis of SIFT data if certain minor modifications are made. The minor modifications to this analysis for application to the SIFT are discussed in Section IV.B.

D. Some Recent Successes of the Flowing Afterglow

Since its inception, the flowing afterglow has been established as a first rate, extremely versatile technique and has been used to study many hundreds of thermal energy, binary, and ternary reactions of both positive and negative ions many of which cannot at present readily be investigated by other than flow tube techniques. The rate coefficients and ionized products so measured are included in compilations by Fehsenfeld *et al.* (1969b), Ferguson (1973), and Albritton (1978). The advantages and versatility of the flowing afterglow technique are also possessed by the more recently developed flow tubes, and therefore it seems only fitting to mention some of its successes which will act as a standard against which the recent advances should be compared. These successes have been reviewed in some detail in the literature (see, e.g., Ferguson *et al.*, 1969; Ferguson, 1972).

A great virtue of the flowing afterglow technique is its ability to produce a wide variety of primary ion species in their ground electronic state and with a Boltzmann distribution of rovibronic states in thermal equilibrium with the background carrier gas. The reactant neutral is also in thermal equilibrium with the background gas. However, in a few favorable cases it has also been possible to determine rate coefficients for ions reacting with excited state neutrals, e.g., vibrationally excited N_2 (Schmeltekopf *et al.*, 1968) and $O_2(^1\Delta_g)$ (Fehsenfeld *et al.*, 1969a). The striking increase in the rate coefficient for the reaction

$$O^+ + N_2 \rightarrow NO^+ + N \qquad \text{(II)}$$

of one and a half orders of magnitude for a factor of six increase in the nitrogen vibrational temperature shows the virtue of knowing the excitation states of the reactants. Another aspect of the versatility is the capability for studying the reactions of ions with chemically unstable neutral species such as H, N, O, OH, and O_3 (see, e.g., Fehsenfeld and Ferguson, 1968; Howard *et al.*, 1974; Fehsenfeld *et al.*, 1975a; Fehsenfeld, 1976). The study of negative-ion reactions at thermal energies has been almost solely the province of the flowing afterglow (Fehsenfeld, 1975b) and indeed it was using this technique that the process of associative detachment was first experimentally observed (Fehsenfeld *et al.*, 1966b). For example, reactions such as

$$Cl^-(^1S) + H(^2S) \rightarrow HCl^{-*} \rightarrow HCl + e \qquad \text{(III)}$$

lead directly to information concerning the short lived, intermediate, negative-ion states. Indeed recently, Bierbaum *et al.* (1977) using a flowing afterglow have observed infrared emission from the vibrationally excited CO_2 product of the associative detachment reaction of O^- with CO.

One major, very successful application of the data accumulated using the flowing afterglow technique has been to the study of the positive- and negative-ion chemistry of the Earth's ionosphere including water-cluster reactions, e.g.,

$$NO^+ + H_2O + M \rightleftarrows NO^+\cdot H_2O + M \tag{IV}$$

and

$$NO^+\cdot(H_2O)_3 + H_2O \rightarrow H_3O^+\cdot(H_2O)_2 + HNO_2 \tag{V}$$

and metal ion reactions, e.g.,

$$Mg^+ + O_3 \rightarrow MgO^+ + O_2. \tag{VI}$$

This application has been extensively reviewed (see e.g., Ferguson, 1975a,b; Thomas, 1974).

In many cases it is essential to study reactions at temperatures appropriate to a particular application. The flowing afterglow has the capability for temperature variation over the range 80–600 K (Dunkin *et al.*, 1968). This has recently been extended up to 900 K, the practical limit for such a system (Lindinger *et al.*, 1974). Such a temperature range is very adequate for many natural plasmas. Temperature dependences of several binary rate coefficients have been measured over the available temperature range (Lindinger *et al.*, 1974) and rate coefficients for some ternary association reactions have been obtained at several temperatures (see, e.g., Bohme *et al.*, 1968; Adams *et al.*, 1970; Fehsenfeld *et al.*, 1974; see also the review by Good, 1975). Reaction equilibria have also been studied as a function of temperature (Fehsenfeld *et al.*, 1976) as well as at room temperature (Bohme, 1975) and have provided much fundamental information concerning reaction energetics.

III. THE FLOW–DRIFT TUBE

The restrictions in the temperature range over which the flowing afterglow technique can be used (which corresponds to a mean energy range from about 0.01 to 0.1 eV) are unfortunate since a detailed understanding of ion–neutral reactive collisions will only be obtained if such interactions can be studied over a wider energy range. In principle, ion beam studies can provide the required data but in practice it is very difficult to operate beam apparatuses below a few electron volt interaction energies. The drift tube technique, which has been greatly exploited over the last few decades to study nonreactive collision processes of electrons and ions, i.e., the determination of diffusion coefficients and mobilities (see Huxley and Crompton, 1974), bridges the energy gap between the flowing afterglow and beam experiments. Its potential to study reactive ion–neutral collisions has been exploited in several laboratories during the last decade notably by the following workers and their colleagues: Biondi (Heimerl *et al.*, 1969; Johnsen *et al.*, 1970; Johnsen and Biondi, 1973), McDaniel (Miller *et al.*, 1968), Hasted (Kosmider and Hasted, 1975), and Phelps and Moruzzi (Moruzzi *et al.*, 1968; Coxon and Moruzzi, 1977). The principle of the drift tube is simple; the average energy of a swarm of ions in a nonreactive buffer gas is increased by the application of a uniform electric field and hence, with the addition of relatively small concentrations of a reactive gas, ion–neutral reaction rates as a function of energy can be determined. However, except at very low values of E/N, the kinetic energy distribution of the drifting ions may depart significantly from Maxwellian and similarly their internal energy states may be non-Boltzmann and so the assignment of a mean interaction energy must be treated with caution. A thorough discussion of these aspects of drift tubes has recently been presented by Albritton *et al.*, (1977) and briefly reviewed by Biondi (1977). The chemical versatility of the flowing afterglow has been married with the energy range of the drift tube by the NOAA group to form the flow–drift tube.

A. General Description of Apparatus and Technique

The details of the apparatus have been given by McFarland *et al.* (1973a) and so will only briefly be presented here with reference to the schematic of Fig. 3. It consists of a four-section stainless steel flow tube of approximate dimensions 8 cm internal diameter and 125 cm long, dimensions typical of most flow tubes in use at present. The upstream section is the *ion production*

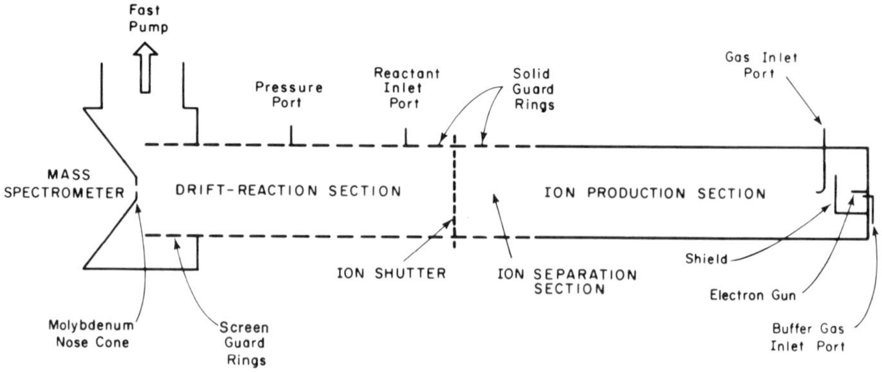

FIG. 3 Schematic diagram of the flow–drift tube. [From McFarland *et al.* (1973a).]

region and is simply a flowing afterglow plasma source of thermalized ions in which a wide variety of positive or negative ions can be synthesized. Downstream of the ion production region, the flow tube walls are segmented into nearly 100 "guard rings." The vacuum seals between the rings are accomplished with O-ring seals and electrical insulation between the rings is effected by Mylar spacers. Thus the application of suitable potentials to the guard rings, readily provided by coupling the rings together with appropriate resistors (external to the flow tube and hence not a source of contamination), establishes the required electric field within the flow tube. The *ion separation section* is utilized to separate the plasma positive ions from the negative ions and electrons and this is achieved by establishing an electric field in the appropriate direction in this region. The magnitude of this field is held constant during an experiment. The *ion shutter* consists of two closely spaced grids (~ 1 mm apart) between which a relatively intense electric field can be established to prevent the flow of ions into the drift-reaction section. To determine drift velocities of ions, the shutter is operated in the pulsed mode so that drift times can be determined. During the determination of reaction rate coefficients, the shutter is held open by shorting the two grids together. The *drift-reaction section* represents about one-half of the length of the flow tube. The thickness and diameter of the guard rings is such as to produce an electric field which is no less than 99% uniform to approximately 83% of the tube radius. A grid in front of the mass spectrometer sampling orifice terminates the field and by pulsing this grid to an appropriate potential, the transit time of the ions through the *ion detection system* can be determined. This parameter is necessary in order to determine the drift times in the gas alone from the total time taken for the ions to transit the region between the ion shutter and the ion detector (Fig. 3; see also McFarland *et al.*, 1973a). The detection system is a conventional mass-filter/particle-multiplier system (Ferguson *et al.*, 1969). Thus, the drift velocities of ions can readily be determined as a function of E/N. Reactant gases are introduced into the drift-reaction section through several small holes equally spaced around the inside of one of the guard rings which is of a hollow construction. A similar approach has been adopted in the SIFT described in Section IV.A.2.

B. Determination of Rate Coefficients as a Function of Interaction Energy

Reaction rate coefficients are determined in a manner similar to that described for the conventional flowing afterglow (Section II.C). An electric field is established in the reaction region corresponding to a particular value of E/N and the decrease in the reactant ion signal and the appearance of the products ions are monitored as a function of the reactant gas flow rate. The analysis of the data is, however, greatly simplified compared to that outlined for the flowing afterglow (Section II.C), essentially due to the fact that the radial profile of the ion velocity for the accelerated ions is uniform, i.e., slug flow prevails rather than the parabolic radial velocity distribution in the flowing plasma. This is a consequence of the ion drift velocity being significantly greater than the gas flow velocity even for small values of E/N (McFarland *et al.*, 1973b). Thus for the generalized reaction

$$A^+ + B \rightarrow C^+ + D, \qquad \text{(VII)}$$

the rate of loss of A^+ in the drift-reaction region is given by

$$v\frac{\partial[A^+]}{\partial z} = -k[A^+][B], \qquad (5)$$

where k is the reaction rate coefficient and $[A^+]$ and $[B]$ are the concentrations of A^+ and the reactant neutral B, respectively, and where v is the sum of the flow velocity and drift velocity of the ions. As in the case of the flowing afterglow, the diffusive loss terms can be eliminated from the analysis (McFarland *et al.*, 1973a). It follows that $[B] = Q/(\pi a^2 v_0)$, where Q is the reactant gas flow rate, a the flow tube radius, and v_0 the bulk flow velocity of the carrier gas. Substitution of $[B]$ into Eq. (5), followed by integration and rearrangement of the resulting expression, yields a value for

$$k = \frac{\pi a^2 v v_0}{QL} \ln\left(\frac{[A^+]_0}{[A^+]}\right), \qquad (6)$$

where $[A^+]_0$ corresponds to $[A^+]$ at $Q = 0$ and L is the distance between the reactant gas inlet port and the sampling orifice (again an end correction has to be included, see Section II.C). As in the case of the flowing afterglow, it is assumed that the diffusive loss of ions is controlled by the carrier gas only, which necessitates that the reactant gas concentration must always be small. Additionally, in the flow–drift experiment, excessive concentrations of the reactant gas will reduce the drift velocity of the ions below that determined in its absence, thus increasing the reaction time. Both of these effects will tend to produce erroneously high rate coefficients but will only be significant for slow reactions ($k \lesssim 10^{-12}$ cm^3 s^{-1}, see below).

Thus, rate coefficients can readily be determined as a function of E/N, but it is necessary to relate them to the center-of-mass kinetic energy of interaction E_{cm}. Recently the simple relationship derived much earlier by Wannier (1953) has been adopted which relates the mean kinetic energy KE_{ion} of an ion drifting through the gas to the drift velocity v_d, viz,

$$KE_{ion} = \tfrac{1}{2}m{v_d}^2 + \tfrac{1}{2}M_b{v_d}^2 + \tfrac{3}{2}kT \qquad (7)$$

for $KE_{ion} \gg \frac{3}{2}kT$, where m is the mass of the ion, M_b the mass of the buffer gas atom or molecule, and T the gas temperature. Users of conventional drift tubes were reluctant to accept the applicability of this expression to a real situation since it is based on a polarization interaction alone and so mobility data were simply plotted as a function of E/N. However, justification for the use of Eq. (7) has come from the recent experimental observations of McFarland *et al.* (1973b) (see Section III.C) and from the theoretical work of Skullerud (1973) who was able to show that when the experimental drift velocity was used in Eq. (7), it reliably predicted KE_{ion} for a wide variety of model interaction potentials. The kinetic theory approach of Viehland *et al.* (1974) showed its general applicability (for a detailed discussion, see Albritton *et al.*, 1977). The center-of-mass kinetic energy for a reacting ion–neutral pair KE_{cm} is given by

$$KE_{cm} = \tfrac{1}{2}[Mm/(M + m)](v_i^2 + v_n^2), \tag{8}$$

where M is the mass of the reactant neutral and v_n^2 its mean square velocity, i.e., $\frac{1}{2}Mv_n^2 = \frac{3}{2}kT$, and v_i^2 the ion mean square velocity, i.e., $\frac{1}{2}mv_i^2 = KE_{ion}$. Combining these well-known relationships,

$$KE_{cm} = [M/(m + M)](KE_{ion} - \tfrac{3}{2}kT) + \tfrac{3}{2}kT \tag{9}$$

so from the appropriate value for KE_{ion}, which is itself obtained from the experimentally determined value of v_d, the required value for KE_{cm} is obtained. A minor complication arises in that the inevitable small pressure gradient along the reaction-drift region results in a gradiant in v_d. In the NOAA apparatus, this results in a $\pm 10\%$ variation in KE_{ion} which needs to be considered in overall error estimates (discussed in detail in McFarland *et al.*, 1973b).

C. *Some Important Results*

Having developed the technique, the NOAA group has exploited it to determine a large number of atomic and molecular positive- and negative-ion mobilities and ion–molecule reaction rate coefficients over a center-of-mass energy range from 0.05 to 3 eV, and at a gas temperature of about 300 K. These data have been reported and briefly reviewed in several papers (see, e.g., McFarland *et al.*, 1973a,b,c; Fehsenfeld, 1975a; Lindinger *et al.*, 1975a; Lindinger and Albritton, 1975; Dotan *et al.*, 1976b) and include data on several reactions of metastable $O_2^+(a^4\pi_u)$ ions (Lindinger *et al.*, 1975b), some product branching ratio studies (Dotan *et al.*, 1976a; McFarland *et al.*, 1974), and isoenergetic reactions (Fehsenfeld *et al.*, 1975b).

The success of the flow–drift technique is well illustrated by the excellent agreement which is so often obtained with the rate coefficients deduced from crossed-beam measurements. The cross sections σ obtained from the beam measurements are converted to rate coefficients k, using the relation $k = \sigma\overline{v}$, where $\overline{v}$ is the mean relative velocity of the reacting ion–molecule pair given by $\overline{v} = (2KE_{em}/M_r)^{1/2}$, M_r being the reduced mass of the ion–molecule pair (this procedure has only been adopted when the energy dependence of σ is not great). This agreement is well illustrated by the data shown in Fig. 4 (from McFarland *et al.*, 1973b) for the fast ion–molecule reaction of N^+ with O_2, and is very significant since its indicates that no gross errors are involved in using the Wannier relation to calculate KE_{cm}. The beam data is that due to Neynaber *et al.* (1971) and Rutherford and Vroom (1971). Many examples of data which support this conclusion are given in the published papers of the NOAA group. The static drift-tube data of Johnsen *et al.* (1970) are also shown. The good agreement between the flow–drift tube and the static drift tube illustrates the good precision between the measurements but cannot be used to justify the accuracy of the calculated energy scale since the Wannier relation is used in both experiments to calculate KE_{cm}.

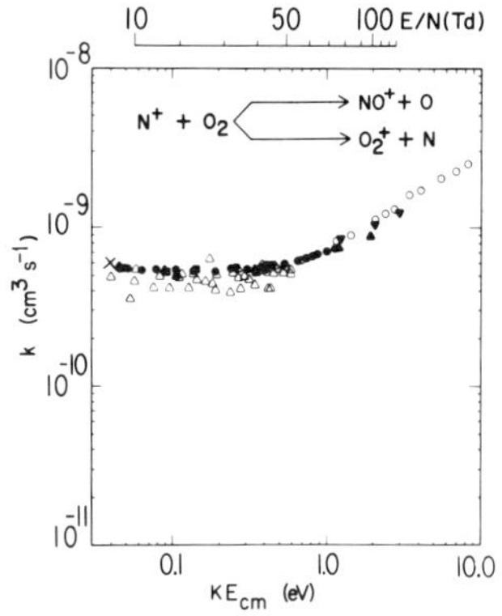

FIG. 4 Rate coefficients for the $N^+ + O_2$ reaction as a function of KE_{cm}. [From McFarland *et al.* (1973b).] The data are solid symbols, flow–drift tube (■, 0.595 torr; ◆, 0.429 torr; ●, 0.386 torr; ▲, 0.283 torr; ▼, 0.208 torr; ×, flowing afterglow [Dunkin *et al.* (1968)]; △, drift tube [Johnsen *et al.* (1970)], and ○, ion beam data [Neynaber *et al.* (1971); Rutherford and Vroom (1971)].

As previously mentioned (Section II.D), a great deal of attention has been given in flow tube studies to the $O^+ + N_2$ reaction (II). The rate coefficient reduces from an already low value of 1.2×10^{-12} $cm^3\ s^{-1}$ at room temperature to about 5×10^{-13} $cm^3\ s^{-1}$ at about 600 K (Dunkin *et al.*, 1968). Further increase in effective temperature results in a rapid increase in the rate coefficient toward the Langevin value ($\sim 10^{-9}$ $cm^3\ s^{-1}$) (McFarland *et al.*, 1973b). The wide ranging values of the rate coefficient and the rapid change with energy make it especially valuable for critical comparisons of the flow tube, drift tube, and beam techniques used for ion–molecule reaction studies. The results of the study of this and other reactions give great credence to the various techniques used (McFarland *et al.*, 1973b; Albritton *et al.*, 1977).

The $O^+ + N_2$ reaction has also been used to demonstrate forcibly an important consideration in the study of slow ion–molecule reactions ($k \lesssim 10^{-12}$ cm s^{-1}) in flow–drift tubes. For such slow reactions, the relatively large amount of reactant gas which must be introduced into the drift-reaction region to produce a significant reduction in the ion signal, also results in an enhancement of the transverse diffusive loss of ions and thus gives an erroneously high rate coefficient for the ion–molecule reaction. This nonreactive ion loss has been demonstrated by adding argon rather than nitrogen to the buffer gas containing the drifting O^+ ions, where a finite "rate coefficient" can be obtained without, as expected, any product ions being observed! This unfortunate problem has been circumvented by monitoring the rate of increase of the product ion signal in slow reactions (in this case NO^+) (Albritton *et al.*, 1977).

A most interesting feature of the flow–drift tube and conventional drift-tube data is the very obvious differences which result in the rate coefficients as a function of center-of-mass kinetic energy for a given reaction when different buffer gases are used. This point is graphically illustrated in Fig. 2 in Chapter 2 for the reaction of O^+ with O_2. These differences are most obvious for reactions whose rate coefficients increase rapidly with increasing kinetic energy. A further discussion is found in Chapter 2.

The most recent development in flow–drift tube techniques is due to Rayment and Moruzzi (1978) and is primarily intended for negative-ion studies. The essential difference in their approach to that of the NOAA group is that the drift electric field is directed along a diameter of the flow tube that is normal to the gas flow direction. Electrons from a thermionic emitter are introduced into the drift field and negative ions are produced by adding small concentrations of attaching gases to the buffer gas flow. Any excited neutral species generated by electron collisions are swept out of the drift region and therefore cannot significantly interfere with the negative-ion–molecule reactions under study. To date, the only reaction studied in detail has been the associative detachment of O^- with ground state N_2, the rate coefficient for which is seen to increase from about 7×10^{-15} cm^3 s^{-1} at low E/N to 10^{-12} cm^3 s^{-1} at higher E/N. The advantage of this crossed-flow–drift approach appears to be that it can be used to study, with reasonable precision, electron-attachment processes in unstable gas mixtures, e.g., those containing atomic oxygen.

IV. THE SELECTED ION FLOW TUBE (SIFT)

As research into ion–neutral interactions progresses, interest is turning to the study of more exotic ions such as those present in planetary atmospheres and in the interstellar medium (see, e.g., Huntress, 1977a), and it is inevitable that the limitations of the flowing afterglow will become more restrictive. The following complications are variously present in flowing afterglows, flow–drift tubes, and ICR cells:

(a) The presence of the source gas from which the primary ions of interest are created can result in concurrent reactions which can compete with the reaction under study and can complicate product identification. Additional complications in the product ion spectrum are produced by secondary reactions with the source gas. Also, for reactions which are almost thermoneutral, the presence of the source gas may permit reverse reaction to occur resulting in a slower decay of the primary ion (see, e.g., Adams *et al.*, 1970; Fehsenfeld *et al.*, 1975b).

(b) The presence of primary ions other than that being studied can sometimes lead to erroneously low rate coefficients if the ion under study is a product of the simultaneous reaction of another primary ion. In addition, the reactions of the other primary ions can complicate the ion spectrum making product ion identification difficult (Govers *et al.*, 1974; Marx *et al.*, 1975; Anicich *et al.*, 1975).

(c) Metastable excited species may react with the gases present producing the primary ion and thus resulting in erroneously low rate coefficients, or in the production of ions which could be wrongly identified as products of the primary ionic reaction (Kim and Huntress, 1975).

(d) The presence of electrons, inevitable in most afterglow plasma experiments, together with adequate concentrations of molecular positive ions can result in deionization by dissociative recombination. This loss mechanism may compete with ion loss by the ion–neutral reaction under study, thus leading to erroneous rate coefficients and distorted product ion distributions. Electrons can also efficiently destroy excited ions and excited neutrals by superelastic collisions making the study of the reactions of such species difficult (Phelps, 1955; Bolden *et al.*, 1970b; Adams *et al.*, 1972; Smith *et al.*, 1975; see also Section IV.C).

(e) Energetic photons from the ionization source may ionize gases in the reaction zone producing a distributed source of the primary ion under study (Ferguson *et al.*, 1965, 1969). Such photons may also enhance the background count rate at the particle multiplier (Heimerl *et al.*, 1969; Farragher, 1970).

The above complications could be effectively avoided by producing the ions of interest in a remote ion source and injecting a single mass selected ion species into a flowing carrier gas. That this apparently obvious approach has not previously been adopted is probably due to the inhibiting experimental factor that the injection of ions from a low pressure mass spectrometer into a high pressure gas requires the ions to move against a back flow of neutral carrier gas with the attendant loss of ion beam intensity. This major problem has been essentially overcome by the authors by the realization that if the fast flowing carrier gas is introduced into the flow tube through a series of small apertures located around the ion injection point and parallel to the ion beam, then the resulting "aspirator effect" (Section IV.A) will produce a much lower pressure at the flow tube side of the injection orifice and consequently a large reduction in the back flow of the carrier gas into the mass filter chamber below that otherwise expected. This principle has been used in the design and construction of the selected ion flow tube (SIFT) at Birmingham (Adams and Smith, 1976a). The details of the ion injector, flow tube, and detection system are described in Section IV.A, following a general description of the SIFT technique. Section IV.B contains a discussion of the method of determining reaction rate coefficients and product ion distributions and examples of some of the many types of reactions which have been studied using this technique are given in Section IV.C.

A. *General Description of Apparatus and Technique*

A schematic diagram of the SIFT apparatus is shown in Fig. 5. Ions are produced in an ion source (Section IV.A.1) remote from the flow tube (Section IV.A.2) and after mass selection in a quadrupole mass filter they are injected through an orifice O_1 into the flow tube. The ions then rapidly thermalize in collision with the carrier gas and the resulting ion swarm is

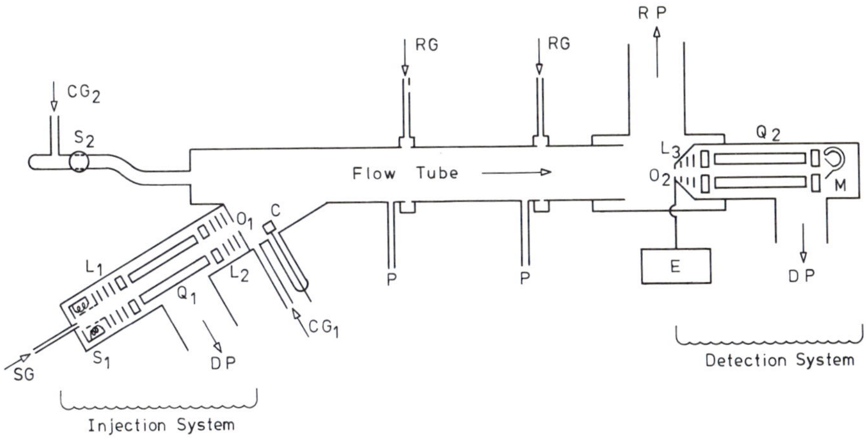

FIG. 5 Schematic diagram of the SIFT apparatus. S_1—electron impact ion source, S_2—microwave discharge source; L_1, L_2, L_3—electrostatic lenses; Q_1, Q_2—quadrupole mass filters; O_1, O_2—orifices; C—ion collector; *SG*—ion source gas; CG_1, CG_2—carrier gas; *RG*—reactant gas; *P*—pressure measurement ports; *E*—electrometer; *M*—channel multiplier; *RP*—roots pump; *DP*—diffusion pumps. CG_2 and S_2 enable the apparatus to be used as a conventional flowing afterglow.

transported down the flow tube and past radial entry ports through which neutral reactant gases are introduced. At the end of the reaction region, the ions are sampled through another orifice O_2, mass selected by a second quadrupole mass filter, and detected with a channel-multiplier/pulse-counting system. This ion sampling and detection system can be calibrated for mass discrimination as described in Section IV.A.3, essential if accurate product ion distributions are to be determined (Adams and Smith, 1976b). The methods of gas flow rate control and measurement, ion detection, and data analysis leading to reaction rate coefficients are very similar to the conventional flowing afterglow.

1. The Principle of Ion Injection

A diagram of the ion injector is shown in Fig. 6. The ions produced in ion source *I* are focused using a three element, electrostatic aperture lens L_1 into the entrance aperture of a quadrupole mass filter Q_1. The mass selected ions are then partially focused by a second lens L_2 through O_1 in a molybdenum plate and then into the flow tube. Molybdenum is used for the construction of the orifice disk since, as discussed earlier this material is resistant to contamination. The carrier gas is introduced into an annular space around

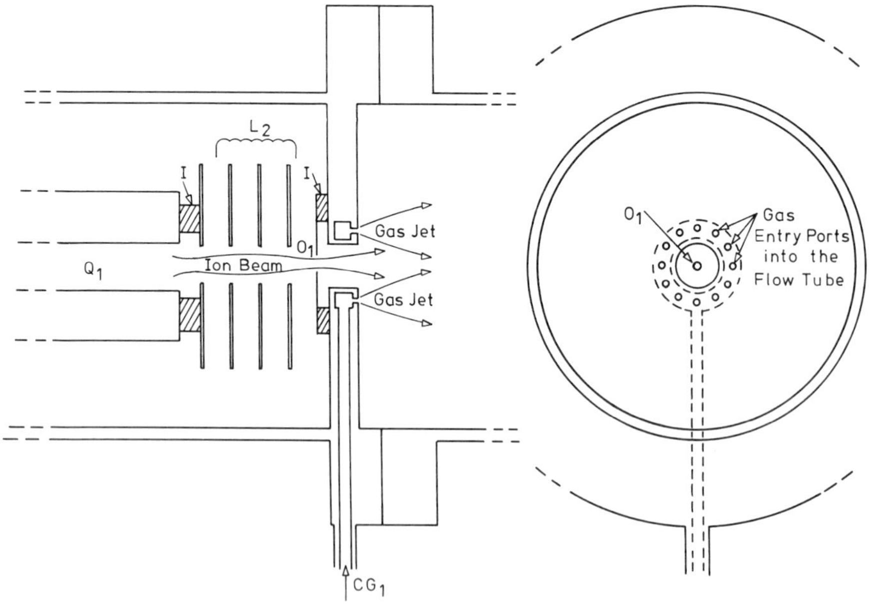

FIG. 6 Diagram of the SIFT ion injector showing both elevation and end views. *I*—electrical insulators; other symbols as in Fig. 5. The dashed lines indicate the tube and cavity through which the carrier gas passes before entry into the flow tube.

the orifice and enters the flow tube through 12 1-mm diameter apertures arranged in a circle with the ion beam passing through its center. These holes direct the gas at high velocity (approaching the speed of sound) parallel to the incoming ion beam and then down the flow tube. This method of gas injection results in the pressure in the vicinity of O_1 being much lower than that downstream in the flow tube. Thus, the back flow of gas through O_1 is considerably reduced and the necessary low pressure is maintained in the mass filter chamber even for relatively large connecting orifices (typically 1 mm diameter). Equally important is the fact that the initial interaction of the ions entering the flow tube is with a relatively tenuous carrier gas which reduces the scattering losses and ensures that most of the ions are convected downstream. It is important to note that the efficiency of ion collection by the flowing gas near to O_1 can to some extent be optimized by varying the gas injection rate. If the gas flow rate is too great, a reduction in collection efficiency results, presumably due to a transition from streamline to turbulent flow in the vicinity of O_1. A detailed appreciation of this phenomenon has not yet been obtained and it is certain that improvements will be made in the coming years. However, the efficiency of the multiaperture injector in the existing apparatus is acceptably high.

The major attenuation in the mass analyzed ion beam is in the region between the quadrupole exit aperture and the orifice, a direct result of the divergence and spatial extent of the ion trajectories exiting the mass filter. This makes it impossible to efficiently focus these ions onto the orifice using electrostatic lenses. Typically in the Birmingham SIFT, the ion current decreases by two orders of magnitude in this region. Ion currents up to 5×10^{-9} A have been measured at the collector C at an injected ion energy of about 10 eV for ion species which are rich in the ion source, for example, for N_2^+ and Ar^+ from N_2 and Ar. All lens potentials are optimized to maximize this current. In a collaborative experiment between the authors and Twiddy at the University College of Wales, Aberystwyth, in which an existing flowing afterglow apparatus has been converted to a SIFT, much higher currents (about 10^{-7} A of Ar^+ at an injected ion energy of 5 eV) have been obtained in the flow tube using Brubaker type, rf quadrupole segments at the entrance and exit apertures of the mass filter (Brubaker, 1968) which tend to concentrate the ion beam along the axis of the mass filter.

The fraction of the ion beam which passes through O_1 is very dependent on ion mass. For example, from an electron impact ion source with a nitrogen source gas and an electron energy of 70 eV, approximately equal currents of N^+ and N_2^+ are obtained at the exit aperture of the quadrupole. However, after the injection through the orifice, the N^+ current is only about 10% of that of N_2^+ (as monitored by collector C), the lenses being optimized for each ion to achieve maximum current. This is thought to be a manifestation of the more divergent trajectories of the lighter ion at the quadrupole exit aperture. In addition, the profiles of the individual mass peaks are not smooth but show multiple peak splitting. This is not due to inhomogeneities in the quadrupole fields (Dawson and Whetten, 1969) since normal smooth peaks are obtained at the exit aperture when the ion current is monitored on the quadrupole side of O_1, and hence it is presumably due to the focusing and defocusing of the ions at O_1 by the electrostatic lens, this being caused by the change in the ion trajectories out of the quadrupole as the magnitudes of the tuned rf and dc field intensities are varied. In principle, these effects could be avoided by replacing the quadrupole with, for example, a magnetic mass filter such as the Wien filter. However, these instruments conventionally operate with much higher ion energies than the quadrupole and thus a much greater deceleration of the ions would be required (with the associated ion beam divergence) before injection into the flow tube.

The energy at which ions are injected into the flow tube is necessarily small since many molecular ions can be fragmented in collision with the carrier gas. For a helium carrier gas, strongly bonded ions such as N_2^+ can be injected at energies as large as 30 eV without significant fragmentation. In practice, such relatively high energies are unnecessary and would not be desirable since considerable collisional excitation of the injected ion and nonthermal reaction with the carrier gas become energetically possible. With more weakly bound molecular ions, collisional fragmentation is more obvious. For example, when injecting CH_4^+ at 15 eV, greater than 50% collisional breakup to CH_3^+ is observed which becomes insignificant at an injection energy of about 5 eV. Thus with weakly bound ions, injection energies of this order have to be used to eliminate fragmentation and the likelihood of incorrect product ion identification in ion–neutral reaction studies. This reduction in ion injection energy results in a considerable but not intolerable reduction in the injected ion current and a consequent decrease in the ion count rate at the channel multiplier (Fig. 5). However, even at an injection energy of a few electron volts, the count rate is still very adequate for the determination of rate coefficients and product ion distributions. The above example of the degree of collisional fragmentation refers to helium as the target/carrier gas. For heavier carrier gases such as argon and nitrogen, fragmentation is more severe, but nevertheless both of these gases have been successfully used (Smith *et al.*, 1977).

Up to the present time, two types of ion source have been used, a gas discharge and a low pressure electron impact source. The former consists of a microwave discharge (10–50 W dissipation) in a quartz tube containing an appropriate gas (pressure typically several tens of microns). The ions so produced effuse into the low pressure mass filter chamber through an orifice (about 1 mm diameter) and are focused by lens L_1 into the quadrupole. This plasma source produced copious positive ions; however it was only suitable for atomic ions (e.g., He^+, C^+, N^+, O^+, Ar^+) and strongly bound molecular ions (e.g., $N_2{}^+$, NO^+). This is because the positive ions, in traversing the accelerating wall sheath between the plasma and the orifice plate, fall through a potential difference amounting to a few tens of volts [for an estimate source electron temperature of 10^5 K (Smith *et al.*, 1974)] and are thus injected at excessively high energies into the flow tube. Since the injection energy is effectively established by the potential at the point of origin of the ions with respect to the flow tube (ground potential), this problem can be overcome by reducing the potential of the whole plasma. However the electron impact ion source has proved to be much more suitable for the weakly bound ions. That used to date is a conventional low pressure Brink ionizer (Brink, 1966) which has the advantage that the potential at the point of production of the ions can readily be adjusted and hence the injection energy at O_1 can readily be varied. Another desirable feature is that the spread in the energy of the ions emerging from the source (and hence into the flow tube) is small. Also from such sources, by using high electron energies (approaching 100 eV) and low source pressures, appreciable currents of metastable excited ions and multiply charged ions can be obtained (Lindholm, 1972). These ions can often survive at the relatively high pressures in the flow tube without collisional deexcitation or appreciable reaction with the carrier gas and this has been used to great effect to study the reactions of these species at thermal energies (Section IV.C). Conversely to produce ground state, singly charged ions only, much lower electron energies are used to inhibit the production of excited states. In many studies, reactions which are endothermic for ground state ions, but which are exothermic and rapid for excited state ions, have been used to confirm that unwanted excited ions are absent from the beam.

Negative ions have also been generated in the electron impact ion source. However because of the small cross sections for the production of negative ions by electron collisions, to date only the readily generated ions Cl^- and F^- have been produced at sufficiently large beam intensities to give adequate negative-ion count rates at the downstream detector. This situation will soon be rectified by the use of better negative-ion sources (Section V) so that a program of negative-ion reaction studies can be pursued vigorously.

2. *The Flow Tube*

Although the flow tube section of the SIFT is similar to that of the flowing afterglow (Section II), there are several important differences which are worthy of note. As previously stated, unlike the flowing afterglow, no electrons are present in the SIFT to neutralize the diffusive wall current of the positive ions when these constitute the injected current. Thus no insulating material can be exposed to the flowing ion swarm, since surface charge buildup would inhibit the passage of any ions down the flow tube. Similar remarks apply to negative-ion injection. Because of this, glass flow tubes are not practical and thus a stainless steel flow tube (7.3 cm internal diameter, about 100 cm long), with its internal surface free of insulating oxide layers, has been used. Typical flow tube pressures are limited in the present apparatus to a maximum of about 1 torr because of pressure rise in the mass filter chambers (Section II.A). This range could be extended to greater pressures by adopting smaller orifices O_1 and O_2 or by using larger diffusion pumps. After injection into the tube, rapid collisional thermalization occurs and the ions undergo free diffusion resulting in a closely fundamental mode, Bessel function, radial density distribution about 20 to 30 cm downstream. The carrier gas bulk velocity is about 10^4 cm s^{-1} and thus for typical injected ion currents of 10^{-11}–10^{-9} A, average ion densities of about 10^2–10^4 cm^{-3} are obtained upstream in the flowing gas. At these densities, space charge mutual repulsion of the ions is insignificant and only normal field-free diffusive loss to the flow tube walls occurs. (However, an upper limit to the number density and hence to the ion current which can be usefully injected, will eventually be reached due to space charge repulsion.) Thus, the diffusive loss of ions along the SIFT is much less than in a positive-ion–electron flowing afterglow plasma, in which the positive ions are lost by ambipolar diffusion with electrons [Eq. (4)], for which the diffusion coefficient D_a for a given ion species is twice the free diffusion coefficient $D_{+,-}$. The net result of this is that in the SIFT much smaller ion number density gradients are established along the flow tube.

On their passage along the flow tube, the ions pass ports *RG* through which the reactant gases are introduced. The availability of entry ports at more than one axial position along the flow tube facilitates the determination of reaction-length end corrections (ϵ, Section II.C) and allows consecutive reactions with different reactant gases to be studied. The entry ports are similar to those employed in the flow–drift tube (Section III.A) consisting of 12 1 mm diameter holes located in the flow tube walls and disposed in a circle normal to the flow tube axis. This port design allows the gas to be introduced fairly uniformly without seriously disturbing the ion density distribution or depleting the ion concentration as do the more conventional axial ports (Smith *et al.*, 1975). It is important to note that for gas injection from the walls of the flow tube, the end correction ϵ is of the opposite sign to that for the conventional axial gas entry ports, meaning that the effective reaction length is less than the physical distance between the port and the downstream sampling orifice.

3. *Ion Sampling and Detection*

The ions are sample through an orifice O_2 (diameter about 0.3 mm) located in a molybdenum disk which is mounted on, but electrically insulated from, the nose cone, then focused by an electrostatic lens L_3, mass selected using a second quadrupole mass filter Q_2, and detected using a channel multiplier M, together with conventional pulse-counting equipment. The channel multiplier is situated on axis viewing directly into the mass filter, a position in which the detected count rate is considerably enhanced. Such positioning, most desirable in the SIFT because of the low signal levels, is not profitable in the flowing afterglow due to the large background count rate produced by photons from the ionization source and in some cases by metastable excited states. These complications are not present in the SIFT, the background count rate being 0.1–0.2 counts s^{-1}, the dark count rate of the channel multiplier.

The ion current reaching the downstream end of the flow tube is sampled by monitoring the current flowing to the orifice disk. It was found necessary to apply a potential within the range ± 1 V to this orifice disk with respect to the surrounding flow tube to optimize the current flow through O_2, presumably to counteract the effects of contact potential differences. The currents to this disk of individually injected ion species are typically 10^{-13}–10^{-11} A and can be directly related to the count rate at the channel multiplier. This procedure can be repeated for several ion species by simply retuning the upstream mass filter, and hence a curve of detection sensitivity against ionic mass can be constructed. Thus the entire detection system (the orifice sampling probe, the lens, the mass filter, and the channel multiplier) can be calibrated for mass discrimination for any resolution setting of the quadrupole which can then be accounted for when determining the product ion distributions of ion–neutral reactions (Section IV.B). Secondary emission of electrons from the orifice disk is not expected to distort the measured ion currents significantly because of the very small emission coefficients for thermal energy positive ions on gas-covered surface (experimental data supporting this conclusion is summarized in McDaniel, 1964). This calibration procedure is not possible in the flowing afterglow because it is not possible to produce a specific ion type in the absence of other ion species, electrons, metastables, and photons.

Mass discrimination is critically dependent on the resolution setting of the mass filter. Very large mass discrimination factors (greater than an order of magnitude) can result when the mass filter is operated at high resolution. However, if only modest resolution is required, e.g., unity resolution up to mass 50 amu, with our mass filter (Extranuclear Laboratories Inc., pole diameter 1.6 cm), a factor of two discrimination between the lowest and highest masses is more appropriate. Clearly it is essential to operate at the lowest resolution possible, consistent with resolving all of the ions of interest, if accurate product distributions are to be obtained. Under these low resolution conditions, count rates approaching the upper limit of the channel multiplier (about 10^5 counts s^{-1}) can be obtained at a helium carrier gas pressure of about 0.5 torr. Sampling of both positive and negative ions is straightforward since there is no plasma sheath, the cause of negative-ion sampling problems in stationary afterglows and to a lesser extent in flowing afterglows (Section II.B).

B. Determination of Rate Coefficients and Product Ion Distributions

The procedure for calculating the reaction rate coefficients from the decay of the primary ion count rate as a function of reactant gas addition is almost identical to that employed in the case of the flowing afterglow (Section II.C). Equations (1)–(4) apply equally well to the SIFT except that D_a (the ambipolar diffusion coefficient) is replaced by $D_{+,-}$ (the free diffusion coefficient) and again the axial diffusion contribution [term B in Eq. (3)] can be neglected. A typical example of SIFT data is given in Fig. 7a for the reaction of N_2^+ with H_2S. The rate coefficient for the reaction is obtained from the exponential loss of the primary ion and the product ion distribution is obtained from an analysis of the product ion curves. Initially, in order to demonstrate the validity of the technique, we studied several miscellaneous reactions whose thermal energy rate coefficients had previously been well established by other techniques (Adams and Smith, 1976a). In all cases the rate coefficients obtained using the SIFT were in good agreement with previous data.

Where the product ions of the primary reaction do not undergo secondary reactions with the reactant gas, the product ion distribution can simply be obtained from the relative product ion count rates, accounting for mass discrimination. However, it is more usual for successive reactions of the product ions to occur and then the product ion distribution is obtained by plotting the percentage of each product ion as a function of the reactant gas flow rate and extrapolating the resulting curves to zero flow rate (Adams and Smith, 1976b). The procedure is illustrated in Fig. 7b for the data of Fig. 7a and it can be seen that the product distribution is HS^+ (75%), S^+ (15%), and H_2S^+ (10%). The product ion distributions for many positive-ion reactions have now been determined using this approach. For the reaction of He^+ with N_2, the only reaction for which the product ion distribution has been measured at thermal energies by several techniques including the SIFT, good consistency is obtained (Adams and Smith, 1976b). However,

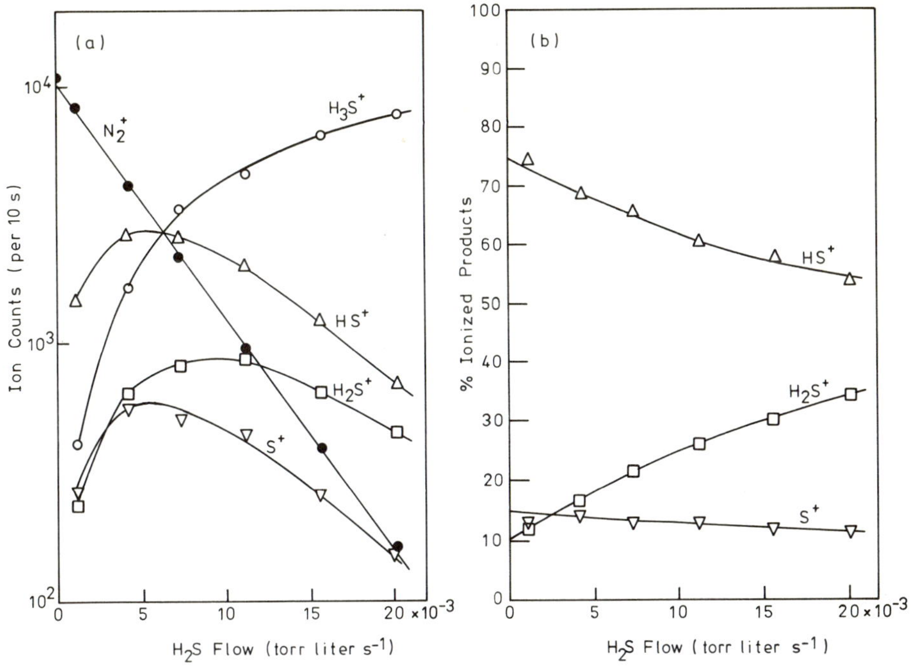

FIG. 7 Data obtained for the reaction of N_2^+ with H_2S at 300 K (from Smith *et al.*, 1978). The rate coefficient is 1.5×10^{-9} cm^3 s^{-1}. (a) indicates HS^+, H_2S^+, and S^+ to be the primary products of the reaction which all react with H_2S at differing rates. H_3S^+ (the terminating ion in the sequence) is the product of secondary and tertiary reactions and is therefore not included in (b) which illustrates the procedure for obtaining product ion distributions. That the curves in (b) are not parallel to the abscissa is a manifestation of the different reaction rates of the primary product ions with H_2S.

comparisons are possible between SIFT data and those obtained using the ICR technique as reported in several papers by Huntress and co-workers (recently reviewed in Huntress, 1977b). Good agreement is generally obtained and where the distributions are significantly different it can usually be ascribed to the differing conditions in the collision dominated SIFT and the low pressure ICR cell (see e.g., Smith and Adams, 1977b).

C. *Some Programmes Illustrating the Versatility of the SIFT*

In the two year period since its inception, we have used the SIFT to measure the thermal energy rate coefficients and product ion distributions for several hundred positive-ion–neutral reactions (Adams and Smith, 1976a,b, 1977, 1978; Smith and Adams, 1977a,b, 1978a; Smith *et al.*, 1977, 1978; Adams *et al.*, 1978, 1979). This work has been undertaken not only as a contribution to fundamental atomic and molecular physics and chemistry but also as a contribution to the challenging problems involved in constructing models for molecular synthesis in the interstellar medium and in planetary atmospheres, particularly the Earth's ionosphere. In this section, the results of some of the studies carried out to date are briefly presented to illustrate the outstanding capabilities of the SIFT technique. These include (1) a detailed survey of the reactions of the CH_n^+ ions ($n = 0$–4) with some 13 different molecules and the reactions of the $C_2H_n^+$ ions ($n = 0$ to 4) with H_2 and CH_4; (2) the reactions of the formaldehyde-derived ions H_nCO^+ ($n = 0$ to 3) with a series of molecules; (3) a detailed study of the reactions of CH_3^+ ions, principally because of their likely significance to interstellar chemistry; (4) reactions of the primary and secondary stratospheric ions N^+, N_2^+, N_3^+, N_4^+, O^+, O_2^+, and NO^+ with several stratospheric neutral constituents including a study of several ternary association reactions of NO^+. (5) several reactions of the metastable excited ions $O^+(^2D, ^2P)$, $NO^+(a^3\Sigma)$, and $O_2^+(a^4\Pi_u)$; and (6) reactions of the doubly charged ions Ar^{2+} and Xe^{2+}, including the reactions of metastable electronically excited states of Xe^{2+}.

(1) The unique capability of the SIFT apparatus to permit the selective injection and detailed study of the reactions of a sequence of ions with successive degrees of saturation is well demonstrated by the reactions of C^+, CH^+, CH_2^+, CH_3^+, and CH_4^+ (generated from methane) with the molecules H_2, N_2, O_2, CO, CO_2, COS, H_2O, H_2S, NH_3, CH_4, H_2CO, CH_3OH, and CH_3NH_2 (Smith and Adams, 1977a,b; Adams and Smith, 1977, 1978). Very great differences in the rate coefficients and product ion distribution are observed with the degree of hydrogenation. These studies have revealed, for example, that in general the ions C^+, CH^+, and CH_2^+ react at rates which are close to the Langevin or ADO values (Su and Bowers, 1973) giving multiple products. The CH_3^+ reactions are varied, including ternary association, which is usually rapid [these are discussed in more detail in (3) below], while CH_4^+ reacts almost exclusively in a binary fashion by charge transfer and/or proton transfer at a gas kinetic rate, the respective rate coefficients for these two reaction channels when they occur simultaneously being inversely related to their relative exoergicities (Adams and Smith, 1978). C_2^+, C_2H^+, $C_2H_2^+$, $C_2H_3^+$, and $C_2H_4^+$ (derived from C_2H_4) react similarly with H_2 to their single carbon analogs, CH_n^+. In both series, the least hydrogenated ions (excluding C^+) react rapidly with H_2 by H-atom abstraction, each series producing a terminating ion CH_3^+ and $C_2H_2^+$, respectively, which only slowly reacts further by ternary association

$$CH^+ \xrightarrow{+H_2} CH_2^+ \xrightarrow{+H_2} CH_3^+ \xrightarrow{+H_2+He} CH_5^+ \nrightarrow, \qquad \text{(VIII)}$$

$$C_2^+ \xrightarrow{+H_2} C_2H^+ \xrightarrow{+H_2} C_2H_2^+ \xrightarrow{+H_2+He} C_2H_4^+ \nrightarrow. \qquad \text{(IX)}$$

A significant binary channel is also observed in the reaction of $C_2H_2^+$ with H_2 at 300 K although for ground state reactants this channel is slightly endoergic, viz.,

$$C_2H_2^+ + H_2 \rightarrow C_2H_3^+ + H - 0.02 \text{ eV}. \qquad \text{(X)}$$

This is a result of the Boltzmann population (about 14%) in the first vibrational state of $C_2H_2^+$ at 300 K, which is sufficient to compensate for the slight endoergicity (Buttrill *et al.*, 1974; Adams and Smith, 1977).

Many similarities are evident between the reactions of CH_4 with the CH_n^+ and $C_2H_n^+$ ion series, carbon addition reactions being favored in both cases and multiple reaction channels being common. Surprisingly, the most hydrogenated ion observed in these reactions is $C_3H_5^+$ (rather than $C_3H_7^+$). These studies illustrate the serious complications in interpretation which would result from the presence of CH_4 or other hydrocarbons as a source gas in the flow tube. Detailed deconvolution of mass spectrometer data would be practically impossible.

Many of the reactions studied in these sequences are thought to be significant in molecular synthesis in interstellar clouds. Using these detailed data on rate coefficients and product ion distributions, several likely channels for the synthesis of observed interstellar species have been recognized and have tentatively explained the relative abundances of some hydrogen–carbon molecules in the gas clouds in Orion A and Sagittarius B2 (Smith and Adams, 1977c).

(2) Among the species most widely observed in the molecular clouds in the galaxy are CO, HCO, H_2CO, and HCO^+ (see, e.g., Dalgarno and Black, 1976). In an attempt to determine how these species have evolved in these regions, we have studied the reactions of the H_nCO^+ ($n = 0$ to 3) ions with several molecules observed (as well as others expected to exist) in these regions, i.e., H_2, N_2, O_2, CO, CO_2, COS, H_2O, NH_3, CH_4, H_2CO, and CH_3OH (Adams *et al.*, 1978). Some of these ions are also thought to be of significance in the chemistry of the stratosphere (Arnold *et al.*, 1977). This is a further example of a sequence of reactions which again illustrates the value of the SIFT technique. These reactions generally proceed at a gas kinetic rate and many distinct classes of reactions are apparent in the data including direct charge transfer, proton transfer, and H-atom abstraction. The application of this data to the problem of interstellar molecular synthesis has led to a likely production mechanism for H_2CO in dense clouds (Smith and Adams, 1978b).

A good example of the value of the SIFT for the separate study of the forward and reverse reactions is the closely thermoneutral reaction

$$H_2O^+ + H_2CO \underset{k_2}{\overset{k_1}{\rightleftharpoons}} H_3CO^+ + H_2O + 0.2 \text{ eV}. \qquad \text{(XI)}$$

k_1 and k_2 have been measured to be 3×10^{-9} and 3×10^{-11} $cm^3\ s^{-1}$, respectively, at 300 K, which immediately provides an equilibrium constant for the reaction. In our new SIFT (Section V) it will be possible to study this type of reaction over an appreciable temperature and energy range and thus obtain critical thermochemical data.

(3) The above studies have revealed that the behavior of certain ions is anomalous. For example, the reactions of the $CH_n{}^+$ ions are generally fast binary in character except, that is, for the $CH_3{}^+$ reactions which are very varied. Three types of $CH_3{}^+$ reactions are apparent which are (a) pure binary, (b) simultaneous binary and ternary, and (c) pure ternary association. In several cases, the ternary association reaction channel is "saturated" at 300 K, a phenomenon also observed for several reactions at this temperature by Spears and Ferguson (1973) and at 80 K by Bohme *et al.* (1968). Because of this unusual behavior and because of the likely significance of $CH_3{}^+$ to interstellar chemistry, we have made a detailed study of its reactions at 300 K in the SIFT. In addition, by cooling the walls of the SIFT with solid CO_2 and by precooling the carrier gas, we have obtained the rate coefficients for ternary association at the lower temperature of 225 K. Such measurements have enabled the lifetimes against unimolecular decomposition of the excited associated complex to be estimated at the two temperatures (Smith and Adams, 1978a). This has allowed us to postulate that radiative association [the radiatively stabilized analog of collisional stabilized (ternary) association] of $CH_3{}^+$ is a likely mechanism by which many of the molecules observed in the interstellar medium are generated (Smith and Adams, 1978b). For example, the rate coefficient for the reaction

$$CH_3{}^+ + H_2 + He \rightarrow CH_5{}^+ + He \qquad \text{(XII)}$$

has been shown to vary approximately as T^{-5} and if this holds down to interstellar temperatures then the radiatively stabilized analog of this reaction could be an important mechanism for the ultimate production of CH_4 in these regions (from $CH_5{}^+ + e \rightarrow CH_4 + H$).

(4) One major application of thermal energy ion–neutral reaction rate data has been to the understanding of the ion chemistry of the Earth's ionosphere (e.g., Ferguson, 1975a,b). Currently the major interest is in the lower atmospheric regions, specifically in the stratosphere (Arnold *et al.*, 1977; Fehsenfeld *et al.*, 1978) and the D-region (Ferguson, 1971; Thomas, 1974). As a contribution to stratospheric chemistry, we have studied the reactions of the primary (N^+, $N_2{}^+$, O^+, and $O_2{}^+$) and secondary ($N_3{}^+$, $N_4{}^+$, and NO^+) stratospheric ions with a series of 13 neutral molecules, including the major atmospheric molecules and several possible "contaminant" molecules (Smith *et al.*, 1978). The $N_3{}^+$, $N_4{}^+$, and NO^+ were produced from the reactions which followed the injection of N^+, $N_2{}^+$, and O^+, respectively, into a helium/

nitrogen carrier gas mixture. This detailed study has revealed many interesting features concerning the relative reactivity of these ions. For example, the reactions of N_2^+ and N_4^+ are very similar, the most common mechanism is apparently direct charge transfer usually resulting in ionized fragmentation of the reactant neutral molecules leaving the N—N bonds in the reactant ions intact. The N^+ and N_3^+ reactions are very different in that they show a greater propensity to form N—X bonds (when X = H, C, O, S, etc.). In several reactions, direct mass spectrometric identification of the product ions was ambiguous since more than one product ion having the same mass but a different chemical composition is energetically possible or a possible product ion had the same mass as the primary ion. These situations were often resolved either by injecting isotopic species or by deduction based on the known reaction rates of the possible product ions. The latter procedure is well demonstrated by the data shown in Fig. 8 for the $N^+ + H_2CO$ reaction.

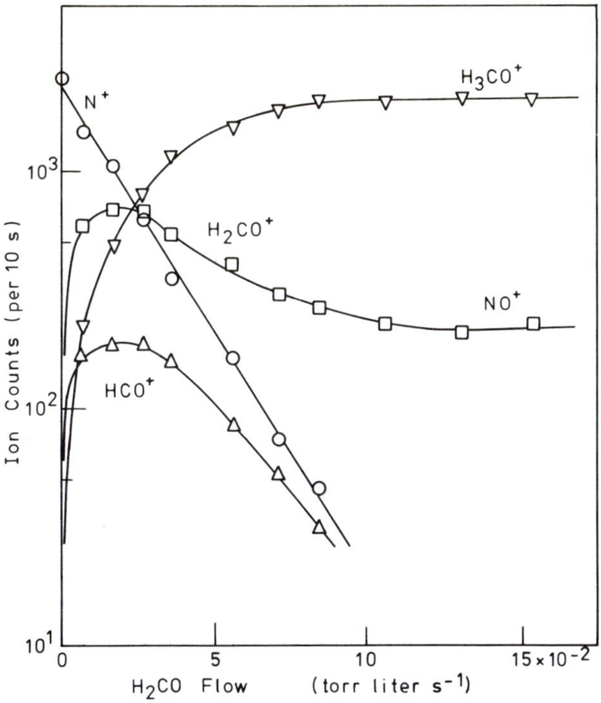

FIG. 8 Data obtained for the reaction of N^+ with H_2CO at 300 K. The rate coefficient is 2.9×10^{-9} cm^3 s^{-1}. Primary product ions result at masses 29 and 30 amu, the former being HCO^+. The ions at mass 30 amu consist of H_2CO^+, which in common with HCO^+ reacts with H_2CO to produce H_3CO^+, and of NO^+ which can be seen to be unreactive with H_2CO. The product ion distribution is H_2CO^+(65%), HCO^+(25%), NO^+(10%).

The product ion at mass 30 amu could either be H_2CO^+ or NO^+. In fact, in this case, both ions are products of the reaction, the rapidly reacting component being H_2CO^+ [known from the H_nCO^+ studies described in (2)] and the relatively unreactive component being NO^+ (known from the detailed studies of NO^+ reactions referred to in what follows). Where data are not available for this kind of analysis, they can rapidly be acquired with the SIFT.

In the binary reactions of the ions of large recombination energy (e.g., N^+, N_2^+, N_4^+) with low ionization potential molecules (e.g., CH_3NH_2, NH_3, H_2S), multiple products generally result and the rate coefficients are close to the gas kinetic limit. Conversely, the low recombination energy ions NO^+ and O_2^+ generally undergo ternary association with the large ionization potential molecules (e.g., H_2, N_2, CO) since binary reactions are effectively prohibited due to the energetics. The ternary association reactions of NO^+ with N_2, O_2, and CO_2, e.g.,

$$NO^+ + CO_2 + N_2 \rightleftharpoons NO^+{\cdot}CO_2 + N_2 \qquad \text{(XIII)}$$

have been the subject of a separate detailed study (Smith *et al.*, 1977) because of the importance of these associations in D-region reaction schemes (e.g., Thomas, 1974). Ternary association rate coefficients have been determined at both 300 and 225 K in order to obtain information concerning their temperature dependence. The study of such reactions is difficult since the NO^+—Y bond (Y = N_2, O_2, or CO_2) is very weak, permitting both collisional breakup in the mass spectrometer sampling system and significant reverse reaction in the flow tube to occur. These difficulties were overcome by the addition to the carrier gas of small concentrations of water vapor as a "scavenger" to rapidly convert the $NO^+{\cdot}Y$ to the strongly bonded $NO^+{\cdot}H_2O$, viz,

$$NO^+{\cdot}Y + H_2O \rightarrow NO^+{\cdot}H_2O + Y \qquad \text{(XIV)}$$

thus ensuring that the decay of NO^+ due to the primary reaction is effectively observed. This work has revealed that reaction (XIII) may be important in the synthesis of water-cluster ions in the D-region (Smith *et al.*, 1977).

(5) Reactions of kinetically relaxed, metastable electronically excited ions have been measured in the Aberystwyth SIFT by using an electron impact ion source operating at suitable electron energies. Both ground state and metastable excited state ions are injected into the flow tube and can only be distinguished at the downstream mass filter by their different reactivities with the reactant gases. When reaction with the ground state is endothermic but reaction occurs for the excited state, the detected primary ion count rate reduces to a constant value (equivalent to the ground state population in the flow tube) with the addition of the reactant gas and it is a straight-

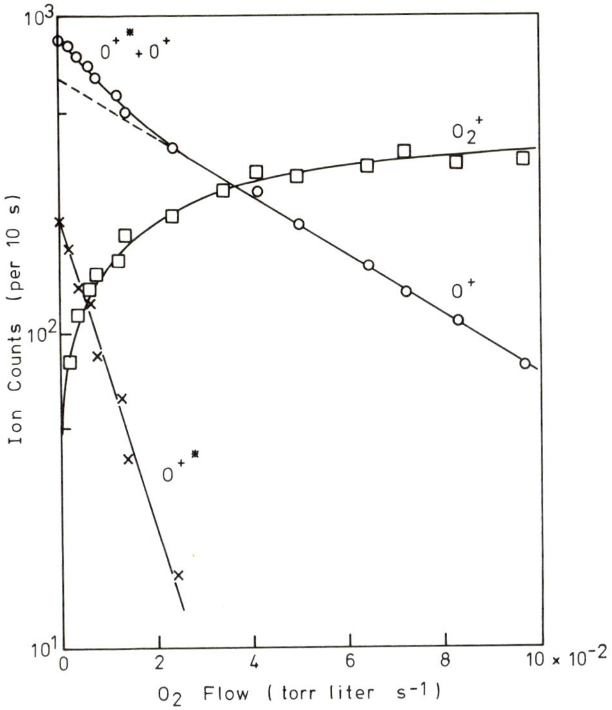

FIG. 9 Data obtained for the reaction of ground state $O^+(^4S)$ ions and excited state $O^+(^2D, ^2P)$ ions with O_2 at 300 K. The rate coefficient for the $O^+(^4S)$ reaction ($k = 2.3 \times 10^{-11}$ cm^3 s^{-1}) is obtained from the linear portion of the decay curve at large O_2 flows; that for the O^{+*} reaction ($k = 1.3 \times 10^{-10}$ cm^3 s^{-1}) is obtained from the data points at small O_2 flows by the extrapolation and subtraction procedure indicated [i.e., $(O^{+*} + O^+) - O^+$ to give O^{+*}].

forward process to deduce the excited ion rate coefficient by subtracting this "background" count rate. Where the ground state ion also reacts, extrapolation of the linear portion of the decay curve obtained at large flow rates is necessary to extract the separate rate coefficients (Fig. 9). When collisional quenching of the excited ion by the reactant gas as well as chemical reaction occurs, as in the case of $NO^+(a^3\Sigma)$ with N_2, viz,

$$\begin{aligned} NO^+(a^3\Sigma) + N_2 &\rightarrow N_2^+ + NO\ (55\%) \quad &\text{chemical reaction} \\ &\rightarrow NO^+(X^1\Sigma^+) + N_2\ (45\%) \quad &\text{quenching,} \end{aligned} \qquad \text{(XV)}$$

the separate contributions can be obtained by the monitor ion method (Bolden *et al.*, 1970b; Glosik *et al.*, 1978). To date, the reactions of $O^+(^2D, ^2P)$, $NO^+(a^3\Sigma)$, and $O_2^+(a^4\Pi_u)$ (30–50% of the injected ions in the excited state) with several molecules have been studied including variously H_2, N_2, O_2,

NO, CO, CO_2, and Ar and rate coefficients and product ion distributions have been obtained (Glosik *et al.*, 1978). The individual rate coefficients have been obtained to an accuracy of $\pm 30\%$, which indicates the quality of the technique. The rate coefficients for the $O_2^+(a^4\Pi_u)$ reactions are in acceptable agreement with those previously obtained with the NOAA flow–drift tube in which the excited ions are produced by the reaction of $He(2^3S)$ metastables with O_2 (Lindinger *et al.*, 1975b). The rate coefficients for a few reactions of $NO^+(a^3\Sigma)$ have also been obtained using the flow–drift tube (Dotan *et al.*, 1977) and again acceptable agreement is obtained, although the separate contributions, due to chemical reaction and quenching which are both often observed for this excited ion, have not been obtained using the flow–drift technique.

For the O^+ ion studies, it was necessary to limit the energy of the ionizing electrons to about 50 eV when O_2 was used as the source gas, since otherwise O_2^{2+} ions will pass through the mass filters with the O^+ ions. The presence of an adequate fraction of O_2^{2+} ($m/e = 16$) in the beam, at an electron energy in the ion source of about 100 eV, has been exploited to determine the rate coefficients for the single charge transfer reactions of O_2^{2+} with both N_2 and CO (Glosik *et al.*, 1978).

(6) The relatively high recombination energy of doubly charged ions often means that reaction with the helium carrier gas is energetically possible. Such is the case for $Ar^{2+}(^3P)$ states, which react relatively rapidly to give Ar^+ and He^+, while the $Ar^{2+}(^1S_0)$ which reacts only slowly and survives the passage down the flow tube. Conversely, in argon carrier gas the $^1S_0$ state reacts relatively rapidly to give $Ar^+ + Ar^+$, whereas the ^3P states survive the passage down the flow tube. Thus, it has been possible to determine the rate coefficients for both $Ar^{2+}(^3P)$ and $Ar^{2+}(^1S_0)$ ions with H_2, N_2, O_2, CO_2, He, Ar, and Xe. The reaction rates for both ionic states with the molecules are almost always gas kinetic, are in general agreement with the data of Lindinger *et al.* (1977) and often proceed by double charge transfer, e.g.,:

$$\begin{array}{l} Ar^{2+} + O_2 \rightarrow (O_2^{2+})^* + Ar \\ \qquad\qquad\quad \llcorner\!\!\longrightarrow O^+ + O^+. \end{array} \qquad \text{(XVI)}$$

The reaction with Xe was surprisingly fast—near Langevin—and indicates an energy resonance in the reaction. These data have been discussed in detail by Smith *et al.* (1979).

The studies of Xe^{2+} were not complicated by reaction with the helium carrier gas, no measurable signal of He^+ being observed by the downstream sampling system. This is to be expected for the ground state $Xe^{2+}(^3P_{2,1,0})$ and the $^1D_2$ metastable state (endoergic to produce He^+ with helium), but the $^1S_0$ metastable state reaction with helium is energetically possible but no significant reaction apparently occurs. Thus it was possible to study the reactions of this ion in more detail and to determine the product ion distributions. This was possible by selectively removing the different states from the Xe^{2+} ion swarm using "gas filters" and then studying the remaining states by exploiting the fact that the $^3P_{2,1,0}$ states rapidly react with argon, whereas the other two states do not and the $^3P_{2,1,0}$ and $^1D_2$ states react rapidly with nitrogen, whereas the $^1S_0$ state does not. Also the $^1D_2$ and $^1S_0$ states can effectively be eliminated from the Xe^{2+} ion beam by reducing the electron energy in the ion source to below their production thresholds, leaving only the $^3P_{2,1,0}$ ground state ions. The rate coefficients for the three different states each with H_2, N_2, O_2, CO_2, He, Ar, and Xe have been determined. Direct single change transfer (leaving the molecular bond intact) is the most common mechanism, although for H_2 and O_2 dissociative charge transfer channels are evident, e.g.,

$$Xe^{2+} + H_2 \rightarrow H^+ + H + Xe^+ \qquad \text{(XVII)}$$
$$\rightarrow H_2^+ + Xe^+.$$

No evidence for varying reactivity of the $^3P_0$, $^3P_1$, and $^3P_2$ states has been obtained, although this has not as yet been looked at in any detail. Actually, the rate coefficients measured for each of the excited states are the total rate coefficients, i.e., the sum of chemical reaction plus any collisional quenching [see (5) above], although we have found no evidence for the latter effect in these studies. These data have been discussed in detail by Adams *et al.* (1979).

V. FUTURE DEVELOPMENTS

Since the inception and development of the temperature variable flowing afterglow to study ion–neutral reactions the advances in flow tube techniques have followed a logical course towards the ultimate objective of obtaining a detailed understanding of ionic reactions at thermal energies. The flow–drift technique has added the vital energy variability and the SIFT technique has extended the flow tube method to the study of a wider range of ionic species in a cleaner reactive medium. While the exploitation of flowing afterglow has provided a wealth of data which has almost solely resulted in the elucidation of the chemistry of the earth's ionosphere, the future now lies with the vigorous exploitation and development of the SIFT technique. In Section IV.C we have briefly described some of the studies we have made to date mostly with the prototype apparatus at Birmingham as well as the somewhat improved version at Aberystwyth. These data simply represent a beginning. Many of the programmes which have been initiated need to be pursued much further, not least those involving excited ions and doubly charged ions for which there exists a sparcity of information. These and much more can be pursued with a basic SIFT apparatus, in particular the technique is eminently suited to the use of isotopic species to determine structures of product ions when rearrangement in reactive collisions is suspected. For example, the injection of selectively deuterated hydrocarbon ions and their subsequent reaction with hydrogen-containing neutral molecules will provide information concerning any "scrambling" in such reactions. The SIFT has not yet been used for the quantitative study of reactions of ions with atoms or radicals, a subject of great interest in interstellar chemistry, although Fehsenfeld 1976 has accomplished some such measurements in the flowing afterglow. The authors have made a start in negative-ion–neutral studies (Section IV.A.1), but the pursuit of this subject depends on the development of suitable negative-ion sources.

How can the basic SIFT technique be improved and developed? Obvious developments are to include temperature and energy variability over as wide a range as is practically possible, thus combining all the best features of the existing flowing afterglows, the flow drift, and SIFT apparatuses. To this end, the authors have designed and constructed the temperature variable SIFT–DRIFT apparatus illustrated diagrammatically in Fig. 10. This new apparatus, which is shortly to be commissioned, will have the following features.

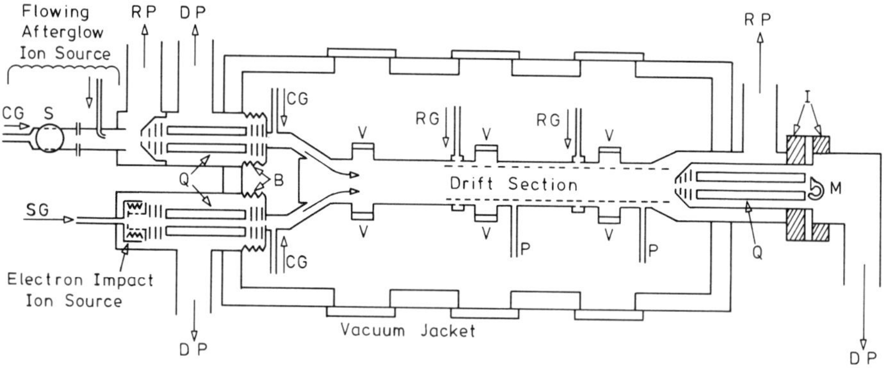

FIG. 10 Schematic diagram of the SIFT–DRIFT apparatus. The major improvements on the basic SIFT illustrated in Fig. 5, include (1) two ion injection limbs, (2) a surrounding vacuum jacket to facilitate temperature variation of the flow tube, (3) a drift section, (4) flowing afterglow ion sources which can be used on both injection limbs simultaneously and (5) optical viewports labeled *V*. *I*—electrical insulators, *B*—stainless steel bellows; all other symbols are as in Fig. 5.

(1) Two ion injection limbs which will enable two different ionic species, either positively or negatively charged, to be introduced into the flow tube simultaneously. This feature clearly has many advantages, but a particularly appealing example will suffice here. In the study of hydrogen atom reactions

it is, of course, necessary to determine the relative concentration of H atom to H_2 molecules in the partially dissociated hydrogen used as the reactant gas for such studies. Thus we intend to exploit the fact that Cl^- ions undergo an associative detachment reaction with H atoms (Fehsenfeld, 1975b; see also Section II.D) whereas they are unreactive with H_2, and so one of the two limbs will be used to inject Cl^- ions as an "hydrogen atom probe," while the other limb will be used to inject the ions whose reaction we wish to study.

(2) The flow tube is surrounded by a vacuum jacket and can be heated to about 600 K or cooled to 80 K, thus allowing temperature dependences of ion–neutral reactions to be studied. Amongst such studies, a great deal of data is required relating to ternary association reactions in order to cast some light on the role of internal degrees of freedom in the intermediate complexes of such reactions (Smith and Adams, 1978a) and to provide a greater understanding of molecular synthesis in the Earth's lower atmosphere (Ferguson, 1975b) and in interstellar gas clouds (Smith and Adams, 1978b). A temperature variable SIFT is admirably suited to such studies.

(3) The inclusion of a drift section and thus the creation of a SIFT–DRIFT system will extend the flow–drift tube technique to a large number of ionic species, and the combination of the ion energy and gas temperature variation offers the exciting prospect of exploring the respective roles of kinetic and internal excitation in determining ionic reaction rates and product distributions (Albritton *et al.*, 1977). We will be particularly interested in studying the several near thermoneutral reactions we have briefly investigated in our previous work (see Section IV.C for references).

(4) Associated with each injection limb is a glass flow tube (approximate dimensions 5 cm internal diameter by 50 cm long) in which flowing afterglow plasmas can be created. A second Roots pump is included in the apparatus for this purpose. These flowing afterglows are to be used as ion sources and will provide ions which are not supplied by low pressure electron collision sources, such as cluster ions, and protonated molecules. It is also expected that they will provide an adequate source of negative ions, particularly from negative-ion dominated plasmas (Smith and Church, 1976). Problems might well arise when attempts are made to inject weakly bound cluster ions and negative ions of low electron affinity. The injection energy will necessarily have to be low to prevent collisional break up and electron detachment, the net result being to severely reduce the injected current. This will need to be compensated for by improvements in ion source intensities and in the efficiency of ion injection through the final aperature into the flow tube. Improvements have been made in the Aberystwyth SIFT (Section IV.A.1) and doubtless many more improvements will be made as the technique is adopted in other laboratories. The prospect of being able to inject adequate currents of cluster ions into a temperature variable flow tube with, for

example, successive numbers of water molecules is particularly exciting in that some of the remaining problems associated with cluster-ion formation in the ionosphere (Thomas, 1976) could be solved, as well as the provision of a good deal of data relating to bond association energies.

(5) With the inclusion of strategical positioned optical viewports, that is near to gas inlet ports, it is hoped to observe the light output from reactions. The success of this part of the experiment will clearly depend on such parameters as oscillator strengths and Franck–Condon factors and on the number density of the emitting atoms or molecules. The latter will depend directly on the injected ion current and our estimates show that with the present SIFT injection currents and efficient optics, such observations are within reach for some reactions. This is an area in which rich scientific returns will be forthcoming in the near future.

These combined features of the SIFT–DRIFT provide the opportunity for a concerted investigation of ion–neutral reactions at thermal energies and will provide a great deal of data on rate coefficients of binary and ternary reactions, product distributions, including information on states of excitation, relative proton affinities, critical thermochemical data, etc. These developments in flow tube experiments also offer the best opportunity to maintain the progress made in recent years in the understanding of the chemistry of natural plasmas such as the ionosphere and interstellar gas clouds. Doubtless further developments and new innovations in flow tubes will be made in the future, but even with the presently available techniques an increased output of data on low energy ionic interactions is assured during the next few years.

ACKNOWLEDGMENT

We are grateful to Dr. D. L. Albritton for his constructive comments on the flow–drift tube section.

REFERENCES

Adams, N. G., and Smith, D. (1976a). *Int. J. Mass Spectrom. Ion Phys.* **21**, 349–359.
Adams, N. G., and Smith, D. (1976b). *J. Phys. B* **9**, 1439–1451.
Adams, N. G., and Smith, D. (1977). *Chem. Phys. Lett.* **47**, 383–387.
Adams, N. G., and Smith, D. (1978). *Chem. Phys. Lett.* **54**, 530–534 (1978)
Adams, N. G., Bohme, D. K., Dunkin, D. B., Fehsenfeld, F. C., and Ferguson, E. E. (1970). *J. Chem. Phys.* **52**, 3133–3140.
Adams, N. G., Dean, A. G., and Smith, D. (1972). *Int. J. Mass Spectrom. Ion Phys.* **10**, 63–76.
Adams, N. G., Church, M. J., and Smith, D. (1975). *J. Phys. D* **8**, 1409–1422.
Adams, N. G., Smith, D., and Grief, D. (1978). *Int. J. Mass Spectrom. Ion Phys.* **26**, 405–415.
Adams, N. G., Smith, D., and Grief, D. (1979) *J. Phys. B* (in press).
Albritton, D. L. (1978). *At. Data Nucl. Data Tables* **22**, 1–101.
Albritton, D. L., Dotan, I., Lindinger, W., McFarland, M., Tellinghuisen, J., and Fehsenfeld, F. C. (1977). *J. Chem. Phys.* **66**, 410–421.

Anicich, V. G., Futrell, J. H., Huntress, W. T., Jr., and Kim, J. K. (1975). *Int. J. Mass Spectrom. Ion Phys.* **18**, 63–64.

Arnold, F., Krankowsky, D., and Marien, K. J. (1977). *Nature* (*London*) **267**, 30–32.

Batey, P. H., Court, G. R., and Sayers, J. (1965). *Planet. Space Sci.* **13**, 911–917.

Bierbaum, V. M., DePuy, C. H., Shapiro, R. H., and Stewart, J. H. (1976). *J. Am. Chem. Soc.* **98**, 4229–4235.

Bierbaum, V. M., Ellison, G. B., Futrell, J. H., and Leone, S. R. (1977). *J. Chem. Phys.* **67**, 2375–2376.

Biondi, M. A. (1977). *Comments At. Mol. Phys.* **6**, 159–161.

Bohme, D. K. (1975). *In* "Interactions Between Ions and Molecules" (P. Ausloos, ed.), pp. 489–504. Plenum, New York.

Bohme, D. K., Dunkin, D. B., Fehsenfeld, F. C., and Ferguson, E. E. (1968). *J. Chem. Phys.* **49**, 5201–5202.

Bohme, D. K., Hemsworth, R. S., Rundle, H. W., and Schiff, H. I. (1973). *J. Chem. Phys.* **58**, 3504–3518.

Bolden, R. C., and Twiddy, N. D. (1972). *Discuss. Faraday Soc.* **53**, 192–200.

Bolden, R. C., Hemsworth, R. S., Shaw, M. J., and Twiddy, N. D. (1970a). *J. Phys. B* **3**, 45–60.

Bolden, R. C., Hemsworth, R. S., Shaw, M. J., and Twiddy, N. D. (1970b). *J. Phys. B* **3**, 61–71.

Brink, G. O. (1966). *Rev. Sci. Instrum.* **37**, 857–860.

Brubaker, W. M. (1968). *Adv. Mass Spectrom.* **4**, 293–299.

Burt, J. A., Dunn, J. L., McEwan, M. J., Sutton, M. M., Roche, A. E., and Schiff, H. I. (1970). *J. Chem. Phys.* **52**, 6062–6075.

Buttrill, S. E., Jr., Kim, J. K., Huntress, W. T., Jr., Le Breton, P., and Williamson, A. (1974). *J. Chem. Phys.* **61**, 2122–2128.

Copsey, M. J., Smith, D., and Sayers, J. (1966). *Planet. Space Sci.* **14**, 1047–1055.

Coxon, P., and Moruzzi, J. L. (1977). *J. Phys. D* **10**, 969–977.

Dalgarno, A., and Black, J. H. (1976). *Rep. Prog. Phys.* **39**, 573–612.

Dawson, P. H., and Whetten, N. R. (1969). *Adv. Electron. Electron Phys.* **27**, 60–185.

Dickinson, P. H. G., and Sayers, J. (1960). *Proc. Phys. Soc., London* **76**, 137–148.

Dotan, I., Albritton, D. L., and Fehsenfeld, F. C. (1976a). *J. Chem. Phys.* **64**, 4334–4336.

Dotan, I., Lindinger, W., and Albritton, D. L. (1976b). *J. Chem. Phys.* **64**, 4544–4547.

Dotan, I., Albritton, D. L., and Fehsenfeld, F. C. (1977). *Gaseous Electron. Conf., 30th, Palo Alto, Calif.*

Dunkin, D. B., Fehsenfeld, F. C., Schmeltekopf, A. L., and Ferguson, E. E. (1968). *J. Chem. Phys.* **49**, 1365–1371.

Farragher, A. L. (1970). *Trans. Faraday Soc.* **66**, 1411–1422.

Farragher, A. L., Peden, J. A., and Fite, W. L. (1969). *J. Chem. Phys.* **50**, 287–293.

Fehsenfeld, F. C. (1975a). *Int. J. Mass Spectrom. Ion Phys.* **16**, 151–166.

Fehsenfeld, F. C. (1975b). *In* "Interactions Between Ions and Molecules" (P. Ausloos, ed.), pp. 387–412. Plenum, New York.

Fehsenfeld, F. C. (1976). *Astrophys. J.* **209**, 638–639.

Fehsenfeld, F. C., and Ferguson, E. E. (1968). *Planet. Space Sci.* **16**, 701–702.

Fehsenfeld, F. C., Ferguson, E. E., and Schmeltekopf, A. L. (1966a). *J. Chem. Phys.* **44**, 3022–3024.

Fehsenfeld, F. C., Ferguson, E. E., and Schmeltekopf, A. L. (1966b). *J. Chem. Phys.* **45**, 1844–1845.

Fehsenfeld, F. C., Albritton, D. L., Burt, J. A., and Schiff, H. I. (1969a). *Can. J. Chem.* **47**, 1793–1795.

Fehsenfeld, F. C., Schmeltekopf, A. L., Dunkin, D. B., and Ferguson, E. E. (1969b). ESSA Tech. Rep. E.R.L. 135-AL3. U.S. Dep. Commer. Res. Lab., Boulder, Colorado.

Fehsenfeld, F. C., Dunkin, D. B., and Ferguson, E. E. (1974). *Astrophys. J.* **188**, 43–44.

Fehsenfeld, F. C., Howard, C. J., Harrop, W. J., and Ferguson, E. E. (1975a). *J. Geophys. Res.* **80**, 2229–2235.

Fehsenfeld, F. C., Lindinger, W., and Albritton, D. L. (1975b). *J. Chem. Phys.* **63**, 443–445.

Fehsenfeld, F. C., Lindinger, W., Schiff, H. I., Hemsworth, R. S., and Bohme, D. K. (1976). *J. Chem. Phys.* **64**, 4887–4891.

Fehsenfeld, F. C., Dotan, I., Albritton, D. L., Howard, C. J., and Ferguson, E. E. (1978). *J. Geophys. Res.* **83**, 1333–1336.

Ferguson, E. E. (1971). *In* "Mesospheric Models and Related Experiments" (G. Fiocco, ed.), pp. 188–197. Reidel Publ., Dordrecht, Netherlands.

Ferguson, E. E. (1972). *In* "Ion–Molecule Reactions" (J. L. Franklin, ed.), Vol. 2, pp. 363–393. Plenum, New York.

Ferguson, E. E. (1973). *At. Data Nucl. Data Tables* **12**, 159–178.

Ferguson, E. E. (1975a). *In* "Atmospheres of Earth and the Planets" (B. M. McCormac, ed.), pp. 197–210. Reidel Publ., Dordrecht, Netherlands.

Ferguson, E. E. (1975b). *In* "Interactions Between Ions and Molecules" (P. Ausloos, ed.), pp. 313–339. Plenum, New York.

Ferguson, E. E., Fehsenfeld, F. C., Goldan, P. D., Schmeltekopf, A. L., and Schiff, H. I. (1965). *Planet. Space Sci.* **13**, 823–827.

Ferguson, E. E., Fehsenfeld, F. C., and Schmeltekopf, A. L. (1969). *Adv. At. Mol. Phys.* **5**, 1–56.

Glosik, J., Rakshit, A. B., Twiddy, N. D., Adams, N. G., and Smith, D. (1978). *J. Phys. B* **11**, 3365–3379.

Good, A. (1975). *Chem. Rev.* **75**, 561–583.

Govers, T. R., Fehsenfeld, F. C., Albritton, D. L., Fournier, P. G., and Fournier, J. (1974). *Chem. Phys. Lett.* **26**, 134–137.

Heimerl, J., Johnsen, R., and Biondi, M. A. (1969). *J. Chem. Phys.* **51**, 5041–5048.

Henchman, M. (1972). *In* "Ion–Molecule Reactions" (J. L. Franklin, ed.), Vol. 1, pp. 101–259. Plenum, New York.

Howard, C. J., Rundle, H. W., and Kaufman, F. (1970). *J. Chem. Phys.* **53**, 3745–3751.

Howard, C. J., Rundle, H. W., and Kaufman, F. (1971). *J. Chem. Phys.* **55**, 4772–4776.

Howard, C. J., Fehsenfeld, F. C., and McFarland, M. (1974). *J. Chem. Phys.* **60**, 5086–5089.

Huntress, W. T., Jr. (1977a). *Chem. Soc. Rev.* **6**, 295–323

Huntress, W. T., Jr. (1977b). *Astrophys. J. Suppl. Ser.* **33**, 495–514.

Huxley, L. G. H., and Crompton, R. W. (1974). "The Drift and Diffusion of Electrons in Gases." Wiley, New York.

Johnsen, R., and Biondi, M. A. (1973). *J. Chem. Phys.* **59**, 3504–3509.

Johnsen, R., Brown, H. L., and Biondi, M. A. (1970). *J. Chem. Phys.* **52**, 5080–5084.

Kebarle, P. (1975). *In* "Interactions Between Ions and Molecules" (P. Ausloos, ed.), pp. 459–487. Plenum, New York.

Kim, J. K., and Huntress, W. T., Jr. (1975). *Int. J. Mass Spectrom. Ion Phys.* **16**, 451–454.

Kosmider, R. G., and Hasted, J. B. (1975). *J. Phys. B* **8**, 273–289.

Liddy, J. P., Freeman, C. G., and McEwan, M. J. (1977). *Int. J. Mass Spectrom. Ion Phys.* **23**, 153–161.

Lindholm, E. (1972). *In* "Ion–Molecule Reactions" (J. L. Franklin, ed.), Vol. 2, pp. 457–484. Plenum, New York.

Lindinger, W., and Albritton, D. L. (1975). *J. Chem. Phys.* **62**, 3517–3522.

Lindinger, W., Fehsenfeld, F. C., Schmeltekopf, A. L., and Ferguson, E. E. (1974). *J. Geophys. Res.* **79**, 4753–4756.

Lindinger, W., Albritton, D. L., McFarland, M., Fehsenfeld, F. C., Schmeltekopf, A. L., and Ferguson, E. E. (1975a). *J. Chem. Phys.* **62**, 4101–4110.

Lindinger, W., Albritton, D. L., Fehsenfeld, F. C., and Ferguson, E. E. (1975b). *J. Geophys. Res.* **80**, 3725–3726.

Lindinger, W., Alge, E., Stori, H., Paul, M., and Varney, R. N. (1977). *J. Chem. Phys.* **67**, 3495–3499.
Lineberger, W. C., and Puckett, L. J. (1969a). *Phys. Rev.* **186**, 116–127.
Lineberger, W. C., and Puckett, L. J. (1969b). *Phys. Rev.* **187**, 286–291.
McDaniel, E. W. (1964). "Collision Phenomena in Ionised Gases." Wiley, New York.
McDaniel, E. W., Cermak, V., Dalgarno, A., Ferguson, E. E., and Friedman, L. (1970). "Ion–Molecule Reactions." Wiley, New York.
McFarland, M., Albritton, D. L., Fehsenfeld, F. C., Ferguson, E. E., and Schmeltekopf, A. L. (1973a). *J. Chem. Phys.* **59**, 6610–6619.
McFarland, M., Albritton, D. L., Fehsenfeld, F. C., Ferguson, E. E., and Schmeltekopf, A. L. (1973b). *J. Chem. Phys.* **59**, 6620–6628.
McFarland, M., Albritton, D. L., Fehsenfeld, F. C., Ferguson, E. E., and Schmeltekopf, A. L. (1973c). *J. Chem. Phys.* **59**, 6629–6635.
McFarland, M., Albritton, D. L., Fehsenfeld, F. C., Ferguson, E. E., and Schmeltekopf, A. L. (1974). *J. Geophys. Res.* **79**, 2925–2926.
Marx, R., Mauclaire, G., and Fenistein, S. (1975). *Chem. Phys. Lett.* **33**, 357–361.
Meisels, G. G. (1975). *In* "Interactions Between Ions and Molecules" (P. Ausloos, ed.), pp. 604–608. Plenum, New York.
Miller, T. M., Moseley, J. T., Martin, D. W., and McDaniel, E. W. (1968). *Phys. Rev.* **173**, 115–123.
Moruzzi, J. L., Ekin, J. W., Jr., and Phelps, A. V. (1968). *J. Chem. Phys.* **48**, 3070–3076.
Neynaber, R. H., Rutherford, J. A., and Vroom, D. A. (1971). Gulf Radiat. Technol. Rep. No. A 12209. San Diego, Calif.
Phelps, A. V. (1955). *Phys. Rev.* **99**, 1307–1313.
Puckett, L. J., and Lineberger, W. C. (1970). *Phys. Rev. A* **1**, 1635–1641.
Rayment, S. W., and Moruzzi, J. L. (1978). *Int. J. Mass Spectrom. Ion Phys.* **26**, 321–326.
Rutherford, J. A., and Vroom, D. A. (1971). *J. Chem. Phys.* **55**, 5622–5624.
Sayers, J., and Smith, D. (1964). *Proc. Int. Conf. Phys. Electron. At. Collisions, 3rd* pp. 871–876.
Schmeltekopf, A. L., and Fehsenfeld, F. C. (1970). *J. Chem. Phys.* **53**, 3173–3177.
Schmeltekopf, A. L., Fehsenfeld, F. C., Gilman, G. I., and Ferguson, E. E. (1967). *Planet. Space Sci.* **15**, 401–406.
Schmeltekopf, A. L., Ferguson, E. E., and Fehsenfeld, F. C. (1968). *J. Chem. Phys.* **48**, 2966–2973.
Skullerud, H. R. (1973). *J. Phys. B* **6**, 728–742.
Smith, D., and Adams, N. G. (1977a). *Int. J. Mass Spectrom. Ion Phys.* **23**, 123–135.
Smith, D., and Adams, N. G. (1977b). *Chem. Phys. Lett.* **47**, 145–149.
Smith, D., and Adams, N. G. (1977c). *Astrophys. J.* **217**, 741–748.
Smith, D., and Adams, N. G. (1978a). *Chem. Phys. Lett.* **54**, 535–540.
Smith, D., and Adams, N. G. (1978b). *Astrophys. J.* **220**, L87–L92.
Smith, D., and Church, M. J. (1976). *Int. J. Mass Spectrom. Ion Phys.* **19**, 185–200.
Smith, D., and Dean, A. G. (1975). *J. Phys. B* **8**, 997–1008.
Smith, D., and Fouracre, R. A. (1968). *Planet. Space Sci.* **16**, 243–252.
Smith, D., and Plumb, I. C. (1973). *J. Phys. D* **6**, 1431–1446.
Smith, D., Dean, A. G., and Plumb, I. C. (1972). *J. Phys. B* **5**, 2134–2142.
Smith, D., Dean, A. G., and Adams, N. G. (1974). *J. Phys. D* **7**, 1944–1962.
Smith, D., Adams, N. G., Dean, A. G., and Church, M. J. (1975). *J. Phys. D* **8**, 141–152.
Smith, D., Adams, N. G., and Church, M. J. (1976). *Planet Space Sci.* **24**, 697–703.
Smith, D., Adams, N. G., and Grief, D. (1977). *J. Atmos. Terr. Phys.* **39**, 513–521.
Smith, D., Adams, N. G., and Miller, T. M. (1978). *J. Chem. Phys.* **69**, 308–318.
Smith, D., Grief, D., and Adams, N. G. (1979). Int. J. Mass Spectrom. Ion Phys. (in press).
Spears, K. G., and Ferguson, E. E. (1973). *J. Chem. Phys.* **59**, 4174–4183.

Su, T., and Bowers, M. T. (1973). *Int. J. Mass Spectrom. Ion Phys.* **12**, 347–356.
Thomas, L. (1974). *Radio Sci.* **9**, 121–136.
Thomas, L. (1976). *J. Atmos. Terr. Phys.* **38**, 61–67.
Twiddy, N. D. (1974). *Contemp. Phys.* **15**, 427–444.
Viehland, L. A., Mason, E. A., and Whealton, J. H. (1974). *J. Phys. B* **7**, 2433–2439.
Wannier, G. H. (1953). *Bell Syst. Tech. J.* **32**, 170–254.

Chapter 2
Ion chemistry of the earth's atmosphere

E. E. FERGUSON, F. C. FEHSENFELD, AND D. L. ALBRITTON

AERONOMY LABORATORY
ENVIRONMENTAL RESEARCH LABORATORIES
NATIONAL OCEANIC AND ATMOSPHERIC ADMINSTRATION
BOULDER, COLORADO

I. INTRODUCTION

Interest in the behavior of the electrons and ions in the earth's atmosphere has provided a major impetus to atomic physics for many decades. Concern with the details of ionization, attachment, recombination, radiation, and

ISBN-0-1208 01-X

other properties has provided the stimulus for numerous theoretical and experimental studies of charged particle production and behavior. Significant deductions about atmospheric ion chemistry were made prior to either laboratory measurements or *in situ* ion composition measurements and were based only on the behavior of the ionospheric electron density observed by ground-based radio sounders. An interesting account of the early development of the understanding of ionospheric behavior has been provided by Bates, one of the pioneering leaders in the field (Bates, 1974). However, the subject of atmospheric ion chemistry, as we now know it, was initiated by the *in situ* ion composition measurements of Johnson *et al.* (1958), who developed a rocket-borne mass-spectrometer sampling technique. The acquisition of atmospheric ion composition data provided a keen stimulus to obtain the laboratory reaction rate constant data necessary to understand the observed ion composition. This required the development of new technology, an ongoing effort continuing to this day.

The atmosphere is divided into altitude regions in different ways according to different disciplines. The region of the atmosphere where free electrons occur has been the domain of radio physicists who called this region the ionosphere, with a subdivision of the ionosphere into layers. The ionosphere above about 120 km is called the F-region, between about 80 and 120 km the E-region, and between about 60 and 80 km the D-region. Below the D-region, the electrons are almost entirely attached to form negative ions and this lower atmosphere does not attract the attention of ionospheric physicists. Much of our concern in this chapter will be with the traditional ionosphere; however, we also will discuss some aspects of the largely unexplored ion chemistry of the atmosphere below the ionosphere. Thus, we will need to bridge a nomenclature gap. The lower atmosphere has been primarily the domain of meteorologists, who label atmospheric regions according to the way in which the temperature varies with altitude. The lowest atmospheric region, in which the temperature decreases with increasing altitude, is called the troposphere. At about 10–15 km, depending on latitude and season, the temperature decrease ceases. This change defines the upper boundary of the troposphere and the beginning of the stratosphere. The temperature in the stratosphere increases with increasing altitude up to about 60 km, roughly the base of the D-region. We will therefore be discussing the ion chemistry of the troposphere, stratosphere, and the D-, E-, and F-regions of the ionosphere.

From a chemical kinetics viewpoint, the atmospheric ion chemistry is rather well divided into simply low-pressure (high-altitude) and high-pressure (low-altitude) regimes, the boundary being roughly 80 km. The low-pressure, high-altitude ion chemistry of the F- and E-regions is dominated by a relatively small number of bimolecular positive ion reactions occurring at temperatures between about 300 and 1500 K. As we shall show, this chemistry is just becoming very well understood at this time. The high-pressure, low-altitude ion chemistry of the D-region, stratosphere, and troposphere involves both negative and positive ion reactions, as well as termolecular and bimolecular reactions. A large number of relevant laboratory reaction studies have been carried out. However, only very limited atmospheric composition measurements have been made and these are rather qualitative and sometimes contradictory, so that we do not yet have a detailed understanding of lower atmospheric ion chemistry.

We shall discuss first the high altitude (ionospheric E- and F-region) chemistry of the major constituents and then turn to the more complex lower atmospheric ion chemistry. There is some metal ion chemistry occurring in the atmosphere, primarily in the E-region around 100 km, initiated by the ablation of metallic meteors in this region. We shall not discuss this topic. There have been both *in situ* observations of the metal ions and laboratory measurements of the related rate constants. However, the metallic ion distributions are determined to a large extent by dynamic, rather than chemical, considerations and they do not provide a well-posed chemical problem. This area of atmospheric ion chemistry has received little attention and there has been little advance in knowledge since an earlier review (Ferguson, 1972).

II. E- AND F-REGION ION CHEMISTRY

A. O^+ *and* N_2^+ *Reactions*

The most important reactions in the upper ionosphere are those that convert O^+ to molecular ions:

$$O^+ + N_2 \rightarrow NO^+ + N \tag{1}$$

and

$$O^+ + O_2 \rightarrow O_2^+ + O. \tag{2}$$

The O^+ ion is the major F-region ion, being produced by solar uv ionization of the dominant neutral species, atomic oxygen. Radiative recombination of electrons with O^+ is extremely slow. Dissociative recombination of electrons with NO^+ and O_2^+ is about 10^5 times faster than radiative recombination; thus, reactions (1) and (2) are the rate-limiting steps for electron loss in the F-region (where the electron density reaches its maximum, about 10^6 electrons cm^{-3}).

Bates and Nicolet (1960) pointed out that the rate constants for both reactions (1) and (2) must be very much less than the collision (or Langevin) rate constant, otherwise the ionosphere below 200 km would be rapidly depleted after sunset, contrary to observations. Norton *et al.* (1963) deduced quite good values for k_1 and k_2 from an analysis of ionospheric composition data obtained from rocket-borne mass spectrometers by Johnson *et al.* (1958). The first successful laboratory measurement of an ionospheric reaction was carried out by Dickinson and Sayers (1960) using a stationary afterglow system, in which they obtained a value $k_2 = 2.6 \times 10^{-11}$ cm^3 s^{-1}, only 30% higher than the presently accepted value. Unfortunately they obtained a rather inaccurate value for k_1 in the same apparatus and Langstroth and Hasted (1962) obtained quite different values for k_1 and k_2 in a very similar stationary afterglow experiment, so that ionospheric physicists were quite skeptical of laboratory results for some time. The development of the flowing afterglow technique for measuring the rate constants of ion–molecule reactions in 1964 rapidly led to a large number of ionospheric rate constants, as described in an early review by Ferguson *et al.* (1965) and a recent review by Albritton (1978). These measurements have had a typical accuracy of 30%, which has been quite adequate for ionospheric analyses, since uncertainties in ionospheric production rates, neutral reactant concentrations, etc., have often been much greater than this. For reactions (1) and (2), several independent laboratory measurements now agree to yield the presently accepted values of $k_1 = 1.2 \times 10^{-12}$ cm^3 s^{-1} and $k_2 = 2.0 \times 10^{-11}$ cm^3 s^{-1} at 300 K.

The F-region attains temperatures as high as 1500–2000 K during solar sunspot maximum. Yet the early laboratory rate constant measurements were all carried out at room temperature. Thus, as the quality of the ionospheric observations improved, it became necessary to consider the energy dependence of the rate constants. Donahue (1966), in a careful ionospheric analysis, deduced that k_1 and k_2 must decrease with increased temperature in order to rationalize the laboratory measurements and the ionospheric observations. Smith and Fouracre (1968) obtained the first reliable temperature measurements for an ionospheric reaction, reaction (2), which they measured between 185 and 576 K using a stationary afterglow system. Temperature measurements were also carried out in a heated flowing afterglow system up to 600 K (Dunkin *et al.*, 1968). These measurements were later extended to 900 K (Lindinger *et al.*, 1974), but because of the small rate constant, it has proven to be very difficult to get definitive data for reaction (1) as a function of temperature over the range needed for ionospheric analyses.

The latest chapter in this story is the measurement of cross sections for reactions (1) and (2) as a function of relative kinetic energy (Albritton *et al.*, 1977a). Figures 1 and 2 show the values (denoted by the circles and triangles)

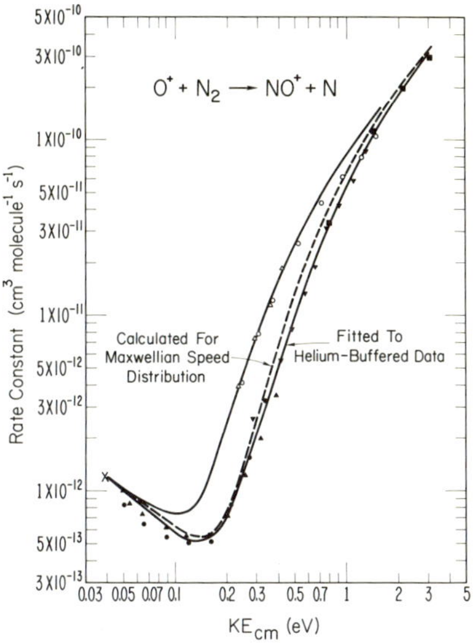

FIG. 1 Comparison of experimental differences between helium- and argon-buffered rate constants for reaction (1) to those predicted from the numerical O^+-in-He and O^+-in-Ar speed distributions of Lin and Bardsley (1977). [From Albritton *et al.* (1977a).] The lower solid curve is the fit to the helium-buffered data (the filled symbols) and the upper solid curve is the prediction for the argon-buffered values (the unfilled symbols) (× denotes flowing afterglow). The dashed curve is the prediction for Maxwellian O^+ speed distributions with $\frac{3}{2}kT = KE_{cm}$, where KE_{cm} is the mean relative kinetic energy.

obtained for the rate constants over the relative kinetic energy range from 0.04 to 3 eV. Each reaction was measured in argon and helium buffer gases separately. As Figs. 1 and 2 show, the argon-buffered and the helium-buffered rate constants do not always agree at the same relative energy. The reason for these differences is that the speed distributions of O^+ ions drifting in argon and helium buffer gases are different, even when the mean energy of the ions is the same. Since the rate constant is a folding of the cross section with the ion–reactant relative speed distribution, it will therefore be different in the two buffer gases.

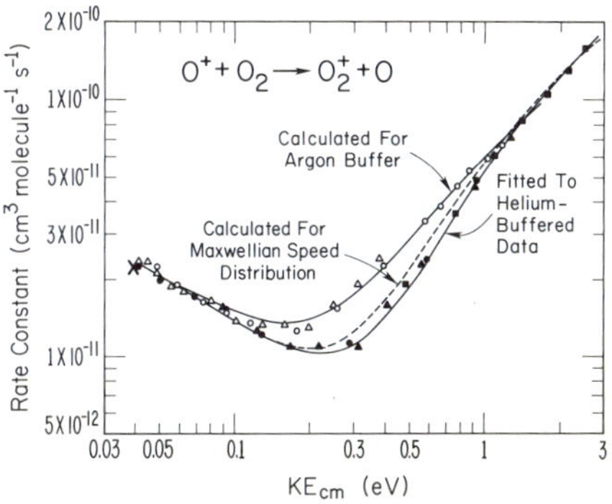

FIG. 2 Comparison of experimental differences between helium- and argon-buffered rate constants for reaction (2) to those predicted from the numerical O^+-in-He and O^+-in-Ar speed distributions of Lin and Bardsley (1977). [From Albritton *et al.* (1977a).] The lower solid curve is the fit to the helium-buffered data (the filled symbols) and the upper solid curve is the prediction for the argon-buffered values (the unfilled symbols) (× denotes flowing afterglow). The dashed curve is the prediction for Maxwellian O^+ speed distributions with $\frac{3}{2}kT = KE_{cm}$, where KE_{cm} is the mean relative kinetic energy.

The speed distribution of ions drifting under the action of a dc electric field is generally non-Maxwellian and unknown. However, the data in Figs. 1 and 2 have been combined with recent theoretical analyses to solve this long-standing problem for the situation of atomic ions drifting in rare gas buffers. Speed distributions for O^+ in helium and O^+ in argon were calculated with Monte Carlo trajectory methods (Lin and Bardsley, 1977). The O^+-in-He speed distributions were used with the helium-buffered rate constant data in Figs. 1 and 2 to unfold the reaction cross sections given in Figs. 3 and 4. The lower solid curve in Figs. 1 and 2 is a refolding of the cross section and the O^+-in-He speed distributions as a check on the inversion procedure. Experiment and theory can then be tested by combining the cross section with the theoretical O^+-in-Ar speed distribution to yield predicted rate constants

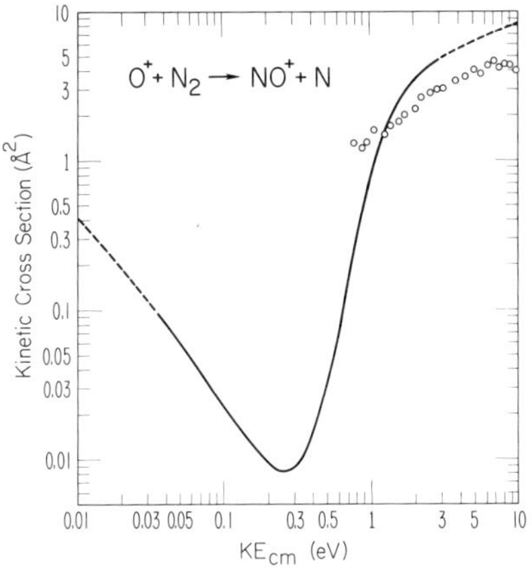

FIG. 3 Kinetic cross section deduced for reaction (1) from the helium-buffered data in Fig. (1) and the O^+-in-He speed distributions of Lin and Bardsley (1977). [From Albritton *et al.* (1977a).] The dashed portions are the more uncertain. The open circles are crossed-beam cross section measurements of Rutherford and Vroom (1971).

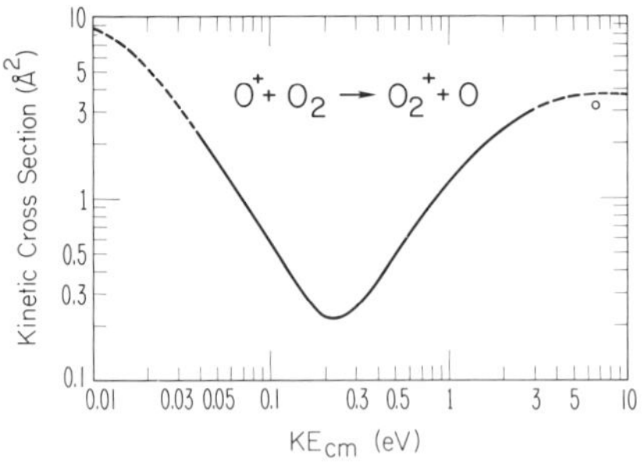

FIG. 4 Kinetic cross section deduced for reaction (2) from the helium-buffered data in Fig. (2) and the O^+-in-He speed distributions of Lin and Bardsley (1977). [From Albritton *et al.* (1977a).] The dashed portions are the more uncertain. The open circle is a crossed-beam cross section measurement of Stebbings *et al.* (1966).

appropriate for an argon buffer gas. These predictions are upper solid curves in Figs. 1 and 2 and are in excellent agreement with the argon-buffered data. The cross sections can then be combined with Maxwellian relative speed distributions to yield rate constants for reactions (1) and (2) as a function of ion kinetic temperature. These results are the dashed lines in Figs. 1 and 2 with $KE_{cm} = 3kT/2$. Quite similar results were obtained from a parallel and independent theoretical approach (Viehland and Mason, 1977) that used a moment solution of the Boltzmann transport equation.

It is to be emphasized that the cross sections in Figs. 3 and 4 express the role of kinetic energy alone, since the neutral reactants N_2 and O_2 were not vibrationally or rotationally excited in the flow–drift tube experiments. Reaction (1) is known to be very sensitive to N_2 vibrational energy for $T_{vib} \geq 1000$ K (Schmeltekopf *et al.*, 1968). In many respects, it is valuable to know the kinetic and vibrational aspects of reaction (1) separately. This permits ionospheric models to explore the effects of different ion and neutral temperatures. Furthermore, the kinetic cross section permits the testing of proposed ionospheric non-Maxwellian relative speed distribution (M. R. Torr *et al.*, 1977; St.-Maurice and Torr, 1978).

The other major reactions of the F-region are those that destroy ${N_2}^+$, which is found to have a relatively low concentration in spite of its large production rate. This is a consequence of the fast reaction

$$ {N_2}^+ + O \rightarrow NO^+ + N \quad (3a)$$

$$ \rightarrow O^+ + N_2 \quad (3b)$$

and to a lesser extent the charge-transfer reaction

$$ {N_2}^+ + O_2 \rightarrow {O_2}^+ + N_2 . \quad (4)$$

There is less O_2 than O and $k_4 \leq k_3$, so that the ${N_2}^+$ loss by O is much more significant than that by O_2 in the F-region. Laboratory measurements of k_3 have yielded $k_3 = (1.4 \pm 0.7) \times 10^{-10}\,(T/300)^{-0.44}$ cm^3 s^{-1} with $k_{3a}/k_{3b} = 0.95$, relatively independent of kinetic energy (McFarland *et al.*, 1974). The minor reactions that occur in the upper ionosphere are reviewed by Banks and Kockarts (1973). Ferguson (1973) has recently summarized the current values for the rate constants.

The close agreement between the present laboratory data and atmospheric ion composition has been shown by several analyses of the data obtained from the satellite Atmosphere Explorer-C (AE-C). (Dalgarno *et al.*, 1973). This satellite was comprehensively instrumented to measure simultaneously the ion composition, the neutral composition, and the solar ionizing flux and temperature, which are the relevant parameters required for a detailed analysis of the ion chemistry. An example of the kind of agreement obtained is

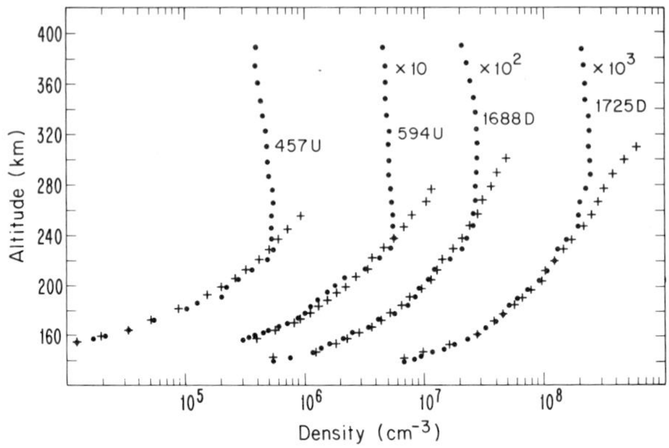

FIG. 5 Comparison of calculated (plus) and observed (circle) O^+ ion densities as a function of altitude for four Atmosphere Explorer-C satellite orbital passes. [From Oppenheimer *et al.* (1977).]

shown in Fig. 5, taken from Oppenheimer *et al.* (1977), showing a comparison of measured O^+ profiles with theory from four different satellite passes, the theory embodying the laboratory rate constant data. Below the O^+ ion density peak (i.e., below about 240 km), a photochemical steady state exists and the agreement is excellent. At higher altitudes diffusion becomes important, leading to an expected divergence between O^+ concentrations and steady-state photochemical theory. Figure 6 is a comparison of observed and theoretical profiles for the major F-region molecular ions $O_2{}^+$, NO^+, and $N_2{}^+$, further illustrating the excellent agreement between laboratory ion–molecule reaction rate measurements and ionospheric requirements. Similar agreement was found in other AE analyses. For example,

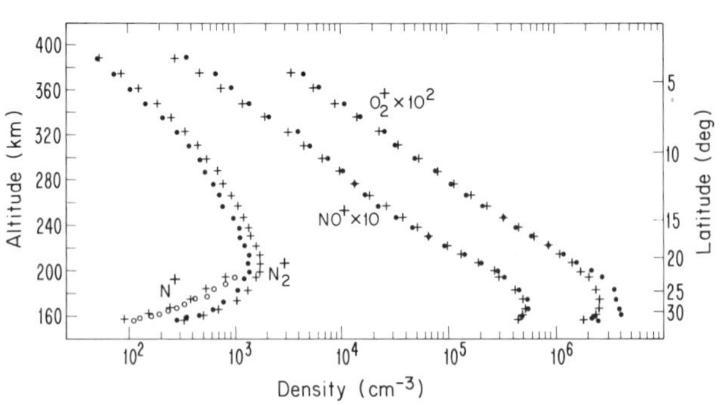

FIG. 6 Comparison of calculated (plus) and observed (circle) N^+, $N_2{}^+$, NO^+, and $O_2{}^+$ ion densities as a function of altitude and latitude for an orbit pass of Atmosphere Explorer-C satellite. [From Oppenheimer *et al.* (1977).]

D. G. Torr *et al.* (1977) deduced a value of $(1.4 \pm 0.2) \times 10^{-10}$ cm^3 s^{-1} for the rate constant at 300 K for reaction (3), using the slight temperature dependence found in the laboratory measurements to extrapolate from the ~ 650 K temperature of the atmospheric observations. This agreement with the laboratory study, and the close correspondence shown in Figs. (5) and (6), is typical of the results of several AE analyses of ion composition and demonstrates that the major features of F-region ion chemistry are now well understood.

B. *Metastable and Doubly Charged Ion Reactions*

Dalgarno (1970) has reviewed the role of metastable species in the ionosphere. Since then, there have been several *in situ* and laboratory studies of the reactive properties of metastable ionic species. Substantial fractions of the primary ions in the upper ionosphere are produced in metastable electronic states. The reactivities of metastable ions are, of course, generally quite different than those of the ground state ions. In particular, the additional energy often opens up new reaction channels that are endothermic for the ground state ions. Metastable ions can, therefore, play a role in atmospheric ion composition. Even when they do not impact significantly on ion composition as such, they may manifest their presence in the atmosphere in other ways. The forbidden optical transitions of metastable ions in some cases lead to detectable airglow emissions. Observations of such radiations are being carried out, particularly with the Atmosphere Explorer satellites, thereby providing valuable information on atmospheric processes. In some cases, as described below, reaction rate constant data are deduced from atmospheric measurements. Table I lists the available reaction rate constant data for important atmospheric metastable ions with major atmospheric neutrals. There is no known case where the metastable ion is not lost efficiently by at least one of the major ionospheric species O, N_2, or O_2, so we do not tabulate data for less abundant neutral reactants such as CO_2. Some of the data have been obtained in the laboratory and some from atmospheric observation.

1. $O^+(^2P)$ *Reactions*

The $O^+(^2P)$ state, lying 5.0 eV above the $O^+(^4S)$ ground state, has a raditive lifetime of 4.57 s. Dalgarno and McElroy (1965) pointed out that about 20% of the O^+ ions produced in the F-region are in the ^2P state. Walker *et al.* (1975), and more recently Rusch *et al.* (1977), have used optical data from the visible airglow experiments on AE-C and -D of the $O^+(^2P) \rightarrow O^+(^2D) + h\nu$ doublet at 7320 and 7330 Å to deduce $O^+(^2P)$ loss rates by atomic oxygen and molecular nitrogen, which are the dominant neutral species between 190 and 330 km where the observations were made. They found that the

TABLE I

F- *and* E-*Region Metastable Ion Reaction Rate Constants* (300 K)

	Reaction	Rate constant[a] ($cm^3\ s^{-1}$)	Method[b]	Source
(1)	$O^+(^2P) + N_2 \rightarrow O^+(^4S) + N_2$	4.0 (−10)	AE	Oppenheimer *et al.* (1976)
	$\rightarrow N_2^+ + O$	5.0 (−11)	AE	Oppenheimer *et al.* (1976)
	→ both channels	4.8 ± 1.4 (−10)	AE	Rusch *et al.* (1977)
(2)	$O^+(^2P) + O \rightarrow O^+(^4S) + O$	1.8 (−10)	AE	Oppenheimer *et al.* (1976)
		5.2 ± 2.5 (−11)	AE	Rusch *et al.* (1977)
(3)	$O^+(^2D) + N_2 \rightarrow N_2^+ + O$	1.0 ± 0.5 (−9)	B	Rutherford and Vroom (1971)
		5.0 ± 1.7 (−10)[c]	AE	Torr and Orsini (1977)
(4)	$O^+(^2D) + O_2 \rightarrow O_2^+ + O$	2.0 ± 1.5 (−9)	AE	Torr *et al.* (1979)
(5)	$O^+(^2D) + O \rightarrow O^+(^4S) + O$	2 ± 1 (−11)[c]	AE	Orsini *et al.* (1977)
(6)	$O_2^+(a^4\Pi_u) + N_2 \rightarrow N_2^+ + O_2$	4.1 ± 1.6 (−10)[c]	FDT	Lindinger *et al.* (1975a)
(7)	$O_2^+(a^4\Pi_u) + O_2 \rightarrow O_2^+(X^2\Pi_g) + O_2$	3.1 ± 1.5 (−10)[c]	FDT	Lindinger *et al.* (1975a)
(8)	$NO^+(a^3\Sigma^+) + N_2 \rightarrow N_2^+ + NO$	3.5 ± 1.3 (−10)[c]	FDT	Dotan *et al.* (1979)
	$\rightarrow NO^+(X^1\Sigma^+) + N_2$	3.5 ± 1.3 (−10)	FDT	Dotan *et al.* (1979)
(9)	$NO^+(a^3\Sigma^+) + O_2 \rightarrow NO^+(X^1\Sigma^+) + O_2$	4.5 ± 2.0 (−10)[c]	FDT	Dotan *et al.* (1979)

[a] 4.8 ± 1.4 (−10) means $(4.8 \pm 1.4) \times 10^{-10}$, etc.

[b] AE denotes an analysis of Atmosphere Explorer satellite data, B denotes laboratory beam experiment, and FDT denotes laboratory flow–drift tube experiment.

[c] See Notes added in proof, (a) for revised and additional measurements.

volume emission rate of 7320–7330 Å photons could be well described by the photochemical equilibrium expression

$$\eta = \frac{0.781 \times 0.218P}{0.218 + k_O[O] + k_{N_2}[N_2] + k_e[e]}, \tag{5}$$

where 0.781 is the fraction of $O^+(^2P)$ ions that radiate the 7320–7330 Å doublet, P the production rate cm $^{-3}$ s^{-1}, 0.218 s^{-1} the $O^+(^2P)$ reciprocal lifetime, and the ks are the appropriate loss rates. P was calculated using the solar uv spectrum obtained with spectrometers on the AE satellites and the appropriate photoionization and photoabsorption cross sections. The electron quenching rate constant had been calculated theoretically. The concentrations [O], [N_2], and [e] and the photon flux η were measured on board the satellite. Since η is measured over a range of altitudes 190–330 km, the separate contributions to the $O^+(^2P)$ loss from O and N_2 can be determined, due to the fact that O and N_2 have different altitude profiles. The results, $k_{N_2} = (4.8 \pm 1.4) \times 10^{-10}$ cm^3 s^{-1} and $k_O = (5.2 \pm 2.5) \times 10^{-11}$ cm^3 s^{-1}, are that $O^+(^2P)$ ions are lost on nearly every collision with N_2, but that atomic oxygen is about an order of magnitude slower. From AE-C N_2^+ measurements, Oppenheimer *et al.* (1976) obtained similar rate constants, as shown in Table I. They found that the $O^+(^2P)$ loss from N_2 is mainly by quenching, rather than by charge transfer. These studies are excellent examples of how detailed and precise AE satellite data can be analyzed for rate constants that are only just now becoming available from the laboratory (Glosik *et al.*, 1978).

2. $O^+(^2D)$ *Reactions*

The $O^+(^2D)$ state lies 3.3 eV above the ground state and has a radiative lifetime of 3.6 hr. Rutherford and Vroom (1971) measured the reaction of $O^+(^2D)$ with N_2 to give N_2^+ in a beam experiment and extrapolated their results to a thermal energy value of $(1.0 \pm 0.5) \times 10^{-9}$ cm^3 s^{-1}, essentially equal to the collision frequency. Some 38% of the O^+ ions in the F-region are produced in the ^2D state. Rather involved analyses of N_2^+ profiles from AE-C allowed Torr and Orsini (1977) to deduce the rate constant for $O^+(^2D)$ charge transfer with N_2, in agreement with the laboratory measurement within the uncertainty involved. They also showed (Orsini *et al.*, 1977) that $O^+(^2D)$ must be quenched with atomic oxygen with a rate constant of 2×10^{-11} cm^3 s^{-1}, within a factor of two. Torr *et al.* (1979) also showed that the reaction of $O^+(^2D)$ with O_2 is a significant O_2^+ source under certain atmospheric conditions and deduced a very large rate constant, 2×10^{-9} cm^3 s^{-1}, with an estimated error of a factor of two. [See Notes added in proof, (a).]

3. $O_2^+(a^4\Pi_u)$ *Reactions*

The $O_2^+(a^4\Pi_u)$ state has an energy 4.0 eV greater than the ground electronic state $O_2^+(X^2\Pi_g)$. The bands of the $O_2^+b^4\Sigma_g^- - a^4\Pi_u$ First-Negative system (Albritton *et al.*, 1977b) are important features of many types of aurora (Gattinger, 1976). The radiative lifetime of the $a^4\Pi_u$ state is unknown, but it is undoubtedly very long. Therefore, the ultimate fate of these energetic ions in the ionosphere will be governed by their reactive properties with the neutrals of that region. Some of the most important of these have been measured (Lindinger *et al.*, 1975a,b; Glosik *et al.*, 1978):

$$O_2^+(a^4\Pi_u) + N_2 \rightarrow N_2^+ + O_2, \tag{6}$$

$$O_2^+(a^4\Pi_u) + O_2 \rightarrow O_2^+ + O_2, \tag{7}$$

and

$$O_2^+(a^4\Pi_u) + NO \rightarrow NO^+ + O_2. \tag{8}$$

All three of these are fast; the rate constants are listed in Table I. In reactions (6) and (8), the charge-transfer mechanism appears to predominate over quenching. In reaction (7), quenching is the only available mechanism. Above 200 km, the reaction

$$O_2^+(a^4\Pi_u) + O \rightarrow O^+ + O_2 \tag{9}$$

would be important. Its rate constant has not been measured.

4. $NO^+(a^3\Sigma^+)$ *Reactions*

Another metastable ion that could play a role during periods of possible NO enhancement (Zipf *et al.*, 1970) is $NO^+(a^3\Sigma^+)$, which has an energy 6.4 eV greater than the ground state ion $NO^+(X^1\Sigma^+)$. The reactions

$$NO^+(a^3\Sigma^+) + N_2 \rightarrow N_2^+ + NO \tag{10a}$$

$$NO^+(a^3\Sigma^+) + N_2 \rightarrow NO^+(X^1\Sigma^+) + N_2 \tag{10b}$$

and

$$NO^+(a^3\Sigma^+) + O_2 \rightarrow NO^+(X^1\Sigma^+) + O_2 \tag{11}$$

have been studied recently in the laboratory (Dotan *et al.*, 1979; Glosik *et al.*, 1978). The rate constants are listed in Table I and are both large. In reaction (10), the reactive and quenching channels are about equal at 300 K and, interestingly enough, reaction (11) proceeds entirely by quenching.

5. O^{2+} *Reactions*

Atmosphere Explorer-C data have been analyzed (Brieg *et al.*, 1977) to yield O^{2+} reaction rate constants that have not yet been measured in the laboratory. These are

$$O^{2+} + O \rightarrow O^+ + O^+, \qquad k_{12} = 1 \times 10^{-11} \quad cm^3\ s^{-1} \tag{12}$$

$$O^{2+} + N_2 \rightarrow O^+ + N_2^+, \qquad k_{13} \leq 10^{-11} \quad cm^3\ s^{-1}. \tag{13}$$

[See Notes added in proof, (b).]

III. D-REGION POSITIVE ION CHEMISTRY

Since the pioneering rocket-borne mass-spectrometer flights of Narcisi and Bailey (1965), there have been numerous *in situ* measurements of the D-region positive ions (see e.g., Goldberg and Aiken, 1971; Narcisi *et al.*, 1972; Arnold and Krankowsky, 1974; Zbinden *et al.*, 1975). Although there is considerable variation among the various flights, these measurements conclusively show that the dominant ions of the lower E-region, O_2^+ and NO^+, are replaced by the water cluster ions, $H_3O^+ \cdot nH_2O$, in the D-region. Considerable laboratory, theoretical, and *in situ* efforts have been directed toward identifying the D-region reaction paths that convert the primary O_2^+ and NO^+ ions into the observed $H_3O^+ \cdot nH_2O$ ions. The reaction scheme that has emerged is given in Fig. 7. Laboratory data are available for the rate constants and/or equilibrium constants for the reactions represented by solid lines and are currently unavailable for those represented by dashed lines. The details of this chemistry have been reviewed recently by Reid (1976).

There has been no problem in explaining the conversion of the primary O_2^+ ions into the water cluster ions via the laboratory-verified paths shown along the top part of Fig. 7 (Fehsenfeld and Ferguson, 1969; Good *et al.*, 1970). However, the major source of ionization in the undisturbed daytime D-region is photoionization of NO by solar Lyman-α radiation. The understanding of the conversion of these NO^+ ions into the water cluster ions has proven to be a more difficult problem. Because of the low mixing ratio of water in the D-region and the relatively low D-region pressures, it is not possible to explain the conversion via the three-body association paths shown in the center of Fig. 7 (Howard *et al.*, 1971; Fehsenfeld *et al.*, 1971; French *et al.*, 1973).

It was suggested by Ferguson (1971) that this conversion process could be significantly shortened if $NO^+ \cdot nH_2O$ would associate with the more abundant atmospheric neutrals such as N_2 and CO_2, i.e.,

$$NO^+ \cdot nH_2O + X + M \rightarrow NO^+ \cdot nH_2O \cdot X + M, \tag{14}$$

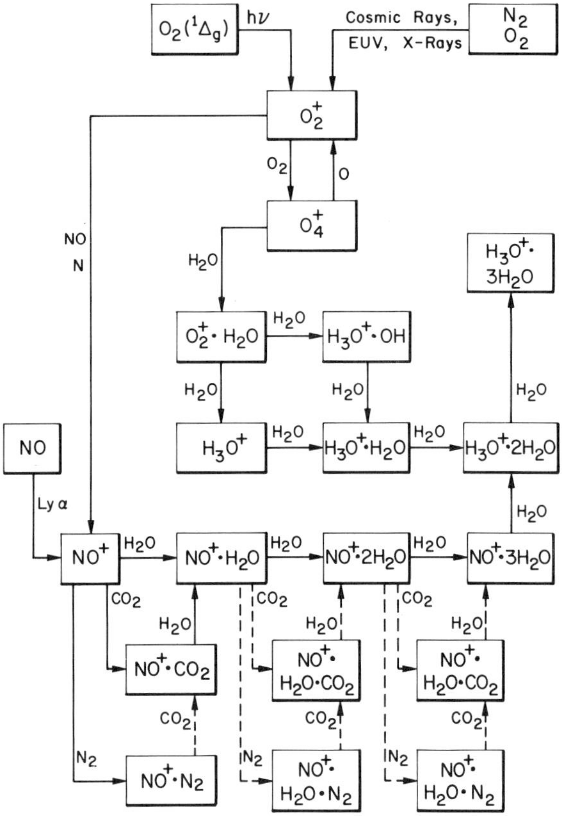

FIG. 7 D-region positive ion chemistry.

where $n = 0, 1, 2$ and $X = N_2, CO_2$. The weakly bound association complexes thus formed would then undergo a series of fast switching reactions,

$$NO^+\cdot nH_2O\cdot N_2 + CO_2 \leftrightarrows NO^+\cdot nH_2O\cdot CO_2 + N_2 \tag{15}$$

$$NO^+\cdot nH_2O\cdot CO_2 + H_2O \rightarrow NO^+\cdot(n+1)H_2O + CO_2 \tag{16}$$

that would greatly accelerate each hydration step and hence the eventual production of the water cluster ions. These reaction paths are shown in the lower part of Fig. 7. However, the verification of this scheme in the laboratory or the detection of the key ions in the atmosphere has presented a formidable experimental problem, since the intermediate $NO^+\cdot nH_2O\cdot X$ ions are extremely fragile.

Arnold and Krankowsky (1974, 1977a) have had considerable success in solving the problem of sampling such weakly bound cluster ions in rocket-borne mass spectrometers. As a result, some key ions in this reaction scheme now have been observed *in situ*, for example $NO^+ \cdot N_2$, $NO^+ \cdot CO_2$, and $NO^+ \cdot H_2O \cdot CO_2$. A sample of these results is shown in Fig. 8. Progress has also been made in the laboratory. The rate constant and thermodynamic data for the clustering of N_2 to NO^+ [reaction (14) with $X = N_2$, $n = 0$] are now available. Heimerl and Vanderhoff (1974) measured k_{14} at room temperature. Studies were later made as a function of temperature: k_{14} and $K_{eq}(14)$ by Johnsen *et al.* (1975); $K_{eq}(14)$ by Turner and Conway (1976); and k_{14} by Smith *et al.* (1977). Arnold and Krankowsky (1977b) were also able to deduce a value for k_{14} from their *in situ* measurements. All of these values are in good agreement.

Although very few of these switching reactions have been measured in the laboratory, such reactions are known to be fast. For example, Arnold and Krankowsky (1977b) find that a value $k_{15} \approx 10^{-9}$ cm^3 s^{-1} ($n = 0$) is consistent with their *in situ* observations. Thus, this sequence of reactions seems to offer an efficient mechanism for the formation of $NO^+ \cdot H_2O$ ions, particularly around 80 km where the D-region temperatures are the lowest. Reid (1977) has devised a reasonable extension of this cycle that can account for most of the observed D-region positive ion features. According to this model, the nonpolar neutrals that are clustered to NO^+ are so weakly bound that these neutrals are collisionally dissociated even at D-region temperatures. This is, in fact, the case for $NO^+ \cdot N_2$, which has a bond dissociation energy of 5 kcal/mole (Johnsen *et al.*, 1975; Turner and Conway, 1976). Accordingly, when the temperature is above about 215 K, the collisional decomposition of $NO^+ \cdot N_2$ will be larger than the switching reaction with CO_2 (Arnold and Krankowsky, 1977b).

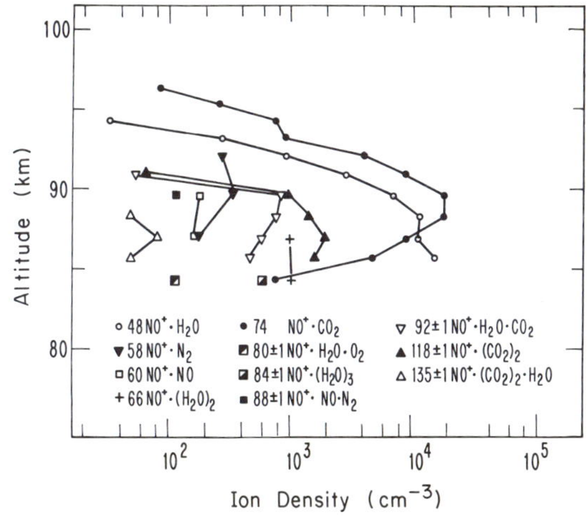

FIG. 8. Altitude profiles of NO^+ cluster ions. [From Arnold and Krankowsky (1974).] The identification of masses 80 ± 1 and 88 ± 1 amu are less certain.

For this reason, the conversion of NO^+ ions to the observed water cluster ions is extremely sensitive to temperature, and the D-region positive ion composition will, therefore, manifest strong seasonal, latitudinal, and even irregular variations as a result of variations in the atmospheric temperature. Moreover, electron–ion recombination coefficients for these weakly bound cluster ions may be considerably larger than those for the unclustered NO^+ and $O_2{}^+$ ions, which may explain, at least in part, the strong variations in the electron density that are observed at about 80 km (Arnold and Krankowsky, 1977a; Reid, 1977). While many of the details still await laboratory verification, the reaction scheme shown in Fig. 7 has been shown (Reid, 1977) to qualitatively solve the long-standing D-region problem of the conversion of the primary NO^+ ions into the observed water cluster ions.

IV. D-REGION NEGATIVE ION CHEMISTRY

In contrast to the D-region positive ion chemistry, where *in situ* measurements discovered the unexpected water cluster ions, the negative ion chemistry of the D-region has been derived predominately from laboratory

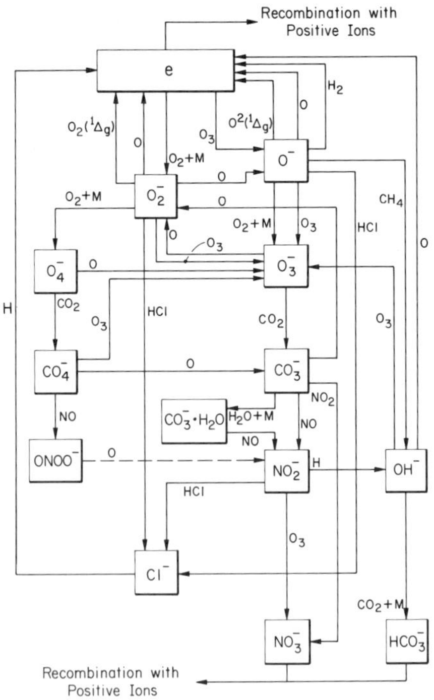

FIG. 9 Reaction scheme for D-region negative ion chemistry.

studies. This chemistry is shown schematically in Fig. 9. The measured rate constants for these reactions are listed in Table II. The table contains improved values for several of the reactions from yet-unpublished studies. Many of the details of D-region negative ion chemistry have been reviewed recently by Reid (1976).

The primary negative ion of the D-region is largely O_2^-, which is formed by the termolecular attachment of electrons to O_2 [see Table II, reactions (1) and (2)]. To a much less extent, O^- is formed by dissociative attachment to O_3 [see Table II, reaction (3)]. The relatively unstable O_2^- ions undergo a series of ion–molecule reactions with the D-region minor *molecular* neutral constituents to form progressively the more stable ions shown toward the bottom in Fig. 9. In competition with this progression at each intermediate step are reactions with the minor D-region *atomic* neutral constitutents that lead back eventually to the release of the electrons. The terminal ions NO_3^- and HCO_3^- are sufficiently stable to resist attack by the atomic species. For example, the electron affinity of NO_3 is 4.0 eV, one of the largest known for simple molecules.

The first few steps in the reaction sequence of Fig. 9 are fast in comparison to the conversion of CO_3^- to NO_2^- by reaction with NO [see Table II, reaction (24)]. The relatively small rate constant for this reaction and low NO concentration implies that this is the limiting step in Fig. 9. However, this CO_3^- bottleneck does not mean that there will be a large CO_3^- concentration in the D-region, but rather it implies an overall low negative ion centration. The rapid conversion of CO_3^- back to O_2^- via reaction with O [see Table II, reaction (23)] and then detachment of the electron by reaction with O [see Table II, reaction (11b)] holds the CO_3^- concentration in check. The rate constants and densities involved in this loop are such that it will occur many times before a CO_3^- ion is converted to an NO_2^- ion. The step from NO_2^- to NO_3^- is fast compared to the preceding one, thus the steady-state NO_3^- concentration is determined largely by the ratio of the NO_2^- production rate, k [NO] [see Table II, reaction (24)] (where [NO] is the NO concentration), and the NO_3^- loss by ion–ion recombination. Hence, this model predicts a rather large $[NO_3^-]/[CO_3^-]$ ratio in the D-region.

Thus, to a large extent the negative ion chemistry of the D-region is controlled by the neutral composition, and its variability is linked directly with the large variations in the minor species O, O_3, NO, and H. It is clear from the upper part of Fig. 9 that higher pressure, hence lower altitude, favors the attachment of electrons to O_2, thereby initiating the negative ion chemistry, and the termolecular reactions that lead toward more stable negative ions. Furthermore, the increasing concentrations of O_3 and NO and the decreasing concentration of O at the lower altitudes all imply a

TABLE II

D-*Region Negative Ion Reaction Rate Constants* (300 K)

	Reaction	Rate constant[a] ($cm^3\ s^{-1}$, $cm^6\ s^{-1}$)	Source
(1)	$e + 2O_2 \rightarrow O_2^- + O_2$	1.9 (−30)	Phelps (1969)
(2)	$e + O_2 + N_2 \rightarrow O_2^- + N_2$	1.0 (−31)	Phelps (1969)
(3)	$e + O_3 \rightarrow O^- + O_2$	9 (−12)	Stelman *et al.* (1972)
(4)	$O^- + O \rightarrow O_2 + e$	1.9 (−10)	Fehsenfeld (1979)
(5)	$O^- + O_2(^1\Delta_g) \rightarrow \rightarrow O_3 + e$	3 (−10)	Fehsenfeld *et al.* (1969a)
(6)	$O^- + H_2 \rightarrow H_2O + e$	6.4 (−10)	McFarland *et al.* (1973)
(7)	$O^- + O_3 \rightarrow O_3^- + O_2$	8.0 (−10)	Dotan *et al.* (1979)
(8)	$O^- + 2O_2 \rightarrow O_3^- + O_2$	1.0 (−30)	Snuggs *et al.* (1971)
(9)	$O^- + HCl \rightarrow Cl^- + OH$	2.0 (−9)	Dotan *et al.* (1978a)
(10)	$O^- + CH_4 \rightarrow OH^- + CH_3$	8.0 (−11)	Lindinger *et al.* (1975c)
(11a)	$O_2^- + O \rightarrow O^- + O_2$	1.5 (−10)	Fehsenfeld (1979)
(11b)	$\rightarrow O_3 + e$	1.5 (−10)	Fehsenfeld (1979)
(12)	$O_2^- + O_2(^1\Delta_g) \rightarrow 2O_2 + e$	2 (−10)	Fehsenfeld *et al.* (1969a)
(13)	$O_2^- + 2O_2 \rightarrow O_4^- + O_2$	3.0 (−31)	Parkes (1971)
(14)	$O_2^- + O_3 \rightarrow O_3^- + O_2$	6.0 (−10)	Dotan *et al.* (1979)
(15)	$O_2^- + HCl \rightarrow Cl^- + HO_2$	1.6 (−9)	Dotan *et al.* (1978a)
(16)	$OH^- + O \rightarrow \rightarrow HO_2 + e$	2 (−10)	Fehsenfeld *et al.* (1966)

(17)	$OH^- + O_3 \rightarrow O_3^- + OH$	9 (−10)	Dotan *et al.* (1977)
(18)	$OH^- + CO_2 + M \rightarrow HCO_3^- + M$	7.6 (−28)	Fehsenfeld and Ferguson (1974)
(19)	$O_3^- + O \rightarrow O_2^- + O_2$	2.5 (−10)	Fehsenfeld (1979)
(20)	$O_3^- + CO_2 \rightarrow CO_3^- + O_2$	5.5 (−10)	Dotan *et al.* (1977)
(21)	$O_4^- + O \rightarrow O_3^- + O_2$	4.0 (−10)	Fehsenfeld *et al.* (1969b)
(22)	$O_4^- + CO_2 \rightarrow CO_4^- + O_2$	4.3 (−10)	Fehsenfeld *et al.* (1969b)
(23)	$CO_3^- + O \rightarrow O_2^- + CO_2$	1.1 (−10)	Fehsenfeld (1979)
(24)	$CO_3^- + NO \rightarrow NO_2^- + CO_2$	1.1 (−11)	Albritton *et al.* (1979)
(25)	$CO_3^- + NO_2 \rightarrow NO_3^- + CO_2$	2.0 (−10)	Fehsenfeld and Ferguson (1974)
(26)	$CO_3^- + H_2O + M \rightarrow CO_3^- \cdot H_2O + M$	1.0 (−28)	Fehsenfeld and Ferguson (1974)
(27a)	$CO_3^- \cdot H_2O + NO \rightarrow NO_2^- + CO_2 + H_2O$	7 (−12)	Fehsenfeld and Ferguson (1974)
(27b)	$\rightarrow NO_2^- \cdot H_2O + CO_2$		
(28)	$CO_4^- + O \rightarrow CO_3^- + O_2$	1.4 (−10)	Fehsenfeld (1979)
(29)	$CO_4^- + NO \rightarrow ONOO^- + CO_2$	4.8 (−11)	Fehsenfeld *et al.* (1969b)
(30)	$CO_4^- + O_3 \rightarrow O_3^- + CO_2 + O_2$	1.3 (−10)	Fehsenfeld and Ferguson (1974)
(31)	$NO_2^- + H \rightarrow OH^- + NO$	3.7 (−10)	Fehsenfeld (1975)
(32)	$NO_2^- + O_3 \rightarrow NO_3^- + O_2$	1.2 (−10)	Dotan *et al.* (1979)
(33)	$NO_2^- + NO_2 \rightarrow NO_3^- + NO$	≤2 (−13)	Fehsenfeld *et al.* (1975)
(34)	$NO_2^- + HCl \rightarrow Cl^- + HNO_2$	1.4 (−9)	Ferguson *et al.* (1972)
(35)	$Cl^- + H \rightarrow HCl + e$	9.6 (−10)	Howard *et al.* (1974)

[a] 1.9 (−30) means 1.9×10^{-30}, etc.

rather sharp upper boundary for the D-region negative ions at 75–80 km, with the free electrons showing a complementary variation with altitude. Diurnal effects are expected to be large, since the O atom concentration, for example, increases throughout the D-region during the day. At the same time, the less-stable negative ions are subject to photodestruction (Huber *et al.*, 1977).

The temperature of the D-region and lower atmosphere is less than 300 K and varies with time, place, and altitude and particularly with season and latitude. It is therefore also necessary to have temperature dependences for the reaction rate constants in order to understand the negative ion chemistry in detail. Figures 10 and 11 show examples of recent measurements of energy dependence of D-region negative ion reactions. The rate constant for the $CO_3^- + NO \rightarrow NO_2^- + CO_2$ reaction shown in Fig. 11 decreases from 1.8×10^{-11} cm^{-3} s^{-1} at 180 K to 6.5×10^{-12} cm^3 s^{-1} at 450 K. Such negative activation energies are characteristic for slow reactions in the thermal-energy range.

Figure 9 does not include the effects of neutral molecules clustering to the negative ions. The rate constants of such termolecular association reactions typically increase with decreasing temperature and the breakup rates of weakly bound cluster ions decrease. Thus, at the relatively low D-region temperatures, all of the negative ions are subject to clustering, which is not very chemically specific. The terminal, i.e., long-lived, ions like NO_3^- are certainly expected to be clustered; hence, $NO_3^- \cdot nH_2O$ ions are predicted to be the dominant ions in the D-region. If the intermediate ions cluster in their shorter lifetimes, the question arises as to whether this would significantly alter their subsequent chemistry. A limited study carried out a few years ago (Fehsenfeld and Ferguson, 1974) indicated that water clusters would not significantly effect the reaction scheme of Fig. 9. For example, one and two water molecules do not significantly alter the rate of O_2^- charge transfer with O_3 [see Table II, reaction (14)]. One water molecule does not significantly effect the rate of O_3^- reaction with CO_2 [see Table II, reaction (20)], but two water molecules quench this reaction, probably by making the reaction effectively endothermic. However, because of the large abundance of CO_2, the O_3^- ions will not have time to cluster twice, so that this should have little effect. Clustering to CO_3^- may somewhat alleviate the CO_3^- bottleneck if it inhibits the reaction of CO_3^- with O relative to that with NO. At D-region temperatures, there could be important negative-ion association reactions with O_2, N_2, CO_2, and H_2O, followed by switching, exactly analogous to the D-region NO^+ association sequence proposed in the lower part of Fig. 7. Only one such D-region negative-ion reaction path has been studied in the laboratory, namely, the one involving $CO_3^- \cdot H_2O$ shown in

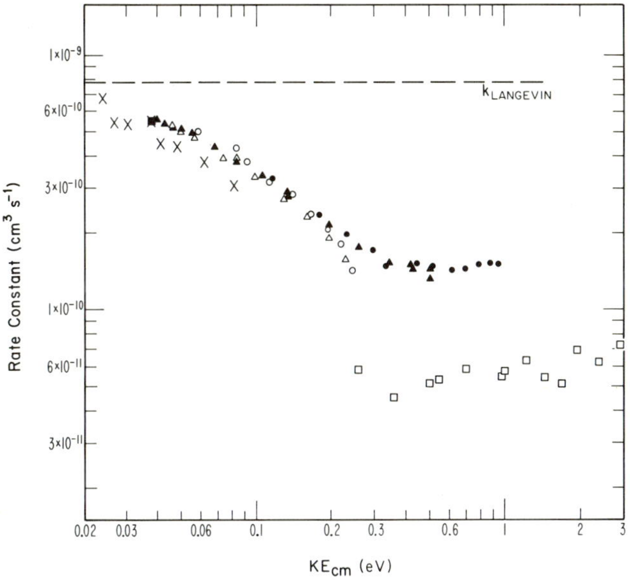

FIG. 10 Rate constants for the reaction $O_3^- + CO_2 \rightarrow CO_3^- + O_2$ as a function of mean relative kinetic energy (flow–drift tube) and temperature (flowing afterglow, using $\frac{3}{2}kT = KE_{cm}$). [From Dotan *et al.* (1977).] These data are compared to static-drift tube [Parkes (1972)] and crossed-beam (Neynaber *et al.*, 1970) measurements and to the collision-limiting value [Gioumousis and Stevenson (1958)]. ×, flowing afterglow; □, crossed beams; and ■, static-drift tube. Flow-drift tube with helium buffer gas: ●, 0.250 torr and ▲, 0.450 torr. Flow-drift tube with argon buffer gas: ○, 0.125 torr and △, 0.200 torr.

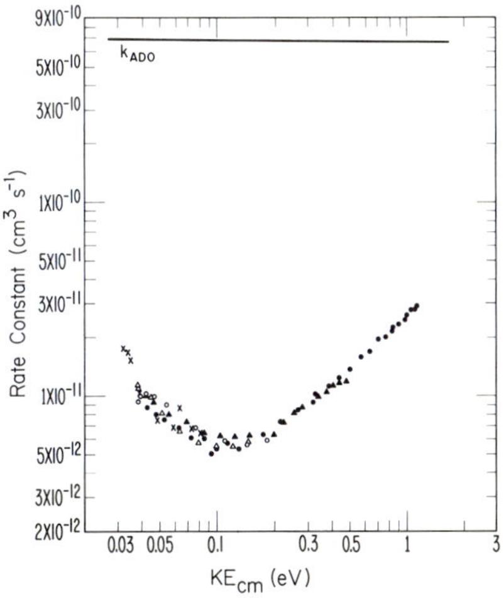

FIG. 11 Rate constants for the reaction $CO_3^- + NO \rightarrow NO_2^- + CO_2$ as a function of mean relative kinetic energy (flow–drift tube) and temperature (flowing afterglow, using $\frac{3}{2}kT = KE_{cm}$). [From Albritton *et al.*, (1979).] The data are compared to the collision-limiting values [Su and Bowers (1973a,b)]. ×, flowing afterglow. Flow–drift tube with helium buffer gas: ●, 0.250 torr and ▲, 0.450 torr. Flow–drift tube with argon buffer gas: ○, 0.125 torr and △, 0.200 torr.

Fig. 9. Other reactions that need to be investigated at D-region temperatures are

$$CO_3^-\cdot H_2O + O \rightarrow O_2^- + CO_2 + H_2O, \quad (17)$$

$$CO_3^- + X + M \rightarrow CO_3^-\cdot X + M, \quad (18)$$

$$CO_3^-\cdot X + H_2O \rightarrow CO_3^-\cdot H_2O + X, \quad (19)$$

$$CO_3^-\cdot O_2 + CO_2 \rightarrow CO_3^-\cdot CO_2 + O_2, \quad (20)$$

$$CO_3^-\cdot O_2 + N_2 \leftrightarrows CO_3^-\cdot N_2 + O_2, \quad (21)$$

$$CO_3^-\cdot N_2 + CO_2 \rightarrow CO_3^-\cdot CO_2 + N_2, \quad (22)$$

where $X = N_2, O_2$, and CO_2. With the improved ability to study individual cluster ions that now exists, more extensive studies of the effects of clustering on chemical reactivity are being undertaken.

The first *in situ* measurements of the D-region negative ion composition were made by Narcisi *et al.* (1971) and Arnold *et al.* (1971). Figure 12 shows the ion concentration recorded in a later flight by Narcisi and co-workers. The dominant ions below 90 km are those with e/m ratios of $62 + n18$ ($n = 1$–5). An interpretation of these ions as $NO_3^-\cdot nH_2O$ is consistent with the model shown in Fig. 9. However, the persistence of e/m ratios of 16 and 32, which are presumably O^- and O_2^-, at the higher altitudes is not in accord with the chemistry of Fig. 9, which predicts an upper boundary for the negative ions at lower altitudes. The flights of Arnold and co-workers found somewhat different results. Namely, they report a rather sharp upper boundary to the D-region negative ions, in accord with the predictions based on laboratory measurements, but the $NO_3^-\cdot nH_2O$ sequence was not ob-

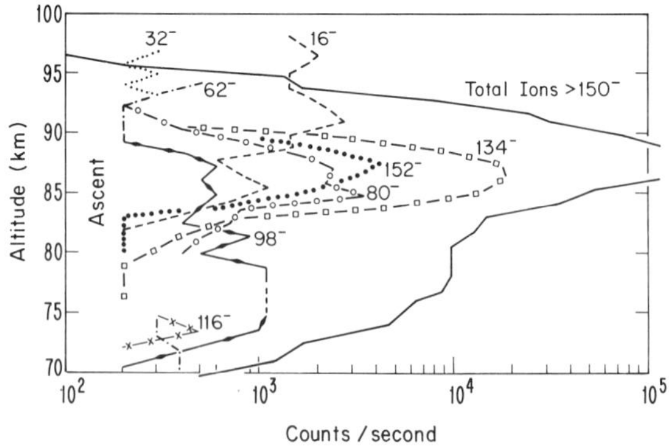

FIG. 12 D-region negative ion composition measurements during total eclipse. [From Narcisi *et al.* (1972).]

served. The major ion species in these flights were identified as CO_3^-, Cl^-, and HCO_3^-.

In view of the sparcity and differences between the atmospheric negative ion measurements, the substantial uncertainties in the concentrations of certain critical minor neutral species, and the need for more laboratory reaction studies, a quantitative understanding of D-region negative ion chemistry cannot be claimed at the present.

V. STRATOSPHERIC AND TROPOSPHERIC POSITIVE ION CHEMISTRY

A positive ion reaction scheme for the stratosphere and troposphere is given in Fig. 13. It is essentially the O_2^+ reaction sequence of the D-region positive ion chemistry (Fig. 7), augmented by reactions that involve some of the minor constituents of the lower atmosphere.

Galactic cosmic rays are the primary source of ionization in the stratosphere and upper troposphere. Radioactive decay is the primary source of ionization near the earth's surface. Cole and Pierce (1965) have given a typical ionization production-rate altitude profile. A value of 10 ion pairs cm^{-3} s^{-1} is fairly typical for most altitudes. The ambient concentration of positive ions is determined by the recombination rate with negative ions. Measurements of this concentration have been typically 10^3 cm^{-3} in the troposphere, increasing to about 5×10^3 cm^{-3} in the lower stratosphere (Cole and Pierce, 1965; Paltridge, 1965, 1966). Positive ion lifetimes are on the order of 200 s.

As shown in Fig. 13, the initial products of the ionization below 60 km are predominately N_2^+ and O_2^+, with lesser amounts of O^+ and N^+. These ions are rapidly converted to O_2^+, as well as an inconsequential amount of NO^+, by well-established reactions. Once formed, the O_2^+ ions associate with O_2:

$$O_2^+ + O_2 + M \leftrightarrows O_2 \cdot O_2^+ + M, \quad (23)$$

(Yang and Conway, 1964; Durden *et al.*, 1969; Good *et al.*, 1970). The formation of $O_2^+ \cdot O_2$ begins a series of fast switching reactions that involve H_2O and that lead to the formation of the water cluster ions, $H_3O^+ \cdot nH_2O$.

In the troposphere, where the H_2O mixing ratio is about 10^{-2}, the conversion of $O_2^+ \cdot O_2$ to $O_2^+ \cdot H_2O$ proceeds so rapidly that there is no conceivable alternative to the path described above that leads to the water cluster ions. However, in the stratosphere, where the H_2O mixing ratio is only on the order of 10^{-6}, other neutral constituents can have comparable abundances. Figure 14 compares the mixing ratios of three minor constituents—CO_2, O_3, and CH_4—with that of H_2O. If $O_2^+ \cdot O_2$ reacts with any of these neutrals, then the ion chemistry outlined in Fig. 13 could be significantly altered.

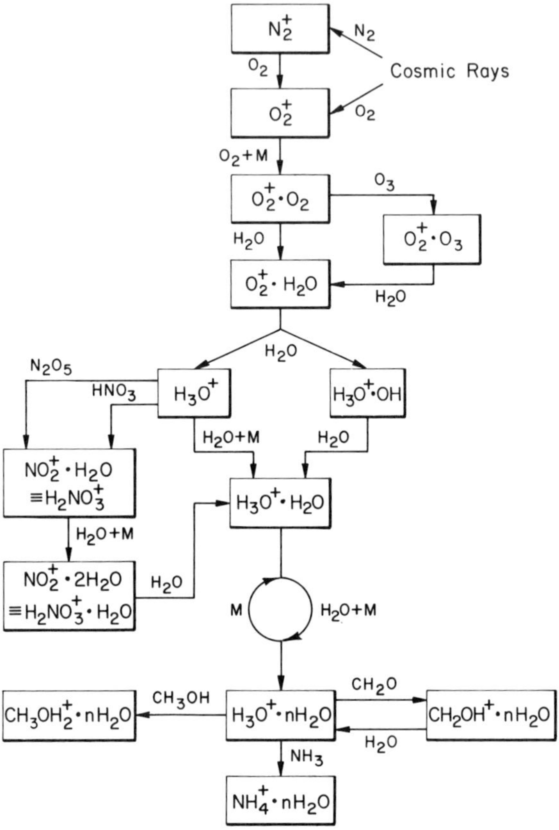

FIG. 13 Stratospheric and tropospheric positive ion chemistry.

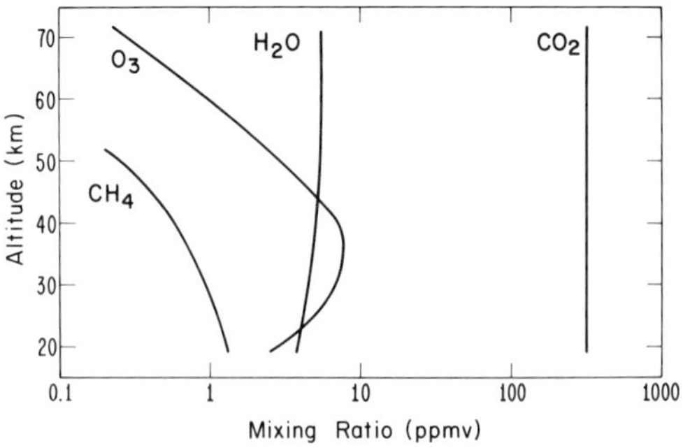

FIG. 14 Mixing ratios of CH_4, O_3, H_2O, and CO_2 as a function of altitude. [From Dotan *et al.* (1978b).] The concentration of CH_4 is from Crutzen *et al.* (1978) and those of O_3, H_2O, and CO_2 are from Dubin *et al.* (1976).

TABLE III

$O_2^+ \cdot O_2$ *Reactions with* H_2O, CH_4, CO_2, *and* O_3 (300 K)

Reactions	Result[a]	Source
(1) $O_2^+ \cdot O_2 + H_2O \rightarrow O_2^+ \cdot H_2O + O_2$	$k = 1.5\,(-9)\ cm^3\ s^{-1}$	Howard *et al.* (1972)
(2) $O_2^+ \cdot O_2 + CH_4 \rightarrow$ products	$k = 3\,(-12)\ cm^3\ s^{-1}$	Dotan *et al.* (1978b)
(3) $O_2^+ \cdot O_2 + CO_2 \leftrightarrows O_2^+ \cdot CO_2 + O_2$	$\Delta H = 0.3 \pm 1.0$ kcal mole^{-1} $\Delta S = 4.3 \pm 2.6$ cal mole^{-1} K^{-1}	Dotan *et al.* (1978b)
(4) $O_2^+ \cdot O_2 + O_3 \leftrightarrows O_2^+ \cdot O_3 + O_2$	$\Delta H = -3.7 \pm 1.0$ kcal mole^{-1} $\Delta S = 4.5 \pm 2.6$ cal mole^{-1} K^{-1}	Dotan *et al.* (1978b)

[a] 1.5 (−9) means 1.5×10^{-9}, etc.

These $O_2^+ \cdot O_2$ reactions have been examined recently (Dotan *et al.*, 1978b) and the results are given in Table III. The first entry is the fast reaction against which the alternative paths must compete. In contrast to it, the reaction of $O_2^+ \cdot O_2$ with CH_4 is very slow. The reactions of $O_2^+ \cdot O_2$ with CO_2 and O_3 were studied in equilibrium and the thermochemical constants are given. For CO_2, the equilibrium constant is about unity. The ratio of O_2 to CO_2 in the atmosphere is about 600, thus the equilibrium will always strongly favor $O_2^+ \cdot O_2$ rather than $O_2^+ \cdot CO_2$. However, the ΔH value for the reaction involving O_3 is much larger; hence, the $O_2^+ \cdot O_2$ and $O_2^+ \cdot O_3$ ion concentrations will be much more comparable. Figure 15 shows the ratio of the $O_2^+ \cdot O_3$ and $O_2^+ \cdot O_2$ concentration as a function of altitude calculated from the atmospheric densities and temperature and the above thermochemical data. The $O_2^+ \cdot O_3$ ion is a fairly strong rival to $O_2^+ \cdot O_2$, never dropping below one-tenth of the $O_2^+ \cdot O_2$ and even equaling it at 30 km. This means that the $O_2^+ \cdot O_3$ chemistry must also be considered.

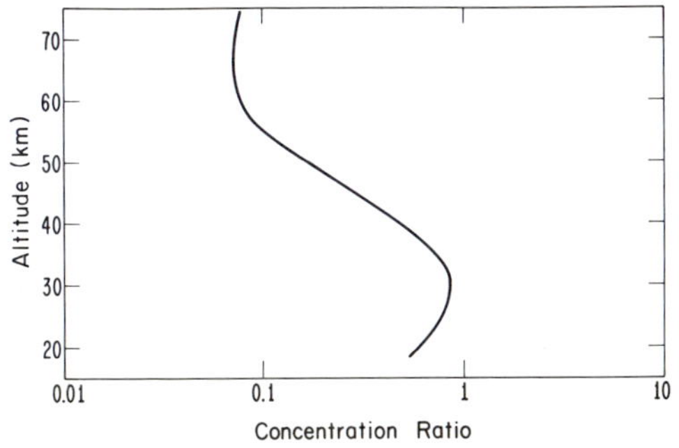

FIG. 15 Calculated equilibrium ratio of $O_2^+ \cdot O_3$ to $O_2^+ \cdot O_2$ as a function of altitude. [From Dotan *et al.* (1978b).]

The reaction of $O_2^+ \cdot O_3$ with CH_4 was found to be slow, $k \leq 5 \times 10^{-13}$ $cm^3\ s^{-1}$ (Dotan *et al.*, 1978b). In contrast, the reaction

$$O_2^+ \cdot O_3 + H_2O \rightarrow O_2^+ \cdot H_2O + O_3 \tag{24}$$

is fast, $k_{24} = 1.2 \times 10^{-9}$ $cm^3\ s^{-1}$. However, the product of this reaction is $O_2^+ \cdot H_2O$, which is the same ion that is produced by the reaction of $O_2^+ \cdot O_2$ with H_2O, the reaction for which competitors were being sought. Therefore, as Fig. 13 shows, the $O_2^+ \cdot O_3$ chemistry simply represents a side excursion that loops the stratospheric positive ion chemistry back on the path involving $O_2^+ \cdot H_2O$.

Lastly, it was observed that $O_2^+ \cdot H_2O$ does not react with CH_4, $k \leq 5 \times 10^{-13}$ $cm^3\ s^{-1}$. Thus, the neutrals CO_2, O_3, and CH_4 do not appear to disrupt the ion chemistry that leads to the water cluster ions.

Turning to neutrals that occur in even smaller concentrations, HNO_3 is of interest here because it is known to exist in both the stratosphere and troposphere. For example, its mixing ratio at 25 km is expected to be greater than 10^{-9} (Crutzen *et al.*, 1978). Furthermore, the reaction

$$H_3O^+ + HNO_3 \rightarrow H_2NO_3^+ + H_2O \tag{25}$$

is fast, $k_{25} = 1.6 \times 10^{-9}$ $cm^3\ s^{-1}$ (Fehsenfeld *et al.*, 1975). However, HNO_3 does not create a new reaction path that leads away from the ultimate formation of the water cluster ions. A rather remarkable situation exists in that protonated nitric acid $H_2NO_3^+$ is chemically equivalent to hydrated NO_2^+, i.e., $NO_2^+ \cdot H_2O$, as has been established by chemical reactivities (Fehsenfeld *et al.*, 1975). When this ion is further hydrated in the atmosphere

$$NO_2^+ \cdot H_2O + H_2O + M \rightarrow NO_2^+ \cdot 2H_2O + M, \tag{26}$$

the product ion does not undergo additional hydration, but instead reacts via a fast binary reaction with H_2O,

$$NO_2^+ \cdot 2H_2O + H_2O \rightarrow H_3O^+ \cdot H_2O + HNO_3. \tag{27}$$

Thus, the $NO_2^+ \cdot H_2O$ chemistry only leads back to the water cluster ions. As is shown in Fig. 13, the same is also true for the possible path involving the trace constituent N_2O_5 (Davidson *et al.*, 1978). At the present, no reasonable path has been found that would disrupt the formation of the water cluster ions in the stratosphere and troposphere.

Because of the relatively long lifetimes of atmospheric ions against ion–ion recombination, one must consider whether the water cluster ions would react with one or more trace atmospheric constituents to form even more stable ions. To predict this chemistry requires a knowledge of the identity and mixing ratios of the trace atmospheric neutrals, which at present is not fully available, either from *in situ* neutral measurements or from neutral-chemistry model calculations. Recently, however, the first *in situ* mass-spectrometric samplings of the stratospheric ion composition have been made. Such data can provide clues as to the identity of the terminal ions of this region and, when coupled with laboratory reaction studies, can then provide information about the trace neutral constituents.

The rocket-borne mass-spectrometer flights of Arnold *et al.* (1977) were the first such observations. Three high-latitude studies sampled the positive ions during descent through the altitudes 55 to 35 km, the lower of which is well into the stratosphere. Above 45 km, the substantial signal levels permitted high-resolution identification of the water cluster ions $H_3O^+ \cdot nH_2O$, with $n = 0$–3, as the dominant species in this region. However, as Fig. 16 shows, Arnold and co-workers reported a rather sharp transition at about 45 km from the water cluster ions to ions with e/m ratios of 29 ± 2, 42 ± 2, 60 ± 2, and 80 ± 2. In the region below this altitude the non-water-cluster ions become the dominant species. In addition to these data, recent balloon-borne mass-spectrometeric measurements have been made. Olson *et al.* (1977) have reported tentative results that suggest $H_3O^+ \cdot nH_2O$ ions with $n = 3$–5 dominate the ion composition down to 28 km. More recently balloon flights by Arijs *et al.* (1978) indicate that the water cluster ions are major species at 35 km. Other ions were observed in appreciable concentration, in particular, an ion with an e/m ratio of 96 ± 2. [See Notes added in proof, (c).]

The observation of the water cluster ions as the dominant species down to altitudes of 45 km or below, having been formed undoubtedly by the chemistry described in Fig. 13, places usefully low upper limits on neutral molecules that react rapidly with $H_3O^+ \cdot nH_2O$ ions. For example, laboratory studies (Fehsenfeld *et al.*, 1978) find that these water cluster ions proton transfer rapidly with methanol with rate constants that are independent of temperature in the stratospheric range. Since Arnold *et al.* (1977) found no masses that could be associated with protonated methanol ions or their hydrates, an

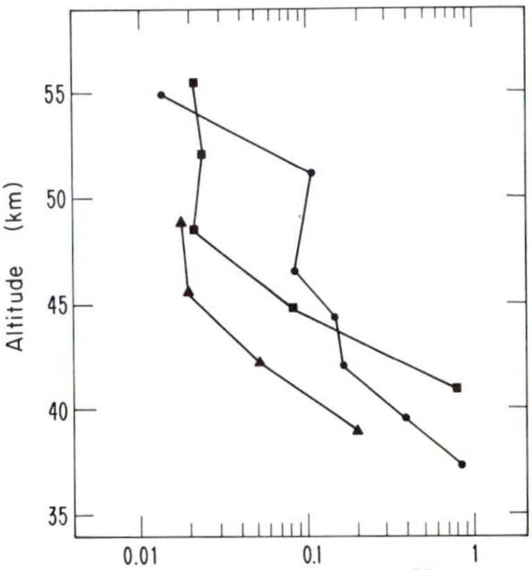

FIG. 16 Altitude profile of observed fractional abundance of non-water-cluster ions. [From Arnold *et al.*, (1977).] ●, F33; ■, F42; ▲, S18.

upper limit of 10^{-11} on the CH_3OH mixing ratio could be deduced for this region of the atmosphere. This upper limit is several orders of magnitude less than that inferred from the methane-oxidation chemistry with its present uncertainties (Crutzen, 1977).

The situation with ammonia appears to be quite similar. The reactions

$$H_3O^+ \cdot nH_2O + NH_3 \rightarrow NH_4{}^+ \cdot mH_2O + (n + 1 - m)H_2O \tag{28}$$

(where $m = 0, 1, \ldots, n$) have been found (Fehsenfeld and Ferguson, 1973) to be very fast for $n = 0$–2. No detailed atmospheric profiles of NH_3 have been obtained. The concentration in the lower troposphere is believed to be significant, but is expected to fall off rapidly with altitude because of the large solubility of ammonia and its consequent rainout of the atmosphere. Presumably, $NH_4{}^+ \cdot nH_2O$ ions do exist in the lower part of the troposphere and indeed may even be the dominant ions, but no ion composition measurements exist yet in this region. It is interesting to note that the measurement of $NH_4{}^+$ cluster ions in the atmosphere has the potential of being an extremely sensitive analytical method for NH_3 detection, having an estimated detection limit of 10^7 NH_3 molecules cm^{-3} (Ferguson and Fehsenfeld, 1973).

Arnold *et al.* (1977) have tentatively identified the non-water-cluster ions that were observed in their flights (Fig. 16) as arising from reactions of the $H_3O^+ \cdot nH_2O$ ions with formaldehyde and used their observations to deduce a CH_2O altitude profile. The reasonable choice of CH_2O as the trace neutral with which the water cluster ions were reacting was based on the fact that the proton affinity of CH_2O is greater than that of H_2O and that CH_2O is expected to be present in the atmosphere from the oxidation of methane. However, recent laboratory studies (Fehsenfeld *et al.*, 1978) of the CH_2O–H_2O ion chemistry do not support this interpretation. The results of this study are shown in Table IV. The first line, with its negative ΔH, shows that the unclustered H_3O^+ ion does indeed proton transfer with CH_2O. This ΔH value is not large in magnitude and the small difference in proton affinity is offset by the fact that H_2O bonds more strongly to H_3O^+ than to CH_3O^+. The reactions of the higher-order water cluster ions with CH_2O become endothermic. Even if the protonated formaldehyde ions were produced in some way in the stratosphere, the data in Table IV show that they will be rapidly converted to $H_3O^+ \cdot nH_2O$ ions by reaction with H_2O. Thus, the identity of the non-water-cluster ions observed by Arnold and co-workers remains an open question. Clearly, these valuable observations are strong stimuli for further *in situ* and laboratory observations. [See Notes added in proof, (d).]

Finally, the question as to whether ion chemistry in the troposphere and stratosphere can significantly perturb the neutral chemistry sometimes arises. So far the answer has been negative; no cases have yet been recognized in

TABLE IV

Measured Thermochemical Constants for the Reactions of Hydrated Hydronium and Protonated Formaldehyde Ions[a]

	Reaction	ΔH_T (kcal mole^{-1})	ΔS_T (cal mole^{-1} K^{-1})	Mean temperature (K)
(1)	$H_3O^+ + CH_2O \leftrightarrows CH_2OH^+ + H_2O$[b]	-5.2 ± 0.8	2.2 ± 3.0	470
(2)	$H_3O^+ \cdot H_2O + CH_2O \leftrightarrows CH_2OH^+ \cdot H_2O + H_2O$	-0.41 ± 0.29	2.6 ± 2.5	400
(3)	$H_3O^+ \cdot 2H_2O + CH_2O \leftrightarrows CH_2OH^+ \cdot 2H_2O + H_2O$	$+0.58 \pm 0.43$	2.3 ± 2.8	400
(4)	$H_3O^+ \cdot 3H_2O + CH_2O \leftrightarrows CH_2OH^+ \cdot 3H_2O + H_2O$	$+1.1 \pm 0.6$	4.1 ± 2.0	325
(5)	$CH_2OH^+ \cdot H_2O + CH_2O \leftrightarrows CH_2OH^+ \cdot CH_2O + H_2O$	-0.80 ± 0.13	-0.71 ± 2.2	400

[a] From Fehsenfeld *et al.* (1978).

[b] $k_1 = 2.2 \times 10^{-9}$ cm^3 s^{-1} (300 K).

which ion chemistry has a significant impact on neutral chemistry. One speculation that received considerable attention was the possibility of destruction of chlorofluoromethanes by ion reactions in the troposphere (before they could rise into the stratosphere where they are photodissociated and lead to ozone destruction). Detailed laboratory measurements (Fehsenfeld *et al.*, 1976b) excluded this possibility. The expected ambient tropospheric ions are simply too stable to have exothermic reaction channels with either $CFCl_3$ or CF_2Cl_2. One surprise that arose, however, was the finding that the reaction

$$H_3O^+ + CFCl_3 \rightarrow CCl_3^+ + H_2O + HF \quad (29)$$

was quite fast, $k_{29} = 4 \times 10^{10}$ cm^3 s^{-1}. The surprise lay in the fact that the generally accepted literature value of $\Delta H(CCl_3^+)$ implied that reaction (29) was quite endothermic! The occurrence of a large rate constant clearly indicates that the reaction is exothermic and the most recent determination (Lias and Ausloos, 1977) of the heat of formation of CCl_3^+ is consistent with this. When H_3O^+ was hydrated even once reaction (29) was found not to occur; the reaction in this case is almost certainly endothermic. Furthermore, H_3O^+ did not react with CF_2Cl_2. More stable ions that might exist in the troposphere such as NH_4^+ did not react. Presumably, none of these had exothermic reaction channels.

Recently some positive and negative ion reactions with N_2O_5 have been measured (Davidson *et al.*, 1978). These reactions include

$$O_2^+ + N_2O_5 \rightarrow NO_2^+ + NO_3 + O_2, \quad (30)$$

$$NO^+ + N_2O_5 \rightarrow \text{products}, \quad (31)$$

$$H_3O^+ + N_2O_5 \rightarrow \text{products}, \quad (32)$$

where $k_{30} = 7.0 \times 10^{-10}$, $k_{31} = 5.9 \times 10^{-10}$, and $k_{32} = 1.3 \times 10^{-9}$ cm^3 s^{-1}. If the ambient positive ions of the stratosphere, which are expected to be $H_3O^+ \cdot nH_2O$ ions with n mostly in the range 2–4, do react with N_2O_5, then two molecules of HNO_3 are produced. This is true in the case of reaction (32) no matter what the initial products are. Any hydrated $H_2NO_3^+$ ions produced will convert back to H_3O^+ hydrates via reaction (27), which produces another nitric acid molecule. This ion chemistry cycle is essentially an ion catalyzed reaction of H_2O and N_2O_5 to produce $2HNO_3$ that may be of some significance as an atmospheric source of HNO_3. At high enough hydrations, reactions such as

$$H_3O^+ \cdot 8H_2O + N_2O_5 \rightarrow H_3O^+ \cdot 7H_2O + 2HNO_3 \quad (33)$$

become exothermic and should be considered. In such a reaction, the $H_3O^+ \cdot 7H_2O$ ion would essentially act as a "surface" on which the $H_2O + N_2O_5 \rightarrow 2HNO_3$ reaction occurs.

VI. STRATOSPHERIC AND TROPOSPHERIC NEGATIVE ION CHEMISTRY

The negative ion chemistry of the stratosphere and troposphere is even more speculative than the positive ion chemistry of these regions. In addition to there being few measurements of the trace neutrals involved, there are no negative ion composition data. Furthermore, the D-region measurements, from which one could draw guidance, are relatively sparse and somewhat ambiguous at present, as pointed out in Section IV. Thus, our present understanding of the negative ion processes in the lower atmosphere stems entirely from laboratory studies. [See Notes added in proof, (e).]

The approach here has been to start with the D-region negative ion chemistry given in Fig. 9 and modify it to be in accord with the expected differences in neutral composition of the two regions. These considerations fall into two classes. First, to look for possible alternatives to the chemistry that leads to the NO_3^- ions and second, to consider whether these ions would react with trace neutral species that are suspected in the lower atmosphere to form even more stable ions.

The first necessary modification of the D-region scheme is to disregard the reactions of atomic oxygen, whose concentration below 50 km is negligible in comparison to that of O_3. In addition, NO will no longer play a role.

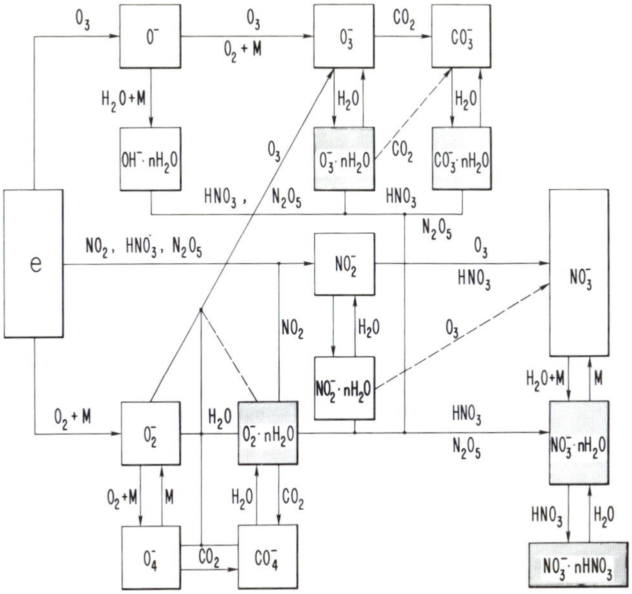

FIG. 17 Stratospheric and tropospheric negative ion chemistry.

Furthermore, the rapid formation of cluster ions can strongly alter the evolution of ion chemistry. For example, the reaction of O_3^- with CO_2 is known to decrease rapidly with O_3^- hydration (Fehsenfeld and Ferguson, 1973). Because H_2O bonds more strongly to O_3^- than to CO_3^-, this reaction may become endothermic when O_3^- becomes heavily hydrated (Dotan *et al.*, 1977). In the lower atmosphere, O_3^- hydration is likely to occur before the reaction of O_3^- with CO_2.

The reaction scheme based on these considerations is shown in Fig. 17. The shaded boxes and dashed lines represent points of considerable uncertainty. For example, it is not clear whether $O_2^- \cdot nH_2O$ ions will react with O_3, as do the unclustered O_2^- ions. The situation with $O_3^- \cdot nH_2O$, O_3^-, and CO_2 is the same; namely, the hydrated ions may not follow the same reaction paths as the unhydrated ions.

However, even if the paths that lead to NO_3^- are somewhat uncertain, its eventual formation does not appear to be in doubt. Both HNO_3 and N_2O_5 provide effective NO_3^- production routes. Some of these reactions are listed in Table V. Furthermore, it has been shown that HNO_3 exothermically displaces H_2O in hydrated negative ions (Fehsenfeld, 1975); therefore, the agent for NO_3^- formation can always enter the cluster.

The high stability of NO_3^- seems to ensure its role in the terminal negative ions of the lower atmosphere. At the present, no neutral species that can be expected to have a significant atmospheric concentration has been found to react with NO_3^-. The ion will, of course, hydrate and it is known that there are several molecules that can displace H_2O from a hydrated NO_3^- cluster. The molecules SO_2 (Fehsenfeld and Ferguson, 1973) and HNO_3 (Fehsenfeld *et al.*, 1975; Davidson *et al.*, 1977) are examples. Thus, it is likely that the terminal negative ions of the lower atmosphere are complex cluster ions, like $NO_3^- \cdot lH_2O \cdot mSO_2 \cdot nHNO_3$, with NO_3^- as the core ion.

TABLE V

Negative Ion Reactions with HNO_3 *and* N_2O_5 (300 K)

	Reactions	Rate constant[a] ($cm^3 s^{-1}$)	Source
(1)	$CO_3^- + HNO_3 \rightarrow NO_3^- + HCO_3$(?)	8.0 (−10)	Fehsenfeld *et al.* (1975)
(2)	$NO_2^- + HNO_3 \rightarrow NO_3^- + HNO_2$	1.6 (−9)	Fehsenfeld *et al.* (1975)
(3)	$Cl^- + HNO_3 \rightarrow NO_3^- + HCl$	1.6 (−9)	Fehsenfeld *et al.* (1975)
(4)	$CO_3^- + N_2O_5 \rightarrow NO_3^- + NO_3 + CO_2$	2.8 (−10)	Davidson *et al.* (1978)
(5)	$NO_2^- + N_2O_5 \rightarrow NO_3^- + 2NO_2$	7.0 (−10)	Davidson *et al.* (1978)
(6)	$Cl^- + N_2O_5 \rightarrow NO_3^- + ClNO_2$	9.4 (−10)	Davidson *et al.* (1978)

[a] 8.0 (−10) means 8.0×10^{-10}, etc.

Lastly, just as in the case for positive ions, negative ion reactions have been postulated as sinks for certain lower-atmosphere neutral constituents. One such case was proposed by Ruderman *et al.* (1976) in which the reaction

$$NO_3^- + O_3 \rightarrow NO_2^- + 2O_2 \tag{34}$$

was suggested as a possible stratospheric sink for ozone. This was an attempt to explain an observed solar cycle variation in the ozone concentration as a consequence of the known solar cycle variation of atmospheric ionization by galactic cosmic rays. However, it was found (Fehsenfeld *et al.*, 1976a) that reaction (34) has a rate constant less than 10^{-13} cm^3 s^{-1}, which is too low for this reaction to have significance in this case.

Another consideration has been that negative ion reactions might remove tropospheric N_2O, which provides the stratospheric NO that is the natural balance for the ozone layer. Such reactions would be of considerable importance because no major tropospheric sink for N_2O has been discovered. Thus, even slow processes would be quite significant, i.e., processes that would yield an N_2O lifetime of less than 100 years. Many possible atmospheric negative ions have exothermic reaction channels with N_2O, e.g., O_2^-, O_3^-, CO_3^-, and NO_2^-. However, all of these reactions are found to have rate constants below the limits of detectability, which range from $\leq 1 \times 10^{-12}$ cm^3 s^{-1} to $\leq 2 \times 10^{-14}$ cm^3 s^{-1} for the various ions (Fehsenfeld and Ferguson, 1976). Therefore, these reactions do not constitute a significant sink for N_2O.

Perhaps the most interesting prospect, and certainly the most controversial, have been attempts to propose chlorofluoromethane loss in the troposphere via negative ion reactions. Since these molecules are detected with extremely high sensitivity ($\sim$ a part per 10^{12}) by electron capture detectors, it was not unreasonable to look for a fast negative ion atmospheric chemistry. However, no chlorofluoromethane can compete with the formation of the primary negative ion O_2^- by termolecular electron attachment to O_2 due to the large abundance of the latter (Fehsenfeld and Albritton, 1977). This primary negative ion does react with $CFCl_3$ and CF_2Cl_2:

$$O_2^- + CFCl_3 \rightarrow Cl^- + CFCl_2 + O_2 \tag{35}$$

and

$$O_2^- + CF_2Cl_2 \rightarrow Cl^- + CF_2Cl + O_2 \tag{36}$$

with $k_{35} = 7.6 \times 10^{-10}$ cm^3 s^{-1} and $k_{36} = 2.1 \times 10^{-10}$ cm^3 s^{-1} (Fehsenfeld *et al.*, 1976b). However, reactions (35) and (36) are both only slightly exothermic and $O_2^- \cdot H_2O$ and higher hydrates, as well as O_3^-, CO_3^-, NO_2^-, and NO_3^- ions, are found to not react. Therefore, just as in the case of positive ions, it has been shown that negative ion reactions are not a significant sink for the chlorofluoromethanes in the lower atmosphere.

VII. SUMMARY

The ion chemistry of the major species of the high-altitude region of the atmosphere, the so-called E- and F-regions of the ionosphere above ~120 km, is now quite well understood. There are precise laboratory measurements of the reaction rate constants over the necessary temperature range (300–1000 K) and the reaction scheme has been critically tested by detailed and precise ionospheric observation. The remaining problems in high-altitude ion chemistry concern minor species such as metastable ion reactions and the reactions of doubly charged ions. There is some laboratory data on the metastable reactions and some ionospheric data on both the metastable and doubly charged ion reactions.

The ion chemistry of the atmosphere below ~80 km remains incompletely understood. There are numerous relevant laboratory measurements, although many more are needed, but direct atmospheric observations are limited and in some altitude regions nonexistent. In addition, the ion chemistry of the lower atmosphere is controlled by trace neutral constituents, whose concentrations are poorly known in many cases. Some of the critical trace neutral constituents have probably not yet been identified.

Notes Added in Proof

(a) Torr and Orsini (1978) have reexamined the AE data in the light of new laboratory measurements of $N_2^+ + e$ recombination rates and have revised downward the rate constants for two $O^+(^2D)$ reactions determined from these satellite data. First, the rate constant for the charge-transfer reaction of $O^+(^2D)$ with N_2 [Table I, reaction (3)] was found to be $(1.75 \pm 1.25) \times 10^{-10}$ cm^3 s^{-1}, which is now considerably smaller than the value extrapolated from Rutherford and Vroom's (1971) beam measurements. This revised value is in agreement with the thermal energy value $(1.5 \pm 0.5) \times 10^{-10}$ cm^3 s^{-1} obtained very recently by Glosik *et al.* (1978) in a selected-ion flow tube for possibly mixed ions $O^+(^1P, ^2D)$ reacting with N_2. Second, Torr and Orsini (1978) have revised the quenching rate constant for the reaction of $O^+(^2D)$ by O atoms [Table I, reaction (5)], now giving only an upper limit: $k \ll 3 \times 10^{-11}$ cm^3 s^{-1}.

The value obtained by Glosik *et al.* (1978) for the reaction of $O^+(^2P, ^2D)$ with O_2 [Table I, reaction (4)], $(1.3 \pm 0.4) \times 10^{-10}$ cm^3 s^{-1}, is an order of magnitude smaller than that obtained by Torr *et al.* (1979) from the AE data. For the $O_2^+(a\ ^4\Pi_u)$ and $NO^+(a\ ^3\Sigma^+)$ reactions [Table I, reactions (6)–(9)], the rate constants of Glosik *et al.* (1978) and those of Lindinger *et al.* (1975a) and Dotan *et al.* (1979) are in good agreement.

(b) Johnsen and Biondi (1978) have recently reported the first laboratory measurements of the rate constants of the reactions of O^{2+} ions with N_2 and O_2 at thermal energies, $(1.3 \pm 0.3) \times 10^{-9}$ and $(1.5 \pm 0.3) \times 10^{-9}$ cm^3 s^{-1}, respectively. This fast rate constant of the reaction of O^{2+} with N_2 differs considerably from the low upper limit, $k_{13} \leq 10^{-11}$ cm^3 s^{-1} given by Brieg *et al.* (1977) from an analysis of AE data. Thus, another conflict between satellite and laboratory data has arisen (see above).

(c) Very recent ballon flights by Arnold *et al.* (1978) at 37 km also found the conversion to non-water-cluster ions, but the three lower masses reported earlier were not observed. An ion at mass 78 ± 1 may be the same ion that was reported earlier at 80 ± 2 amu. Several ions with masses ≥ 96 amu were identified as non-water-cluster ions.

(d) Ferguson (1978) has recently interpreted the stratospheric non-water-cluster ions as protonated sodium hydroxide ions $NaOH_2{}^+$ and related clusters $NaOH_2{}^+ \cdot nH_2O \cdot mNaOH$. These ions have mass numbers that match almost all of the numerous observed ions within the uncertainties of the observations. This interpretation argues that the well-known atmospheric sodium layer, which arises from the ablation of meteors, will give rise to NaOH, which, because of its exceedingly high proton affinity, will react with the water-cluster ions.

(e) The first *in situ* measurements of stratospheric negative ions were reported recently by Arnold and Henschen (1978), resulting from balloon-borne observations at 33–37 km. The range of large mass numbers, 125–295 amu, suggested $NO_3{}^-$ as the core ion of a variety of clusters.

REFERENCES

Albritton, D. L. (1978). *At. Data Nucl. Data Tables* **22**, 1–101.

Albritton, D. L., Dotan, I., Lindinger, W., McFarland, M. Tellinghuisen, J., and Fehsenfeld, F. C. (1977a). *J. Chem. Phys.* **66**, 410–421.

Albritton, D. L., Schmeltekopf, A. L., Harrop, W. J., Zare, R. N., and Czarny, J. (1977b). *J. Mol. Spectrosc.* **67**, 157–184.

Albritton, D. L., Dotan, I., Fehsenfeld, F. C., Streit, G., and Ferguson, E. E. (1979). *Planet Space Sci.* (to be submitted).

Arijs, E., Ingels, J., and Nevejans, D. (1978). *Nature* (*London*) **271**, 642–644.

Arnold, F., and Krankowsky, D. (1974). *Int. COSPAR Symp. Sol.-Terr. Phys., 4th, Sao Paulo.*

Arnold, F., and Krankowsky, D. (1977a) *In* "Dynamical and Chemical Coupling" (B. Grandal and J. A. Holtet, eds.), pp. 93–127. Riedel Publ., Dordrecht, Netherlands.

Arnold, F., and Krankowsky, D. (1977b). *J. Atmos. Terr. Phys.* **39**, 625–629.

Arnold, F., and Henschen, G. (1978). *Nature* (*London*) **275**, 521–522.

Arnold, F., Kissel, J., Krankowsky, D., Wieder, H., and Zähringer, J. (1971). *J. Atmos. Terr. Phys.* **33**, 1169–1175.

Arnold, F., Krankowsky, D., and Marien, K. H. (1977). *Nature (London)* **267**, 30–32.
Arnold, F., Böhringer, H., and Henschen, G. (1978). *Geophys. Res. Lett.* **5**, 653–656.
Banks, P. M., and Kockarts, G. (1973). "Aeronomy." Academic Press, New York.
Bates, D. R. (1974). *J. Atmos. Terr. Phys.* **36**, 2287–2307.
Bates, D. R., and Nicolet, M. (1960). *J. Atmos. Terr. Phys.* **18**, 65–70.
Brieg, E. L., Torr, M. R., Torr, D. G., Hanson, W. B., Hoffman, J. H., Walker, J. C. G., and Nier, A. O. (1977). *J. Geophys. Res.* **82**, 1008–1012.
Cole, R. K., and Pierce, E. T. (1965). *J. Geophys. Res.* **70**, 2735–2749.
Crutzen, P. J. (1977). Personal communication.
Crutzen, P. J., Isaksen, I. S. A., and McAfee, J. R. (1978). *J. Geophys. Res.* **83**, 345–363.
Dalgarno, A. (1970). *Ann. Geophys.* **26**, 601–607.
Dalgarno, A., and McElroy, M. B. (1965). *Planet. Space Sci.* **13**, 947–957.
Dalgarno, A., Hanson, W. B., Spencer, N. W., and Schmerling, E. R. (1973). *Radio Sci.* **8**, 263–266.
Davidson, J. A., Viggiano, A. A., Howard, C. J., Dotan, I., Fehsenfeld, F. C., Albritton, D. L., and Ferguson, E. E. (1978). *J. Chem. Phys.* **68**, 2085–2087.
Dickinson, P. H. G., and Sayers, J. (1960). *Proc. Phys. Soc., London* **76**, 137–148.
Donahue, T. M. (1966). *Planet. Space Sci.* **14**, 33–48.
Dotan, I., Davidson, J. A., Streit, G. E., Albritton, D. L., and Fehsenfeld, F. C. (1977). *J. Phys. Chem.* **67**, 2874–2879.
Dotan, I., Albritton, D. L., Fehsenfeld, F. C., Streit, G. E., and Ferguson, E. E. (1978a). *J. Chem. Phys.* **68**, 5414–5416.
Dotan, I., Davidson, J. A., Fehsenfeld, F. C., and Albritton, D. L. (1978b). *J. Geophys. Res.* **83**, 4036–4038.
Dotan, I., Albritton, D. L., and Fehsenfeld, F. C. (1979). *J. Chem. Phys.* (to be submitted).
Dubin, M., Hull, A. R., and Champion, K. S. W. (1976). "U.S. Standard Atmosphere, 1976." U.S. Gov. Print. Off., Washington, D.C.
Dunkin, D. B., Fehsenfeld, F. C., Schmeltekopf, A. L., and Ferguson, E. E. (1968). *J. Chem. Phys.* **49**, 1365–1371.
Durden, D. A., Kebarle, P., and Good, A. (1969). *J. Chem. Phys.* **50**, 805–813.
Fehsenfeld, F. C. (1975). *J. Chem. Phys.* **63**, 1686–1687.
Fehsenfeld, F. C. (1979). *J. Chem. Phys.* (to be published).
Fehsenfeld, F. C., and Albritton, D. L. (1977). *Geophys. Res. Lett.* **4**, 61–63.
Fehsenfeld, F. C., and Ferguson, E. E. (1969). *J. Geophys. Res.* **74**, 2217–2222.
Fehsenfeld, F. C., and Ferguson, E. E. (1973). *J. Chem. Phys.* **59**, 6272–6276.
Fehsenfeld, F. C., and Ferguson, E. E. (1974). *J. Chem. Phys.* **61**, 3181–3193.
Fehsenfeld, F. C., and Ferguson, E. E. (1976). *J. Chem. Phys.* **64**, 1853–1854.
Fehsenfeld, F. C., Albritton, D. L., Burt, J. A., and Schiff, H. I. (1969a). *Can. J. Chem.* **47**, 1793–1795.
Fehsenfeld, F. C., Ferguson, E. E., and Bohme, D. K. (1969b). *Planet. Space Sci.* **17**, 1759–1762.
Fehsenfeld, F. C., Mosesman, M., and Ferguson, E. E. (1971). *J. Chem. Phys.* **55**, 2120–2125.
Fehsenfeld, F. C., Howard, C. J., and Schmeltekopf, A. L. (1975). *J. Chem. Phys.* **63**, 2835–2841.
Fehsenfeld, F. C., Ferguson, E. E., Streit, G. E., and Albritton, D. L. (1976a). Science **194**, 544–545.
Fehsenfeld, F. C., Crutzen, P. J., Schmeltekopf, A. L., Howard, C. J., Albritton, D. L., Ferguson, E. E., Davidson, J. A., and Schiff, H. I. (1976b). *J. Geophys. Res.* **81**, 4454–4460.
Fehsenfeld, F. C., Dotan, I., Albritton, D. L., Howard, C. J., and Ferguson, E. E. (1978). *J. Geophys. Res.* **83**, 1333–1336.
Ferguson, E. E. (1971). *Rev. Geophys. Space Phys.* **9**, 997–1008.
Ferguson, E. E. (1972). *Radio Sci.* **7**, 397–401.
Ferguson, E. E. (1973). *At. Data Nucl. Data Tables* **12**, 159–178.

Ferguson, E. E. (1978). *Geophys. Res. Lett.* **5**, 1035–1038.
Ferguson, E. E., Fehsenfeld, F. C., Goldan, P. D., and Schmeltekopf, A. L. (1965). *J. Geophys. Res.* **70**, 4323–4329.
French, M. A., Hills, L. P., and Kebarle, P. (1973). *Can. J. Chem.* **51**, 456–461.
Gattinger, R. L. (1976). *Planet. Space Sci.* **24**, 999–1001.
Gioumousis, G., and Stevenson, D. P. (1958). *J. Chem. Phys.* **29**, 294–299.
Glosik, J., Rakshit, A. B., Twiddy, N. D., Adams, N. G., and Smith, D. (1978). *J. Phys. B* **11**, 3365–3379.
Goldberg, R. A., and Aiken, A. C. (1971). *J. Geophys. Res.* **76**, 8352–8364.
Good, A., Durden, D. A., and Kebarle, P. (1970). *J. Chem. Phys.* **52**, 222–229.
Hiemerl, J. M., and Vanderhoff, J. A. (1974). *J. Chem. Phys.* **60**, 4362–4368.
Howard, C. J., Rundle, H. W., and Kaufman, F. (1971). *J. Chem. Phys.* **55**, 4772–4776.
Howard, C. J., Bierbaum, V. M., Rundle, H. W., and Kaufman, F. (1972). *J. Chem. Phys.* **57**, 3491–3497.
Howard, C. J., Fehsenfeld, F. C., and McFarland, M. (1974). *J. Chem. Phys.* **60**, 5086–5089.
Huber, B. A., Cosby, P. C., Peterson, J. R., and Moseley, J. T. (1977). *J. Chem. Phys.* **66**, 4520–4526.
Johnsen, R., and Biondi, M. A. (1978). *Geophys. Res. Lett.* **5**, 847–848.
Johnsen, R., Huang, C.-M., and Biondi, M. A. (1975). *J. Chem. Phys.* **63**, 3374–3378.
Johnson, C. Y., Meadows, E. B., and Holmes, J. C. (1958). *J. Geophys. Res.* **63**, 443–444.
Langstroth, G. F. O., and Hasted, J. B. (1962). *Discuss. Faraday Soc.* **33**, 298.
Lias, S. G., and Ausloos, P. (1977). *Int. J. Mass Spectrom. Ion Phys.* **23**, 273–292.
Lin, S. L., and Bardsley, J. N. (1977). *J. Chem. Phys.* **66**, 435–445.
Lindinger, W., Fehsenfeld, F. C., Schmeltekopf, A. L., and Ferguson, E. E. (1974). *J. Geophys. Res.* **79**, 4753–4756.
Lindinger, W., Albritton, D. L., McFarland, M., Fehsenfeld, F. C., Schmeltekopf, A. L., and Ferguson, E. E. (1975a). *J. Chem. Phys.* **62**, 4101–4110.
Lindinger, W., Albritton, D. L., Fehsenfeld, F. C., and Ferguson, E. E. (1975b). *J. Geophys. Res.* **80**, 3725–3726.
Lindinger, W., Albritton, D. L., Fehsenfeld, F. C., and Ferguson, E. E. (1975c). *J. Chem. Phys.* **63**, 3238–3242.
McFarland, M., Albritton, D. L., Fehsenfeld, F. C., Ferguson, E. E., and Schmeltekopf, A. L. (1973). *J. Chem. Phys.* **59**, 6629–6635.
McFarland, M., Albritton, D. L., Fehsenfeld, F. C., Ferguson, E. E., and Schmeltekopf, A. L. (1974). *J. Geophys. Res.* **79**, 2925–2926.
Narcisi, R. S., and Bailey, A. D. (1965). *J. Geophys. Res.* **70**, 3687–3700.
Narcisi, R. S., Bailey, A. D., Della Lucca, L., Sherman, C., and Thomas, D. M. (1971). *J. Atmos. Terr. Phys.* **33**, 1147–1159.
Narcisi, R. S., Bailey, A. D., Wlodyka, L. E., and Philbrick, C. R. (1972). *J. Atmos. Terr. Phys.* **34**, 647–658.
Neynaber, R. A., Layton, J. K., Rutherford, J. A., and Vroom, D. A. (1970). Gulf Radiat. Technol. Rep. GA-10286. (IRT Corporation, Box. 80817, San Diego, CA 92138).
Norton, R. B., Van Zandt, T. E., and Denison, J. S. (1963). *Proc. Int. Conf. Ionosphere, 1962, London* p. 26.
Olson, J. R., Amme, R. C., Brooks, J. N., Murcray, D. G., and Keller, G. E. (1977). *Trans. Am. Geophys. Union* **58**, 1201.
Oppenheimer, M., Dalgarno, A., and Brinton, H. C. (1976). *J. Geophys. Res.* **81**, 3762–3766.
Oppenheimer, M., Constantinides, E. R., Kirby-Docken, K., Victor, G. A., Dalgarno, A., and Hoffman, J. H. (1977). *J. Geophys. Res.* **82**, 5485–5492.
Orsini, N., Torr, D. G., Torr, M. R., Brinton, H. C., Brace, L. H., Nier, A. O., and Walker, J. C. G. (1977). *J. Geophys. Res.* **82**, 4829–4833.

Paltridge, G. W. (1965). *J. Geophys. Res.* **70**, 2751–2761.
Paltridge, G. W. (1966). *J. Geophys. Res.* **71**, 1945–1952.
Parkes, D. A. (1971). *Trans. Faraday Soc.* **67**, 711–729.
Parkes, D. A. (1972). *J. Chem. Soc., Faraday Trans. I* **68**, 627–640.
Phelps, A. V. (1969). *Can. J. Chem.* **47**, 1783–1793.
Reid, G. C. (1976). *Adv. At. Mol. Phys.* **63**, 375–413.
Reid, G. C. (1977). *Planet. Space Sci.* **25**, 275–290.
Ruderman, M. A., Foley, H. M., and Chamberlain, J. W. (1976). *Science* **192**, 555–557.
Rusch, D. W., Torr, D. G., Hays, P. B., and Walker, J. C. G. (1977). *J. Geophys. Res* **82**, 719–722.
Rutherford, J. A., and Vroom, D. A. (1971). *J. Chem. Phys.* **55**, 5622–5624.
St.-Maurice, J,-P., and Torr, D. G. (1978). *J. Geophys. Res.* **83**, 969–977.
Schmeltekopf, A. L., Ferguson, E. E., and Fehsenfeld, F. C. (1968). *J. Chem. Phys.* **48**, 2966–2973.
Smith, D., and Fouracre, R. A. (1968). *Planet. Space Sci.* **16**, 243–252.
Smith, D., Adams, N. G., and Grief, D. (1977). *J. Atmos. Terr. Phys.* **39**, 513–521.
Snuggs, R. M., Volz, D. J., Gatland, I. R., Schummers, J. H., Martin, D. W., and McDaniel, E. W. (1971). *Phys. Rev.* **3**, 487–493.
Stebbings, R. F., Turner, B. R., and Rutherford, J. A. (1966). *J. Geophys. Res.* **71**, 771–784.
Stelman, D., Moruzzi, J. L., and Phelps, A. V. (1972). *J. Chem. Phys.* **56**, 4183–4189.
Su, T., and Bowers, M. T. (1973a). *J. Chem. Phys.* **58**, 3027–3037.
Su. T., and Bowers, M. T. (1973b). *Int. J. Mass Spectrom. Ion Phys.* **12**, 347–356.
Torr, D. G., and Orsini, N. (1977). *Planet. Space Sci.* **25**, 1171–1176.
Torr, D. G., and Orsini, N. (1978). *Geophys. Res. Lett.* **5**, 657–659.
Torr, D. G., Orsini, N., Torr, M. R., Hanson, W. B., Hoffman, J. H., and Walker, J. C. G. (1977). *J. Geophys. Res.* **12**, 1631–1634.
Torr, D. G., Donahue, K., Rusch, D. W., Torr, M. R., Nier, A. O., Kayser, D., Hanson, W. B., and Hoffman, J. H. (1979). *J. Geophys. Res.* **84**, 387–392.
Torr, M. R., St.-Maurice, J.-P., and Torr, D. G. (1977). *J. Geophys. Res.* **82**, 3287–3290.
Turner, D. L., and Conway, D. C. (1976). *J. Chem. Phys.* **65**, 3944–3947.
Viehland, L. A., and Mason, E. A. (1977). *J. Chem. Phys.* **66**, 422–434.
Walker, J. C. G., Torr, D. G., Hays, P. B., Rusch, D. W., Docken, K., Victor, G., and Oppenheimer, M. (1975). *J. Geophys. Res.* **80**, 1026–1029.
Yang, J.-H., and Conway, D. C. (1964). *J. Chem. Phys.* **40**, 1729–1735.
Zbinden, P. A., Hidalgo, M. A., Eberhardt, P., and Geiss, J. (1975). *Planet. Space Sci.* **23**, 1621–1642.
Zipf, E. C., Borst, W. L., and Donahue, T. M. (1970). *J. Geophys. Res.* **75**, 6371–6376.

Chapter 3

Classical ion–molecule collision theory

TIMOTHY SU

DEPARTMENT OF CHEMISTRY
SOUTHEASTERN MASSACHUSETTS UNIVERSITY
NORTH DARTMOUTH, MASSACHUSETTS

MICHAEL T. BOWERS

DEPARTMENT OF CHEMISTRY
UNIVERSITY OF CALIFORNIA
SANTA BARBARA, CALIFORNIA

ISBN-0-12-120801-X

I. INTRODUCTION

Theoretical treatments of ion–molecule reactions are important. They not only provide rational explanations to experimental observations, but can also be used in conjunction with experimental results to investigate reaction mechanisms. In this chapter, we will concentrate on the low energy (less than a few electron volts) ion–molecule collision theories. The following sections will begin with the Langevin ion-induced dipole theory for nonpolar molecules. The various ion–dipole theories will then be presented. Comparison of these theories with experimental results will be discussed. Following this, the theoretical studies of the effect of the ion–quadrupole potential and the induced dipole-induced dipole potential will be given. Finally, the effect of the anisotropy of polarizability of molecules on ion–molecule collisions will be discussed.

II. PURE POLARIZATION THEORY

The classical capture collision cross section for a structureless point charge and a point polarizable molecule was first solved by Eyring *et al.* (1936) for the reaction of H_2^+ with H_2. Vogt and Wannier (1954) then derived a general form of the collision cross section based on a model first developed by Langevin (1905). This model was later elaborated by Gioumousis and Stevenson (1958). It was subsequently reformulated and extensively discussed by a number of authors (McDaniel, 1964; Futrell and Tiernan, 1968; Henglein, 1970). In this section, a summary of the theoretical treatment will be given.

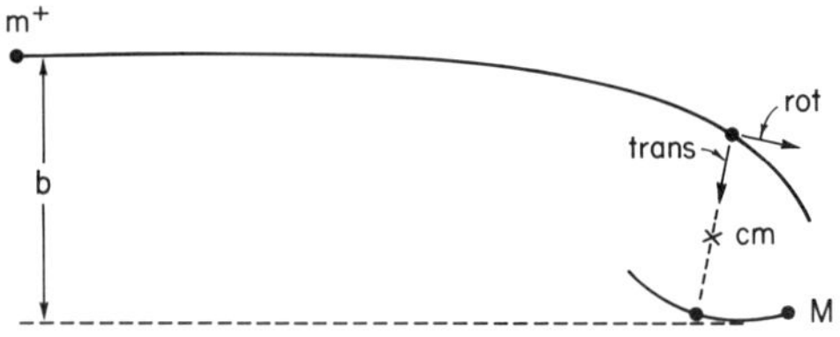

FIG. 1 Schematic ion–molecule collisions.

The pure polarization theory calculates the collision cross section of an ion–molecule pair with a given relative velocity. Both the ion and the neutral molecule are assumed to be point particles with no internal energy. Consider an ion and molecule approaching each other with relative velocity v and impact parameter b, as shown in Fig. 1. The classical potential at an ion–molecule separation r is given by

$$V(r) = -\int_r^\infty F(r)\,dr = -\int_r^\infty \frac{2\alpha q^2}{r^5}\,dr = -\frac{\alpha q^2}{2r^4}, \tag{1}$$

where α is the polarizability of the neutral and q the charge on the ion. Hence for $r < \infty$, the relative energy of the system E_r is a sum of the instantaneous kinetic energy $E_\mathrm{kin}(r)$ and the potential energy

$$E_\mathrm{r} = \tfrac{1}{2}\mu v^2 = E_\mathrm{kin}(r) + V(r), \tag{2}$$

where μ is the reduced mass. From Fig. 1, it is apparent that there are two components to the kinetic energy term

$$E_\mathrm{kin}(r) = E_\mathrm{rot}(r) + E_\mathrm{trans}(r), \tag{3}$$

where $E_\mathrm{trans}(r)$ is the translational energy along the line of centers of the collision and $E_\mathrm{rot}(r)$ is the energy of relative rotation of the particles. The rotational energy of the system is given by

$$E_\mathrm{rot}(r) = L^2/2\mu r^2 = \mu v^2 b^2/2r^2 = E_\mathrm{r} b^2/r^2, \tag{4}$$

where L is the classical orbiting angular momentum of the two particles. This rotational energy is associated with an outwardly directed centrifugal force. Thus, the effective potential of the ion–molecule system is the sum of the central potential energy $V(r)$ and the centrifugal potential energy $E_\mathrm{rot}(r)$;

$$V_\mathrm{eff}(r) = -(q^2\alpha/2r^4) + (L^2/2\mu r^2). \tag{5}$$

The total relative energy of the system is then

$$E_\mathrm{r} = E_\mathrm{trans}(r) + V_\mathrm{eff}(r). \tag{6}$$

A plot of $V_\mathrm{eff}(r)$ versus r at constant E_r for several values of the impact parameter is given in Fig. 2. For $b = 0$, there is no contribution from the centrifugal potential and $V_\mathrm{eff}(r)$ is attractive for all values of r. Hence, the system passes through the scattering center (i.e., $r = 0$). When $b > 0$, the centrifugal potential terms in $V_\mathrm{eff}(r)$ creates a so-called "centrifugal barrier" to a capture collision. For the special case in which the centrifugal barrier height equals E_r ($b = b_\mathrm{c}$ in Fig. 2), V_eff at r_c is equal to E_r. Thus from Eq. (6), $E_\mathrm{trans}(r) = 0$, and the particles will orbit the scattering center with a constant ion–molecule separation r_c (Fig. 3). For all impact parameters $< b_\mathrm{c}$, a

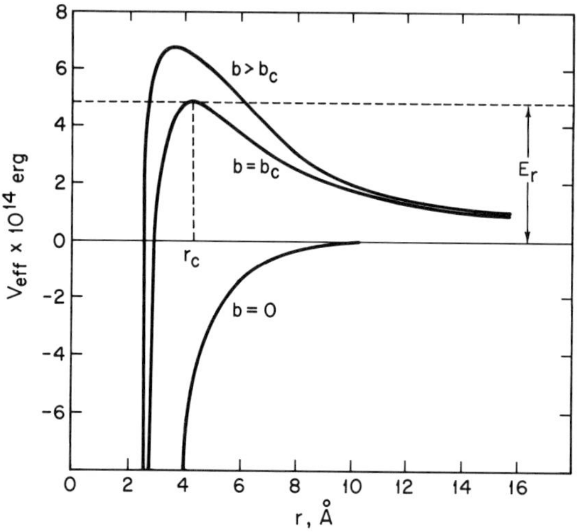

FIG. 2 Plot of V_{eff} versus r from Eq. (5) for N_2^+ colliding with N_2.

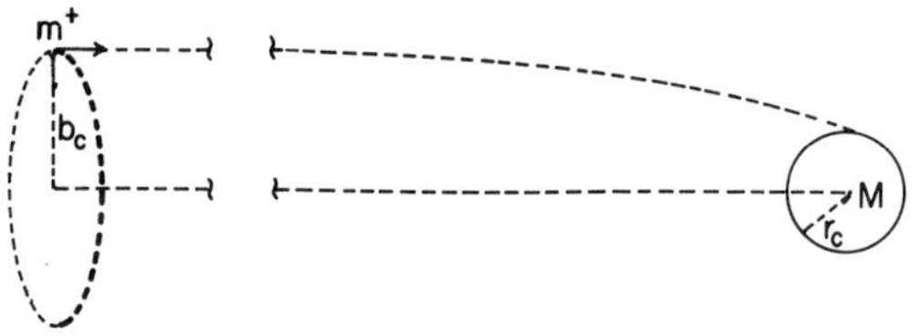

FIG. 3 Capture cross section of ion–molecule collisions.

capture collision will occur. (A capture collision is defined as one in which the particles have appropriate energy and impact parameters to pass through $r = 0$.) For $b > b_c$, the centrifugal barrier prevents capture and the particles are merely scattered at large r. Imagine a circle with radius b_c perpendicular to the line of collision at infinite ion–molecule separation (see Fig. 3). All ions that pass through the circle when approaching the neutral molecule will pass through $r = 0$ (i.e., capture collision will occur). The area of the circle is called the capture collision cross section σ for a given relative velocity v:

$$\sigma(v) = \pi b_c^{\,2}(v). \tag{7}$$

To determine the capture cross section at a given relative velocity, one has to calculate the critical impact parameter such that the maximum value of V_{eff} is equal to E_r (see Fig. 2). This condition occurs when $\partial V_{eff}(r)/\partial(r) = 0$

and $E_r = V_{eff}(r)$ at $r = r_c$:

$$\frac{\partial V_{eff}(r)}{\partial r} = 0 = -\frac{L^2}{\mu r^3} + \frac{2q^2\alpha}{r^5} \tag{8}$$

$$V_{eff}(r) = \frac{L^2}{2\mu r^2} - \frac{q^2\alpha}{2r^4} = E_r. \tag{9}$$

Equations (8) and (9) give

$$r_c = (q/b_c)(2\alpha/E_r)^{1/2}, \tag{10}$$

$$b_c = (2q^2\alpha/E_r)^{1/4}, \tag{11}$$

and

$$r_c = b_c/\sqrt{2} \tag{12}$$

Finally, the capture cross section and rate constant are given by

$$\sigma_c(v) = \pi b_c^{\,2} = \pi q(2\alpha/E_r)^{1/2}, \tag{13}$$

$$k_c = v\sigma_c = v\pi q(2\alpha/E_r)^{1/2} = 2\pi q(\alpha/\mu)^{1/2}. \tag{14}$$

Hence, the pure polarization theory predicts that the microscopic capture cross section is inversely proportional to v (or $E_r^{1/2}$) and that the capture rate constant is independent of v. The above derivation is for low energy ion–molecule collisions. The high energy limit of Eq. (13) is zero according to the Langevin theory. However, since the collision partners are not actually point charges, the higher energy limit should be $\pi r_k^{\,2}$, where r_k is the radius of the hard core gas kinetic cross section. This limit gives an inverse v^2 dependence of the cross section at energies above a few electron volts (Boelnijk and Hamill, 1962). The pure polarization theory agrees quite well with some simple low energy ion–molecule reaction rate constants but underestimates the rate constants of most of ion–polar molecule collisions. The various theoretical treatments of ion–molecule collisions in which the neutral molecule has a permanent dipole moment will be reviewed in the next section.

III. ION–DIPOLE THEORY

In this section the current state of the theory of ion–dipole collisions will be reviewed. Emphasis will be on the "macroscopic" average dipole orientation (ADO) theory that predicts low energy rate constants and cross sections and their dependence on energy and temperature.

A. The Locked Dipole Approximation

It was determined early (Moran and Hamill, 1963) that the contribution of the ion–dipole potential to the collision cross section could not be ignored for target molecules with significant permanent dipole moments. Theard and Hamill (1962) and Moran and Hamill (1963) treated the reaction between an ion and a polar molecule by introducing the ion–point dipole potential (15),

$$\phi = -(q\mu_D/r^2)\cos\theta, \tag{15}$$

where μ_D is the dipole moment of the molecule and θ the angle the dipole makes with the line of centers of the collision. These authors used the simplifying assumption that the dipole "locks in" on the ion and set $\theta = 0°$. Thus the effective potential of the system becomes

$$V_{\text{eff}}(r) = \frac{L^2}{2\mu r^2} - \frac{\alpha q^2}{2r^4} - \frac{\mu_D q}{r^2}. \tag{16}$$

Using an approach similar to the pure polarization theory, the capture collision rate constant becomes

$$k_{\text{LD}}(v) = \frac{2\pi q}{\mu^{1/2}}\left[\alpha^{1/2} + \frac{\mu_D}{v}\right]. \tag{17}$$

For thermal ion velocities, the rate constant becomes (Gupta *et al.*, 1967)

$$\bar{k}_{\text{LD}}(\text{therm}) = 2\pi q\left(\frac{\alpha}{\mu}\right)^{1/2} + \frac{2\pi q\mu_D}{\mu}\int_0^\infty \frac{P(v)}{v}\,dv, \tag{18}$$

where $P(v)$ is the Maxwell–Boltzmann distribution of relative velocities. Integrating Eq. (18) gives

$$\bar{k}_{\text{LD}}(\text{therm}) = \frac{2\pi q}{\mu^{1/2}}\left[\alpha^{1/2} + \mu_D\left(\frac{2}{\pi KT}\right)^{1/2}\right], \tag{19}$$

where K is Boltzmann's constant and T the absolute temperature. Equations (17) and (18) provide useful upper limits to the ion–dipole capture rate constants. However, they seriously overestimate the dipole effect (Su and Bowers, 1973a–c, 1974).

B. The Frozen Rotor Approximation

In this model (Dugan and Magee, 1966), it was assumed that the polar rotor molecule is fixed at angle θ with respect to the line of centers of the collision. The capture collision cross section determined by this approximation is given by

$$\sigma_\theta = \pi(\mu_D q/E_r)\cos\theta + (2\alpha q^2/E_r)^{1/2}. \tag{20}$$

The average cross section is obtained by integrating over all orientations under the constraint that σ_θ is always positive. The result is given by

$$\bar{\sigma}_\theta = \pi\frac{\mu_D q}{4E_r}\left(1 - \frac{E_r}{\epsilon_c}\right) + \left(\frac{\alpha q^2}{2E_r}\right)^{1/2} + \frac{\alpha q}{\mu_D}, \tag{21}$$

where $\epsilon_c = \mu_D{}^2/2\alpha$. The thermal average rate constant is given by (Dugan, 1973)

$$\bar{k}_\theta = \left(\frac{8\pi}{\mu}\right)^{1/2}(KT)^{-3/2}\left[\frac{\mu_D q}{4E_r}(KT) + \left(\frac{\alpha q}{\mu} - \frac{\mu q}{4\epsilon_c}\right)(KT)^2 + \frac{3}{2}\left(\frac{\alpha q^2}{2E_r}\right)^{1/2}(KT)^{3/2}\right]. \tag{22}$$

Rate constants calculated from this approximation are usually higher than experimental results. Comparison with experimental rate constants will be presented later.

C. *Ion–Dipole Trajectory Calculations*

Dugan and Magee (1967) have solved numerically the equations of motion for the collision of an ion with a rotating polar molecule. A summary of this elegant microscopic theory will be given here. Their method used the equations of motion in Lagrangian form:

$$\frac{d}{dt}\left(\frac{\partial L}{\partial \dot{q}_i}\right) = \frac{\partial L}{\partial q_i}, \tag{23}$$

where L is the Lagrangian function, the q_i are generalized coordinates, and the $\dot{q}_i$ are generalized velocities. The appropriate coordinate system for a "diatomic" polar molecule and a positive ion is given in Fig. 4, and the Lagrangian function in Eq. (24)

$$L = \tfrac{1}{2}\mu(\dot{X}^2 + \dot{Y}^2 + \dot{Z}^2) + \tfrac{1}{2}I(\dot{x}_1{}^2 + \dot{x}_2{}^2 + \dot{x}_3{}^2) + \frac{q\mu_D}{r^2}\cos\theta + \frac{q^2\alpha}{2r^4}, \tag{24}$$

where I is the moment of inertia of the polar molecule. For a diatomic rotor, the rotation about the molecular axis can be ignored. Hence, x_3 can be

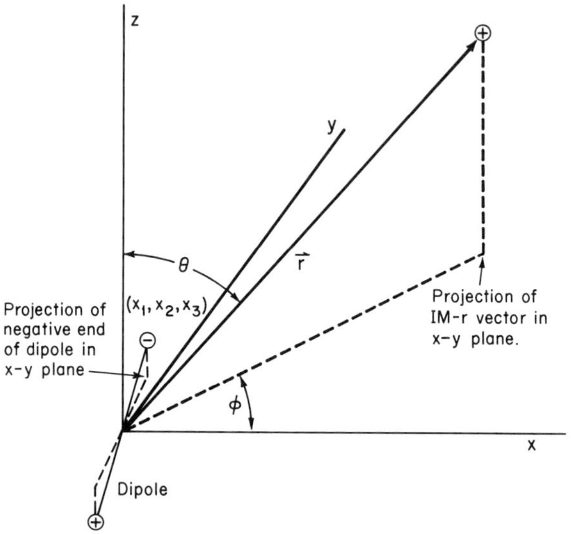

FIG. 4 Coordinate system for an ion–diatomic rotor collision.

expressed in terms of x_1 and x_2. The Lagrangian equations of motion can then be solved numerically to obtain $X(t)$, $Y(t)$, $Z(t)$, $x_1(t)$, and $x_2(t)$. Thus, the microscopic trajectories (time–history plot) can be traced out for a given initial velocity, impact parameter, dipole orientation, and rotational energy of the dipole rotor. Trajectories were calculated for many impact parameters at a given rotational energy and relative translational energy. For each impact parameter, many trajectories were calculated with random initial orientations for the rotating polar molecule. To find the average capture collision cross section, the fraction of "capture" type collisions is calculated for a given impact parameter. A "capture" type collision is assumed to occur when the ion and polar molecule approached a predetermined separation (usually 2–3 Å). The average capture cross section is defined as

$$\sigma_c(v) = \pi \int_0^{b_m 2} C_R(b)\, d(b^2) \tag{25}$$

where C_R is the fraction of collisions which tend to capture and b_m is the impact parameter at which C_R becomes zero. From a series of trajectory calculations, it was estimated that the numerical capture cross section depends on relative translational energy E_r approximately as $E_r^{-0.60}$ to $E_r^{-0.75}$ at 300–500 K (Dugan and Canright, 1972). If the cross section is assumed to be

$$\sigma_c(v) = (\sigma_0)_c E_r^{-n} \tag{26}$$

the average thermal rate constant is then given by (Dugan, 1973)

$$k_c(\text{scaled}) = \frac{1.956 \times 10^{-4}}{T^{3/2}\mu^{1/2}} (KT)^{2-n}(\sigma_0)\Gamma(2-n), \tag{27}$$

where T is in degrees Kelvin, μ in atomic mass units, KT in electron volts, and $(\sigma_0)_c$ in angstroms squared (electron volts)^n. Comparison of these calculations with experimental results will be given later.

D. *The Average Dipole Orientation (ADO) Theory*

It is generally agreed that a permanent dipole moment has a much smaller effect on ion–molecule capture collision cross sections than that predicted by the "locked dipole" model because the ion field does not completely quench the rotational angular momentum of the polar molecule and only partial "locking" can occur. Bowers and Laudenslager (1972) made the first attempt to estimate experimentally the degree of locking of the dipole. They measured thermal energy charge transfer rate constants from various rare gas ions to the three geometric isomers of difluoroethylene by ion cyclotron resonance spectrometry. All of these isomers have essentially the same angular average polarizability and ionization potential and similar photoelectron spectra, but they have considerably different dipole moments: trans, 0.0 D; cis, 2.42 D; and 1,1, 1.38 D. Thus, an incremental increase in rate constant with increasing polarity of the molecule presumably reflects the degree of locking of the dipole. If a parameter c with values between 0 and 1 represents the effectiveness of locking of the dipole, the rate constant can be expressed as

$$k_{\text{PLD}} = \frac{2\pi q}{\mu^{1/2}}\left[\alpha^{1/2} + c\mu_{\text{D}}\left(\frac{2}{\pi k T}\right)^{1/2}\right]. \tag{28}$$

The dipole locking constant c can be determined from the experimental data using

$$c = \Delta k (kT\mu/8\pi q^2 \mu_{\text{D}})^{1/2}, \tag{29}$$

where Δk is the experimental incremental increase in rate constant for the polar isomer relative to the nonpolar isomer. Experimental c values were determined to be $c_{1,1} = 0.093 \pm 0.05$ and $c_{\text{cis}} = 0.111 \pm 0.04$. The c value can be considered qualitatively as $\cos\langle\theta\rangle$, where $\langle\theta\rangle$ is the average orientation angle of the dipole. Thus, $\langle\theta\rangle_{1,1} = 84°6'$ and $\langle\theta\rangle_{\text{cis}} = 83°6'$. This result demonstrated that the extent of locking of the dipole is not high and it increases as the dipole moment of the molecule increases. These experimental results stimulated the development of the average dipole orientation (ADO) theory presented below.

1. The cos $\bar{\theta}$ *Model**

The ADO theory calculates an overall average orientation angle of the dipole molecule in a given ion–molecule collision process. The primary assumptions of the theory are as follows.

(1) At a given ion–molecule separation r, an average angle between the dipole axis and the line of collision can be calculated. This average angle can then replace the microscopic angle in the expression for the effective potential.

(2) There is no net angular momentum transfer between the rotating polar molecule and the system as a whole. From the trajectory calculations of Dugan and co-workers, it was shown that on the microscopic level there is an oscillatory momentum transfer between the rotor and the system; as the negative end of the dipole approaches the positive ion, the rotational motion accelerates, resulting in a net angular momentum transfer from the system to the rotor, and as the negative end moves away from the ion, the angular velocity decreases resulting in a transfer of angular momentum from the rotor to the system. From the trajectory calculations, it was shown that this oscillating angular momentum transfer almost exactly cancels except for small ion–molecule separations.

(3) The polar molecule can be approximated by a two-dimensional rotor in the plane of collision. The third degree of freedom is assumed to be aligned along the dipole axis and does not contribute to the change of θ. This assumption is rigorously true for diatomic molecules and introduces an error of less than 5% in polyatomic molecules. The ADO approach is summarized below.

Suppose $\bar{\theta}(r)$ is the orientation angle of the dipole with respect to the line of centers of collision at an ion–molecule separation r. The effective potential energy at r is given by

$$V_{\mathrm{eff}}(r)_{\mathrm{ADO}} = \frac{L^2}{2\mu r^2} - \frac{\alpha q}{2r^4} - \frac{q\mu_{\mathrm{D}}}{r^2}\cos\bar{\theta}(r). \tag{30}$$

* See Su and Bowers (1973a,b).

Using an approach similar to the Langevin treatment [i.e., setting $d[V_{\mathrm{eff}}(r)]/dr = 0$ and $E_r = V_{\mathrm{eff}}(r)$], the following equations are derived:

$$\sigma(v) = \pi r_c^{\,2} + \frac{\pi q^2 \alpha}{r_c^{\,2} \mu v^2} + \frac{2\pi q \mu_D}{\mu v^2} \cos \overline{\theta}(r = r_c) \tag{31}$$

and

$$v^2 = \frac{q^2 \alpha}{\mu r_c^{\,4}} + \frac{q \mu_D}{r_c \mu} \sin \overline{\theta}(r = r_c) \left(\frac{d\overline{\theta}}{dr} \right)_{r_c}, \tag{32}$$

where r_c is the value of r at the top of the centrifugal barrier. The function $\overline{\theta}(r)$ is calculated using

$$\overline{\theta}(r) = \int \theta P(\theta)\, d\theta \bigg/ \int P(\theta)\, d\theta, \tag{33}$$

where $P(\theta)$ is the probability that the dipole is at an orientation angle θ,

$$P(\theta) \propto \sin \theta / \dot{\theta} = \sin \theta / [E_{\mathrm{rot}} + q\mu_D / r^2]^{1/2}, \tag{34}$$

where $\dot{\theta}$ is the angular velocity of the dipole in the plane of collision, E_{rot} the thermal rotational energy of the polar molecule in the plane of collision, and $\sin \theta$ a spatial factor accounting for the number of ways the dipole can be oriented with an angle θ. The method of evaluating Eq. (33) and its derivative $d\overline{\theta}/dr$ is given in Appendix A.

Once $\overline{\theta}(r)$ and $d\overline{\theta}/dr$ are calculated, $\sigma(v)$ can be plotted against v from Eqs. (31) and (32) by specifying r_c. For thermal velocities, the macroscopic rate constant is obtained by integrating

$$\overline{k}_{\mathrm{ADO}}(\mathrm{therm}) = \int_0^\infty v \langle \sigma(v) \rangle P(v)\, dv, \tag{35a}$$

where $P(v)$ is the normalized Maxwell–Boltzmann velocity distribution function. Thermal energy momentum transfer rate constants can be calculated using

$$\xi / n = [M/(m + M)] k_{\mathrm{ADO}}(\mathrm{therm}), \tag{35b}$$

where ξ is the collision frequency for momentum transfer, n the concentration of the neutral, M the mass of the neutral, and m the mass of the ion.

2. *The $\overline{\cos \theta}$ Model**

This model calculates an average value of $\cos \theta$ as a function of (r). This approach is theoretically more justified because it calculates directly the

* See Bass *et al.* (1975).

average effective potential of the ion–polar molecule system:

$$V_{\text{eff}}(r, \overline{\cos\theta}) = \frac{L^2}{2\mu r^2} - \frac{\alpha q^2}{2r^4} - \frac{q\mu_{\text{D}}}{r^2}\overline{\cos\theta}(r). \tag{36}$$

From Eq. (36), the following equations are obtained for capture rate constant as a function of v

$$\langle\sigma(v)\rangle_{\overline{\cos\theta}} = \pi r_{\text{c}}^{\,2} + \frac{\pi q^2\alpha}{r_{\text{c}}^{\,2}\mu v} + \frac{2\pi q\mu_{\text{D}}}{\mu v^2}\overline{\cos\theta}(r = r_{\text{c}}) \tag{37}$$

$$v^2 = \frac{q^2\alpha}{\mu r_{\text{c}}^{\,4}} - \frac{q\mu_{\text{D}}}{r_{\text{c}}\mu}\left[\frac{d\,\overline{\cos\theta}}{dr}\right]_{r_{\text{c}}}. \tag{38}$$

Calculations of $\overline{\cos\theta}$ are given in Appendix B. It turns out that the theoretical capture rate constants calculated from both $\bar{\theta}$ and $\overline{\cos\theta}$ models are very close to each other (Bass *et al.*, 1975).

3. *Parametrization of the* ADO *Theory*

Equation (28) can be used to parametrize the ADO theory to permit a simple semiempirical calculation of the ADO capture rate constants (Su and Bowers, 1973b). It turns out that at constant temperature, the theoretical dipole locking constant c is a function of $\mu_{\text{D}}/\alpha^{1/2}$ only. Theoretical c values are calculated by substituting the theoretical rate constants in Eq. (28) for specific values of α and μ_{D}. Figure 5 shows a plot of c values (calculated

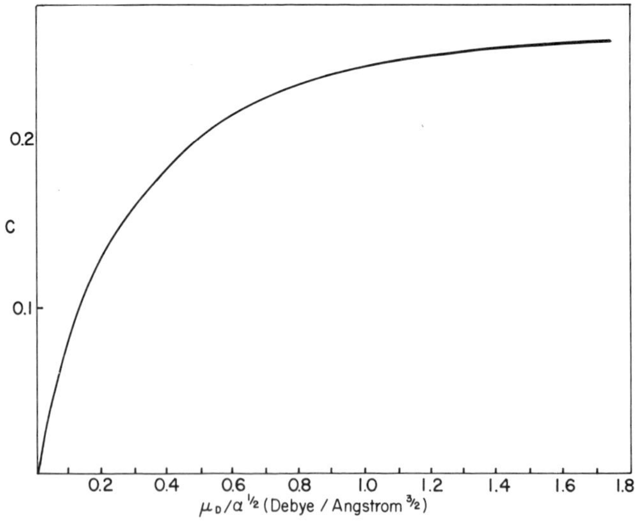

FIG. 5 A plot of the dipole "locking" constant c versus $\mu_{\text{D}}/\alpha^{1/2}$.

from the $\overline{\cos\theta}$ model) versus $\mu_D/\alpha^{1/2}$ for $T = 300$ K. To calculate a capture rate constant, determine $\mu_D/\alpha^{1/2}$ for the polar molecule of interest, determine c from Fig. 5, and then calculate $\bar{k}_{ADO}$(therm) from Eq. (28).

E. *Comparison of Various Theories with Experimental Results*

To test the degree of accuracy of the various ion–dipole collision theories mentioned above, a number of comparisons should be made with experimental results: (1) comparison of the incremental effects of the dipole moment on the rate constants, (2) comparison of the absolute rate constants at thermal energies, (3) comparison of the relative kinetic energy dependences of the rate constants, and (4) comparison of the temperature dependences of the rate constants.

The incremental effect of the dipole on rate constant can be approximated experimentally by the technique described previously in Section III.B. Table I is a comparison of average $\Delta k_{exp}/\Delta k_{theory}$ from a collection of experimental data for the ADO theory, the frozen rotor approximation of Dugan and Magee [Eq. (22)], and the scaled numerical results of Dugan [Eq. (27)]; Δk is the difference between the polar and nonpolar rate constants. The ADO theory, on the average, gives results that are incrementally ~10–40% larger than experiment for the geometric isomer substrates with a scatter of 50–60% positive and 30% negative. The frozen rotor approximation predicts incremental increases in the rate constant that are ~150% larger than those of experiment, while the scaled numerical results give no predictable pattern with large deviations from experimental values. It is not surprising that the scaled numerical model gives erratic results because it assumes the same scaling law for all systems ($E_r^{0.7}$ dependence). The experiment–theory comparison for the locked dipole approximation is not included

TABLE I

Average Values of $\Delta k_{exp}/\Delta k_{theory}$ *for Various Theoretical Models*

No. of experiments	ADO	Frozen rotor	Dugan[a] scaled
40[b]	0.69	0.41	(±5.0)
52[c]	0.82	⟨—⟩	⟨—⟩

[a] Dugan (1973).

[b] Experimental charge transfer and proton transfer to the geometric isomers of $C_6H_4F_2$, $C_2H_2Cl_2$, and $C_2F_2Cl_2$. Experimental data from Su and Bowers (1973a,c).

[c] All UCSB experiments. From Su and Bowers (1973a,c, 1974) and Bowers and Laudenslager (1972).

TABLE II

Comparison of Theoretical Predictions with Experiment for Absolute Proton Transfer Rate Constants[a]

	Theoretical Average Percent Deviations[b]			
No. of experiments	ADO[c] (absolute)	ADO[d]	Langevin	Locked dipole
63	11	−5.2	−43	+170

[a] All thermal energy positive ion proton transfer reactions to polar substrate that have been done at UCSB; experimental data from Hemsworth *et al.* (1974) and from Mackay *et al.* (1976).

[b] The percent deviation is obtained from $100(k_{\text{theory}} - k_{\text{exp}})/k_{\text{exp}}$.

[c] Average total deviation regardless of sign.

[d] Average deviation, taking the sign of deviation into account.

in the table, but it should be noted that it overestimates the incremental effect of the dipole on the rate constant by more than an order of magnitude.

To compare the absolute theoretical rate constant with experiment, it is advisable to choose those reactions with unit or near-unit reaction efficiency. The best suited reactions are exothermic proton transfer reactions for which there is only one reaction channel. Most exothermic proton transfer reactions appear to occur on virtually every capture collision. Table II summarizes the average deviation of theoretical predicted rate constants from experimental observations for the ADO, Langevin, and locked dipole calculations. Predictably, the Langevin theory gives results that are much too low and the locked dipole approximation is much too high. The ADO theory, on the other hand, predicts capture rate constants to a reasonable accuracy.

It is also important to test the theories for their predictions of kinetic energy dependence capture rate constants. Bowers and Su (1975) and Dugan (1973) have made comparisons of various theories with the experimental kinetic energy dependence of rate constants measured by Bowers *et al.* (1973) for

$$CH_3OH^+ + CH_3OH \rightarrow CH_3OH_2^+ + CH_3O \tag{39}$$

and by Huntress *et al.* (1971) for

$$NH_3^+ + NH_3 \rightarrow NH_4^+ + NH_2 \tag{40}$$

The comparisons are reproduced in Figs. 6 and 7. The ADO theory agrees best with the experimental results, both in magnitude and in the shape of the curve. The scaled numerical curve has about the right magnitude but

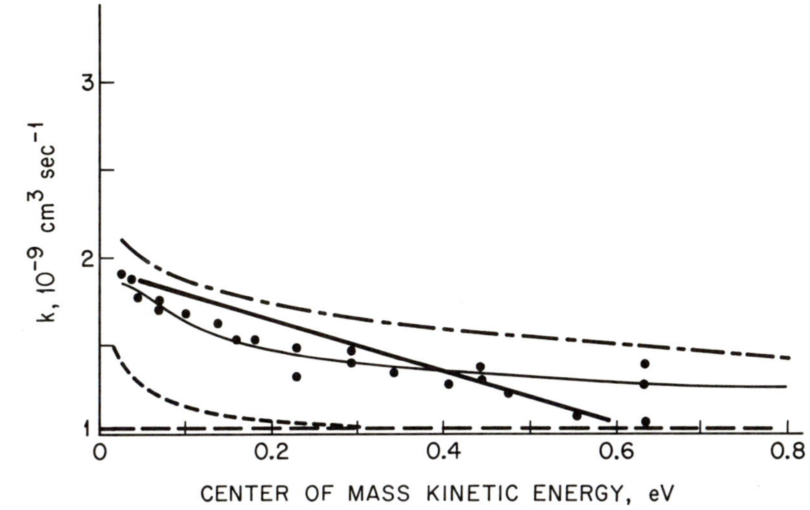

FIG. 6 The rate constant versus relative kinetic energy for the reaction $CH_3OH^+ + CH_3OH \rightarrow CH_3OH_2^+ + CH_3O$. The experimental data——are from Bowers *et al.* (1973). The numerical prediction (---) is from Dugan and Magee (1967) type trajectory calculations as reported by Dugan (1973). The scaled numerical (═══) and frozen rotor (-----) predictions are from Dugan (1973). The (——) prediction is from Gioumousis and Stevenson (1958).

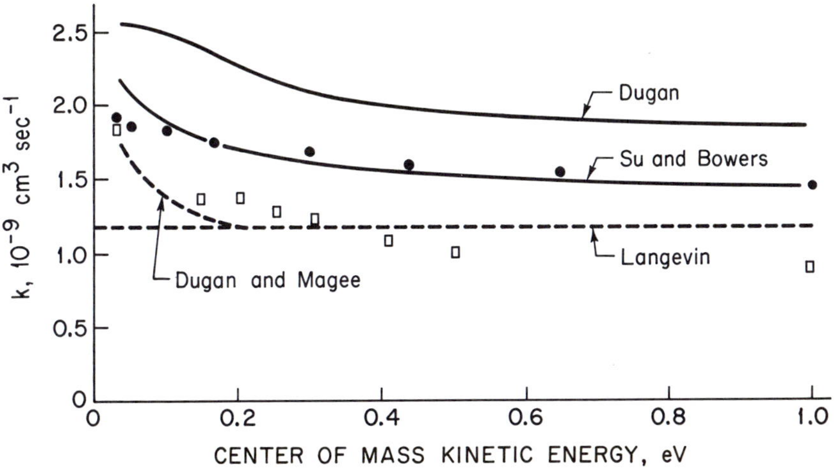

FIG. 7 The rate constant versus relative kinetic energy for the reaction $NH_3^+ + NH_3 \rightarrow NH_4^+ + NH_2$. The results of Dugan (———) are from numerical trajectory calculations of the Dugan and Magee (1967) type. The results labeled Dugan and Magee (-----) are the frozen rotor results of Eq. (26) as reported by Dugan. [The data are from Huntress *et al.* (1971).]

the wrong shape, the "exact" trajectory calculation results are about 30% higher but have a shape similar to experiment. The frozen rotor model clearly has the wrong energy dependence when compared with experiment.

Lindinger *et al.* (1975) have measured the kinetic energy dependence of proton transfer rate constants from several ions on NH_3 [including reaction (40)] in a flow–drift tube. It was observed that thermal energy rate constants are in good agreement with the ADO predictions. However, at higher relative kinetic energies, the experimental rate constants are considerably higher than those obtained from ADO calculations, except for reaction (40), in which the experimental rate constants at higher relative energies are lower than those of the ADO theoretical predictions. The relatively high reaction rate constants at nonthermal kinetic energies may be caused in part by the induced dipole–induced dipole effect discussed later in this chapter. The strong negative kinetic energy dependence of reaction (40), measured in a flow–drift tube, does not agree well with the ICR measurement (Huntress *et al.*, 1971) or with the ADO prediction. The flow–drift tube measurement may be more reliable since it measures rate constants at ground state internal energies. The sharp decrease of the rate constant may be an indication of a decrease in reaction efficiency as relative kinetic energy increases. This matter needs further investigation.

From the above comparisons, it seems that the ADO theory gives the best prediction of rate constants in most cases. This suggests that the ADO theoretical approach is on the right track. Dugan's numerical approach is very useful in the investigation of details of trajectories of nonreactive ion–polar molecule phenomena, including energy transfer studies, but does not provide a clearcut criterion for defining capture collision.

The ADO theory predicts a substantial temperature dependence of capture rate constants for ion–polar molecule collisions (Bowers and Su, 1973; Su and Bowers, 1975a). Several tests of the ADO theory temperature dependence have been made (Meot-Ner and Field, 1975) with mixed results. Some of the ion–dipole rate constants varied with temperature as predicted, some not. More experimental data should be obtained.

F. *Barker–Ridge Model*

Barker and Ridge (BR) (1976) have developed a theory of ion–polar molecule collisions using a simple statistical approach. They used

$$\bar{\phi}_{BR} = \int_0^\pi \phi \exp(-\phi/KT_R) \sin\theta \, d\theta \Big/ \int_0^\pi \exp(-\phi/KT_R) \sin\theta \, d\theta \qquad (41a)$$

$$= -(q\mu_D/r^2)\mathscr{L}(q\mu_D/r^2 KT_R), \qquad (41b)$$

to calculate the average ion–premanent dipole potential energy, where $\mathscr{L}$ is the Langevin function and T_R the rotational temperature of the dipole at infinite separation. The effective potential of the system is then given by

$$V_{\text{eff}}(r)_{\text{BR}} = \frac{L^2}{2\mu r^2} - \frac{\alpha q^2}{2r^4} + \bar{\phi}_{\text{BR}}. \tag{42}$$

When Eq. (42) is solved using the Langevin criteria for capture collisions, an expression for the capture cross section is obtained:

$$\sigma_{\text{BR}} = \pi b_0^{\,2} \tag{43}$$

$$b_0^{\,2} = q\mu_{\text{D}}\left(\frac{1}{KT_{\text{R}}\eta} + \frac{1}{E_{\text{r}}\tanh\eta} - \frac{1}{E_{\text{r}}\eta}\right) + \frac{\alpha q K T_{\text{R}}\eta}{2\mu_{\text{D}}E_{\text{r}}}. \tag{44}$$

The value of η is obtained from

$$\eta^2\left(\frac{\alpha K T_{\text{R}}}{2\mu_{\text{D}}^{\,2}} - \frac{1}{\sinh^2\eta}\right) + 1 = \frac{E_{\text{r}}}{KT_{\text{R}}}. \tag{45}$$

Thermal energy capture rate constants are obtained by integrating

$$k_{\text{BR}}(\text{therm}) = \int_0^\infty v\sigma_{\text{BR}}(v)P(v)\,dv, \tag{46}$$

where $P(v)$ has been defined previously.

G. *Comparison of* ADO *Theory with Barker and Ridge* (BR) *Theory*

The BR method has the great virtue of being computationally simple and gives good agreement with absolute values of some momentum transfer rate constants of ions colliding with polar molecules (Barker and Ridge, 1976). It does not agree well with ADO calculations, however, nor does it appear correctly to account for the dipole dependence of the momentum transfer rate constant. This is illustrated by the example of momentum transfer rate constants between Cl^- and geometric isomers of dichloroethylene (DCE) and difluorobenzene (DFB). These momentum transfer rate constants are plotted in Fig. 8 versus $\mu_D/\alpha^{1/2}$ for the two sets of isomers. Also included are other experimental rate constants from the paper of Barker and Ridge (1976), and the predictions of the BR and ADO theories. Clearly, the BR theory agrees very well with experiment for the absolute magnitude of the polar isomers, while the ADO theory predictions are too low. However, the experimental rate constants for the nonpolar isomers are much larger than those obtained from Langevin theory, which is the limit of both the BR and ADO theories as

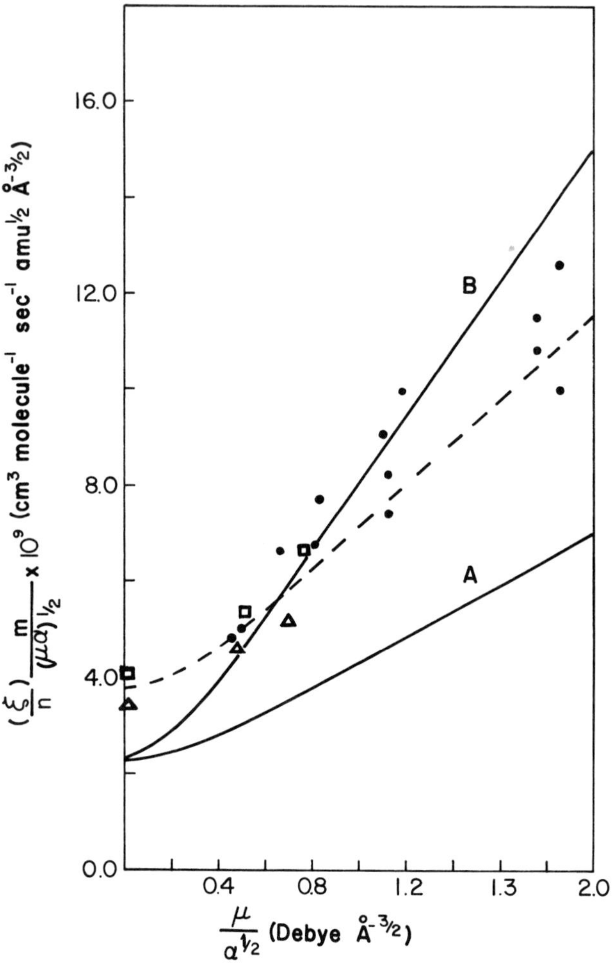

FIG. 8 The variation of ion–neutral momentum transfer collision frequency ξ/η in cm^3 $molecule^{-1}$ sec^{-1} with $\mu_D/\alpha^{1/2}$. Here μ is the reduced mass and m is the mass of the ion. Line A is the ADO theory, line B is the BR theory, and the dashed line is ADO theory multiplied by 1.65. The triangles and squares represent experimental results for $C_2H_2Cl_2$ isomers and $C_6H_4F_2$ isomers, respectively; the solid circles are other data from Barker and Ridge (1976, Fig. 1).

$\mu_D \to 0$. Since BR theory is based on the Langevin model, an agreement with experiment for the polar isomers but lack of agreement for the nonpolar isomers is inconsistent. Interestingly, if the ADO theoretical results are multiplied by a factor of 1.65, very good agreement with experiment is obtained for all values of $\mu_D/\alpha^{1/2}$ (see Fig. 8). While there is no theoretical justification for doing this, it does suggest that ADO theory adequately explains the dipole effect, whereas the BR theory overestimates the dipole effect

TABLE III

Average Values of $\Delta k_{\text{theory}}/\Delta k_{\text{exp}}$ *for* ADO *and* BR *Models*

No. of experiments	ADO	BR
24[a]	1.17	4.17
24[b]	2.31	8.08
4[c]	0.64	2.00

[a] Charge transfer to geometric isomers of $C_6H_4F_2$, $C_2H_2F_2$, and $C_2H_2Cl_2$. Experimental data from Su and Bowers (1973a) and Bowers and Laudenslager (1972).

[b] Proton transfer to geometric isomers of $C_6H_4F_2$, $C_2H_2F_2$, and $C_2H_2Cl_2$. Experimental data from Su and Bowers (1973c).

[c] Momentum transfer to geometric isomers of $C_2H_2Cl_2$ and $C_6H_4F_2$. Experimental data from Su and Bowers (1974).

by nearly a factor of 2. Similar comparisons between ADO theory and experiment for charge transfer (Su and Bowers, 1973a) and proton transfer (Su and Bowers, 1973c) reactions confirm the agreement found for momentum transfer reactions. In Table III, a comparison of the average theory–experiment incremental dipole effect $\Delta k_{\text{theory}}/\Delta k_{\text{exp}}$ is given for the ADO and BR theories. Thus, the BR model seriously overestimates the dipole effect for charge transfer and proton transfer, as well as momentum transfer collisions. It is interesting that both charge transfer and momentum transfer rate constants for nonpolar molecules are sometimes significantly larger than Langevin predictions. Theories based on the Langevin criteria are capture theories; that is, an intimate collision is required. Such collisions are required for transfer of massive particles and Langevin-type theories should apply best to these types of reaction. Since electron and momentum transfer reactions do not require transfer to massive particles, it is not surprising that Langevin theories have not yet been generally successful in predicting rate constants for them. The good agreement between ADO theoretical and experimental dipole effects in momentum transfer collisions suggests that a modified Langevin and ADO approach may be successful in predicting momentum transfer rate constants, however.

The question remains as to why the BR and ADO theories differ so substantially. We suggest the following reasons. In the BR calculation of $\bar{\phi}_{\text{BR}}$, it was assumed that the rotational temperature of the dipole molecules at infinite ion–molecule separation remains unchanged throughout the course of the collision. This assumption may not be valid. The rotational temperature should increase as the ion and the polar molecule approach each other. To illustrate this, consider the limit $T_R \to 0$; in this case the BR theory approaches the locked dipole limit. In reality, however, dipole locking never occurs. Since the initial orientation of the dipole relative to the line of centers of the collision is random, virtually all collisions will occur with θ other than zero even if there is no rotational energy in the rotor at infinite separation. As the dipole approaches the ion, a torque is exerted on it by the ion field and it starts to rotate such that θ will initially decrease. As a result, the rotational energy increases. From ADO calculations and from the trajectory calculations of Dugan and co-workers (see e.g., Dugan and Magee, 1967), it is apparent that the average rotational energy of the polar rotor increases as r decreases. It will be shown shortly that this increase in rotational energy, thus T_R, is significant in calculating $\bar{\phi}_{BR}$ and should not be neglected. In fact, the rotational energy distribution is no longer Boltzmann. Hence, the averaging method used by Barker and Ridge to obtain $\bar{\phi}_{BR}$ should be modified. One method of doing this is to obtain an expression for an "effective" temperature T_{eff} as a function of the ion–dipole separation r. A crude method of calculating an approximate T_{eff} is as follows: As $r \to \infty$, $\bar{\phi} \to 0$ and the average one-dimensional rotational energy of the polar rotor in the plane of collision is $\bar{E}_R = \frac{1}{2}KT$. At any finite value of r, the maximum value of the rotational energy will then be $\frac{1}{2}KT + q\mu_D/r^2$. Since the ion–dipole interaction can either oppose or reinforce the thermal rotational motion the dipole, the following limits apply*: $0 < E_R < \frac{1}{2}KT + q\mu_D/r^2$. For ease of calculation, it will be assumed that the average energy is near the middle of this range. Hence,

$$\bar{E}_R \simeq \tfrac{1}{2}(\tfrac{1}{2}KT + q\mu_D/r^2). \tag{47}$$

If T_{eff} is defined as

$$\bar{E}_R = \tfrac{1}{2}KT_{eff}, \tag{48}$$

then combining Eqs. (47) and (48) yields

$$T_{eff} = \tfrac{1}{2}T + q\mu_D/Kr^2. \tag{49}$$

This T_{eff} can thus be substituted in Eq. (41a) to obtain a modified value of $\bar{\phi}_{BR}$.

* A better way to estimate T_{eff} is as follows. Since the ion–dipole interaction can either oppose or reinforce the thermal motion of the polar molecule, the limits on the rotational energy are $\frac{1}{2}kT - q\mu_D/r^2 \leq E_R \leq \frac{1}{2}kT + q\mu_D/r^2$. These limits hold for large values of r where $\frac{1}{2}kT \geq q\mu_D/r^2$. Since E_R cannot be negative, the limits $0 \leq E_R \leq \frac{1}{2}kT + q\mu_D/r^2$ hold when $\frac{1}{2}kT < q\mu_D/r^2$. This is the limit used in the calculations discussed in this paper. A more correct analysis would have used both sets of limits. Since a qualitative point only was being made, the cruder (and much simpler) approximation was used.

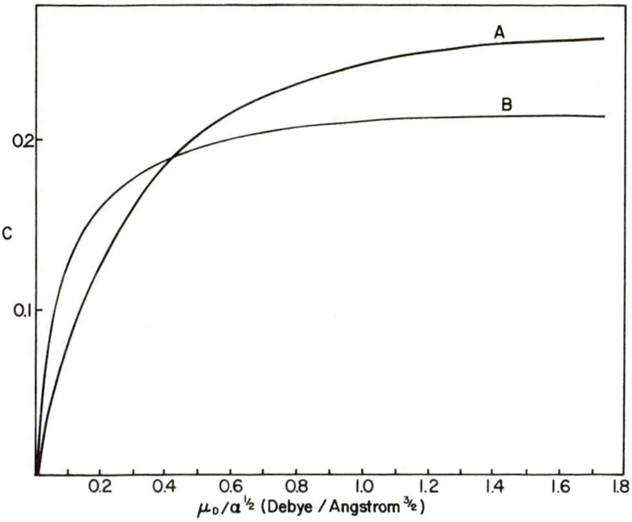

FIG. 9 A plot of the dipole "locking" constant c plotted against $\mu_D/\alpha^{1/2}$. Curve A is the ADO theory and curve B is the modified BR theory.

Thus, an approximate expression of the average ion–polar interaction energy using the modified BR method is

$$\bar{\phi}_{BR} = \int \phi \exp(-\phi/KT_{eff})\, d\phi \bigg/ \int \exp(-\phi/KT_{eff})\, d\phi \qquad (50a)$$

$$= \int_0^{\pi} \phi \exp(-\phi/KT_{eff}) \sin\theta\, d\theta \bigg/ \int_0^{\pi} \exp(-\phi/KT_{eff}) \sin\theta\, d\theta \qquad (50b)$$

When Eq. (50) is used to calculate the Langevin criteria rate constant, the results agree with ADO calculations to within 10–15% for all values of μ_D and α. Figure 9 compares the locking constant c of the ADO and the modified BR theories. The agreement is not exact but, considering the crude method of estimating T_{eff}, the comparison is satisfactory. This treatment suggests that the BR approach should give about the same theoretical rate constant as do the ADO calculations if a correct rotational temperature of the dipole is used.

H. The Average Dipole Orientation Theory with Conservation of Angular Momentum: The AADO *Theory**

The major assumption in the original formulation of ADO theory is that there is no net angular momentum transfer between the rotating molecule and the ion–molecule orbital motion. While this assumption is quite good at

* See Su *et al.* (1978a).

large ion–molecule separations, it becomes less valid as the separation decreases toward the capture limit. It was shown above that as the ion–molecule separation decreases, the rotational energy of the dipole rotor increases. This increase in rotational energy will cause a subsequent decrease in orbiting angular momentum to conserve the total angular momentum. Mackay *et al.* (1976) have measured some thermal energy proton transfer rate constants for various ions on HCN ($\mu_D = 2.98$ D) and CH_3CN ($\mu_D = 3.92$ D). The experimental rate constants are significantly higher than those predicted by ADO. Since these molecules have very large dipole moments, it is expected that angular momentum effects will be large and may contribute in a significant way to the capture rate constant

The approximate inclusion of conservation of angular momentum into the ADO theory has been developed by Su and Bowers (Su *et al.*, 1978a) and termed the AADO theory. The theory is described as follows.

There are two approaches that can be taken. The BR model can be modified to accurately account for the r dependence of the rotational energy of the polar rotor. This is a complicated matter and is not the method of choice here. The approach that will be pursued is to modify the ADO theory to include angular momentum conservation. It should be noted that in the original ADO theory, the r dependence of the rotational energy (and hence rotational angular momentum) of the polar rotor was implicitly accounted for. The decrease in average orbital angular momentum as r decreases will be calculated and this $L(r)$ will be used in the ion–polar molecule effective potential to generate Langevin criteria capture rate constants. In this ap-

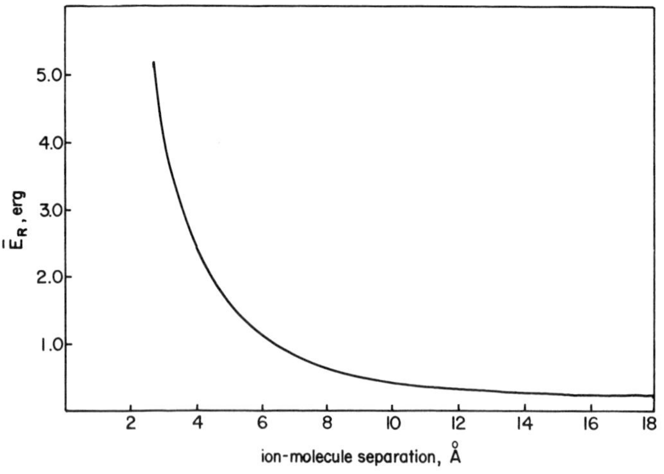

FIG. 10 Average rotational energy E_R of polar molecule at 300 K versus ion–molecule separation r. $\alpha = 3.0$ Å^3, $\mu_D = 2.4$ D, and $\mu = 1.41 \times 10^{-23}$ g.

proach, it is assumed that the angular momenta of the polar rotor and of the ion–molecule orbital motion are collinear. This is not an unreasonable assumption since the collision between the ion and the dipole is considered to occur in a plane. Hence, as the ion approaches the dipole, the torque exerted on the dipole by the ion occurs in the plane of collision and the angular momentum vector is pointed in a direction normal to the plane. In the same way, the direction of the orbital angular momentum vector is also normal to the plane of collision. Hence the two vectors, to a good approximation, are collinear. Added to these angular momenta will be the thermal angular momenta of the ion and of the dipolar molecule. This treatment assumes that these motions are decoupled from the orbital motion and ion-induced rotation of the polar molecule. Finally, the rotation of the plane of collision is ignored.

The calculation of the r dependence of the average rotational energy of the rotor in the plane of collision $\bar{E}_R$ is given in Appendix C. Figure 10 shows a typical plot of $\bar{E}_R$ versus r. At large r, $\bar{E}_R$ stays quite close to the thermal rotational energy, but $\bar{E}_R$ increases rapidly at small r. The r-dependent average angular momentum of the rotor can be calculated using

$$I\dot{\theta}(r) = (2I\bar{E}_R)^{1/2}, \tag{51}$$

where I is the moment of inertia of the polar rotor. Here the relation $E_R = \frac{1}{2}I[\dot{\theta}(r)]^2$ has been used. The angular momentum at $r = \infty$ is

$$I\dot{\theta}_\infty = (IkT)^{1/2}. \tag{52}$$

Hence, the increase in angular momentum C_L as r decreases is given by (53)

$$C_L = (2I\bar{E}_R)^{1/2} - (IkT)^{1/2}. \tag{53}$$

The orbital (i.e., translational) angular momentum at any separation r is then

$$L(r) = \mu bv - C_L \tag{54}$$

and the resulting effective potential is

$$V_{\text{eff}}(r) = \frac{(\mu bv - C_L)^2}{2\mu r^2} - \frac{\alpha q^2}{2r^4} - \frac{q\mu_D}{r^2}\cos\theta. \tag{55}$$

When the Langevin criteria are applied to Eq. (55), the following two parametric equations are obtained:

$$\langle\sigma(v)\rangle = \pi\left\{\left[r_c^{\,2} + \frac{\alpha q^2}{\mu v^2 r_c^{\,2}} + \frac{2q\mu_D}{\mu v^2}\,\overline{\cos\theta(r = r_c)}\right]^{1/2} + \frac{C_L}{\mu v}\right\}^2, \tag{56}$$

$$v = \left[\frac{B + (B^2 - 4D)^{1/2}}{2}\right]^{1/2}, \tag{57}$$

where

$$B = \frac{2q\mu_D}{\mu r_c}\left[\frac{d\overline{\cos\theta}}{dr}\right]_{r_c} - \frac{2\alpha q^2}{\mu r_c^{\,4}} - \frac{1}{\mu^2}\left(\frac{dC_L}{dr}\right)_{r_c}^2,$$

$$D = \frac{\alpha q^2}{\mu^2 r_c^{\,4}}\left\{\frac{\alpha q^2}{r_c^{\,4}} - \frac{2\mu_D q}{r_c}\left[\frac{d(\overline{\cos\theta})}{dr}\right]_{r_c} - \frac{1}{\mu}\left(\frac{dC_L}{dr}\right)_{r_c}^2\right\},$$

$$+ \frac{q\mu_D}{\mu^2 r_c^{\,2}}\left\{q\mu_D\left[\frac{d\overline{\cos\theta}}{dr}\right]_{r_c}^2 - \frac{2}{\mu}\overline{\cos\theta}(r = r_c)\left(\frac{dC_L}{dr}\right)_{r_c}^2\right\},$$

Thermal energy rate constants are then obtained by an equation analogous to Eq. (35). Notice that when C_L and dC_L/dr are both zero, Eqs. (56) and (57) reduce to Eqs. (31) and (32) of the original ADO theory.

The AADO theory has been parametrized to provide easy calculations of thermal energy capture rate constants. Equation (58) is the parametrized equation for 300 K.

$$k_{\text{therm}}(\text{AADO}) = \frac{2q}{\mu^{1/2}}\left[\alpha^{1/2} + c\mu_D\left(\frac{2}{\pi KT}\right)^{1/2} + \left(\frac{0.790I + 1.40 \times 10^{37}I^2}{0.106 \times 10^{-37} + BI}\right)^{1/2}\frac{\mu_D}{\alpha^{1/2}}\right], \tag{58}$$

TABLE IV

Experiment and Theoretical Rate Constants for Reactions of the Type $XH^+ + NH_3 \rightarrow NH_4^+ + X$[a]

XH^+	k_{exp}[b]	k_{ADO}	k_{AADO}	k_{exp}/k_{ADO}	k_{exp}/k_{AADO}
H_3^+	4.2	3.98	4.52	1.06	0.93
NH_3^+	2.5	2.18	2.48	1.15	1.01
CH_5^+	2.5	2.18	2.48	1.15	1.01
H_3O^+	2.4	2.12	2.41	1.13	1.00
COH^+	2.4	1.94	2.21	1.24	1.09
N_2H^+	2.3	1.94	2.21	1.19	1.04
$C_2H_5^+$	2.1	1.94	2.21	1.08	0.95
C_2H^+	2.0	1.92	2.18	1.04	0.92
N_2OH^+	2.1	1.81	2.05	1.16	1.04
$C_4H_8^+$	1.9	1.75	1.99	1.09	0.95
$C_3H_7^+$	1.9	1.82	2.07	1.04	0.92
				⟨1.12⟩	⟨0.99⟩

[a] All rate constants × 10^9 cm^3 molecule sec^{-1}.
[b] Data from Hemsworth *et al.* (1974).

where

$$B = \frac{1.39}{\alpha \times 10^{24} + 2.64} + 0.375$$

and all units are in the cgs system.

The AADO calculation yields capture rate constants larger than the ADO theory for a given reaction system. This result is predictable because the changes to ADO theory outlined above result in a lower effective potential [Eq. (55)], and hence enhance the possibility for capture. From Eq. (58), the AADO theory predicts that as μ_D and/or I increase, the capture rate constant will also increase over those predicted by ADO theory. Tables IV–VI compare the AADO prediction with ADO theory and experiment for thermal energy proton transfer reactions of various ions with NH_3, HCN, and CH_3CN. The properties of interest of the neutral molecules are given in Table VII. From Tables IV–VI, it is shown that when the conservation of angular momentum is taken into account, the theoretical rate constants increase, on the average, 13, 22, and 31% for the reactions with NH_3, HCN, and CH_3CN, respectively, bringing the theory into better agreement with experiment.

More comparison with experiment should be made for various temperatures and with reactions involving polar molecules with various moments of inertia. The AADO theory presented above approximately takes into account the conservation of angular momentum. However, it is not obvious

TABLE V

Experimental and Theoretical Rate Constants for Reactions of the Type $XH^+ + HCN \rightarrow H_2CN^+ + X$ or $X^- + HCN \rightarrow CN^- + XH$[a]

XH^+, X^-	k_{exp}[b]	k_{ADO}	k_{AADO}	k_{exp}/k_{ADO}	k_{exp}/k_{AADO}
H_3^+	7.4	6.48	7.50	1.14	0.94
H_3O^+	3.5	3.20	3.88	1.09	0.90
N_2H^+	3.2	2.86	3.47	1.12	0.92
HCO^+	3.0	2.86	3.47	1.05	0.86
H^-	15	10.8	13.11	1.39	1.09
D^-	9.9	7.8	9.45	1.27	1.05
NH_2^-	4.8	3.37	4.08	1.42	1.18
OH^-	4.1	3.31	4.00	1.24	1.03
C_2H^-	3.9	2.97	3.60	1.31	1.08
SH^-	2.9	2.77	3.34	1.05	0.87
			Average	⟨1.21⟩	⟨0.99⟩

[a] All rate constants × 10^9 cm^3 $molecule^{-1}$ sec^{-1}.
[b] Data from Mackay *et al.* (1976).

TABLE VI

Experimental and Theoretical Rate Constants for Reactions of the Type $XH^+ + CH_3CN \rightarrow CH_3CNH^+ + X^a$

XH^+	k_{exp} [b]	k_{ADO}	k_{AADO}	k_{exp}/k_{ADO}	k_{exp}/k_{AADO}
H_3^+	10	8.40	10.64	1.19	0.94
H_3O^+	4.7	3.91	5.36	1.20	0.88
N_2H^+	4.1	3.42	4.69	1.20	0.87
HCO^+	4.1	3.42	4.69	1.20	0.87
N_2OH^+	3.8	3.05	4.18	1.25	0.99
CO_2H^+	4.1	3.05	4.18	1.34	0.98
			Average	⟨1.23⟩	⟨0.92⟩

[b] Data from Mackay *et al.* (1976).
[a] All rate constants $\times 10^9$ cm^3 molecule^{-1} sec^{-1}.

TABLE VII

Properties of Neutral Molecules in Tables IV–VI

Neutral molecule	Polarizability α (Å^3)	Dipole moment μ_D (Debye)	Average moment of inertia (g cm^2)
NH_3	2.26	1.47	2.82×10^{-40}
HCN	2.56	2.98	1.92×10^{-39}
CH_3CN	4.56	3.92	1.05×10^{-38}

that energy is conserved as well. Further theoretical work should be done to clarify this.

IV. ION–QUADRUPOLE THEORY

This section gives the theoretical treatment of ion–quadrupole molecule interactions for molecules that belong to the $D_{\infty h}$ point group. Emphasis will be on the average quadrupole orientation (AQO) theory by Su and Bowers (1975b).

The effective potential for an ion interacting with a quadrupole molecule of D_∞ symmetry is given by

$$V_{eff}(r) = (L^2/2\mu r^2) - (\alpha^2 q/2r^4) + V_Q(r), \tag{59}$$

$$V_Q(r) = Qq(3\cos^2\theta - 1)/2r^3, \tag{60}$$

where Q is the static quadrupole moment of the molecule and θ is the angle the quadrupolar axis makes with r. Dugan and Palmer (1971) have calculated numerically the classical trajectories of ion–quadrupole molecule collisions similarly to those calculated for ion–dipole molecule collisions (Dugan and Magee, 1967). The calculation suggested that partial locking of the quadrupole will occur.

Su and Bowers (1975c) have developed a classical theory called the average quadrupole orientation (AQO) theory for ion–quadrupole molecule interactions utilizing a framework similar to that of the ADO theory. The theory is for those molecules that belong to the $D_{\infty h}$ point group, which includes those molecules whose quadrupole moments transform under the $D_{\infty h}$ point group, e.g., C_6H_6, C_6F_6, even though the molecules themselves belong to different point groups.

The ion–quadrupole interaction potential is given in Eq. (60). There are two cases to be considered: *even* and *odd*. In the even case, q and Q have the same sign, and V_Q is a minimum (i.e., attractive phase) when $\theta = 90°$ and a maximum when $\theta = 0°$. In the odd case, q and Q have different signs, and V_Q is minimum when $\theta = 0°$ and maximum at $\theta = 90°$. These two cases will be treated separately.

Odd case. The angle-averaged effective radial potential of the ion–quadrupole pair is

$$V_{\text{eff}} = \frac{L^2}{2\mu r^2} - \frac{q^2\alpha}{2r^4} + \frac{Qq(3\cos^2\overline{\theta}(r) - 1)}{2r^3}, \tag{61}$$

where $\overline{\theta}(r)$ is the average orientation angle at r. Solving Eq. (61) for Langevin criteria capture cross sections, the following equations are obtained for the relation between the angle-averaged capture cross section $\langle\sigma\rangle$ and the ion–molecule relative velocity v:

$$\langle\sigma(v)\rangle = \pi\left[r_c^{\,2} + \frac{q^2\alpha}{r_c^{\,2}\mu v^2} - \frac{Qq(3\cos^2\overline{\theta}(r_c) - 1)}{r_c\mu v^2}\right], \tag{62}$$

$$v^2 = \frac{q^2\alpha}{\mu r_c^{\,4}} - \frac{Qq(3\cos^2\overline{\theta}(r_c) - 1}{2\mu r_c^{\,3}} - \frac{3qQ\cos\overline{\theta}(r_c)\sin\overline{\theta}(r_c)}{\mu r_c^{\,2}}\left(\frac{d\overline{\theta}}{dr}\right)_{r_c}, \tag{63}$$

where r_c is the distance of closest approach that leads to capture and $\overline{\theta}(r)$ and $d\overline{\theta}(r)/dr$ are calculated as in ADO theory.

Even case. The parametric equations for Langevin criteria capture cross section and relative velocities are

$$\langle\sigma(v)\rangle = \pi\left[r_c^{\,2} + \frac{q^2 d}{\mu v^2 r_c^{\,2}} - \frac{Qq(3\sin^2\beta(r_c) - 1)}{\mu v^2 r_c}\right], \tag{64}$$

$$v^2 = \frac{q^2\alpha}{\mu r_c^{\,4}} - \frac{Qq(3\sin^2\beta(r_c) - 1)}{2\mu r_c^{\,3}} + \frac{3Qq\cos\beta(r_c)\sin\beta(r_c)(d\beta/dr)}{2\mu r_c^{\,2}}, \tag{65}$$

where $\beta = \frac{1}{2}\pi - \theta$. For thermal velocities, the macroscopic rate constant is obtained by an equation similar to Eq. (35).

To provide general use, two equations are obtained for *odd* and *even* cases by an empirical fit of the theoretical thermal energy rate constants at 300 K.

Odd case

$$\bar{k}_{\text{therm}} = \begin{cases} 2\pi q\left(\dfrac{\alpha}{\mu}\right)^{1/2} + \dfrac{2.17Q^{2/3}}{\mu^{1/2}}\left[\dfrac{Q^{2/3}}{\alpha^{1/2}} - 10.01 \times 10^{-6}\right], \\ \dfrac{Q^{2/3}}{\alpha^{1/2}} \geq 8.73 \times 10^{-6}, \\ 2\pi q\left(\dfrac{\alpha}{\mu}\right)^{1/2} + \dfrac{Q^{2/3}}{\mu^{1/2}}\left[\dfrac{((Q^{2/3}/\alpha^{1/2}) - 4.29 \times 10^{-6})^2}{2.08 \times 10^{-6}} - 7.58 \times 10^{-6}\right], \\ \dfrac{Q^{2/3}}{\alpha^{1/2}} \leq 8.73 \times 10^{-6}. \end{cases} \tag{66}$$

Even case

$$\bar{k}_{\text{therm}} = 2\pi q\left(\frac{\alpha}{\mu}\right)^{1/2} + \frac{3.97}{\mu^{1/2}}\left(\frac{Q^{2/3}}{\alpha^{1/2}} - 1.59 \times 10^{-6}\right). \tag{67}$$

A theoretically more justified approach is to calculate the average V_Q or average $\cos^2\theta$. The result is expected to be very close to the $\bar{\theta}$ calculation. Table VIII compares the AQO calculations with Langevin theory and some experimental results (Su and Bowers, 1975c). It is shown from this table that the AQO calculation predicts significant quadrupole effects for those

TABLE VIII

Experimental and Theoretical Thermal Energy Proton Transfer Rate Constants at 300 K for D_3^+ to Neutral Molecules[a]

Neutral	$Q \times 10^{26}$	α (Å^2)	Exp	Langevin	Exp/Langevin	AQO	Exp/AQO
CH_4	0.00	2.60	1.87	1.81	1.03	1.81	1.03
C_2H_2	+0.72	3.33	2.40	1.93	1.24	2.26	1.06
C_2N_2	−9.0	5.01	2.40	2.26	1.06	2.36	1.02
CO_2	−4.4	2.65	1.74	1.06	1.05	1.69	1.03
C_6F_6	+17.2	9.48	3.68	2.99	1.23	3.56	1.03
c-C_6H_{12}	+13.2	10.87	4.14	3.26	1.27	3.63	1.14
C_6H_6	−5.6	10.32	3.37	3.18	1.05	3.15	1.07
CS_2	0.1	8.74	3.10	2.93	1.06	3.07	1.01
CO	+0.4	1.95	1.49	1.47	1.01	1.48	1.01

[a] All rate constants $\times 10^9$ cm^3 molecule^{-1} sec^{-1}.

molecules with high quadrupole moments, especially in the even case, and improves the agreement between theory and experiment.

V. EFFECT OF INDUCED DIPOLE–INDUCED DIPOLE FORCES ON ION–MOLECULE COLLISIONS

It was suggested (Su and Bowers, 1974) that the r^{-6} term arising from the induced dipole–induced dipole interaction may be important in some ion–molecule collisions when the polarizability of the ion is large. (Some negative ions could have very large polarizability.) Empirical treatment of the r^{-6} term has been carried out by Mason and Schamp (1958), Dymerski *et al.* (1974), and Ridge and Beauchamp (1976) for ion–molecule collisions and by Bell *et al.* (1968) for neutral systems. Su and Bowers (Su *et al.*, 1978b) have made an attempt to calculate theoretically the effect of the r^{-6} term on ion–molecule collisions by using both r^{-4} and r^{-6} potentials and solving the effective potential

$$V_{\text{eff}}(r) = \frac{L^2}{2\mu r^2} - \frac{q^2\alpha_1}{2r^4} - \frac{C}{r^6} \tag{68}$$

for Langevin criteria capture cross sections. Here C is a constant for an ion–molecule system and is given by (Rothe and Bernstein, 1959)

$$C = \tfrac{3}{2}a^{1/2}q^2\left[\frac{\alpha_1\alpha_2}{(\alpha_1/N_1)^{1/2} + (\alpha_2/N_2)^{1/2}}\right], \tag{69}$$

where a is the Bohr radius, α_1 and α_2 are polarizabilities of the neutral and the ion, respectively, and N_1 and N_2 are the number of outer shell electrons of the neutral and the ion, respectively. The following equations are obtained for calculating the capture cross section as a function of relative velocity:

$$\sigma(v) = \frac{2\pi}{v^2\mu r^2}\left(q^2\alpha_1 + \frac{3C}{r^4}\right), \tag{70}$$

$$v^2 = \frac{q^2\alpha}{\mu r^4} + \frac{4C}{\mu r^6}. \tag{71}$$

The inclusion of the r^{-6} term slightly increases the theoretical thermal energy capture rate constant. Table IX compares the theoretical thermal energy capture collision rate constants using Eqs. (70) and (71) with Langevin rate constants for some collisions of nonpolar molecules with Cl^- [$\alpha_2 = 2.96$ Å^3, Tessman and Kahn (1953)] and H^- [$\alpha_2 = 2.68$ Å^3, Schwartz (1961)]. Interestingly, when the r^{-6} term is included, the theoretical capture rate constant increases with increasing relative translational energy. Figure 11

TABKE IX

Theoretical Capture Rate Constants[a]

Ion–molecule system	Langevin	r^{-6} term included
$Cl^- + trans\text{-}C_2H_2Cl_2$	1.28	1.34
$Cl^- p\text{-}C_6H_4F_2$	1.41	1.47
$H^- + C_2H_2$	4.35	4.55

[a] All rate constant × 10^9 cm^3 molecule^{-1} sec^{-1}.

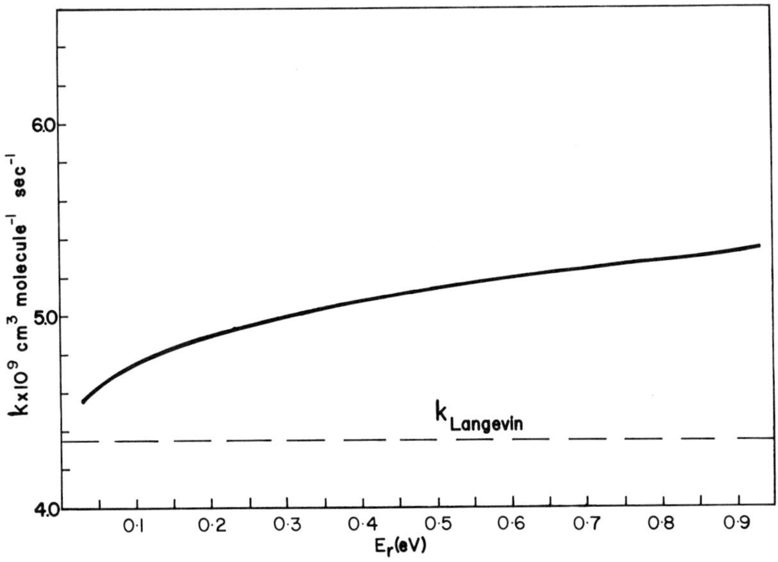

FIG. 11 A plot of theoretical capture collision rate constant with r^{-4} and r^{-6} potentials as a function of relative translational energy E_r for the collision of H^- with C_2H_2.

gives a plot of theoretical capture rate constant versus relative translational energy for the collision of H^- ($\alpha = 2.68$ Å^3) with C_2H_2. The figure shows a considerable increase of capture rate constant with increasing relative kinetic energy. Lindinger *et al.* (1975) have observed rate constants of some ion–polar molecule proton transfer reactions as a function of the kinetic energy of the ion. Some rate constants are essentially unchanged with energy and one even increases with increasing kinetic energy. The ADO theory predicts that collision rate constants should decrease with increasing energy. The induced dipole–induced dipole interaction may be one of the factors causing the higher rate constants.

VI. OTHER THEORIES

The anisotropy in the polarizability has also received some attention. Since most molecules are anisotropic, the rotationally averaged polarizability usually used in Langevin theory calculations is always smaller than the maximum component. It has been suggested (Smith and Futrell, 1972–1973; Dymeski and Dunbar, 1972; Miasek and Beauchamp, 1974) that the ion might "lock in" on the maximum component of the polarizability, thus explaining rate constants larger than Langevin predictions. Bass *et al.* (1978) have developed a Langevin treatment for anisotropic molecules. The results show that the anisotropy has little effect on the capture collision rate constant. It was pointed out that rate constants larger than those of Langevin predictions are better explained by the quadrupole effect (Su and Bowers, 1975c).

Finally, Gentry (1974) has derived a classical theory for scattering of ions by dipole molecules at high translational energy and small scattering angles. The theory compares quantitatively with experiment (Udseth *et al.*, 1974) for collisions of protons with CO, HCl, and HF, and has been shown to have some diagnostic importance.

VII. SUMMARY AND PROGNOSIS

As this chapter makes clear, there has been a tremendous amount of progress made in the area of ion–molecule capture theory in the past 10 years. With theories like the ADO theory, one can now estimate simply the capture rate constant fairly accurately for most ion–neutral encounters. However, one worrisome point remains. There has not yet been a rigorous capture theory published that both conserves energy and angular momentum and simultaneously defines a capture collision using the Langevin criteria. The ADO and ADO-type theories do not rigorously account for all of the coordinates in the colliding system, and hence the theoretical validity of their predictions is uncertain. In the trajectory calculations of Dugan and Magee, a unique capture criterion was not defined and the numerical results were obtained by assigning a rather arbitrary distance of closest approach in the ion–molecule pair to sort capture collisions from noncapture collisions.

Both problems appear to be near solution. We have recently applied transition state concepts to the ion–dipole capture problem and derived a theory that rigorously conserves energy and angular momentum (Chesnavich *et al.*, 1978). This theory provides an upper limit to the capture rate constant for ion–dipole collisions. Preliminary results indicate that this rigorous transition state theory predicts capture rate constants that are somewhat higher (20–30%) than those of the ADO theory for identical systems. Furthermore, we are nearly ready to perform classically exact trajectory calculations on the ion–dipole potential for comparison with the transition state theory. These calculations will be an improvement on those of Dugan and Magee in that a Langevin-type capture criterion will be built in. It appears that both of these theories may lend themselves to including multiple terms in the potential (e.g., dipole and quadrupole), an extremely tedious problem using the ADO procedure. Also, accurate modeling, taking into account the relative importance of the various terms in the potential, will be possible with these theories and determinations of reaction mechanisms by comparison of theory with experiment facilitated.

APPENDIX A. CALCULATION OF $\bar{\theta}(r)$ AND $d\bar{\theta}(r)/dr$ FOR THE ADO $\cos\bar{\theta}$ MODEL

The term $\bar{\theta}(r)$ is calculated by evaluating Eq. (33). To do this, there are two cases to be considered. (1) $E_{\text{rot}} \leq E_m = q\mu_D/r^2$; the motion is an oscillating one. (2) $E_{\text{rot}} \geq E_m$; the motion is nonoscillatory; the dipole molecule has enough energy to swing around in a complete circle.

Case (1)

Suppose E_1 is the average rotational energy in the plane of collision of all those molecules such that E_{rot} is less than E_m; then the maximum value of θ, G, is given by

$$E_1 = -(q\mu_D/r^2)\cos G. \tag{A1}$$

The average orientation angle θ is then

$$\bar{\theta}_1(r) = \int_0^G \frac{\theta \sin\theta \, d\theta}{[\cos\theta - \cos G]^{1/2}} \bigg/ \int_0^G \frac{\sin\theta \, d\theta}{(\cos\theta - \cos G)^{1/2}} = \frac{N}{D}, \tag{A2}$$

where D can be integrated directly and N is integrated numerically.

Case (2)

Suppose E_2 is the average E_{rot} in this case; then the average orientation angle $\bar{\theta}_2$ is

$$\bar{\theta}_2(r) = \int_0^\pi \frac{\theta \sin\theta \, d\theta}{[E_2 + (q\mu_D/r^2)\cos\theta]^{1/2}} \bigg/ \int_0^\pi \frac{\sin\theta}{[E_2 + (q\mu_D/r^2)\cos\theta]^{1/2}} = \frac{B}{C}, \tag{A3}$$

where C can be integrated directly and B is integrated numerically.

The overall average θ at a given r is given by

$$\bar{\theta}(r) = \bar{\theta}_1(r)F_1(r) + \bar{\theta}_2(r)F_2(r), \tag{A4}$$

where F_1 and F_2 are the fractions of molecules that have E_{rot} less than and greater than E_m, respectively.

To evaluate E_1, E_2, F_1, and F_2, the one-dimensional energy distribution in the plane of collision is used since this is the component of interest. Thus,

$$F_1 = \frac{1}{\sqrt{\pi KT}} \int_0^{E_m} \frac{1}{\sqrt{E}} \exp(-E/KT)\, dE, \tag{A5}$$

$$F_2 = 1 - F_1 = \frac{1}{\sqrt{\pi KT}} \int_{E_m}^{\infty} \frac{1}{\sqrt{E}} \exp(-E/KT)\, dE, \tag{A6}$$

$$E_1 = \frac{1}{\sqrt{\pi KT}} \int_0^{E_m} \sqrt{E} \exp(-E/KT)\, dE/F_1, \tag{A7}$$

and

$$E_2 = \frac{1}{\sqrt{\pi KT}} \int_{E_m}^{\infty} \sqrt{E} \exp(-E/KT)\, dE/F_2. \tag{A8}$$

$d\bar{\theta}(r)/dr$ is calculated by

$$\frac{d\bar{\theta}(r)}{dr} = \bar{\theta}_1 \frac{dF_1}{dr} + F_2 \frac{d\bar{\theta}_1}{dr} + \bar{\theta}_2 \frac{dF_2}{dr} + F_2 \frac{d\bar{\theta}_2}{dr^2}. \tag{A9}$$

All derivatives in Eq. (A9) can be evaluated by applying the Leibniz rule and numerical integration.

APPENDIX B: CALCULATION OF $\overline{\cos\theta}(r)$ FOR THE ADO $\overline{\cos\theta}$ MODEL

To evaluate Eq. (37) for $\overline{\cos\theta(r)}$, again, it is separated into two cases as in the $\cos\bar{\theta}$ model (Appendix A)

Case (1). $E_{\text{rot}} < E_m$

$$\overline{\cos\theta_1} = \int_0^G \frac{\cos\theta \sin\theta\, d\theta}{(\cos\theta - \cos G)^{1/2}} \Bigg/ \int_0^G \frac{\sin\theta\, d\theta}{(\cos\theta - \cos G)^{1/2}}$$

or

$$\overline{\cos\theta_1} = \tfrac{1}{3}(1 - 2E_1 r^2/q\mu_D), \tag{B1}$$

where $G = \cos^{-1}(-E_1 r^2/q\mu_D)$, and E_1 is defined in Eq. (A7).

Case (2). $E_{\text{rot}} \geq E_m$

$$\overline{\cos\theta_2} = \int_0^{\pi} \frac{\cos\theta \sin\theta\, d\theta}{[E_2 + (q_D/r^2)\cos\theta]^{1/2}} \Bigg/ \int_0^{\pi} \frac{\sin\theta\, d\theta}{[E_2 + (q\mu_D/r^2)\cos\theta]^{1/2}}$$

or

$$\overline{\cos\theta_2} = \frac{1}{3}\left[\left(\frac{E_2 r^2}{q\mu_D} - 1\right)^{1/2} - \frac{E_2 r^2}{q\mu_D}\right], \tag{B2}$$

where E_2 is defined in Eq. (A8). Thus

$$\overline{\cos\theta}(r) = F_1\,\overline{\cos\theta_1} + F_2\,\overline{\cos\theta_2},$$

where F_1 and F_2 are defined in Appendix A.

APPENDIX C. CALCULATION OF THE r-DEPENDENT AVERAGE ROTATIONAL ENERGY IN THE PLANE OF COLLISION IN THE AADO THEORY

As in the ADO theory, the rotors are separated into two groups: those with sufficient rotational energy to overcome the energy of the ion field, and those with insufficient energy to overcome the field.

Case (1). $E_{rot} \leq E_m$

The instantaneous rotational energy in the plane of collision at an angle θ ($\leq$ the maximum angle G) and a separation r is

$$E_{R1} = E_1 + (q\mu_D/r^2)\cos\theta, \tag{C1}$$

where E_1 is calculated from Eq. (A7) in Appendix A. The average rotational energy for these molecules is then

$$\bar{E}_{R1} = \int_0^G E_{R1} P(\theta)\,d\theta \bigg/ \int_0^G P(\theta)\,d\theta, \tag{C2}$$

where $P(\theta)$ is defined in Eq. (33). Integrating Eq. (C2) yields

$$\bar{E}_{R1} = (q\mu_D/3r^2)(1 - \cos G) \tag{C3}$$

Case (2). $E_{rot} \geq E_m$

In this case, the instantaneous rotational energy in the plane of collision is

$$E_{R2} = E_2 + (q\mu_D/r^2)\cos\theta, \tag{C4}$$

where E_2 is defined in Eq. (A8). Hence

$$\bar{E}_{R2} = \int_0^\pi E_{R2} P(\theta)\,d\theta \bigg/ \int_0^\pi P(\theta)\,d\theta = \tfrac{1}{3}[2E_2 + (E_2{}^2 + q^2\mu_D{}^2/r^4)^{1/2}]. \tag{C5}$$

The average rotational energy in the plane of collision is then

$$\bar{E}_R = F_1\bar{E}_{R1} + F_2\bar{E}_{R2}, \tag{C6}$$

where F_1 and F_2 are defined in Eqs. (A5) and (A6), respectively.

REFERENCES

Barker, R. A., and Ridge, D. P. (1976). *J. Chem. Phys.* **64,** 4411.

Bass, L., Su, T., Chesnavich, W. J., and Bowers, M. T. (1975). *Chem. Phys. Lett.* **34**, 119.

Bass, L., Su, T., and Bowers, M. T. (1978). *Int. J. Mass Spectrom. Ion Phys.* **28**, 389.

Bell, K. L., Dalgarno, A., and Kingston, A. E. (1968). *J. Phys. B.* **1**, 18.

Boelnijk, N., and Hamill, W. H. (1962). *J. Am. Chem. Soc.* **84**, 529.

Bowers, M. T., and Laudenslager, J. B. (1972). *J. Chem. Phys.* **56**, 4711.

Bowers, M. T., and Su, T. (1973). *Adv. Electron. Electron Phys.* **34**, 223.

Bowers, M. T., and Su, T. (1975). *In* "Interactions Between Ions and Molecules" (P. Ausloos, ed.), p. 163. Plenum, New York.

Bowers, M. T., Su, T., and Anicich, V. G. (1973). *J. Chem. Phys.* **58**, 5175.

Chesnavich, W. J., Su, T., and Bowers, M. (1979). *In* "Kinetics of Ion–Molecule Reactions" (P. Ausloos, ed.), Plenum, New York.

Dugan, J. V. (1973). *Chem. Phys. Lett.* **21**, 476.

Dugan, J. V., and Canright, R. B. (1972). *J. Chem. Phys.* **56**, 3623.

Dugan, J. V., and Magee, J. L. (1966). *NASA Tech. Note* **NASA TN-D-3229.**

Dugan, J. V., and Magee, J. L. (1967). *J. Chem. Phys.* **47**, 3103.

Dugan, J. V., and Palmer, R. W. (1971). *Chem. Phys. Lett.* **13**, 144.

Dymerski, P. P., and Dunbar, R. C. (1972). *J. Chem. Phys.* **57**, 4049.

Dymerski, P. P., Dunbar, R. C., and Dugan, J. V. (1974). *J. Chem. Phys.* **61**, 298.

Eyring, H., Hirschfelder, J. O., and Taylor, H. S. (1936). *J. Chem. Phys.* **4**, 479.

Futrell, J. H., and Tiernan, T. O. (1968). *In* "Fundamental Process in Radiation Chemistry" (P. Ausloos, ed.), p. 171. Wiley (Interscience), New York.

Gentry, W. R. (1974). *J. Chem. Phys.* **60**, 2547.

Gioumousis, G., and Stevenson, D. P. (1958). *J. Chem. Phys.* **29**, 294.

Gupta, S. K., Jones, E. G., Harrison, A. G., and Myher, J. J. (1967). *Can. J. Chem.* **45**, 3107.

Hemsworth, R. S., Payzant, J. D., Schiff, H. I., and Bohme, D. K. (1974). *Chem. Phys. Lett.* **26**, 417

Henglein, A. (1970). *In* "Molecular Beams and Reaction Kinetics" (C. Schlier, ed.). Academic Press, New York.

Huntress, W. T., Mosesman, M. M., and Elleman, D. E. (1971), *J. Chem. Phys.* **54**, 843.

Langevin, P. M. (1905). *Ann. Chim. Phys.* **5**, 245.

Lindinger, W., Albritton, D. L., Felsenfeld, F. C., Schmeltekopf, A. L., and Ferguson, E. E. (1975). *J. Chem. Phys.* **62**, 3549.

McDaniel, E. W. (1964), "Collision Phenomena in Ionized Gases," Wiley, New York.

Mackay, G. I., Betowski, L. D., Payzant, J. D., Schiff, H. I., and Bohme, D. K. (1976). *J. Phys. Chem.* **80**, 2919.

Mason, E. W., and Schamp, H. W. (1958). *Ann. Phys.* (*N.Y.*) **4**, 233.

Meot-Ner, M., and Field, F. H. (1975). *J. Am. Chem. Soc.* **97**, 2014.

Miasek, P. G., and Beauchamp, J. C. (1974). *Int. J. Mass Spectrom. Ion Phys.* **15**, 49.

Moran, T. F., and Hamill, W. H. (1963). *J. Chem. Phys.* **39**, 1413.

Ridge, D. P., and Beauchamp, J. L. (1976). *J. Chem. Phys.* **64**, 2735.

Rothe, E. W., and Bernstein, R. B. (1959). *J. Chem. Phys.* **31**, 1619.

Schwartz, C. (1961). *Phys. Rev.* **123**, 1700.
Smith, D. L., and Futrell, J. H. (1972–1973). *Int. J. Mass Spectrom. Ion Phys.* **10**, 405.
Su, T., and Bowers, M. T. (1973a). *J. Chem. Phys.* **58**, 3027.
Su. T., and Bowers, M. T. (1973b). *Int. J. Mass Spectrom. Ion Phys.* **12**, 347.
Su, T., and Bowers, M. T. (1973c). *J. Am. Chem. Soc.* **95**, 1370.
Su, T., and Bowers, M. T. (1974). *J. Chem. Phys.* **60**, 4897.
Su, T., and Bowers, M. T. (1975a). *Int. J. Mass Spectrom. Ion Phys.* **17**, 211.
Su, T., and Bowers, M. T. (1975b). *Int. J. Mass Spectrom. Ion Phys.* **25**, 1.
Su, T., and Bowers, M. T. (1975c). *Int. J. Mass Spectrom. Ion Phys.* **17**, 309.
Su, T., Su, E. C. F., and Bowers, M. T. (1978a). *J. Chem. Phys.* **69**, 2243.
Su, T., Su, E. C. F., and Bowers, M. T. (1978b). *Int. J. Mass Spectrom. Ion Phys.* **28**, 285.
Tessman, J. R., and Kahn, A. H. (1953). *Phys. Rev.* **92**, 89.
Theard, L. P., and Hamill, W. H. (1962). *J. Am. Chem. Soc.* **84**, 1134.
Udseth, H., Giese, C. F., and Gentry, W. R. (1974). *J. Chem. Phys.* **60**, 3051.
Vogt, E., and Wannier, G. H. (1954). *Phys. Rev.* **95**, 1190.

Chapter 4
Statistical methods in reaction dynamics

WALTER J. CHESNAVICH AND MICHAEL T. BOWERS

DEPARTMENT OF CHEMISTRY
UNIVERSITY OF CALIFORNIA
SANTA BARBARA, CALIFORNIA

I. INTRODUCTION

For nearly half a century statistical methods have played an important role in the attempt to understand reaction dynamics. Theoretical research

ISBN-0-12-120801-X

in the last decade has been particularly active, focusing on the validity of the transition state and quasi-equilibrium assumptions, the role these assumptions play in quantum theory, and the development of variational techniques and new transition state models. Parallel to the advances in theory has been the development of experimental techniques, particularly in ion–molecule chemistry, which are capable of obtaining near-monoenergetic data on absolute reaction rates and product energy disposal, thus providing the most precise experimental tests to date of the statistical hypotheses.

In this chapter many of the recent advances in statistical methods are reviewed. The chapter is divided into three main parts covering (1) statistical theory, (2) the orbiting transition state, and (3) the application of statistical theory to chemical systems. Specifically, Section II discusses transition state theory, variational theory, RRKM–QET unimolecular theory with emphasis on lifetime distributions, and complex formation theory. In Section III recent research on the orbiting transition state is summarized. Section IV summarizes the results obtained in the application of statistical theory to ionic systems for which monoenergetic data have been obtained for absolute rate constants, cross sections, and/or translational energy disposal. Section V provides a brief conclusion.

It is impossible to cover all of statistical theory in one chapter. The interested reader can also consult with profit the articles by Vestal (1968), Hase (1976), Pechukas (1976), and Lifshitz (1978), and the books by Robinson and Holbrook (1972) and Forst (1973).

II. STATISTICAL THEORY

This section is divided into three subsections. The division is based on the three major categories of chemical reactions for which statistical methods have been most fully developed. Schematic potential surfaces for these categories are given in Fig. 1. The first subsection discusses direct bimolecular reactions (Fig. 1a), the second subsection pure unimolecular reactions (Fig. 1b), and the third subsection unimolecular reactions in which the reactant is an intermediate complex created by a bimolecular reaction (Fig. 1c).

A. *Transition State Theory*

The statistical theory of direct bimolecular reactions is transition state theory. Most textbook presentations of transition state theory are based on a comparison between an equilibrium distribution of transition state complexes (in which the reactants and products are also at equilibrium) and the nonequilibrium reactive distribution created by a thermal Boltzmann distribution of reactants [which is also often called an equilibrium distribution; see, e.g., Laidler (1969, Chapter 3)]. In recent years, however, the original

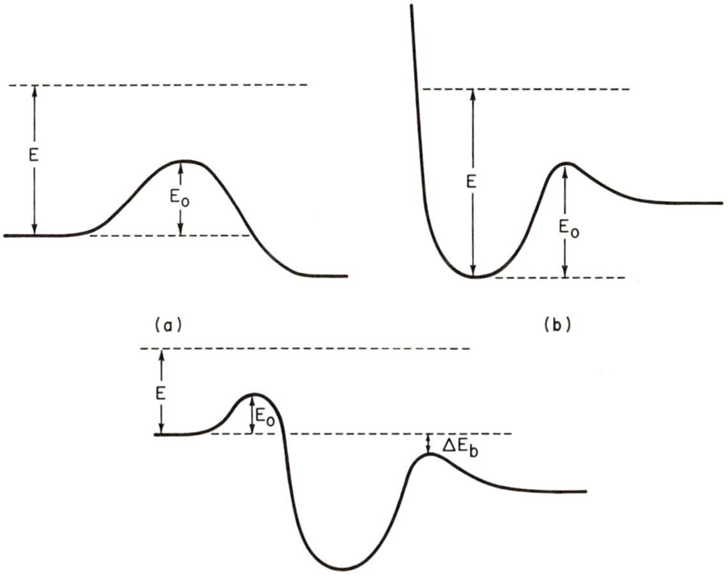

FIG. 1 Schematic potential surfaces for (a) direct bimolecular, (b) unimolecular, and (c) complex-formation reactions.

language of transition state theory (Pelzer and Wigner, 1932; Evans and Polanyi, 1935; Eyring, 1935; Wigner, 1938) has been revived (Pechukas and McLafferty, 1972; Miller, 1974, 1976a,b; Pechukas, 1976) since this viewpoint states more precisely the differences between transition state theory and exact dynamics and thus leads to a clearer understanding of the conditions under which the theory may fail. According to the early authors (see, e.g., Wigner, 1938), the three assumptions basic to the transition state theory approach are the following:

(1) The Born–Oppenheimer approximation may be used to separate electronic and nuclear motion.

(2) The motion of the nuclei can be described by classical mechanics (with quantum corrections when necessary).

(3) A surface exists in the system configuration space, completely separating reactants and products, with the property that all systems crossing this surface in the direction from reactants to products originated as reactants, cross *once and only once*, and will terminate as products.

Application of these three assumptions to the general bimolecular reaction

$$A + B \rightarrow X^{\ddagger} \rightarrow C + D \tag{1}$$

produces the well known canonical transition state theory rate constant

$$k(T) = \frac{kT}{h}\frac{Q^{\ddagger}}{Q}\exp(-E_0/kT), \tag{2}$$

where h, T, k, and E_0 are the Planck constant, the system temperature, the Boltzmann constant, and the activation energy, respectively. The internal partition function for the transition state complex $X^{\ddagger}$ is $Q^{\ddagger}$ and

$$Q = Q_A Q_B Q_t, \tag{3}$$

where Q_A and Q_B are the internal partition functions of A and B and Q_t is the partition function per unit volume for relative translational motion. The microcanonical version of Eq. (2) is

$$k(E) = W^{\ddagger}(E - E_0)/h\rho(E), \tag{4}$$

where $W^{\ddagger}(E - E_0)$ is the total sum of states (vibrational and rotational) of the transition state complex with nonfixed energy less than or equal to $E - E_0$, where E is the total system energy, and $\rho(E)$ is the total density of states of the reactants (a convolution of the vibrational and rotational densities and the relative translational density per unit volume). Symmetry numbers are assumed to be included in $Q^{\ddagger}$, $W^{\ddagger}$, Q, and ρ.

Microcanonical transition state theory rate constants such as Eq. (4) are often of limited practical use since it is usually not possible to determine which distribution in the reactants of the total system energy is responsible for the formation of a given transition state complex. Stated in another way, the transition state Hamiltonian usually cannot be expressed in terms of the coordinates for vibrational, rotational, and relative translational motion of the reactants (or of the products if product state distributions are desired). There are a number of methods by which one can attempt to bypass this problem. One method is to "place" the transition state in a region where the reactants retain their original identity. An example of this type is the orbiting transition state discussed in Section III. Another method is to estimate the dependence of the rate constant on the energy distribution in the reactants using information gleaned from the potential surface. This approach has been taken by Marcus in a series of papers on the translational energy dependence of rate constants and is discussed further in Sections III.B and IV.C.

1. The Exact Nature of the Fundamental Assumption

The third assumption stated above is simply a detailed statement of the concept that the transition state is the point of no return. It defines precisely the difference between transition state theory and exact dynamics. For this reason the third assumption is often called the fundamental assumption of transition state theory (Miller, 1976a).

The following is an important property the fundamental assumption imparts to the transition state theory rate constant. If the classical versions of the sums and densities in Eq. (4) are rigorously calculated (i.e., without harmonic oscillator–rigid rotor approximations) for a chosen potential surface, the resulting transition state theory rate constant will be a rigorous upper limit to the true classical dynamical rate constant for that surface. Stated symbolically,

$$k_{\mathrm{TST}} \geq k_{\mathrm{CLDYN}}, \tag{5}$$

where k_{CLDYN} is the exact classical rate constant.

For most potential surfaces, the total system energy determines whether the equality holds in Eq. (5). However, for certain classes of potentials, classical transition state theory is always rigorously exact. Included in this class are central potentials since motion in a central potential can always be reduced to motion in a corresponding one-dimensional system and transition state theory is exact for one-dimensional systems (see, e.g., Pechukas, 1976). An example is the charge-induced dipole capture rate constant which can be obtained from transition state theory (Eyring *et al.*, 1936) and from dynamical arguments (Vogt and Wannier, 1954; Gioumousis and Stevenson, 1958).

Several efforts have been made to determine the conditions for which transition state theory is exact (Pechukas and McLafferty, 1972; Chapman *et al.*, 1975; Grimmelmann and Lohr, 1977; Chesnavich, 1978). The results of these studies suggest that transition state theory is exact in the threshold region and that the magnitude of the failure above threshold depends in part on the number of degrees of freedom the system possesses (Miller, 1976a). One interesting point is that the failure of transition state theory indicates the onset of chattering or statistical collisions in which the system passes through the transition state many times before reacting. An example of a system in which chattering may influence the energy dependence of the cross section is the proton transfer reaction $H_2O^+(H_2, OH)H_3^+$, which is discussed in Section IV.C.

2. *Variational Theory*

Lifshitz (1978) has recently suggested that a variational criterion for the location of the transition state may be particularly well suited for ionic systems since many ionic dissociations have a negligible reverse activation energy. The basis of the variational approach is that since transition state theory provides a rigorous upper limit to the true rate constant for a given potential surface, the best choice of transition state is the dividing surface between reactants and products that produces the smallest transition state theory rate constant.

The first detailed application of variational theory to chemical systems was made by Keck (1960, 1967), using an approach related to earlier presentations by Wigner (1937) and Horiuti (1938). The essential features of the theory can be understood by considering a microcanonical system with s center of mass degrees of freedom, one of which is a separable reaction coordinate r. In this case the best location of the transition state is obtained by solving

$$\frac{\partial}{\partial r} W_{s-1}(E - V(r); r) = 0, \tag{6}$$

where $W_{s-1}(E - V(r); r)$ is the sum of states of the $s - 1$ degrees of freedom perpendicular to r at energy less than or equal to $E - V(r)$, where $V(r)$ is the potential energy at r. The variational procedure given by Eq. (6) is equivalent to varying the total rate constant or reactive flux as a function of r. Wong and Marcus (1971) and Quack and Troe (1977) have shown that this procedure is equivalent to a maximum free energy criterion.

An alternative variational approach was developed by Bunker and Pattengill (1968). Instead of varying the sum of states as in Eq. (6), they proposed to vary the convoluted product of the density of states in the reaction coordinate and the density of the rest of the complex. This approach, however, does not determine the minimum rate since it neglects the frequency at which reacting systems cross the dividing surface (see, e.g., Kay, 1976).

B. Unimolecular Theory

The present form of the statistical theory of the unimolecular reaction of an isolated energized molecule was developed independently by Marcus and Rice (Marcus and Rice, 1951; Marcus, 1952) and by Rosenstock *et al.* (1952). The theory has two aliases, the RRKM theory (after Rice, Ramsperger, Kassel, and Marcus) and the quasi-equilibrium theory or QET. The basic assumptions used to construct RRKM–QET theory are (1) the assumptions of transition state theory (Section II.A; see also Kay, 1976) and (2) the quasi-equilibrium hypothesis. The quasi-equilibrium hypothesis can be stated in the following way:

Subject to known conservation laws the density in the phase space of energized molecules prior to their reaction is uniform and independent of the method of initial preparation.

When the quasi-equilibrium hypothesis is applied, the resulting unimolecular rate constant is similar to the microcanonical transition state theory rate constant in Eq. (4). However, since it is possible in principle to measure rates as a function of the total angular momentum $\mathscr{J}$, it is useful to express the rate constant as a function of $\mathscr{J}$. Hence,

$$k(E, \mathscr{J}) = \frac{W^{\ddagger}(E - E_0, \mathscr{J})}{h\rho(E, \mathscr{J})}, \tag{7}$$

where k, $W^{\ddagger}$, and ρ are functions of E, $\mathscr{J}$, and $\mathscr{J}_z$, $\mathscr{J}_z$ is the projection of $\mathscr{J}$ on a space-fixed axis, and $\rho(E, \mathscr{J})$ the vibrational–rotational density of the energized molecule.

1. *Quasi-Equilibrium in Perspective*

In an ideal statistical molecule, the coupling provided by the intramolecular forces between the internal modes is strong enough so that no matter how unusual the pattern of initial energization, the intramode coupling would render it equivalent to a random one in a negligibly short time. An immediate consequence of this fact is that the reacting molecules will produce a random lifetime distribution given by (Robinson and Holbrook, 1972; Forst, 1973; Hase, 1976)

$$P(\tau) = ke^{-k\tau}. \tag{8}$$

Bunker and Hase (1973) have suggested that this ideal is never quite attained in real molecules. They define two catagories of nonstatistical behavior, *apparent* and *intrinsic*.

A molecule that exhibits apparent nonstatistical behavior possesses reasonably strong internal coupling. If the excess energy is initially deposited in a random pattern, for example by collisional excitation, a quasi-equilibrium will occur producing a random distribution of lifetimes. If however the excess energy is initially deposited in a specific location within the molecule, for example by chemical activation, the initial internal energy distribution may not relax before reaction occurs leading to an increase in the probability of a long (Fig. 2a) or short (Fig. 2b) lifetime with respect to reaction and an apparent failure of the quasi-equilibrium hypothesis.

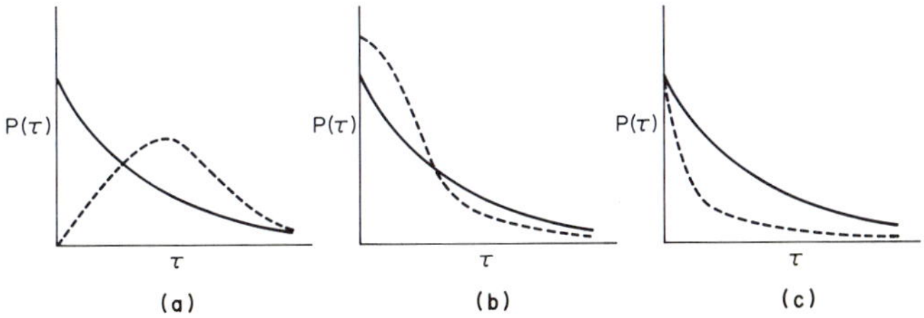

FIG. 2 Unimolecular lifetime distributions for (a) apparent nonstatistical behavior biased toward long lifetimes, (b) apparent nonstatistical behavior baised toward short lifetimes, and (c) intrinsic nonstatistical behavior. The solid line is a random lifetime distribution.

Experimental and theoretical evidence suggests that the time scale for apparent nonstatistical behavior is usually quite short. Internal relaxation times on the order of 10^{-12} sec have been obtained by Rabinovitch and co-workers in chemical activation experiments on neutral systems (Ryndbrandt and Rabinovitch, 1970, 1971a,b; Meogher *et al.*, 1974; Ko *et al.*, 1977; see also Forst, 1973, Chapter 2). Similar results have been obtained from trajectory calculations on model systems (Bunker, 1962, 1964; Hase and Feng, 1974, 1976; Hase and Sloane, 1976; McDonald and Marcus, 1976; Sloane and Hase, 1977). On the other hand, charge exchange and photoelectron–photoion coincidence experiments (Andlauer and Ottinger, 1971, 1972; Eland and Schulte, 1975; Werner and Baer, 1975; Tsai *et al.*, 1975; Baer *et al.*, 1975, 1976), which measure lifetimes of excited ions in the 10^{-4}–10^{-8} sec range, have uncovered no evidence of apparent nonstatistical behavior. These results appear to be consistent with the fact that a relaxation time of 10^{-12} sec corresponds to approximately 10 vibrations per mode in the energized reagent, whereas the lifetime range in the ionic experiments corresponds to about 10^5–10^9 vibrations per mode.

Intrinsic nonstatistical behavior occurs in molecules possessing internal coupling weak enough to produce transitions between two or more groups of states that are less probable than transitions leading to products. When the internal coupling between different groups is vanishingly small, the groups are said to be isolated. If the different groups are all accessible to the excitation process, and if each group has a pathway leading to the same products, then intrinsic behavior will produce a lifetime distribution which has the form of a sum of exponentials (Fig. 2c),

$$P(\tau) = \sum_i c_i e^{-k_i \tau}. \tag{9}$$

If, however, each group produces different products, or if only one group is accessible to the excitation process, the resulting lifetime distribution will be that of a single exponential.

Intrinsic nonstatistical behavior falls within the domain of the charge exchange and coincidence experiments. Systems which exhibit this type of behavior include the decompositions of $C_2H_5Cl^+$, $CH_2ClCH_2Cl^+$, $CHCCH_2Cl^+$, $CHCCH_2Br^+$ (Baer *et al.*, 1975; Tsai *et al.*, 1975), $C_2F_6{}^+$ (Simm *et al.*, 1973, 1974, 1976; Simm and Danby, 1976), and benzene (Andlauer and Ottinger, 1971, 1972; Eland and Schlute, 1975). The halocarbons studied by Baer and co-workers produce lifetime distributions which have the form of a sum of exponentials suggesting that these ions possess weakly coupled groups of internal states (either geometric isomers or electronic states) which are all accessible to the ionization process and produce the same products. Although lifetime distributions have not been reported for

$C_2F_6^+$, the breakdown pattern and translational energy release suggest that these ions possess isolated electronic states, of which the ground state yields exclusively CF_3^+ and the first excited state $C_2F_5^+$. The lifetime distributions of benzene yield single exponentials for the formation of $C_6H_5^+$ and $C_4H_4^+$, however the experiments demonstrate that these decompositions are not in competition with each other, indicating that the reactions occur from isolated electronic states or weakly coupled isomeric forms of $C_6H_6^+$ (Rosenstock *et al.*, 1973, 1977; Eland, 1974). The decompositions of benzene is discussed in greater detail in Section IV.A.

C. Complex Formation Theory

Statistical approaches to reactions which form a long-lived complex have been developed by Keck (1958), Light (1964, 1967), and Nikitin (1965a,b,c). Essential to these approaches is the assumption that the intermediate complex undergoes quasi-equilibrium and decays into each available channel at a rate proportional to the flux through the transition state for that channel. The resulting rate constant can be written as

$$k = \int \frac{\partial k_c}{\partial \mathscr{J}} P(\mathscr{J})\, d\mathscr{J}, \tag{10}$$

where $\partial k_c/\partial \mathscr{J}$ is the differential capture rate constant to form the complex and $P(\mathscr{J})$ is the decomposition probability of the complex into the final channel of interest.

Consider for simplicity a two channel reaction where the subscript a refers to the incoming channel (i.e., the left-hand side of Fig. 1c) and the subscript b to the product channel (i.e., the right-hand side of Fig. 1c). To obtain $\partial k_c/\partial \mathscr{J}$, Eq. (4) must be written as a function of the total angular momentum of the collision complex:

$$k_c(E) = \frac{\int W_a^{\ddagger}(E - E_0, \mathscr{J}) 2\mathscr{J}\, d\mathscr{J}}{h\rho(E)}. \tag{11}$$

Taking the derivative of Eq. (11) and substituting into Eq. (10) gives

$$k(E) = \frac{1}{h\rho(E)} \int W_a^{\ddagger}(E - E_0, \mathscr{J}) P(\mathscr{J}) 2\mathscr{J}\, d\mathscr{J}. \tag{12}$$

Using Eq. (7) to obtain the unimolecular rate constants for decomposition of the complex into each channel gives

$$P(\mathscr{J}) = k_b(E + \Delta E_b; \mathscr{J})/(k_a(E - E_0, \mathscr{J}) + k_b(E + \Delta E_b, \mathscr{J})), \tag{13a}$$

or

$$P(\mathscr{J}) = W_b^{\ddagger}/(W_a^{\ddagger} + W_b^{\ddagger}), \tag{13b}$$

where the arguments have been deleted from the sums of states for notational brevity.

Miller (1976b) has suggested that Eq. (13b) is not a general result since, for example, as the potential surface in Fig. 1c is distorted to look like Fig. 1a, $P(\mathscr{J})$ does not approach the limit $P(\mathscr{J}) \rightarrow 1$ as it should if Eq. (12) is to reduce to Eq. (11). To remedy this defect, Miller used variational arguments to obtain the following general result for $P(\mathscr{J})$:

$$P(\mathscr{J}) = W_b^{\ddagger}/(W_a^{\ddagger} + W_b^{\ddagger} - W_a^{\ddagger}W_b^{\ddagger}/W_d). \tag{14}$$

In this equation W_d is the flux (i.e., sum of states) of the intermediate complex through a surface located at the bottom of the well. For deep wells W_d will be much larger than either $W_a^{\ddagger}$ or $W_b^{\ddagger}$ and Eq. (14) reduces to Eq. (13b). As the well is distorted so that Fig. 1c approaches Fig. 1a, W_d approaches $W_b^{\ddagger}$ and $P(\mathscr{J}) \rightarrow 1$.

Unimolecular reagents generated by a complex formation reaction are expected to have an angular momentum distribution peaked at higher values of $\mathscr{J}$ than the corresponding thermal distribution. This is particularly true of ion–molecule reactions due to the strength of the long range attractive force. In a high $\mathscr{J}$ complex most of the potential energy released upon formation of the complex will be consumed as kinetic energy of rotation resulting in a shallow effective potential well. In this situation the conditions leading to Eq. (14) will occur even if the thermodynamic well is deep.

III. THE ORBITING TRANSITION STATE

An orbiting transition state is located at the maximum of the long range effective potential. An important property of this location is that the transition state Hamiltonian can be determined from the Hamiltonians of the individual molecules plus a potential energy term describing their mutual long range interaction. This choice removes a certain arbitrariness usually associated with the selection of transition state parameters. Also, if the long range intermolecular potential is spherically symmetric, and if vibrational–rotational coupling and electronic transitions can be ignored, the transition state Hamiltonian is separable and can be written as

$$H = H_{\mathrm{v}_1} + H_{\mathrm{v}_2} + H_{\mathrm{r}_1} + H_{\mathrm{r}_2} + H_{\mathrm{t}}, \tag{15}$$

where the subscripts v, r, and t signify vibrational, rotational, and relative translational motion. Separability allows one to calculate state to state rate constants and product energy disposal using the methods of transition state theory without making additional dynamical assumptions. For these reasons an orbiting model is a desirable transition state candidate. However, it must be kept in mind that its appropriateness as a candidate depends on how well the reaction dynamics are governed by the long range part of the intermolccular potential.

The remainder of this section discusses in greater detail the orbiting transition state. The section is divided into two parts. In the first part the angular momentum conserved sums and densities of rotational states are introduced and the orbiting expressions for bimolecular and unimolecular rate constants are presented. It is shown that the rate constant for arbitrary $\mathscr{J}$ for a pair of rigid rotors is simply a generlization of the result obtained by Pechukas and Light (1965) for an atom–diatom pair. The second part discusses the application of the orbiting model to the prediction of product translational energy disposal.

A. *Orbiting Rate Constants*

1. *Sums and Densities of Rotational States*

Since there is presently little experimental data available on the rotational energy dependence of rate constants and on product vibrational and rotational distributions, it is often convenient to discuss orbiting transition states for separable Hamiltonians in terms of the total vibrational–rotational contribution to the reactive flux at the transition state. If one does not have to treat angular momentum conservation explicitly, this can be accomplished through a convolution of the total vibrational and total rotational density of states of the molecular pair. However, when angular momentum is explicitly conserved, the rotational density of states must be amended to include the angular momentum associated with the rotational motion of the molecular pair. The desired quantity therefore is the J_r, J_{rz}-conserved rotational density of states of the molecular pair, where J_r is the total rotational angular momentum and J_{rz} is the space-fixed projection (Chesnavich and Bowers, 1976, 1977a). Table I lists classical expressions for $\Gamma(E_r, J_r)$, the corresponding sum of states at total rotational energy less than or equal to E_r, for various pairs of rigid rotors (although classical mechanics is used for rotational and orbital motion, all angular momenta are expressed in units of $\hbar$ to avoid excessive recurrence of this quantity in the equations). The density of states is given by

$$\rho(E_r, J_r) = \frac{d}{dE_r}\Gamma(E_r, J_r). \tag{16}$$

TABLE I

Rotational Sums of States[a]

System	$\Gamma(E_r)$	$\Gamma_z(E_r)$	$\Gamma(E_r, J_r)$	Boundaries[b]
Linear–atom	E_r/B_l	$(E_r/B_l)^{1/2}$	1	$E_r \geq B_l J_r^2$
Sphere–atom	$4E_r^{3/2}/3B_s^{3/2}$	E_r/B_s	$2J_r$	$E_r \geq B_s J_r^2$
Linear–linear[c]			$q \sin^{-1}\left\{\frac{J_r}{\omega q} z\right\} - \frac{J_r}{\omega} z$	$B_1 J_r^2 \geq E_r \geq B_r J_r^2$
	$E_r^2/B_1 B_2$	$\frac{2}{3}[(\sqrt{B_1} + \sqrt{B_2} - \sqrt{\omega})/B_1 B_2]E_r^{3/2}$	$q \sin^{-1}\left\{2 \frac{\sqrt{B_1 B_2}}{\omega}\right\} - \frac{B_1}{\omega} J_r^2$	$B_2 J_r^2 \geq E_r \geq B_1 J_r^2$
			$q \sin^{-1}\left\{\frac{J_r}{\omega q} z\right\} + \frac{J_r}{\omega} z - J_r^2$	$E_r \geq B_2 J_r^2$
Sphere–linear[d]	$8E_r^{5/2}/15B_s^{3/2}B_l$	$E_r^2 \sin^{-1}(B_s/\omega)^{1/2}/2B_s(B_s B_l)^{1/2}$	$(4/3B_s\omega^2)z^3$ $(2J_r/\omega^2)[\omega E_r - B_s(B_l + B_s/3)J_r^2]$	$B_s J_r^2 \geq E_r \geq B_r J_r^2$ $E_r \geq B_s J_r^2$
Sphere–sphere[e]	$\pi E_r^3/6(B_1 B_2)^{3/2}$	$8E_r^{5/2}/15B_1 B_2 \omega^{1/2}$	$(8J_r/3\omega^3)z^3$	$E_r \geq B_r J_r^2$

[a] B_s and B_l are spherical top and linear rotational constants. B_r is the reduced rotational constant of the rigid rotor pair. For symmetric and asymmetric tops use $B_s = (ABC)^{1/3}$.

[b] These boundaries refer only to $\Gamma(E_r, J_r)$.

[c] For the linear–linear system $q = E_r/2\sqrt{B_2 B_2}$, $z = (\omega E_r - B_1 B_2 J_r^2)^{1/2}$, $\omega = B_1 + B_2$ and $B_1 < B_2$.

[d] For the sphere–linear system $z = (\omega E_r - B_s B_l J_r^2)^{1/2}$ and $\omega = B_s + B_l$.

[e] For the sphere–sphere system $z = (\omega E_r - B_1 B_2 J_r^2)^{1/2}$ and $\omega = B_1 + B_2$.

Note that $\Gamma(E_r, J_r)$ and $\rho(E_r, J_r)$ are zero for $E_r \leq B_r J_r^2$, where B_r is the reduced rotational constant of the rigid rotor pair.

Also included in Table I are expressions for atom–rigid rotor pairs. For these systems, rotational energy and angular momentum are related by $E_r = BJ_r^2$. Therefore, $\Gamma(E_r, J_r)$ is simply the rotational degeneracy of the rigid rotor. In the rate constant expressions summarized below, dE_r is eliminated and $\rho(E_r, J_r)$ is replaced by $\Gamma(E_r, J_r)$ for these systems.

Two important properties of $\Gamma(E_r, J_r)$ [and $\rho(E_r, J_r)$] are the following. First, the total rotational sum of states (see e.g., Forst, 1973, Chapter 6) can be obtained by integrating over all J_{rz} and J_r,

$$\Gamma(E_r) = \int_0^{(E_r/B_r)^{1/2}} \int_{-J_r}^{+J_r} \Gamma(E_r, J_r)\, dJ_{rz}\, dJ_r. \tag{17}$$

Second, when the space-fixed degeneracy is neglected, the integration of $\Gamma(E_r, J_r)$ over J_r produces the sum of states obtained by Klots (1971, 1972),

$$\Gamma_z(E_r) = \int_0^{(E_r/B_r)^{1/2}} \Gamma(E_r, J_r)\, dJ_r. \tag{18}$$

Expressions for $\Gamma(E_r)$ and $\Gamma_z(E_r)$ are given in Table I. Note that $\Gamma(E_r) \propto E_r^{s/2}$ and $\Gamma_z(E_r) \propto E_r^{(s-1)/2}$, where s is the total number of rotational degrees of freedom of the rigid rotor pair

The expressions in Table I consider only spherical top three-dimensional rotors. Most real molecules, however, are symmetric or asymmetric tops. By equating the partition functions for a spherical top and symmetric or asymmetric top, one finds that $B_s = (ABC)^{1/3}$ where A, B, and C are the rotational constants of the asymmetric top. This value of B_s can be used in the spherical top expressions given in Table I. Comparison of this approximation to exact expressions for $\Gamma(E_r J_r)$ for sphere–oblate and sphere–prolate pairs (Chesnavich and Bowers, 1977a) shows it to be an excellent approximation for these systems.

2. *The Bimolecular Rate Constant*

The orbiting rate constant for a bimolecular reaction can be obtained from standard transition state theory arguments (Miller, 1976b), from the Liouville equation for time independent phase space density (Keck, 1967), or from detailed balance (Light, 1967). If the system total angular momentum is not explicitly conserved, the general microcanonical result for a direct reaction (i.e., a capture collision) and separable Hamiltonian can be written for rigid rotor pairs as

$$k(E) = \int \rho_v(E_v)\rho(E_r, J_r) 2J_r 2L\, d\tau / h\rho(E), \tag{19}$$

where $\rho_v(E_v)$ is the density of vibrational states of the reactants, L is the orbital angular momentum, and the $2J_r$ and $2L$ degeneracies arise from projections on a space-fixed axis. The total density of states per unit volume of the reactants is defined by

$$\rho(E) = \iint \rho_v(E_v)\rho_r(E_r)\rho_t(E_t)\, dE_v\, dE_r, \tag{20}$$

where $\rho_r(E_r) = d\Gamma(E_r)/dE_r$, $\rho_t(E_t)$ is the density of states for relative translational motion per unit volume at relative translational energy E_t, and $E = E_v + E_r + E_t$. The integration in Eq. (19) is over all variables in the differential volume element $d\tau$, where

$$d\tau = dJ_r\, dL\, dE_v\, dE_r. \tag{21}$$

The boundaries of the integration depend on the long range potential. For a single term potential $V(r) = -c/r^n$, the translational Hamiltonian is

$$H_t = P_{rc}^2/2\mu + L^2\hbar^2/2\mu r^2 - c/r^n, \tag{22}$$

where P_{rc} is the momentum in the reaction coordinate and μ is the reduced mass of the reactants. The location of the transition state is obtained by determining the value of r for which $\partial H_t/\partial r = 0$. When this result is substituted back into the Hamiltonian and the reaction coordinate momentum is allowed to span its range of values, one obtains

$$0 \le E_v + E_r + L^{2n/(n-2)}/\lambda \le E, \tag{23a}$$

where λ is a collection of constants that arises in the solution of $\partial H_t/\partial r = 0$ (for the charge-induced dipole potential $n = 4$ and $\lambda = 8\mu^2 q^2 \alpha/\hbar^4$, where q is the total charge on the ion and α the polarizability of the neutral). Equation (23a) and

$$0 \le B_r J_r^{\,2} \le E_r \tag{23b}$$

constitute the boundaries in the integration of Eq. (19) for a single term long range potential. If the potential is expanded in powers of $1/r$ (Fullerton and Moran, 1971), Eqs. (22) and (23a) are modified.

Equations (19)–(23b) determine $k(E)$ for a direct bimolecular reaction. For separable Hamiltonians, one can also obtain the rate constant as a function of E_v, E_r, and E_t or of other energy and/or angular momentum combinations. For example, for k as a function of E_v, E_r, and E_t, the integrations in Eq. (20) are eliminated, dE_v and dE_r are dropped from Eq. (21), and Eq. (23a) becomes $E_t \ge L^{2n/(n-2)}/\lambda \ge 0$. The vibrational densities of states cancel from the numerator and denominator of Eq. (19) and when Eq. (17) is used to integrate over dJ_r, the rotational densities cancel also.

The final integration over dL can now be performed giving

$$k = (\lambda E_t)^{(n-2)/n}/h\rho_t(E_t). \tag{24}$$

By substituting the appropriate results for the charge-induced dipole potential, and using $\rho_t(E_t) = 4\pi\mu(2\mu E_t)^{1/2}/h^3$, one sees that Eq. (24) gives the charge-induced dipole rate constant $k = 2\pi q(\alpha/\mu)^{1/2}$.

Equation (19) describes only the capture rate constant for a direct reaction. For reactions which form an intermediate complex the general results of Section II.C can be used. To do so, Eq. (19) must first be rewritten so that $\mathcal{J}$ is explicitly conserved. This can be accomplished by replacing $2J_r 2L$ by $2\mathcal{J}$ (which accounts for the degeneracy of $\mathcal{J}$ along a space-fixed axis), changing $d\tau$ to

$$d\tau = dJ_r\, dL\, d\mathcal{J}\, dE_v\, dE_r, \tag{25}$$

and integrating over $\mathcal{J}$ (Light, 1964; Pechukas and Light, 1965; Keck, 1967). An additional boundary, due to conservation of total angular momentum, now arises, given by

$$|L - J_r| \leq \mathcal{J} \leq L + J_r. \tag{26}$$

The bimolecular rate constant for a complex forming reaction can now be written. For example, for the rate constant as a function of E_v, E_r, and E_t one has

$$k = \frac{1}{h\rho_r(E_r)\rho_t(E_t)} \int \rho(E_r, J_r)\, dJ_r\, dL P(\mathcal{J}) 2\mathcal{J}\, d\mathcal{J}, \tag{27}$$

where the boundaries for L, J_r, and $\mathcal{J}$ are given by Eq. (26) and Eqs. (23) with E_v and E_r separately conserved. Comparing Eq. (27) to Eq. (12), one sees that the sum of states at the orbiting transition state for the incoming channel is given by the integral of $\rho(E_r, J_r)$ over J_r and L. Note also that Eq. (27) is really a general result independent of the exact form of $P(\mathcal{J})$. For example if $P(\mathcal{J}) = 1$ for all $\mathcal{J}$, Eq. (27) reduces to Eq. (24).

If the translational energy of the reactants is large, Eq. (27) reduces to a familiar result. In this case one has $L \to \mathcal{J}$ for the reactants. Therefore, the integration over L can be approximated as $2J_r$ times the value at $L = \mathcal{J}$. For rigid rotor pairs, the integration over J_r can be performed by using Eq. (17) to obtain $\rho(E_r)$, which cancels with the rotational density in the denominator. Finally, making the substitution $\mathcal{J}\hbar = \mu v b$, where b is the impact parameter and v is the initial relative velocity of the reactants, Eq. (17) becomes

$$k = v \int_0^{b_{\max}} 2\pi b P(b) db, \tag{28}$$

where $b_{\max} = (\lambda E_t)^{(n-2)/2n}\hbar/\mu v$.

3. The Unimolecular Rate Constant

The unimolecular rate constant for decomposition of an energized reagent is similar to Eq. (19) except that the terms in the numerator refer to the products of the decomposition and the density of states in the denominator refers to the unimolecular reagent. Since the $\mathscr{J}$ distribution of the reagent is a well defined quantity, at least in principle, it is desirable to express the rate constant as a function of E, $\mathscr{J}$, and $\mathscr{J}_z$. By defining the total translational–rotational energy as $E_{tr} = E - E_v$ and defining E as the energy above the zero point potential surface of the products, the rate constant as a function of E, $\mathscr{J}$, and $\mathscr{J}_z$ can be written as (Chesnavich and Bowers, 1977b)

$$k(E, \mathscr{J}) = S \int \rho_v(E - E_{tr})\Gamma_{ro}(E_{tr}, \mathscr{J})\, dE_{tr}/h\rho(E, \mathscr{J}), \tag{29}$$

where

$$\Gamma_{ro}(E_{tr}, \mathscr{J}) = \iint \Gamma(E_r^*, J_r)\, dJ_r\, dL \tag{30}$$

and S is the symmetry number ratio of reactant to products. Equation (30) defines the $\mathscr{J}, \mathscr{J}_z$-conserved sum of rotational–orbital states with rotational–orbital energy less than or equal to E_{tr} [for atom–rigid rotor pairs no alterations are necessary in Eq. (30)]. The sum of states $\Gamma(E_r^*, J_r)$ arises from the integration of $\rho(E_r, J_r)$ over E_r within the range defined by Eq. (23a). The energy E_r^* is given by

$$E_r^* = E_{tr} - L^{2n/(n-2)}/\lambda. \tag{31a}$$

The boundaries on J_r and L in Eq. (30) are given by Eq. (26) and by

$$L^{2n/(n-2)}/\lambda + B_r J_r^2 \leq E_{tr} \tag{31b}$$

The denominator of Eq. (29) depends on the molecular structure of the reagent. For example, if the reagent is a spherical top with rotational constant B_0, $\rho(E, \mathscr{J})$ is given by

$$\rho(E, \mathscr{J}) = 2\mathscr{J}\rho(E + E_0 - B_0\mathscr{J}^2), \tag{32}$$

where the $2\mathscr{J}$ term arises from the classical internal degeneracy of a spherical top rigid rotor. If the reagent is linear, the $2\mathscr{J}$ factor is deleted from Eq. (32).

Equation (29) is a rigorous definition of $k(E, \mathscr{J})$ for the orbiting transition state for arbitrary $\mathscr{J}$. For an atom–diatom pair $\Gamma(E_r^*, J_r) = 1$ and Eq. (30) is identical to the result obtained by Pechukas and Light (1965) (in classical language).

There are two points which should be kept in mind in the evaluation of Eqs. (29) and (30). First, inspection of $\Gamma(E_r, J_r)$ for the systems listed in Table I shows that the integration over J_r can be evaluated analytically for all systems. Therefore, only the integrations over E_{tr} and L must be performed numerically. Second, for sphere–atom and sphere–sphere pairs $\Gamma(E_r^*, J_r)$ can be convoluted with $\rho_v(E - E_{tr})$ before the integrations over J_r and L are performed. By noting that for these systems

$$\Gamma(E_r, J_r)\, dJ_r = \rho_z(E_r - E_J)\, dE_J, \tag{33}$$

where $E_J = B_r J_r^2$ and $\rho_z(E_r) = d\Gamma_z(E_r)/dE_r$, the result can be viewed as the convolution of $\rho_v(E_v)$ with $\Gamma_z(E_r^*)$. Therefore, only the integration over L has to be evaluated numerically for these systems to obtain $k(E, \mathscr{J})$ for arbitrary $\mathscr{J}$.

In thermal systems the maximum in the $\mathscr{J}$ distribution is often near zero. In this case an important simplification can be made to Eq. (29) (Klots, 1971, 1972). For small $\mathscr{J}$, provided that $(\lambda E)^{(n-2)/2n} > (E/B_r)^{1/2}$, the integration of Eq. (30) over L can be approximated as $2\mathscr{J}$ times $\Gamma(E_{tr}, J_r)$. The integration over J_r can then be performed in the limits $0 \leq J_r \leq (E_{tr}/B_r)^{1/2}$, using Eq. (18) and giving

$$\lim_{\mathscr{J}\to 0} k(E, \mathscr{J}) = 2\mathscr{J} S \int_0^E \Gamma_z(E_{tr})\rho_v(E - E_{tr})\, dE_{tr}/h\rho(E, \mathscr{J}). \tag{34}$$

Equation (34) is readily evaluated by noting that the numerator is simply a vibrational–rotational sum of states (Klots, 1971, 1972).

For small $\mathscr{J}$, $\Gamma_{ro}(E_{tr}, J)$ is proportional to $2\mathscr{J}$. As $\mathscr{J}$ increases, $\Gamma_{ro}(E_{tr}, \mathscr{J})$ peaks and then declines (see, e.g., Chesnavich and Bowers, 1977a, Fig. 7). For large $\mathscr{J}$, provided that $(\lambda E)^{(n-2)/2n} > (E_{tr}/B_r)^{1/2}$, the integration of Eq. (30) over L can be approximated as $2J_r$ times $\Gamma(E_r^*, J_r)$ evaluated at $L = \mathscr{J}$. The integration over J_r can then be performed using Eq. (17) and giving

$$k(E, \mathscr{J}) \sim S \int_{\mathscr{J}^{2n/(2-n)}/\lambda}^{E} \Gamma(E_{tr} - \mathscr{J}^{2n/(n-2)}/\lambda)\rho_v(E - E_{tr})\, dE_{tr}/h\rho(E, \mathscr{J}). \tag{35}$$

B. *Translational Energy Disposal*

For spherically symmetric long range potentials, the orbiting model result for the probability of a given translational energy release is (Chesnavich and Bowers, 1976, 1977a)

$$P_{\mathscr{J}}(E; E_t) \propto \int \rho_v(E - E_t - E_r)\rho(E_r, J_r)\, dJ_r\, dL\, dE_r, \tag{36}$$

where the boundaries are given by Eq. (26) and by $L \leq (\lambda E_t)^{(n-2)/2n}$. It can be shown that in the limit $\mathscr{J} \to 0$ the average translational energy release estimated from Eq. (36) can be written as

$$\bar{E}_t = \frac{2}{s+1} \frac{\int_0^E \rho_v(E_v)(E - E_v)^{(s+1)/2}\, dE_v}{\int_0^E \rho_v(E_v)(E - E_v)^{(s-1)/2}\, dE_v}, \tag{37}$$

where s is the total number of rotational degrees of freedom of the products. Equation (37) was originally reported by Klots (1972).

Equations (36) and (37) can be compared to

$$P(E, E_{rc}) \propto \rho^{\ddagger}(E - E_{rc}), \tag{38}$$

which is often used to estimate translational energy release. In this equation a saddle point transition state is assumed and E_{rc} is the kinetic energy of the reaction coordinate. In the application of Eq. (38) it is assumed that the translational energy of the products arises entirely from the kinetic energy of the reaction coordinate.

Although both Eqs. (36) and (38) have been refered to as statistical predictions, they are definitely based on different assumptions. The orbiting model produces a transition state in which the various degrees of freedom of the products can be identified unambiguously and are all accounted for. In the saddle point model, however, the degrees of freedom of the transition state moiety do not directly correspond to product vibrations, rotations, and translations. Marcus (1973) has emphasized this point strongly (see also Chesnavich and Bowers, 1977b). Within the context of transition state theory, rate constants are governed only by system motion on the reactant side of the transition state, whereas product energy disposal is governed only by system motion on the products side. Therefore, only when the transition state is located in the products region of configuration space can product energy disposal be obtained from transition state theory without the need for additional assumptions to describe the evolution of reacting systems from the transition state to products.

Carter (1976) has noted that the average translational energy release calculated from Eqs. (37) and (38) will in many instances differ only by a small amount with the difference approaching zero for large molecules. Carter's comparison suggests that *average* translational energies may, under certain conditions, be insensitive to the exact nature of the mechanism by which translational energy release occurs. This insensitivity can partially be overcome by examining translational energy *probabilities*. For example, Eq. (38) predicts a monotonically decreasing probability peaked at $E_{rc} = 0$. Equation (36) predicts a probability which rises smoothly from zero for $\mathcal{J} \leq (E/B_r)^{1/2}$ (or from some energy above zero for $\mathcal{J} \geq (E/B_r)^{1/2}$), peaks, and then decays to zero as $E_t \rightarrow E$. Klots (1976) has shown that when averaged over a distribution in $\mathcal{J}$ the probability is zero at $E_t = 0$ and has the general form of a Boltzmann distribution for $E_t > 0$.

Marcus (1966, 1967, 1970, 1975, 1977) has discussed in detail methods by which translational energy release can be obtained from saddle point transition states. He considered explicitly late barrier reactions with reverse activation energies. If one assumes that the probability of a given translational energy in excess of the reverse activation energy is a function of the excess energy in the reaction coordinate, the assumptions of transition state theory lead to (Worry and Marcus, 1977)

$$W^{\ddagger}(E - E_0, \mathscr{J}) = \int dE_{tr}\rho_v(E - E_0 - E_{tr}) \iint dE_t\, dL P(E_t - B_L L^2) \int dJ_r \rho(E_r, J_r), \tag{39}$$

where $W^{\ddagger}(E - E_0, \mathscr{J})$ is the sum of states at the transition state and $P(E_t - B_L L^2)$ is the probability that the products have translational energy E_t with orbital component $B_L L^2$, where B_L is a rotational constant associated with the transition state. Equation (39) is to be solved for P (usually by applying Laplace transforms). By appeal to detailed balance, it is also possible to obtain reaction cross sections from Eq. (39) as a function of the translational energy of the reactants. An example is given in Section IV.C.3.

IV. APPLICATION TO CHEMICAL SYSTEMS

This section summarizes the results obtained in the application of statistical methods to certain chemical systems. The systems discussed include only those for which near monoenergetic data have been obtained for absolute rate constants, cross sections, and/or product translational energy release.

A. Decomposition of Benzene and Substituted Benzene Ions

The following unimolecular decompositions of benzene and substituted benzene ions have been investigated:

$$C_6H_6^+ \rightarrow C_6H_5^+ + H \tag{40}$$

$$C_6H_6^+ \rightarrow C_6H_4^+ + H_2 \tag{41}$$

$$C_6H_6^+ \rightarrow C_4H_4^+ + C_2H_2 \tag{42}$$

$$C_6H_6^+ \rightarrow C_3H_3^+ + C_3H_3 \tag{43}$$

$$C_6H_5CN^+ \longrightarrow C_6H_4^+ + HCN \tag{44}$$

$$C_6H_5X^+ \longrightarrow C_6H_5^+ + X. \tag{45}$$

Andlauer and Ottinger (1971, 1972) reported the first monoenergetic $k(E)$ versus E data for metastable ions in an investigation of reactions (40), (42), and (44). A more accurate energy scale for these reactions was established by Eland and Schulte (1975), who also obtained rate constant data for reaction (43). Rate constant data for the decompositions of the halobenzenes, reaction (45) (X = C, Br, I), have been reported by Baer *et al.* (1976). In addition, the metastable kinetic energy release has been reported for reaction (40) (Terwilliger, Beynon, and Cooks, unpublished results, reported in Klots, 1976) and for reaction (45) (Terwilliger *et al.*, 1974). Data have not yet been obtained for the rate of reaction (41) or for the product translational energy distributions of reactions (41), (43), and (45).

The decomposition of $C_6H_6^+$ is an interesting example of intrinsic non-statistical behavior (see Section II.B.1). Reactions (40), (42), and (43) produce single exponential lifetime distributions. Reactions (42) and (43) are in competition with each other but not with reaction (40) (Andlauer and Ottinger, 1972; Eland and Schulte, 1975). Also, current evidence indicates that reaction (41) is in competition with (40) but not with (42) and (43) (Rosenstock *et al.*, 1973; Eland and Schulte, 1975). These results, when combined with photoionization studies (Rosenstock *et al.*, 1973, 1977), indicates that reactions (40) and (41) occur from the ground state of $C_6H_6^+$, whereas reactions (42) and (43) occur from an excited state lying ≈ 2 eV above the ground state (see also Eland, 1974). Furthermore, the photoionization behavior of a benzene isomer, 1,5-hexadiene, suggests that the $k(E)$ versus E data for this molecule will be similar to benzene (Rosenstock *et al.*, 1977).

The lifetime distributions of reactions (44) and (45) are single exponentials. These reactions also show no other evidence of isolated states. In addition, although reaction (44) involves a rearrangement, the product translational energy distribution indicates that this reaction does not have an appreciable reverse activation energy.

A number of statistical theory calculations of $k(E)$ versus E have been reported for reactions (40)–(45). The results obtained, using orbiting transition states, are as follows. Klots (1972) calculated $k(E)$ versus E for reactions (40), (42), and (44). The thermochemical data available to him, however, were inaccurate. Calculations based on more recent appearance potential measurements indicate that the theoretical rate constants are much higher than the experimental results throughout the energy range investigated [for a recalculation of reaction (44), see Chesnavich and Bowers (1977b)]. The same trend is obtained for the halobenzene decompositions. Results for the decomposition of bromobenzene are given in Fig. 3. If these reactions are simple bond cleavages involving no rearrangement of the phenyl group, one would expect better performance from an orbiting model than is obtained.

Statistical theory calculations, using saddle point transition states (with no reverse activation energy), gave much better agreement with experiment. Rosenstock *et al.* (1973) duplicated $k(E)$ versus E for reaction (42) by assuming that decomposition occurs from an excited state of $C_6H_6^+$ lying ~ 2 eV above the ground state. Chesnavich and Bowers (1977b) and Baer *et al.* (1976) obtained agreement between experiment and theory for reactions (44) and (45), respectively. However, no statistical theory calculation to date has been able to mimic the slow rise of $k(E)$ versus E for reaction (40).

An interesting point concerning reaction (40) is that the experimental result appears to be higher than that predicted for the orbiting model near

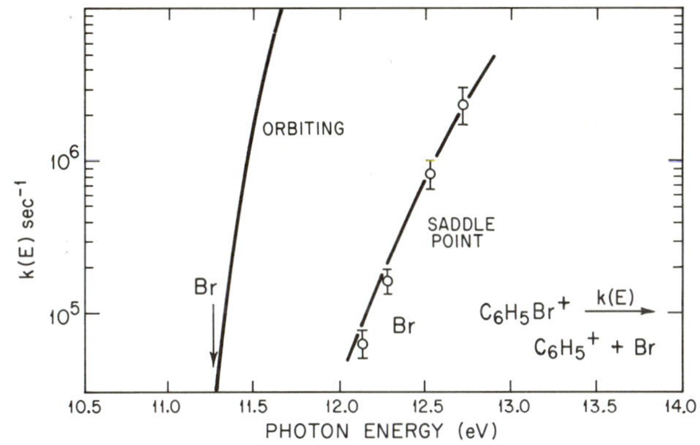

FIG. 3 $k(E)$ vs. E for the decomposition of bromobenzene ions. Experimental data (○) and saddle point curve as reported by Baer *et al.* (1976); orbiting curve calculated using Eq. (29). [Adapted from Baer *et al.* (1976).]

threshold. Similar behavior occurs in the $C_4H_6^+$ system (Section IV.B) and has been interpreted in terms of a two-step mechanism in which a rate determining isomerization precedes decomposition (Chesnavich and Bowers, 1977b).

Klots (1976) was able to match the experimental product translational energy distribution for reaction (40), but only by letting $n \to \infty$ in the long range potential $V(r) = -c/r^n$ (i.e., a square well potential). The experimental product translational energy distribution for reaction (44) was easily reproduced by using the charge-induced dipole long range potential (Klots, 1976; Chesnavich and Bowers, 1977b).

The agreement obtained between the experimental product translational energy distributions and the results obtained, using orbiting models, is satisfying, especially for the benzonitrile decomposition. However, it should be kept in mind that, although the metastable experiments select reactant ions with fairly well defined lifetimes, the internal energy distributions are not well defined. Product translational energy distributions obtained from monoenergetic $C_4H_6^+$ ions (see Section IV.B) suggest that deviations from theoretical predictions may occur as the internal energy of the parent ion is increased.

The ability of a square well potential to predict the product translational energy distribution for reaction (40) suggests that perhaps a square well potential produces a more realistic transition state than the charge-induced dipole potential for this reaction. The square well prediction of $k(E)$ versus E can be determined from the data of Table I by recalling that the orbital term of the translational Hamiltonian at the square well transition state is given by $L^2\hbar^2/2\mu r^2$, where r is a constant. Therefore, the orbital contribution to the system phase space at the transition state is equivalent to the contribution obtained from a linear rigid rotor with rotational constant $\hbar^2/2\mu r^2$. Since the products of reaction (40) are an atom and a three-dimensional rotor, the square well rate constant can therefore be obtained from Eq. (29) by replacing $\Gamma_{ro}(E_{tr}, \mathscr{J})$ by $\Gamma(E_r, J_r)$ for the sphere–linear system and making the substitutions $B_l \to \hbar^2/2\mu r^2$, $E_r \to E_{tr}$, and $J_r \to \mathscr{J}$.

A facile comparison of the square well rate constant with the calculations reported by Klots (1972) can also be made. To do this, note that Klots performed his calculations in the limit $\mathscr{J} \to 0$; that is, he used Eq. (34). To obtain $\Gamma_z(E_{tr})$ for the square well potential, the value of $\Gamma_{ro}(E_{tr}, \mathscr{J})$ in the limit, $\mathscr{J} \to 0$ must be obtained. Using $\Gamma(E_r, J_r)$ for the sphere–linear system, making the appropriate substitutions, and taking the limit gives

$$\lim_{\mathscr{J} \to 0} \Gamma_{ro}(E_{tr}, \mathscr{J}) = 2\mathscr{J} E_{tr}/\omega, \tag{46a}$$

where

$$\omega = B + \hbar^2/2\mu r^2 \tag{46b}$$

and B is the geometric mean of the three rotational constants of $C_6H_5^+$. Therefore, $\Gamma_z(E_{tr})$ for the square well potential differs from $\Gamma_z(E_{tr})$ used by Klots only by the replacement $B \to \omega$. The ratio of the square well rate constant (k_{sw}) to the charge-induced dipole rate constant (k_{cd}) in the limit $\mathscr{J} \to 0$ can now be obtained as

$$\lim_{\mathscr{J} \to 0} \frac{k_{sw}}{k_{cd}} = \frac{B}{B + \hbar^2/2\mu r^2}. \tag{47}$$

This result will not accommodate the slow rise of $k(E)$ versus E for reaction (40).

The fact that the square well potential does not explain the behavior of $k(E)$ versus E for the loss of H atoms from benzene ions does not necessarily imply that the description obtained by Klots for the product translational energy release is invalid. It must be kept in mind that transition state theory suggests that the behavior of the rate constant and translational energy release are determined by different stages of the dynamical process (see Section III.B).

B. *Decomposition of* $C_4H_6^+$ *and* $C_4H_8^+$

1. $C_4H_6^+$

Rate constant data for the reaction

$$C_4H_6^+ \to C_3H_3^+ + CH_3 \tag{48}$$

have been obtained by preparing the parent ion from the following neutral isomers: 1,3-butadiene, 1,2-butadiene, cyclobutene, 2-butyne, 1-butyne (Werner and Baer, 1975), and methyl cycloprop-2-ene (Baer, 1977). The product translational energy distribution has been measured by photoelectron–photoion coincidence (Klots *et al.*, 1977) for ions prepared from 1,3-butadiene and by crossed beams (Herman and Birkinshaw, 1973) for ions prepared through collisions of $C_2H_2^+$ and C_2H_4. In addition, the product translational energy distribution for metastable $C_4H_6^+$ ions prepared from 1,3-butadiene has also been reported [Elder, Beynon, and Cooks, unpublished results, reproduced in Klots *et al.* (1977) and Chesnavich and Bowers (1977b)].

The rate constant data reproduced in Fig. 4 show that all of the isomers decompose with the same rate. This result suggests that all $C_4H_6^+$ isomers sample a common phase space (Werner and Baer, 1975). Also, the lifetime distributions of all isomers are single exponentials.

The average translational energy release obtained from the coincidence and beam experiments is given in Fig. 5. In plotting the beam data, it was assumed that the initial vibrational energy of $C_2H_2^+$ was 0.25 eV and that the bimolecular reaction is 1.76 eV exoergic. These results are interesting in that the $C_4H_6^+$ ions created in the coincidence experiments have a near-thermal distribution of angular momentum, whereas those created in the beam experiments have a distribution which is expected to be quite large. Although the energy ranges investigated by the two methods don't overlap,

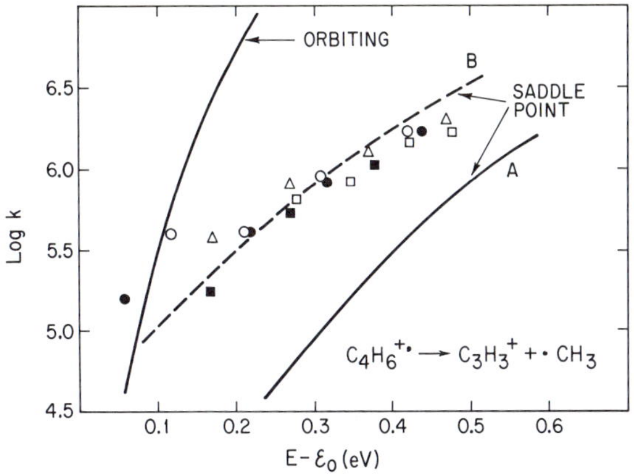

FIG. 4 $k(E)$ versus E for the decomposition of $C_4H_6^+$. Experimental data (Werner and Baer, 1975): (●) 1,3-butadiene, (□) 1,2-butadiene, (○) cyclobutene, (■) 2-butyne, and (▲) 1-butyne. Saddle point curve A is that obtained by Werner and Baer (1975), assuming 2-methylcyclopropene to be the transition state structure. Saddle point curve B is that obtained by assuming reaction mechanism (49). Orbiting curve calculated using Eq. (29). [Adapted from Chesnavich and Bowers (1977b).]

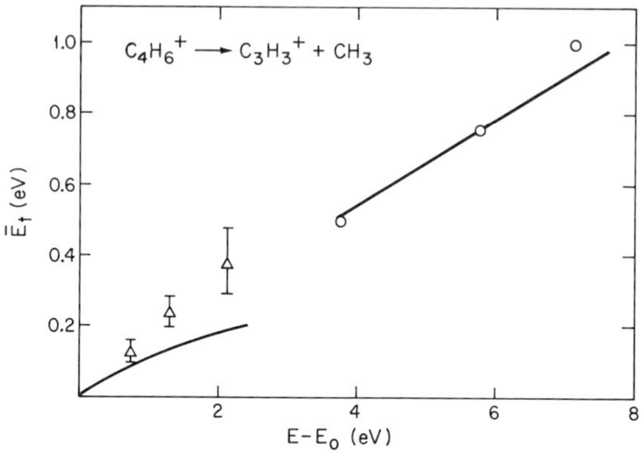

FIG. 5 Product average translational energy release for the decomposition of $C_4H_6^+$. Molecular beam data (●) from Herman and Birkinshaw (1973). Photoelectron–Photoion coincidence (PIPECO) data (△) and statistical theory calculations (———) as reported in Klots *et al.* (1977).

the data in Fig. 5 suggest that at about 3 eV above the appearance potential the thermal and high angular momentum reactions may yield similar average translational energies. This result is in contrast to the measured product translational energy distributions (cf. the various figures given in Klots *et al.*, 1977) and is indicative of the fact that translational energy distributions are more sensitive to reaction dynamics than are average translational energies (see Section III.B).

Statistical theory calculations of $k(E)$ versus E have been performed using a number of different transition states. Calculations with linear and cyclic transition state structures, using the appearance potential as the activation energy, produced curve A of Fig. 4 (Werner and Baer, 1975). When the $C_4H_6^+$ parent ion structure was loosened, the resulting curve was 2.8 log units lower than the experimental data. Agreement between experiment and theory was obtained however when the following two-step mechanism was assumed (Chesnavich and Bowers, 1977b):

$C_4H_6^+ \xrightarrow{k_1}$ 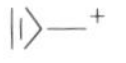 $\xrightarrow{k_2} C_3H_3^+ + CH_3.$ (49)

In reaction (49), the $C_4H_6^+$ parent ion undergoes an initial isomerization to produce the methyl-cycloprop-2-enium ion which then fragments to products. The isomerization rate constant k_1 was assumed to be rate determining and the activation energy of the isomerization was treated as a variable parameter. The resulting curve, using an isomerization activation energy 0.2 eV lower than the appearance potential, is given by curve B in Fig. 4.

Also shown in Fig. 4 is the result obtained by using an orbiting transition state (Chesnavich and Bowers, 1977b). At threshold, the theoretical rate constant is smaller than the experimental result, whereas at higher energies it is orders of magnitude larger. This result is consistent with the mechanism of reaction (49) since it implies that the actual activation energy is lower than the zero point energy difference used in the theoretical calculations.

The orbiting result for average translational energy release is plotted in Fig. 5 (Klots *et al.*, 1977). For $C_4H_6^+$ ions with a thermal distribution of angular momentum, it is seen that experiment and theory coincide at lower parent ion energies and diverge as the energy is increased. However, for $C_4H_6^+$ ions with a collisionally generated distribution of angular momentum, the agreement between experiment and theory is consistently good. The same trends are evident in the distributions of product translational energy. The results suggest that as the angular momentum of the $C_4H_6^+$ ions is increased, centrifugal effects begin to dominate the potential surface.

A comparison of the metastable translational energy distribution with orbiting model predictions gives excellent agreement (Klots *et al.*, 1977; Chesnavich and Bowers, 1977b). This agreement is consistent with the results summarized in Fig. 5 since the range of internal energies corresponding to metastable transitions (see, e.g., Fig. 4) is approximately the same range for which the coincidence experiments and theory coincide. In the comparison of Figs. 4 and 5, it is useful to note that the average metastable translational energy release is ≈ 0.025 eV and that the energy deposition function for 1,3-butadiene can be approximated by $P(E - E_0) = \text{const}$ for $0 \leq E - E_0 \leq 0.1$ eV and $P(E - E_0) = \text{const}(\frac{3}{2} - 5(E - E_0))$ for $0.1 \leq E - E_0 \leq 0.3$ eV (Klots *et al.*, 1977).

2. $C_4H_8^+$

The translational energy distribution of $C_3H_5^+$, CH_3 fragments produced by the decomposition of energized $C_4H_8^+$ has been measured by Herman *et al.* (1969). They created the butene ions from the reaction

$$C_2H_4^+ + C_2H_4 \rightleftarrows [C_4H_8^+]^* \rightarrow C_3H_5^+ + CH_3, \quad (50)$$

using crossed molecular beams. These experiments stimulated the first attempts to apply the orbiting model to polyatomic bimolecular reactions (Safron *et al.*, 1972; Lee *et al.*, 1972). Unfortunately, no measurements of $k(E)$ versus E or of product translational energy distributions have been reported for the $C_4H_8^+$ ions formed directly from C_4H_8 neutrals. These results would be interesting because a comparison of ICR (Huntress, 1972) and photoionization (Le Breton *et al.*, 1975) studies of this reaction with statistical theory calculations suggest that an isomerization may be the rate determining step (Su and Bowers, 1979).

The first attempt to calculate the translational energy distribution for reaction (50) (Lee *et al.*, 1972) used the approach of Safron *et al.* (1972), which was shown by Klots (1976) to be based on an invalid approximation. The calculation has been repeated at least three times using more rigorous methods. Klots (1976) began with the derivative of Eq. (35) and added a first-order perturbation correction based on Eq. (31b). Carter (1976) integrated

$$P(E, E_t, \mathscr{J}) \propto \int dK_1\, dK_2\, dJ_1\, dJ_2\, dJ_r\, dL \rho_v(E_v)\delta(E - H)\delta(E_t - H_t) \quad (51)$$

numerically. (The delta functions for angular momentum conservation in this equation have been omitted for clarity.) By recalling that for two spherical top rigid rotors,

$$\rho(E_r, J_r) = \int_{|J_1 - J_2| \leq J_r \leq J_1 + J_2} dK_1\, dK_2\, dJ_1\, dJ_2 \delta(E_r - H_r), \quad (52)$$

is obtained, Eq. (51) is seen to be identical to Eq. (36), which was integrated numerically by Chesnavich and Bowers (1977a). All three calculations indicate that the experimental translational energy release is in satisfactory agreement with the results obtained using orbiting transition states.

C. *Threshold Behavior of Endoergic Bimolecular Reactions*

1. *Theory*

If an endoergic bimolecular reaction forms an intermediate complex, and if an orbiting transition state is a valid approximation, the threshold behavior of the cross section can be determined from Eqs. (13), (34), and (35) in the following manner. At threshold $k_a \gg k_b$, where k_a and k_b are the unimolecular rate constants for decomposition of the complex into the reactant and product channels, respectively, and a two channel reaction has been assumed for simplicity. Therefore, the decomposition probability [Eq. (13)] becomes

$$P(\mathscr{J}) \sim k_b/k_a. \quad (53)$$

Also E_b, the available energy in the product channel, is small at threshold. Since orbiting transition states for potential surfaces more attractive than a square well are L-dominated at low energy, one has $\mathscr{J} \to L$ in the product channel. Therefore, the overall rotational motion of the individual molecules is largely unhindered by angular momentum conservation requirements and k_b is given by Eq. (35). For small E_b, vibrations remain unexcited giving

$$\lim_{E_b \to 0} k_b \propto (E_b - \mathscr{J}^{(n-2)/2n}/\lambda)^{s/2}, \quad (54)$$

where s is the number of rotational degrees of freedom of the products.

The rate constant for the reverse decomposition into the reactant channel is governed, when $E_b \geq 0$, by the following considerations at threshold. First, since E_b is small, only low $\mathscr{J}$ collisions contribute to the forward reaction. Second, the available energy in the reactant channel is given by $E_b + \Delta E_0$, where ΔE_0 is the reaction endoergicity. Therefore, k_a is governed by low $\mathscr{J}$ and moderate to large available energy. In this situation, k_a is given by Eq. (34). Provided that the endoergicity is large enough so that $\Delta E_0 \gg E_b$ for the energy range of interest, the sums and densities of states in the numerator and denominator of Eq. (34) remain effectively constant and one has

$$\lim_{E_b \to 0} k_a \propto 2\mathscr{J}. \tag{55}$$

Equations (53)–(55) can now be used to obtain the desired result. If the reactants are translationally excited, one can use Eq. (28), $\mathscr{J} = \mu b v$, and $k = \sigma v$. When these equations are combined with Eqs. (53)–(55) and the integrations are performed, one obtains

$$\lim_{E_b \to 0} \sigma \propto E_b^{(n-2)/2n + (s/2)}. \tag{56}$$

Note that for an atom–diatom pair $s = 2$ and $\sigma \propto E_b^{(3n-2)/2n}$, which is the result obtained by Nikitin (1965a) and Pechukas and Light (1965).

As the system energy increases, the maximum $\mathscr{J}$ available to the product channel increases rapidly. Therefore, high $\mathscr{J}$ collisions begin to contribute to the forward reaction. When this occurs, the orbital angular momentum in the reactant channel must play a greater role in the accommodation of $\mathscr{J}$ and k_a, when $E_b \geq 0$, begins to take the form given by Eq. (35). Making the substitution $k_a \sim$ const in place of Eq. (55) leads to

$$\sigma(E_b > 0) \propto E_b^{(n-2)/n + (s/2)}. \tag{57}$$

The range of validity of Eqs. (56) and (57) depends on the reaction endoergicity and also on the rotational constants, parameters in the expression for the long range potential and, to a lesser extent, the vibrational frequencies of each channel. It is usually not possible to state precisely when these equations are valid. This must be determined individually for each system.

2. *Applications*

A. $C^+ + D_2 \rightarrow CD^+ + D$. The behavior of the cross section for this deuterium atom transfer reaction in the threshold region has been measured by Maier (1967). Truhlar (1969) has performed extensive statistical theory calculations on this reaction including threshold behavior. Truhlar's exact calculations agreed qualitatively with experiment, however the calculated $\sigma(E)$ versus E curve rose at about twice the rate as the experimental result to approximately 1 eV above threshold. At that point, inclusion of electronically excited $^3\Pi$ CD^+ caused the calculated cross section to rise rapidly in comparison with experiment. The endoergicity used by Truhlar was 0.4 eV. When the calculations were repeated, using an endoergicity based on known heats of formation (0.825 eV), the calculated $\sigma(E)$ versus E curve rose at about one-half the rate of that of the experimental result (Chesnavich and Bowers, 1978).

Truhlar performed a detailed investigation of the threshold region in his calculations. He found that the $E_b^{5/4}$ threshold law agreed with the exact calculations only for the first 0.03 eV above threshold. At this point, the exact calculations began to climb at a slower rate than the threshold law. It can be shown (Chesnavich and Bowers, 1978) that the early failure of the threshold law for this system occurs because the reaction endoergicity is overcome by the rotational constant difference between D_2 and CD^+. This difference makes the rate constant for decomposition of the complex into the CD^+–D channel competitive with the rate constant for decomposition into the C^+–D_2 channel soon after threshold and Eq. (53) becomes an invalid approximation.

The lack of quantitative agreement between experiment and exact theory for this reaction may be due to uncertainties in the heats of formation, failure of the quasi-equilibrium hypothesis for the $CD_2{}^+$ complex, or to failure of the orbiting model for one or both transition states. An uncertainty in the thermochemical data of 0.2 eV would be needed to bring experiment and theory into agreement. The fact that the $CD_2{}^+$ intermediate is stable by 4.25 eV suggests complex formation theory may be applicable, at least in the threshold region. This suggestion is supported by beam scattering experiments (Iden *et al.*, 1971) which show that the CD^+–D product distribution is isotropic with respect to the center of mass for energies up to at least 2.5 eV above threshold. It appears, therefore, that the orbiting model provides only a qualitative description of the transition states for this reaction.

B. $H_2O^+ + H_2 \rightarrow H_3{}^+ + OH$. The proton transfer reaction

$$H_2O^+ + H_2 \rightarrow H_3{}^+ + OH \tag{58}$$

is of interest because accurate determination of the threshold energy provides an upper limit to the proton affinity of H_2. The translational energy dependence of the cross section near threshold has been investigated by Cotter *et al.* (1972). Also studied was the fully deuterated counterpart of the reaction. In a related study Cotter and Koski (1973) studied the threshold behavior of

$$D_3O^+ + D_2 \rightarrow D_2O + D_3{}^+. \tag{59}$$

The experimental results for reaction (58) and the theoretical predictions based on complex formation and orbiting transition states are given in

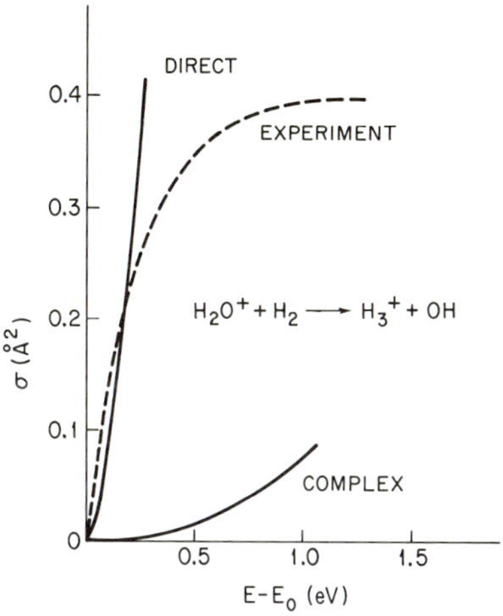

FIG. 6 Reaction cross section versus relative translational of H_2O^+, H_2 reactants (E_0 = 1.82 eV). Experimental curve from Cotter *et al.* (1972). Complex orbiting curve calculated from Eq. (27) [Chesnavich and Bowers (1978)]. Direct-saddle point curve calculated from Eq. (61).

Fig. 6. Analogous results were obtained for the fully deuterated counterpart and for reaction (59).

It is clear from the comparison of experiment and theory that complex formation and the orbiting model are not applicable to these reactions. This is not surprising since neither H_4O^+ nor D_5O^+ have been experimentally observed, indicating that these ions cannot be stable to dissociation by more than a few kilocalories per mole. The results suggest, therefore, that these reactions are direct rather than complex forming.

Direct reactions, driven by translational energy at threshold, provide exactly the conditions for which the transition state theory of Marcus for translational energy disposal was developed (see Section III.B). An estimate of the predictions of the Marcus theory can be obtained in the following manner. If it is assumed that four of the rotational degrees of freedom of the reactants pass directly to bending vibrations at the transition state, the reaction probability in the limit $L \rightarrow \mathscr{J}$ is given by (Marcus, 1975)

$$P(E_t, \mathscr{J}) = B_s B_l \frac{d}{dE_t} \rho_4^{\ddagger}(E_t - E_0 - \mathscr{J}^2\hbar^2/2\mu r^2), \tag{60}$$

where B_s and B_l are the rotational constants of the spherical top and linear reactants, r is the distance of closest approach at the transition state, $\rho_4^{\ddagger}(E)$ the density of states of the four transition state vibrations created from the rotational degrees of freedom of the reactants, and it has been assumed that two rotations of the spherical top pass to vibrations. When Eq. (60) is integrated over impact parameters [see, e.g., Eq. (28)], one obtains

$$\sigma = \pi r^2 (B_s B_l / E_t) \rho_4{}^{\ddagger}(E_t - E_0). \tag{61}$$

The result obtained from Eq. (61) for reaction (58) is plotted in Fig. 6 (labeled direct). The transition state bending vibrations were taken to be 400 cm^{-1}, a value of 3 Å was used for r, and a reaction path degeneracy of two was assumed. Although the comparison between experiment and theory is not exact, the results suggest that at threshold reaction (58) is direct and is governed by a saddle point transition state.

A similar result is obtained for the deuterated version of reaction (58). In particular, the isotope effect is predicted correctly. This latter result is important since it is independent of the r^2 term in Eq. (61) and is only mildly dependent on the choice of vibrational frequencies.

The relative behavior of the experimental results and Eq. (61) is almost identical to the relative behavior of the exact computer calculations and the transition state theory result for the cross section for the three-dimensional $H + H_2$ reactions (Marcus, 1967). In particular, the computer calculations and transition state theory show the same divergence at higher energy that is evident in Fig. 6. Marcus (1967) has suggested that vibrational nonadiabaticity is the primary cause of this divergence. Nonadiabaticity produces "chattering" collisions which pass through the transition state many times before reacting (see, e.g., Miller, 1976b). If the correspondence between the $H + H_2$ reaction and reaction (58) is valid, the plateau reached by the experimental cross section at higher energy (Fig. 6) may be viewed as evidence of the failure of the fundamental assumption of transition state theory.

V. CONCLUSIONS

Although it is too early to predict how the recent developments in statistical theory and the results obtained from monoenergetic experiments will ultimately intertwine, some patterns do appear to be emerging. Evidence already suggests that orbiting unimolecular rate constants are unrealistically high. However, the orbiting bimolecular capture rate constant $k = 2\pi q(\alpha/\mu)^{1/2}$ often agrees quite well with experiment at thermal energies. A possible explanation of this behavior may be the different energy and angular momentum distributions in the unimolecular and bimolecular reactions. Another interesting trend is the ability of the orbiting model to qualitatively, and often quantitatively, predict translational energy disposal or reactions with no reverse activation energy. Also, the investigation of the $C_4H_6{}^+$ system, which is the only system for which translational energy release data have been obtained for reagent ions with both thermal and collisionally generated $\mathcal{J}$ distributions, suggests that angular momentum can play a major role in product energy disposal.

There are still many unsettled questions in statistical theory. Most research on the concept of ergoticity, which is related to but not identical to the concept of quasi-equilibrium, has focused on the classical dynamics of simple but "unchemical" Hamiltonians. The role of quantum mechanics in ergoticity (see e.g., Nordholm and Rice, 1975) and the behavior of more realistic Hamiltonians are expected to receive more thorough scrutiny in the future. Unanswered questions in variational theory, such as the energy and angular momentum dependence of the optimal transition state location, and the role of nonseparability also await the interested researcher.

In conclusion, the research summarized in this chapter supports the contention that statistical methods will continue to play an important role in the understanding of reaction dynamics.

REFERENCES

Andlauer, B., and Ottinger, C. (1971). *J. Chem. Phys.* **55**, 1471.

Andlauer, B., and Ottinger, C. (1972). *Z. Naturforsch., Teil A* **27**, 293.

Baer, T. (1977). Personal communication.

Baer, T., Werner, A. S., and Tsai, B. P. (1975). *J. Chem. Phys.* **62**, 2497.

Baer, T., Tsai, B. P., Smith, D., and Murray, P. T. (1976). *J. Chem. Phys.* **64**, 2460.

Bunker, D. L. (1962). *J. Chem. Phys.* **37**, 393.

Bunker, D. L. (1964). *J. Chem. Phys.* **40**, 1946.

Bunker, D. L., and Hase, W. L. (1973). *J. Chem. Phys.* **59**, 4621.

Bunker, D. L., and Pattengill, M. P. (1968). *J. Chem. Phys.* **48**, 772.

Carter, D. E. (1976). *J. Chem. Phys.* **65**, 2584.

Chapman, S., Hornstein, S. M., and Miller, W. H. (1975). *J. Am. Chem. Soc.* **97**, 892.

Chesnavich, W. J. (1978). *Chem. Phys. Lett.* **53**, 300.

Chesnavich, W. J., and Bowers, M. T. (1976). *J. Am. Chem. Soc.* **97**, 892.

Chesnavich, W. J., and Bowers, M. T. (1977a). *J. Chem. Phys.* **66**, 2306.

Chesnavich, W. J., and Bowers, M. T. (1977b). *J. Am. Chem. Soc.* **99**, 1705.

Chesnavich, W. J., and Bowers, M. T. (1978). *J. Chem. Phys.* **68**, 901.

Cotter, R. J., and Koski, W. S. (1973). *J. Chem. Phys.* **59**, 784.

Cotter, R. J., Rogett, R. W., and Koski, W. S. (1972). *J. Chem. Phys.* **57**, 4100.

Eland, J. H. D. (1974). *Int. J. Mass. Spectrom. Ion Phys.* **13**, 457.

Eland, J. H. D., and Schulte, H. (1975). *J. Chem. Phys.* **62**, 3835.

Evans, M. C., and Polanyi, M. (1935). *Trans. Faraday Soc.* **31**, 875.

Eyring, H. (1935). *J. Chem. Phys.* **3**, 107.

Eyring, H., Hirschfelder, J. O., and Taylor, H. S. (1936). *J. Chem. Phys.* **4**, 479.

Forst, W. (1973). "Theory of Unimolecular Reactions". Academic Press, New York.

Fullerton, D. C., and Moran, T. F. (1971). *Chem. Phys. Lett.* **10**, 626.

Gioumousis, G., and Stevenson, D. P. (1958). *J. Chem. Phys.* **29**, 294.

Grimmelmann, E. K., and Lohr, L. L., Jr., (1977). *Chem. Phys. Lett.* **48**, 487.

Hase, W. L. (1976). *In* "Dynamics of Molecular Collisions" (W. H. Miller, ed.), Part B, p. 121. Plenum, New York.

Hase, W. L., and Feng, D. (1974). *J. Chem. Phys.* **61**, 4690.
Hase, W. L., and Feng, D. (1976). *J. Chem. Phys.* **64**, 651.
Hase, W. L., and Sloane, C. S. (1976). *J. Chem. Phys.* **64**, 2256.
Herman, Z., and Birkinshaw, K. (1973). *Ber. Bunsenges. Phys. Chem.* **7**, 566.
Herman, Z., Lee, A., and Wolfgang, R. (1969). *J. Chem. Phys.* **51**, 452.
Horiuti, J. (1938). *Bull. Chem. Soc. Jpn.* **13**, 210.
Huntress, W. T. (1972). *J. Chem. Phys.* **56**, 5111.
Iden, C. R., Liardon, R., and Koski, W. S. (1971). *J. Chem. Phys.* **54**, 2757.
Kay, K. G. (1976). *J. Chem. Phys.* **65**, 3813.
Keck, J. C. (1958). *J. Chem. Phys.* **29**, 410.
Keck, J. C. (1960). *J. Chem. Phys.* **32**, 1035.
Keck, J. C. (1967). *Adv. Chem. Phys.* **13**, 85.
Klots, C. E. (1971). *J. Phys. Chem.* **75**, 1526.
Klots, C. E. (1972). *Z. Naturforsch., Teil A* **27**, 553.
Klots, C. E. (1976). *J. Chem. Phys.* **64**, 4269.
Klots, C. E., Mintz, D., and Baer, T. (1977). *J. Chem. Phys.* **66**, 5100.
Ko, A. N., Rabinovitch, B. S., and Chao, K. J. (1977). *J. Chem. Phys.* **66**, 1374.
Laidler, K. J. (1969). "Theories of Chemical Reaction Rates." McGraw-Hill, New York.
Le Breton, P. R., Williamson, A. D., and Beauchamp, J. L. (1975). *J. Chem. Phys.* **62**, 1623.
Lee, A., Leroy, R. L., Herman, Z., Wolfgang, R., and Tully, J. C. (1972). *Chem. Phys. Lett.* **12**, 596.
Lifshitz, C. (1978). *Adv. Mass Spectrom.* **7A**, 3.
Light, J. C. (1964). *J. Chem. Phys.* **40**, 3221.
Light, J. C. (1967). *Discuss. Faraday Soc.* **44**, 14.
McDonald, J. D., and Marcus, R. A. (1976). *J. Chem. Phys.* **65**, 2180.
Maier, W. B., II (1967). *J. Chem. Phys.* **46**, 4991.
Marcus, R. A. (1952). *J. Chem. Phys.* **20**, 4658.
Marcus, R. A. (1966). *J. Chem. Phys.* **45**, 2630.
Marcus, R. A. (1967). *J. Chem. Phys.* **46**, 959.
Marcus, R. A. (1970). *J. Chem. Phys.* **53**, 604.
Marcus, R. A. (1973). *Discuss. Faraday Soc.* **55**, 381.
Marcus, R. A. (1975). *J. Chem. Phys.* **62**, 1372.
Marcus, R. A. (1977). *Ber. Bunsenges. Phys. Chem.* **81**, 190.
Marcus, R. A., and Rice, O. K. (1951). *J. Phys. Colloid Chem.* **55**, 894.
Meogher, J. F., Chao, J. K., Barker, J. R., and Rabinovitch, B. S. (1974). *J. Phys. Chem.* **78**, 2535.
Miller, W. H. (1974). *J. Chem. Phys.* **61**, 1823.
Miller, W. H. (1976a). *Acc. Chem. Res.* **9**, 306.
Miller, W. H. (1976b). *J. Chem. Phys.* **65**, 2216.
Nikitin, E. (1965a). *Theor. Exp. Chem. (USSR)* **1**, 83.
Nikitin, E. (1965b). *Theor. Exp. Chem. (USSR)* **1**, 90.
Nikitin, E. (1965c). *Theor. Exp. Chem. (USSR)* **1**, 275.
Nordholm, S., and Rice, S. A. (1975). *J. Chem. Phys.* **62**, 157.
Pechukas, P. (1976). *In* "Dynamics of Molecular Collisions" (W. H. Miller, ed.), Part B, p. 269. Plenum, New York.
Pechukas, P., and Light, J. C. (1965). *J. Chem. Phys.* **42**, 3281.
Pechukas, P., and McLafferty, F. J. (1972). *J. Chem. Phys.* **58**, 1622.
Pelzer, H., and Wigner, E. (1932). *Z. Phys. Chem., Abt. B* **15**, 445.
Quack, M., and Troe, J. (1977). *Ber. Bunsenges. Phys. Chem.* **81**, 329.
Robinson, J. P., and Holbrook, K. A. (1972). "Unimolecular Reactions." Wiley (Interscience), New York.

Rosenstock, H. M., Wallenstein, M. B., Wahrhaftig, A. L., and Eyring, H. (1952). *Proc. Natl. Acad. Sci. U.S.A.* **38**, 667.

Rosenstock, H. M., Larkins, J. T., and Walker, J. A. (1973). *Int. J. Mass. Spectrom. Ion Phys.* **11**, 309.

Rosenstock, H. M., McColloh, K. E., and Lossing, F. P. (1977). *Int. J. Mass. Spectrom. Ion Phys.* **25**, 327.

Ryndbrandt, J. D., and Rabinovitch, B. S. (1970). *J. Phys. Chem.* **74**, 4175.

Ryndbrandt, J. D., and Rabinovitch, B. S. (1971a). *J. Phys. Chem.* **75**, 2164.

Ryndbrandt, J. D., and Rabinovitch, B. S. (1971b). *J. Chem. Phys.* **54**, 2275.

Safron, S. A., Weinstein, N. D., Herschbach, D. R., and Tully, J. C. (1972). *Chem. Phys. Lett.* **12**, 564.

Simm, I. G., and Danby, C. J. (1976). *J. Chem. Soc., Faraday Trans. II* **72**, 860.

Simm, I. G., Danby, C. J., and Eland, J. H. D. (1973). *Chem. Commum.* p. 832.

Simm, I. G., Danby, C. J., and Eland, J. H. D. (1974). *Int. J. Mass Spectrom. Ion Phys.* **11**, 309.

Simm, I. G., Danby, C. J., Eland, J. H. D., and Mansell, P. I. (1976). *Chem. Soc., Faraday Trans. II* **72**, 426.

Sloane, C. S., and Hase, W. L. (1977). *J. Chem. Phys.* **66**, 1523.

Su, T., and Bowers, M. T. (1979). Unpublished data.

Terwilliger, D. T., Beynon, J. H., and Cooks, R. G. (1974). *Proc. R. Soc. London, Ser. A* **341**, 135.

Truhlar, D. G. (1969). *J. Chem. Phys.* **51**, 4991.

Tsai, B. P., Werner, A. S., and Baer, T. (1975). *J. Chem. Phys.* **63**, 4384.

Vestal, M. (1968). *In* "Fundamental Processes in Radiation Chemistry" (P. Ausloos, ed.), p. 59. Wiley (Interscience), New York.

Vogt, E., and Wannier, H. (1954). *Phys. Rev.* **95**, 1190.

Werner, A. S., and Baer, T. (1975). *J. Chem. Phys.* **62**, 2900.

Wigner, E. (1937). *J. Chem. Phys.* **5**, 720.

Wigner, E. (1938). *Trans. Faraday Soc.* **34**, 29.

Wong, W. H., and Marcus, R. A. (1971). *J. Chem. Phys.* **55**, 5625.

Worry, G., and Marcus, R. A. (1977). *J. Chem. Phys.* **67**, 1636.

Chapter 5

State selection by photoion–photoelectron coincidence

TOMAS BAER

DEPARTMENT OF CHEMISTRY
UNIVERSITY OF NORTH CAROLINA
CHAPEL HILL, NORTH CAROLINA

I. INTRODUCTION

Photoion–photoelectron coincidence (PIPECO) is a method by which ions can be prepared in selected internal energy states, and their subsequent

ISBN-0-120801-X

reactions investigated. The goal of this chapter is to point out, wherever appropriate, the information obtained from PIPECO studies which is relevant to ion–molecule interactions. Because most of the coincidence studies have dealt with unimolecular reactions, the connection to bimolecular reactions might at first glance appear tenuous. However, unimolecular reactions can be considered as the latter half of bimolecular reactions and the results from such investigations can therefore be applied to ion–molecule reactions which go through long lived intermediate collision complexes. It is often important to know on which potential surface a reaction proceeds. Preparation of an ion in a specific state and the study of its subsequent decay can lead to a better understanding of the ion–molecule reaction. In addition to the investigation of unimolecular decay, PIPECO has also been used to investigate ion–molecule reactions. Its ability to state select with energy resolutions as low as 5 meV, and the possibility of incorporating this in a crossed beam experiment, makes this approach an extremely attractive and powerful one.

The first coincidence study of ion dynamics was an electron impact investigation by McCulloh *et al.* (1965) in which double ionization was investigated by measuring the two positive singly charged ion fragments in coincidence. The first PIPECO experiment was reported by Brehm and von Puttkamer (1967).

The basis of PIPECO is that when a molecule is ionized by a photon of energy $h\nu$, ejecting an electron of energy E_e, the ion internal energy E_i is determined by conservation of energy and momentum as given by

$$E_i = h\nu - E_e(1 + m_e/m_i), \tag{1}$$

in which m_e and m_i are the electron and ion masses, respectively. Therefore, if ions are detected in delayed coincidence with electrons of energy E_e, those ions with internal energy E_i are selected from all the ions formed. By measuring the ion flight time and/or its kinetic energy, it is possible to extract a great deal of dynamical information, such as the kinetic energy released in the dissociation, the ion lifetime with respect to dissociation, or the cross section for a particular reaction which the ion might experience before its detection.

The only review of the PIPECO field was one which covered the literature through the early part of 1975 (Gellender and Baker, 1977). Since that time, the PIPECO literature has more than tripled and the applications of coincidence studies have been considerably expanded.

II. EXPERIMENTAL CONSIDERATIONS

Two types of photoion-photoelectron coincidence experiments are currently being pursued. Their differences are related entirely to the type of light source used. In the historically first approach, a rare gas resonance lamp of fixed energy, usually He(I), is employed using a dispersive type analyzer to energy analyze the electrons. By collecting electrons of different energies, ions of varying internal energy can be studied in coincidence with their corresponding electrons. The other approach is to use a continuum light source in conjunction with a photon monochromator. In this experiment, the energy of the detected electron is held fixed while the photon energy is varied. Although the two techniques differ profoundly in many respects, there are a number of common characteristics and problems.

A. The Experiment and the Information Derived

Figure 1 illustrates the arrangement of a typical PIPECO experiment in which the ion time of flight (TOF) is determined by measuring the time difference between the arrivals of the energy analyzed electron and ion at their respective detectors. The electron and ion flight times are on the order of 0.10 and 10 μsec, respectively. When total ionization rates are below 10,000 cps, the detection of the electron and ion pairs is essentially unambiguous. For experiments with such low total ionization rates, the most common approach is to use the electron and ion signals as start and stop inputs, respectively, to a time-to-pulse height converter whose output is fed into a multichannel pulse height analyzer. In this manner, the ion TOF distribution is obtained. While the delayed coincidence requirement between an ion and its energy analyzed electron insures the detection of only state selected ions, the TOF distribution contains all the dynamical information.

If an ion dissociates with release of kinetic energy, the fragment ion TOF will be symmetrically broadened. The precise shape depends on whether

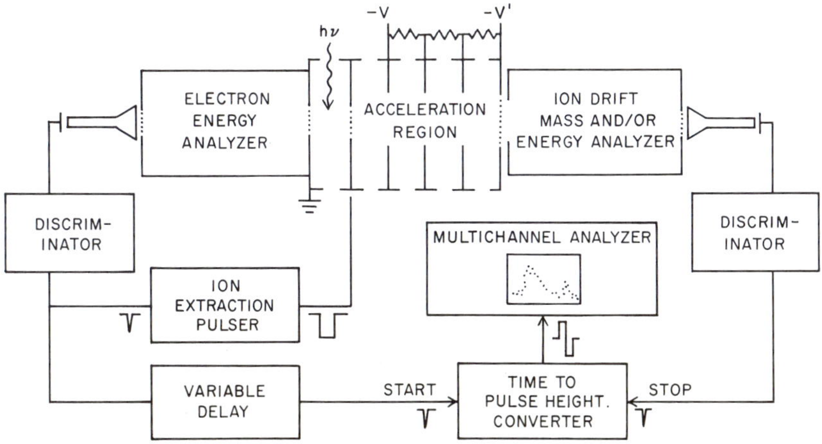

FIG. 1 Block diagram of a photoion–photoelectron coincidence experiment.

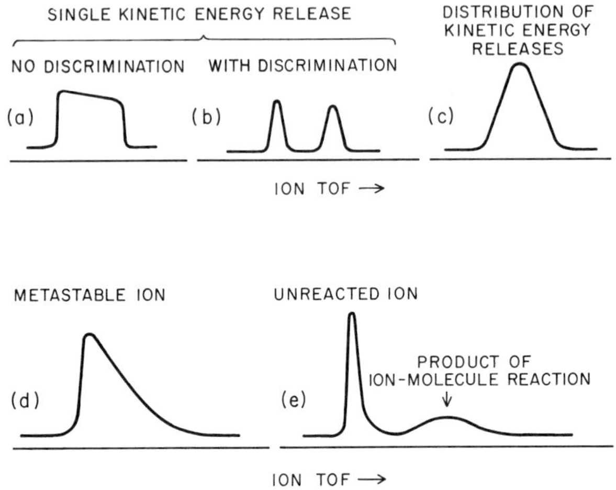

FIG. 2 Schematic ion time-of-flight (TOF) distributions obtained in PIPECO experiments. These distributions are not mutually exclusive. For instance, some metastable ions dissociate with considerable kinetic energy release.

the apparatus does or does not discriminate against ions with initial velocity components perpendicular to the experimental axis (Fig. 2a,2b). Polyatomic ions usually dissociate with a distribution of kinetic energy releases, resulting in TOF distributions as shown in Fig. 2c. The detection of fragments from metastable precursor ions with lifetimes in the 10^{-6} sec range results in asymmetric fragment ion TOF distributions as shown in Fig. 2d. The distribution is asymmetric because the precursor ions dissociate after having been accelerated some distance in the acceleration region. Finally, ions which are the products of ion–molecule interactions occurring in the acceleration region have flight times greater than unreacted ions as shown in Fig. 2e. The product ion TOF is a function of both the collision dynamics and the position of reaction in the acceleration region.

In order to analyze the data, the ion TOF is calculated taking into account the particular dynamical event which is being investigated. For instance, in the lifetime analysis of metastable ions which dissociate while being accelerated in a single acceleration plus drift region, the total TOF is given by

$$\mathrm{TOF} = T_{\mathrm{a}}(\text{precursor ion}) + T_{\mathrm{a}}'(\text{fragment ion}) + T_{\mathrm{d}}(\text{fragment ion}), \quad (2)$$

where T_{a} and T_{d} are the flight times in the acceleration and drift regions, respectively. For some mean lifetime τ, the distribution in T_{a} is given by $\exp(-T_{\mathrm{a}}/\tau)$. In this manner the expected TOF distributions can be calculated for various τs and compared with the experimental TOF distribution.

B. *Coincidences, False Coincidences, and Paralysis in Nonpulsed Experiments*

Some of the statistical considerations involved in coincidence experiments have been presented by Eland (1972a) and Gellender and Baker (1975). The major points are summarized here. The ion, electron, and coincidence signals (I, E, and C, respectively) can be expressed in terms of the total ionization rate N and the electron and ion collection efficiencies if the detected electrons and ions originate from the same photoionization volume. Thus,

$$\text{ion signal} = I = f_i N, \tag{3}$$

$$\text{electron signal} = E = f_e g_e N, \tag{4}$$

$$\text{coincidence signal} = C = E f_i = I f_e g_e = N f_i f_e g_e, \tag{5}$$

where f_i is the ion collection efficiency, f_e the electron collection efficiency, g_e the fraction of electrons in energy analyzer band pass, and N the total photoionization rate. It is apparent from Eq. (5) that the collection efficiencies f_i and $f_e g_e$, as well as the total photoionization rate N, can be determined from the three measurable quantities I, E, and C. The ability to measure the ion collection efficiency is of some importance in various kinds of coincidence experiments, because it yields knowledge about the fraction of ions which are *not* collected.

One of the problems encountered in coincidence experiments is that of false coincidences. For the case of a steady ionization rate and nonpulsed ion extraction, the false coincidence rate C_f per time interval Δt is given by

$$C_f = EI\,\Delta t = NC\,\Delta t \qquad \text{for} \qquad I\,\Delta t < 1. \tag{6}$$

These false coincidences, which increase with the square of the total ionization rate, add a relatively structureless background to the coincidence signal. The resulting signal-to-noise ratio is given by

$$\frac{\text{signal}}{\text{noise}} = \frac{C}{\sqrt{(C + C_f)^{1/2}}} = \sqrt{\left(\frac{N f_i f_e g_e}{1 + N\,\Delta t}\right)^{1/2}}. \tag{7}$$

It is evident from Eq. (7) that the best approach to increasing the signal-to-noise ratio is to increase f_i,f_e, and g_e. Up to a limit it is also possible to maximize this quantity by increasing N. Gellender and Baker (1975) have analyzed this in considerable detail and have suggested optimum conditions for coincidence experiments.

When using a time-to-pulse height converter as shown in Fig. 1 (but without the ion extraction pulser), the effect of the false coincidences goes far beyond that of simply increasing the noise. Because the first stop signal prevents further detection of additional stop signals, the false coincidences can paralyze the system. Because false coincidences are random events when timed from some electron start, the probability of observing the first false coincidence signal in a time t' is

$$\text{prob} = Nf_i \int_0^{t'} \exp(-tNf_i)\,dt = 1 - \exp(-t'Nf_i). \tag{8}$$

In turn, the probability of observing a true coincidence in a time t' after an electron start is an exponentially decreasing function of time. Such a paralysis of the system leads to decreased signals for ions of high mass (long TOF) relative to those of lower mass. This problem can be solved experimentally by discarding the simple time-to-pulse height converter system in favor of a timing unit which accepts several stops for each start and subsequently deposits all of the signals in the multichannel pulse height analyzer.

C. *False Coincidences in Pulsed Experiments*

When either the light source is pulsed (e.g., in the He Hopfield continuum) or the ions are pulsed out of the ionization region as illustrated in Fig. 1, the problems encountered with false coincidences are even more severe than in nonpulsed experiments. This is because the false coincidences are no longer uncorrelated but in fact are detected at nearly the same time of flight as the true coincidences of similar mass. This problem can be minimized in the pulsed light source experiment by operating at high repetition rates, while keeping the probability for an ionization event per light pulse significantly less than unity. In the case of a continuously operating light source and a pulsed ion extraction, it is possible to correct the coincidence spectra by subtracting a "false coincidence time of flight mass spectrum." The latter can be obtained by randomly pulsing the ions.

D. *Energy Analysis of Threshold Electrons*

1. *Angular Discrimination*

The methods of energy analyzing energetic electrons by dispersive analyzers need no further elaboration as this topic has been treated thoroughly in numerous publications (see, for example, Carlson, 1975; Shirley, 1972). The principle of threshold electron detection has been less widely disseminated, particularly with regard to its relation to coincidence experiments. The basis of threshold electron detection is the large collection efficiency of zero energy electrons produced in the presence of small electric fields. The small solid angle in which electrons are collected efficiently stops all energetic, or hot, electrons except those whose initial velocity vector is directed toward the detector.

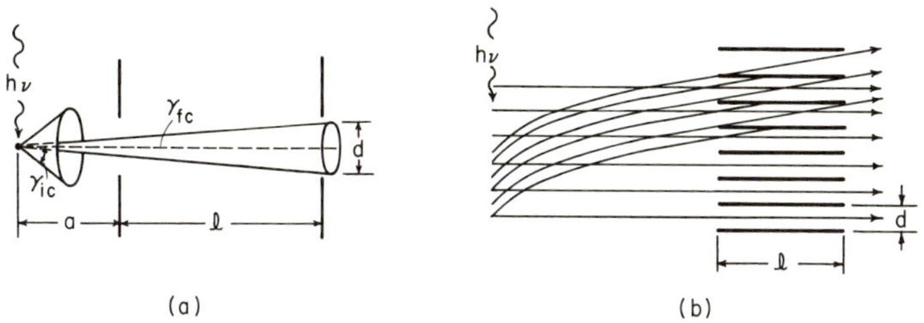

FIG. 3 Two types of threshold electron energy analyzers. An idealized analyzer with a point ionization source is shown in (a) in which γ_{ic} is the initial planar angle of electron ejection and γ_{fc} the final planar angle after acceleration. An analyzer useful for extended ionization sources is shown in (b). The long pipes, available commercially as collimated hole structures, discriminate against electrons with significant perpendicular velocity components.

The principle of this analyzer was proposed by Baer *et al.* (1969), and calculations for two types of geometry (Fig. 3) were performed by Spohr *et al.* (1971). In the idealized angular discriminator shown in Fig. 3a, electrons originate at a point ionization source and are accelerated toward the second aperture which determines the final critical planar angle γ_{fc}, into which electrons must be accelerated in order to be detected. The expression for the fraction F of electrons transmitted as a function of the initial electron energy E_i in an electric field, in which the electrons gain energy E_g, is given by

$$\begin{aligned} F &= 1 - \left[1 - \left(\frac{E_g}{E_i} + 1\right)\sin^2\gamma_{fc}\right]^{1/2}, \qquad & E_i \geq E_g \tan^2\gamma_{fc} \\ F &= 1, & E_i \leq E_g \tan^2\gamma_{fc} \end{aligned} \tag{9}$$

For the analyzer shown in Fig. 3a, $\gamma_{fc} = 0.5\, d/(l + 2a)$.

Figure 3b shows an analyzer constructed of a commercially available* set of collimated stainless steel holes whose length is on the order of a millimeter with length to diameter ratios (l/d) between 10 and 50. This analyzer has the advantage that large ionization regions can be used. In addition, the large l/d ratio allows the use of relatively large electric fields without sacrificing electron energy resolution. A serious disadvantage is that the collection efficiency is greatly reduced.

2. *Eliminating the Hot Electron Tail*

The major problem with a purely angular discriminator is that the transmission function F in Eq. (9) never goes completely to zero. That is, those electrons whose initial velocity vectors are parallel to the applied electric field will be collected regardless of their energy. These hot electrons

* Brunswick Corporation, 5223 West Imperial Highway, Los Angeles, California. The continued production of these collimated hole structures as of April 1978 is in doubt.

can be minimized by properly designed lenses and low electric fields as has been done by Frey *et al.* (1977), who obtained a resolution of 4 meV on the $^2P_{3/2}$ argon peak. But even in that experiment, hot electrons are not totally eliminated. In large ions at high photon energies, the integrated transition amplitudes producing hot electrons will cause a serious buildup of background electrons. In order to overcome this, Stockbauer (1973) used angular discrimination in conjunction with a dispersive analyzer which truncated the high energy tail. Even for the relatively poor resolution ($\approx$200 meV) needed for this purpose, ionization must be carried out in low electric field conditions. Ions must therefore be extracted by means of an applied electrical pulse. In this manner, good resolution ($\approx$20 meV) with little contribution from hot electrons can be achieved. Its major drawback is the delay of approximately 400 nsec in extracting the ions caused by the long electron flight path. A different approach has been taken by Peatman *et al.* (1975) who have developed a threshold electron energy analyzer with excellent resolution and good rejection of hot electrons. However, its collection efficiency is low and its geometry is not well suited for coincidence experiments.

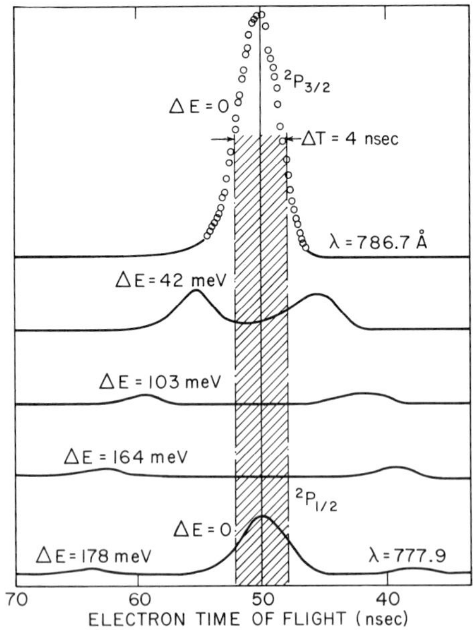

FIG. 4 Time-of-flight distribution of photoelectrons from the photoionization of argon with pulsed synchroton radiation. The distributions at each photon energy are normalized to the total ion signal. The shaded region is the time window in which electron signals are accepted for the PIPECO studies. [Data from Baer *et al.* (1979).]

The ideal threshold electron detector combines angular and time-of-flight electron discrimination. The reason for this is that both angular and TOF analysis are optimum for low energy electrons, and because of electron TOF space focusing (Wiley and McLaren, 1955), a wide or unfocused photon beam will not cause the electron resolution to deteriorate (Tsai *et al.*, 1974). In addition, apertures on the order of 3 mm diameter with drift distances of 1–2 cm allow for every high collection efficiencies. Finally, because the total TOF is on the order of 50 nsec, the delay in ion extraction is minimized. Synchrotron radiation from an electron storage ring which has photon pulses with less than 2 nsec FWHM and pulse repetition times of the order of 100 nsec is ideally suited for such experiments. A delayed signal, synchronous with the photon pulses, can be used to gate the electron signal. Such coincidence experiments have recently been done by Guyon *et al.* (1978a) and Baer *et al.* (1979) at the LURE synchrotron facility (Guyon *et al.*, 1976) in Orsay, France. In that experiment the dispersed light from a 1.5 m monochromator was refocused at the center of a 1 cm wide ionization/acceleration region with a field of 1 V/cm. A 1 cm long drift region designed for space focusing was terminated by a 4 mm diameter aperture. Typical electron TOF distributions obtained by photoionizing argon at various photon energies are shown in Fig. 4. By collecting only those electrons in the shaded region, it was possible to obtain a resolution for threshold electrons of 20 meV with a photon resolution of 12 meV. The collection efficiency for threshold electrons was estimated to be 50 %.

E. *Comparison of Threshold and Energetic Electron Detection for Coincidence Studies*

1. *Electron Collection Efficiencies*

Typical collection efficiencies for threshold electrons are between 2 and 6% and by careful design this can be increased to 50%. In contrast, most dispersive analyzers operate with efficiencies of less than 0.1%. An exception is the cylindrical mirror analyzer which has an efficiency close to 1% (Citrin *et al.*, 1972), but whose geometry is ill suited for coincidence studies with ions. It is therefore evident that all other things being equal, the threshold electron technique will result in 50 to 500 times higher signal-to-noise ratio than is possible with the energetic electron coincidence approach.

Another term in Eq. (7) which affects the signal quality is g_e. In threshold electron detection, this factor is always optimum being equal to 1 at the ionization potential and decreasing with increasing photon energy. On the other hand, in the energetic electron coincidence experiment, the factor g_e has its minimum value throughout the range of ion energies being investigated.

From this discussion it is clear that it is best to utilize the lowest photon energy possible in order to maximize g_e.

2. Total Ion Flux

The threshold electron technique must be associated with a dispersed light source and therefore the total photon flux is orders of magnitude less than available from a typical He(I) resonance source. This has the effect of canceling the gain in electron collection efficiency, so that the total coincidence rates in the two types of experiment are comparable. However, the ratio of true to false coincidences is greater by a factor of about 1000 in the threshold technique.

3. Accessible Available States

Of major interest in coincidence work is the ability to prepare ions in selected internal energy states. By He(I) photoionization, ions are prepared in states which can be reached via allowed one-electron transitions with favorable Franck–Condon factors. On the other hand, the threshold approach can be used to prepare ions in a considerably wider variety of states because autoionization aids in the preparation of ions not accessible via direct ionization (Murray and Baer, 1979). An example of this effect is shown in Fig. 5 which demonstrates the possibility of state selecting NO^+ ions in

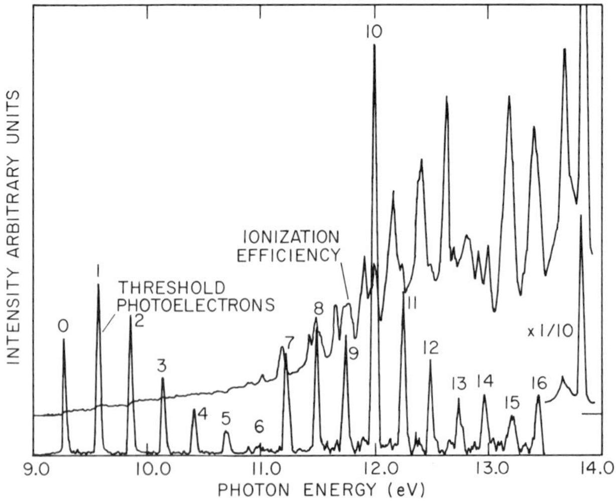

FIG. 5 Photoionization efficiency and threshold photoelectron spectrum (TPES) of NO. The complete rejection of energetic electrons in the TPES was achieved by a combination of angular and time-of-flight discrimination. [For details, see Murray and Baer (1979).]

vibrational states up to $v = 18$ and beyond. The corresponding PES obtained with the He(I) source exhibits only the first five vibrational states with relative intensities similar to those in Fig. 5 (Turner *et al.*, 1970).

A similar enhancement of the upper vibrational state production in NO^+ has been reported by Natalis *et al.* (1977), who observed the production of NO^+ (v) up to $v = 32$ by photoionizing with a Ne(I) resonance lamp. Other studies with the Ne(I) lamp on small ions have revealed similar enhancements which have been attributed to the participation of autoionizing channels (Carlson, 1971; Collin *et al.*, 1972). Although the lower energy of the Ne(I) light source and the ability to prepare ions in a wider variety of vibrational states makes it an attractive light source for PIPECO studies of ions below 16 eV, it has not yet been used for such experiments.

III. APPLICATIONS

A. Predissociation in Di- and Triatomics

Oxygen is the only diatomic whose ionic predissociation has been investigated by PIPECO. Most of the other stable diatomics such as $H_2{}^+$, $N_2{}^+$, CO^+, and NO^+ have dissociation limits at energies in excess of 21 eV which are inaccessible by He(I) radiation or the He Hopfield continuum. These ions will certainly be investigated in the future by coincidence techniques with He(II) light sources and synchrotron radiation. The only state of $O_2{}^+$ studied by PIPECO with He(I) radiation is the $\tilde{B}^2\Sigma_g^-$ state which Danby and Eland (1972) found to be fully predissociated with a lifetime of less than 10^{-7} sec. The $\tilde{b}^4\Sigma_g^-$, $\tilde{B}^2\Sigma_g^-$, $\tilde{c}^4\Sigma_u^-$ as well as the very diffuse $^2\Pi_u$ states just below the $\tilde{c}^4\Sigma_u^-$ states of $O_2{}^+$ have recently been investigated by Guyon *et al.* (1978a), using synchrotron radiation and threshold electron detection with electron TOF analysis. The $\tilde{b}^4\Sigma_g^-$, $v = 5$ and 6, states were found to be 90% predissociated with lifetime less than 50 nsec. Recent experiments (see Moseley *et al.*, 1979) have indicated that the lifetimes of these levels are in the 0.5 nsec range. These results are contrary to the 1 μsec lifetimes claimed by Erman and Larsson (1977). However, the latter data are being reinterpreted.

The diffuse $^2\Pi_u$ region at an energy between 23.5 and 24.5 eV is particularly interesting because, although its integrated signal is appreciable, its intensity within the usual spectroscopic band pass is exceedingly small. It is therefore difficult to study experimentally. Examples of O^+ fragment TOF distributions obtained by preparing $O_2{}^+$ ions with various energies within the $^2\Pi_u$ state are shown in Fig. 6. Because the ionization region was designed to transmit all fragment ions with up to 10 eV kinetic energy, the TOF distributions are thermally broadened trapezoids (see Fig. 2a). The width of the

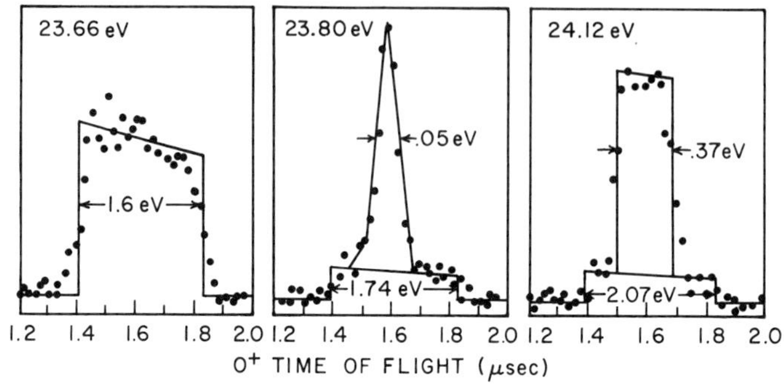

FIG. 6 Time-of-flight distributions of O^+ fragment ions from $O_2^+(^2\Pi_u)$ ions at the photon energies indicated in the figures. The solid lines are calculated TOF distributions obtained by neglecting the thermal spread of O_2 molecules in the ionization region. [Data taken in part from Guyon *et al.* (1978a).]

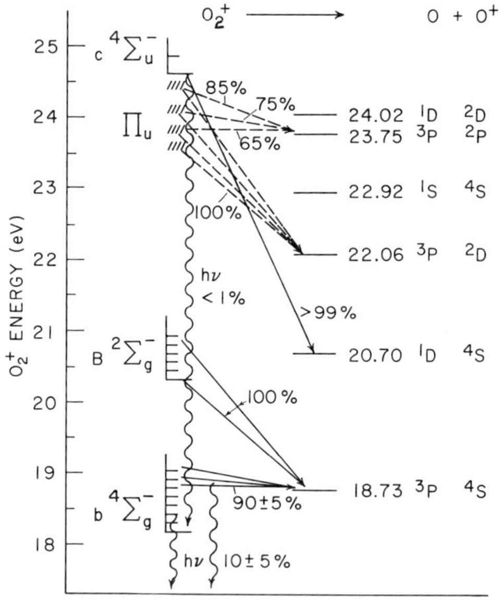

FIG. 7 A summary of the fate of excited O_2^+ ions as determined by PIPECO.

TOF distributions are consistent with dissociation to just two atomic limits. It suggests that there are two accessible dissociative states in this energy region. A summary of all the information obtained on the O_2^+ dissociation is presented in Fig. 7.

The results on triatomics are summarized in Table I. The kinetic energy release data obtained directly from the spread in the TOF distribution is

TABLE I

Dissociation Pathways for Energy Selected Triatomics

Ion energy (eV)	State	Products	Thermochemical dissociation Limit (eV)	Branching ratios (%)	Average KE release (eV)	Comments
	$N_2O^+(\tilde{X}^2\pi_{3/2}$ 12.888 ± 0.002 eV[a,b])					
15.17	$\tilde{X}^2\Pi$	$NO^+(^1\Sigma^+) + N(^4S)$	14.19	100[c]	0.8 ± 0.1[c]	The highly vibrationally excited states of $N_2O^+(\tilde{X}^2\Pi)$ are reached via autoionizing transitions[d].
15.55	$\tilde{X}^2\Pi$	$NO^+(^1\Sigma^+) + N(^4S)$		11[c]		
		$O^+(^4S) + N_2(^1\Sigma)$	15.29	89[c]	0.28 ± 0.05[c]	
16.11	$\tilde{X}^2\Pi$	$NO^+(^1\Sigma^+) + N(^4S)$		22[c]		
		$O^+(^4S) + N_2(^1\Sigma^+)$		78[c]	0.20 ± 0.05[c]	
16.388 ± 0.002[a,b]	$\tilde{A}^2\Sigma^+(0, 0, 0)$	$N_2O^+(\tilde{X}^2\Pi)$		98[c]		$N_2O^+(\tilde{X}^2\Pi)$ is formed by fluorescence stabilization. The $\tilde{A}^2\Sigma^+(0, 0, 0)$ lifetime is is 225 ± 10 nsec[a,e]
		$NO^+(^1\Sigma^+) + N(^4S)$		2[c]		
16.557[a]	$\tilde{A}^2\Sigma^+(1, 0, 0)$	$N_2O^+(\tilde{X}^2\Pi)$		67 ± 1[a,c]		The $N(^2D)$ channel is energetically accessible (16.57 eV) for rotationally excited N_2O^+ ions, but is not observed[f]
16.545[b]		$NO^+(^1\Sigma^+) + N(^4S)$		32[a,c]	1.6 ± 0.2[c]; 1.2[f]	
		$O^+(^4S) + N(^1\Sigma^+)$		1[c]		
16.693 ± 0.002[a,b]	$\tilde{A}^2\Sigma^+(0, 0, 1)$	$N_2O^+(\tilde{X}^2\Pi)$		42[c]		Brehm *et al.* (1974) report that 5% of the $NO^+ + N$ channel produces $N(^2D)$
16.724[a]	$\tilde{A}^2\Sigma^+(2, 0, 0)$	$NO^+(^1\Sigma^+) + N(^4S)$		56[c]	1.6 ± 0.2[c]; 1.0[f]	
		$O^+(^4S) + N_2(^1\Sigma^+)$		2[c]		

TABLE ↑ (*Continued*)

Ion energy (eV)	State	Products	Thermochemical dissociation Limit (eV)	Branching ratios (%)	Average KE release (eV)	Comments
18.10	$\tilde{B}(^2\Pi)$	$NO^+(^1\Sigma^+) + N(^2D)$	16.57		0.53^f	It seems most probable that $NO^+(v \approx 3) + N(^2D)$ is produced because this dissociation is spin allowed, and it is unlikely that $NO^+(v \approx 12) + N(^4S)$ is produced
		$N_2{}^+(^2\Sigma^+) + O(^3P)$	17.25		0.33^f	
20.105^b	$\tilde{C}^2\Sigma^+(0, 0, 0)$	$NO^+(^1\Sigma^+) + N(^2D)$				The branching ratios are obtained by combining photoionization[b] and PIPECO[c] data
		$NO^+(^1\Sigma^+) + N(^2P)$	17.76	$60^{b,c}$	1.4^f	
		$N^+(^3P) + NO(^2\Pi)$	19.46	16^c	0.45^c	
		$N_2{}^+(^2\Sigma^+) + O(^3P)$		$22^{b,c}$	1.1^f	
		$O^+(^4S) + N_2(^1\Sigma^+)$				
		$O^+(^2D) + N_2(^1\Sigma^+)$	18.61	2^c		
20.26	$\tilde{C}^2\Sigma^+(1, 0, 0)$	$NO^+(^1\Sigma^+) + N(^2D)$				Because the TOF experiment did not distinguish between NO^+ and $N_2{}^+$, no kinetic energy release data for these two ions were obtained
		$NO^+(^1\Sigma^+) + N(^2P)$		90^c		
		$N_2{}^+(^2\Sigma^+) + O(^3P)$				
		$N_2{}^+(^2\Sigma^+) + O(^1D)$	19.22			
		$N^+(^3P) + NO(^2\Pi)$			0.50 ± 0.1^c	In extracting the kinetic energy release for the N^+ channel, the O^+ channel was assumed to be negligibly small
		$O^+(^2D) + N_2(^1\Sigma^+)$		10^c		
		$O^+(^2P) + N_2(^1\Sigma^+)$	20.31			

	$CO_2^+(\tilde{X}^2\Pi_{3/2}$ 13.776 ± 0.002 eV[g,h])					
19.388 ± 0.004[g,j]	$\tilde{C}^2\Sigma_g^+(0, 0, 0)$	$O^+(^4S) + CO(^1\Sigma^+, v = 0)$	19.06	15 ± 7[i] 23 ± 10[g]	0.32	The values for the release energies were assumed in fitting the coincidence data[i]
		$O^+(^4S) + CO(^1\Sigma^+, v = 1)$	19.33	85 ± 7[i] 77 ± 10[g]	0.05	
19.469 ± 0.004[j]	$\tilde{C}^2\Sigma_g^+(0, 1, 0)$	$O^+(^4S) + CO(^1\Sigma^+)$ $CO^+(^2\Sigma^+) + O(^3P)$	 19.46	≈ O[j] ≈ 100[j]		These results are determined from photoionization data[j]
	$COS^+(\tilde{X}^2\Pi_{3/2}$ 11.190 eV[k])					
15.075[k]	$\tilde{A}^2\Pi(0, 0, 0)$	$S^+(^4S) + CO(^1\Sigma^+)$ $COS^+(^2\Pi)$	13.52	92.4[e,l,m] 7.6[e,l,m]	0.73[f]	The $\tilde{A}^2\Pi$ state lifetime is < 30 nsec[e]
15.58	$\tilde{A}^2\Pi(2, 0, 1)$ and others	$S^+(^4S) + CO(^1\Sigma^+)$ $S^+(^2D) + CO(^1\Sigma^+)$	 15.34	70 ± 15[n] 30 ± 15[n]	0.80[f] 0.20[f]	
16.04[k]	$\tilde{B}^2\Sigma^+(0, 0, 0)$	$S^+(^4S) + CO(^1\Sigma^+)$ $S^+(^2D) + CO(^1\Sigma^+)$			0.22[f]	
	$CS_2^+(\tilde{X}^2\Pi_{3/2}$ 10.077 eV[k])					
16.192[k]	$\tilde{C}^2\Sigma_g^+(0, 0, 0)$	$S^+(^4S) + CS(^1\Sigma^+)$	14.81	42[o]	0.38[o]	CS is produced in a broad distribution of vibrational states
		$CS^+(^2\Sigma^+) + S(^3P)$	15.78	58[o]	0.12[0]	CS^+ is produced in $v = 0$, and 1 in approximately equal intensities

TABLE 1 (*Continued*)

Ion energy (eV)	State	Products	Thermochemical dissociation Limit (eV)	Branching ratios (%)	Average KE release (eV)	Comments
17	$\tilde{D}^2\Sigma_u^+$	$S^+(^4S) + CS(^1\Sigma^+)$		31°	0.25^o	
		$S^+(^2D) + CS(^1\Sigma^+)$	16.65			
		$CS^+(^2\Sigma^+) + S(^3P)$				
		$CS^+(^2\Sigma^+) + S(^1D)$	16.92	60°	0.35^o	
		$CS^+(^2\Pi) + S(^3P)$	17.21			
		$S_2^+(^2\Pi) + C(^3P)$	16.91	9°	O^o	
	$D_2O^+(\tilde{X}^2B_1$ 12.624 eV[p])					
19.0[q]	$\tilde{B}^2B_2$	$OD^+(^3\Sigma^-) + D(^2S)$	18.22	87^r	0.22^r	
		$D^+(^1S) + OD(^2\Pi)$	18.82	13^r	0.01^r	Only two of the numerous energies reported by Eland (1975) between 18 and 20 eV are shown here
19.5	$\tilde{B}^2B_2$	$OD^+(^3\Sigma^-) + D(^2S)$		67^r	0.30^r	
		$D^+(^1S) + OD(^2\Pi)$		33^r	0.05^r	

	$SO_2^+(\tilde{X}\,^2A_1$ 12.32 eV[s])					
16.26	$^2B_2$[t]	$SO^+(^2\Pi) + O(^3P)$	16.02 ± 0.05[u]	≈100[v,w]	0.17[v]	The kinetic energy release distributions are broad.
		$S^+(^4S) + O_2(^3\Sigma^-)$	16.26	≈0[v]		
16.40	$^2B_2$; $^2A_1$	$SO^+(^2\Pi) + O(^3P)$		≈100[v]	0.29[v]	Approximately 50% of the available energy is released as translational energy.
16.81	$^2A_1$; $^2B_1$	$SO^+(^2\Pi) + O(^3P)$		≈100[v]	0.46[v]	Coincidence data were taken at 6 energies from 16.26 to 16.93 eV[v].

[a] Frey *et al.* (1978b).
[b] Berkowitz and Eland (1977).
[c] Guyon *et al.* (1978b).
[d] Baer *et al.* (1979).
[e] Eland *et al.* (1976c).
[f] Brehm *et al.* (1974).
[g] Frey *et al.* (1977).
[h] Dibeler and Walker (1968).
[i] Eland (1972b).
[j] Eland and Berkowitz (1977).
[k] Frey *et al.* (1978a).
[l] Eland (1974b).
[m] Eland (1973).
[n] The percentages are the author's estimates from the data of Brehm *et al.* (1974).
[o] Brehm *et al.* (1973b).
[p] Asbrink and Rabalais (1971).
[q] The origin of the $^2B_2$ band is probably at 17.301 eV (Bergmark *et al.* 1974). 19.0 eV is about in the middle of this broad band.
[r] Eland (1975).
[s] Dibeler and Liston (1968)
[t] Assignment is based on PES analysis of Lloyd and Roberts (1973). The IPs for the states are 15.99($^2B_2$), 16.32($^2A_1$), and 16.50($^2B_1$).
[u] This onset is not firmly established (see text).
[v] Brehm *et al.* (1973a).
[w] Wu and Yencha (1977) have reported the observation of fluorescence from the $v = 0$, 1, and 2 levels of the $^2B_2$ state.

characterized by an energy resolution which varies inversely with the release energy. In addition, in most of the experiments there was discrimination against the energetic fragments. There is a considerable uncertainty associated with the data, and in general it has not been possible to obtain precise information about the final vibrational energy of the diatomic fragments. Table I is a summary of the average kinetic energy released in the dissociation from which the average vibrational energy of the product diatomic fragment can be extracted. In a number of cases, the kinetic energy release distribution was reported and the reader can consult the original reference for such information.

Among the triatomics, N_2O^+ is probably the most studied ion. Recent investigations include work in photoionization (Berkowitz and Eland, 1977), photodissociation (Orth and Dunbar, 1977; Thomas *et al.*, 1977), PIPECO (Eland, 1973; Brehm *et al.*, 1974; Baer *et al.*, 1979; Guyon *et al.*, 1978b), and extensive calculations of the $^4A''$ surface (Hopper, 1975, and 1978). It is the only triatomic in which the kinetic energy released upon fragmentation has been measured by two different PIPECO techniques in different laboratories. The agreement between the two experiments for the energy releases associated with NO^+ formation from the N_2O^+ $\tilde{A}\Sigma^+$ state cannot be considered satisfactory. A problem with the earlier work of Brehm *et al.* (1974) may be that the discrimination against energetic NO^+ ions caused the calculated average value of the energy release to be too low.

One of the motivations behind the study of N_2O^+ has been the importance of the $O^+(^4S) + N_2(^1\Sigma) \rightarrow NO^+(^1\Sigma) + N(^4S)$ ion–molecule reaction which has an unusual temperature dependence in that the rate first decreases and then sharply increases with increasing temperature (McFarland *et al.*, 1973; Lindinger *et al.*, 1974). A long lived N_2O^+ complex was proposed to account for the low temperature behavior. In the low energy range, the $^2\Sigma^+$ state of N_2O^+, which is about 1 eV above the $O^+ + N_2$ limit, cannot take part in the reaction. At higher energies its participation can be ruled out for dynamical reasons. This can be deduced directly from the branching ratios in Table I and the measured fluorescence lifetime of the $^2\Sigma^+$ (1, 0, 0) state which was found by Smith (1969) and, more recently, by Frey *et al.* (1978b) to be 165 nsec. Predissociation to form O^+ and NO^+ is in competition with fluorescence, and the predissociation rates for O^+ and NO^+ formation can thus be calculated to be 6×10^4 and 2×10^6 sec^{-1}, respectively. Not only are these slow rates consistent with the spin forbidden nature of this predissociation, but more importantly they indicate that the cross section for surface crossing from the $O^+ + N_2$ quartet surface to the bound N_2O^+ $^2\Sigma^+$ surface is negligibly small. It thus appears that the reaction $O^+ + N_2$ remains on the quartet surface which, according to Hopper (1975, 1978), is the bent and mostly repulsive $^4A''$ state in the region of the N_2O^+ transition state.

This surface is reached by photoionization at photon energies below the $^2\Sigma^+$ state via autoionizing transitions to vibrationally excited $N_2O^+ \tilde{X}^2\Pi$ states and subsequent curve crossing to the $^4A''$ state (Baer *et al.*, 1979). Guyon *et al.* (1978b) measured the branching ratios and the kinetic energy released in the dissociation at these energies. NO^+ was found to be ejected with considerable kinetic energy consistent with the crossed beam data for the $O^+ + N_2 \rightarrow NO^+ + N$ reaction (Smith and Cross, 1974). In addition it was found that when the total energy is above that necessary to produce $O^+ + N_2$, this channel dominates by as much as a factor of 10. This is also consistent with the low cross section for the ion–molecule reaction.

The PIPECO results are consistent with all the kinetic results and in fact contribute some important information to the whole problem, yet it is unlikely that the definitive explanation for the energy dependence of the $O^+ + N_2$ reaction will be forthcoming from such data. This is because by preparing N_2O^+ via photoionization, we sample only a small part of the $^4A''$ surface, a section which may have very little relevance to the ion–molecule reaction. Hopper (1978) has rationalized most of the reaction characteristics such as the apparent reaction threshold of 0.15 eV on the basis of his calculated surface. It may be that trajectory calculations on this surface will answer some of the remaining questions.

Another interesting and well studied dissociation is that of the $\tilde{C}^2\Sigma^+$ state of CO_2^+. The dissociation limit to O^+ is just 0.40 eV below that of CO^+. Because the CO^+ onset is 75 meV above the $^2\Sigma^+(0,0,0)$ level and because the upper vibrational levels are populated only weakly, the study of the CO^+ dissociation was difficult and has not been done by PIPECO. Thus, all the conclusions are based on photoionization results. On the basis of these data, Eland and Berkowitz (1977) concluded that the CO^+ channel predominates when it is energetically accessible. In fact, they found that rotationally excited $^2\Sigma^+(0,0,0)$ ions predissociate to form CO^+. Using the rotational constant of the neutral CO_2 (0.39 cm^{-1}), one can calculate that those $^2\Sigma^+(0,0,0)$ ions with $J > 40$ are above the CO^+ dissociation limit. According to the photoionization data, essentially 100% of these ions dissociate to $CO^+ + O$. In order to conserve angular momentum the impact parameters for dissociation from $J = 40$, 42, 44, and 46 are 8.5, 5.3, 4.2, and 3.7 Å, the first one of which is larger than the Langevin impact parameter. This is not the first case in which rotational energy has been found to contribute to dissociation (Chupka, 1968; McCulloh and Dibeler, 1976), but it is rather striking that the CO^+ channel is preferred over the O^+ channel in a dissociation involving such large impact parameters.

A less well understood dissociation is that of SO_2^+. The dissociation limit to $SO^+(^2\Pi) + O(^3P_g)$ is still in question. The value of 16.02 eV listed in Table I is based on the ionization potential of SO measured by Jonathan *et al.* (1971) and the heats of formation of SO_2, SO, and O (Rosenstock *et al.*, 1977b). On the other hand, a direct measurement by photoionization of the SO^+ fragment from SO_2 by Dibeler and Liston (1968) yielded a value of 15.81 eV. Brehm *et al.* (1973a) measured PIPECO mass spectra of SO^+ at various energies between 16.26 and 16.93 eV. They found no evidence of $SO_2{}^+$ ions at these energies indicating that within the experimental error, $SO_2{}^+$ is fully predissociated. However, recently Wu and Yencha (1977) observed fluorescence when SO_2 reacted with helium metastables. They assigned the fluorescing level as $SO_2{}^+[\tilde{B}^2A_1(v_1, 0, 0)]$, the origin of which is at 15.97 eV. This appears to be the same as the $\tilde{B}^2B_2$ state at 15.99 eV assigned by Lloyd and Roberts (1973) on the basis of PES. Wu and Yencha (1977) observed fluorescence from $v_1 = 0, 1$, and 2. In the Lloyd and Roberts assignment the $\tilde{B}^2B_2(2, 0, 0)$ level is at 16.189 eV, just 70 meV below the energy investigated by Brehm *et al.* (1973a). This leaves open the possibility that SO_2 between the dissociation limit of 16.02 and 16.19 eV decays via fluorescence, while above the latter energy, the ion decays 100% via predissociation. Because the Wu and Yencha experiment provided neither quantum yields nor lifetimes, an alternative explanation for these results is also possible. This is because the fluorescence channel is less than about 5%, in which case the PIPECO experiment would not have had enough sensitivity to detect the fluorescence stabilized $SO_2{}^+$ ions.

B. *Breakdown Diagrams of Polyatomics*

A breakdown diagram is a plot of the ratio of the various decomposition products to the total ionization yield as a function of the parent ion internal energy. One of the first goals of PIPECO studies was to measure breakdown diagrams directly rather than differentiating photoionization curves or even doubly differentiating electron impact ionization scans. The first application of PIPECO was the breakdown diagram of CH_4 (Brehm and von Puttkamer, 1967). This, as well as the breakdown diagrams of $CD_4{}^+$, $C_2H_6{}^+$, and $C_2D_6{}^+$, were measured by Stockbauer (1973) with a significantly improved experiment. The breakdown diagrams of $C_2H_4{}^+$ and $C_2D_4{}^+$ (Stockbauer and Inghram, 1975), $C_3H_8{}^+$ (Stockbauer and Inghram, 1976; Brehm *et al.*, 1976), $C_5H_{12}{}^+$ (Brehm *et al.*, 1976), $C_3H_4{}^+$ (Parr *et al.*, 1978; Dannacher and Vogt, 1978), $CH_3COCH_3{}^+$ (Cant *et al.*, 1975; Golovin *et al.*, 1977), $C_6H_6{}^+$ (Eland *et al.*, 1976a), 2,4 hexadiyne (Dannacher, 1978), CH_3I^+ (Eland *et al.*, 1976b), and BF_3 (Batten *et al.*, 1978) have also been measured by PIPECO. The coincidence technique is uniquely suited for such investigations because by computing ratios of fragment ion signals, such variables (in time) as photon intensity, electron collection efficiency, sample gas pressure, etc., cancel out. The only unknown is the collection efficiency for the various fragments which

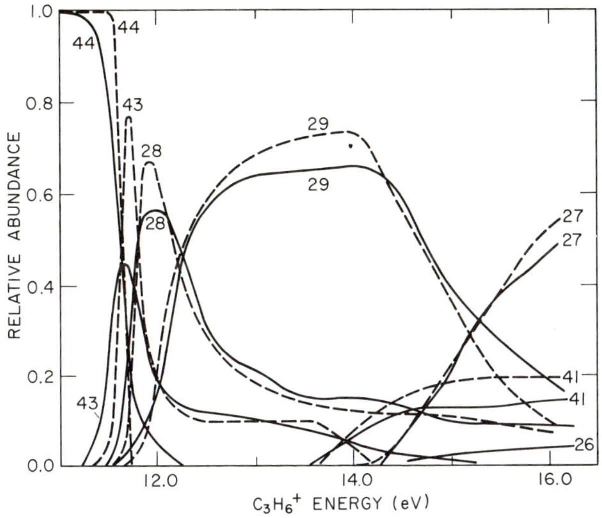

FIG. 8 Breakdown diagram of propane. [Data taken from Brehm *et al.* (1976) (solid lines); Stockbauer and Inghram, (1976) (dashed lines).]

may depend on their mass, usually a small effect, and their kinetic energy. In large ions, the energy released as translational energy is generally small so that even the kinetic energy discrimination is not usually a major source of error. In contrast, the method of differentiating integral data, such as photoionization or electron impact scans, is subject to serious problems because autoionization often is a dominant channel contributing to the total signal. An example, which also serves to illustrate the reproducibility of PIPECO results, is the breakdown diagram of propane (Fig. 8) recently reported by two different laboratories, one using the He(I) resonant source (Brehm *et al.*, 1976) and the other using a dispersed continuous light source with threshold electron detection (Stockbauer and Inghram, 1976).

A major aim of these studies has been to determine whether ionic dissociations can be described by the statistical theory of unimolecular decay known either as the RRKM (Rice and Ramsperger, 1927; Kassel, 1928; Marcus and Rice, 1951) theory or the quasi-equilibrium theory (QET) (Rosenstock *et al.*, 1952). All the hydrocarbons so far investigated have breakdown diagrams consistent with the statistical theory. On the other hand, Eland *et al.* (1976b) have pointed out that the dissociation of CH_3I^+ produces more I^+ than CH_2I^+, even though the latter should dominate the dissociation at all energies on statistical grounds. However, the CH_3I^+ dissociation appears to be consistent with the statistical theory from other viewpoints, such as the kinetic energy release. The latter aspect of the CH_3I^+

dissociation, as well as a possible explanation for the nonstatistical breakdown diagram, is discussed more fully in the following section.

More dramatic examples of nonstatistical behavior have been found for the dissociation of the other methyl halides on the basis of the products formed and the kinetic energy released (Eland *et al.*, 1976b). Results from the photodissociation of CH_3Cl^+ confirmed these findings (Dunbar and Kramer, 1973). Similar results have been reported for the $C_2F_6{}^+$ dissociation (Simm *et al.*, 1973, 1974). These fragmentations are nonstatistical in the sense that some of the ionic states dissociate directly and specifically to certain fragments without first exploring all the available phase space. It is significant that this is not simply a case of failure to undergo a radiationless transition to the ground electronic state, which is at times confused with nonstatistical behavior.

Although breakdown diagrams are extremely useful in assessing the statistical or nonstatistical nature of the dissociation of polyatomic ions, their utility goes beyond this function. These diagrams are of considerable value because they can represent in a compact fashion the dissociation paths of large ions as a function of the internal energy content. In addition, it is possible to extract fragment ion appearance potentials with considerably higher precision than in photoionization scans, because the onsets are sharper in the coincidence scans (Tsai *et al.*, 1975a).

C. *Kinetic Energy Released in Polyatomic Dissociations*

1. *The Kinetic Energy Release Distribution*

For the reasons previously mentioned, kinetic energy release determined from TOF distributions is one of the most difficult pieces of information to extract. In polyatomics, the problem is particularly acute because the energy released is generally a rather broad distribution. In addition, the ion fragment is often the more massive one and because its energy (E_i) is shared with that of the neutral fragment (E_n) such that $E_i M_i = E_n M_n$, where M_i and M_n are fragment ion and neutral masses, respectively, the precision is often poor.

Various methods have been used to obtain not only the average energy released but also the kinetic energy release distribution (KERD). The approach is largely determined by the particular ion investigated. In the case of the dissociation of propargyl chloride

$C_3H_3Cl^+ \longrightarrow$ 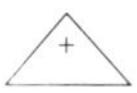 $+ \text{Cl} \qquad \text{KE} = 0.58\ \text{eV},$ (10a)

$C_3H_3Cl^+ \longrightarrow$  $+ \text{Cl} \qquad \text{KE} \approx 0\ \text{eV},$ (10b)

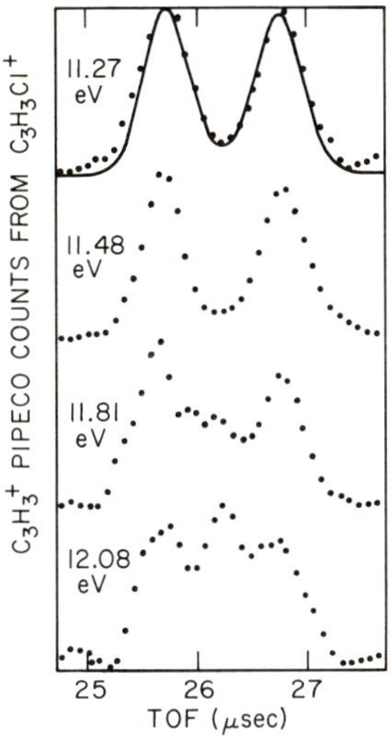

FIG. 9 $C_3H_3{}^+$ fragment ion TOF distributions from the photoionization of propargyl chloride (CH_2ClCCH) at the indicated photon energies. The drift tube was a quadrupole mass spectrometer set to transmit ions with an *m/e* of 39. The solid line is a calculated TOF distribution for $C_3H_3{}^+$ ions formed in a fragmentation accompanied by a total release energy of .58 eV. [Data taken from Tsai *et al.* (1975b).]

in which the cyclic $C_3H_3{}^+$ is formed with the release of nearly all of the total available energy of 0.58 eV, the best approach is to fit the ion TOF distribution (Fig. 9) with calculated distributions, taking into account the initial thermal kinetic energy of the parent $C_3H_3Cl^+$ and the effect of apertures on the collection efficiency of energetic fragment ions. Once the most probable release energy is determined, a distribution about this most probable energy can be assumed to obtain a final fit. For the best fit, shown in Fig. 9 as a solid line, Tsai *et al.* (1975*b*) assumed that the energy release is characterized by a 100 meV wide gaussian function centered at 580 meV. As is evident from Fig. 9, at higher photon energies, a new dissociation channel, characterized by a single peak at 26.2 μsec, becomes energetically possible. This channel releases little energy and has been associated with Eq. (10b). The two forms of $C_3H_3{}^+$ have been identified on the basis of electron impact data (Lossing, 1972) and molecular orbital calculations (Radom *et al.*, 1976).

For the case of a smoothly varying KERD, it is possible to vary an assumed distribution of energy releases until a good fit is obtained (Cant *et al.*, 1975), or to utilize a less biased approach such as a deconvolution in discrete steps, as was reported by Mintz and Baer (1976) for the reaction.

$$CH_3I^+ \rightarrow CH_3{}^+ + I. \tag{11}$$

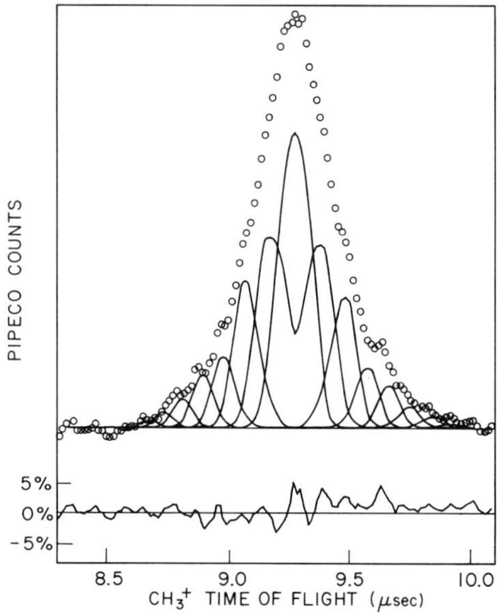

FIG. 10 $CH_3{}^+$ fragment ion TOF distribution from the photoionization of CH_3I^+ at 13.42 eV photon energy ($h\nu$ − AP = 1.18 eV). The solid lines are calculated TOF curves for various total single release energies. The release energies for the calculated curves are 6, 24, 54, 150, 216, 294, 384, and 486 meV. The solid line near the bottom is the difference between the sum of the calculated distributions and the experimental curve (points). [Data taken from Mintz and Baer (1976).]

A deconvolution in discrete steps was carried out by first calculating expected TOF flight distributions (basis functions) for a set of single kinetic energy releases. These basis functions, shown in Fig. 10, appear as double humped peaks originating from the forward and backward ejected $CH_3{}^+$. A set of release energies ($E_t = \epsilon n^2, n = 1, 2, \ldots$) was chosen such that adjacent basis functions were equally spaced in the TOF distribution. This prescription was adopted in order to minimize the linear dependence of adjacent basis functions. The data at each flight time were then expressed as contributions from each basis function B_n weighted by the coefficients C_n. The resulting set of linear equations (there were approximately 10 times as many equations as coefficients) was inverted by a standard linear regression analysis, yielding the coefficients C_n along with their uncertainty. The value of this method is that it guarantees that the maximum information is obtained from the data without extracting more information than should be obtainable.

In the case of $CH_3{}^+$ from CH_3I^+, the spacing of the basis functions could be made very small ($\epsilon = 6$ meV) because of the favorable fragment to parent

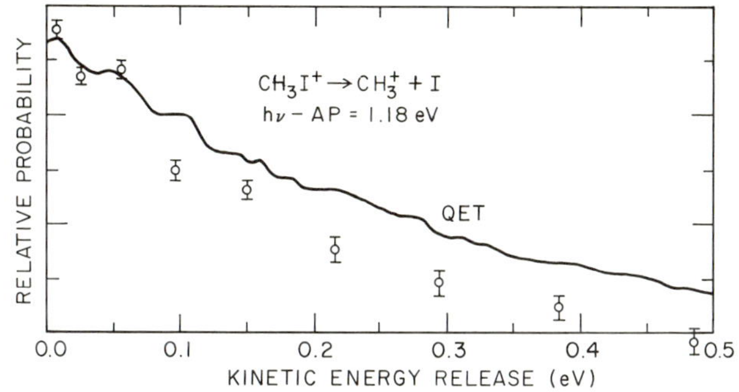

FIG. 11 The points are the derived kinetic energy release distributions (KERD) from the data of Fig. 10. The solid line is a KERD calculated by the statistical theory.

ion mass ratio. The derived KERD is shown as open circles in Fig. 11 along with the calculated KERD based on the statistical theory as formulated by Klots (1976). In the calculation, angular momentum was conserved and the results summed over the distribution of the initial rotational states of CH_3I^+. It is evident that for this ion energy, the experimental results lie below the theoretical ones. On the other hand, the lower ion energy KERDs were reproduced accurately by the theory. When the average release energy was plotted against the CH_3I^+ internal energy, a break at 0.65 eV ion internal energy was observed (Mintz and Baer, 1976). Above this ion energy, the kinetic energy released is less than the calculated energy. It is intriguing that 0.65 eV is just the energy at which the $I^+ + CH_3$ channel becomes energetically accessible. The question arises; How does the I^+ fragmentation channel affect the energy release of the $CH_3{}^+$ fragmentation channel? One explanation is that above 0.65 eV internal energy, the ion may dissociate along the $I^+ + CH_3$ channel. As the two fragments move apart, a charge transfer can take place yielding $I + CH_3{}^+$ which leaves the $CH_3{}^+$ vibrationally excited because of the very different geometries of CH_3 and $CH_3{}^+$. It would seem only reasonable that the inverse process could also take place, i.e., initial dissociation along the $CH_3{}^+ + I$ channel and a subsequent charge transfer to form $CH_3 + I^+$. Such a mechanism could explain the anomalously large I^+ yield noted by Eland *et al.* (1976b). At 0.95 eV above the dissociation onset, the $CH_3{}^+ + I(^2P_{1/2})$ dissociation channel becomes energetically accessible. This will further decrease the kinetic energy released.

In contrast to the CH_3I^+ dissociation, the KERDs of the other methyl halide ion fragmentations (Eland *et al.*, 1976b) are far from consistent with the statistical theory. These data support the conclusion, based on the fragmentation patterns, that the dissociations of CH_3X^+ (X = F, Cl, Br) are too rapid to allow internal energy equilibration prior to reaction.

The KERDs of some larger ions have also been measured and calculated. These include CH_3CO^+ from acetone (Cant *et al.*, 1975; Mintz and Baer, 1977) and $C_3H_3{}^+$ from butadiene (Klots *et al.*, 1977). The latter study is particularly interesting because the data from a crossed beam study (Herman and Birkinshaw, 1973; Herman, 1975) of the reaction

$$C_2H_2{}^+ + C_2H_4 \rightarrow \text{complex} \rightarrow C_3H_3{}^+ + CH_3 \tag{12}$$

are compared to the PIPECO results. A major difference between the two methods of preparing the $C_4H_6{}^+$ is that additional angular momentum is introduced into the $C_4H_6{}^+$ complex as a result of the collision. The KERDs obtained for the low angular momentum case, as measured by PIPECO, are close to exponentially decreasing functions, while the KERDs for the high angular momentum case are best described as Maxwellian $[E^{1/2}\exp(-BE)]$. That is, the latter KERDs go to zero in the limit of both high and low release energies. The KERDs for the low and high angular momentum cases are both described reasonably well by the statistical theory (Klots *et al.*, 1977).

2. *The Average Release Energy*

From an experimental viewpoint, it is considerably simpler to measure the average release energy than the KERD because a single parameter such as the width of the fragment ion TOF distribution can be related to the average release energy. However, this relation involves an assumption about the KERD; hence average release energies based on a single parameter determination can be in error. Stockbauer (1977a) made the assumption of Maxwellian KERDs and derived average release energies for CH_3 loss from acetone ions. These data along with other PIPECO results for the acetone dissociation are plotted in Fig. 12. The slightly higher values obtained by Stockbauer may be a reflection of the Maxwellian assumption because Mintz and Baer (1977) found that the KERDs for the acetone ion dissociation are rather better described by exponentials ($\exp -BE$) than by Maxwellian type functions.

Although accurate calculations of energy release by the statistical theory are not difficult in principle, there is continued interest in using simpler relations for predicting average release energies. A remarkably successful one is the empirical relationship of Haney and Franklin (1968) given by

$$\langle \epsilon_t \rangle = E^*/0.44N \tag{13}$$

in which E^* is the parent ion internal energy in excess of the dissociation limit, while N is the number of vibrational degrees of freedom in the parent ion. The average energy released in the dissociation of acetone to yield

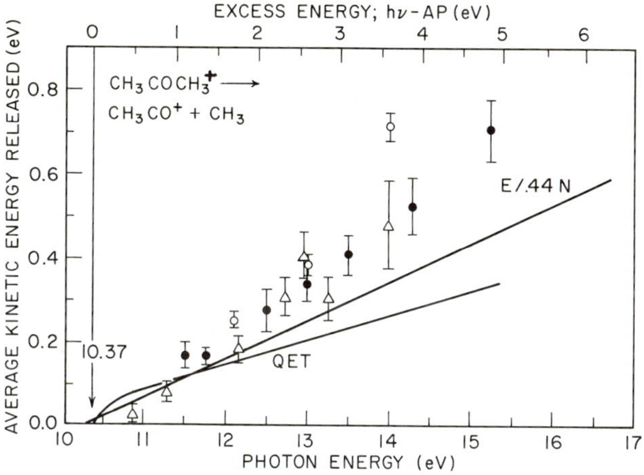

FIG. 12 The average release energy for the dissociation of acetone as a function of the acetone ion internal energy. [Data from: ●, Cant *et al.* (1975); ○, Stockbauer (1977a); △, Mintz and Baer (1977).

CH_3CO^+ is compared to predicted release energies in Fig. 12. The statistical theory calculations (QET) were carried out by Klots (see Mintz and Baer, 1977). Although there is considerable scatter in the experimental results, the data lie clearly above the QET calculation. A tight activated complex associated with a reverse activation energy cannot be blamed for this discrepancy, since there is no reverse activation energy for this dissociation. It may be significant that the serious deviation from statistical behavior begins at about 12 eV, just at the onset of a series of at least three broad bands in the PE spectrum (Brundle *et al.*, 1972). The fact that the KERDs are smooth (Mintz and Baer, 1977) and similar in shape to that of Fig. 11 indicates that considerable energy equilibration is taking place, but evidently not enough to yield the statistically expected result. The better agreement between the empirical relation of Eq. (13) and the data is very likely fortuitous. The constant 0.44 has been chosen to give the best limiting slope at low ion energies for a wide variety of dissociations. Because it is a linear equation, it will always predict a higher release energy at the higher ion energies than will the statistical theory which tends toward the equipartition limit of approximately E^*/N.

Similar minor deviations between experiment and the statistical theory have been found for H loss from CH_4^+ (Stockbauer, 1977b) and CH_3 loss from $C_4H_6^+$ (Klots *et al.*, 1977). Gross deviations from statistical behavior have been found for F loss from CF_4^+ (Brehm *et al.*, 1974) and the previously mentioned X loss from CH_3X^+, X = F, Cl, Br (Eland *et al.*, 1976b).

D. *Lifetime of Metastable Ions*

One of the most sensitive probes for the study of unimolecular reactions is the lifetime measurement of ions in selected internal energy states. The first such measurement was done not by PIPECO but by preparing ions in selected energy states by charge transfer (Andlauer and Ottinger, 1971, 1972). The fragmention of benzonitrile ion ($C_6H_5CN^+$) to $C_6H_4^+$ and the fragmentation of benzene to $C_6H_5^+$ and $C_4H_4^+$ were investigated in some detail. The results for the benzene dissociation revealed an interesting and remarkable result. It was found that the rates for $C_6H_5^+$ and $C_4H_4^+$ formation were very different, indicating that the dissociation was proceeding via two noncompeting reactions. Although the benzene dissociation since then has also been investigated by PIPECO (Eland, 1974a; Eland and Schulte, 1975; Eland *et al.*, 1976a; Sergeev *et al.*, 1977), the precise mechanism is still under discussion (Rosenstock *et al.*, 1973, 1977a). The basic problem centers around the mechanism for the dissociation. The decay of an excited ion A* can be represented in a general way by

$$A^* \begin{cases} \nearrow A' \longrightarrow P_i \\ \searrow A'' \longrightarrow P_i \end{cases} \qquad i = 1, 2, \cdots. \tag{14}$$

In this mechanism A′ and A″ are isomers of A* (either geometric or electronic), while P_i are the various products. If the densities of states of A′ and A″ are significantly bigger than that of A*, the likelihood of back reaction to A* is very small. If in addition there is no coupling between A′ and A″ except via A*, the two isomers will decay with different rates. In the case of benzene, the Andlauer and Ottinger results imply that A′ and A″ dissociate to different product ions, at least within the limits of detectability. However, a number of problems with the interpretation of these data remain. One is that the statistical theory cannot account for the low energy dependence of the H loss channel (Rosenstock *et al.*, 1973, 1977a). Photoionization (Rosenstock *et al.*, 1977a) and PIPECO (Dannacher, 1978) studies on other $C_6H_6^+$ isomers are beginning to shed some light on this question. Soon to be published PIPECO results from our laboratory on 1,5 and 2,4 hexadiyne, in fact, suggest that the $C_6H_6^+$ dissociation channels compete with each other, implying that the original data of Andlauer and Ottinger on H loss may be in error.

The Andlauer and Ottinger results on $C_6H_5CN^+$ have been compared to theoretically predicted rates based on the statistical theory by Klots (1972) and more recently by Chesnavich and Bowers (1977). Two formulations of the theory are currently being used. They are the transition state treatment in the forms of the RRKM theory and the QET; and the phase space theory which has been applied to unimolecular dissociations of ions primarily by Klots (1971, 1972). Chapter 4 in this volume discusses and compares these two approaches, and therefore it will not be done here. The important point is that the two formulations predict quite different rates for the $C_6H_5CN^+$ dissociation. The phase space theory, with the use of the ion-induced dipole interaction potential for the departing fragments, predicts rates which are considerably higher than is experimentally observed, while the transition state approach results in acceptable agreement (Chesnavich and Bowers, 1977). This is in accord with Klots' observation that the phase space theory with the ion-induced dipole interaction sets an upper limit to the dissociation rate and is equivalent to the transition state theory with the most loose of all possible transition states (Klots, 1972). Given the rather severe rearrangement necessary to produce $C_6H_4{}^+$ from $C_6H_5CN^+$, these results are not entirely unexpected.

The dissociation rates of a number of energy selected ions have been measured by PIPECO. The major results, along with those of Andlauer and Ottinger, are summarized in Table II. Of the reactions listed in Table II, halogen loss from the halobenzenes and HCN loss from $C_6H_5CN^+$ are most easily interpreted in terms of the statistical theory. As in the $C_6H_5CN^+$ reaction, the RRKM–QET formalism predicts rates for the halobenzene reacion which are in excellent agreement with the experiment (Baer *et al.*, 1976). This agreement lends support to two of the assumptions made, which are (a) that a rapid radiationless transition converts electronic energy into vibrational energy of the ground electronic state and (b) that all the vibrational degrees of freedom participate fully in distributing energy around the ion. In the halobenzene dissociations, the minimum rate of about 10^4 sec^{-1}, which results in significant ion signal on the time scale of the PIPECO experiment, is at an energy considerably above the dissociation threshold. This energy shift, commonly referred to as the kinetic shift, can be calculated from the data of Table II. Recently, Stockbauer and Rosenstock (1978) have carefully measured the kinetic shift in allene and methane ion dissociations by initiating the ion extraction pulse at variable times after the electron signal.

The case of the $C_4H_6{}^+$ dissociation illustrates some of the difficulties involved in predicting the rates by theoretical approaches. A number of C_4H_6 isomers were photoionized, producing ions with various activation energies with respect to dissociation to $C_3H_3{}^+$ and CH_3. The observation of identical rates for $C_4H_6{}^+$ at equal energies above the dissociation limit indicated that the various isomers were isomerizing to a common intermediate prior to dissociation. The only other two series of isomers (Table II) studied exhibited similar evidence for isomerization. The occurrence of ionic rearrangement prior to fragmentation makes the application of a theory difficult because of our lack of knowledge concerning the structure of the isomerized

TABLE II

Rate Studies on Energy Selected Metastable Ions

Molecule and IP (eV)	Ion fragment; thermochemical onset (eV)	Energy range investigated	Isomerization	1-component decay	Reference(s)
C_6H_6 (9.25)	$C_6H_5^+$ (12.9)[a]	14–16	?	See text	*e–g*
	$C_4H_4^+$ (13.5)[b]	14.0–14.7	Yes	See text	*e, f*
	$C_3H_3^+$ (13.6)[c]	14.4–14.6	Yes	See text	*g*
C_4H_4S (8.95)	$C_2H_2S^+$ (?)	12.1–13.4	?	Yes	*f*
C_6H_5CN (9.70)	$C_6H_4^+$ (?)	14–17	?	Yes	*e–g*
C_6H_5Cl (9.07)	$C_6H_5^+$ (12.3)[a]	13.1–13.6	?	Yes	*h*
C_6H_5Br (8.97)	$C_6H_5^+$ (11.7)[a]	12.1–12.7	?	Yes	*h*
C_6H_5I (8.77)	$C_6H_5^+$ (11.0)[a]	11.3–11.7	?	Yes	*h*
1,3 butadiene (9.08)	$C_3H_3^+$ (11.37)	11.4–11.8	Yes	Yes	*i*
1,2 butadiene (9.33)	$C_3H_3^+$ (10.80)	11.1–11.2	Yes	Yes	*i*
Cyclobutene (9.40)	$C_3H_3^+$ (10.85)	11.0–11.3	Yes	Yes	*i*
2 butyne (9.57)	$C_3H_3^+$ (11.05)	11.2–11.5	Yes	Yes	*i*
1 butyne (10.19)	$C_3H_3^+$ (10.70)	11.0–11.3	Yes	Yes	*i*
Methyl-2-cyclopropene (?)	$C_3H_3^+$ (9.9)	10.0–10.2	Yes	Yes	*j*
2 butene (trans) (9.11)	$C_3H_5^+$ (11.37)	11.4–11.7	Yes	No	*k*
2 butene (cis) (9.12)	$C_3H_5^+$ (11.33)	11.4–11.7	Yes	No	*k*

Isobutene (9.22)	$C_3H_5^+$ (11.45)	11.5–11.8	Yes	No	*k*
1-butene (9.61)	$C_3H_5^+$ (11.28)	11.3–11.6	Yes	No	*k*
Methyl cyclopropane (9.64)	$C_3H_5^+$ (11.02)	11.1–11.4	Yes	No	*k*
Styrene (8.4)	$C_6H_6^+$ (10.93)	12.5–13.0	Yes	Yes	*l*
Cyclooctatetraene (8.1)	$C_6H_6^+$ (9.38)	10.8–11.4	Yes	Yes	*l*
CH_3CH_2Cl (10.96)	$C_2H_4^+$ (11.18)[d]	11.3–11.5	?	No	*m*
CH_2ClCH_2Cl (11.15)	$C_2H_3Cl^+$ (10.76)[d]	11.3–11.4	?	No	*m*
$CHCCH_2Cl$ (10.68)	$C_3H_3^+$ (10.44)[d]	11.0–11.3	?	No	*m, n*

[a] These values are based on a $\Delta H_f^\circ(C_6H_5^+) = 266$ kcal/mole.
[b] This value is based on a $\Delta H_f^\circ(C_4H_4^+) = 276$ kcal/mole.
[c] This value is based on a $\Delta H_f^\circ(C_3H_3) = 80$ kcal/mole.
[d] These dissociations have considerable reverse activation energies.
[e] Andlauer and Ottinger (1971).
[f] Andlauer and Ottinger (1972).
[g] Eland and Schulte (1975).
[h] Baer *et al.* (1976).
[i] Werner and Baer (1975).
[j] Baer *et al.* (1977).
[k] Baer *et al.* (1978b).
[l] Smith *et al.* (1979).
[m] Tsai *et al.* (1975b).
[n] Baer *et al.* (1975).

ion. More serious still is the fact that neither the phase space nor the RRKM–QET versions of the statistical theory are able to predict the dependence of the fragmentation rates of $C_4H_6^+$ over the energy range investigated when the activation energy is taken as the difference between the energies of precursor ions and products. Werner and Baer (1975) suggested that this discrepancy of over two orders of magnitude indicates nonstatistical energy flow, particularly at energies close to the dissociation onset. Chesnavich and Bowers (1977) have proposed an alternative explanation which makes the interpretation of the butadiene data consistent with the statistical theory. They propose that the real transition state, or bottleneck in the phase space, lies at an energy *below* the dissociation limit. With reasonable assumptions about the energies and structure for such a transition state, they are able to fit the experimental results. What is particularly disturbing is that there seems to be no way that this hypothesis can be disproven. It may be possible to verify the existence of such a "prebottleneck," possibly followed by a shallow well, if the structure of the ion in that well corresponds to a molecule which can be synthesized and photoionized. To date, it has been verified that methyl-2-cyclopropene is not the appropriate structure (Baer *et al.*, 1977).

Two other ions, whose dissociation rates suggest isomerization prior to fragmentation to the products, $C_6H_6^+ + C_2H_2$, are styrene and cyclooctatretraene (Smith *et al.*, 1979). According to the preliminary analysis, the rates determined by RRKM–QET calculations agree well with the experimental results. The indications are that the cyclooctatretraene ion isomerizes to the styrene structure prior to dissociation. One of the unsolved problems is the development of a method for determining the activation energy of the isomerization reactions in energized ionic systems. A particularly promising approach is one utilized by Wilkins and Gross (1970) who determined on the basis of the different reactivities of cyclooctatetraene and styrene ions with the styrene molecule that no isomerization takes place. Although the energy range sampled by these workers was broad (it was not a PIPECO experiment), these results along with the ion lifetime studies of Smith *et al.* (1979) place lower and upper limits of 0.9 and 1.2 eV, respectively, on the activation energy for the cyclooctatetraene isomerization.

Evidence for the complex nature of unimolecular decay mechanisms comes not only from isomerization of precursor ions but also from the observation of nonexponential decay in a number of fragmentations. An example of such a decay is the HCl loss from $C_2H_5Cl^+$ (Tsai *et al.*, 1975b) shown in Fig. 13a. It is contrasted with the single exponential decay found for the I loss from $C_6H_5I^+$ (Fig. 13b). Clearly, no single exponential decay curve is consistent with the $C_2H_5Cl^+$ data. However, the data are consistent with two component decay, i.e., a rate which is expressed as the sum of two exponential

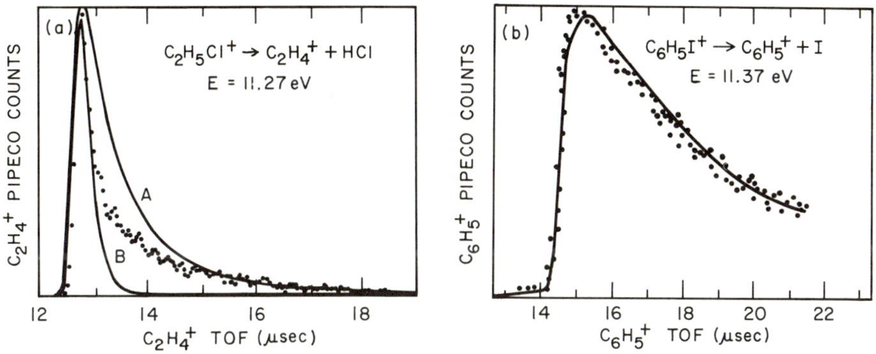

FIG. 13 Time-of-flight distributions of fragments from metastable ion dissociations. (a) An example of (at least) a two component decay; photon energy = 11.27 eV. Curves A and B are calculated one component TOF distributions with assumed rates of 0.8 μsec^{-1} and 3.3 μsec^{-1}, respectively. (b) An example of a one component decay; photon energy = 11.37 eV. The solid line is a calculated TOF distribution with an assumed rate of 0.20 μsec^{-1}. [Data for (a) and (b) taken, respectively, from Tsai *et al.* (1975b); Baer *et al.* (1976).]

decay rates. The results for this particular reaction are most reasonably explained by assuming that initial photoionization is to two noncommunicating electronic states, possibly those formed by removing either of two orthogonal chlorine lone pair electrons. The slow dissociation rate appears to be a case of electronic predissociation.

Two component decay can also be explained by the mechanism of Eq. (14). One possibility is that, A′ and A″ produce different products. However, if, as appears to be the case in the propargyl chloride dissociation ($C_3H_3Cl^+ \rightarrow C_3H_3^+ + Cl$), the two isomeric forms produce the same product, the resultant rate will appear as a two component decay rate. Another fragmentation which exhibits nonexponential decay is the dissociation of $C_4H_8^+$ to $C_3H_5^+$ and CH_3. In addition, this reaction shows evidence of isomerization prior to dissociation for all the isomers investigated (Baer *et al.*, 1978b). Mechanisms which are consistent with these results are complicated and are to a certain extent, arbitrary unless additional information about the reaction is available. In the absence of a unique mechanism, it is impossible to calculate rates theoretically.

The general conclusions from these studies is that although the statistical theory can be used to calculate rates which are in agreement with the experimental rates, it is not yet possible to *predict* the rate of a reaction. The reason has less to do with the theory than with the fact that ionic dissociations are often complex sequences of reactions, and before the theory can be applied to the reaction in question, the sequence of steps must be known with a considerable degree of certainty.

E. Cross Sections of Ion–Molecule Reactions

The study of bimolecular reactions is considerably more difficult than the study of unimolecular reactions because the necessity to limit the process to the single collision domain reduces the signal by an order of magnitude relative to the unimolecular case. Only a few reactions have been investigated. However, it is certainly a field which will receive increased attention in the next decade.

The reactions so far investigated have been carried out in a single chamber experiment (Fig. 1) in which both photoionization and subsequent reaction take place. By the use of collimated hole structures at both exits of the ionization/acceleration region, it is possible to limit reaction to this region. The time of flight of an ion of mass m_2 which is the product of a reactive collision between an ion of mass m_1 and some neutral is given by

$$\mathrm{TOF} = \left(\frac{2d_1}{a_1}\right)^{1/2} + \frac{1}{a_2}\left((2d_2a_2 + v_0{}^2)^{1/2} - v_0\right) + \frac{d_3}{(2d_2a_2 + v_0{}^2)^{1/2}}. \quad (15)$$

The three terms correspond to (1) the TOF of the reactant ion traveling a distance d_1 in an electric field E with an acceleration $a_1 = qE/m_1$, (2) the TOF of the product ion traveling over the remaining distance d_2 in the acceleration region with the acceleration $a_2 = qE/m_2$, and (3) the TOF of the product ion in the drift region traveling a distance d_3. The quantity v_0 is the projection of the initial product ion velocity in the electric field direction. The kinetic energy of the reactant ion is just d_1qE and the cross section in the limit of small P is given by $\sigma = P/n\Delta$, where P is the probability for reaction, n the gas density, and Δ the path length over which σ is determined.

The path length Δ is the distance the reactant ion travels in the time increment ΔT, a quantity which is related but not equal to the final product ion TOF increments. Because the data are collected in units of constant TOF increments, the path length increment Δ varies as a function of the position in the acceleration region. The relation between the final TOF increment and the kinetic energy of the reactant ion, as well as Δ, is given by Eq. (15). However, in order to solve this equation, some assumptions about the collision dynamics must be made. In particular, we must guess v_0. In the absence of any angular dependent scattering data, it is indeed difficult to establish a unique model. This in turn makes the transformation of TOF → KE and product ion signal → cross section somewhat uncertain. However, there are ways of eliminating poor models. As can be seen from Eq. (15), the electric field contained in a_1 and a_2 does not factor out unless $v_0 = 0$. Hence, data taken at different draw-out voltages (electric fields) will not reduce to identical σ versus KE plots for all assumed models. Failure of a model to cast data taken at different draw out voltages into identical σ versus KE plots can thus be used as a criterion for rejecting that model.

Such an analysis was carried out for the collisional dissociation of $CH_2Br_2^+$ (Baer *et al.*, 1974). A model in which the product CH_2Br^+ was backscattered was found to be far more consistent with the data than a forward scattering model, indicating that the positive charge remained on the reactant ion. Although such information can be extracted from a single chamber experiment, it is clearly a far less satisfactory approach than a crossed beam experiment in which the model can be determined from a two dimensional experiment. However, in cases where a particular reaction has already been investigated by crossed beam techniques, a single chamber experiment can give not only internal energy dependence but also very accurate total cross sections.

A class of reactions which is ideally suited for the single chamber arrangement is symmetric charge transfer. In that case, it can be assumed that no momentum is transferred in the collision so that $v_0 = 0$. This has been tested both experimentally (Paulson *et al.*, 1970; Kaiser *et al.*, 1974; Hierl *et al.*, 1977) and theoretically (McCann *et al.*, 1975). An example of the TOF distribution of NO^+ ions in an ionization/acceleration region under a pressure of 5×10^{-4} torr is shown in Fig. 14. The collection time was approximately 5 hr.

Transformation of such TOF data to charge transfer cross sections yields reduced data such as shown in Fig. 15 for Xe symmetric charge transfer. Similar results have been obtained for the $NO^+(X^1\Sigma^+, v)/NO(v' = 0)$ reaction $v = 0, \ldots, 5$ (Squires and Baer, 1976) and the $O_2^+(X^2\Pi_g, v)/O_2(v' = 0)$

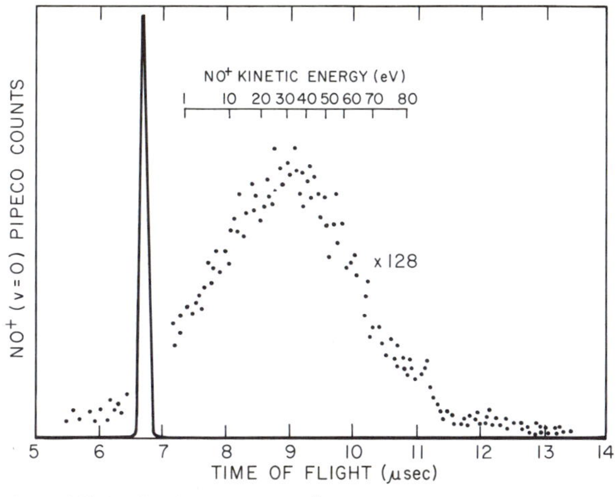

FIG. 14 Time-of-flight distribution of NO^+ ions originating in an ionization/acceleration region at a pressure of 5×10^{-4} torr. The sharp peak at 6.7 μsec is due to unreacted $NO^+(v = 0)$ ions, while the broad peak between 7 and 12 μsec is due to NO^+ product ions from a charge transfer reaction with NO. [Data from Squires and Baer (1976).]

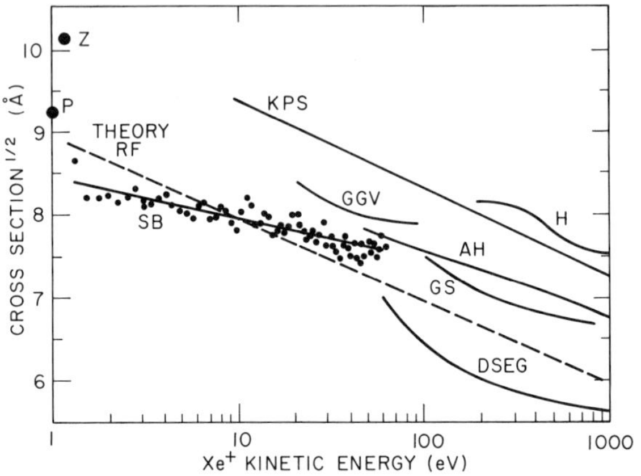

FIG. 15 A summary of published symmetric charge transfer cross sections for Xe^+. The points with a least squares line through them are the PIPECO results taken from Squires and Baer (1976). Other data which are identified by the authors' initials are taken from Amme and Haugsjaa (1968), Dillon *et al.* (1955), Galli *et al.* (1962), Ghosh and Sheridan (1957b), Hishinuma (1972), Kushnir *et al.* (1959), Popescu-Iovitsu (1959), and Ziegler (1953). The dashed line is the calculated cross-section based on the theory of Rapp and Francis (1962).

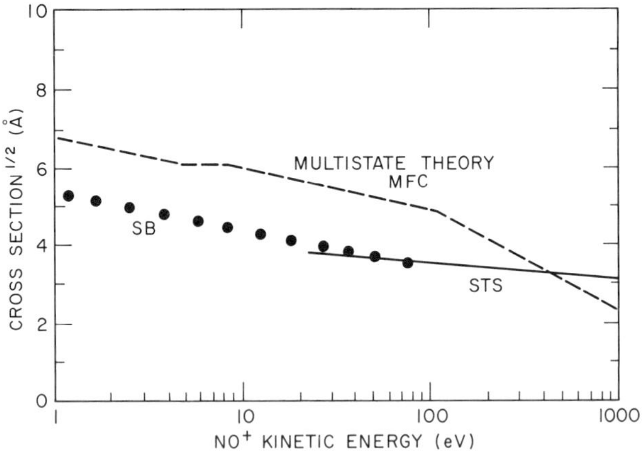

FIG. 16 A summary of published symmetric charge transfer cross sections for NO^+ (11 eV electron impact). The points are taken from the PIPECO results of Squires and Baer (1976). The solid line is the result from Stebbings *et al.* (1963). The dashed line is a theoretical result from Moran *et al.* (1974). The PIPECO cross sections for the $v = 0$–5 were averaged to simulate the internal energy distribution expected for 11 eV electron impact ionization of NO.

$v = 0, \ldots, 7$ (Baer *et* al., 1978a). The dependence of the cross section on the kinetic energy was not found to be affected significantly by the vibrational energy, although the actual magnitude exhibited significant variations. In order to compare the results with cross sections obtained by electron impact

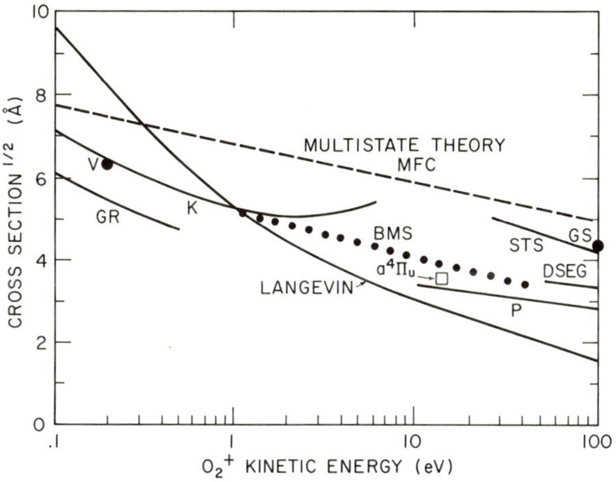

FIG. 17 A summary of published symmetric charge transfer cross sections for $O_2^+(X^2\Pi_g)$. The points between 1 and 60 eV O_2^+ kinetic energy are the PIPECO results from Baer *et al.* (1978a). The individual cross sections for $v = 0$–7 were averaged to simulate the internal energy distribution expected for 14 eV electron impact ionization. The open square is the cross section for O_2^+ in the long lived $^4\Pi_u$ state, also taken from Baer *et al.* (1978a). Other data which are identified by the authors' initials are taken from Dillon *et al.* (1955), Ghosh and Sheridan (1957a), Gray and Rees (1972), Kobayashi (1975), Potter (1954), Stebbings *et al.* (1963), and Varney (1970). The two theoretical curves are the Langevin collision cross section and the results of the multistate theory taken from Moran *et al.* (1974).

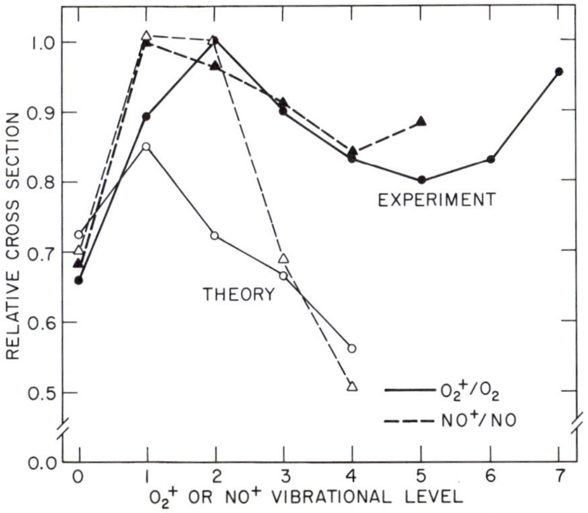

FIG. 18 The dependence of the symmetric charge transfer cross section on vibrational energy level at a reactant ion laboratory frame kinetic energy of 50 eV. The experimental results are taken from Squires and Baer (1976) and from Baer *et al.* (1978a). The theoretical results are taken from the multistate treatment of Moran *et al.* (1974).

ionization in which the ions were prepared in distributions of internal energy, the coincidence results were averaged over the various vibrational states. These data are shown in Figs. 16 and 17 along with theoretical results based on the multistate impact parameter theory (Bates and Reid, 1969; Moran *et al.*, 1974).

The major aim of these studies was to determine the dependence of the charge transfer cross section on the ion vibrational level. Relative cross sections versus the vibrational level are summarized in Fig. 18. The precision of these relative cross sections is approximately $\pm 5\%$. The trend of an increasing cross section with the vibrational level for the higher energy states is particularly interesting and unexpected. Preliminary results of NO^+ ($^1\Sigma, v = 10$) indicate that the relative cross section is about 0.8 ± 0.1, i.e., the cross sections remain high even for the very high vibrational levels. A similar dependence of cross section on vibrational levels was found by Chupka (1972) for symmetric charge transfer in $H_2{}^+$.

The theoretical treatment of charge transfer reactions is based on the impact parameter theory (Massey and Smith, 1933; Dalgarno and Yadav, 1953; Gurnee and Magee, 1957; Rapp and Francis, 1962) which was originally applied to atoms. It treats the motion of the reactants classically and assumes that the relative velocity is constant during the "collision" so that the trajectory is rectilinear. Thus the theory neglects the effect of the ion-induced dipole forces which would be expected to dominate the interaction at low translational energies.

The application of the impact parameter theory to molecules was suggested by Gurnee and Magee (1957) and carried out by Bates and Reid (1969) for the $H_2{}^+$ symmetric charge transfer. In this theory, the multitude of reaction channels accessible to charge transfer between $AB^+(v) + AB(v = 0)$,

$$AB^+(v) + AB(v = 0) \rightarrow AB(v') + AB^+(v''), \tag{16}$$

is taken into account. Three types of reaction channels can be distinguished on the basis of the product ion and molecule vibrational states. The case of $v' = 0$ and $v'' = v$ is the resonant channel. The cases which satisfy the relation $v' + v'' = v$ are the near resonant channels. All other combinations of v' and v'' constitute the nonresonant channels in which a considerable amount of energy must be transferred between internal and translational energy resulting in reduced cross sections. In the multistate treatment, the vibrational and electronic wave functions are separated via the Born–Oppenheimer approximation. The resulting cross sections are sums over the many channels, each term in the sum being a product of an electronic part and Franck–Condon factors connecting the ion and neutral species.

The symmetric charge transfer cross sections have been calculated for a variety of diatomics as a function of the ion vibrational states (Flannery *et al.*,

1973; Moran *et al.*, 1974, 1975). The results of these calculations for O_2^+ and NO^+ are illustrated in Figs. 16–18.

The agreement between theory and experiment is reasonably good. However it is evident that the predicted cross sections are too high by about a factor of two. This discrepancy cannot be explained by the theory's neglect of attractive forces because, as Bates and Reid (1969) point out, including this attraction would have the effect of increasing the calculated cross section still more. There is, however, another approximation which may be responsible for the discrepancy. In calculating the charge transfer cross sections certain Coulomb, exchange, and overlap integrals are evaluated. It is assumed that these integrals are independent of the relative orientation of the reacting pair internuclear axes. If in fact there are some unfavorable orientations of the internuclear axes, the calculated cross sections would be too high. One notes from Fig. 16 that the discrepancy between theory and the experimental results lessens slightly with decreasing translational energy. This fact would support such an orientation effect since, at lower translational energies, the reacting species would have more opportunity to rearrange their relative orientation into ones more favorable for charge transfer. Because the multistate treatment overestimates the O_2^+/O_2 and NO^+/NO cross sections by roughly the same factor, we may rule out the heteronuclear nature of NO^+ as the origin of such an orientation effect. The discrepancy between the experimentally observed dependence of the cross section on reactant ion vibrational energy and the theoretically predicted dependence is illustrated in Fig. 18. It appears significant that the relative cross sections for the first few vibrational levels follow the theory in a qualitative manner but that the cross sections for reactant ions in the higher vibrational states actually increase with increasing vibrational level and show no signs of decreasing. The calculations become long and expensive for the higher vibrational states because of the increased number of accessible channels, and therefore they have not been carried out to high enough vibrational levels to indicate whether the high cross sections are consistent with the multistate theory.

IV. FUTURE DEVELOPMENTS

This chapter has reviewed the application to the study of ion dynamics of only one type of coincidence experiment, that between an energy analyzed photoelectron and its ion. There are other coincidence events which are being investigated. These include coincidences between photoelectrons and fluorescence photons (Schlag *et al.*, 1977; Frey *et al.*, 1978b) and ions and fluorescence photons (Eland *et al.*, 1976c). Ion fluorescence lifetimes from selected ion states can be determined by such experiments.

During the next ten years coincidence studies will certainly be affected dramatically by two recent experimental advances. These are the increased availability of pulsed synchrotron radiation and pulsed tunable lasers. As has already been demonstrated (Guyon *et al.*, 1978a), extremely high collection efficiencies for threshold electrons with excellent energy resolution are made possible by the use of pulsed synchrotron radiation. The improved signal levels will in turn make possible more difficult experiments, such as the measurement of differential cross sections in crossed molecular and ion beams with internal energy selected ions. In addition, it should be possible to use laser-induced fluorescence on products of ion–molecule reactions in order to probe the final states. Thus state to state study of chemical reactions will be possible to achieve.

High efficiencies in the initial preparation of energy selected ions should also make possible triple coincidence experiments. A study which appears possible is the investigation of chemical reactions in high radiation fields. An energy selected ion beam could be crossed with a molecular beam, while a high powered pulse laser is triggered by a delayed signal from the photoelectron pulse. This approach would insure the coincidence of at least the burst of photons (radiation field) and the energy selected ion, and should provide for sufficient coincidences of the photons, ion, and a neutral reactant to allow the investigation of such three particle events. These types of process have been the focus of a great deal of recent theoretical work (George *et al.*, 1977; Zimmerman *et al.*, 1977).

A different extension of the PIPECO technique is the use of pulsed lasers to photodissociate ions formed by photoionization to the ground ionic state. Triggering the laser by the ejected photoelectron insures the presence of an ion at a specific place and time, thereby reducing wasted laser shots which invariably accompany laser excitation of very low density particles. The advantage of such an approach is the ability to state select ions with excellent resolution in very high electric fields, which in turn will allow the investigation of ion fragmentation lifetimes on the 10^{-9} sec time scale. In addition, the kinetic energy released upon dissociation can be studied with an energy resolution orders of magnitude better than is now possible.

ACKNOWLEDGMENTS

I would like to thank the Guggenheim Foundation for a fellowship (1976–1977) which allowed me to spend an exciting year in Paris, carrying out PIPECO experiments on the synchrotron with the research group of P. M. Guyon.

The work presented in this chapter from our laboratory in Chapel Hill has been supported by grants from the National Science Foundation, the Army Research Office, and the Research Corporation.

REFERENCES

Amme, R. C., and Haugsjaa, P. O. (1968). *Phys. Rev.* **165**, 63.

Andlauer, B., and Ottinger, C. (1971). *J. Chem. Phys.* **55**, 1471.

Andlauer, B., and Ottinger, C. (1972). *Z. Naturforsch., Teil A* **27**, 293.

Asbrink, L., and Rabalais, J. W. (1971). *Chem. Phys. Lett.* **12**, 182.

Baer, T., Peatman, W. B., and Schlag, E. W. (1969). *Chem. Phys. Lett.* **4**, 243.

Baer, T., Squires, L., and Werner, A. S. (1974). *Chem. Phys.* **6**, 325.

Baer, T., Tsai, B. P., and Werner, A. S. (1975). *J. Chem. Phys.* **62**, 2497.

Baer, T., Tsai, B. P., Smith, D., and Murray, P. T. (1976). *J. Chem. Phys.* **64**, 2460.

Baer, T., Ormerod, R. C., Smith, D., Aue, D., and Bowers, M. T. (1977). Unpublished observations.

Baer, T., Murray, P. T., and Squires, L. (1978a). *J. Chem. Phys.* **68**, 4901.

Baer, T., Smith, D., Tsai, B. P., and Werner, A. S. (1978b). *Adv. Mass Spectrom. A***7**, 56.

Baer, T., Guyon, P. M., Nenner, I., Govers, T. R., Tabché-Fouhailé, A., Botter, R., and Ferreira, L. F. A. (1979). *J. Chem. Phys.* **70**, 1585.

Bates, D. R., and Reid, R. H. G. (1969). *Proc. R. Soc. London, Ser. A* **310**, 1.

Batten, C. F., Taylor, J. A., Tsai, B. P., and Meisels, G. G. (1978). *J. Chem. Phys.* **69**, 2547.

Bergmark, T., Karlsson, Jadrny, R., Mattsson, L., Albridge, R. G., and Siegbahn (1974). *J. Electron Spectrosc. Relat. Phenom.* **4**, 85.

Berkowitz, J., and Eland, J. H. D. (1977). *J. Chem. Phys.* **67**, 2740.

Brehm, B., and von Puttkamer, E. (1967). *Z. Naturforsch., Teil A* **22,** 8.

Brehm, B., Eland, J. H. D., Frey, R., and Küstler, A. (1973a). *Int. J. Mass Spectrom. Ion Phys.* **12**, 197.

Brehm, B., Eland, J. H. D., Frey, R., and Küstler, A. (1973b). *Int. J. Mass Spectrom. Ion Phys.* **12**, 213.

Brehm, B., Frey, R., Küstler, A., and Eland, J. H. D. (1974). *Int. J. Mass Spectrom. Ion Phys.* **13**, 251.

Brehm, B., Eland, J. H. D., Frey, R., and Schulte, H. (1976). *Int. J. Mass Spectrom. Ion Phys.* **21**, 373.

Brundle, C. R., Robin, M. B., Kuebler, N. A., and Basch, N. A. (1972). *J. Am. Chem. Soc.* **94**, 1451.

Cant, C. S. T., Danby, C. J., and Eland, J. H. D. (1975). *J. Chem. Soc., Faraday Trans. II* **71**, 1015.

Carlson, T. A. (1971). *Chem. Phys. Lett.* **9**, 23.

Carlson, T. A. (1975). "Photoelectron and Auger Spectroscopy." Plenum, New York.

Chesnavich, W. J. and Bowers, M. T. (1977). *J. Am. Chem. Soc.* **99**, 1705.

Chupka, W. A. (1968). *J. Chem. Phys.* **48**, 2337.

Chupka, W. A. (1972). *In* "Ion-Molecule Reactions" (J. L. Franklin, ed.), pp. 33–76. Plenum, New York.

Citrin, P. H., Shaw, R. W., Jr., and Thomas, T. D. (1972). *In* "Electron Spectroscopy" (D. A. Shirley, ed.), pp. 105–120. North-Holland Publ., Amsterdam.

Collin, J. E., Delwiche, J., and Natalis, P. (1972). *In* "Electron Spectroscopy" (D. A. Shirley, ed.), pp. 401–411. North-Holland Publ., Amsterdam.

Dalgarno, A., and Yadav, H. N. (1953). *Proc. Phys. Soc., London, Sect. A* **66**, 173.

Danby, C. J., and Eland, J. H. D. (1972). *Int. J. Mass Spectrom. Ion Phys.* **8**, 153.

Dannacher, J. (1978). *Chem. Phys.* **29**, 339.

Dannacher, J., and Vogt, J. (1978). *Helv. Chim. Acta* **61**, 361.

Dibeler, V. H., and Liston, S. K. (1968). *J. Chem. Phys.* **49**, 482.

Dibeler, V. H., and Walker, J. A. (1968). *Adv. Mass Spectrom.* **4**, 767.

Dillon, J. A., Sheridan, W. F., Edwards, H. D., and Ghosh, S. N. (1955). *J. Chem. Phys.* **23**, 776.

Dunbar, R. C., and Kramer, J. M. (1973). *J. Chem. Phys.* **58**, 1266.
Eland, J. H. D. (1972a). *Int. J. Mass Spectrom. Ion Phys.* **8**, 143.
Eland, J. H. D. (1972b). *Int. J. Mass Spectrom. Ion Phys.* **9**, 397.
Eland, J. H. D. (1973). *Int. J. Mass Spectrom. Ion Phys.* **12**, 389.
Eland, J. H. D. (1974a). *Int. J. Mass Spectrom. Ion Phys.* **13**, 457.
Eland, J. H. D. (1974b). *Adv. Mass Spectrom.* **6**, 917.
Eland, J. H. D. (1975). *Chem. Phys.* **11**, 41.
Eland, J. H. D., and Berkowitz, J. (1977), *J. Chem. Phys.* **67**, 2782.
Eland, J. H. D., and Schulte, H. (1975). *J. Chem. Phys.* **62**, 3835.
Eland, J. H. D., Frey, R., Schulte, H., and Brehm, B. (1976a). *Int. J. Mass Spectrom. Ion Phys.* **21**, 209.
Eland, J. H. D., Frey, R., Kuestler, A., Schulte, H., and Brehm, B. (1976b). *Int. J. Mass Spectrom. Ion Phys.* **22**, 155.
Eland, J. H. D., Devoret, M., and Leach, S. (1976c). *Chem. Phys. Lett.* **43**, 97.
Erman, P., and Larsson, M. (1977). *Phys. Scr.* **15**, 335.
Flannery, M. R., Cosby, P. C., and Moran, T. F. (1973). *J. Chem. Phys.* **59,** 5494.
Frey, R., Gotchev, B., Kalman, O. F., Peatman, W. B., Pollak, H., and Schlag, E. W. (1977). *Chem. Phys.* **21**, 89.
Frey, R., Gotchev, B., Peatman, W. B., Pollak, H., and Schlag, E. W. (1978a). *Int. J. Mass Spectrom. Ion Phys.* **26**, 137.
Frey, R., Gotchev, B., Peatman, W. B., Pollak, H., and Schlag, E. W. (1978b). *Chem. Phys. Lett.* **54,** 411.
Galli, A., Giardini-Guidoni, A., and Volpi, G. G. (1962). *Nuovo Cimento* **26**, 845.
Gellender, M. E., and Baker, A. D. (1975). *Int. J. Mass Spectrom. Ion Phys.* **17**, 1.
Gellender, M. E., and Baker, A. D. (1977). *In* "Electron Spectroscopy" (C. R. Brundle and A. D. Baker, eds.), Vol. 1, pp. 435–455.
George, T. F., Zimmerman, I. H., Yuan, J. M., Laing, J. R., and DeVries, P. L. (1977). *Acc. Chem. Res.* **10**, 449.
Ghosh, S. N., and Sheridan, W. F. (1957a). *J. Chem. Phys.* **27**, 1436.
Ghosh, S. N., and Sheridan, W. F. (1957b). *Indian J. Phys.* **31**, 337.
Golovin, A. V., Sergeev, U. L., Akopyan, M. E., and Vilesov, F. I. (1977). *High Energy Chem. (USSR)* **11**, 325.
Gray, D. R., and Rees, J. A. (1972). *J. Phys. B.* **5**, 1048.
Gurnee, E. F., and Magee, J. L. (1957). *J. Chem. Phys.* **26**, 1237.
Guyon, P. M., Depautex, C., and Morel, G. (1976). *Rev. Sci. Instrum.* **47**, 1347.
Guyon, P. M., Baer, T., Ferreira, L. F. A., Nenner, I., Tabché-Fouhailé, A., Botter, R., and Govers, T. R. (1978a). *J. Phys. B.* **11,** L141.
Guyon, P. M., Nenner, I., Baer, T., Tabché-Fouhailé, A., Govers, T., Ferreira, L. F. A., and Botter, R. (1978b). To be published.
Haney, M. A., and Franklin, J. L. (1968). *J. Chem. Phys.* **48**, 4093.
Herman, Z. (1975). *Abstr., Annu. Conf., Am. Soc. Mass Spectrom., 23rd, Houston* pp. 285–291.
Herman, Z., and Birkinshaw, K. (1973). *Ber. Bunsenges. Phys, Chem.* **77**, 566.
Hierl, P. M., Pacak, V., and Herman, Z. (1977). *J. Chem. Phys.* **67**, 2678.
Hishinuma, N. (1972). *J. Phys. Soc. Jpn.* **32**, 1452.
Hopper, D. G. (1975). *Chem. Phys. Lett.* **31**, 446.
Hopper, D. G. (1978). *J. Am. Chem. Soc.* **100**, 1019.
Jonathan, N., Smith, D. J., and Ross, K. J. (1971). *Chem. Phys. Lett.* **9**, 217.
Kassel, L. S. (1928). *J. Phys. Chem.* **32**, 225.
Kaiser, E. W., Crowe, A., and Falconer, W. E. (1974). *J. Chem. Phys.* **61**, 2720.
Klots, C. E. (1971). *J. Phys. Chem.* **75**, 1526.
Klots, C. E. (1972). *Z. Naturforsch., Teil A* **27**, 553.

Klots, C. E. (1976). *J. Chem. Phys.* **64**, 4269.
Klots, C. E., Mintz, D. M., and Baer, T. (1977). *J. Chem. Phys.* **66**, 5100.
Kobayashi, N. (1975). *J. Phys. Soc. Jpn.* **38**, 519.
Kushnir, R. M., Palyukh, B. M., and Sena, L. A. (1959). *Bull. Acad. Sci. USSR* **23**, 995.
Lindinger, W., Fehsenfeld, F. C., Schmeltekopf, A. L., and Ferguson, E. E. (1974). *J. Geophys. Res*, **79**, 4753.
Lloyd, D. R., and Roberts, P. J. (1973). *Mol. Phys.* **26**, 225.
Lossing, F. P. (1972). *Can. J. Chem.* **50**, 3973.
McCann, K. J., Flannery, M. R., Hornstein, J. V., and Moran, T. F. (1975). *J. Chem. Phys.* **63**, 4998.
McCulloh, K. E., and Dibeler, V. H. (1976). *J. Chem. Phys.* **64**, 4445.
McCulloh, K. E., Sharp, T. E., and Rosenstock, H. M. (1965). *J. Chem. Phys.* **42**, 3501.
McFarland, M., Albritton, D. L., Fehsenfeld, F. C., Ferguson, E. E., and Schmeltekopf, A. L. (1973). *J. Chem. Phys.* **59**, 6620.
Marcus, R. A., and Rice, O. K. (1951). *J. Phys. Colloid Chem.* **55**, 894.
Massey, H. S. W., and Smith, R. A. (1933). *Proc. Soc. London, Ser. A* **142**, 142.
Mintz, D. M., and Baer, T. (1976). *J. Chem. Phys.* **65**, 2407.
Mintz, D. M., and Baer, T. (1977). *Int. J. Mass Spectrom. Ion Phys.* **25**, 39.
Moran, T. F., Flannery, M. R., and Cosby, P. C. (1974). *J. Chem. Phys.* **61**, 1261.
Moran, T. F., Flannery, M. R., and Albritton, D. L. (1975). *J. Chem. Phys.* **62**, 2869.
Moseley, J. T., Cosby, P. C., Ozenne, J. B., and Durup, J. (1979). *J. Chem. Phys.* **70**, 1474.
Murray, P. T., and Baer, T. (1979). *Int. J. Mass Spectrom. Ion Phys.* **29**.
Natalis, P., Delwiche, J., Collin, J. E., Caprace, G., and Praet, M. T. (1977). *Chem. Phys. Lett.* **49**, 177.
Orth, R. G., and Dunbar, R. C. (1977). *J. Chem. Phys.* **66**, 1616.
Parr, A. C., Jason, A. J., and Stockbauer, R. (1978). *Int. J. Mass Spectrom. Ion Phys.* **26**, 23.
Paulson, J. F., Dale, F., and Studniarz, S. A. (1970). *Int. J. Mass Spectrom. Ion Phys.* **5**, 113.
Peatman, W. B., Kasting, G. B., and Wilson, D. J. (1975). *J. Electron Spectrosc. Relat. Phenom.* **7**, 233.
Popescu-Iovitsu, I. (1959). *Rev. Roum. Phys.* **4**, 199.
Potter, R. F. (1954). *J. Chem. Phys.* **22**, 974.
Radom, L., Hariharan, P. C., Pople, J. A., and Schleyer, P. von R. (1976). *J. Am. Chem. Soc.* **98**, 10.
Rapp, D., and Francis, W. E. (1962). *J. Chem. Phys.* **37**, 2631.
Rice, O. K., and Ramsperger, H. C. (1927). *J. Am. Chem. Soc.* **49**, 1617.
Rosenstock, H. B., Wallenstein, M. B., Wahrhaftig, A. L., and Eyring, H. (1952). *Proc. Natl. Acad. Sci. USA.* **38**, 667.
Rosenstock, H. M., Larkins, J. T., and Walker, J. A. (1973). *Int. J. Mass Spectrom. Ion Phys.* **11**, 309.
Rosenstock, H. M., McCulloh, K. E., and Lossing, F. P. (1977a). *Int. J. Mass Spectrom. Ion Phys.* **25**, 327.
Rosenstock, H. M., Draxl, K., Steiner, B. W., and Herron, J. T. (1977b). "Energetics of Gaseous Ions." Natl. Bur. Stand., Washington, D.C.
Schlag, E. W., Frey, R., Gotchev, B., Peatman, W. B., and Pollak, H. (1977). *Chem. Phys. Lett.* **51**, 406.
Sergeev, Y. L., Golovin, A. V., Akopyan, M. E., and Vilesov, F. I. (1977). *High Energy Chem. (USSR)* **11**, 88.
Shirley, D. A., ed. (1972). "Electron Spectroscopy." North-Holland Publ., Amsterdam.
Simm, I. G., Danby, C. J., and Eland, J. H. D. (1973). *Chem. Commun.* p. 832.
Simm, I. G., Danby, C. J., and Eland, J. H. D. (1974). *Int. J. Mass Spectrom. Ion Phys.* **14**, 285.

Smith, D., Baer, T., Willett, G. D., and Ormerod, R. C. (1979) *Int. J. Mass Spectrom. Ion Phys.* **29**.
Smith, G. P. K., and Cross, R. J. (1974). *J. Chem. Phys.* **60**, 2125.
Smith, W. H. (1969). *J. Chem. Phys.* **51**, 3410.
Spohr, R., Guyon, P. M., Chupka, W. A., and Berkowitz, J. (1971). *Rev. Sci. Instrum.* **42**, 1872.
Squires, L., and Baer, T. (1976). *J. Chem. Phys.* **66**, 4001.
Stebbings, R. F., Turner, B. R., and Smith, A. C. H. (1963). *J. Chem. Phys.* **38**, 2277.
Stockbauer, R. (1973). *J. Chem. Phys.* **58**, 3800.
Stockbauer, R. (1977a). *Int. J. Mass Spectrom. Ion Phys.* **25**, 89.
Stockbauer, R. (1977b). *Int. J. Mass Spectrom. Ion Phys.* **25**, 401.
Stockbauer, R., and Inghram, M. G. (1975). *J. Chem. Phys.* **62**, 4862.
Stockbauer, R., and Inghram, M. G. (1976). *J. Chem. Phys.* **65**, 4081.
Stockbauer, R., and Rosenstock, H. M. (1978). *Int. J. Mass Spectrom. Ion Phys.* **27**, 185.
Thomas, T. F., Dale, F., and Paulson, J. F. (1977). *J. Chem. Phys.* **67**, 793.
Tsai, B. P., Baer, T., and Horovitz, M. L. (1974). *Rev. Sci. Instrum.* **45**, 494.
Tsai, B. P., Baer, T., Werner, A. S., and Lin, S. F. (1975a). *J. Phys. Chem.* **79**, 570.
Tsai, B. P., Werner A. S., and Baer, T. (1975b). *J. Chem. Phys.* **63**, 4384.
Turner, D. W., Baker, C., Baker, A. D., and Brundle, C. R. (1970). "Molecular Photoelectron Spectroscopy." Wiley (Interscience), New York.
Varney, R. N. (1970). *Phys. Rev. A* **2**, 370.
Werner, A. S., and Baer, T. (1975). *J. Chem. Phys.* **62**, 2900.
Wiley, W. C., and McLaren, I. H. (1955). *Rev. Sci. Instrum.* **26**, 1156.
Wilkins, C. L., and Gross, M. L. (1970). *J. Am. Chem. Soc.* **93**, 895.
Wu, K. T., and Yencha, A. J. (1977). *Can. J. Phys.* **55**, 767.
Ziegler, B. (1953). *Z. Phys.* **136**, 108.
Zimmerman, I. H., Yuan, J. M., and George, T. F. (1977). *J. Chem. Phys.* **66**, 2638.

Chapter 6

Temperature and pressure effects in the kinetics of ion–molecule reactions

MICHAEL MEOT-NER (MAUTNER)

THE ROCKEFELLER UNIVERSITY
NEW YORK, NEW YORK

ISBN-0-120801-X

I. INTRODUCTION

A. General

This chapter deals with the effects of temperature and pressure on rate constants and mechanisms of ion–molecule (IM) reactions and on the thermal pyrolysis of gaseous ions. The discussion will be restricted to reaction systems in which both the ionic and neutral reactants are in thermal equilibrium, i.e., where a Maxwell–Boltzmann distribution in all degrees of freedom of the reactants may be assumed.

The temperatures and pressures at which ion–molecule reactions occur in nature or technology are extremely diverse. At one extreme, ion–molecule reactions are believed to be prominent in dark interstellar clouds at temperatures of $T < 100$ K and particle densities of 10–10^6 cm^{-3} (Dalgarno, 1975; Watson, 1976). At the other extreme, ion–molecule reactions will occur in flames and electric discharges at atmospheric pressures and thousands of degrees Kelvin (Calcote, 1972). It will be observed in this chapter that the rates and sometimes the mechanisms of most types of gaseous ionic processes are effected, in some cases quite extensively by the ambient temperature and pressure. For example, rate constants of fast ion–polar molecule reactions may increase by nearly an order of magnitude, and the rate constants of three body association reactions of astrochemical interest may increase by over six orders of magnitude as the temperature decreases from 1000 to 50 K (Meot-Ner and Field, 1975a). Evidently the temperature dependence (TD) of IM reactions must be taken into account for realistic modeling of ionizing planetary and astrochemical environments. Indeed, the realization of the potential significance of temperature effects in ionospheric chemistry motivated the first systematic studies of the TD of IM reactions by Ferguson, Fehsenfeld, and co-workers (Dunkin *et al.*, 1968) and by Kebarle and co-workers (Durden *et al.*, 1969).

Significant temperature effects are also observed in chemical ionization mass spectrometry (CIMS) (Field, 1969a). Ion source temperatures in analytical instruments usually range from 20 to 300°C. The relative yields of fragment versus quasi-molecular ions may vary extensively over this range of temperature. The nature of the chemical ionization agent may also be affected: For example, low temperatures will increase the rate of formation of the cluster ions $(H_2O)_nH^+$ and $(NH_3)_nH^+$ in water and ammonia CIMS, respectively; cluster ions will in turn be more moderate proton transfer reagents than the corresponding monomeric ions. Temperature effects may thus have useful analytical applications.

B. Temperature and Pressure Studies as a Probe of the Reaction Mechanism

From the fundamental viewpoint, the temperature and pressure dependences of IM reaction kinetics serve as a probe of the properties of the reaction complexes. In general, the mechanism of an IM reaction may be written as

$$A^+ + B \underset{b}{\overset{a}{\rightleftarrows}} (AB)^{+*} \underset{p}{\rightarrow} C^+ + D$$
$$\underset{s}{\overset{M}{\rightarrow}} AB^+$$
$$\underset{r}{\overset{b}{\rightarrow}} AB^+$$

The outcome of the reaction is determined by the relative values of k_b, k_p, k_s, and k_r. The latter, radiative stabilization, is usually insignificant relative to the other processes under accessible experimental conditions, although it may lead to condensation products in interstellar clouds (Watson, 1976; Herbst, 1976). If k_r is negligible, the overall forward rate constant k_f is given by

$$k_f = k_a[k_p + k_s(M)]/[k_b + k_p + k_s(M)]. \tag{1}$$

Under high pressure conditions $k_s(M) \gg k_b$. Under such conditions, and if only association products are observed, i.e., $k_p = 0$, then $k_f = k_a$ and k_a is obtained directly by measuring k_f.

At intermediate pressures, the variation of k_f with (M) may be used to obtain k_a and k_b, using Eq. (1). Examples of such pressure studies will be discussed later. The important point to note here is that pressure studies may be used to obtain k_a, the rate constant for the formation of a long-lived intermediate; k_a may be written as

$$k_a = p_a k_{c(A^+,B)}, \tag{2}$$

where $k_{c(A^+,B)}$ is the rate constant for capture collisions between A^+ and B. Since $k_{c(A^+,B)}$ can be calculated, measurement of k_a yields p_a, the probability that a capture collision will lead to a long-lived excited intermediate complex AB^{+*}. p_a is of fundamental interest from the viewpoint of collision dynamics. The assumption that $p_a \approx 1$ was made in several studies of association kinetics (Bohme *et al.*, 1969; Meot-Ner and Field, 1975b; Porter, 1975; Olmstead *et al.*, 1977). However, there are indications that in some reactions p_a may be much smaller than unity and display a negative TD (Bohme *et al.*, 1968; Neilson *et al.*, 1978; Meot-Ner, 1978). The probability that a collision will lead to a long-lived intermediate which may be described by statistical theory is one of the more important unresolved questions in IM reaction mechanisms.

The rate constant for the dissociation of the excited reaction complexes k_b is also of fundamental interest. If a stable association product AB^+ is observed, k_b may be obtained from pressure studies, using Eq. (1). If a stable association product is not formed but a transfer reaction occurs, and if p_a is known or assumed to be unity, then k_p/k_b may still be obtained from the experimental value of k_f and the applicable terms in Eq. (1)

$$\frac{k_p}{k_b + k_p} = \frac{k_f}{p_a k_{c(A^+,B)}}. \quad (3)$$

In some reactions a structure may be assigned to the reaction complex with reasonable confidence. Unimolecular rate theory may then be used to calculate k_b and k_p or k_p/k_b as a function of temperature (Olmstead *et al.*, 1977). Experimental measurements of the temperature dependence (TD) of k_p and k_b or of k_b/k_p serve as a test of the theoretical calculations, and thus, indirectly, as a test of the assumed structure of the reaction intermediate.

C. *Experimental Requirements for the Generation of Thermal Reactant Ions*

As was noted above, the discussion of this chapter will be restricted to thermal systems. To satisfy this requirement, the excess energy of the ionic reactant, which is usually generated by electron impact or exothermic IM reactions, must be removed prior to the occurrence of the reactions of interest. Direct spectroscopic evidence on the achievement of thermal ion population is generally not available. Thus, collisional thermalization of excited reactants must be defined in an operational sense as the removal of excess energy from the excited ion to the extent that further collisions of the ion with M will not measurably change the rate constant for the reaction of interest.

To evaluate the conditions necessary to yield thermalized ion populations, vibrational energy transfer rates must be considered. Quantitative information on the rate of collisional deactivation of vibrationally excited ions is available for the deactivation of H_3^{+*} in the sequence

$$H_2^+ + H_2 \rightarrow H_3^{+*} + H \quad (4)$$

$$H_3^{+*} + H_2 \xrightarrow{d} H_3^+ + H_2^*. \quad (5)$$

Kim *et al.* (1974) investigated the deactivation of H_3^{+*}, as determined from the variation of the ratio of dissociative to nondissociative H^+ transfer with increasing P_{H_2} in the reactions of $H_3{}^+$ with CH_4, C_2H_4, C_2H_2, and C_2H_6. A value of $k_d = 2.7 \pm 0.6 \times 10^{-10}$ cm^3 sec^{-1} was measured. It was also observed that k_d was approximately independent of the excess energy (E^*) of H_3^{+*} from $E^* \approx 12$ kcal mole^{-1} to $E^* \approx 59$ kcal mole^{-1}. Since k_d is smaller than $k_{c(H_3{}^+,H_2)}$ by a factor of about four, the results imply that about four collisions with H_2 are necessary for the deactivation of H_3^{+*}. The deactivation of $D_3{}^+$ by H_2 was also studied by Smith and Futrell (1976) who found that about eight collisions were required to decrease the rate constant for the reaction of D_3^{+*} with H_2 to thermal values.

The collisional deactivation of excited ions was reviewed by Harrison (1975). It was observed that H_2 is an inefficient third body and that third-body efficiencies, for example, for the deactivation of $C_5H_9^{+*}$, increase with complexity of M; although third-body efficiencies increase by not more than a factor of five in going from simple third bodies like He and H_2 to third bodies as complex as SF_6 and C_2H_6. As noted above, even highly excited H_3^{+*} is deactivated by the inefficient third-body H_2 in less than ten collisions. It may be reasonable to assume that, in the operational sense defined above, vibrationally excited ions will be thermalized in ten collisions, and that Maxwell–Boltzmann distribution will be well approximated after 100 collisions with M.

The great majority of the temperature studies available to date were performed by flowing afterglow or high pressure mass spectrometry. In both techniques the reactant is an additive of 1% or less in a buffer gas. The reactant ions will therefore collide on the average 100 times or more with buffer gas molecules before colliding with a neutral reactant, and thermalization of the ion will be achieved. In some cases, when reactions with low efficiencies are investigated, the buffer itself is the neutral reactant. In these cases thermal kinetic data may be obtained by observing the reaction after at least 100 collisions of I^+ with M occurred. At a typical value of $(M) = 3 \times 10^{16}$ cm^{-3}, this condition will be satisfied after 30 μsec. Indeed, reactions are usually observed to reaction times significantly longer than 30 μsec in high pressure measurements. In ICR measurements, the low pressures may be compensated for by proportionally longer residence times, and ion energies may be very close to thermal (Meisels, 1975). The data quoted in this chapter thus represents thermal rate constants unless otherwise indicated.

Nonthermal energies of excited reactants will be reflected in the observed rate constants or product distribution. However, the variation of rate constants with the excitation of selected degrees of freedom is in general different from variation of rate constants with increasing temperature. For example, the effects of increasing the electric field in a flow drift afterglow apparatus on the reaction $N_2OH^+ + CO \rightarrow COH^+ + N_2O$ were found to be different from the kinetic effects of an equivalent variation of temperature (Lindinger *et al.*, 1975a). Nonthermal energy effects supply information complementary to temperature effects. The subject has been reviewed by Chupka (1975).

Temperature effects in IM *equilibria* have led to an extensive and rapidly growing body of thermochemical data on heats of formation of ions, intrinsic acid–base properties, ion solvent interaction, and ion clustering. Thermochemical data from IM equilibria was reviewed recently by Kebarle (1977). Thermochemical data will be considered in this chapter only in relation to kinetic phenomena.

The precision of the rate constants measured for IM reaction is usually about ±15%. Amongst the factors that contribute to errors in kinetic measurements are errors in pressure determination, errors in the composition of the reaction mixture, side reactions which are not taken into account, and scattering of points in plots of ion intensity versus reaction time or versus pressure. The nature and magnitudes of errors depend on the technique used, and the reader is referred to the original sources for their evaluation. Reported error limits significantly greater than ±15% will be listed in the tables. Errors in the rate constants listed in the tables in this chapter also result from the fact that in most instances the data were read from graphic representations of temperature dependence of rate constants. Errors of about ±15% in k and ±10 K in T may have resulted from extracting data from published plots for tabulation here.

II. TEMPERATURE EFFECTS IN BIMOLECULAR EXCHANGE REACTIONS

For the present purposes, bimolecular exchange reactions will be divided into two groups: fast reactions which proceed near unit collision efficiency and slow processes with collision efficiencies substantially below unity. This division is somewhat arbitrary, in that a given reaction may belong to either of the two groups depending on the temperature itself, as will be shown below. As a practical rule, reactions will be assigned to one of these categories according to the rate constants at 300 K.

A. *Fast Reactions*

1. *Reactions Involving Apolar Neutral Reactants*

By definition, the rate constants of fast IM reactions, and therefore the TDs thereof, are determined by k_c, the rate constant of the capture collisions between the reactants. For reactions of apolar neutral reactants, k_c is determined by the well-known Langevin rate constant (Giomousis and Stevenson, 1958)

$$k_c = 2\pi q(\propto/\mu)^{1/2}. \tag{6}$$

This expression is derived from a pure polarization theory which predicts that the capture cross section k_c is inversely proportional to the relative velocity (v) of the collision partners. Consequently, $k_c = v\sigma_c$ is independent of v. The result is that k_c is independent of temperature for reactions which proceed at unit collision efficiency, and in which the neutral reactant has no permanent dipole moment. This prediction was verified in several fast exothermic exchange reactions. Meot-Ner and Field (1975c) found that for the fast proton transfer reaction

$$N_2H^+ + CH_4 \rightarrow CH_5^+ + N_2,$$

$k = 1.3 \pm 0.2 \times 10^{-9}$ cm^3 sec^{-1}, without a measurable TD between 130 and 640 K. This observation compares well with Eq. (6), which predicts $k_c = 1.1 \times 10^{-9}$ cm^3 sec^{-1} through this temperature range. Pierce and Porter (1974) observed that the rate constant for fast D^+ transfer from D_3^+ to CH_4 was 24×10^{-10} cm^3 sec^{-1} at 80 K, similar to values reported earlier at 300 K. H^+ transfer from N_2H^+ to CO_2 is also independent of T between 300 and 700 K (Fehsenfeld *et al.*, 1976).

Fast processes with rate constants independent of temperature were also observed in dissociative charge transfer reactions (CT). Po and Porter (1977) measured the rate constant for $D_3^+ + Mg \rightarrow Mg^+ + D_2 + D$ as $16 \pm 4 \times 10^{-10}$ cm^3 sec^{-1}, without significant TD between 324 and 374°C. Dunkin *et al.* (1968) found that the rate constant for $He^+ + N_2 \rightarrow$ products was 12×10^{-10} cm^3 sec^{-1} between 300 and 600 K; indeed, the rate constant did not change when the vibrational temperature of N_2 was increased up to 6000 K, although the fraction of dissociative CT yielding N^+, compared with nondissociative CT yielding N_2^+ increased with $T_v(N_2)$ (Schmeltekopf *et al.*, 1968). Several other fast CT reactions with slight TD or without experimentally significant TD, observed by flowing afterglow techniques by Ferguson's group, are shown in Table I.

Fast rate constants associated with temperature independent behavior were also observed by Meot-Ner and Field (1976a) in H^- transfer reactions of $C_2H_5^+$ with 2-methyl butane between 320 and 700 K and with 2,4-dimethylbutane between 320 and 620 K. The reaction of $C_2H_5^+$ with propane and $C_3H_7^+$ with isobutane also exhibited fast kinetics without TD between 320 and 480 K and 130 and 210 K, respectively. However, the collision

TABLE I

Temperature Dependence of Ion–Molecule Exchange and Dissociative Exchange Reactions

Reaction	Temp. (K)	k ($cm^3\ sec^{-1}$)	Temp. coef. (n in $k = CT^n$)	Comments	Reference
$He^+ + N_2 \rightarrow N^+ + N + He$ $\rightarrow N_2^+ + He$	300–6000	1.5 (−9)	0	Only N_2 vibrational temperature was varied. The fractional yield of N^+, ($N^+/N^+ + N_2^+$), increases from 0.68 to 0.82 between $T_v(N_2) = 300$–6000 K	Schmeltekopf *et al.* (1968)
$He^+ + N_2 \rightarrow$ Products	90 380	1.7 (−9) 1.5 (−9)	0		Bowers *et al.* (1977)
$He^+ + N_2 \rightarrow N^+ + N + He$ $\rightarrow N_2^+ + He$	300–600	1.2 (−9)	0	k is independent of T	Dunkin *et al.* (1968)
$He^+ + N_2 \rightarrow He + N^+ + N$	195 503	1.8 (−9) 1.0 (−9)	0.6		Sayers and Smith (1964)
$He^+ + NH_3 \rightarrow$ Products	300 770	1.6 (−9) 1.2 (−9)	Negative TD	k is smaller than ADO rate constant, but TD is in good agreement with ADO predictions	Lindinger *et al.* (1975b)
$Ar^+ + H_2 \rightarrow ArH^+ + H$	333–614			Reaction cross section is constant at $259 \pm 13 \times 10^{-16}\ cm^2$	Stevenson and Schissler (1958)

$Ar^+ + H_2 \rightarrow$ Products	80 300 550	5.5 (−10) 9.5 (−10) 1.3 (−9)	0.5–1.0	At $T > 400$ K, k increases with decreasing pressure between $0.4 < P < 2.0$ torr	Adams *et al.* (1970)
$Ar^+ + O_2 \rightarrow O_2^+ + Ar$	82 280 400 500	9 ± 3 (−11) 4.5 (−11) 3.5 (−11) 3.0 (−11)	−0.5	$k = 7.8 \times 10^{-10} T^{-0.5}$ $cm^3\ sec^{-1}$	Adams *et al.* (1970)
$Kr^+ + H_2 \rightarrow KrH^+ + H$	380–616			Reaction cross section is constant $123 \pm 2.6 \times 10^{-16}\ cm^2$	Stevenson and Schissler (1958)
$Kr^+ + H_2 \rightarrow KrH^+ + H$	90–380	1.8 (−10)	0	k is constant for H_2, D_2 but decreases somewhat with T for HD	Kemper *et al.* (1978)
$N^+ + O_2 \rightarrow O_2^+ + N$ $NO^+ + O$	300–600	6 (−10)	0	k is independent of T	Dunkin *et al.* (1968)
$O^+ + O_2 \rightarrow O_2^+ + O$	185–576		0.48 ± 0.05	$k = (3.4 \pm 0.5) \times 10^{-15} T^{(-0.48 \pm 0.05)}$ $cm^3\ sec^{-1}$	Smith and Fouracre (1968)
$O^+ + O_2 \rightarrow O_2^+ + O$	80 300 600	3.5 (−11) 2.0 (−11) 1.5 (−11)	−0.4		Ferguson *et al.* (1969)

TABLE I (*Continued*)

Reaction	Temp. (K)	k ($cm^3\ sec^{-1}$)	Temp. coef. (n in $k = CT^n$)	Comments	Reference
$O^+ + N_2 \rightarrow NO^+ + N$	300	1.3 (−12)	Positive TD	Only N_2 vibrational temperature was varied	Schmeltekopf *et al.* (1968)
	1000	1.3 (−12)			
	3000	9.0 (−12)			
	6000	5.0 (−11)			
$O^+ + N_2 \rightarrow NO^+ + N$	80	2 (−12)	n varies with T	k decreases between 80 and 500 K, and is constant from 500 to 600 K. k is proportional to third body (He) pressure at 82 K, $0.2 < P_{He} <$ 1.1 torr, but is independent of P_{He} at higher T	Ferguson *et al.* (1969); Lindinger *et al.* (1974)
	300	1 (−12)			
	500	7 (−13)			
	600	7 (−13)			
	900	8 (−13)			
$O^+ + CO_2 \rightarrow O_2{}^+ + CO$	300	1.1 (−9)	≈ 0	Slight negative TD	Dunkin *et al.* (1968)
	600	8. 5(−10)			
$N_2{}^+ + O_2 \rightarrow O^+{}_2 + N_2$	80	9 (−10)	≈ -0.5		Ferguson *et al.* (1969); Lindinger *et al.* (1974)
	300	5 (−10)	n varies slightly with T		
	550–900	2.5 (−10)			
$N_2{}^+ + O_2 \rightarrow O_2{}^+ + N_2$	300	4.7 (−11)	Slightly negative TD		Dunkin *et al.* (1968)
	520	3.0 (−11)			
$D_2{}^+ + D_2 \rightarrow D_3{}^+ + D$	380–615			Reaction cross section is constant $79 \pm 7 \times 10^{-16}\ cm^2$	Stevenson and Schissler (1958)

$ArH^+ + H_2 \rightarrow$ Products	80–550	5 ± 1 (−10)	0		Adams *et al.* (1970)
$CO_2^+ + O_2 \rightarrow O_2^+ + CO_2$	150 270 740 900	5 (−11) 4 (−11) 1.6 (−11) 2.5 (−11)	−0.8 (150 − 740 K)	TD is negative from 150 to 740 K, then k increases up to 900 K	Lindinger *et al.* (1975a)
$COH^+ + N_2O \rightarrow N_2OH^+ + CO$	300 620 870	2.6 (−12) 7.0 (−11) 1.4 (−10)	Positive TD	This is an endothermic reaction with $k = 1.58 \times 10^{-9} \exp(-3830/RT)$ $cm^3\ sec^{-1}$	Lindinger *et al.* (1975a)
$N_2H^+ + CO_2 \rightarrow CO_2H^+ + N_2$	298 700	9.8 ± 2.0 (−10) 8.2 ± 2.5 (−10)	≈0		Fehsenfeld *et al.* (1976)
$D_3^+ + Mg \rightarrow Mg^+ + D_2 + D$	597–647	1.4 (−9)	0		Po and Porter (1977)
$N_2H^+ + NH_3 \rightarrow NH_4^+ + N_2$	320 640	2.0 (−9) 1.6 (−9)	Slightly negative TD	The TD of this fast ion–polar molecule reaction agrees well with ADO theory prediction	Meot-Ner and Field (1975c)
$N_2H^+ + CH_4 \rightarrow CH_5^+ + N_2$	130–650	1.3 (−9)	0		Meot-Ner and Field (1975c)
$N_2H^+ + CH_3CHO \rightarrow CH_3CHOH^+ + N_2$	270 410 640	6.0 (−9) 5.5 (−9) 4.2 (−9)	Slightly negative TD	TD agrees with collisional (ADO) theory	Meot-Ner and Field (1975c)

TABLE I (*Continued*)

Reaction	Temp. (K)	k ($cm^3\ sec^{-1}$)	Temp. coef. (n in $k = CT^n$)	Comments	Reference
$N_2OH^+ + CO \rightarrow COH^+ + N_2O$	277–505	5 (−10)	0	k is independent of T	Hemsworth *et al.* (1973)
$N_2OH^+ + CO \rightarrow COH^+ + N_2O$	300 620 870	5.0 (−10) 3.2 (−10) 2.8 (−10)	−0.54	TD is given by authors as $k = 1.74 \times 10^{-10} \exp(728/RT)$ $cm^3\ sec^{-1}$	Lindinger *et al.* (1975a)
$NH_3^+ + H_2 \rightarrow NH_4^+ + H$	300 800	4 (−13) 4 (−12)	Positive TD	This reaction involves a positive activation energy: $k(T) = 2.7 \times 10^{-11} \exp(-2.1/RT)$ $cm^3\ sec^{-1}$	Fehsenfeld *et al.* (1975)
$CH_4^+ + CH_4 \rightarrow CH_5^+ + CH_3$	90 400	9 (−10) 12 (−10)		Small positive TD corresponds to activation energy of 80 cal $mole^{-1}$	Bowers *et al.* (1977)
$CH_4D^+ + CH_4 \rightarrow CH_5^+ + CH_3D$	90 110 150 167	3.8 (−9) 4.1 (−9) 4.2 (−9) 3.5 (−9)		No significant TD between 90 and 150 K. Slight decrease from 150 to 170 K (Pierce and Porter); much lower rates are reported about 300 K by other workers (next two references)	Pierce and Porter (1974)

$CH_4D^+ + CH_4 \rightarrow CH_5^+ + CH_3D$	195 298	$\approx 1\,(-12)$ $\approx 1\,(-12)$			Lawrence and Firestone (1965)
$CH_4D^+ + CH_4 \rightarrow CH_5^+ + CH_3D$	300	3 (−11)			Abramson and Futtrell (1966)
$s\text{-}C_3H_7^+ + NH_3 \rightarrow NH_4^+ + C_3H_6$	220 400 550	2.2 (−9) 1.7 (−9) 1.5 (−9)	Slightly negative TD	TD agrees with collisional (ADO) theory	Meot-Ner (1979a)
$t\text{-}C_4H_9^+ + (C_2H_5)_3N \rightarrow (C_2H_5)_3NH^+ + C_4H_8$	310–640	1.6 (−9)	≈ 0	TD agrees with collisional (ADO) theory	Meot-Ner and Field (1975c)
$t\text{-}C_4H_9^+ + C_6H_5CH_2OC(O)CH_3$ $\rightarrow C_6H_5CH_2^+$ + Neutrals $\rightarrow C_6H_5CH_2OC(O)CH_3{\cdot}H^+$ + Neutrals	323 440 520	2.4 (−9) 2.8 (−10) 6.1 (−10)	n varies from −6 at 320 K to −10 at 520 K	Branching ratio a function of T, with $C_6H_5OC(O)CH_3{\cdot}H^+$ major product at 320 K, and $C_6H_5CH_2^+$ fragment major product at $T > 400$ K	Meot-Ner *et al.* (1977a)
$NO^- + He \rightarrow NO + e^- + He$	193 506	6.1 (−14) 1.0 (−12)	Positive TD	$k(T) = 5.23 \times 10^{-12} \exp(1.7/RT)$	McFarland *et al.* (1972)
$NO^- + He \rightarrow NO + e^- + H_2$	193 506	4.2 (−14) 9.4 (−13)	Positive TD	$k(T) = 6.29 \times 10^{-12} \exp(1.9/RT)$	McFarland *et al.* (1972)

TABLE I (*Continued*)

Reaction	Temp. (K)	k ($cm^3\ sec^{-1}$)	Temp. coef. (n in $k = CT^n$)	Comments	Reference
$NO^- + NO \rightarrow NO + e^- + NO$	285 506	5.0 (−12) 3.0 (−11)	Positive TD	$k(T) = 3.28 \times 10^{-10} \exp(2.4/RT)$ Arrhenius parameters for collisional detachment of NO^- with NH_3, N_2O, CO_2, CO, and Ne are also given in this ref	McFarland *et al.* (1972)
$H_3^+ + e^- \rightarrow$ Products	205 300 450	2.9 (−7) 2.3 (−7) 2.0 (−7)	−0.5	Dissociative electron recombination. TD is due to variation of electron temperature	Leu *et al.* (1973)

TABLE II

Temperature Dependences of Some Nonreactive Momentum Transfer Collision Rate Constants

Reactants	Temp. (°K)	k ($cm^3\ sec^{-1}$)	Temp. coef.	Reference
$CH_2F^+ + CH_3F$	200	1.8 (−9)	$\approx T^{-0.5}$	Buttrill (1973)
	400	1.2 (−9)		
$H_3S^+ + H_2S$	200	1.1 (−9)	Small negative TD	Buttrill (1973)
	430	9.2 (−10)		
$CH_5^+ + CH_4$	150	9.7 (−10)	Small negative TD	Buttrill (1973)
	430	6.5 (−10)		
$N_2^+ + N_2$	180	7.4 (−10)	0	Buttrill (1973)
	410	7.4 (−10)		
$(CH_3)_3NH^+ + (CH_3)_3N$	200	1.4 (−9)	Small negative TD in agreement	Neilson *et al.* (1978)
	370	9.0 (−10)	with ADO theory	
$((CH_3)_3N)_2H^+ + (CH_3)_3N$	200	6.8 (−10)	Small negative TD in agreement	Neilson *et al.* (1978)
	370	5.2 (−10)	with ADO theory	
$N_2^+ + N_2$	80	5.9 (−10)	No TD at $150 < T < 380$ K;	Bowers *et al.* (1977)
	220	7.1 (−10)	small positive TD at $T < 150$ K	
	380	7.2 (−10)		
$Ne^+ + Ne$	90	3.5 (−10)	Small positive TD	Bowers *et al.* (1977)
	380	4.2 (−10)		
$N_2^+ + Ne$	80	2.0 (−10)	Small negative TD	Bowers *et al* (1977)
	380	1.5 (−10)		
$CH_5^+ + CH_4$	80	8.2 (−10)	≈ 0	Bowers *et al.* (1977)
	380	7.9 (−10)		
$CF_3^+ + CF_4$	80	5.9 (−10)	Small negative TD	Bowers *et al.* (1977)
	380	4.5 (−10)		

efficiencies of the latter reactions decreased below unity and assumed negative TD at higher temperatures.

Rate constants of exchange reactions provide only an indirect probe of the rate constant for collisions between the reactants, since the reaction rate constants are obviously affected by reaction probabilities. Direct information on the TD of ion–molecule collision rate constants is obtained by measurements of momentum transfer rate constants. Such data are derived from ICR resonance linewidth (Table II). Buttrill (1973) and Bowers *et al.* (1977) used variable temperature ICR apparatus to examine the TD of momentum transfer rate constants in nonreactive collisions of ions with nonpolar molecules. Both workers found that k_c for the $N_2{}^+/N_2$ system was higher than $k_{LANGEVIN}$, which may be attributed to symmetric charge transfer reactions. While Buttrill found no TD, Bowers *et al.* found a small positive TD, particularly below 150 K (Fig. 1), indicating that the cross section for symmetric charge transfer decreases only slowly with increasing velocity of the reactants. Bowers *et al.* (1977) also found a small positive TD for the He^+/He symmetric charge transfer system, but a small negative TD for the completely unreactive $N_2{}^+/Ne$, $CF_3{}^+/CF_4$, and $CH_5{}^+/CH_4$ systems; in the latter systems the Langevin rate constant was 20–60% lower than the experimental rate constant. Following Ridge and Beauchamp (1976), Bowers *et al.* suggested that the latter observations could be accounted for by using the 12-6-4 potential to describe momentum transfer in these systems, and that the 12-6-4 potential, unlike the pure polarization potential, is temperature dependent.

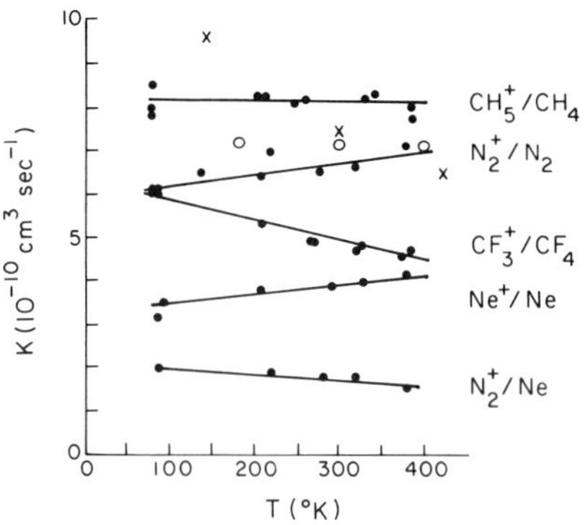

FIG. 1 Momentum transfer rate constants for nonreactive collisions of the ion–molecule pairs indicated. Replotted from the data of Bowers *et al.* (1977). ×, points for the $CH_5{}^+/CH_4$ system and ○, points for the $N_2{}^+/N_2$ system measured by Buttrill (1973). (Reprinted with permission from Elsevier, Amsterdam.)

2. *Reactions Involving Polar Neutral Reactants*

A theoretical equation for the rate constant of capture collisions between ions and polar molecules was given by Bowers and Laudenslager (1972):

$$k_c = \frac{2\pi q}{\mu^{1/2}}\left[\alpha^{1/2} + c\mu_D\left(\frac{2}{\pi kT}\right)^{1/2}\right], \tag{7}$$

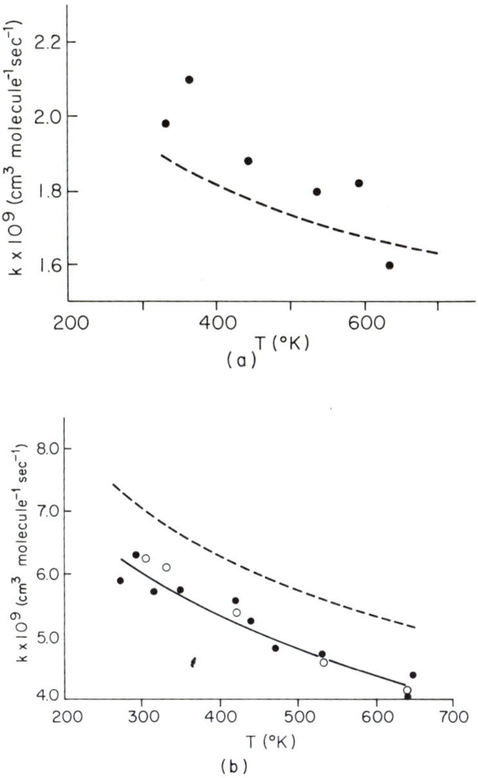

FIG. 2 (a) The temperature dependence of the rate constant of the reaction $N_2H^+ + NH_3 \rightarrow NH_4 + N_2$ between 320 and 640 K. The broken line represents the theoretical rate constants calculated from ADO theory with the dipole locking constant $c = 0.21 = c_{ADO}^{300}$. (b) The temperature dependence of the rate constant for the reaction $N_2H^+ + CH_3CHO \rightarrow CH_3CHOH^+ + N_2$ between 270 and 640 K. Solid circles represent experimental values and the solid line shows the experimental temperature dependence. The broken line represents the theoretical (locked dipole) rate constants. Open circles are experimental rate constants for the reaction $N_2D^+ + CH_3CHO \rightarrow CH_3CHOD^+ + N_2$. [From Meot-Ner and Field (1975c). Reprinted with permission from *J. Am. Chem. Soc.* Copyright by the American Chemical Society.]

where μ is the reduced mass, α and μ_D the polarizability and dipole moment of the neutral reactant, and the other symbols have their usual meaning; c is a "locking constant" which depends on the degree that the dipole is aligned with the ionic charge during the collision. In the "locked-dipole" model (Moran and Hamill, 1963; Gupta *et al.*, 1967), $c = 1$. The second term in the parentheses in Eq. (7) introduces a slight negative TD of $T^{-1/2}$. The weight of this term in k_c will be dependent on the relative contributions of the permanent and induced dipoles to k_c. According to the more advanced average dipole orientation theory (Su and Bowers, 1973, 1975), c is determined by the average locking of the dipole in the electric field of the ion. The degree of locking decreases as the rotational energy of the reactant molecule increases. Thus c decreases with T, and the TD of the second term in Eq. (7) will be more negative than $T^{-1/2}$. However, since $0 < c < 1$, the contribution of the second term to k_c will be smaller than in the locked dipole model, and the overall TD predicted by the two models remains comparable. The measurement of the absolute values of the rate constants is thus a more sensitive comparative test of the two models than the measurement of TDs.

The predictions of collision theory concerning the TD of collision rates may be tested by measurements on fast exchange reactions. Exothermic H^+ transfer reactions should be especially suitable, since Bohme (1975) showed that the majority of exothermic H^+ transfer reactions proceeds at unit collision efficiency at 300 K. Thus, the TD of fast H^+ transfer reactions should reflect the TD of the collision rates. The TDs of H^+ transfer from N_2H^+ to NH_3 and to CH_3CHO (Fig. 2) were measured by Meot-Ner and Field (1975c). The TD of H^+ transfer from s-$C_3H_7^+$ to NH_3 were measured by Meot-Ner (1979a) (Fig. 3). In the case of the reactions of NH_3, both the

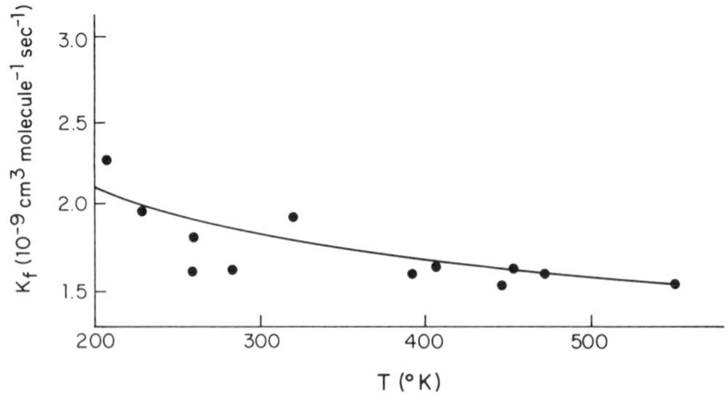

FIG. 3 The temperature dependence of the rate constant for s-$C_3H_7^+ + NH_3 \rightarrow NH_4^+ + C_3H_6$ between 200 and 550 K. Solid line represents rate constant calculated from ADO theory. [From Meot-Ner (1979a).]

measured rate constants and their TDs are in good agreement with ADO predictions. In the case of the reaction of N_2H^+ with CH_3CHO, the measured rate constant is higher than ADO predictions, and the TD is consistent with the locked-dipole prediction. However, as noted above, the TDs predicted by ADO and the locked-dipole model are similar and the experimental TD would fit both predictions within experimental accuracy. Meot-Ner and Field (1975c) also showed that in H^+ transfer from $t\text{-}C_4H_9{}^+$ to $(C_2H_5)_3N$, a molecule with small μ_D and large α, the TD was too small to be experimentally observable, also in agreement with ADO predictions.

Concerning the results plotted in Figs. 2 and 3, it should be noted that the TDs correspond to a change of 20–30% in k in the experimentally accessible temperature range. This change is of the same order of magnitude as the precision of the measurements. Thus, although the agreement with collision theory predictions is encouraging, further comparison of predicted and experimental TDs would be necessary to verify the predictions of collision theory with confidence.

The temperature dependences of momentum transfer rate constants for collisions of ions with polar molecules were measured by Buttrill (1973) and by Neilson *et al.* (1978). Buttrill found that for nonreactive collisions of H_3S^+ with H_2S and of CH_2F^+ with CH_3F the measured rate constants exceeded the predictions of the locked dipole model (and thus also the ADO model). In both cases a negative TD was observed in qualitative but not quantitative agreement with collision theory. Neilson *et al.* (1978) measured momentum transfer rate constants for collisions of $(CH_3)_3NH^+$ and of $((CH_3)_3N)_2H^+$ with $(CH_3)_3N$ between 170 and 370 K. A negative TD of about the magnitude predicted by collision theory was observed.

In summary, data on the TD of ion–polar molecule collision rate constants are scarce. The existing data is generally in qualitative agreement with the predictions of collision theory, in that negative TDs are observed. The existing data are not sufficient to judge the degree of quantitative agreement between experimental and theoretical TDs.

B. *Slow Bimolecular Exchange Reactions*

The available experimental data concerning the TD of slow bimolecular exchange reactions may be divided to two sets. One is a set of data concerning reactions of atmospheric interest involving mono-, di-, and triatomic reactants. The TD of these reactions was investigated mainly by Fehsenfeld and Ferguson and their associates, using the flowing afterglow technique. A second set of data concerning the TD of H^- transfer reactions involving saturated hydrocarbons and carbonium ions were obtained by Meot-Ner and Field using pulsed high pressure MS.

The results of the temperature studies reveal the mechanisms leading to the low collision efficiencies of slow bimolecular reactions. Specifically, the low collision efficiency of an exchange reaction may result from an energy barrier which is higher than the energy of the separated reactants. If such a barrier is present, the reaction will require thermal activation which will result in a positive TD. Alternatively, reactions that do not involve positive activation energies may proceed with low efficiency as a result of competition between the back dissociation of the ion–molecule complex into reactants or dissociation to products as described in Scheme I. As will be shown below, the qualitative treatments applied to slow bimolecular reactions to date predict that dissociation into exothermic channels leading to products will become increasingly favorable as the temperature decreases, resulting in a negative TD. At the lowest order of theoretical sophistication, determination of the sign of the TD of slow reactions distinguishes reactions with positive activation energies from reactions whose slow rates result from other factors. The data in Table I may be inspected in this perspective.

Table I contains a list of seven charge transfer (CT) reactions. From the viewpoint of rate constants and TDs, these CT reactions divide as follows: four slow reactions ($k_{300} < 5 \times 10^{-10}$ cm^3 sec^{-1}) with negative TD and three fast reactions ($k_{300} > 5 \times 10^{-10}$ cm sec^{-1}) without TD or with slightly negative TDs compatible with collision theory. It is evident that the following generalization may be applied for this set of reactions: in the absence of an activation energy slow reaction exhibiting negative TD between 80 and 600 K, fast reactions exhibit no TD. The division between fast reactions with zero or slightly positive TD and slow reactions with negative TD is illustrated by plots of the rate constants of several ionospheric reactions between 80 and 900 K (Fig. 4). No exothermic charge transfer reaction examined to date showed a positive activation energy.

Negative TD was also observed in slow reactions involving the exchange of massive particles. Three slightly exothermic proton transfer reactions listed in Table I behave in this manner. The reaction of N_2OH^+ with CO exhibits a small negative TD, according to a recent study by Lindinger *et al.* 1975a). although a previous study by Hemsworth *et al.* (1973) found no observable TD. For this reaction $\Delta H^\circ = 3.4$ kcal mole^{-1} (Bohme, 1975). Proton transfer from t-$C_4H_9{}^+$ to benzyl acetate displays the largest negative temperature coefficient observed for a bimolecular reaction so far. ($k \propto T^{-6}$–T^{-10}). Since the proton affinity of i-C_4H_8 is 194 kcal mole^{-1}, and the proton affinities of esters are 192–199 kcal mole^{-1} (Kebarle, 1977), the exothermicity of this reaction is probably not more than 5 kcal mole^{-1}. The temperature coefficient of H^+ transfer from CH_4D^+ to CH_4, at the temperature range where such negative TD exists, i.e., at $167 < T < 300$ K,

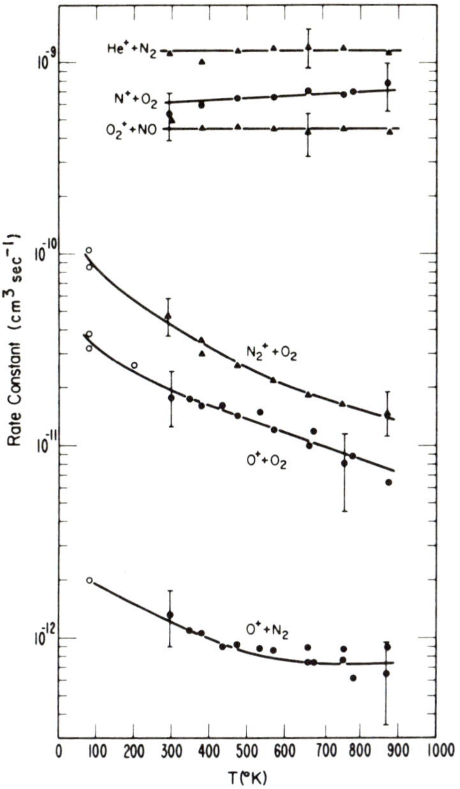

FIG. 4 Rate constants for several ionospheric reactions from 300 K (in some cases from 80 K) to 900 K. [From Lindinger *et al.* (1974). Reprinted with permission from the American Geophysical Union.]

is calculated from the data of Pierce and Porter (1974) and of Abramson and Futrell (1966) as $k = CT^{-8}$. Obviously, this is a very nearly thermoneutral reaction. Thus all three H^+ transfer reactions for which negative TD was observed were exothermic by not more than 5 kcal mole^{-1}. In contrast, all H^+ transfer reactions with exothermicities greater than 10 kcal mole^{-1} exhibited fast kinetics and TD as determined by collision dynamics. These observations correlate well with the observations by Bohme (1975) that only H^+ transfer reactions with exothermicities of 5 kcal mole^{-1} or less proceed with collision efficiencies significantly less than unity.

The temperature measurements show that negative TD is associated with low efficiency and low exothermicity; a more extensive and quantitative relationship between these parameters will be shown for H^- transfer reactions.

Hydrogen atom transfer reactions are the only class of exchange reaction for which positive TD was observed to date. Examples are $Ar^+(H_2, H)ArH^+$ and $NH_3^+(H_2, H)NH_4^+$. The cross sections of H abstraction from H_2 by Kr^+ and of the reaction $D_2^+(D_2, D)D_3^+$ were found to be independent of temperature (Table I), indicating that the rate constants of these reactions may possess a positive TD. However, recent experiments by Kemper *et al.* (1978) indicate very little TD in the reaction rate of Kr^+ with H_2. The reaction $CH_4^+(CH_4, CH_3)CH_5^+$ also showed a slight positive TD, corresponding to a small activation energy of 80 cal mole^{-1}. (Note that D^+ or H^+ transfer may also contribute to the latter two reactions.) It will be of interest whether the requirement for an energy threshold, leading to positive TD, will prove a general feature of H abstraction reactions.

C. *Temperature Dependence of H^- Transfer Reactions. Correlations between Rate Constants, Temperature Coefficients, and Exothermicity*

In radiation chemistry experiments Ausloos and Lias (1970) observed that the hydride transfer reaction $t\text{-}C_4H_9^+(i\text{-}C_5H_{12}, i\text{-}C_4H_{10})t\text{-}C_5H_{11}^+$ is a slow reaction at 298 K, with $k_{298} = 2.5 \times 10^{-11}$ cm^3 sec^{-1}. Following the suggestion of Ausloos and Lias that the slow rate may be due to an activation energy, Solomon *et al.* (1974) measured the TD of k, but found a negative TD of $k = CT^{-3}$ rather than a positive activation energy. Subsequently, Meot-Ner and Field (1976a) measured the TD of a number of reactions of $C_2H_5^+$, $s\text{-}C_3H_7^+$, and $t\text{-}C_4H_9^+$ with C_3-C_8 hydrocarbons (Table III). Most of the reactions exhibited slow rates and negative TD and temperature coefficients of $n = -1.5$ to -5.5 were observed.[1] However, in the reactions of $s\text{-}C_3H_7^+$ with $i\text{-}C_4H_{10}$ and with several other higher tertiary hydrocarbons, it was observed that at sufficiently low temperatures the rate constants reached the collision limit. Since below these transition temperatures the rate was collisionally limited and since IM collision rates are independent of temperature, these reactions assumed fast kinetics without TD at the low temperature range (Fig. 5). Similar behavior was observed in the reaction of $C_2H_5^+$ with $i\text{-}C_4H_{10}$, with a transition temperature at 480 K. It is interesting to note that a similar transition from slow kinetics with negative TD at higher temperatures to fast kinetics at low temperatures

[1] Negative TD in the $s\text{-}C_3H_7^+ + i\text{-}C_4H_{10}$ system was confirmed recently in low pressure ICR experiments (Lias, 1978).

TABLE III

Temperature Dependences of Hydride Transfer Reactions $R_1^+ + R_2H \rightarrow R_2^+ + R_1H$

R_1^+	R_2	Temp. (K)	k ($cm^3\ sec^{-1}$)	Temp. Coef. (n in $k = CT^n$)	$-\Delta H^\circ$ ($kcal\ mole^{-1}$)	Comments	Reference
(1) $C_2H_5^+$		320–480 580 680	1.2 (−9) 9.0 (−10) 7.2 (−10)	0 −1	40	Collisionally limited fast kinetics without TD between 320 and 480 K; slow kinetics with negative TD between 480 and 680 K	Meot-Ner and Field (1976a)
(2) $C_2H_5^+$		320–700	2.7 (−9)	0	42		Meot-Ner and Field (1976a)
(3) $C_2H_5^+$		320–620	3.0 (−9)	0	45		Meot-Ner and Field (1976a)
(4) 		130–210 300 600	1.1 (−9) 5.8 (−10) 1.4 (−10)	0 −2	18	Fast kinetics without TD below 210 K; slow kinetics with negative TD between 210 and 600 K	Meot-Ner and Field (1976a)
(5) 		300 600	8.6 (−10) 2.3 (−10)	−2	21		Meot-Ner and Field (1976a)
(6) 		300–400 450 600	1.0 (−9) 8.2 (−10) 3.9 (−10)	0 −2.5	23	Fast kinetics without TD below 300 K; slow kinetics with negative TD between 400 and 600 K	Meot-Ner and Field (1976a)
(7) 		280–380 500 600	1.3 (−9) 6.7 (−10) 4.5 (−10)	0 −2.5	22	Fast kinetics without TD below 380 K; slow kinetics with negative TD between 380 and 600 K	Meot-Ner and Field (1976a)

TABLE III (*Continued*)

R_1^+	R_2	Temp. (K)	k ($cm^3\ sec^{-1}$)	Temp. Coef. (n in $k = CT^n$)	$-\Delta H^\circ$ ($kcal\ mole^{-1}$)	Comments	Reference
(8) 		300–400 500 600	1.3 (−9) 1.0 (−9) 6.0 (−10)		0	Fast kinetics without TD below 400 K; slow kinetics with negative TD between 400 and 600 K	Meot-Ner and Field (1976a)
(9) 		190 570	5.2 (−11) 2.3 (−12)	−3.2	3.4		Solomon *et al.* (1974)
(10) 		320 410	5.5 (−11) 1.4 (−11)	−5.0	5.4		Meot-Ner and Field (1976a)
(11) 		320 440	2.0 (−11) 6.0 (−12)	−4.4	4.7		Meot-Ner and Field (1976a)
(12) 		320 450	9.5 (−12) 6.5 (−12)	−1.5	6.9		Meot-Ner and Field (1978)
(13) 		450	2.7 (−11)	−2.7	2.2		Meot-Ner and Field (1978)
(14) 		320 450	6.7 (−12) 1.2 (−12)	−5.4	1.3		Meot-Ner and Field (1978)
(15) 		450	8.6 (−12)	−4.1			Meot-Ner and Field (1978)
(16) 		320 420	4.5 (−12) 8.6 (−13)	−6.0			Meot-Ner and Field (1978)
(17) 		450	1.6 (−13)	−6.8			Meot-Ner and Field (1978)

(18) 	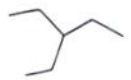	450	7.0 (−13)	−6.0	1.5	Meot-Ner and Field (1978)
(19) 		380	2.0 (−12)	−6.0		Meot-Ner and Field (1978)
		440	8.2 (−13)			
(20) 	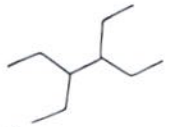	320	1.8 (−12)	−4.5		Meot-Ner and Field (1978)
		390	7.0 (−13)			
(21) 	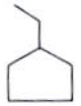	450	1.1 (−12)	−3.7	1.8	Meot-Ner and Field (1978)
(22) 	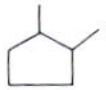	370	1.1 (−11)	−4.2	2.2	Meot-Ner and Field (1978)
		520				
(23) $NO^+ + CH_3CHO$		320	8.0 (−10)			Meot-Ner and Field (1975c)
		480	4.5 (−10)			
		645	3.3 (−10)			

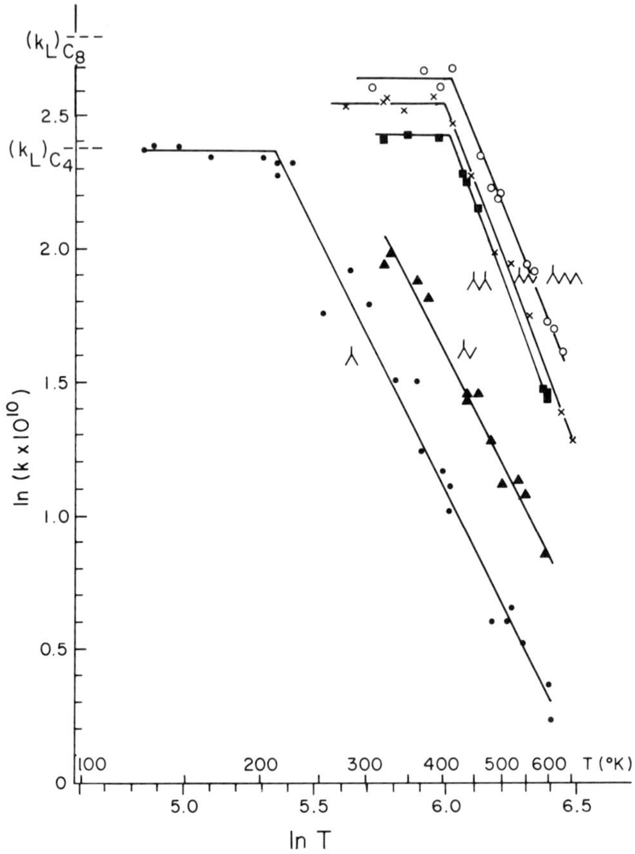

FIG. 5 Temperature dependence of H^- transfer reactions $s\text{-}C_3H_7^+ + RH \rightarrow R^+ + C_3H_8$. R is the hydrocarbon indicated on the figure [From Meot-Ner and Field (1976a).]

was observed by Pierce and Porter (1974) in the proton transfer reaction $CH_4D^+(CH_4, CH_3D)CH_5^+$ (Fig. 6).

Two hydride transfer reactions: the reactions of $C_2H_5^+$ with isopentane and with 2,4-dimethylpentane exhibited fast kinetics up to 700 K. If a transition to slow kinetics with negative TD for these reactions occurs, it must take place above 700 K. Such high temperatures were not accessible experimentally.

Temperature studies of H^- transfer reactions were extended to a total of 22 reactions (Meot-Ner and Field, 1978). The collision efficiencies of the reactions vary between 1 and 10^{-4}, the temperature coefficients between $T^{-1.5}$ and $T^{-6.8}$, and exothermicities between -1 and -45 kcal mole^{-1}.

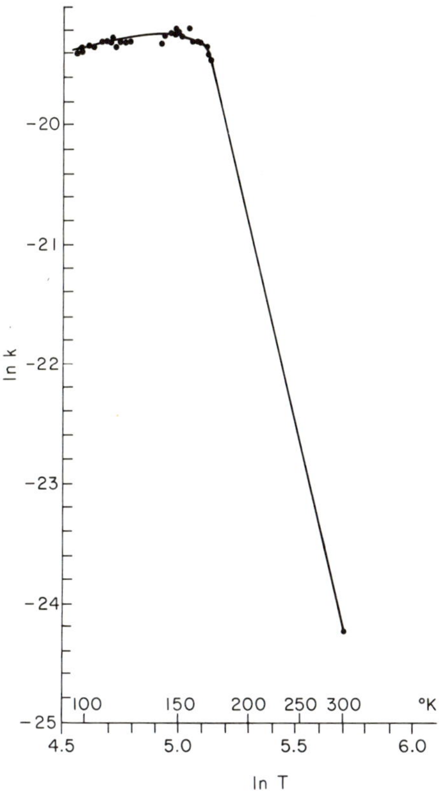

FIG. 6 Temperature dependence of the rate constant for $CH_4D^+ + CH_4 \rightarrow CH_5^+ + CH_3D$ [From Pierce and Porter. (1974). Reprinted with permission from *J. Phys. Chem.* Copyright by the American Chemical Society.] The point at 300 K ($10^3/T = 3.33$) was obtained by Abramson and Futrell (1966).

The following relationships are observed:

(a) A negative correlation is noted between the absolute magnitude of the rate constants and their temperature coefficients (Fig. 7).

(b) Increasing rate constants are associated with increasing exothermicity. The data suggests an empirical relationship:

$$k_{450} = Ce^{-\beta \Delta H_\circ}, \tag{8}$$

where C and β are constants which depend upon the nature of the reactant ion (primary, secondary, or tertiary).

(c) Relations (a) and (b) imply a correlation of increasingly negative temperature coefficients with decreasing exothermicity.

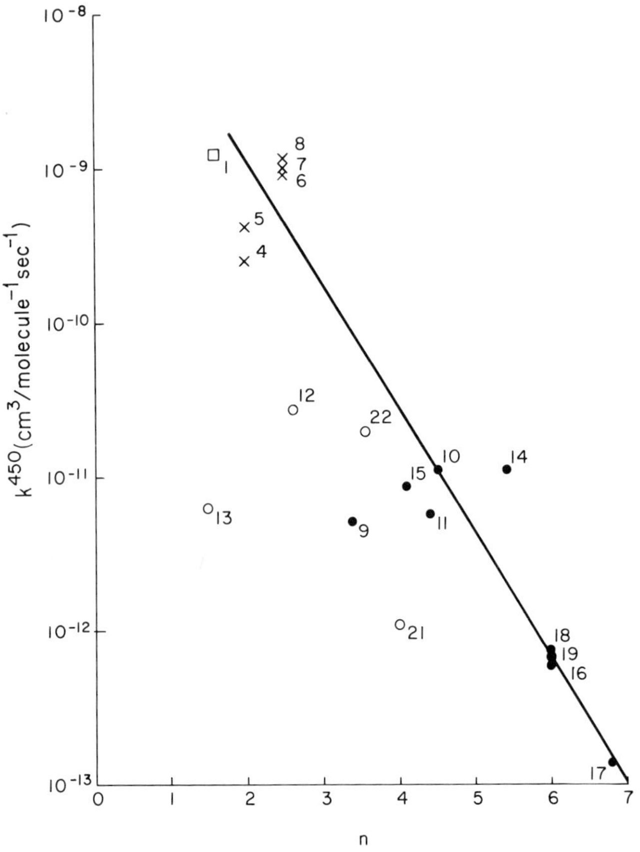

FIG. 7 Correlation between temperature coefficients, i.e., n in $k = CT^{-n}$ and the rate constants for H^- transfer reactions. Numbers indicate reactions as listed in Table III. [From Meot-Ner and Field (1978).]

(d) Increasing complexity of the reactants is correlated with decreasing rates and increasingly negative temperature coefficients.

(e) The temperature coefficients for reactions of cyclic reactants are less negative than for reactions of acyclic reactants with comparable rates and exothermicities.

Solomon *et al.* (1974), Meot-Ner *et al.* (1974), and Meot-Ner and Field (1978) used transition state theory (TST) to account for the negative TDs and for some of the kinetic correlations noted above for H^- transfer reactions. The general TST equation for the forward rate constant is given by

$$k_f = \kappa \frac{kT}{h} \frac{Q^{\ddagger}}{\pi Q_i} \exp\left(\frac{-E_a}{kT}\right), \tag{9}$$

where κ is a transmission coefficient, $Q^\ddagger$ and Q_i are the partition functions for the transition complex and the reactants, respectively, and the other symbols have their usual meaning. If the contribution of the electronic and vibrational partition functions to $Q^\ddagger$ and πQ_i cancel, i.e., $Q^\ddagger_{\mathrm{EL}} Q^\ddagger_{\mathrm{VIB}} / \pi (Q_{\mathrm{EL}} Q_{\mathrm{VIB}})_i \approx 1$ at the moderate temperatures of interest, and the temperature independent terms are collected, and if the activation energy is negligible, the temperature dependence of k_{f} is given by

$$k_{\mathrm{f}} = CT \frac{(T^\ddagger_{\mathrm{trans}})^{3/2} (T^\ddagger_{\mathrm{rot}})^{3/2}}{\pi (T_{\mathrm{trans}})^{3/2} (T_{\mathrm{rot}})^{3/2}} T^{-r/2}_{\mathrm{int,\,rot}} \qquad (10)$$

$$= CT^{-(2+r/2)},$$

where r is the total number of internal rotations in the reactants which become hindered upon the formation of the complex. (r may be negative if less internal rotors of the reactants are hindered upon the formation of the complex than there are new internal rotors formed in the process.)

The applicability of TST to IM reactions should be commented upon. In most applications of the theory the activated or transition complex is chosen at a maximum in the potential energy surface of the reaction. In reactions with zero activation energy such a maximum is absent. However, it has been pointed out by Johnston (1966) that the coordinates used to describe the reacting system and the location of the surface dividing reactants and products are arbitrary and may be located with no reference to saddle points. It is only necessary that the motion along the reaction coordinate be separable from the other degrees of freedom of the complex. The other basic requirements for TST are that the reactants will be distributed among their possible internal energy states as they would be at thermal equilibrium; and that the activated complexes will be in equilibrium with the reactants in the sense that the forward flux of the activated complexes through the dividing surface from reactants to products will be the same as the forward flux at complete thermal and chemical equilibrium. The latter condition can be satisfied in ion–molecule reactions, even though the intermediates are chemically activated, since they will be equally activated in the presence of overall equilibrium. Also, the latter condition may be satisfied in the absence of chemical equilibrium, i.e., in the absence of the inverse reaction (Laidler, 1969).

Equation (10) obviously satisfies the functional form $k_{\mathrm{f}} = CT^n$ which is observed in the temperature dependent regimes of the plots on Fig. 6. The correlation between k_{f} and n [correlation (a) noted above] is also accounted for by Eq. (10) since increasing r increases n but decreases the absolute value of $T^{-(2+r/2)}$. Correlation (d) may be also accounted for since increasingly more complex reactants permit the hindrance of increasingly more rotors upon the formation of an $(R_1 \cdots \mathrm{H{-}R_2})^+$ complex; thus r increases, n

increases, and k_f decreases with increasing complexity. Correlation (e) may also be understood by considering the effects of reactant structure on r: In reactions of cyclic reactants there are fewer free rotors in the reactants than in the reactions of open-chain reactants. Fewer rotors are therefore subject to hindrance upon the formation of the transition complex from cyclic than from acyclic reactants; r should be smaller and n less negative in the former than in the latter case, as is indeed observed.

The negative TDs of H^- transfer reactions, the functional form of the temperature coefficients, and the relations between reactant structure, the magnitude of the rate constants and their temperature coefficients, may thus be adequately explained by simple TST arguments. On the other hand, the observed relation between exothermicity and the kinetic parameters cannot be explained on this bases. An alternative approach to the negative TD of IM reactions is based on RRK or RRKM considerations of the decomposition of the reaction complex. This approach to bimolecular reactions will be discussed in the last section of this chapter.

D. *The Pressure Dependence of Exchange Reactions*

In bimolecular reactions of ions which are produced with nonthermal energy distributions, increasing pressure may effect the rate constant by collisional deactivation of the excited reactant, as was discussed earlier. However, once all the degrees of freedom of the reactants are in thermal equilibrium, no further pressure effects by this mechanism should be observed. Indeed, the thermal rate constants of numerous fast exchange reactions were measured both by ICR techniques at pressures of 10^{-5}–10^{-3} torr, and by high pressure MS or flowing afterglow methods in the pressure range 1–4 torr. Rate constants of exchange reactions measured in these widely different pressure regimes are usually in good agreement, which proves that for fast reactions of thermal ions no significant pressure effects exist, as may be expected.

A very interesting pressure effect in the kinetics of a slow exchange reaction was observed by Bohme *et al.* (1969). These workers studied the effects of third-body (He) pressure on the reaction $O^+(N_2, N)NO^+$ at 82 K. The rate constant was observed to increase linearly with P_{He} at $0.2 < P_{He} < 1.0$ torr. The rate constant for the same reaction at 300 K was independent of P_{He} in the same pressure range. Ferguson (1972) interpreted these observations as evidence for the collisional deactivation of excited vibrational levels of the N_2O^{+*} reaction intermediate, since the lower N_2O^+ vibrational levels can dissociate only back to reactants but not to the products of this reaction. The fact that the N_2O^{+*} ion can be deactivated collisionally shows that the lifetime of this reaction intermediate is comparable to or longer than the time between collisions with He. From this observation Ferguson deduced that at 82 K $\tau(N_2O^{+*}) \geq 10^{-10}$ sec. At a higher temperature $\tau(N_2O^{+*})$ should be shorter and back dissociation to reactants should occur before a collision with the third body can take place; this explains the absence of a pressure effect at 300 K.

It should also be noted that the temperature dependence of this reaction and the contribution of the excited states of the intermediate were given a detailed quantum chemical treatment by Kaufman (1973).

The observations of Bohme *et al.* are of much interest, since similar pressure effects may occur in other slow bimolecular reactions which exhibit negative TDs. The negative TD of such reactions indicates that decomposition of the reaction complex to products becomes more favorable as the internal energy of the complex decreases. All ion–molecule reaction complexes $(A^+ \cdot B)^*$ are chemically activated to an energy equal to the dissociation energy of the complex to reactants, i.e., $D°(A^+ \cdot B)$. Collisions with M should dissipate some of the excitation energy of $(A^+ \cdot B)^*$. Dissipation of the internal energy of $(A^+ \cdot B)^*$ should, in turn, increase the probability of dissociation to products compared with back dissociation to reactants, as is demonstrated by the negative TD of these reactions. In a manner similar to the $O^+ + N_2$ reaction, observation of such pressure effects could yield valuable information on the lifetimes of the excited reaction intermediates.

III. TEMPERATURE AND PRESSURE EFFECTS IN ASSOCIATION REACTIONS

A. Temperature Dependence of Clustering Reactions

1. Experimental Results

Negative temperature dependence has been a common feature of IM association reactions since the first observation of the TD of the association of He^+ with He (Niles and Robertson, 1965) and the first systematic studies of temperature effects on the rate constants of IM association reactions (Bohme *et al.*, 1969; Durden *et al.*, 1969). Experimental results on the TD of clustering reactions are summarized in Table IV.

2. The Energy Transfer Mechanism. Decomposition Rates of Excited Reaction Complexes

The energy transfer mechanism, first proposed by Rabinowitz (1937) for atom recombination reactions, is generally considered to be applicable also to IM association reactions (e.g., Bohme *et al.*, 1968; Kebarle, 1972; Anicich

TABLE IV

Temperature and Pressure Effects in Clustering Reactions

Reaction	Temp. (K)	k ($cm^6 sec^{-1}$)	Temp. Coef. (n in $k = CT^n$)	$(M)^a$ (cm^{-3})	Comments	Reference
$He^+ + 2He \rightarrow He_2^+ + He$	80 410	4.4 (−31) 8.8 (−32)	−1	1.4 (17)–3.5 (18)	Third-order kinetics confirmed at 300 K, $4 > P_{He} > 10$ torr	Niles and Robertson (1965)
$Ar^+ + Ar + He \rightarrow Ar_2^+ + He$	82 290	1.6 (−30) 1.0 (−31)				Bohme *et al.* (1969)
$N^+ + N_2 + He \rightarrow N_3^+ + He$	82 280	7.2 (−29) 8.6 (−30)	−1.7	1.2 (16)–1.2 (17)	Third-order kinetics confirmed at 82 K, $0.1 > P_{He} > 1.0$ torr	Bohme *et al.* (1969)
$N^+ + 2N_2 \rightarrow N_3^+ + N_2$	300 380	5 (−29) 2.8 (−29)	−2.5	2.0 (16)–8.5 (15)	Third-order kinetics could not be confirmed unambigously	Good *et al.* (1970)
$O_2^+ + 2O_2 \rightarrow O_4^+ + O_2$	300 340	3 (−30) 2 (−30)	−2.8	1.0–1.3 (17)		Durden *et al.* (1969)
$O_2^+ + 2O_2 \rightarrow O_4^+ + O_2$	80 290	1.2 (−28) 2 (−30)	−3.2	4.1 (16)–9.5 (16)	Third-order kinetics confirmed at $T = 143$ K, $0.6 > P_{O_2} > 1.4$ torr	Payzant *et al.* (1973)
$O_4^+ + 2O_2 \rightarrow O_6^+ + O_2$	80 180	4 (−29) 2 (−30)	−3.7	4.1 (16)–9.5 (16)	Third-order kinetics confirmed at $T = 143$ K, $0.6 < P_{O_2} < 1.4$ torr	Payzant *et al.* (1973)
$O_2^+ + O_2 + He \rightarrow O_4^+ + He$	82	3.1 (−29)			Third-order kinetics confirmed at 82 K, $0.1 < P_{He} < 1.2$ torr; second-order kinetics at 82 K, $1.2 < P_{He} < 2.7$ torr, with $k = 5 \times 10^{-12}$ $cm^3 sec^{-1}$	Bohme *et al.* (1968)

$N_2^+ + 2N_2 \rightarrow N_4^+ + N_2$	300 380	8 (−29) 3 (−29)	−4	3.5 (16)–1.4 (17)	Third-order kinetics confirmed at 380 K, $0.5 < P < 3.5$ torr	Good *et al.* (1970)
$N_2^+ + 2N_2 \rightarrow N_4^+ + N_2$	270 650	1.2 (−28) 1.6 (−29)	−1.7			Meot-Ner and Field (1974)
$N_2^+ + N_2 + He \rightarrow N_4^+ + He$	82 280	1.2 (−28) 1.9 (−29)	−1.5	1.2 (16)–1.7 (17)	Third-order kinetics at 280 K, $0.6 < P_{He} <$ 1.3 torr, at 82 K, kinetics is third order at $0.1 < P_{He} <$ 0.5 torr, but becomes second order at $0.5 < P_{He} < 1.5$ torr, with $k = 7 \times 10^{-12}$ cm^3 sec^{-1}	Bohme *et al.* (1969)
$CO^+ + 2CO \rightarrow C_2O_2^+ + CO$	120 165 650	2.8 (−28) 2.8 (−28) 5.6 (−29)	−1.5	1.6 (16)	Third-order kinetics confirmed at 120 K, $P =$ 0.2 − 0.8 torr. TD deviates from $k = CT^{-1.5}$ at low T, possibly due to small positive E_a of $\approx$0.2 kcal mole^{-1}	Meot-Ner and Field (1974)
$CO^+ + CO + He \rightarrow C_2O_2^+ + He$	120 650	7.9 (−28) 6.3 (−29)	−1.5	5.9 (15)–2.0 (16)	Third-order kinetics confirmed at 122 K, $0.075 < P_{He} < 0.25$ torr	Meot-Ner and Field (1974)
$H_3^+ + 2H_2 \rightarrow H_5^+ + H_2$	130 270	1.5 (−28) 1.3 (−29)	−2.3	2.3 (16) 3.5 (16)	Third-order kinetics confirmed at high temperature range of this work	Hiraoka and Kebarle (1975a)

TABLE IV (*Continued*)

Reaction	Temp. (K)	k ($cm^6\ sec^{-1}$)	Temp. Coef. (n in $k = CT^n$)	$(M)^a$ (cm^{-3})	Comments	Reference
$NO^+ + H_2O + NO \rightarrow$ $(NO^+ \cdot H_2O) + NO$	298 330	1.1 (−28) 9.0 (−29)	−3.4			Su and Bone (1975)
$NO^+ + H_2O + SF_6 \rightarrow$ $(NO \cdot H_2O)^+ + SF_6$	298 348	4.2 (−28) 2.3 (−28)	−4.1			Su and Bone (1975)
$(NO \cdot H_2O)^+ + H_2O + N_2 \rightarrow$ $(NO \cdot 2H_2O)^+ + N_2$	300 350	1.0 (−27) 4.5 (−28)	−4.7			French *et al.* (1973)
$HCO^+ + CO + H_2 \rightarrow$ $(HCO \cdot CO)^+ + H_2$	120 370	7.9 (−29) 4.0 (−30)	−3.0	4.0 (16)–1.6 (17)	Third-order kinetics confirmed at 120 K, $0.5 < P_{H_2} < 2.0$ torr; TD slightly deviates from $k = CT^{-3}$ below 160 K	Meot-Ner and Field (1974)
$HCO^+ + 2CO \rightarrow$ $(HCO \cdot CO)^+ + CO$	120 200 370	2.2 (−29) 1.6 (−29) 1.0 (−30)	−3.0	1.6 (16) 6.4 (17)	Third-order kinetics confirmed at 120 K, $0.2 < P < 0.8$ torr; TD deviates from $k = CT^{-3}$ below 200 K	Meot-Ner and Field (1974)
$H_3O^+ + H_2O + CH_4 \rightarrow$ $(H_2O)_2H^+ + CH_4$	398 805	1.3 (−27) 7.2 (−29)	−4.1	5 (16) 1.4 (17)		Cunningham *et al.* (1972)
$H_3O^+ + H_2O + CH_4 \rightarrow$ $(H_2O)_2H^+ + CH_4$	305 570	4.0 (−27) 7.4 (−28)	−3.1	2.7 (16)	Third-order kinetics confirmed at 359 K, $0.5 < P_{CH_4} < 1.2$ torr	Meot-Ner and Field (1977)

$(H_2O)_2H^+ + H_2O + CH_4 \rightarrow$ $(H_2O)_3H^+ + CH_4$	398 496	9.8 (−28) 2.2 (−28)	−6.5	5 (16) 1.4 (17)		Cunningham *et al.* (1972)
$H_3S^+ + H_2S + CH_4 \rightarrow$ $(HCN)_2H^+ + CH_4$	310 400	9.0 (−29) 2.4 (−29)	−4.8	2.7 (16)	Third-order kinetics confirmed at 329 K, $1.0 < P_{CH_4} < 2.2$ torr	Meot-Ner and Field (1977)
$HCNH^+ + HCN + CH_4 \rightarrow$ $(HCN)_2H^+ + CH_4$	300 400 520	3.6 (−27) 2.7 (−27) 7.4 (−28)	−4.9 at $400 < T < 520$ K	1.5 (16)	Third-order kinetics at $250 < T < 600$ K, $2 < N_{CH_4} < 8 \times 10^{16}$ cm^{-3}; TD somewhat less negative then $T^{-4.9}$ at $T <$ 400 K. Positive TD at $580 < T < 640$ K	Meot-Ner (1978)
$NH_4^+ + NH_3 + CH_4 \rightarrow$ $(NH_3)_2H^+ + CH_4$	250 350	3.9 (−27) 1.3 (−27)	−3.2	2.5 (16)	Third-order kinetics assumed	Meot-Ner and Field (1975b)
$NH_4^+ + 2NH_3 \rightarrow$ $(NH_3)_2^+ + NH_3$	213 302	3.3 (−26) 5.1 (−27)	−5.3	2 (13)–1.2 (14)	Transition from third- toward second-order kinetics through pressure range of study	Neilson *et al.* (1978)
$CH_3NH_3^+ + CH_3NH_2 + CH_4 \rightarrow$ $(CH_3NH_2)_2H^+ + CH_4$	314 364	1.3 (−26) 8.3 (−27)	3.0	1.6 (16)	Approx. third-order kinetics at (M) < 1.5(16) cm^{-3}, transition toward second order at higher (M)	Meot-Ner and Field (1975b)
$CH_3NH_3^+ + 2CH_3NH_2 \rightarrow$ $(CH_3NH_2)_2H^+ + CH_3NH_2$	220 302 420	4.4 (−25) 1.5 (−25) 2 (−26)	-5 ± 2	5 (12)–6 (13)	Transition from third- toward second-order kinetics through pressure range of study	Neilson *et al.* (1978)

TABLE IV (*Continued*)

Reaction	Temp. (K)	k ($cm^6\ sec^{-1}$)	Temp. Coef. (n in $k = CT^n$)	$(M)^a$ (cm^{-3})	Comments	Reference
$(CH_3)_2NH_2^+ + (CH_3)_2NH + i\text{-}C_4H_{10} \rightarrow ((CH_3)_2NH)_2H^+ + i\text{-}C_4H_{10}$	314 388	2.9 (−26) 1.2 (−26)	−4.2	1.5–5 (16)	Approx. third-order kinetics at (M) < 1.5(16) cm^{-3}, transition toward second order at higher (M)	Meot-Ner and Field (1975b)
$(CH_3)_2NH_2^+ + 2(CH_3)_2NH \rightarrow ((CH_3)_2NH)_2H^+ + (CH_3)_2NH$	280 302 370	2 (−25) 9.6 (−26) 2 (−26)	−7 ± 1	3 (12)–4 (13)	Transition from third- toward second-order kinetics through pressure range of study	Neilson *et al.* (1978)
$(CH_3)_3NH^+ + 2(CH_3)_3N \rightarrow ((CH_3)_3N)_2H^+ + (CH_3)_3N$	302	1.1 (−24)	−7 ± 1	3 (12)–4 (13)	Transition from third- toward second-order kinetics through pressure range of study	Neilson *et al.* (1978)

[a] Listed values of (M) indicate range of third-body densities at which temperature and/or pressure studies were performed.

and Bowers, 1974; Porter, 1975; Meot-Ner and Field, 1975b)

$$A^+ + B \underset{k_b}{\overset{k_a}{\rightleftarrows}} AB^{+*}, \qquad AB^{+*} + M \underset{k_{ac}}{\overset{k_s}{\rightleftarrows}} AB^+ + M,$$

Scheme II

where Scheme II is a special case of Scheme I with $k_p = 0$ and $k_r = 0$.

In the absence of other information k_a, the rate constant for the formation of an excited association complex, is often equated with the collision rate constant $k_{c(A^+,B)}$ (e.g., Anicich and Bowers, 1974; Ferguson, 1972; Meot-Ner and Field, 1975b; Olmstead *et al.*, 1977). (Some reservations concerning this assumption will be discussed later.) It is also generally assumed that k_s may be approximated by the rate constant for the collision between AB^{+*} and M. The latter assumption is justified on the basis that the removal of a fraction of the excitation energy from AB^{+*} by a collision with M makes back dissociation to $A^+ + B$ sufficiently slow to allow for the eventual stabilization of AB^{+*}. This assumption is not always accurate since different third bodies usually vary in stabilization efficiency. However, the third-body efficiencies even of monoatomic species usually differ by no more than a factor of five from those of more complex molecules such as CH_4 or C_2H_6, for which the strong collision assumption is probably correct (Anicich and Bowers, 1974; Miasek and Harrison, 1975).

Assuming the energy transfer mechanism, the rate constant for the forward reaction leading to AB^+ may be written as

$$k_f = \frac{k_a k_s(M)}{k_b + k_s(M)}. \tag{11}$$

At low pressures, when $k_s(M) \ll k_b$,

$$k_f^{(2)} = (k_a k_s/k_b)(M). \tag{12}$$

Thus, the apparent second-order rate constant $k_f^{(2)}$ increases linearly with (M) at the low pressure limit, i.e., the reaction exhibits third-order behavior with $k_f^{(3)} = k_a k_s/k_b$. Most clustering reactions whose kinetic order was investigated by pressure studies exhibited third-order behavior as summarized in Table IV.

If k_a and k_s may indeed be represented by the corresponding collision rates, their TD should not be significant. Thus, the TD of $k_f^{(3)}$ for a third-order reaction should be determined mainly by the TD of k_b^{-1}. Several workers (e.g., Ferguson, 1972; Hiraoka and Kebarle, 1975a; Meot-Ner and Field, 1975b) used the RRK formula for the decomposition of a collection of s coupled classical oscillators of frequency (Johnston, 1966) for a qualitative description of k_b:

$$k_b = [(E - E_c)/E]^{s-1}, \tag{13}$$

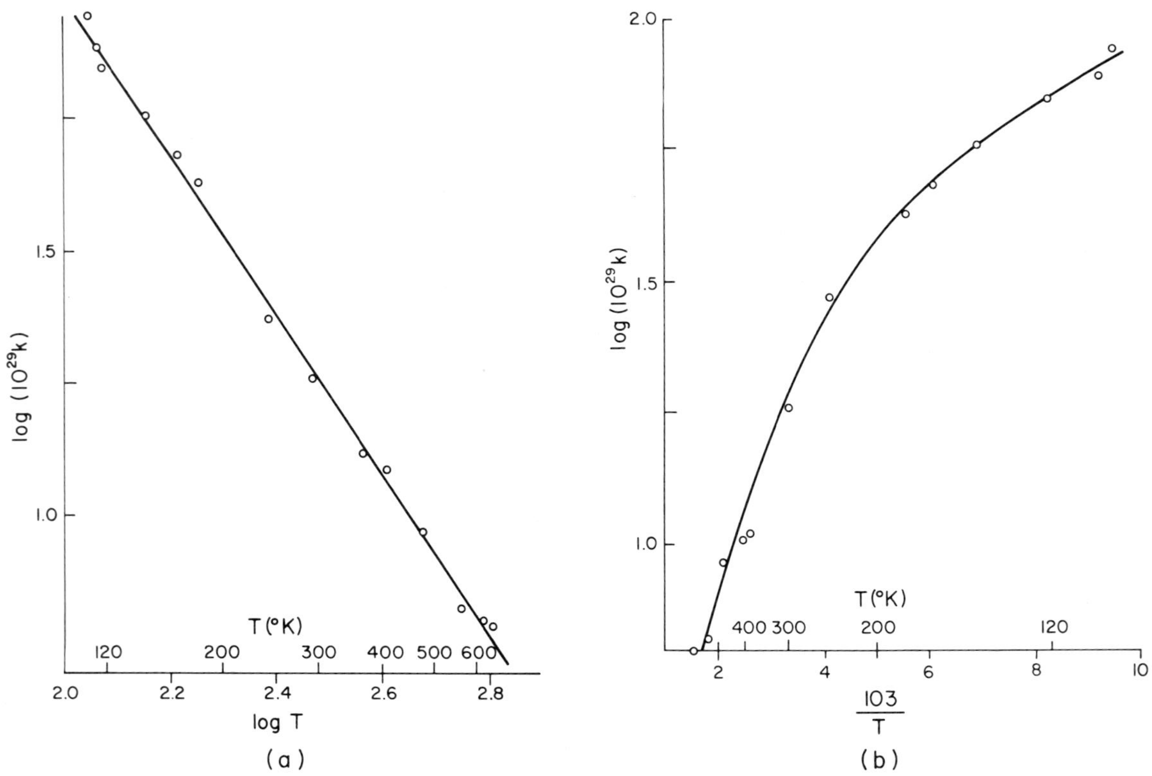

FIG. 8 Temperature dependence of the third-order rate constant for the reaction $CO^+ + CO + He \rightarrow (CO)_2{}^+ + He$. (a) A plot of log *k* versus log *T*. (b) A plot of log *k* versus $10^3/T$. The absolute temperature is also indicated on the abscissa. [From Meot-Ner and Field (1974).]

where E is the total vibrational energy of the system and E_c the critical energy for decomposition. Adapted for the decomposition of the ion–molecule complex, this expression becomes

$$k_b = [rRT/(D^\circ + rRT)]^{s-1}, \tag{14}$$

where D° is the dissociation energy and r is the number of square terms that contribute to the internal energy. If $D^\circ \gg rRT$, which may be a reasonable approximation for small molecules at low temperatures, Eq. (14) predicts that

$$k_b \propto T^{s-1} = T^n \tag{15}$$

and thus $k_f \propto T^{-n}$. Indeed, it was observed that plots of ln k_f versus ln T were linear for several reactions involving simple reactants, such as the formation of Ar_2^+ (Bohme *et al.*, 1969), O_4^+,O_6^+ (Payzant *et al.*, 1973), and H_5^+ Hiraoka and Kebarle, 1975a). The functional form of the TD of association reactions over a wide temperature range was investigated by Meot-Ner and Field (1974). They found that the plot of ln k versus T for the reaction $CO^+ + CO + He \rightarrow (CO)_2^+ + He$ was linear between 120 and 620 K (Fig. 8a). On the other hand, a plot of ln k versus $10^3/T$ for the same reaction clearly deviated from linearity (Fig. 8b), which shows that the concept

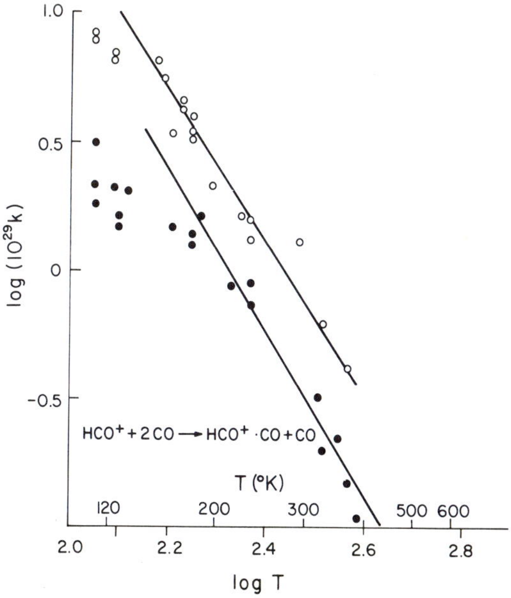

FIG. 9 The temperature dependence of the rate constant for the reactions $HCO^+ + 2CO \rightarrow HCO^+{\cdot}CO + CO$ (●) and $HCO^+ + CO + H_2 \rightarrow CHO^+{\cdot}CO + H_2$ (○). [From Meot-Ner and Field (1974).]

of negative activation energy cannot be applied to describe the negative TD of association reactions.

While the plots of $\ln k$ versus $\ln T$ are linear in many cases, deviation from linearity were observed at $T < 150$ K in the plots of $\ln k$ versus $\ln T$ for reactions shown in Figs. 9 and 10. Some evidence for a similar behavior was observed in the reaction $H_3^+ + 2H_2 \rightarrow H_5^+ + H_2$ (Hiraoka and Kebarle, 1975a). The deviations from linearity observed at low T could be caused by activation energies as small as 0.2 kcal mole^{-1} for these association processes (Fig. 10).

In addition to the functional form of the TD, Eq. (14) also predicts that k_b should be inversely related to $D°$, and thus k_f should be directly related to $D°$. Such relationship was indeed observed. For example, in the following

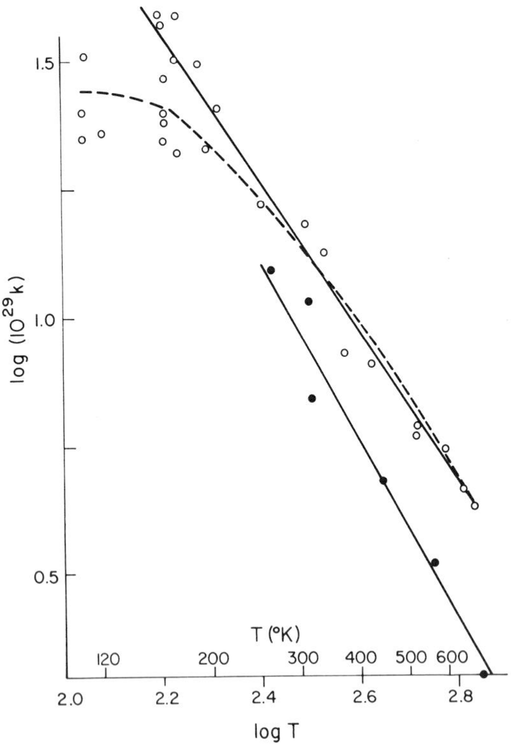

FIG. 10 The temperature dependence of the rate constants of the reactions $CO^{+\cdot} + 2CO \leftrightarrows C_2O_2^{+\cdot} + CO$ (○) and $N_2^+ + 2N_2 \rightarrow N_4^+ + N_2$ (●). The broken line is the plot the logarithm (base 10) of the hypothetical rate constant $k_n = CT^{-2}\exp(-E_a/RT)$ for an activation energy of 0.2 kcal/mole. C is arbitrarily chosen such than k_n equals the experimental rate constant at 328 K. [From Meot-Ner and Field (1974).]

set (Meot-Ner and Field, 1974) of radical and analogous protonated dimers, the dissociation enthalpy is clearly correlated with the rate constants for reactions leading to the dimers: $CO^+{\cdot}CO$, $D^\circ = 25.4$ kcal mole^{-1}, $k_{328} = 19.8 \times 10^{-29}$ cm^6 sec^{-1}; $HCO^+{\cdot}CO$, $D^\circ = 11.7$, $k_{328} = 0.29 \times 10^{-29}$ (in both cases, M = CO). Also, $N_2{}^+{\cdot}N_2$, $D^\circ = 22.8$, $k_{328} = 7.9 \times 10^{-29}$; $N_2H^+{\cdot}N_2$, $D^\circ = 14.5$, $k_{328} = 0.54 \times 10^{-29}$. (In these cases $M = N_2$.) A similar relation between D° and k_f is also observed in $(H_2O)_2H^+$, $D^\circ = 33.0$, $k_{320} = 37 \times 10^{-28}$ and $(H_2S)_2H^+$, $D^\circ = 16.0$ kcal mole^{-1}, $k_{320} = 0.81 \times 10^{-28}$ cm^6 sec^{-1} (Meot-Ner and Field, 1977).

Given the energy transfer mechanism and the approximations $k_a = k_{c(A+,B)}$ and $k_s = k_{c(AB^{+*},M)}$, k_b may be calculated using Eq. (11) if k_f has been measured. Meot-Ner and Field (1975b) used this method to calculate the TD of k_b for the decomposition of the excited dimers $(NH_3)_2H^{+*}$, $k_b^{350} = 2.2 \times 10^9$ sec^{-1}; $(CH_3NH_2)_2N^{+*}$, $k_b^{350} = 1.2 \times 10^8$ sec^{-1} and $((CH_3)_2NH)_2H^{+*}$, $k_b^{350} = 6.0 \times 10^7$ sec^{-1}. The magnitude of k_b was observed to decrease and the TD of k_b to increase with increasing number of degrees of freedom in the excited complex, as may be expected from RRK considerations. However, plots of $\ln k_b$ versus $\ln T$ for these reactions were not linear (Fig. 11), which indicates that the classical RRK equation (Eq. 13) is not a good approximation of the quantum coupled oscillator RRK model (let alone of physical reality). Meot-Ner and Field (1975b) used the quantum degenerate oscillator version of the RRK model (Johnston, 1966) to approximate the TD of k_b for these excited complexes. Good fits of the TD were obtained, even with this simplistic model, when $\nu = 650$ cm^{-1} was used for the ammonia complex and $\nu = 850$ cm^{-1} for the methylamine and dimethylamine complexes (Fig. 11). However, other values of ν were needed to approximate the absolute values of k_b.

Olmstead *et al.* (1977) used more sophisticated RRKM calculations to find k_b and its TD for the $(NH_3)_2H^{+*}$, $(CH_3NH_2)_2H^{+*}$, and $((CH_3)_2NH)_2H^{+*}$ complexes. In these calculations the energy was assumed to be randomized in the complexes; stretching and bending frequencies were approximated from corresponding data of the neutrals; N–H–N bending frequencies, the only adjustable parameter, were selected to fit the entropy of the complex. Both the calculated values of k_b and their TD reproduced well the experimental results of Meot-Ner and Field (1975b).

Equating k_a and k_s with the corresponding calculated collision rates, k_b may be found for all association reactions for which k_f has been measured. For example, k_b^{320} $((H_2O)_2H^{+*} = 5.2 \times 10^8$ sec^{-1} and k_b^{320} $((H_2S)_2H^{+*}) = 2.3 \times 10^{10}$ sec^{-1} were calculated by Meot-Ner and Field (1977). It should be noted again that the assumption $k_a = k_c$, which is used in these calculations, is subject to question, as will be discussed in the last section.

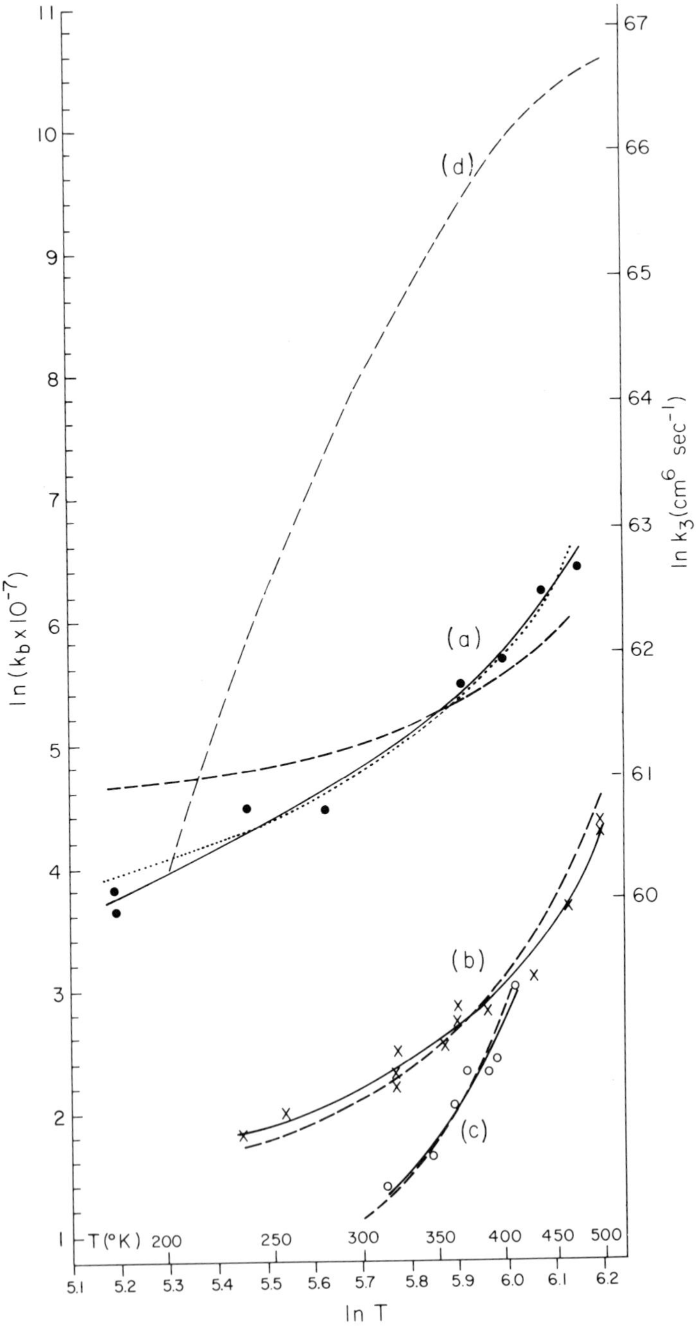
ln (k_b x 10^-7)
ln k_3 (cm^6 sec^-1)
ln T
T (°K) 200 250 300 350 400 450 500
5.1 5.2 5.3 5.4 5.5 5.6 5.7 5.8 5.9 6.0 6.1 6.2
1 2 3 4 5 6 7 8 9 10 11
60 61 62 63 64 65 66 67
(a)
(b)
(c)
(d)

3. *The Ligand Switching Mechanism*

Two major features of clustering reactions, the third-order kinetics and the negative TD, may also be explained by the ligand switching mechanism

$$A^+ + M \rightleftarrows A^+M^*$$
$$A^+M^* + M \rightleftarrows AM^+ + M$$
$$AM^+ + B \xrightarrow{e} AB^+ + M.$$

Scheme III

In this case the forward rate will be given by $k_e(AM^+)(B)$. In many cases AM^+ is a weakly bound dimer held by electrostatic interaction between A^+ and M. Such is the case, for example, when A^+ is a protonated ion and M does not significantly interact by hydrogen bounding, such as CH_4, $i\text{-}C_4H_{10}$, N_2, O_2. In such reactions at temperatures above 300 K, the equilibrium concentration of AM^+ may be small such that AM^+ is not observable experimentally. If the equilibrium concentration of AM^+ is small, i.e., $((AM^+)/(A^+))_{eq} \ll 1$, and if the concentration of B is much smaller than of M, the fast equilibrium

$$A^+ + M \rightleftarrows AM^+ \tag{16}$$

will precede the ligand transfer step. Under favorable circumstances such fast equilibria may be observed. For example, in reaction systems of trace H_2O or NH_3 in CH_4 at $0.5 < P < 1.0$ torr at $200 < T < 250$ K, the ratios $(H_3O^+{\cdot}CH_4)/(H_3O^+)$ and $(NH_4{}^+{\cdot}CH_4)/(NH_4{}^+)$ are constant with reaction time, even as all these ions decay due to the formation of the more stable dimers $(H_2O)_2H^+$ or $(NH_3)_2H^+$, respectively (Meot-Ner *et al.*, 1977b). When the concentration of AM^+ is determined by equilibrium (16), the overall forward rate constant will be given by

$$k_f^{(3)} = k_e K_{16}(M). \tag{17}$$

FIG. 11 The temperature dependences of the rate constants k_b (in units of 10^7 sec^{-1}) for the dissociation of the excited complexes: (a) $(NH_3)_2H^{+*}$, (b) $(CH_3NH)_2H^{+*}$, (c) $((CH_3)_2NH)_2H^{+*}$. The broken lines give theoretical temperature dependences as calculated from the RRK quantum degenerate oscillator model with $s = 3n - 6$ and $\tilde{\nu} = 850$ cm^{-1}. For the dotted line in plot (a) $\tilde{\nu} = 650$ cm^{-1} was used. For the sake of clarity the absolute values of the theoretical rate constants were multiplied by a constant such that they coincide with the experimental plots at 350 K. The original calculated values of k_b^{350} were 4.4×10^6, 5.3×10^5, and 3.7×10^4 sec^{-1} for the rate constants whose temperature dependences are represented by the broken lines (a), (b), (c). respectively, and 3.2×10^6 sec^{-1} for the rate constant represented by the dotted line in plot (a). For comparison (see text), plot (d) represents the temperature dependence of $1/k_f^{(3)}$ calculated on the basis of the ligand switching mechanism for the reaction $NH_4{}^+ + NH_3 + CH_4 \rightarrow (NH_3)_2H^+ + CH_4$, using $\Delta H^\circ = -4.8$ kcal $mole^{-1}$ for this reaction. [From Meot-Ner and Field (1975b).] Reprinted with permission from *J. Am. Chem. Soc.* Copyright by the American Chemical Society.)

The TD of k_1, the rate constant for an exothermic exchange reaction, may be negligible compared with that of K_{16}. Therefore, if ligand switching is the predominant mechanism of association, the TD of $k_f^{(3)}$ will be determined primarily by the TD of K_{16}

$$R\frac{d \ln k_f^{(3)}}{d(1/T)} = -\Delta H_{16}^\circ = D_{(AM)^+}^\circ. \tag{18}$$

Thus, if the ligand switching mechanism applies, $\ln k$ versus $1/T$ should be linear rather than $\ln k$ versus $\ln T$. If D_{AM^+} is known, the TD of $k_f^{(3)}$ may be calculated from Eq. (17) and the result may be compared with experimental TDs. For example, D° of $NH_4^+ \cdot CH_4$ was recently measured by Meot-Ner *et al.* (1977b) as 4.8 kcal $mole^{-1}$. Using this value and neglecting the TD of k_e, $k_f^{(3)}$ was calculated for $NH_4^+ + NH_3 + CH_4 \rightarrow (NH_3)_2H^+ + CH_4$ by using Eq. (17). The results, normalized to the experimental value of $k_f^{(3)}$ at 200 K, are shown in plot (d) in Fig. 11. The calculated TD may be compared with the TD obtained from $k_{b(obs)} = k_a k_s / k_{(obs)}^{(3)}$ using $k_a = k_{c(NH_4^+, CH_4)}$ and $k_s = k_{c(NH_4^+ . CH_4, NH_3)}$. Since k_a and k_ϵ are approximately independent of temperature, $\ln k_{(obs)}^{(3)} \propto -\ln k_{b(obs)}$; $\ln k_{(obs)}^{(3)}$ is represented by plot (a) and the right-hand side ordinate of Fig. 11. It is obvious that the TD of $k_{(obs)}^{(3)}$ and $k_f^{(3)}$ calculated from the ligand exchange mechanism are quite different both in magnitude and in functional form.*

If the concentration of AM^+ is determined by the fast preequilibrium (16), the third-body efficiencies for a given clustering reaction will be determined by the equilibrium constants for the formation of AM^+. For electrostatically bound AM^+ association ions, the equilibrium constants may vary by orders of magnitude depending on the polarizability of M. However, where data exists, third-body efficiencies usually differ by no more than a factor of five, as was discussed above.

In summary, under conditions where the fast preequilibrium (16) obtains, temperature dependence and third-body studies may be used to determine whether the energy transfer mechanism or the ligand switching mechanism contribute most significantly to a given clustering reaction. Although these tests are indirect, they may be useful since the two mechanisms cannot be distinguished by pressure studies. Although for the example quoted, the evidence favors the energy transfer mechanism, this conclusion cannot be generalized and the dominant mechanism must be evaluated individually for each reaction system.

* The ligand switching mechanism gives a much more pronounced negative TD of $k_f(3)$ than the energy transfer mechanism, also for neutral atom recombination reaction (Clarke and Burns, 1972). For example, it was found that the experimental TD of the reaction Br + Br + M is determined by the TD of $K_{Br \cdot M}$, in agreement with the ligand switching mechanism and the arguments presented above.

B. Pressure Dependence of Clustering Reactions

As we noted earlier, most clustering reactions display third-order kinetics, in agreement with the low pressure limit of the energy transfer or ligand exchange mechanisms. However, at the high pressure limit where $k_b \ll k_s(M)$, Eq. (11) reduces to second-order kinetics with $k_f = k_a$. In intermediate cases when k_b and $k_s(M)$ are comparable, k_f should be intermediate between second and third order and approach k_a asymptotically with increasing (M).

An early observation of transition from second- to third-order kinetics was reported by Bohme *et al.* (1968). These workers observed that the apparent second-order rate constant for $N_2^+ + N_2 + He \rightarrow N_4^+ + He$ at $T = 82$ K leveled off at $P_{He} = 0.4$ torr, $k = 7 \times 10^{-12}$ cm^3 sec^{-1}; and the rate constant for $O_2^+ + O_2 + He \rightarrow O_4^+ + He$ leveled off at 1.7 torr, $k = 5 \times 10^{-12}$ cm^3 sec^{-1}. However, Ferguson (1972) noted that the transition from second to third order was much sharper than that predicted by the energy transfer mechanism, and the limiting high pressure second-order rate constants were much smaller than k_c. Ferguson (1972) suggested that the formation of higher clusters might have contributed to these results.

Transition from third- toward second-order kinetics in clustering reactions of protonated amines was investigated by Meot-Ner and Field (1975b). The variation of the second-order rate constants for the formation of $(CH_3NH_2)_2H^+$ and $((CH_3)_2NH)_2H^+$ with the density of the third body (i-C_4H_{10}) was investigated at several temperatures between 210 and 400 K and at $0.8 \times 10^{16} < N_{i-C_4H_{10}} < 9 \times 10^{16}$ cm^{-3}. The observed pressure dependences were generally in good agreement with the energy transfer mechanism at the higher temperatures applied in each study (Figs. 12 and 13). In pressure studies at the lowest temperatures shown in Figs. 12 and 13, the increase in k_f with (M) was slightly slower than predicted by the energy transfer mechanism. In the clustering of the sterically hindered amines $C_2H_5(CH_3)_2N$ and $(C_3H_7)_2NH$, k_f leveled off to values significantly below k_c. The high pressure limiting values for $(C_3H_7)_2NH$ were 2.9×10^{-10} cm^3 sec^{-1} at 364 K and 1.1×10^{-10} cm^3 sec^{-1} at 376 K. The results for $C_2H_5(CH_3)_2N$ are shown in Fig. 14. The limiting high-pressure second-order rate constant for the formation of $(C_2H_5(CH_3)_2N)_2H^+$ showed a very large negative TD of $k = CT^{-9.3}$ (Fig. 15). The mechanistic interpretation of these observations will be discussed later.

The pressure and temperature dependence of the clustering reactions leading to the formation of $(NH_3)_2H^+$, $(CH_3NH_2)_2H^+$, $((CH_3)_2NH)_2H^+$, and $((CH_3)_3N)_2H^+$ were also investigated by Bowers and co-workers (Neilson *et al.*, 1978). They used a variable temperature ICR instrument and worked with neat amines, using the amine both as reactant and third body. The studies were conducted in the pressure range $1 \times 10^{-4} < P \leq 3 \times 10^{-3}$ torr and at temperatures of $195 < T < 400$ K. Continuous transition from

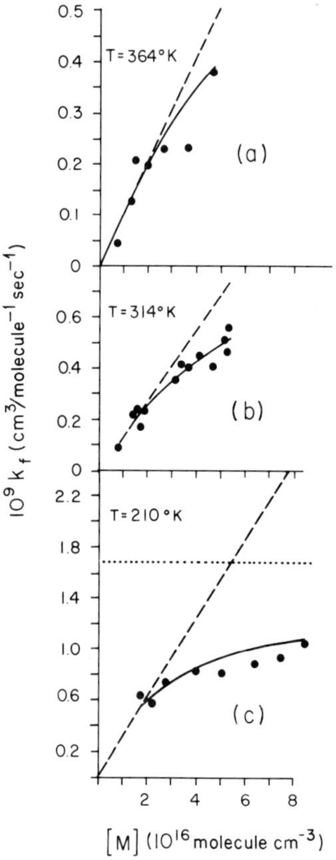

FIG. 12

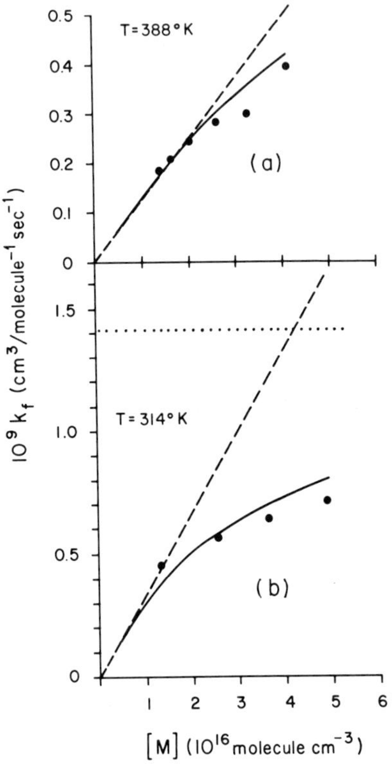

FIG. 13

FIG. 12 The dependence of the apparent second-order rate constant k_f on third-body density (M) for the reaction $CH_3NH_3^+ + CH_3NH_2 \xrightarrow{M} (CH_3NH_3)_2H^+$ at 364, 314, and 210 K. The solid circles (●) are experimental values. The solid lines (———) show the dependence of k_f on (M) as predicted by the energy transfer mechanism, with the experimental point at (M) = 2×10^{16} molecule cm^{-3} as reference. The broken lines (---) show the calculated third-order behavior of k_f based on this reference point. The dotted line in (c) represents the collisional (ADO) second-order capture limit for k_f. M = i-C_4H_{10} in (a) and (b); M = CH_4 in (c). [From Meot-Ner and Field (1975b).]

FIG. 13 The dependence of the apparent second-order rate constant k_f on third-body (i-C_4H_{10}) density for the reaction $(CH_3)_2NH_2^+ + (CH_3)_2NH + M \rightarrow ((CH_3)_2NH)_2H^+ + M$. The meaning of the symbols is as in Fig. 12. The experimental point at (M) = 1.2×10^{16} molecule cm^{-3} was used as reference for the broken line in (b). [From Meot-Ner and Field (1975b).]

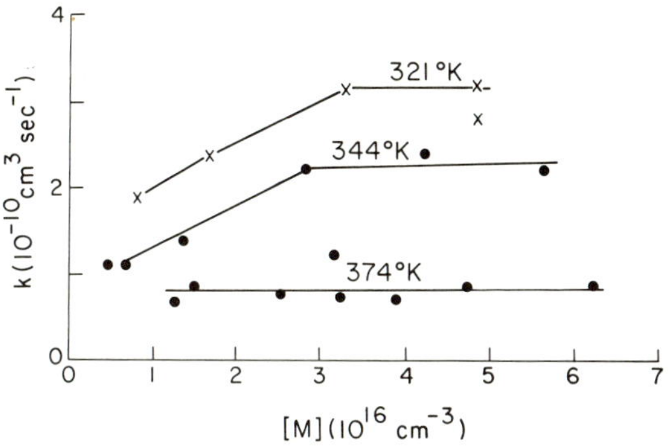

FIG. 14 The dependence of the apparent second-order rate constant for

$$C_2H_5(CH_3)_2NH^+ + C_2H_5(CH_3)_2N \rightarrow (C_2H_5(CH_3)_2N)_2H^+$$

on the third-body (i-C_4H_{10}) number density at the temperatures indicated. [From Meot-Ner (1975).]

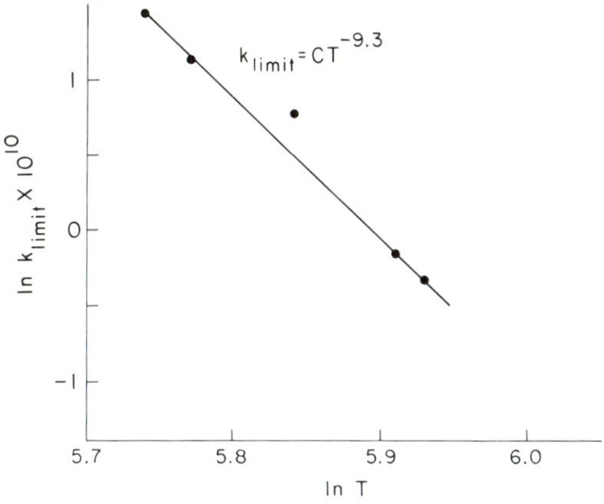

FIG. 15 Temperature dependence of the second-order high pressure limiting rate constant $k_{\text{lim}}^{(2)}$ for $C_2H_5(CH_3)_2NH^+ + C_2H_5(CH_3)_2N \rightarrow (C_2H_5(CH_3)_2N)_2H^+$. [From Meot-Ner (1975).]

third- toward second-order kinetics was observed in all the reactions over the entire pressure range, in good agreement with predictions of the energy transfer mechanism (Fig. 16). Neilson *et al.* inverted Eq. (11), which yields

$$\frac{1}{k_f} = \frac{1}{k_a} + \left(\frac{k_b}{k_a k_s}\right)\left(\frac{1}{(M)}\right). \tag{19}$$

According to Eq. (19), the intercept of a plot of $1/k_f$ versus $1/(M)$ should yield k_a and the slope should yield $k_b/(k_a k_s) = k_f^{(3)}$ (third-order rate constant). With k_a known and with the assumption of strong collisions for deactivation, i.e., $k_s = k_c$, k_b may also be calculated.

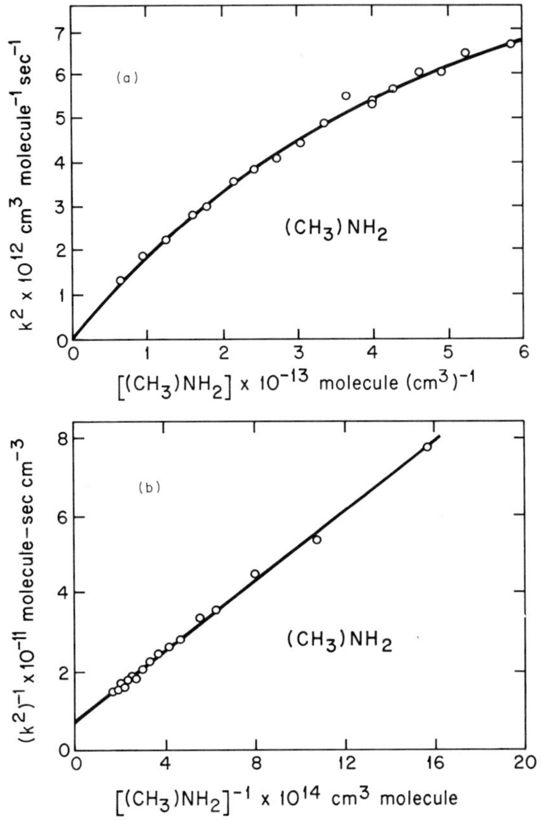

FIG. 16 (a) Plot of the apparent second-order dimerization rate constant k_f for $CH_3NH_3^+ + CH_3NH_2 \rightarrow (CH_3NH_2)_2H^+$ versus $N_{CH_3NH_2}$ at 302 K. [From Neilson *et al.* (1978).] (b) Plot of the inverse of the rate constant shown in (a), i.e., $1/k_f$ versus $1/N_{CH_3NH_2}$ at 302 K. The straight line is a least square fit of the data points. The values of k_a and k_3 determined from the intercept and slope of this line produce the solid line in curve (a) when substituted in the kinetic equation for the energy transfer mechanism. [From Neilson *et al.* (1978).]

Neilson *et al.* used the results of their pressure studies to construct plots of $1/k_f$ versus 1/(M) (Fig. 16b). Values of k_a obtained from the intercepts had the following properties: (i) For all reactions k_a was significantly smaller than the collision rate constant calculated from ADO theory. (ii) k_a/k_{ADO} increased with increasing complexity of the reactants. For example, at 302 K for NH_3, $k_a = 0.13 \times 10^{-11}$ cm^3 sec^{-1}, $k_a/k_{ADO} = 6.5 \times 10^{-4}$; for CH_3NH_2, $k_a = 1.7 \times 10^{-11}$ cm^3 sec^{-1}, $k_a/k_{ADO} = 1.1 \times 10^{-2}$; for $(CH_3)_2NH$, $k_a = 4.3 \times 10^{-11}$ cm^3 sec^{-1}, $k_a/k_{ADO} = 3.3 \times 10^{-2}$. However, for sterically hindered amines, k_a was smaller than would be expected from the above trend, for $(CH_3)_3N$, $k_a = 2.8 \times 10^{-11}$ cm^3 sec^{-1}, $k_a/k_{ADO} = 2.6 \times 10^{-2}$. (iii) k_a exhibits large negative TD. In the dimerizations of mono-, di-, and trimethylamine, $k_a = CT^{-4}$.

These results for k_a are quite different from the above-quoted results of Meot-Ner and Field (1975b), who observed a pressure dependence for the dimerization of NH_3, CH_3NH_2, and $(CH_3)_2NH$ in good agreement with approach to the collision rate at the high pressure limit, i.e., $k_a/k_{ADO} \approx 1$. Indeed, Meot-Ner and Field directly measured values of k_f higher than the limiting k_a values obtained by Neilson *et al.*, using Eq. (19).

Neilson *et al.* calculated k_b from

$$k_b = k_a k_s / k_f^{(3)} \tag{20}$$

using k_a obtained from the plots of $1/k_f$ versus $1/(M)$. The values of k_b obtained in this manner were smaller by several orders of magnitude than the k_b values obtained in the high pressure measurements of Meot-Ner and Field. For example, for the decomposition of $(CH_3NH)_2H^{+*}$ and of $((CH_3)_2N)_2H^{+*}$ at 345 K, Neilson *et al.* measured $k_b = 2.1 \times 10^5$ and 12×10^5 sec^{-1}, as compared with $k_b = 1.2 \times 10^8$ and 6×10^7 sec^{-1}, respectively, obtained by Meot-Ner and Field.

The low values of k_a obtained from the intercept plots of $1/k_f$ versus $1/M$ and the negative TDs of these k_a values obtained in the low pressure measurements were unexpected. Neilson *et al.* offered several hypotheses to explain these results and the difference between these results and the high pressure results of Meot-Ner and Field. One possibility is that, as suggested by Olmstead *et al.* (1977), k_b in fact increases with pressure for the following reason. At low pressure all complexes with energy higher than the critical energy decompose. However, at high pressure, only the more energetic, short lived complexes dissociate, since low-energy, longer lived complexes will be stabilized by collision. Thus, k_b at high pressure reflects only the decomposition of the high energy, short lived activated complexes. As a result, k_b may increase by several orders of magnitude in going from ICR measurements at $P \approx 10^{-3}$ torr to high pressure MS at $P \approx 1$ torr. Such a change in k_b would explain the difference between the ICR and MS results. It would also invalidate the extrapolation used to obtain k_a values from plots of $1/k_f$ versus $1/(M)$, since in this extrapolation k_b is considered constant from measurements at 10^{-2} torr until the intercept at $(M) = \infty$. Indeed, when Neilson *et al.* used $k_a = k_{ADO}$ instead of the extrapolated k_a values, and $k_f^{(3)}$ as obtained from slopes of the plots of $1/k_f$ versus $1/M$, they obtained from Fig. 16 $k_b = 2.5 \times 10^7$ and 4.1×10^6 sec^{-1} for the decomposition of $(CH_3NH_2)_2H^{+*}$ at 345 K. These values are much closer to the high pressure values of Meot-Ner and Field than those obtained using extrapolated k_a values.

If the change in k_b with P is not large, the extrapolated k_a values should be meaningful. Small values k_a/k_{ADO} could be then explained (Neilson *et al.*, 1978) if most collisions between the dimerization partners result in back dissociation within one vibrational period. According to this mechanism, only in a small fraction of collision will the relative translational energy of the reactants be distributed into internal degrees of freedom. Only this small fraction of collisions will form a long lived complex which could be stabilized under ICR conditions and to which statistical RRKM theory may be applied. Such a dynamic model could also explain the increase of k_a with reactant complexity, but it would not easily explain the negative TD of k_a.

Despite quantitative differences in the kinetic results, the results of Neilson *et al.* and Meot-Ner and Field agree on some qualitative trends: In both works, the pressure dependence agrees with the energy transfer mechanism; a steric effect decreases k_a for the association of tertiary and other hindered amines; and the limiting high-pressure value of the second-order rate constant, when different from k_c, exhibits a large negative TD.

At the present time there are not sufficient data to decide whether differences between the high and low pressure measurements are due to a real pressure dependence of k_b or same other factors. Measurements at intermediate pressures would be useful to resolve this problem.

C. *Temperature and Pressure Effects in Condensation Reactions*

Some IM association reactions lead to the formation of new covalent bonds (Table V). Thus Hiraoka and Kebarle (1975b, 1976, 1977a,b) observed condensation reactions of $C_2H_5^+$ and s-$C_3H_7^+$ with CH_4 to yield higher carbonium ions and with H_2 to yield protonated alkanes; also with H_2O and $CO + H_2O$ to yield protonated alcohols and protonated carboxylic acids. Meot-Ner (1978) investigated the condensation of s-$C_3H_7^+$ and t-$C_4H_9^+$ with HCN which yields protonated isocyanides. Meot-Ner (1979a) also investigated reactions of carbonium ions with ammonia, in which condensation yielding protonated alkyl amines competes with proton transfer.

Hiraoka and Kebarle (1977a) pointed out that condensation reactions in methane which yield carbonium ions, and the further condensation of these carbonium ions to yield protonated alcohols and acids, may take place in reducing planetary atmospheres. The synthesis of amines and nitriles by atmospheric ion–molecule reactions and possible role in prebiotic synthesis was discussed by Meot-Ner (1979b). Planetary conditions encompass a wide variety of pressures and temperatures, and the dependence of the kinetics of the condensation reactions on these variables must be understood

TABLE V

Temperature and Pressure Effects in Ion–Molecule Condensation and Eliminative Condensation Reactions

Reaction	Temp. (K)	k ($cm^3\ sec^{-1}$ for Bimolecular or $cm^6\ sec^{-1}$ for Trimolecular Reactions)	Comments	Reference
$C_2H_5^+ + H_2 \rightarrow C_2H_7^+$	175 294	2 (−15) 1 (−14)	$k(T) = 7.3 \times 10^{-14} \exp(-1.2/RT)\ cm^3\ sec^{-1}$ at $T > 175$ K; at this high T range reaction is second order, product is protonated ethane; at $T < 140$ K association leading to $C_2H_5^+ \cdot H_2$ occurs	Hiraoka and Kebarle (1976)
$C_2H_5^+ + CH_4 \rightarrow C_3H_7^+ + H_2$	250 632	4 (−15) 9 (−14)	$k(T) = 6 \times 10^{-13} \exp(-2.5/RT)\ cm^3\ sec^{-1}$; product is *s*-propyl ion at $T > 250$ K; at lower T the addition product $C_3H_9^+$ is formed; at $T > 250$ K second-order kinetics is confirmed by P study	Hiraoka and Kebarle (1975b)

TABLE V (*Continued*)

Reaction	Temp. (K)	k ($cm^3\ sec^{-1}$ for Bimolecular or $cm^6\ sec^{-1}$ for Trimolecular Reactions)	Comments	Reference
$s\text{-}C_3H_7^+ + HCN(+CH_4) \rightarrow s\text{-}C_3H_7NCH^+$ (or $s\text{-}C_3H_7^+ \cdot HCN$) $(+CH_4)$	250 280 280 395 640	1.5 (−10) 4.3 (−11) 7 (−28) 4 (−28) 1 (−11)	Three body reaction, reaching high pressure second-order limit at 250 K, $1.7 < N_{CH_4} < 7.5 \times 10^{-16}\ cm^{-3}$, and at 280 K, $6.5 < N_{CH_4} < 10.5 \times 10^{16}\ cm^{-3}$; limiting second-order rate constant shows negative TD; kinetics intermediate between third and second order at 280 K, $2.5 < N_{CH_4} < 4.5 \times 10^{16}\ cm^{-3}$ and at 395 K, $1.5 < N_{CH_4} < 7.5 \times 10^{16}\ cm^{-3}$; at 580 K second-order kinetics with positive TD is observed; product at high T is protonated propyl isocyanide; at low T, structure is unknown	Meot-Ner (1978)
$t\text{-}C_4H_9^+ + HCN(+CH_4) \rightarrow t\text{-}C_4H_9NCH^+$ (or $t\text{-}C_4H_9^+ \cdot HCN$) $(+CH_4)$	270 310 360 420	1.0 (−27) 5.6 (−28) 2.5 (−11) 6.0 (−11)	Third-order kinetics at 277 K, $2.8 < N_{CH_4} < 9.5 \times 10^{16}\ cm^{-3}$; at $T < 320$ TD is negative, structure of product uncertain; at $T > 320$ K TD is positive, kinetics is second order, product is protonated t-butyl isocyanide	Meot-Ner (1978)

$s\text{-}C_3H_7^+ + NH_3 + CH_4 \rightarrow s\text{-}C_3H_7NH_3^+ + CH_4$	200–370 420 550	8 (−27) 4 (−27) 0	In this reaction condensation, yielding a protonated amine, competes with bimolecular H^+ transfer yielding NH_4^+; branching ratio (condensation/H^+ transfer) is proportional to N_{CH_4} between $2 < N_{CH_4} < 9 \times 10^{16}$ cm^{-3}, and independent of T in range indicated (200–370 K); condensation decreases above 400 K	Meot-Ner (1978)
$s\text{-}C_3H_7^+ + n\text{-}C_3H_7NH_2 + CH_4 \rightarrow (s\text{-}C_3H_7)(n\text{-}C_3H_7)NH_2^+ + CH_4$	230–370	1 (−26)	See comments for preceding reaction	Meot-Ner (1979a)
$t\text{-}C_4H_9^+ + NH_3 + CH_4 \rightarrow t\text{-}C_4H_9NH_3^+ + CH_4$	200–400 450 >500	1 (−27) 5 (−28) ≈0	See comments for preceding reaction	Meot-Ner (1979a)
$t\text{-}C_4H_9^+ + n\text{-}C_3H_7NH_2 + CH_4 \rightarrow (t\text{-}C_4H_9)(n\text{-}C_3H_7)NH_2^+ + CH_4$	230–470	4 (−27)	See comments for preceding reaction	Meot-Ner (1979a)
$C_6H_5CH_2^+ + C_6H_5OC(O)CH_3 \rightarrow C_6H_5CH_2C_6H_5CH_2^+ + CH_3COOH$	410 520 572	7.1 (−10) 4.6 (−10) 2.3 (−10)	Temperature coefficient increasingly negative with increasing temperature	Meot-Ner *et al.* (1977a)

in order to evaluate the contribution of these reactions to prebiotic organic synthesis. The complex pressure and temperature dependence of condensation reactions will be the subject of this section.

1. Condensation of Carbonium Ions with H_2

These reactions were investigated by Hiraoka and Kebarle (1976). Equilibrium studies of the reactions $C_2H_5^+ + H_2 \rightleftarrows C_2H_7^+$ at low temperatures (113–143 K) yielded $\Delta H^\circ = -4 \pm 0.5$ kcal mole^{-1}. Using this value, the authors obtained $\Delta H_f^\circ = 215$ kcal mole^{-1} for the product ion, and concluded that the low temperature reaction yields a cluster-type ion in which $C_2H_5^+$ interacts weakly with the covalent bond of H_2

$$\left(CH_3CH_2\!-\!\!\!<\begin{matrix}H\\H\end{matrix}\right)^+. \quad \text{(a)}$$

Qualitative observations indicated that at low temperature the reaction was of third order with a negative TD and involving no activation energy.

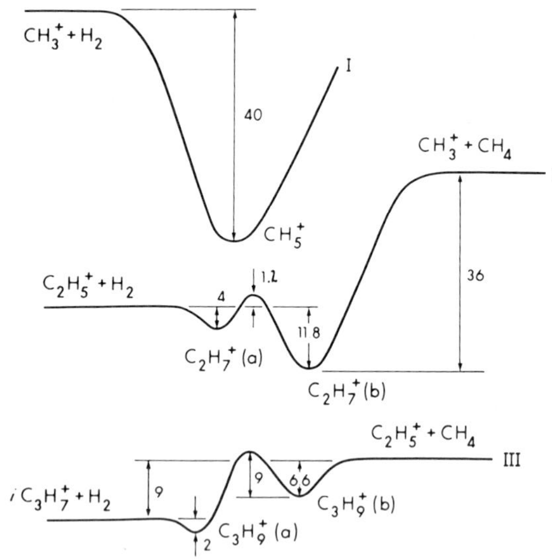

FIG. 17 Schematic potential energy diagrams for some isomeric protonated alkanes. Diagram I is based on literature data for CH_5^+. Diagram II is based on reaction $C_2H_5^+ + H_2 = C_2H_7^+$ (a), $\Delta H = -4.0$ kcal mole^{-1} occurring at low temperature; and $C_2H_5^+ + H_2 = C_2H_7^+$ (b) occurring at higher temperature with an activation energy $E = 1.2$ kcal mole^{-1} and $\Delta H = -11.8$ kcal mole^{-1}. Species $C_2H_7^+$ (a) and $C_2H_7^+$ (b) are identified in the text as having structures a and b. Diagram III is for $C_3H_9^+$. [From Hiraoka and Kebarle (1976). Reprinted with permission from *J. Am. Chem. Soc.* Copyright by the American Chemical Society.]

The association of $C_2H_5^+$ with H_2 was also investigated at higher temperatures. Equilibrium measurements at 358–473 K yielded $\Delta H^\circ = -11.8$ kcal mole^{-1}, and ΔH_f° of the association product at high temperature was 207.2 kcal mole^{-1}. The authors assigned to this product the C—C protonated ethane structure

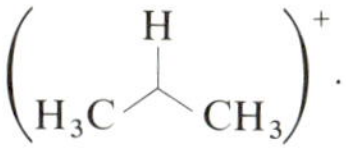

(b)

A pressure study at 226 K, $2.2 < P_{H_2} < 6.8$ torr, revealed that the association reaction at high temperature displays second-order kinetics. In the range $173 < T < 313$ K, a positive TD was observed with an activation energy of 1.2 ± 0.1 kcal mole^{-1}, with $k = 7.3 \pm 0.9 \times 10^{-14} \exp(-1.2/RT)$ cm^3 sec^{-1}. Hiraoka and Kebarle used thermochemical data obtained from equilibrium studies combined with the activation energy measured in the kinetic study to construct the diagram shown in Fig. 17.

2. *The Eliminative Condensation* $C_2H_5^+ + CH_4 \rightarrow C_3H_7^+ + H_2$

This reaction too was studied by Hiraoka and Kebarle (1975b) in the temperature range 250–650 K. Pressure studies at $2 < P < 7$ torr showed that the kinetics is of second order. The rate constant exhibited a positive TD with $k = 6 \times 10^{-13} \exp(-2.5/RT)$ cm^3 sec^{-1}. At temperatures below 250 K, association to form $C_3H_9^+$ is observed. The authors proposed the following mechanism:

$$C_2H_5^+ + CH_4 \rightleftarrows (C_3H_9)^{+*} \xrightarrow{P} C_3H_9^+ \rightarrow s\text{-}C_3H_7^+ + H_2,$$

Scheme IV

wherein at high temperature, decomposition to product occurs, while at low temperature, stabilization predominates. The activation energy was considered to be due to the transition from the low energy structure (b) of the intermediate $C_3H_9^+$ to structure (a)

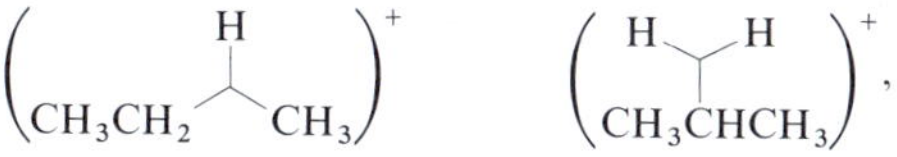

(a) (b)

which is the higher energy precursor of dissociation to s-$C_3H_7^+$ and H_2. Using the measured activation energy and other themical data, the potential energy diagram for this system could also be constructed (Fig. 17).

The diagrams in Fig. 17 constitute an elegant example of the use of temperature studies to obtain detailed quantitative energy diagrams for interacting ion–molecule systems.

3. Condensation of Carbonium Ions with HCN

The condensation of s-$C_3H_7^+$ and t-$C_4H_9^+$ with HCN was investigated recently by Meot-Ner (1978).

Equilibrium studies on the reactions

$$\text{s-}C_3H_7^+ + HCN \rightleftarrows \text{s-}C_3H_7NCH^+ \tag{21}$$

and

$$\text{t-}C_4H_9^+ + HCN \rightleftarrows \text{t-}C_4H_9NCH^+, \tag{22}$$

in the range $570 < T < 645$ and $320 < T < 420$ K, respectively, yielded the results of $\Delta H^\circ = -30.8$ and $\Delta H^\circ = -16.3$ kcal mole^{-1}, respectively. These results yielded $\Delta H_f^\circ = 193$ kcal mole^{-1} for the product of reaction (21) and $\Delta H_f^\circ = 186$ kcal mole^{-1} for the product of reaction (22). These values are higher by about 15 kcal mole^{-1} than ΔH_f° of the respective protonated cyanides s-$C_3H_7CNH^+$ and t-$C_4H_9CNH^+$. The thermodynamic data indicates the formation of protonated isocyanides in these condensation reactions. In a pressure study the condensation of t-$C_4H_9^+$ with HCN showed second-order kinetics at temperatures of $320 < T < 420$ K where equilibrium could be observed. In this temperature range the reaction displayed a positive TD with an activation energy of ~5.5 kcal mole^{-1}. At lower temperatures third-order kinetics with a negative TD was observed (Fig. 18).

The kinetics of reaction (21) shows even more complex changes with temperature and pressure (Fig. 19). At 280 K a transition from third- to second-order kinetics is observed; at 250 K the second-order limit is reached even at the lowest third-body density available ($N_{CH_4} = 1.5 \times 10^{16}$ cm^{-3}). It is interesting that these high-pressure limiting second-order rate constants are significantly lower than k_c and exhibit a negative TD. At a higher temperature (395 K), transition from third- toward second-order kinetics is seen. The pressure studies thus indicate that at low temperature the reaction is a three-body process and, correspondingly, negative TD of $k_f^{(3)}$ observed. At a high temperature, second-order kinetics is observed and positive TD commences (Fig. 18).

The kinetic changes with temperature and pressure in the condensation of s-$C_3H_7^+$ and t-$C_4H_9^+$ with HCN are similar to the observations of Hiraoka and Kebarle (1976) concerning the condensation of $C_2H_5^+$ with H_2: third-order kinetics with negative TD at low temperatures; second-order kinetics with positive TD at high temperature. Indeed, this seems to be a general feature of the condensation reactions investigated to date. In the case of $C_2H_5^+ + H_2$, this behavior was elegantly demonstrated to result from the formation of a weakly bound cluster ion at low T and rearrangement thereof

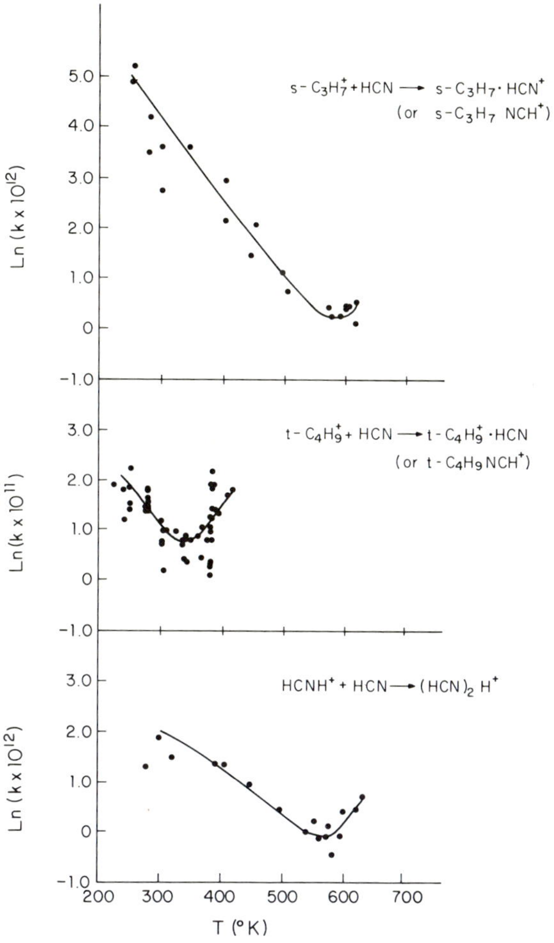

FIG. 18 Apparent second-order rate constants for the association reactions of $HCNH^+$, s-$C_3H_7{}^+$, and t-$C_4H_9{}^+$ with HCN as a function of temperature. All the measurements for the dimerization of $HCNH^+$ were performed at a third-body density within $\pm 10\%$ of $N_{CH_4} = 1.5 \times 10^{16}$ cm^{-3}; for the reaction of s-$C_3H_7{}^+$ at $N_{CH_4} = 4.8 \times 10^{16}$ cm^{-3} and the reactions of t-$C_4H_9{}^+$ at $N_{CH_4} = 4.4 \times 10^{16}$ cm^{-3}. Precision of the rate constants is estimated as $\pm 20\%$. [From Meot-Ner (1978).]

to a more stable condensation product at high T. In the case of s-$C_3H_7{}^+$ + HCN and t-$C_4H_9{}^+$ + HCN, no thermodynamic data is available on the low T product. It is possible that this is a cluster ion. This should be surprising, however, since attack on the positive charge center of the carbonium

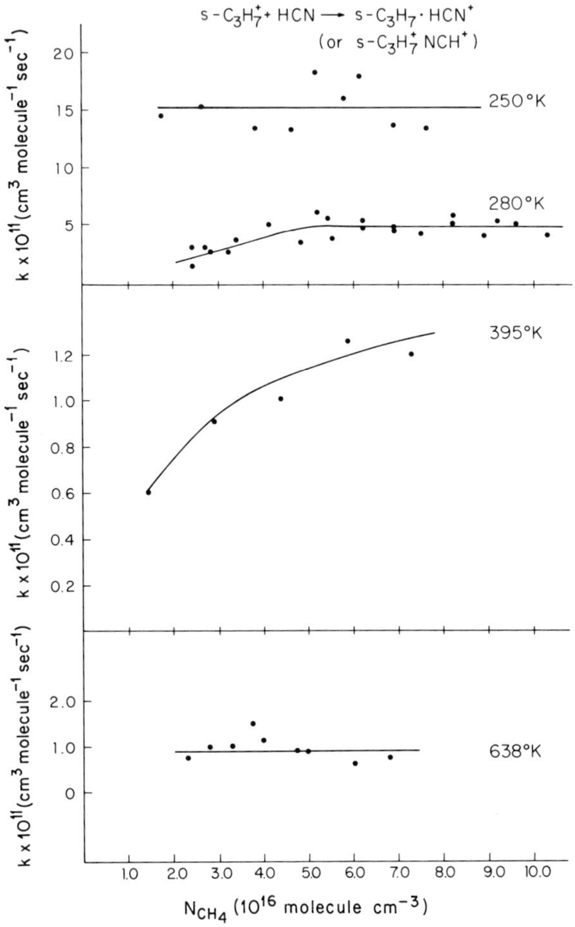

FIG. 19 The dependence of the apparent second-order rate constant for the association of *s*-$C_3H_7^+$ with HCN on the number density of the third-body CH_4 at four temperatures between 250 and 640 K. [From Meot-Ner (1978).]

ions by the cyanide nitrogen which carries a large fractional negative charge should lead straightforward to a covalently bonded protonated isocyanide.

4. *Competitive Condensation–Proton Transfer Reactions*

In reactions of carbonium ions (R^+) with amines (A) condensation can compete with proton transfer if the pressure is sufficiently high (Meot-Ner, 1979a). Pressure and temperature studies were preformed to investigate whether an excited σ-bonded complex, which is similar to the condensation

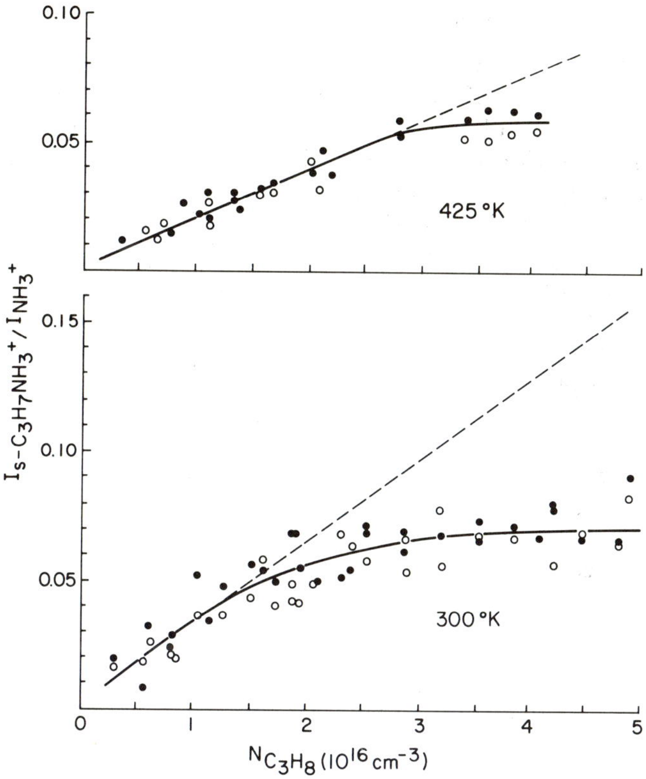

FIG. 20 Effects of the third-body density $N_{C_3H_8}$ on the ratio (condensation product/transfer product) in the reaction

$$s\text{-}C_3H_7^+ + NH_3 \longrightarrow NH_4^+ + C_3H_6$$
$$\hookrightarrow s\text{-}C_3H_7NH_3^+.$$

[From Meot-Ner (1979a).]

product, is the intermediate for proton transfer, i.e., if the following mechanism is adequate:

$$R^+ + A \rightarrow (RA^{+*}) \begin{cases} \xrightarrow{p} AH^+ + (R - H) \\ \xrightarrow[(M)]{s} RA^+ \end{cases}$$

Scheme V

RA^{+*} represents the σ-bonded complex, for example, $RA^{+*} = s\text{-}C_3H_7NH_3^{+*}$ in the $s\text{-}C_3H_7^+ + NH_3$ system.

The following results were obtained:

(a) In the reaction system of $s\text{-}C_3H_7^+ + NH_3$, the product distribution expressed as the ratio of I_{RA^+}/I_{AH^+}, i.e., $I_{s\text{-}C_3H_7NH_3^+}/I_{NH_4^+}$, was measured as a function of third-body (C_3H_8) density. It was found that I_{RA^+}/I_{AH^+} increased linearly with $N_{C_3H_8}$ at low third-body densities, but the ratio leveled off to 0.06 at high densities (Fig. 20). Similar results were obtained in the system $s\text{-}C_3H_7^+ + C_2H_5NH_3$ (Fig. 21).

(b) In both systems, the high pressure limiting value $(I_{RA^+}/I_{AH^+})_{HP}$ was reached at higher values of $N_{C_3H_8}$ as the temperatures increased (Figs 20, 21).

(c) $(I_{RA^+}/I_{AH^+})_{HP}$ was independent of temperature in all four reaction systems where the TD of $(I_{RA^+}/I_{AH^+})_{HP}$ was investigated (Fig. 22).

(d) The displacement reaction

$$s\text{-}C_3H_7^+ + C_2H_5NH_2 \rightarrow s\text{-}C_3H_7NH_3^+ + C_2H_4$$

is insignificant even though it is thermodynamically favored over proton transfer. Similar observations were made in the $t\text{-}C_4H_9^+ + C_2H_5NH_2$ system.

(e) Little condensation is observed when A is a highly hindered tertiary amine.

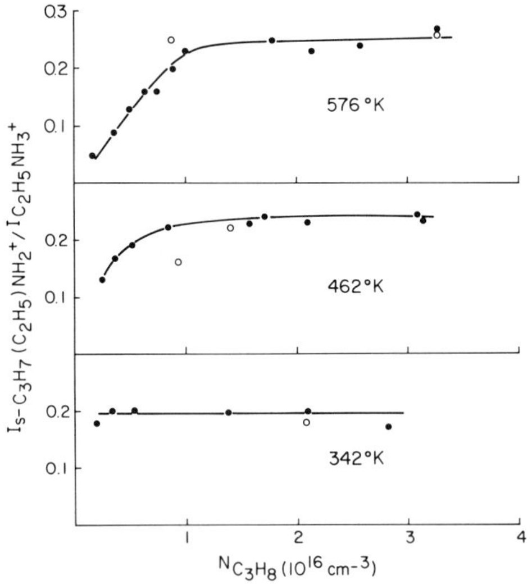

FIG. 21 Effects of the third-body density $N_{C_3H_8}$ on the ratio (condensation product/transfer product) in the reaction

$$s\text{-}C_3H_7^+ + C_2H_5NH_2 \longrightarrow C_2H_5NH_3^+ + C_3H_6$$

$$\hookrightarrow s\text{-}C_3H_7(C_2H_5)NH_2^+.$$

[From Meot-Ner (1979a).]

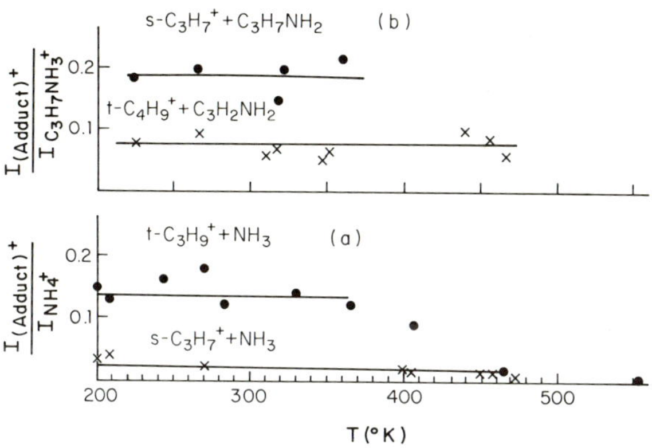

FIG. 22 Ratios of adduct ions to protonated ammonium ions formed in the reactions of s-$C_3H_7^+$ and t-$C_4H_9^+$ with NH_3 and n-$C_3H_7NH_2$ as a function of temperature. Experiments were conducted in mixtures of 1% C_3H_8 or i-C_4H_{10} + 0.05% NH_3 or n-$C_3H_7NH_2$ in CH_4 at total particles densities within 20% of 3×10^{16} cm^{-3}. [From Meot-Ner (1979a).]

Observation (d) is not consistent with a σ-bonded, statistical intermediate

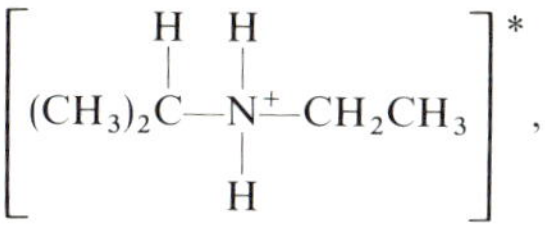

since all exothermic reaction channels of such a species should be observed. Observation (a) is also inconsistent with Scheme V, since according to this scheme I_{RA^+}/I_{AH^+} should increase with (M) until I_{RA^+} is the only product. The temperature effects are also inconsistent with Scheme V, since it is expected that unimolecular rate constant k_p should increase with T, but k_s should not be significantly affected. Thus, according to Scheme V, $I_{RA^+}/I_{AH^+} = k_s(M)/k_p$ should be inversely related to T, but according to observation (c) this is not so.

Since all the results indicate that Scheme V is not adequate, the following mechanism is proposed:

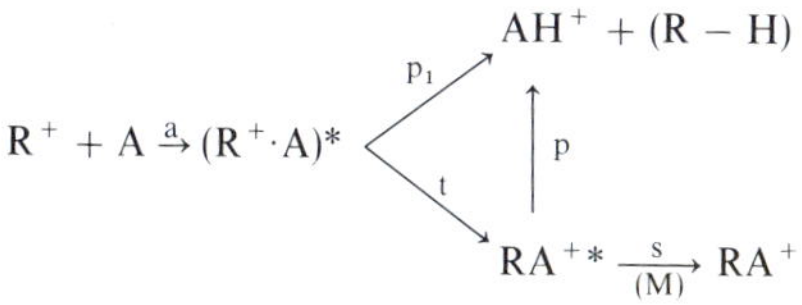

Scheme VI

The essential new feature is that the loose complex $(R^+ \cdot A)^*$ plays an important role as the common precursor of RA^{+*} and of the proton transfer product AH^+. In other words, the major route to AH^+ is via proton transfer in the loose complex. Mathematical analysis of Scheme VI shows that at low pressures

$$(I_{RA^+}/I_{AH^+})_{LP} = k_t k_s(M)/(k_t + k_{p_1})k_p, \tag{23}$$

i.e., the product ratio is proportional to (M). At high pressures

$$\left(\frac{I_{RA^+}}{I_{AH^+}}\right)_{HP} = \frac{k_t}{k_{p_1}} \tag{24}$$

and the product ratio is finite and independent of (M). These predictions are consistent with result (a).

According to Scheme VI, the transition between low and high pressure behavior depends on the relative values of k_p and $k_s(M)$. If k_p increases with T, the high pressure limit will be reached at higher (M) as T increases; this is consistent with (b). Since RA^{+*} is a minor reaction intermediate, displacement will be at most a minor channel; this is consistent with observation (d).

It was also proposed that $(R^+ \cdot A)^*$ is a loose complex in which R^+ and A tumble freely and undergo multiple intracomplex collisions. The occurrence of processes p_1 or t depends only on the relative orientations of R^+ and A in the intracomplex collisions, and the ratio k_t/k_{p_1} is determined only by the relative magnitudes of the steric factors which are conducive to the respective processes. Thus, $k_t/k_{p_1} = (I_{RA^+}/I_{AH^+})_{HP}$ will be independent of T, consistent with observation (c). The observation that steric factors effect $(I_{RA^+}/I_{AH^+})_{HP}$ [observation (e)] is also consistent with this model.

IV. TEMPERATURE AND PRESSURE EFFECTS IN THE THERMAL DECOMPOSITION OF GASEOUS IONS

Although the decomposition of energetic ions has been thoroughly investigated in relation to electron ionization mass spectrometry, there is very little information on thermal decomposition of gaseous ions. The data available to date on temperature and pressure effects relates to the thermal decomposition of carbonium ions and of cluster ions. Rate data and Arrhenius parameters were also reported for the decomposition of protonated benzyl acetate (Field, 1969a,b), amino acid derivatives (Meot-Ner and Field, 1973), and other protonated ions formed in CIMS (Weeks and Field, 1970, 1973; Field and Weeks, 1970; Laurie and Field, 1972a,b). In all of these systems, the formation of fragment ions as compared with protonated molecular ions increases substantially with temperature. A subsequent investigation using pulsed CIMS (Meot-Ner *et al.*, 1977a) showed however that the fragment $C_7H_7^+$ is formed not by decomposition of the protonated benzyl acetate ion $BzOAcH^+$, but probably by the competitive decomposition of the BzOA ⊙ c-t-$C_4H_9^+$ intermediate to $BzOAcH^+$ or to $C_7H_7^+$. While the pronounced enhancement of fragementation is due to the thermal activation of the BzOAC ⊙ t-$C_4H_9^+$ intermediate, the thermal decomposition of $BzOAcH^+$ assumed in the calculation of the kinetic data by Field and co-workers is thus subject to question, and it will not be reviewed here.

A. Thermal Decomposition of Carbonium Ions

The data on the thermal pyrolysis of C_7 carbonium ions in CH_4 carrier gas is summarized in Table VI. At the accessible source pressures of $0.5 < P_{CH_4} < 3.0$ torr at $T = 537$ K, the reactions are in the bimolecular falloff region, i.e., the unimolecular rate constants increase with increasing P_{CH_4} (Fig. 23). Similar results, although somewhat closer approach to the unimolecular high-pressure limit, were observed at a lower temperature, 503 K.

Activation energies were obtained from temperature studies at two constant CH_4 densities. The results are also summarized in Table VI.

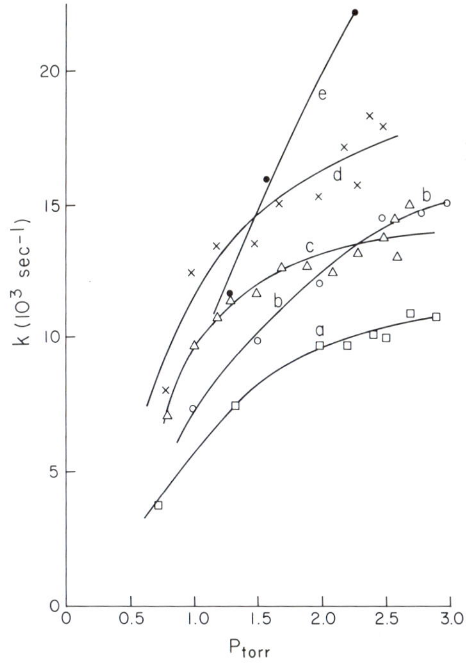

FIG. 23 Pressure dependences of the decomposition rate constants of $C_7H_{15}^+$ ions formed in 1% heptane in propane at 537 K for the following heptanes: (a) 2-methylhexane, (b) 2,2,3-trimethylbutane, (c) 3-ethylpentane, (d) n-heptane, (e) 2,4-dimethylpentane. [From Meot-Ner and Field (1976b). Reprinted with permission from *J. Phys. Chem.* Copyright by the American Chemical Society.]

TABLE VI

Temperature Dependences of Thermal Decomposition Reactions of Gaseous Ions

Reaction	Temp. (°K)	k^a	A^a	E_a (kcal $mole^{-1}$)	$(M)^b$ (10^{16} cm^{-3})	$D°$ (kcal $mole^{-1}$)	Reference
		$k(T) = A \exp(-E_a/RT)$					
$H_5^+(+H_2) \rightarrow H_3^+ + H_2(+H_2)$	150	4 (−18)	8.7 (−6)	8.4		9.7	Hiraoka and Kebarle (1975a)
	270	2 (−12)					
$C_2H_7^+(+CH_4) \rightarrow C_2H_5^+ + H_2(+CH_4)$	330	2.2 (−15)	8.3 (−8)	10.5			French and Kebarle (1975)
	485	1.2 (−12)					
$H_3O^+ \cdot H_2O(+CH_4) \rightarrow H_3O^+ + H_2O(+CH_4)$	400	8.2 (−23)	2.0 (−2)	29.8		33.0	Meot-Ner and Field (1977)
$H_3S^+ \cdot H_2S(+CH_4) \rightarrow H_3S^+ + H_2S(+CH_4)$	300	5.0 (−12)	1.3 (−4)	10.2		12.6	Meot-Ner and Field (1977)
	400	3.6 (−10)					
$NH_4^+ \cdot NH_3(+CH_4) \rightarrow NH_4^+ + NH_3(+CH_4)$	400	3.1 (−15)	2.0 (−4)	19.8		24.8	Meot-Ner and Field (1977)
$CH_3NH_3^+ \cdot CH_3NH_2(+i\text{-}C_4H_{10}) \rightarrow CH_3NH_3^+ + CH_3NH_2 + (i\text{-}C_4H_{10})$	400	1.7 (−14)	1.6 (−4)	18.3		21.7	Meot-Ner and Field (1977)
$(CH_3)_2NH_2^+ \cdot (CH_3)_2NH(+i\text{-}C_4H_{10}) \rightarrow (CH_3)_2NH_2^+ + (CH_3)_2NH(+i\text{-}C_4H_{10})$	400	1.1 (−11)	4.0 (−2)	17.5		20.8	Meot-Ner and Field (1977)
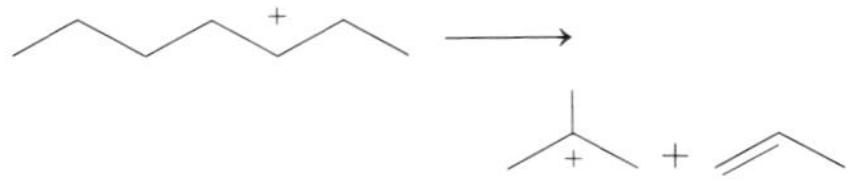	503	2.2 (+3)	2.0 (+9)	13.9	1.6		Meot-Ner and Field (1976b)
	503	6.2 (+3)			4.4		
	537	1.2 (+4)			1.6		
	537	1.8 (+4)			4.4		
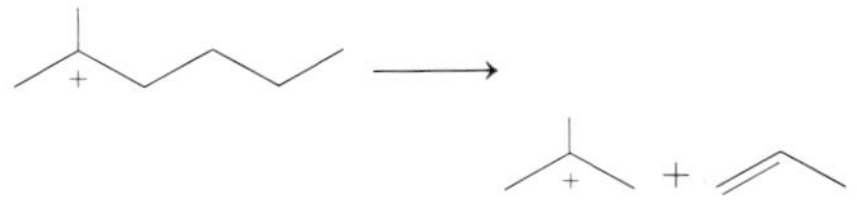	503	2.8 (+3)	1.0 (+9)	12.9	1.6	23.0	Meot-Ner and Field (1976b)
	503	4.2 (+3)	2.3 (+10)	15.6	4.4		
	537	6.0 (+3)			1.6		
	537	1.6 (+4)			4.4		

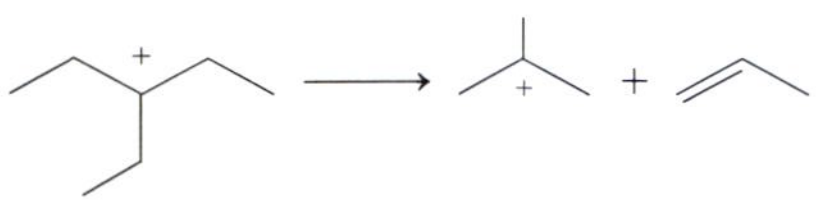	503	3.8 (+3)	1.4 (+9)	12.9	1.6	26.0	Meot-Ner and Field (1976b)
	503	5.5 (+3)	4.4 (+10)	16.0	4.4		
	537	9.5 (+3)			1.6		
	537	1.4 (+3)			4.4		
	503	5.2 (+3)	7.9 (+8)	12.0	1.6	25.6	Meot-Ner and Field (1976b)
	503	1.2 (+3)			4.4		
	537	1.1 (+4)			1.6		
	537	2.4 (+4)			4.4		
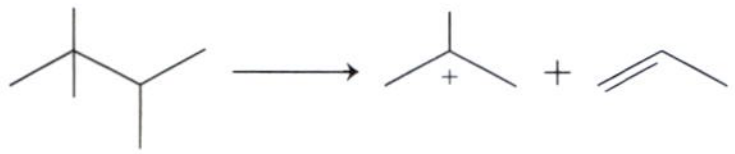	503	0.9 (+3)			1.6	29.5	Meot-Ner and Field (1976b)
	503	3.0 (+3)	2.4 (+10)	16.0	4.4		
	537	8.8 (+3)			4.4		

[a] Rate constants and Arrhenius A factors are reported in units of $cm^3\ sec^{-1}$ for the first seven reactions, since these reactions are second order in the pressure range where kinetic data were obtained (1–3 torr) (see text).

[b] Pressure studies for the decomposition of the carbonium ions listed indicate "falloff" behavior, i.e., kinetics intermediate between first and second order. For these reactions, k and A are reported in units of sec^{-1}. Since these reactions are approximately second order at the lower density reported ($1.6 \times 10^{15}\ cm^{-3}$), second-order rate constants and A factors can be calculated by $k^{(2)} = k^{(1)}/(M)$ and $A^{(2)} = A^{(1)}/(M)$, using $k^{(1)}$ and $A^{(1)}$ values reported at $(M) = 1.6 \times 10^{16}\ cm^{-3}$.

We note that activation energies obtained at $N_{CH_4} = 4.4 \times 10^{16}\ cm^{-3}$ are higher than those obtained at $N_{CH_4} = 1.6 \times 10^{16}\ cm^{-3}$. In all cases the measured activation energies are in the range 12–16 kcal mole^{-1}, which is significantly below the endothermicities of the decomposition reactions (23–30 kcal mole^{-1}), and therefore necessarily below the energy barriers for the decomposition reactions. All of the features observed: transition from second- to first-order kinetics with increasing P_{CH_4}, increase in activation energies with P_{CH_4}, and activation energies in the bimolecular region which are lower than the energy barrier, are common features of unimolecular reactions and are dealt with in detail by the RRKM theory (Robinson and Holbrook, 1972). The limiting low pressure activation energies E_a° are related to the limiting high pressure activation energy E_a^{∞} in terms of the RRK coupled degenerate classical oscillator model (Wayne, 1969)

$$E_a^{\circ} = E_a^{\infty} - (s - 1)RT. \tag{25}$$

At the low pressure limit second-order rate constants may be derived from the first-order constants by $k^{(2)} = k^{(1)}/N_{CH_4}$. Similarly, second-order Arrhenius A factors $A^{(2)}$ may be calculated from the first-order A factors $A^{(1)}$ listed in Table VI by $A^{(2)} = A^{(1)}/N_{CH_4}$. For example, for the decomposition of the 2-methylhexyl ion at $N_{CH_4} = 1.6 \times 10^{16}\ cm^{-3}$, we find $A^{(2)} = 1 \times 10^9/(1.6 \times 10^{16}) = 6.3 \times 10^{-8}\ cm^3\ sec^{-1}$. $A^{(2)}$ is larger by approximately a factor of 60 than the collision rate between the decomposing ions and CH_4. Indeed, for all the carbonium ion decompositions listed in Table VI, $A^{(2)}$ is greater by one to two orders of magnitude than the rate constant for collision of the ions with the third body. The large $A^{(2)}$ values reflect the fact that even collisions in which the relative kinetic energy is smaller than the critical energy may lead to decomposition due to contribution from internal degrees of freedom of the decomposing ion.

B. *Thermal Dissociation of Cluster Ions*

All information on the kinetics of decomposition of cluster ions results from calculation of decomposition rate constants using the measured association equilibrium constants (K) and the relation

$$k_d = k_f K_{assoc}^{-1}, \tag{26}$$

where k_d is the rate constant for dissociation and k_f is the overall rate constant for association. In all cases listed in Table VI, k_f was measured as a third-order rate constant. Since the dimension of K is atomsphere^{-1}, Eq. (26) implies that if k_f is of third order, k_d must be of second order, i.e., the dissociation reaction is at the low pressure bimolecular limit. Inspection of Table VI shows that the activation energies are lower than the dissociation energies D°, similar to what was observed in the decomposition if carbonium ions. Table VI also shows that the preexponential factors for the decomposition rate constants of cluster ions are larger by several orders of magnitude than the typical preexponential factor for activating collisions, i.e., $\sim 10^{-9}\ \mathrm{cm^3\ sec^{-1}}$. As noted in the previous section, this can be explained on the basis of activation from internal degrees of freedom.

C. *The Relation between Activation Energies of Dissociation and Negative Temperature Dependences of the Reverse Association Reactions*

Activation energies for decomposition which are lower than D° are related to the negative temperature coefficients of the inverse association reaction as follows. The TD of the association reactions is usually of the form $k_f = CT^{-n}$. Substituting for k_f, Eq. (24) yields

$$k_d = \mathrm{C}T^{-n}K_{\mathrm{assoc}}^{-1} = \mathrm{C}T^{-n}\exp(\Delta G^\circ_{\mathrm{assoc}}/RT) \tag{27}$$

and differentiation yields:

$$E_a{}^\circ \equiv -R\frac{d(\ln k_d)}{d(1/T)} = D^\circ - nRT. \tag{28}$$

If the association reaction has no energy barrier, then for the corresponding dissociation reaction $E_a{}^\infty = D^\circ$. Thus, comparing Eqs. (26) and (28), one obtains $n = s - 1$, which is consistent with Eq. (15).

Physically, the lowering of the activation energy for dissociation from $E_a{}^\infty = D^\circ$ at high pressure to $E_a{}^\circ$ at low pressure and the negative temperature coefficient of k_{assoc} are related since they both result from the dissociation of excited complexes at low pressures due to contribution of energy from internal degrees of freedom.

V. CONCLUSIONS

A. *A Summary of Empirical Observations*

In conclusion we shall list some general trends which appear to be emerging from temperature and pressure studies of IM reaction kinetics.

(i) Fast reactions exhibit no TD or small negative TD which may be accounted for on the basis of collision theory.

(ii) Most slow exchange reactions and association reactions exhibit negative TD, which is often of the functional form $k = CT^n$ over a wide range of temperatures. Temperature coefficients of $n = -0.5$ to -10 were reported.

(iii) Slow reactions with negative TD may reach the collision limit at sufficiently low temperatures, and assume fast kinetics without TD below this transition temperature.

(iv) Some slow exchange reactions, in particular H atom transfer reactions, show positive activation energies.

(v) Clustering reactions exhibit third-order kinetics at low pressure, and change to second-order kinetics with increasing pressure, in agreement with the energy transfer mechanism.

(vi) The limiting high pressure rate constant $k_{\text{lim}}^{(2)}$ of association reactions may reach the collision rate k_c of the association partners. However, in some reactions $k_{\text{lim}}^{(2)}$ is up to several order of magnitude smaller than k_c. In such cases $k_{\text{lim}}^{(2)}$ exhibits significantly negative TD.

(vii) Some association reactions produce weakly bound cluster ions at low temperatures and covalently bound condensation products at high temperatures. In these reactions the kinetic behavior at low temperatures is as described in (v) and (vi) above; however, the high temperature condensation reactions exhibit second-order kinetics with positive activation energies.

(viii) When fast proton transfer reactions between carbonium ions (R^+) and amines (A) compete with condensation between the same reactants, the branching ratio (condensation/proton transfer $= I_{RA^+}/I_{AH^+}$) increases with third-body density (M) at low values of (M), and levels off to finite values, $(I_{RA^+}/I_{AH^+})_{HP}$, at high values of (M). $(I_{RA^+}/I_{AH^+})_{HP}$ is independent of temperature. Little condensation occurs when A is a highly hindered amine.

(ix) Positive TD corresponding to small energy barriers of 0.2–1.0 kcal mole^{-1}, were observed in H^+ transfer reactions and clustering reactions. The effects are observed as positive TDs or positive deviation from negative TDs at $T < 150$ K.

(x) Thermal decomposition reactions of gaseous ions at the low pressure limit were observed to proceed with activation energies lower than the reaction endothermicity by 2–15 kcal mole^{-1}.

B. *A Detailed Scheme of Reactions of the Reaction Complex*

Temperature and pressure effects in slow reactions are determined by the competitive reactions of the reaction complex. In Scheme VII several processes which were proposed to account for some of the observations listed above are unified into one general mechanism. Figure 24 presents one possible diagram of the potential energies of the various reaction intermediates in Scheme VII. The relative significance of the reactions in Scheme VII, as well as the absence or presence of the various energy maxima and minima, and their relative heights may vary extensively depending on the

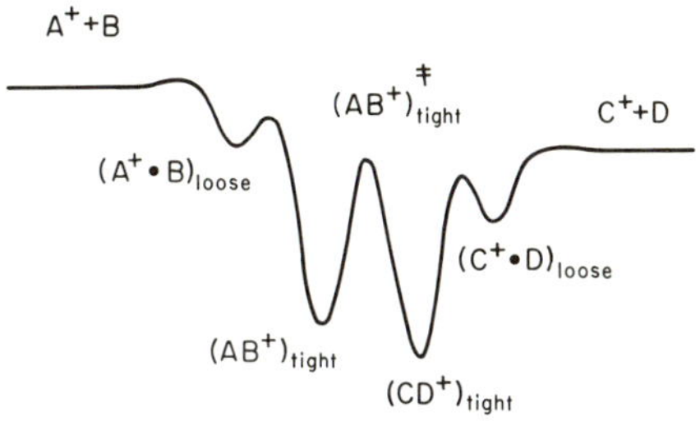

FIG. 24 A schematic potential energy diagram for the generalized ion molecule reaction $A^+ + B \rightarrow C^+ + D$.

individual reaction. It should be noted that Scheme VII shows only the forward reaction in detail; similar sequences may be postulated for the reactions leading from the products to the reactants. Also, the possibility of direct reactions which proceed without a long lived complex is omitted from Scheme VII.

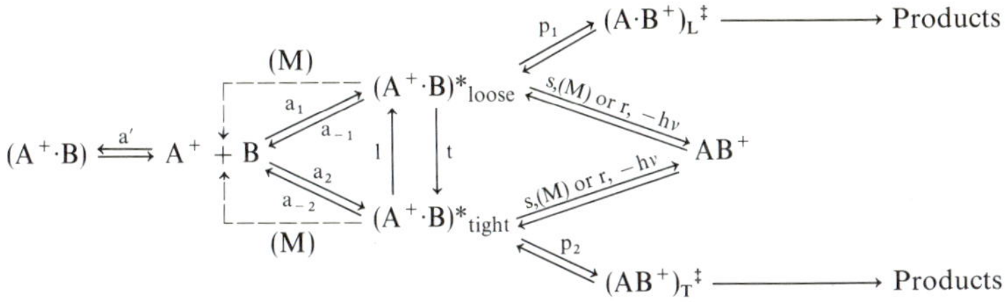

The important feature of Scheme VII is that it postulates a long lived loose complex which is formed directly following the capture collision between A^+ and B and preceding the formation of a tight, specifically bound complex when that is possible. Competitive reactions of the loose complex may result in some of the less well understood aspects of association, transfer, and competitive association–transfer reactions [observations (ii), (iii), (vi), and (viii) above], as will be explained presently. The model was mentioned in the discussion of competitive condensation–transfer reactions.

The reaction sequence between A^+ and B starts with a capture collision. This results in the spiraling orbit of A^+ and B, leading to a collision at the repulsive radii. The collision may result in reflection and prompt back dissociation to reactants within one vibrational period if the relative translational energy of A^+ and B is not distributed into the internal degrees of freedom during the collision. Direct reflection and dissociation process a' in Scheme VII was proposed by Neilson *et al.* (1978) as one possible explanation for limiting high pressure rate constant $k_{\text{lim}}^{(2)}$ which are smaller than $k_{c(A^+,B)}$ in association reactions [observation (vi) above]. According to these authors, the rate constant for the formation of long lived complexes, represented by $(k_{a_1} + k_{a_2})$ in Scheme VI, is given as

$$(k_{a_1} + k_{a_2}) = k_{c(A^+,B)} \frac{k_{ET}}{k_{ET} + \tau_c^{-1}}. \tag{29}$$

Here k_{ET} is the rate constant for intramolecular transfer of energy from the translational to internal degrees of freedom; τ_c the collisional duration. The fraction $k_{ET}/(k_{ET} + \tau_c^{-1})$ may be orders of magnitude smaller than unity.

If the translational energy of the collision is converted into internal degrees of freedom, a long lived complex will be formed. Ion-induced dipole forces are always sufficient to create a shallow potential well, usually 10 kcal mole^{-1} or less deep, for this complex. In this loosely bound complex $(A^+\cdot B)^*_{loose}$, the reactants may be freely tumbling and they may undergo multiple collisions as A^+ and B oscillate between close approach and separation to a distance where reflection will occur due to a maximum in the combined potential + rotational energy surface. $(A^+\cdot B)^*_{loose}$ has a finite lifetime against back dissociation because of the distribution of the electrostatic association energy into internal degrees of freedom. If the mutual orientation of the reactants is favorable in an intracomplex collision, intracomplex transfer reactions in $(A^+\cdot B)^*_{loose}$ may yield a transfer product (process p_1). Alternatively, proper orientation in an intracomplex collision may lead to the specifically bonded (hydrogen or electronically bonded) complex $(A^+\cdot B)^*_{tight}$ (process t). The formation of $(A^+\cdot B)^*_{tight}$ may, however, be slow due to an energy barrier or a low entropy transition state between $(A^+\cdot B)^*_{loose}$ and $(A^+\cdot B)^*_{tight}$. When entropy or steric factors are determinant, k_p and k_t will depend only on the frequency of intracomplex collisions and on the steric factors. The rate constant for back dissociation (k_{a-1}) will depend on the factors that usually effect unimolecular decomposition. The relative values of k_{a-1}, k_t, and k_p will determine if the overall reaction is a fast or slow process. When the reaction proceeds through $(A^+\cdot B)^*_{tight}$, competitive reactions of the specifically bonded complex will also effect the overall kinetics.

As we noted in Section III.C.4, the kinetics of competitive condensation–H^+ transfer reactions strongly implied H^+ transfer directly through the loose complex. Mathematical analysis of Scheme VI, which is an applicable version of Scheme VII, showed that the pressure and temperature effects could be explained by competitive reactions of the loose complex.

Since early stages of the reaction should be independent of the eventual outcome, it is plausible that a loose complex may precede the tight complex even in pure association and transfer reactions. The applicable portions of Scheme VII for association reactions is shown in Scheme VIII.

$$A^+ + B \underset{a-1}{\overset{a}{\rightleftharpoons}} (A^+\cdot B)^*_{loose} \underset{1}{\overset{t}{\rightleftharpoons}} (A^+\cdot B)^*_{tight} \xrightarrow[M]{s} AB^+$$

Scheme VIII

Using the steady state assumption for $(A^+ \cdot B)^*_{loose}$ and $(A^+ \cdot B)^*_{tight}$, the overall rate constant is given by

$$k_f = \frac{k_{a_1} k_t k_s(M)}{k_1 + k_s(M)(k_{a-1} + k_t) - k_1 k_t}. \tag{30}$$

At low pressures

$$(k_f)_{LP} = \frac{k_{a_1} k_t k_s(M)}{k_1 k_{a-1}}. \tag{31}$$

Thus, $(k_f)_{LP}$ is proportional to (M) and the reaction is of third order. At high pressures

$$(k_f)_{HP} = k_{a_1} k_t / (k_{a-1} + k_t) < k_{a_1}. \tag{32}$$

Thus, Scheme VIII predicts that at high pressures the reaction will be second order; but the high pressure limiting rate constant $(k_f)_{HP}$ can be significantly smaller than the collision rate k_{a_1}, as observed in some reactions. This will happen when $k_{a-1} \gtrsim k_t$.

Moreover, since k_{a-1} is a unimolecular dissociation rate constant, k_{a-1} will have a positive TD. On the other hand k_t, according to the above model, depends only on the frequency of intracomplex collisions and on steric factors. Thus, when $k_{a-1} \gtrsim k_t$ and $(k_f)_{HP} < k_{a_1}$, then $(k_f)_{HP}$ will have a negative TD. The value of $(k_f)_{HP}$ can also be lowered by steric crowding, which will decrease k_t.

The small values of the limiting high pressure rate constants of some association reactions, the negative TD of these rate constants, and the observed effects of steric hindrance, as described in Section III.B, can therefore be all explained by Scheme VIII.

Slow transfer reactions may also be treated by assuming the initial formation of a loose complex and the subsequent formation of a tight complex in competition with back dissociation. It is assumed that transfer takes place only in the specifically bonded complex in these reactions. The applicable scheme is similar to Scheme VIII, and it is obtained by substituting process p_2 for s in Scheme VIII. The overall rate constant k_f is obtained by substituting k_{p_2} for $k_s(M)$ in Eq. (30). Arguments similar to those used for the association reactions will predict that k_f should have a negative TD, and that its value will decrease with steric crowding of the reactants. Both trends were observed in H^- transfer reactions, as described in Section II.C.

It must be noted that some features of the slow transfer reactions may also be explained by RRKM treatment of the tight complex alone. Thus, if a tight complex is required for a transfer reaction, a potential surface with two deep wells, corresponding to the reactantlike $(AB^+)_{tight}$ and the productlike $(CD^+)_{tight}$ may apply (Fig. 24). An energy barrier corresponding to the transition complex $(AB^+)^{\neq}_{tight}$ and unfavorable entropy factors for the forward reaction (i.e., a low density of states for $(AB^+)^{\neq}_{tight}$ could account for the low efficiencies of slow transfer reactions. This was shown by Olmstead *et al.* (1977) who applied RRKM theory to processes equivalent to p_2 and a_{-2} in Scheme VII for S_{N^2} nucleophilic displacement reactions and found that k_f was critically dependent on the energy barrier corresponding to $(AB^+)^{\neq}_{tight}$.

In summary, a direct role for the loose complex in the overall mechanism is strongly indicated in the case of competitive condensation–H^+ transfer reactions between carbonium ions and amines. By generalization it is considered that a long lived loose complex may also precede the tight complex in association and slow transfer reactions. Some unexpected kinetic phenomena may be explained by this mechanism.

If the reactions of loose complex affect significantly the overall reaction, RRKM treatment of the tight complex alone is not sufficient for a complete theoretical treatment. A theoretical treatment of the complex dynamics of motion and steric factors in the loose complex may be very involved. In such cases a complete TST treatment may still be useful, since with this approach only the equilibrium properties of A^+, B, and $(AB^+)^{\neq}_{tight}$ must be known.

The mechanistic implications of temperature and pressure effects illustrate the use of temperature and pressure studies to gain insight into subtle details in the sequence of events in ion–molecule collisions. The body of systematic data on which mechanistic arguments can be based is still quite limited, but the investigation of temperature and pressure effects is gaining momentum. The application of the resultant data to IM reactions in extreme environments and in CIMS adds further value to these studies.

ACKNOWLEDGMENTS

This work was supported by a grant from the National Science Foundation. I thank Dr. H. Gershinowitz for a critical review of the manuscript.

REFERENCES

Abramson, F. P., and Futrell, J. H. (1966). *J. Chem. Phys.* **45**, 1925–1931.

Adams, N. G., Bohme, D. K., Dunkin, D. B., and Fehsenfeld, F. C. (1970). *J. Chem. Phys.* **52**, 1951–1955.

Anicich, V. G., and Bowers, M. T. (1974). *J. Am. Chem. Soc.* **96**, 1279–1284.

Ausloos, P., and Lias, S. G. (1970). *J. Am. Chem. Soc.* **92**, 5037–5045.

Bohme, D. K. (1975). *In* "Interactions Between Ions and Molecules" (P. Ausloos, ed.), pp. 489–504. Plenum, New York.

Bohme, D. K., Dunkin, D. B., Fehsenfeld, F. C., and Ferguson, E. E. (1968). *J. Chem. Phys.* **49**, 5201–5202.

Bohme, D. K., Dunkin, D. B., Fehsenfeld, F. C., and Ferguson, E. E. (1969). *J. Chem. Phys.* **51**, 863–872.

Bohme, D. K., Dunkin, D. B., Fehsenfeld, F. C., and Ferguson, E. E. (1970). *J. Chem. Phys.* **52**, 3133–3140.
Bowers, M. T., and Laudenslager, J. B. (1972). *J. Chem. Phys.* **56**, 4711–4720.
Bowers, M. T., Neilson, P. V., Kemper, P. R., and Wren, A. G. (1977). *Int. J. Mass Spectrom. Ion Phys.* **25**, 103–116.
Buttrill, S. E. (1973). *J. Chem. Phys.* **58**, 656–659.
Calcote, H. F. (1972). *In* "Ion–Molecule Reactions" (J. L. Franklin, ed.), Vol. 2, pp. 673–706. Plenum, New York.
Chau, M., and Bowers, M. T. (1976). *Chem. Phys. Lett.* **44**, 490–494.
Chupka, W. A. (1975). *In* "Interactions between Ions and Moleculess" (P. Ausloos, ed.), pp. 249–262. Plenum, New York.
Clarke, A. G., and Burns, G. (1972). *J. Chem. Phys.* **56**, 4636–4645.
Cunningham, A. J., Payzant, J. D., and Kebarle, P. (1972). *J. Am. Chem. Soc.* **94**, 7627–7632.
Dalgarno, A. (1975). *In* "Interactions Between Ions and Moleculess" (P. Ausloos, ed.), pp. 341–353. Plenum, New York.
Dunkin, D. B., Fehsenfeld, F. C., Schmeltekopf, A. L., and Ferguson, E. E. (1968). *J. Chem. Phys.* **49**, 1365–1371.
Durden, D. A., Kebarle, P., and Good, A. (1969). *J. Chem. Phys.* **50**, 805–813.
Fehsenfeld, F. C., Lindinger, W., Schmeltekopf, A. L., Albritton, D. L., and Ferguson, E. E. (1975). *J. Chem. Phys.* **62**, 2001–2003.
Fehsenfeld, F. C., Lindinger, W., Schiff, H. I., Hemsworth, R. S., and Bohme, D. K. (1976). *J. Chem. Phys.* **64**, 4887–4891.
Ferguson, E. E. (1972). *In* "Ion Molecule Reactions" (J. L. Franklin, ed.), Vol. 2, pp. 363–394. Plenum, New York.
Ferguson, E. E., Bohme, D. K., Fehsenfeld, F. C., and Dunkin, D. B. (1969). *J. Chem. Phys.* **50**, 5039–5040.
Field, F. H. (1969a). *J. Am. Chem. Soc.* **91**, 2827–2839.
Field, F. H. (1969b). *J. Am. Chem. Soc.* **91**, 6334–6341.
Field, F. H., and Weeks, D. P. (1970). *J. Am. Chem. Soc.* **92**, 6521–6525.
French, M. A., and Kebarle, P. (1975). *Can. J. Chem.* **53**, 2668–2674.
French, M. A., Hills, L. P., and Kebarle, P. (1973). *Can. J. Chem.* **51**, 456–461.
Gioumoulis, G., and Stevenson, D. P. (1958). *J. Chem. Phys.* **29**, 294–299.
Good, A., Durden, D. A., and Kebarle, P. (1970). *J. Chem. Phys.* **52**, 212–221.
Gupta, S. K., Jones, E. G., Harrison, A. G., and Myher, J. J. (1967). *Can. J. Chem.* **45**, 3107–3117.
Harrison, A. G. (1975). *In* "Interactions Between Ions and Molecule" (P. Ausloos, ed.), pp. 263–277. Plenum, New York.
Hemsworth, R. S., Rundle, H. W., Bohme, D. K., Schiff, H. I., Dunkin, D. B., and Fehsenfeld, F. C. (1973). *J. Chem. Phys.* **59**, 61–69.
Herbst, E. (1976). *Astrophys. J.* **205**, 94–102.
Hiraoka, K., and Kebarle, P. (1975a). *J. Chem. Phys.* **63**, 736–739.
Hiraoka, K., and Kebarle, P. (1975b). *J. Chem. Phys.* **63**, 394–397.
Hiraoka, K., and Kebarle, P. (1976). *J. Am. Chem. Soc.* **98**, 6119–6125.
Hiraoka, K., and Kebarle, P. (1977a). *J. Am. Chem. Soc.*, **99**, 360–368.
Hiraoka, K., and Kebarle, P. (1977b). *J. Am. Chem. Soc.* **99**, 366–370.
Johnston, H. S. (1966). "Gas Phase Reaction Rate Theory," p. 283. Ronald Press, New York.
Kaufman, J. J. (1973). *In* "Wave Mechanics: The First Fifty Years" (W. C. Price, S. S. Chissick, and T. Ravendale, eds.), pp. 208–238. Wiley, New York.
Kebarle, P. (1972). *In* "Ion–Molecule Reactions" (J. L. Franklin, ed.), Vol. 1, pp. 315–362. Plenum, New York.
Kebarle, P. (1977). *Annu. Rev. Phys. Chem.* **28**, 445–476.
Kemper, P. R., Nilson, P. V., Parent, D., and Bowers, M. T. (1978). *J. Chem. Phys.* **68**, 322.

Kim, J. K., Theard, L. P., and Huntress, W. T. (1974). *Int. J. Mass Spectrom. Ion Phys.* **15**, 223–244.
Laidler, K. J. (1969). "Theories of Chemical Reaction Rates," Ch. 3. McGraw-Hill, New York.
Laurie, W. A., and Field, F. H. (1972a). *J. Am. Chem. Soc.* **94**, 2913–2919.
Laurie, W. A., and Field, F. H. (1972b). *J. Am. Chem. Soc.* **94**, 3359–3360.
Lawrence, R. H., and Firestone, R. F. (1965). *J. Am. Chem. Soc.* **87**, 2288–2290.
Leu, M. T., Biondi, M. A., and Johnsen, R. (1973). *Phys. Rev.* A **8**, 413–419.
Lias, S. (1978). Personal communication.
Lindinger, W., Fehsenfeld, F. C., Schmeltekopf, A. L., and Ferguson, E. E. (1974). *J. Geophys. Res.* **78**, 4753–4756.
Lindinger, W., McFarland, M., Fehsenfeld, F. C., Albritton, D. L., Schmeltekopf, A. L., and Ferguson, E. E. (1975a). *J. Chem. Phys.* **63**, 2175–2181.
Lindinger, W., Albritton, D. L., and Fehsenfeld, F. C. (1975b). *J. Chem. Phys.* **62**, 4957–4958.
McFarland, M., Dunkin, D. B., Fehsenfeld, F. C., Schmeltefopf, A. L., and Ferguson, E. E. (1972). *J. Chem. Phys.* **56**, 2358–2364.
Meisels, G. G. (1975). *In* "Interactions Between Ions and Moleculess" (P. Ausloos, ed.), pp. 595–617. Plenum, New York.
Meot-Ner, M. (1975). Unpublished observation.
Meot-Ner (Mautner), M. (1978). *J. Am. Chem. Soc.* **100**, 4694–4699.
Meot-Ner (Mautner), M. (1979a). *J. Am. Chem. Soc.* (in press).
Meot-Ner (Mautner), M. (1979b). *Origins Life* **9**, 115–131.
Meot-Ner, M., and Field, F. H. (1973). *J. Am. Chem. Soc.* **95**, 7202–7211.
Meot-Ner, M., and Field, F. H. (1974). *J. Chem. Phys.* **61**, 3742–3749.
Meot-Ner, M., and Field, F. H. (1975a). *Origins Life* **3**, 377–394.
Meot-Ner, M., and Field, F. H. (1975b). *J. Am. Chem. Soc.* **97**, 5339–5345.
Meot-Ner, M., and Field, F. H. (1975c). *J. Am. Chem. Soc.* **97**, 2014–2017.
Meot-Ner, M., and Field, F. H. (1976a). *J. Chem. Phys.* **64**, 277–281.
Meot-Ner, M., and Field, F. H. (1976b). *J. Phys. Chem.* **80**, 2865–2869.
Meot-Ner, M., and Field, F. H. (1977). *J. Am. Chem. Soc.* **99**, 998–1003.
Meot-Ner, M., and Field, F. H. (1978). *J. Am. Chem. Soc.* **100**, 1356–1359.
Meot-Ner, M., Solomon, J. J., Field, F. H., and Gershinowitz, H. (1974). *J. Phys. Chem.* **78**, 1773–1774.
Meot-Ner, M., Hunter, E. P., and Field, F. H. (1977a). *J. Am. Chem. Soc.* **99**, 5576–5583.
Meot-Ner, M., Hamlet, P., and Field, F. H. (1977b). Unpublished observations.
Miasek, P. G., and Harrison, A. G. (1975). *J. Am. Chem. Soc.* **97**, 714–721.
Moran, T. F., and Hamill, W. H. (1963). *J. Chem. Phys.* **39**, 1413–1422.
Neilson, P. V., Bowers, M. T., Chau, M., Davidson, W. R, and Aue, D. H. (1978). *J. Am. Chem. Soc.* **100**, 3649–3658.
Niles, F. E., and Robertson, W. W. (1965). *J. Chem. Phys.* **42**, 3277–3280.
Olmstead, W. N., Lev-on, M., Golden, D. M., and Brauman, J. I. (1977). *J. Am. Chem. Soc.* **99**, 992–998.
Payzant, J. D., Cunningham, A. J., and Kebarle, P. (1973). *J. Chem. Phys.* **59**, 5615–5619.
Pierce, R., and Porter, R. F. (1974). *J. Phys. Chem.* **78**, 93–97.
Po, P. L., and Porter, R. F. (1977). *J. Am. Chem. Soc.* **99**, 4922–4930.
Porter, R. F. (1975). *In* "Interactions Between Ions and Molecules" (P. Ausloos, ed.), pp. 231–247, Plenum, New York.
Rabinowitz, E. (1937). *Trans. Faraday Soc.* **33**, 283–293.
Ridge, D. P., and Beauchamp, J. L. (1976). *J. Chem. Phys.* **64**, 2735–2746.
Robinson, P. J., and Holbrook, K. A. (1972). "Unimolecular Reactions," Wiley (Interscience), New York.
Sayers, J., and Smith, D. (1964). *Discuss. Faraday Soc.* **37**, 167.

Schmeltekopf, A. L., Ferguson, E. E., and Fehsenfeld, F. C. (1968). *J. Chem. Phys.* **48**, 2966–2973.

Smith, D., and Fouracre, B. L., (1968). *Planet. Space Sci.* **16**, 231–232.

Smith, D. L., and Futrell, J. H. (1976). *Chem. Phys. Lett.* **40**, 229–233.

Solomon, J. J., Meot-Ner, M., and Field, F. H. (1974). *J. Am. Chem. Soc.* **96**, 3727–3732.

Stevenson, D. P., and Schissler, D. O. (1958). *J. Chem. Phys.* **29**, 282–294.

Su, L., and Bone, L. I. (1976). constant as a function of the complexity of the third body. *Proc. Annu. Conf. Mass Spectrom. Allied Top., 24th, San Diego* pp. 383–384.

Su, T., and Bowers, M. T. (1973). *Int. J. Mass Spectrom. Ion Phys.* **12**, 347–356.

Su, T., and Bowers, M. T., (1975). *Int. J. Mass Spectrom. Ion. Phys.* **17**, 211–212.

Watson, W. D. (1976). *Acc. Chem. Res.* **10**, 221–226.

Wayne, R. P. (1969). *In* "Comprehensive Chemical Kinetics" (C. H. Bamford and D. F. H. Tipper, eds.), Vol. 2, Chap. 3, p. 189–301. Elsevier, Amsterdam.

Weeks, D. P., and Field, F. H. (1970). *J. Am. Chem. Soc.* **92**, 1600–1601.

Weeks, D. P., and Field, F. H. (1973). *J. Org. Chem.* **38**, 3380–3382.

Chapter 7
Energy distribution in the unimolecular decomposition of ions

J. L. FRANKLIN

CHEMISTRY DEPARTMENT, RICE UNIVERSITY
HOUSTON, TEXAS

I. INTRODUCTION

For an elementary unimolecular reaction to occur, energy must be introduced in sufficient amounts to break the appropriate bonds and, in

ISBN-0-12-120801-X

some instances, to surmount the energy barrier to activation. Photolysis can be employed for this purpose but with neutral species the problem of identifying and measuring the products is usually beyond the sensitivity limits of available instruments. On the other hand, the electric charge on an ion makes it very easily detectable and measured. Consequently, the ionic products of a fragmentation process can be readily measured in a time shorter than that required for the ion to collide with a neutral. Thus, the elementary reaction can be studied readily. Ionic decomposition processes then are well suited to such studies and have been employed extensively for this purpose.

There are two aspects of unimolecular reactions that are of major interest and importance. These are (1) the rate of the reaction and (2) the distribution of energy in the products. Both have been the subject of extensive investigations, both theoretical and experimental, for many years, but while progress has been made neither problem can be considered as solved at this time. However, considerably more progress appears to have been made in understanding the rates than the energy distributions of unimolecular reactions. Two rather comprehensive books on the rates of unimolecular reactions have appeared in recent years by Forst (1973) and Robinson and Holbrook (1972), so the state of our present understanding of this aspect is readily available. Consequently, I will devote very little attention to rates of reaction in this paper and will concentrate upon the problem of energy distributions.

II. THEORETICAL CONSIDERATIONS

When a polyatomic ion fragments, the energy in excess of the thermodynamic heat of reaction may be distributed among the fragments as vibrational, rotational, translational, and electronic energy. How will it be distributed? The problem is usually approached by making certain apparently reasonable assumptions and deducing from them mathematical relationships that can be tested experimentally. The most widely used approaches are based on statistical methods (see, e.g., Chesnavich and Bowers, Chapter 4, in this volume). In the calculation of product translational energy release, the main variations of statistical methods include (1) use of the standard saddle point RRKM–QET unimolecular rate expression with the added assumption that the product translational energy arises entirely from the kinetic energy of the transition state reaction coordinate, (2) use of the Langevin or orbiting transition state (i.e., phase space theory), and (3) use of the saddle point transition state with added dynamical assumptions to describe the passage of the reacting system from the saddle point region of the potential surface to the products region. The remainder of this section discusses these three variations in greater detail.

A. *Quasi-Equilibrium Treatment*

The standard saddle point RRKM–QET unimolecular rate expression has been widely used by ion chemists to estimate product translational energy release. The main assumption of this approach, in addition to the usual statistical assumptions is that the relative translational energy of the products arises entirely from the kinetic energy located in the reaction coordinate of the decomposing complex. A further assumption is that rotational motion of the transition state complex can usually be ignored. With these assumptions we can write the distribution function for vibrational energy, $P(E^*, v)$,

$$P(E^*_{\mathrm{v,t}}, v) = \frac{\rho^\ddagger(E^*_{\mathrm{v,t}} - \epsilon_{\mathrm{t}})}{\int_0^{E^*} \rho^\ddagger(E^*_{\mathrm{v,t}} - \epsilon_{\mathrm{t}})\, d\epsilon_{\mathrm{t}}}, \tag{1}$$

where $\rho^\ddagger$ is the density of vibrational states of the transition state complex, v and t refer to vibration and translation, respectively (rotation being neglected), ϵ_{t} is translational energy in the center of mass and $E^*_{\mathrm{v,t}}$ the excess vibrational and translational energy. $E^*_{\mathrm{v,t}}$ is the amount of energy in excess of the adiabatic energy required to raise the molecule to the level of the dissociation assymptote for the state in question. Since rotation is neglected, $\epsilon_{\mathrm{v}} = E^*_{\mathrm{v,t}} - \epsilon_{\mathrm{t}}$ as given in (1). The average value of ϵ_{t} is then

$$\bar{\epsilon}_{\mathrm{t}} = \frac{\int_0^{E^*_{\mathrm{v,t}}} \epsilon_{\mathrm{t}} \rho^\ddagger(E^*_{\mathrm{v,t}} - \epsilon_{\mathrm{t}})\, d\epsilon_{\mathrm{t}}}{\int_0^{E^*_{\mathrm{v,t}}} \rho^\ddagger(E^*_{\mathrm{v,t}} - \epsilon_{\mathrm{t}})\, d\epsilon_{\mathrm{t}}}. \tag{2}$$

This is the quasi-equilibrium treatment of Marcus and Rice (Marcus and Rice, 1951; Marcus, 1952) and of Rosenstock *et al.* (1952) which extends the earlier formalism of Rice and Ramsperger (1927, 1928) and of Kassel (1928, 1932). In the simplest solution of (2) one substitutes the classical expression for the total number of vibrational states

$$W(E^*_{\mathrm{v,t}}) = \frac{(E^*_{\mathrm{v,t}} - \epsilon_{\mathrm{t}})^{n-1}}{(n-1)! \prod_i h\nu_i} \tag{3}$$

and the density of vibrational states $dW/dE^*_{\mathrm{v,t}}$

$$\rho_\ddagger = \frac{(E^*_{\mathrm{v,t}} - \epsilon_{\mathrm{t}})^{n-2}}{(n-2)! \prod_i h\nu_i} \tag{4}$$

into (2) and integrates. Substitution yields

$$\bar{\epsilon}_{\mathrm{t}} = \frac{\int_0^{E^*_{\mathrm{v,t}}} \epsilon_{\mathrm{t}} (E^*_{\mathrm{v,t}} - \epsilon_{\mathrm{t}})^{n-2}\, d\epsilon_{\mathrm{t}}}{\int_0^{E^*_{\mathrm{v,t}}} (E^*_{\mathrm{v,t}} - \epsilon_{\mathrm{t}})^{n-2}\, d\epsilon_{\mathrm{t}}}. \tag{5}$$

In (3)–(5), $n - 1$ is the number of vibrational modes in the transition state, n that of the molecule, and ν_i the frequencies of the various vibrational modes. Integration yields

$$\bar{\epsilon}_t = \frac{E^*_{v,t}}{n}. \tag{6}$$

Obviously, $\bar{\epsilon}_t$ is the average value of the energy in a vibrational mode, the reaction coordinate, which is assumed to become translational energy in the fragments.

Since the classical vibrational density of states is only an approximation, Eq. (6) is an over simplification. A more valid solution is to employ a semi-classical approximation or, better yet, a detailed summation of all vibrational states within the energy range in question (see, e.g., Forst, 1973, Chapter 6). Spotz *et al.* (1969) have done this for several positive ion dissociations. They write

$$\bar{\epsilon}_t = E^* - \sum_i \Omega_i \epsilon_i \Big/ \sum_i \Omega_i. \tag{7}$$

The summation is over all states with energy E_i lying between ΔH_r and $\Delta H_r + E^*$. Since ΔH_r, the heat of reaction, is taken as the zero level of energy, the energy available for the ith quantum is $\epsilon_i = E_i - \Delta H_r$. The energy of the ith state is given by the sum of the energy of each of the oscillators for the quantum assigned; thus,

$$\epsilon_i = \sum_j n_{ij} \nu_j, \qquad \epsilon_i < E^*, \tag{8}$$

where n_{ij} is the quantum number of the jth frequency in the ith state. As a means of simplifying the computations, frequencies of like value were grouped together using the geometric mean ν_j as the frequency of the group. The degeneracy of the ith state is the product of the degeneracies of each of the groups

$$\Omega_i = \prod_j [(n_{ij} + x - 1)!/n_{ij}!(x_j - 1)!]. \tag{9}$$

B. Phase Space Theory

A serious shortcoming of the quasi-equilibrium approach summarized in the previous section is that it fails to adequately account for the passage of the various degrees of freedom from the transition state complex to the products. For example, three degrees of freedom are needed to describe the relative translational motion of the separating products, whereas the reaction coordinate of the activated complex is described by only one degree of freedom. One transition state that does not suffer this problem is the one derived from the Langevin or orbiting model used in phase space theory (Pechukas and Light, 1965; Light, 1967). The orbiting model is discussed in detail in Chesnavich and Bowers (Chapter 4, in this volume). In this chapter it is sufficient to rely on the result obtained by Klots (1971, 1972) for small total system angular momentum

$$\bar{\epsilon}_t = \frac{2}{s+1} \frac{\int_0^{E^*} \rho(\epsilon_v)(E^* - \epsilon_v)^{(s+1)/2}\, d\epsilon_v}{\int_0^{E^*} \rho(\epsilon_v)(E^* - \epsilon_v)^{(s-1)/2}\, d\epsilon_v}. \tag{10}$$

Here ρ is the density of vibrational states of the products and s the total number of rotational degrees of freedom of the products.

If the classical density of states is employed with Eq. (10) we obtain

$$\bar{\epsilon}_t = \frac{E^*}{n'}, \tag{11}$$

where n' is a constant for a given molecule and differs from n in Eq. (6) by not more than two. Thus, both (6) and (11) predict a linear dependence of $\bar{\epsilon}_t$ upon E^*. The difference between n and n' becomes less significant as n increases and thus the predicted result by the two treatments approach each other as the complexity of the parent ion increases.

In both the quasi-equilibrium theory and the phase space theory, the effect of zero point vibrational energy was ignored. As discussed by Whitten and Rabinovitch (1963), zero point energy should be included and when it is, the average translational energy becomes

$$\bar{\epsilon}_t \lessapprox \frac{E^* + E_z}{n} \qquad \text{(QET)} \tag{12}$$

$$\bar{\epsilon}_t \lessapprox \frac{E^* + E_z'}{n'} \qquad \text{(phase space)}, \tag{13}$$

where E_z is the zero point energy of the transition state and E_z' is the sum of the zero point energies of the products. Equations (12) and (13) make relatively minor corrections to Eqs. (6) and (11).

C. The Statistical–Dynamical Method of Marcus

Within the context of statistical theory, reaction rate constants are governed by the region of the potential surface that lies between the reactants and transition state, whereas the distribution of energy in the products is governed by the region between the transition state and products. Therefore, it is possible for a reacting system to exhibit statistical behavior in $k(E)$

versus E, but nonstatistical behavior in the distribution of energy in the products. The exact form of the product energy distribution in these cases will depend on specific details of the potential surface between the transition state and products, such as the location of the transition state (i.e., early versus late barriers) and the magnitude of the reverse activation energy. Marcus, in a series of papers (see Worry and Marcus, 1977, and references therein), had discussed in detail methods by which statistical and dynamical approximations can be combined to estimate product energy distributions. Using detailed balance arguments, these methods can also be applied to the energy dependence of the rate for the reverse reaction. The approach of Marcus is conceptually superior to the naive assumption presented in Section II, that product translational energy arises entirely from the kinetic energy of the reaction coordinate. However, this approach is also technically more difficult to implement and has, to date, only been applied to the neutral system $C_4H_8F \rightarrow CH_3 + C_3H_5F$. For this reason it is not considered further in this chapter.

III. EXPERIMENTAL METHOD

A. *Measurement of Excess Energy*

From the above it appears that both the QET and phase space theories predict a linear dependence of translational energy upon excess energy. Both are statistical treatments that assume that energy is equilibrated rapidly among all vibrational modes.

The measurement of energy distribution in the products of a dissociation reaction is almost entirely the measurement of excess energy and of translational energy of the fragment ion. The estimation of vibrational, rotational, and electronic energies, in most instances, rests upon these measurements plus inference from the nature of the molecule in question.

Measurement of excess energy E^* must rest upon a knowledge of ΔH_r for the dissociation reaction. If heats of formation of the parent neutral and of the fragments are known, the heat of reaction is readily computed. The appearance potential can often be accurately measured and $E^*_{AP} = AP - \Delta H_R$. The appearance potential is normally obtained by comparison to a known ionization potential. With dissociative resonance capture processes, the ionization efficiency curve is usually deconvoluted by a computer technique (MacNeill and Thynne, 1969; Morrison, 1963) to remove the broadening of the resonance curve due to the energy spread in the electron beam. This can also be accomplished by using the retarding potential difference technique (Fox *et al.*, 1951, 1955). With a number of dissociative resonance capture processes, we have found the translational energy of the fragment ion to increase linearly with electron energy, so that extrapolation to thermal energy gives the electron energy that corresponds to the heat of reaction. Electron energies in excess of that can be taken as the excess energy. This, of course, does not apply to pair production or dissociative ionization processes resulting from electron impact, unless the energy of the leaving electron(s) can be determined. This is not possible by ordinary mass spectrometry; however, in the last few years the development of photoion–photoelectron coincidence spectrometry (PIPECO) and threshold photoelectron spectrometry (TPES) has solved this problem and has resulted in important progress in the measurement of energy distributions in positive ion fragmentation measurements. These methods will be discussed later.

B. *Measurement of Translational Energy*

1. *General*

Measurement of the translational energy of ions can be done in several ways, namely, by electrostatic analysis, time of flight, mass spectrometer peak shape analysis, electrostatic beam deflection, and retarding potential. Of these, the first three have been the most frequently employed. Electrostatic analysis consists of passing a beam of ions through an electrostatic field of a carefully designed configuration and variable field strength which separates them according to energy. Cylindrical or hemispherical configurations are most frequently used. Electrostatic beam deflection is a variation of the electrostatic analysis method and employs a variable electric field between parallel plates in a sector field mass spectrometer as a means of measuring the kinetic energy with which the ions are formed in the ionization chamber. The method was originally devised by Taubert (1959, 1964) and Bracher *et al.* (1963) and has recently been employed by Sen Sharma and Franklin (1974).

2. *Peak Shape Analysis*

Time of flight mass analysis depends upon the fact that the time required for an ion of a given mass to travel a known distance will depend upon its velocity and hence upon its energy if the mass is known. Peak shape analysis was originally proposed by Berry (1950) and was later developed by Franklin *et al.* (1967) for use in a time of flight mass spectrometer. The method, briefly, is as follows: consider two ions formed with equal energies in the ionization chamber of a time of flight mass spectrometer. Suppose that the path of one of these, A, is directed initially toward the collector; that of the other, B, away from the collector. When the draw out pulse is applied, A will be accelerated toward the collector; B will be decelerated, stopped, and then reaccelerated toward the collector. B will reach the exit port from the ion chamber a finite time after A. Subsequently, both ions will travel to the collector at the same velocity and hence B will reach the collector a finite but small time after A. This difference in time is dependent upon the initial energies of the two molecules, and of course upon their mass. Of course, there will actually be an assemblage of ions of the same mass but having a distribution of energies, and thus the mass peak will normally appear as approximately a Gaussian curve reflecting the energy distribution. The peak width at half-height will, of course, yield the average translational energy. The mean time required to reach the collector will be a measure of the mass, since the total energy of the ions is the same for all.

3. *Retarding Potential Measurements*

Retarding potential measurements of ion energies have been employed in several studies but the method suffers from relatively poor energy resolution and has not been employed with molecules with more than three or four atoms. It consists of measuring the change in intensity of ions of a given mass with varying retarding potential. In order to extract an energy distribution, the retarding curve must be differentiated with respect to retarding potential.

4. *Photoionization–Photoelectron Coincidence Spectrometry*

As was mentioned above, the problem of studying the energy distribution in positive ion fragmentation is to know the amount of excess energy corresponding to the measured translational energy. This has been solved by photoion–photoelectron coincidence spectroscopy (PIPECO) (Eland, 1972) and by the related threshold photoelectron spectroscopy (TPES) (Stockbauer, 1977a; Tsai *et al.*, 1974). In the former (PIPECO), light of a fixed frequency, usually the 584 Å line of helium, ionizes the gas in question, forming the parent ion in various states which can break up in characteristic ways. The kinetic energy of the electron is measured by electrostatic analysis or time of flight and the time of flight of the fragment ion formed in coincidence with the electron is determined. The latter is a measure of the translational energy of the ion. Since light of a fixed energy is employed, the relation (14) is relevant;

$$h\nu = I_p + \epsilon_{int} + \epsilon_e, \tag{14}$$

where I_p is the ionization potential of the molecule, ϵ_{int} the internal energy of the parent ion before fragmentation, and ϵ_e the energy of the electron. Threshold photoelectron spectroscopy employs light of variable frequency and detects only electrons of very low energy corresponding to the threshold for ionization. A thorough discussion of this technique is given in Baer (Chapter 5, in this volume).

IV. RESULTS AND DISCUSSIONS

A. *Peak Shape Analysis with Diatomic Molecules*

Several early workers have measured the translational energies of ions and employed the results to obtain improved thermochemical values of one of the products (Blewett, 1936; Hagstrum and Tate, 1940; Hagstrum, 1951; Fox and Hipple, 1948; Kandel, 1953, 1954; McDowell and Warren, 1952; Waldron, 1954; Hierl and Franklin, 1967). However, it has been only in the past 10 or 12 years that serious experimental efforts have been directed toward understanding the distribution of energy in unimolecular fragmentation.

Laboratory studies of positive ions from several organic compounds have been made employing the beam deflection method, and of both positive and negative ions from a number of both organic and inorganic compounds have been made employing peak shape analysis, in conjunction with a time of flight mass spectrometer.

As has been discussed above, studies of energy distribution in dissociative ionization are meaningful only at or very near onset. Hierl and Franklin (1967) studied the fragment ions from N_2, CO, and NO and, after correction for kinetic energy, obtained generally very good values for the heats of formation of the various ions. Some corrections were quite large, notably a correction of some 10 eV for the formation of N^{2+} from N_2. It was also possible by measuring the ion kinetic energy with increasing electron energy to detect, in a few instances, the onset of a second process. Thus, the translational energy of N^+ from N_2 shows a sharp increase at an electron energy of 47 eV and the formation of N^+ ions with 8 eV energy in the center of mass. This corresponds to the dissociation of an excited N_2^{2+} ion and, with the correction for kinetic energy, gives N^+ ions in their ground state in good agreement with spectroscopic values. Similar onsets of higher energy processes were detected in ions from CO from sharp changes in the kinetic energies of the fragment ions.

In a similar manner, DeCorpo *et al.* (1970) measured the translational energy of F^-/F_2 at onset and thus were able to determine $D(F_2)$ in good agreement with results obtained by others.

B. *Correlation of Translational and Excess Energies*

Diatomic molecules yield products with only translational and electronic energies, and to extend our understanding of energy distributions, Haney and Franklin (1968) undertook studies of translational energy from polyatomic compounds. For this purpose they selected several dissociative ionization process for which ΔH_r is well known and for which the appearance

TABLE I

Reactions in which an Average Value ($\alpha = 0.44$) was Used

Compound	Ion	Radical	Ap (eV)	$\bar{\epsilon}_t$ (kcal/mole)	ΔH_f (kcal/mole) Ion	Radical	Ref.
HCOOH	HCO^+	OH	13.03	1.1	199		*a*
CH_3CHO	HCO^+	CH_3	12.73	3.5	197		*a*
C_2H_3CHO	HCO^+	C_2H_3	13.30	3.0	196		*a*
C_2H_5OH	CH_2OH^+	CH_3	11.65	0.9	170		*a*
$(CH_3)_2O$	CH_3O^+	CH_3	11.95	4.6	164		*a*
CH_3ONO	CH_3O^+	NO	10.9	6.1	177		*a*
$CH_3OC_2H_5$	$C_2H_5O^+$	CH_3	11.30	2.4	141		*a*
CH_3OCH_2Cl	$C_2H_5O^+$	Cl	10.79	3.2	136		*a*
$(CH_3)_2CO$	CH_3CO^+	CH_3	11.32	2.5	148		*a*
CH_3COCl	CH_3CO^+	Cl	11.25	3.6	148		*a*
CH_3COOH	CH_3CO^+	OH	11.75	2.3	142		*a*
CH_3COOH	$COOH^+$	CH_3	12.27	0	145		*a*
HCO_2CH_3	CH_3O^+	HCO	12.23	2.7		7	*a*
$HCO_2C_2H_5$	$C_2H_5O^+$	HCO	11.50	2.2		9	*a*
C_2H_3CHO	$C_2H_3{}^+$	HCO	13.64	2.1		8	*a*
HCO_2CH_3	HCO^+	CH_3O	13.47	3.9		-2	*a*
$CH_3CO_2CH_3$	CH_3CO^+	CH_3O	11.37	1.6		-3	*a*
CH_3ONO	NO^+	CH_3O	11.15	1.5		-4	*a*
C_2H_5ONO	NO^+	C_2H_5O	11.69	1.6		-6	*a*
$CH_3OC_2H_5$	$CH_3{}^+$	CH_3OCH_2	15.02	2.5		0	*a*
$(CH_3)_2CO$	$CH_3{}^+$	CH_3CO	15.36	3.9		-2	*a*
CH_3COOH	$CH_3{}^+$	COOH	14.0	2.0		-58	*a*
HCO_2CH_3	$CH_3{}^+$	HCO_2	13.71	<2.2		$\overline{>}-46$	*a*

potentials exceed ΔH_r by more than about 0.5 eV. They then determined the translational energy of the ion at several electron voltages at or near the appearance potential and, where necessary, extrapolated the curve to the appearance potential. From this, the average translational energy in the center of mass $\bar{\epsilon}_t$ is obtained from the relationship

$$\bar{\epsilon}_i = \frac{m_i}{m_i + m_n}\left(\tfrac{3}{2}RT\right) + \frac{m_n}{m_i + m_n}\bar{\epsilon}_t, \tag{15}$$

where m_i and m_n are, respectively, the masses of the ion and neutral fragments. In no case did $\bar{\epsilon}_t$, determined in this way, agree with the results predicted by either Eq. (6) or (11), the measured value of $\bar{\epsilon}_t$ being two to three times the predicted value in all cases. By plotting E^*/N against $\bar{\epsilon}_t$, a straight line relation was found with very small deviation of individual points from the line. DeCorpo *et al.* (1971) performed a similar study on several dissociative resonance capture processes with similar results; the points, in fact, falling on the

TABLE I (*Continued*)

Compound	Ion	Radical	Ap (eV)	$\bar{\epsilon}_t$ (kcal/mole)	ΔH_f (kcal/mole) Ion	ΔH_f (kcal/mole) Radical	Ref.
C_2H_5OH	H_3O^+	$H + C_2H_2$[b]	14.30	$1.6(H_3O)^+$	143		c
C_2H_5SH	H_3S^+	C_2H_3	12.41	1.3	193		c
$(CH_3)_2S$	H_3S^+	$H + C_2H_2$[b]	14.14	$1.4(H_3S^+)$	190		c
$C_2H_5NH_2$	NH_4^+	$H + C_2H_2$[b]	12.72	$1.9(NH_4^+)$	148		c
$1\text{-}C_4H_8$	CH_3^+	C_3H_5	14.17	2.0		40	d
$(CH_3)_3CCH{=}CH_2$	CH_3^+	C_5H_9	15.35	2.9		19	d
1-butyne	CH_3^+	C_3H_3	15.09	4.4		82	d
$(CH_3)_3CC{\equiv}CH$	CH_3^+	C_5H_7	14.74	2.5		59	d
$(CH_3)_2C{=}CH_2$	CH_3^+	C_3H_5	16.39	4.6		53	d
$1\text{-}C_4H_8$	$C_3H_5^+$	CH_3	11.75	1.2	222		d
$i\text{-}C_4H_8$	$C_3H_5^+$	CH_3	11.80	1.6	214		d
$1,3\text{-}C_4H_6$	$C_3H_3^+$	CH_3	11.71	1.3	263		d
CH_3NH_2	CH_3^+	NH_2	14.48	4.3		41	d
CH_3NH_2	NH_2^+	CH_3	15.92	3.8	304		d
$(CH_3)_2NH$	CH_3^+	CH_3NH	14.79	3.1		43.6	d
$(CH_3)_3N$	CH_3^+	$(CH_3)_2N$	14.87	2.6		39	d
$C_2H_5NH_2$	CH_3^+	CH_2NH_2	15.61	4.4		43	d
$(C_2H_5)_2NH$	CH_3^+	$CH_2N(C_2H_5)H$	15.39	2.0		37	d
$(C_2H_5)_3N$	CH_3^+	$(C_2H_5)_2NCH_2$	16.74	3.0		23	d
BF_3	F_2^-	BF	10.2	23.2	−67		e
CCl_4	Cl_2^-	CCl_2	2.3	7.8	−58		e
CBr_4	Br_2^-	CBr_2	0.0	0.7	−59		e
CHI_3	I_2^-	CHI	0.0	0.0	−55		e

[a] Haney and Franklin (1969b).
[b] two step process.
[c] Haney and Franklin (1969a).
[d] Sen Sharma and Franklin (1973).
[e] DeCorpo and Franklin (1971a).

same line. If Eq. (6) had been obeyed, the slope of the line would have been unity but the actual slope was found to be 0.44. When this correction factor α is employed, Eq. (6) becomes

$$E^* = \alpha N \bar{\epsilon}_t, \tag{16}$$

where $\alpha = 0.44$ and has the property of being the fraction of the vibrational modes that are active.

I recognize that Eq. (6) was based upon a quite naive theory, and I do not think that Eq. (16) has more than empirical significance. However, we have found it to be a useful way of correlating experimental results and therefore have continued to employ it for this purpose.

C. *Determination of the Heats of Formation of Ions and Neutrals*

In view of the good empirical correlation obtained with Eq. (16), several dissociative ionization and dissociative resonance capture processes were studied for the purpose of determining thermochemical properties of either ionic or neutral products (Haney and Franklin, 1969a,b; DeCorpo and Franklin, 1971). In these studies the appearance potential of the ion in question and its translational energy at onset were determined. E^* was obtained by means of Eq. (16) and the heat of reaction obtained by deducting E^* from the appearance potential. The results were generally satisfactory and are given in Table I.

D. *Dissociative Resonance Capture Processes*

1. *Linear Dependence of Translational Energy upon Excess Energy*

The limitation of our measurements of translational energies to the region of the appearance potential was applicable to dissociative ionization processes because of uncertainty due to the energy lost with the leaving electrons. This limitation does not apply to dissociative resonance capture processes since the electron and its energy remain with the system. So, with better techniques available, studies of the translational energy of ions over the range of the resonance were undertaken (Harland *et al.* 1973; Harland and Franklin, 1974). Measurements of ϵ_t of O^-/CO and of O^-/NO gave linear dependence on electron energy, as shown in Figs. 1 and 2. The excess electron energy above onset appears as translational energy, as is required by conservation of energy. Both processes appear to yield thermal kinetic energies at onset although, with O^-/NO, the very weak signal mode is necessary to extrapolate the curve to onset. It will be noted that the resonance for O^-/NO corresponds to the process

$$\mathrm{NO} + \mathrm{e} \rightarrow \mathrm{O}^- + \mathrm{N}(^2\mathrm{D}). \tag{17}$$

We found no evidence of processes involving the formation of N in either the ground ^4S or the ^2P state. Presumably, the lower state of NO^-, which is isoelectronic with O_2, is a triplet corresponding to the $A^3\Sigma u^+$ state of O_2, and so the decomposition should not be forbidden. Presumably, then, the decomposition asymptote must be at a higher energy than that at which the Franck–Condon region intersects the potential energy diagram. Possibly similar considerations apply to the highest energy process, which also was not observed. Chantry (1968) employed an instrument having a Wien filter between the ionization chamber and the analyzer to measure the kinetic energy of O^- from CO and NO. He found a linear dependence of kinetic

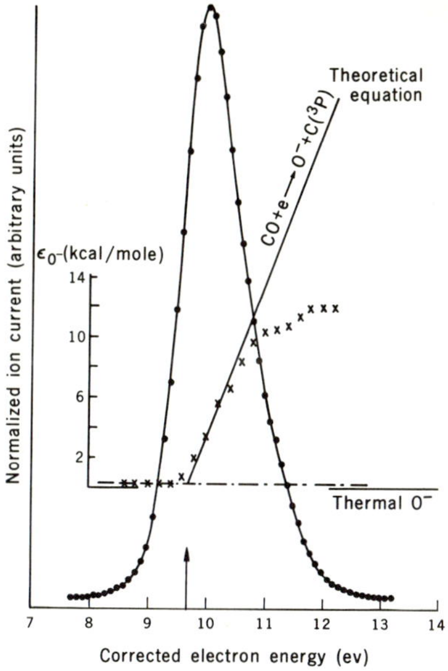

FIG. 1 Resonance and translational energy of O^- from CO. [From Franklin (1976). Copyright 1976 by the American Association for the Advancement of Science.]

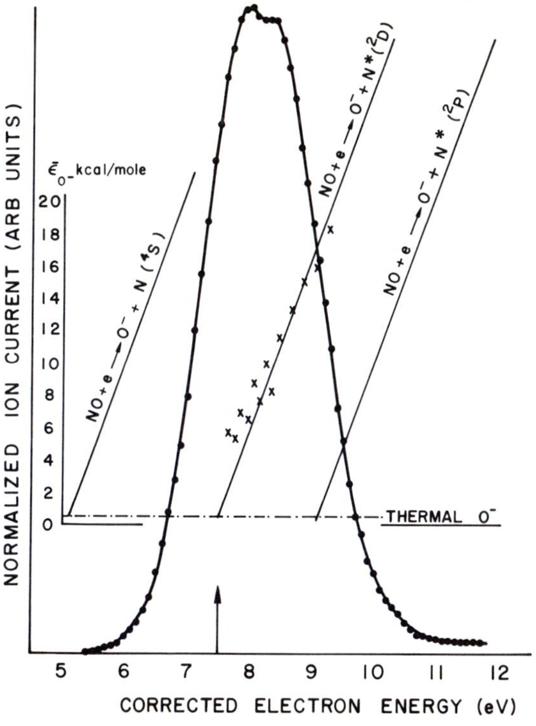

FIG. 2 Resonance and translational energy of O^- from NO. The straight solid lines are the theoretical translational energies for the three processes yielding N in three different electronic states. [From Harland *et al.* (1973).]

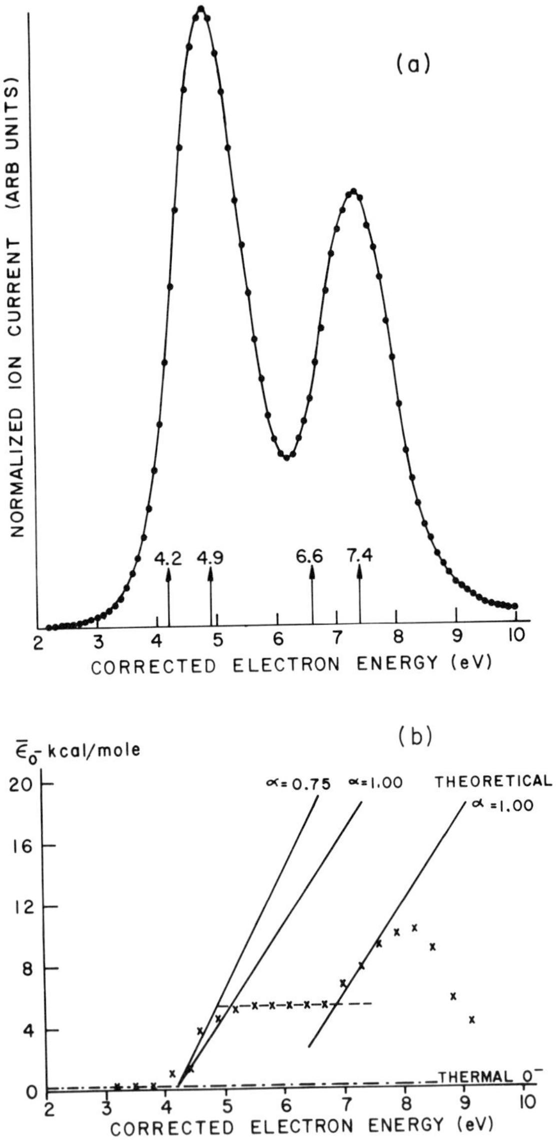

FIG. 3 Resonance and translational energies of O^- from SO_2. The arrows at the base of (a) correspond to the deconvoluted appearance potentials and to the maxima of the two resonances. [From Harland *et al.* (1973).]

energy upon electron voltage which extrapolated to the theoretical value of the heat of reaction at the electron energy axis. Schulz (1962), using a modified Lozier apparatus, also obtained a linear dependence of the kinetic energy of O^- from O_2. A plot of his data had the proper slope but was displaced toward lower electron energy by about 0.5 eV. In a later paper, Chantry and Schulz (1967) employing the mass spectrometer equipped with Wien filter again studied O^-/O_2 and were able to obtain a plot of kinetic energy against electron energy that followed very closely the theoretical line.

Schulz (1962) and Stamatovic and Schulz (1973) have studied O^-/CO_2 by dissociative resonance capture. The earlier study employed the modified Lozier tube and resulted in a linear dependence of ϵ_i upon E^* with a slope of 28/44; i.e., all of the excess energy appeared in translation. In the later study, an electron beam having a very narrow energy spread was prepared by passing the electrons through a trochoidal monochrometer. This beam was then passed through CO_2 gas. The resonance curve was found to have several peaks with spacings corresponding to the first four vibrational levels of the CO product. This is a very interesting result but it is obviously incompatible with the earlier study by Schulz (1962), in which all of the excess energy appeared in translation. Measurements of the energy distribution of O^-/CO_2 in this laboratory, employing both the peak shape analysis and deflection methods, have given values of $\bar{\epsilon}_t/E^*$ of 0.52 and 0.57 and thus show that some energy was available for vibrational excitation of CO.

The process O^-/SO_2 was also studied (Harland *et al.*, 1973) with results shown in Fig. 3. It is noteworthy that there are two resonances separated at their maximum intensities by about 2.5 eV but overlapping slightly. Although the slopes of the translational energy curves are not well defined, especially in the lower energy resonance, it appears that both lines approach having slopes corresponding to values of α = unity. Thus, the energy distributions for both resonances are in near agreement with the naive QET treatment.

Schulz (1961) and Chantry (1969) studied the kinetic energy of O^-/N_2O. Schulz used the modified Lozier tube and Chantry the mass spectrometer equipped with Wien filter, as mentioned above. Both found the kinetic energy to rise with electron energy, but all of the excess energy did not go into translation. Both found that above about 1.5 eV the translational energy curve became constant with a value of 0.4–0.5 eV. Apparently in this region more and more energy is going into vibration. In the energy range below about 1 eV, around 50–60% of the excess energy appears in translation as compared to 33% if the system obeyed the classical quasi-equilibrium theory. The phase space theory predicts that about 40% of the energy will go into translation. Evidently, the behavior of this system is considerably more complex than is predicted by statistical theories.

2. *Heats of Reaction*

Harland and Franklin (1974) extended these studies to somewhat more complex molecules including NF_3, BF_3, CF_4, C_2F_6, C_3F_8, and *cyclo*-C_4F_8. With F^- from both NF_3 and BF_3 and F_2^-/BF_3, the translational energy is

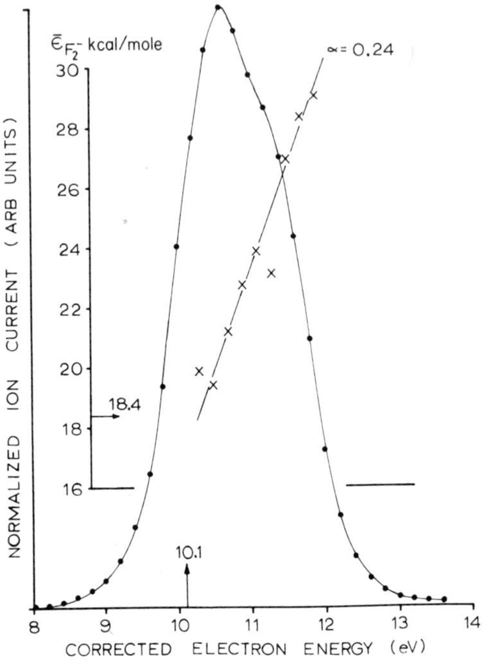

FIG. 4 Resonance and translational energy of F_2^- from BF_3. The arrow represents the deconvoluted appearance potential. The large translational energy at onset should be noted. [From Harland and Franklin (1974).]

quite large at onset and rises linearly with electron energy over the greater part of the resonance. Figure 4 shows the curve for F_2^-/BF_3, which is typical. In all of the processes, the translational energy rose linearly with electron energy and thus from the slope of each line a value of α could be obtained. These varied from 0.24 to 0.78, with, by far, the majority being less than 0.5. Thus, the behavior of all of these processes departed considerably and some drastically from the theoretical value; i.e., all showed significantly more kinetic energy than the statistical treatment predicted. Further, the spread in values of α shows that the early average value of 0.44 is only an approximation. However, at least for dissociative resonance capture processes, it is now possible to measure the dependence of ϵ_i upon electron energy directly for each process so that dependence upon an average value is not necessary. Further, the curve of ϵ_i against electron energy can be extrapolated to the thermal value to obtain the heat of reaction. The thermal value of onset is obtained when $\bar{\epsilon}_t$ in Eq. (15) is zero. In this way it was possible to determine the heats of formation of several ions and free radicals. Some of these are given in Tables II–V.

TABLE II

Energies in the Formation of MX_3^- *from* MX_4

Process	AP (eV)	ΔH_R (kcpm)	ϵ_i(AP) (kcpm)	α	$\epsilon_t/E^*_{v,t}$	EA MX_3 (eV)	Ref.
$SiCl_3^-/SiCl_4$	6.5	—	—	—	—	> 2.3	*a*
$SiBr_3^-/SiBr_4$	4.7	—	—	—	—	> 1.3 or 0.9	*a*
$GeCl_3^-/GeCl_4$		27.6	—	0.26	0.43	1.8, 1.9	*a*
$GeBr_3^-/GeBr_4$	3.7	—	—	—	—	> 1.3 or 0.9	*a*
SnF_3^-/SnF_4	5.0	—	—	—	—	$\gtrsim$1.2	*b*
$SnCl_3^-/SnCl_4$	—	2.3	—	0.11	1.0	2.53	*b*
$SnBr_3^-/SnBr_4$	—	−13.8	—	0.2	0.56	3.08 or 3.3	*b*
SnI_3^-/SnI_4	—	−19.6	—	0.11	1.0	3.21	*b*
$TiCl_3^-/TiCl_4$	3.1	—	1.3	0.12	0.92	0.6	*c*
$TiBr_3^-/TiBr_4$	2.4	—	1.6	0.14	0.80	0.8	*c*
TiI_3^-/TiI_4	1.5	—	—	—	—	>0.9	*c*
CF_3^-/CF_4	5.4	79	4.0	0.33	0.34	2.2	*d*

[a] Pabst *et al.* (1977b).
[b] Pabst *et al.* (1977a).
[c] Bennett *et al.* (1974b).
[d] Harland and Franklin (1974).

TABLE III

Energies in the Formation of X^- *from* MX_4

Process	AP (eV)	ΔH_R (kcpm)	ϵ_i(AP) (kcpm)	α	$\bar{\epsilon}_t/E^*_{v,t}$	Electronic Trans. (MX_3) (eV)	Ref.
$Cl^-/SiCl_4$	—	11.5		0.43	0.26	−89.6[a]	*b*
		131.5		0.46	0.24	5.2	*b*
$Br^-/SiBr_4$	—	99.0		0.4	0.28	4.3	*b*
$Cl^-/GeCl_4$	—	64.5		1.0	0.11	3.5	*b*
$Br^-/GeBr_4$	—	50.7		0.80	0.14	3.2	*b*
F^-/SnF_4	—	124.3		—	—	4.0	*c*
$Cl^-/SnCl_4$	—	40.4		0.91	0.12	2.85	*c*
$Br^-/SnBr_4$	—	23.0		0.63	0.18	2.11	*c*
I^-/SnI_4	—	−13.8		0.67	0.17	1.05	*c*
$Cl^-/TiCl_4$	5.0	107.0	1.2	0.72	0.15	4.8 ± 0.2	*d*
$Br^-/TiBr_4$	3.4	68.0	1.5	0.70	0.16	3.3 ± 0.3	*d*
I^-/TiI_4	2.2	35.7	2.0	0.73	0.15	2.2 ± 0.5	*d*

[a] $\Delta H_f(SiCl_3)$, kcal/mole.
[b] Pabst *et al.* (1977b).
[c] Pabst *et al.* (1977a).
[d] Bennett *et al.* (1974b).

TABLE IV

Energies in the Formation of MX_2^- *from* MX_4

Process	ΔH_R (kcpm)	α	$\bar{\epsilon}_t/E^*_{v,t}$	$EA(MX_2)$ (eV)	Ref.
$SiCl_2^-/SiCl_4$	154	0.30	0.37[a]	0.64[a]	*c*
			0.44[b]	0.90[b]	*c*
$SiBr_2^-/SiBr_4$	97	0.59	0.19[a]	1.7[a]	*c*
			0.35[b]	1.9[b]	*c*
$GeCl_2^-/GeCl_4$	71.6	0.93	0.12[a]	0.3[a]	*c*
			0.25[b]	2.56[a]	
				2.9[b]	*c*
$GeBr_2^-/GeBr_4$	69.2	0.37	0.26[a]	1.6[a]	*c*
			0.3[b]	1.8[b]	*c*
$SnCl_2^-/SnCl_4$	90.0	0.3	0.37[a]	1.04	*d*
$SnBr_2^-/SnBr_4$	62.4	0.22	0.51[a]	1.33	*d*
SnI_2^-/SnI_4	33.4	0.22	0.51[a]	1.74	*d*

[a] assumes neutral is X_2 molecule.
[b] assume neutral is 2X atoms.
[c] Pabst *et al.* (1977b).
[d] Pabst *et al.* (1977a).

TABLE V

Energies in the Formation of Ions from Several Trihalides

Process	AP (eV)	ΔH_r (kcpm)	ϵ_i(AP) (kcpm)	α	$\bar{\epsilon}_t/E^*_{v,t}$	Deduced Energy (eV)	Ref.
$AsF_3 \rightarrow AsF_2^- + F$	5.5	77	5.3	0.26	0.64	$EA(AsF_2)$ 0.8	*a*
$AsF_3 \rightarrow AsF_2^- + F$	8.6	74	13.7	0.23	0.72	$EA(AsF_2)$ 0.9	*a*
$AsCl_3 \rightarrow AsCl_2^- + Cl$	4.2	27	6.6	0.38	0.44	$EA(AsCl_2)$ 2.2	*a*
$AsBr_3 \rightarrow AsBr_2^- + Br$	3.0	−11	11.2	0.32	0.52	$EA(AsBr_2)$ 3.5	*a*
$AsCl_3 \rightarrow AsCl^- + Cl_2{}^*$	5.6	111.4	2.6	0.6	0.28	EA(AsCl) 1.3	*a*
$AsBr_3 \rightarrow AsBr^- + Br_2{}^*$	4.2	79.7	1.9	1.0	0.17	EA(AsBr) 1.3	*a*
$AsF_3 \rightarrow F^- + AsF_2{}^*$	8.8	182	4.5	0.62	0.27	$E^*_{el}(AsF_2)$ 7.3	*a*
$AsCl_3 \rightarrow Cl^- + AsCl_2{}^*$	4.1	92	0.9	0.43	0.39	$E^*_{el}(AsCl_2)$ 4.3	*a*
$AsBr_3 \rightarrow Br^- + AsBr_2{}^*$	3.5	49	4.6	0.88	0.19	$E^*_{el}(AsBr_2)$ 2.6	*a*
$NF_3 \rightarrow F^- + NF_2$	0.6	−20.3	9.3	0.43	0.39	$\Delta H_f(NF_2)$ = 8.8 kcal/mole	*b*
$BF_3 \rightarrow F^- + BF_2$	10.4	192	9.4	0.61	0.27	$E^*BF_2(^2B_2) = 5.3$	*b*

[a] Pabst *et al.* (1976).
[b] Harland and Franklin (1974).

3. *Electronic Excitation*

Studies of several of the group IV tetrahalides and the group V trihalides have also been made (Bennett *et al.*, 1974b; Pabst *et al.*, 1976, 1977a,b). Titanium tetrabromide (Bennett *et al.*, 1974b) gives strong peaks for both Br^- and $TiBr_3$, as shown in Fig. 5. The resonances of the two ions are separated at their maxima by about 1 eV and, although the resonances overlap, this will largely disappear if the curves are deconvoluted to remove the spread in energy of the electron beam. The formation of $TiBr_3^-$ must occur by the reaction

$$TiBr_4 + e \rightarrow TiBr_3^- + Br \tag{18}$$

It is noteworthy that for this process $\alpha = 0.14$ and, since there are nine vibrational modes in $TiBr_4$, $\alpha N \approx 1$, so that nearly all of the excess energy appears in translation.

The formation of Br^-, if it is accompanied by that of $TiBr_3$ in the ground state, leaves some 3.3 eV of excess energy unaccounted for at onset. We must assume then that either $TiBr_3$ is electronically excited or that more than one

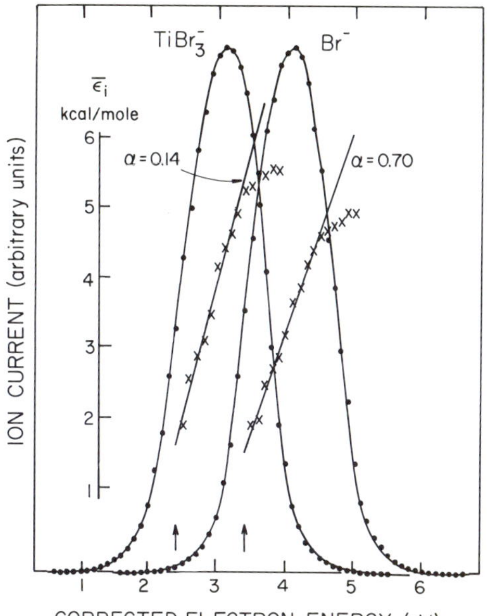

FIG. 5 Resonance and translational energies for $TiBr_3^-$ and Br^- from $TiBr_4$. The arrows at the base are the deconvoluted appearance potentials for the two processes. [From Bennett *et al.* (1974b).]

neutral particle is formed, i.e., that the neutral accompanyment is $TiBr_2$ + Br. If the latter is true, it seems reasonable that the dissociation would not occur in a single step but rather that it would proceed in a sequence of two steps. There are two ways this could occur:

$$TiBr_4 + e \rightarrow TiBr_3^{-*} + Br, \tag{19a}$$

$$TiBr_3^{-*} \rightarrow Br^- + TiBr_2, \tag{19b}$$

or

$$TiBr_4 \rightarrow Br_2^{-*} + TiBr_2, \tag{20a}$$

$$Br_2^{-*} \rightarrow Br^- + Br. \tag{20b}$$

Reaction (19) seems improbable since $TiBr_3^-$ is not formed in the same energy range as Br^-. But if we suppose, nevertheless, that reaction (19a) does occur, nearly all of the excess energy will appear in translation and only 6 kcal/mole will appear in vibration in the $TiBr_3^-$. Since some 14 kcal/mole of internal energy are required for Eq. (19b) to occur, it is evident that this mechanism cannot account for the formation of Br^-. If mechanism (20) occurs, some 42 kcal/mole excess energy will occur in Eq. (20a). We do not know how this will be distributed between translation and vibration, but α can reasonably be taken as 0.5, which is found with processes such as $Cl_2^-/GeCl_4$. This will result in $\bar{\epsilon}_t = 9$ kcal/mole and $\bar{\epsilon}_{Br_2^-} = 5.5$ kcal/mole, leaving 33 kcal/mole for distribution among the various vibrational modes. If Eq. (20b) is to occur, at least 28 kcal/mole of this must go into Br_2^-, which seems improbable. However, assuming that (20b) has just enough energy to take place, the resulting Br^- will have a kinetic energy of 2.8 kcal/mole, which is significantly greater than the energy actually measured for Br^- at onset. We conclude therefore that Br^- cannot be formed by reaction (20). The only remaining mechanism is the formation of $TiBr_3$ in an excited electronic state which we conclude is the most probable one. The electronic transition energy is computed to be 3.3 ± 0.3 eV.

Although it sometimes happens that an X^- ion from MX_4 will have a resonance that yields both products in ground electronic states, this occurance is quite the exception. In Fig. 3 O^-/SO_2 is seen to occur in two resonances separated by about 2.5 eV, the lower of which corresponds to products in the ground state; whereas the higher energy resonance corresponds to SO being formed in an excited electronic state. In Fig. 6 the ionization efficiency curve for F^-/CF_4 appears to involve two extensively overlapping resonances. The peripheral figures at several points around the resonance are reproductions of the mass peaks at those points. It is evident from the peak widths that, on the lower energy side of the resonance, F^- is formed with very large translational energy. On the higher energy side of the resonance, a second peak grows on top of the broad one. This peak is much narrower and thus

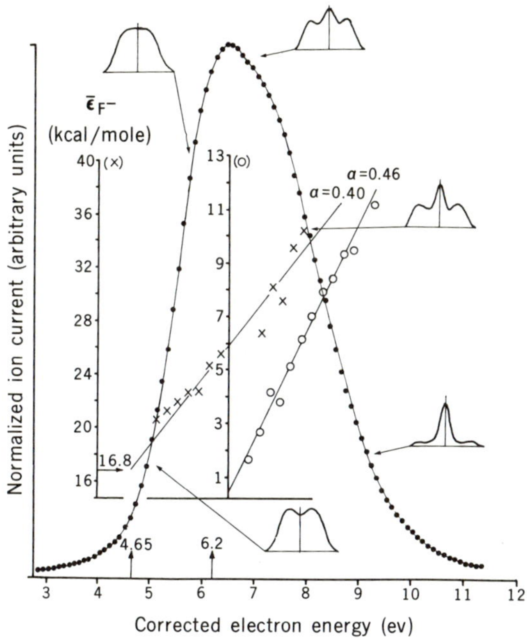

FIG. 6 Resonance and translational energy of F^- from CF_4. The peripheral figures shown at various points on the resonance are mass peaks. Note the continuous growth of a narrow peak on top of the broad one which becomes less intense at higher energies. The broad peak results from high kinetic energy F^- ions formed along with ground state CF_3. The narrow peak shows the growth of F^- of low kinetic energy along with electronically excited CF_3. The straight lines are estimates of the translational energy attributable to the two processes. [From Franklin (1976). Copyright 1976 by the American Association for the Advancement of Science.]

represents a process in which F^- is formed with relatively small kinetic energy. Because of the overlap the individual translational energies could not be measured precisely but in a qualitative sense, it appears that the process at the smaller electron energies probably corresponds to the formation of products in their ground states, whereas the process occurring at the higher electron energies probably corresponds to the formation of electronically excited CF_3. Aside from a few such examples, all of the processes leading to X^- from either MX_4 or MX_3 compounds appear to yield the neutral MX_3 or MX_2 in excited electronic states.

Although there are only two examples in Table V, the process yielding $AsCl^-$ and $AsBr^-$ also seem to involve the formation of the neutral Cl_2 and Br_2 molecules in excited electronic states. Unfortunately, not enough measurements have been made on such reactions to know whether this is typical behavior.

4. Energy Distribution and Molecular Geometry

In Tables II–V a column headed $\bar{\epsilon}_t/E^*_{v,t}$ is given. This is the fraction of the total excess nonelectronic energy that is released in translation upon dissociation and is independent of any theory concerning the dissociation. As seen in Table II, the process MX_3^-/MX_4 releases 43–100% of its excess energy in translation. The classical QET would predict that 11% of the excess energy would appear in translation. In Table V the process AsX_2^-/AsX_3 releases 44–72% of its excess energy in translation. Classical QET predicts that 17% of the excess energy would be released.

On the other hand, Table III shows that the process X^-/MX_4 releases from 11 to 26% of the excess energy in translation. Table V shows that the process X^-/MX_3 releases from 17 to 39% of its excess energy in translation. Classical QET would again predict the fractional release to be 11 and 17%, respectively.

Although there is considerable variation, it is quite evident that there is a wide difference in the fraction of energy released in translation by the processes forming MX_3^- and MX_2^- plus X from those forming X^- plus MX_3 and MX_2. This behavior can be understood in terms of the geometry of the molecules. Consider the reaction.

$$e + MX_4 \rightarrow MX_3^- + X. \tag{21}$$

As the fragments separate, the additional electron will tend to form an unshared pair with the remaining electron of the bond being broken. Thus, the electron pair in MX_3^- will replace the leaving X atom, and MX_3^- will tend to retain its tatrahedral structure. To the extent that this is true, there will be no need for the configuration of the molecule to change and hence no need for energy to be redistributed among the various bonds (and hence vibrational modes of the molecules). Consequently, the excess energy will be located mainly in the breaking bond and thus will appear in translation. Similar arguments can be made for the formation of MX_2^- ions from MX_3.

On the other hand, for the reaction

$$e + MX_4 \rightarrow X^- + MX_3, \tag{22}$$

the configuration of the MX_3 will change from pyramidal to planar as the X^- ion separates. This will involve some redistribution of energy among the bonds (and hence the vibrational modes) of the molecule. Thus, it would be expected that more energy would appear in vibration and less in translation for Eq. (22) than for Eq. (21). Again, similar arguments can be made for the formation of X^- ions from MX_3 as is seen in Table V.

5. *Comparison of Measured and Theoretical Results*

The results given above for dissociative resonance capture processes fail to agree with the results of the classical QET treatment of most systems. The classical QET treatment is generally recognized as unsatisfactory, however, so we performed detailed counts of states in accordance with both the QET, Eq. (2), and "Phase Space Theory," Eq. (10), for two systems studied experimentally in this laboratory. The results for P_2^- obtained experimentally by Bennett *et al.* (1974a) are compared with calculated values in Fig. 7. Several matters of interest are to be seen. (1) Neither theoretical treatment agrees with the experimental results, but the phase space treatment comes closer than the QET. (2) Values calculated by the direct count of states agree very closely with the limiting results of Eqs. (12) and (13), but deviate slightly from the limiting lines for a short distance above onset. This curvature is attributable to the effect of rotation which becomes significant at low energies.

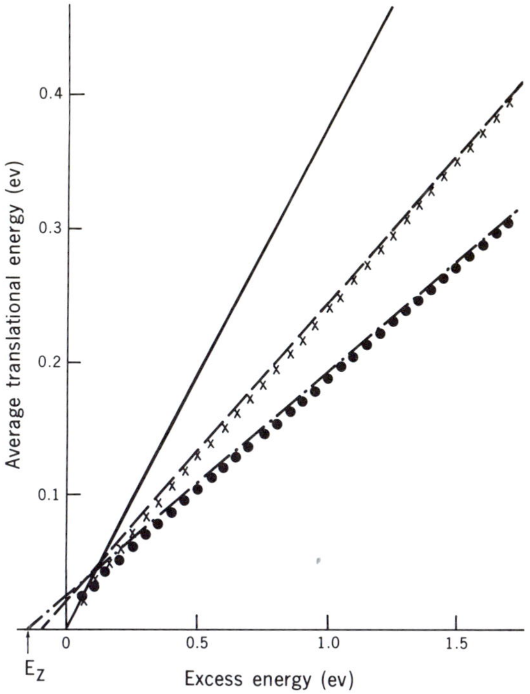

FIG. 7 Average CM translational energy for the process. $P_4 + e \rightarrow P_2^- + P_2$ (Franklin, 1976). Copyright 1976 by the American Association for the Advancement of Science.) The solid line is the experimental result (Bennett *et al.*, 1974a). · · · Direct count of states employing QET, Eq. (2) [From Carter, 1976], Direct count of states employing "phase space treatment," Eq. (10) (Carter, 1976). × × × Upper limits calculated from Eqs. (12) and (13) [Franklin (1976)].

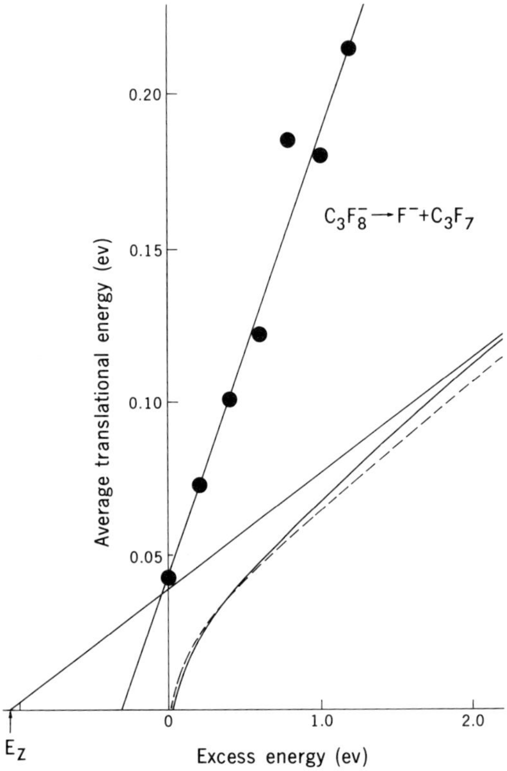

FIG. 8 Average CM translational energy for the process $C_3F_8 + e \rightarrow F^- + C_3F_7$ (Franklin, 1976) Copyright 1976 by the American Association for the Advancement of Science.) · · · Experimental data (Harland and Franklin, 1974). --- Direct count of states employing QET, Eq. (2) (Carter, 1976). Solid curved line, direct count of states employing phase space, Eq. (10) (Carter, 1976). Solid straight line is upper limit of classical QET treatment, Eq. (12) [Franklin (1976)].

Somewhat similar results are found in Fig. 8 for F^-/C_3F_8. The experimental results are from Harland and Franklin (1974). The theoretical results shown in Figs. 7 and 8 were calculated by Carter (1976). The conclusion for this system is again that neither theoretical result agrees with the experimental measurements.

E. Positive Ion Processes

The development of photoionization–photoelectron coincidence spectroscopy (PIPECO) and of threshold photoelectron spectrometry (TPES) made it possible to measure the internal energy of the parent ion and of the corresponding fragment ion, and consequently to relate translational energy to excess energy. Baer (Chapter 5, in this volume), has discussed this point in some detail. A few examples will be given here.

1. *Ions from Hydrocarbons*

Stockbauer (1977b), employing TPES, has measured the translational energy release for the processes CH_3^+/CH_4, CH_2^+/CH_4, CD_3^+/CD_4, CD_2^+/CD_4, and CH_2^+/CH_2CO. For the two methyl ions, $\bar{\epsilon}_t/E^*$ were 0.23 and 0.21 in excellent agreement and definitely greater than the 0.12 calculated by Vestal (1968) and by Klots (1972) and the 0.11 calculated by the classical QET. The $\bar{\epsilon}_t/E^*$ determined for the two methylene ions were also in good agreement at 0.46 and 0.41. This is much greater than the 0.13 computed by Vestal (1968) or 0.11 computed by classical QET. On the other hand, the $\bar{\epsilon}/E^*$ for CH_2^+/CH_2CO was 0.13 and thus is quite close to the statistical value of 0.11 computed by classical QET.

The formation of $C_3H_3^+$ from C_4H_6 is an interesting case. It is discussed in some detail in Baer (Chapter 5, in this volume) and Chesnavich and Bowers (Chapter 4, in this volume) and will not be repeated here.

Eland *et al.* (1976b) have studied the kinetic energy released in the formation of $C_4H_4^+$ and $C_3H_3^+$ ions from benzene, using the PIPECO technique. Both processes gave greater kinetic energies than would be predicted by the classical QET.

Brehm *et al.* (1976), using the PIPECO technique, have studied the fragmentation of propane and neopentane and have measured the kinetic energy released in the formation of $C_2H_5^+$, $C_2H_4^+$, and $C_2H_3^+$ from propane and $C_4H_9^+$, $C_3H_5^+$, and $C_2H_5^+$ from neopentane. In all cases the kinetic energy release was small. The kinetic energy distributions of the three ions from propane agreed reasonably well with values calculated by Vestal (1968), employing the QET formalism. Although there were no calculated values with which to compare the measured kinetic energies of the ions from neopentane, the shape of the kinetic energy distribution and the average energy released were both in agreement with expected results from QET treatment of neopentane.

2. *Ions from Compounds Containing Halogens*

Simm *et al.* (1976) studied the loss of a fluorine atom from SF_6^+, CF_4^+, and $C_2F_6^+$ and found in all cases that a large fraction of the available excess energy was released in translation with a distribution having a preponderance of high energy ions. These results suggest a rapid dissociation of the parent ion and are not in accord with the predictions of any statistical theory. It was shown by Simm *et al.* (1974) that $C_2F_5^+$ is formed directly from the $\tilde{A}$ state without initial conversion to the ground state. However, the loss of fluorine atoms from SF_6^+ and CF_4^+ may occur after initial conversion to the respective ground states. On the other hand, Simm and Danby (1976) found that the energy released in translation by the formation of CF_3^+ from $C_2F_6^+$ in the ground state was in agreement with the prediction of statistical theories. The kinetic energy release distribution from higher states was quite broad and change slowly with excess energy. The forms of the kinetic energy distribution suggest that during dissociation the energy is partially but not fully randomized. However, the slow change in kinetic energy release with excited $C_2F_6^+$ is not compatible with statistical theories.

Eland *et al.* (1976a) have studied ions from the four methyl halides, employing the PIPECO technique. The formation of CH_3^+ from the $\tilde{A}$, $\tilde{B}$ states of CH_3F^+ occurs with a release of about 90% of the excess energy in translation. The formation of CH_3^+ from CH_3Cl^+ ($\tilde{A}$) occurs with the release of a single large kinetic energy that changes very little with excess energy. With the onset of the $CH_3Cl^+(\tilde{B})$, the kinetic energy distribution becomes much broader, but the kinetic energy release characteristic of the $\tilde{A}$ state persists. The kinetic energy released with the formation of CH_3^+ from CH_3Br and CH_3I and I^+ from CH_3I is small and rises slowly with increasing excess energy. A more thorough discussion of the CH_3I system is found in Baer (Chapter 5, in this volume).

The authors also found that with methyl fluoride the loss of H atom occurred only from the ground electronic state and not from states of higher energy, although this should be energetically possible. Loss of H from higher electronic states of the other methyl halides was not observed, although it was energetically possible.

Brehm *et al.* (1974) studied CF_4^+ by PIPECO. When CF_4 is ionized by 584 Å light it gives 99% CF_3^+. It is formed from the $\tilde{X}^2T_1$ state of CF_4^+ with a single kinetic energy release of 1.05–1.10 eV over the entire range of the band. The $\tilde{A}^2T_2$ band similarly gives a single kinetic energy release of 1.3 eV. The $\tilde{B}^2E$ state gives a broad distribution of similar average kinetic energy. The kinetic energy release and the lack of vibrational structure in the $\tilde{X}^2T_1$ and $\tilde{A}^2T_2$ bands suggests an internal conversion of the latter to the former state of CF_4^+ prior to dissociation. Although the $\tilde{B}^2E$ state does exhibit some vibrational structure, its energy release of 1.3 eV suggests that it too probably undergoes at least, in part, an internal conversion to one of the lower states. In view of the rather slight increase in kinetic energy above the directly formed $\tilde{X}^2T_1$ state, most of the added energy probably goes into vibration. This would be expected as the molecular geometry changed during the internal conversion. However, it appears that the statistical theory of energy distribution is not followed.

V. SUMMARY

To recapitulate the results of energy distribution studies of both positive and negative ions, we find that many of the systems examined so far fail to agree with the predictions of either QET or the phase space treatment. Certain dissociative resonance capture processes appear to be understandable in terms of the geometry of the molecules, but similar behavior of dissociative ionization processes has not been observed.

It might be appropriate at this time to consider the appropriateness of some of the assumptions upon which the theories are based. One for which a clear cut decision can be made is the original idea that there are enough crossings of potential energy surfaces to render the ground state accessible to all higher states. In general, this is clearly not true, as has been shown in several of the studies of positive ions discussed above. The two resonances for $TiBr_4$, given in Fig. 5, are evidently completely separate; otherwise, $TiBr_3^-$, which is formed in the lower energy resonance, would be formed in both; and Br^-, which is formed along with electronically excited $TiBr_3$, should also occur at much lower energies. Pabst *et al.* (1977b) have found two resonances for $Cl^-/SiCl_4$ with maxima separated by about 5 eV and thermal onsets at 0.5 and 5.7 eV. It is evident that these two states do not communicate. The same is true of O^-/SO_2, shown in Fig. 3. As a further example, SF_6 attaches an electron having an energy between zero and 0.1 eV of energy. A considerable fraction of the SF_6^- dissociates to SF_5^- in less than 0.3 μsec. However, it is also well known that the lifetime of the SF_6^- is 30 μsec or longer, but if these states communicated there could be only about 10^{-7} of the SF_6^- remaining after 30 μsec. Obviously, the two lifetimes must refer to two separate and noncommunicating states. It is because of these and many similar phenomena that we now assume that many electronic states often do not communicate. Consequently, one should treat internal (vibrational) energy as pertaining only to the electronic state corresponding to the states of the dissociation products.

The assumption that energy will be distributed among the various vibrational modes in a time short compared to the time of dissociation seems to fail for some systems, although proof is not easy to achieve. Many of the dissociative resonance capture processes appear to occur by direct dissociations involving only a single vibration and these generally would probably not have time to distribute their vibrational energy internally. This, for example, might be the case with the $MX_4 + e \rightarrow MX_3^- + X$ reactions which release most of their energy in translation. The obverse reactions $MX_4 + e \rightarrow X^- + MX_3$ tend to release a much smaller part of their internal energy in translation, but whether these processes require a longer time to dissociate is not clear. The behavior can easily be justified in terms of molecular geometry but admittedly this is not proof. It would be helpful if some metastable dissociations similar to those often observed with positive ions could be found. In my opinion, such metastables should be sought.

The present state of energy distribution in unimolecular dissociations suggests that various approaches, both theoretical and experimental, are needed. I would like to suggest a few.

(1) Experimental measurements of ion kinetic energy need to be improved, especially those of negative ion processes. Electron beams of considerably narrower energy width are needed and, along with this, the kinetic energies of ions need to be measured with greater resolution. If these could be accomplished, even with relatively simple molecules, they would constitute important advances. Nevertheless, the one experiment reported by Stamatovic and Schulz (1973) accomplished only one of these but showed how important an improvement could be made.

(2) Very little is known about the electronic states of negative ions and knowledge of the states involved would be of great help in interpreting experimental results. Theoretical studies could be of considerable help. Not many have been made on ions beyond diatomics and a very few triatomics. Some experimental studies may well be useful in identifying the states of parent negative ions that undergo dissociative resonance capture reactions. One approach to this is being attempted in several laboratories by measuring the angular distributions of ions from various processes. With dissociative ionization processes there are so many states spaced so closely that interpretable results are achievable only with very simple diatomics. But dissociative resonance capture processes often appear to involve states that are well separated and so may be expected to give interpretable results. Van Brunt and Kieffer have studied O^-/O_2 by dissociative resonance capture (Van Brunt and Kieffer, 1970) and by pair production (Van Brunt and Kieffer, 1974), but the literature on this subject is not large. In our laboratory an instrument to make such measurements has been constructed and will soon be employed for the measurement of angular distributions of negative ion processes.

(3) It would be of great value if emission spectra of apparently electronically excited products of dissociative resonance capture could be detected and analyzed. If this can be accomplished, it would show conclusively whether our inferences concerning the nature of some reactions is correct and, if so,

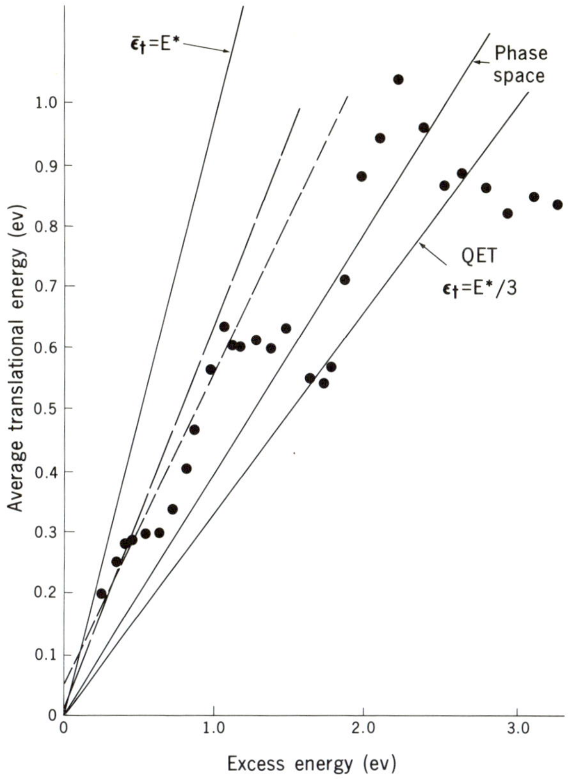

FIG. 9 Average CM translational energy for the process O^- from CO_2. Solid lines are shown for values calculated by the phase space and quasi-equilibrium theories and for the limiting case of $\bar{\epsilon}_1 = E^*$. The circles are the results of trajectory calculations [Carter (1976)]. The lines made up of long and of short dashes represent unpublished experimental results obtained respectively by Carter using the deflection method [Sen Sharma and Franklin (1974)] and by peak shape analysis [Hadley and Franklin (1975)].

would help identify the state of the product in question. We hope to develop an instrument to accomplish this.

(4) Finally, it appears that some new departures in the theory of unimolecular dissociation reactions is called for. Carter (1976) in this laboratory has made a classical trajectory study of the dissociative resonance capture formation of O^- from CO_2. The results, while not perfect, gave reasonably good agreement with experiment over about one electron volt of excess energy, as shown in Fig. 9. Other studies would be of interest but the problem becomes extremely complex and the cost of compution very great with larger molecules.

ACKNOWLEDGMENTS

I wish to express my gratitude for support from the Robert A. Welch Foundation and from the Office of Naval Research (Contract N00 14-75-C-0415). I also wish to express my appreciation to my co-authors.

REFERENCES

Bennett, S. L., Margrave, J. L., and Franklin, J. L. (1974a). *J. Chem. Phys.* **61**, 1647.
Bennett, S. L., Pabst, R. E., Margrave, J. L., and Franklin, J. L. (1974b). *Int. J. Mass Spectrom. Ion Phys.* **15**, 451.
Berry, C. E. (1950). *Phys. Rev.* **78**, 597.
Blewett, J. P. (1936). *Phys. Rev.* **49**, 1900.
Bracher, J., Ehrhardt, H., Fuchs, H., Osberghaus, R., and Taubert, R. (1963). *Adv. Mass Spectrom.* **2**, 285.
Brehm, B., Frey, R., Küstler, A., and Eland, J. H. D. (1974). *Int. J. Mass Spectrom. Ion Phys.* **13**, 251.
Brehm, B., Eland, J. H. D., Frey, R., and Schulte, H. (1976). *Int. J. Mass Spectrom. Ion Phys.* **21**, 373.
Carter, D. E. (1976). *J. Chem. Phys.* **65**, 2584.
Carter, D. E. (1977). *J. Chem. Phys.* **67**, 841.
Chantry, P. J. (1968). *Phys. Rev.* **172**, 125.
Chantry, P. J. (1969). *J. Chem. Phys.* **51**, 3369.
Chantry, P. J., and Schulz, G. J. (1967). *Phys. Rev.* **156**, 134.
DeCorpo, J. J., and Franklin, J. L. (1971a). *J. Chem. Phys.* **54**, 1885.
DeCorpo, J. J., Steiger, R. P., Franklin, J. L., and Margrave, J. L. (1970). *J. Chem. Phys.* **53**, 936.
DeCorpo, J. J., Bafus, D. A., and Franklin, J. L. (1971b). *J. Chem. Phys.* **54**, 1592.
Eland, J. H. D. (1972). *Int. J. Mass Spectrom. Ion Phys.* **9**, 397.
Eland, J. H. D., Frey, R., Kuestler, A., Schulte, H., and Brehm, B. (1976a). *Int. J. Mass Spectrom. Ion Phys.* **22**, 155.
Eland, J. H. D., Frey, R., Schulte, H., and Brehm, B. (1976b). *Int. J. Mass Spectrom. Ion Phys.* **21**, 209.
Forst, W. (1973). "Theory of Unimolecular Reactions." Academic Press, New York.
Fox, R. E., and Hipple, J. A. (1948). *Rev. Sci. Instrum.* **19**, 462.
Fox, R. E., Hickam, W. M., Grove, D. J., and Kjeldaas, T. (1951). *Phys. Rev.* **84**, 859.
Fox, R. E., Hickam, W. M., Grove, D. J., and Kjeldaas, T. (1955). *Rev. Sci Instrum.* **26**. 1101.
Franklin, J. L. (1976). *Science* **193**, 725.
Franklin, J. L., Hierl, P. M., and Whan, D. A. (1967). *J. Chem. Phys.* **47**, 3148.
Hadley, F. J., and Franklin, J. L. (1975). *Int. J. Mass Spectrom. Ion Phys.* **18**, 249.
Hagstrum, H. D. (1951). *Rev. Mod. Phys.* **23**, 185.
Hagstrum, H. D., and Tate, J. T. (1940). *Phys. Rev. B* **59**, 54.
Haney, M. A., and Franklin, J. L. (1968). *J. Chem. Phys.* **48**, 4093.
Haney, M. A., and Franklin, J. L. (1969a). *J. Chem. Phys.* **50**, 2028.
Haney, M. A., and Franklin, J. L. (1969b). *Trans. Faraday Soc.* **65**, 1794.
Harland, P. W., and Franklin, J. L. (1974). *J. Chem. Phys.* **61**, 1621.
Harland, P. W., Franklin, J. L., and Carter, D. E. (1973). *J. Chem. Phys.* **58**, 1430.
Hierl, P. M., and Franklin, J. L. (1967). *J. Chem. Phys.* **47**, 3154.
Kandel, R. J. (1953). *Phys. Rev.* **91**, 436.
Kandel, R. J. (1954). *J. Chem. Phys.* **22**, 1496.
Kassel, L. S. (1928). *J. Phys. Chem.* **32**, 225, 1065.

Kassel, L. S. (1932). "Kinetics of Homogeneous Gas Reactions." Chem. Catalog Co. (Tudor), New York.
Klots, C. E. (1971). *J. Phys. Chem.* **75**, 1526.
Klots, C. E. (1972). *Z. Naturforsch., Teil A* **27**, 553.
Light, J. C. (1967). *Discuss. Faraday Soc.* **44**, 15.
McDowell, C. A., and Warren, J. W. (1952). *Discuss Faraday Soc.* **10**, 53.
MacNeill, K. A. G., and Thynne, J. C. J. (1969). *Int. J. Mass Spectrom. Ion Phys.* **3**, 35.
Marcus, R. A. (1952) *J. Chem. Phys.* **21**, 359.
Marcus, R. A., and Rice, O. K. (1951). *J. Phys. Col. Chem.* **55**, 894.
Morrison, J. D. (1963). *J. Chem. Phys.* **39**, 200.
Pabst, R. E., Bennett, S. L., Margrave, J. L., and Franklin, J. L. (1976). *J. Chem. Phys.* **65**, 1550.
Pabst, R. E., Perry, D. L., Margrave, J. L., and Franklin, J. L. (1977a). *Int. J. Mass Spectrom. Ion Phys.* **24**, 323.
Pabst, R. E., Margrave, J. L., and Franklin, J. L. (1977b). *Int. J. Mass Spectrom. Ion Phys.* **25**, 361.
Pechukas, P., and Light, J. C. (1965). *J. Chem. Phys.* **42**, 328
Rice, O. K., and Ramsperger, H. G. (1927). *J. Am. Chem. Soc.* **49**, 1672.
Rice, O. K., and Ramsperger, H. G. (1928). *J. Am. Chem. Soc.* **50**, 617.
Robinson, P. J., and Holbrook, K. A. (1972). "Unimolecular Reactions." Wiley (Interscience), New York.
Rosenstock, H. M., Wallenstein, M. B., Wahrhaftig, A. L., and Eyring, H. (1952). *Proc. Natl. Acad. Sci. U.S.A.* **38**, 667.
Schulz, G. J. (1961). *J. Chem. Phys.* **34**, 1778.
Schulz, G. J. (1962). *Phys. Rev.* **128**, 178.
Sen Sharma, D. K., and Franklin, J. L. (1973). *J. Am. Chem. Soc.* **95**, 6562.
Sen Sharma, D. K., and Franklin, J. L. (1974). *Int. J. Mass Spectrom. Ion Phys.* **13**, 139.
Simm, I. G., and Danby, C. J. (1976). *J. Chem. Soc., Faraday Trans. II* **72**, 860.
Simm, I. G., Danby, C. J., and Eland, J. H. D (1974). *Int. J. Mass Spectrom, Ion Phys.* **14**, 285.
Simm, I. G., Canby, C. J., Eland, J. H. D., and Mansell, P. I. (1976). *J. Chem. Soc., Faraday Trans. II* **72**, 426.
Spotz, E. L., Seitz, W. A., and Franklin, J. L. (1969). *J. Chem. Phys.* **51**, 5142.
Stamatovic, A., and Schulz, G. J. (1973). *Phys. Rev. A* **7**, 589.
Stockbauer, R. (1977a). *Int. J. Mass Spectrom. Ion Phys.* **25**, 89.
Stockbauer, R. (1977b). *Int. J. Mass Spectrom. Ion Phys.* **25**, 401.
Taubert, R. (1959). *Adv. Mass Spectrom.* **1**, 489.
Taubert, R. (1964). *Z. Naturforsch., Teil A* **19**, 484.
Tsai, B., Baer, T., and Horovitz, M. L. (1974). *Rev. Sci. Instrum.* **45**, 494.
Van Brunt, R. J., and Kieffer, L. J. (1970). *Phys. Rev. A* **2**, 1899.
Van Brunt, R. J., and Kieffer, L. J. (1974). *J. Chem. Phys.* **60**, 3057.
Vestal, M. (1968). *In* "Fundamental Processes in Radiation Chemistry" (P. Ausloos, ed.), p. 59. Wiley, New York.
Waldron, J. D. (1954). *Trans. Faraday Soc.* **50**, 102.
Whitten, G. L., and Rabinovitch, B. S. (1963). *J. Chem. Phys.* **38**, 2466.
Worry, G., and Marcus, R. A. (1977). *J. Chem. Phys.* **67**, 1636.

Chapter 8
On the rapidity of ion–molecule reactions

V. L. TALROSE, P. S. VINOGRADOV, AND I. K. LARIN

INSTITUTE OF CHEMICAL PHYSICS OF THE ACADEMY OF SCIENCES OF THE USSR, 117977 GSP-1 MOSCOW V-334, U.S.S.R.

I. INTRODUCTION

In the present paper we have chosen for consideration three problems which we think denote kinetic aspects of basic investigations of ion–molecule

ISBN-0-12-1208-01-X

reactions. They are the problem of the rapidity (anomalous high rate) of ion–molecule reactions; the problem of the "solvated" shell effect on this rate; and the problem of vibronic–translational (v–t) relaxation. As examples in the chapter we have used data on exothermic and thermoneutral ion–molecule reactions. Nevertheless, it is natural that the principle of detailed equilibrium will allow the reader to extend conclusions to endothermic reactions as well. The study of ion–molecule reactions in gases was begun by one of the authors early in the fifties (Talrose and Ljubimova, 1952). The most interesting and characteristic feature of these reactions is the fact that they are extremely fast processes which can, in many cases, go practically at each collision by one or several possible exothermic or thermoneutral channels. At that time this fact was rather surprising since kineticists knew that even in fast reactions of neutral particles, e.g., the majority of free radicals + molecules reactions, there was an activation energy of about 10 kcal/mole. Such a difference between the reactions of neutral particles and ion–molecule reactions is qualitatively connected with the presence of the electric charge in the system.

By the beginning of the sixties a widespread interest in these reactions had developed and we refer to Talrose (1967) as one of the attempts at systematization of the kinetic data.

As information has been accumulated on ion–molecule reactions, comparatively slow ion–molecule reactions have been discovered in a small percentage of cases. These reactions are exothermic and have no parallel competative exothermic reaction channel. This means that there are cases when the ion–molecule reaction rate is determined by factors other than the reaction thermochemistry. In some cases an approach based on orbital symmetry conservation rules has turned out to be of extraordinary use for qualitative estimation of ion–molecule reactions rate. One of the sections of the chapter is devoted to such an approach.

The processes of isotope exchange of ion–molecule collisions are given special attention. The thermoneutrality of such processes makes the factors connected with energy change unimportant in these systems. The equivalence of reagent and product molecular orbitals allows the application of orbital symmetry conservation rules to ion–molecule reactions in the simplest and most concise form.

Another important problem associated with ion–molecule reaction rates is the influence of a shell of neutral molecules on ion reactivity. Radiolytic deuterium–hydrogen exchange with participation of hydrogen ion clusters in mixtures of H_2 and D_2 is considered as an example in the chapter.

Besides the information on rate constants of elementary ion–molecule reactions, the data on the probability of deactivation of vibrational excitation of the ions are of great significance in kinetics. At present, vibronic–translation relaxation of ions is not well understood. However, the experimental data available on vibration–translational relaxation of diatomic ions show that these relaxation processes are much faster than in the case of neutral particle collisions.

II. SOME CHARACTERISTIC FEATURES OF ION–MOLECULE REACTIONS

Exothermic and thermoneutral ion–molecule reactions are usually extremely fast processes. Based on early experiments on ion–molecule reactions in hydrocarbon gas mixtures, one of the authors (Talrose and Ljubimova, 1952) suggested exothermic and thermoneutral reactions, in most cases, would be extremely fast processes, would have no noticeable activation energy, and would proceed on each collision. Subsequent experiments by other authors confirmed the accuracy of this suggestion for many types of ion–molecule reactions.

Characteristic values of the rate constants of such reactions are often in the range 10^{-9}–10^{-10} $cm^3 sec^{-1}$ and are close to the collision rate constant (Talrose, 1952; Schissler and Stevenson, 1956). The latter is calculated by the Langevin formula in the case of ion–nonpolar molecule collisions, or by more complicated treatments (Moran and Hamill, 1963; Su and Bowers, 1973) if the molecule is polar.

The classical Langevin type treatments rationalize only large values of cross sections of ion–molecule collisions; they have nothing to do with the problem of chemical bond rearrangement, i.e., with the experimental rate of chemical reaction. The reason for the low activation energy in ion–molecule reactions has not been clearly discussed up to now. One has to stress this generally accepted theoretical consideration, since the erroneous idea that a large value of the collision cross section, in itself, determines a large rate for an ion–molecule reaction sometimes slips through in the work of the most competent authors. Further, large values of reaction rate constants of ion–molecule reactions cannot be fully explained by the statistical theory of reactions (Talrose, 1967; Nikitin, 1970; Light, 1962, 1964; see also Chesnavich and Bowers, Chapter 2, in this volume). In fact, statistical theory usually limits itself to explaining relative values of the rate constants for cases where the states are evenly mixed on a multidimensional surface of constant energy of a phase space of the system, $H(P, q) = E$.

In summary, we would like to stress once more that neither Langevin theory nor statistical theory can explain extremely large values of reaction rate constants as well as the absence of activation energy (below we would call it "rapidity") of most exothermic and thermoneutral ion–molecule reactions.

In general, the rapidity of ion–molecule reactions may be said to be bound up with the particular properties of potential energy surfaces of the system: ion + molecule in a quick action zone. Knowing the surfaces of potential energy as well as the probabilities of transition from one surface to another, one could solve the problem of rapid rate constants of ion/molecule reactions. But the calculations of such surfaces and the transition probabilities is a rather complicated problem. It has been accomplished accurately only for simple systems; for example, $H^+ + H_2(D_2, HD)$ (Krenos *et al.*, 1974); $D_2^+ + H$ (Preston and Kross, 1973); $H_2^+ + He$ (Cooper *et al.*, 1974; Brown and Hayer, 1971); $Ar^+ + H_2$ (Chapman and Preston, 1974); $HeH^+ + H_2$ (McLaughlin and Tompson, 1973). At present the calculation of multidimensional potential energy surfaces for more complex polyatomic systems is still a complicated problem. Nevertheless, one may try to give qualitative considerations which could explain why ion–molecule reactions usually have no activation energy, while neutral particles reactions have it.

We have already mentioned above that such a specific character of ion–molecule reactions is apparently connected with the presence of an electric charge in the system of colliding particles. If one compares a fast ion–molecule reaction with a reaction of a neutral radical or atom with a molecule (for which the characteristics activation energy is known to be $E_a \leq 10$ kcal/mole), then the ion–molecule reaction should have approximately the same activation energy as the neutral if the effect of electric charge is not taken into account. But owing to the polarization interaction, $U \approx -\alpha e/2r^4$, the activation barrier which is characteristic for reaction of neutrals as seen in Fig. 1 may be already inside a potential well for IM collisions at distances equal to the interatomic ones.

It is clear that such a simple picture does not take into consideration many details of the interaction at small distances. But even so it follows that if an activation barrier with characteristic value of 10 kcal/mole is arising on a surface of potential energy in a valley corresponding to the reagents approach,

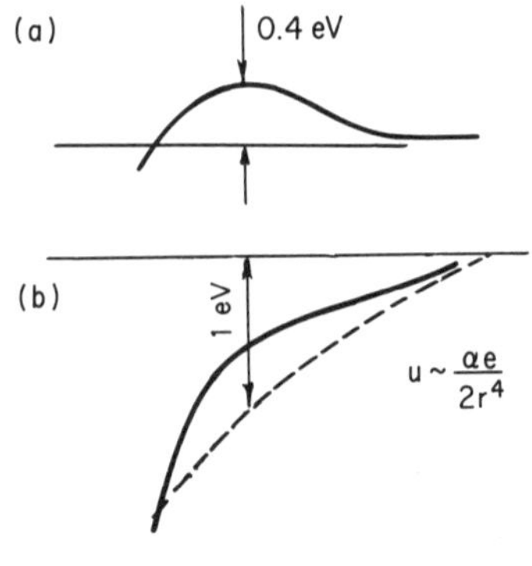

FIG. 1 The effect of the ion–molecule polarization interaction on the potential energy surface (PES). (a) The PES section along the line corresponding to particle approach for reactions of neutrals; (b) the same section of the PES distorted by the polarization interaction.

then it may be comparatively easy to overcome since colliding particles gain considerable kinetic energy due to polarization attraction at small distances.

From the viewpoint of the reaction dynamics, a reaction coordinate is being introduced. It may either correspond to the motion along one of the degrees of freedom of the system X_k, or be expressed as a combination of motions along several coordinates

$$\xi_i = F_i(X_j), \tag{1}$$

i.e., equation system (1) sets the reaction coordinate.

We may try to express the potential energy surface for ion–molecule reaction as follows:

$$U \approx U_{\text{neut}}(X_j) + U_{\text{pol}}(X_j). \tag{2}$$

In expression (2) the potential energy of the system is represented as two parts: U_{neut} corresponds to a potential energy surface which is typical for an equivalent system of neutral particles and U_{pol} is the polarization interaction energy. Of course this approximation is rather rough since potential energy surfaces may have different shapes, but such an equivalent surface is introduced only to demonstrate the influence of the part which is connected with the electric charge in the system.

For the present it is of importance that there is no component of a relatively strong polarization interaction in U_{neut}. Now consider U along the reaction coordinate ξ:

$$U(\xi) = U_{\text{neut}}(\xi) + U_{\text{pol}}(\xi). \tag{3}$$

If $U_{\text{neut}}(\xi)$ has a maximum along the reaction coordinate (i.e., there is an activation barrier), then the polarization attraction may promote the overcoming of this barrier if coordinate projections corresponding to the ion and molecule approach and their coming apart on the reaction coordinate are substantial. In other words, a barrier on the corresponding curve is compensated by U_{pol} in the case if U_{pol} is large enough within the corresponding range of ξ values. A corresponding projection of approach coordinates on the ξ coordinate is determined through the system (1). Perhaps it is the first factor explaining the fact that ion–molecule reactions are usually fast.

Another factor explaining the rapidity of ion–molecule reactions is connected with the fact that a polarization complex AB^+ may be formed, owing to a strong mutual attraction of the A^+ and B colliding particles. Particles A^+ and B will move rather quickly relative to one another, due to the presence of the potential well caused by the polarization interaction. Relative motion energy for even small initial energies of colliding partners will be considerable at short range (about 1 eV). Since a polarization complex may be long lived, an effective exchange between the system degrees of freedom is possible. Now if we compare neutral particles reactions with ion–molecule ones, we may suppose that the lifetime of the AB^+ ionic complex will be much greater than that in the case of a neutral complex. Further, since the energy, $-E_c$, of relative motion of A^+ and B in the AB^+ complex is greater than that in the case of a neutral collision, the energy exchange rate with other degrees of freedom should be greater as well since dE_c/dt may be considered to be proportional to E_c. That is,

$$-\frac{dE_c}{dt} \sim E_c. \tag{4}$$

The above arguments have the following important consequence. It has been suggested (Kondratiyev *et al.*, 1976) that the degrees of freedom of reacting molecules are divided into active and inactive. In this model, for the reaction to go, it is necessary that required energy should be concentrated in active degrees of freedom. One may expect that such a transition of energy into active degrees of freedom will be efficient for ion–molecule reactions for the following reasons. The energy of the particles relative motion in the complex is rather high, 1 eV or even higher. Energy exchange between degrees of freedom may be enhanced by the long lifetime of the complex and fast energy exchange rate [as follows from (4)]. One of the most striking examples of the role of complex lifetimes in the rate of ion–molecule reactions is the reactions (Meot-Ner and Field, 1976) in which the rate sharply increases with the decrease of temperature—approaching Langevin values at temperatures considerably lower than 300 K.

In the case when a reaction is going through a reaction complex, the energy is usually considered to be distributed statistically between degrees of freedom (Kondratiyev and Nikitin, 1974), (of course taking into account conservation laws). In a number of cases the branching ratios generated from polarization complex decay and calculated by statistical theory for a number of ion–molecule reaction are in good agreement with experimental data (Talrose and Karatchevtsev, 1971).

In the case considered here, a possibility of effective energy exchange between degrees of freedom in an ionic complex is important. The experiments on energy exchange and the processes of vibronic relaxation in ion–molecule collisions explain the efficiency of such an exchange. These processes will be considered in details below.

Electric field effects on energy levels and electron wave functions in the system, ion + molecule, is another feature of ion–molecule reactions and is also connected with the presence of an electric charge in the system of colliding particles. As the ion and molecule approach, the ion field intensity reaches a value of 10^8–10^7 V/cm. Such an electric field must not be considered a small disturbance: quantitative calculation of the changes in energy levels and wave functions will be rather complicated in such field. But from a purely qualitative viewpoint, one may say that in the case of ion–molecule reactions, the effects on the molecular orbitals of the system are greater than in the reactions of neutrals. That is, the ion–molecule collision can effectively rebuild initial orbitals into the orbitals of the products resulting in a reaction with a greater probability.

In summary, one may conclude that in a purely qualitative consideration of the features of ion–molecule reactions, three main effects lead to the fact that ion–molecule reactions should be fast processes:

(1) The presence of a potential well on the potential energy surface which is caused by the polarization attraction. The depth of such a well is approximately 1 eV and is greater than typical values for activation barriers in the reactions of neutral molecules and radicals.

(2) The formation of a polarization complex with effective energy exchange between degrees of freedom is possible owing to the strong attraction. The kinetic energy of relative motion of the colliding particles, which is high enough at small distances, facilitates energy transfer into the active degrees of freedom.

(3) The reaction occurs in a strong electric field. This electric field provides a major pertubation which may facilitate the transformation of orbitals corresponding to the reagents into the orbitals of products.

III. APPLICATION OF ORBITAL SYMMETRY CONSERVATION RULES AND CORRELATION DIAGRAMS TO ION–MOLECULE REACTIONS

Despite the fact that exothermic and thermoneutral ion–molecule reactions are as a rule fast processes, slow processes have been found among them which are considered the exceptions to the general rule.

For example, reaction (5) is a relatively slow process

$$He^+ + H_2 \xrightarrow{k} HeH^+ + H_2 \qquad (k \leq 10^{-13}\ \mathrm{cm^3/sec})\ \text{(Giese and Maier, 1961)} \tag{5}$$

despite substantial exothermicity. Other examples of slow reactions are

$$O^+ + N_2 \xrightarrow{k} NO^+ + N, \qquad \text{this process was shown to be slow (Talrose } et\ al. \text{ (1962)} \tag{6}$$

$$k = 10^{-12}\ \mathrm{cm^3/sec},\ T = 300\ \mathrm{K}\ \text{(Ferguson } et\ al.\text{, 1965)}$$

$$N_2^+ + O_2 \xrightarrow{k} NO^+ + NO, \qquad k \leq 2.10^{-13}\ \mathrm{cm^3/sec},\ T = 300\ \mathrm{K}\ \text{(Ferguson } et\ al.\text{, 1965).} \tag{7}$$

Useful qualitative information on potential energy surfaces of these reactions may be obtained with the help of correlation diagrams and orbital symmetry conservation rules.

Let us consider reaction (6) as an example. The ground electronic state of O^+ is $^4S_{3/2}(^4S_u)$, and that of the molecule N_2 is $^1\Sigma_g{}^+$. Suppose that the reacting particles approach in such a way that all the three atoms are situated on the same straight line (such approach corresponds to the $C_{\infty v}$ symmetry point group). Now let us find the allowed electronic states of the quasi-molecule O^+N_2 formed from the states of the reacting particles. We shall follow the following procedure for this (Herzberg, 1966). First, find resolutions of representations which correspond to the reactant particles states. For ^4S_u and $^1\Sigma_g{}^+$ states, these resolutions will be Σ^- and Σ^+. To obtain the resulting electronic state, it is necessary to find the direct product of these resolutions. Thus, $\Sigma^- \times \Sigma^+ = \Sigma^-$ state.

On the other hand, it is known that a $^2\Pi$ state must be the ground state for the linear ion $(ON_2)^+$ (Herzberg, 1971). This means that the reactant particle states do not correlate with the ground state of a linear N_2O^+ ion and therefore the reaction may be expected to be relatively slow if it proceeds via a linear transition state. This is illustrated in Fig. 2a which gives a schematic plot of energy versus distance between the particles.

Consider another reaction:

$$^{14}N^+ + {}^{15}N_2 \rightarrow {}^{15}N^+ + {}^{14}N^{15}N. \tag{8}$$

The system of reacting particles in reaction (8) differs from reaction (6) because the system (8) has one electron less. In the latter case, the combination of ground states of the particles $N^+(^3P_o)$ and $N_2(^1\Sigma_g{}^+)$ may give a $^3\Sigma_g{}^-$ state for the complex ion $N_3{}^+$ (Herzberg, 1971). From correlation between the ground state of the $N_3{}^+$ ion and the ground states of N^+ and N_2 (Fig. 2b), we suspect that reaction (8) should be fast. In fact (Fehsenfeld *et al.*, 1974; Vinogradov *et al.*, 1977) the rate constant is of the order of 10^{-10} cm^3/sec.

Correlation diagrams were considered for the reactions $N^+ + O_2$ (Tully, 1971), $O^+ + N_2$ (Kaufman and Koski, 1969), $H^+ + H_2$ (Holliday *et al.*, 1971).

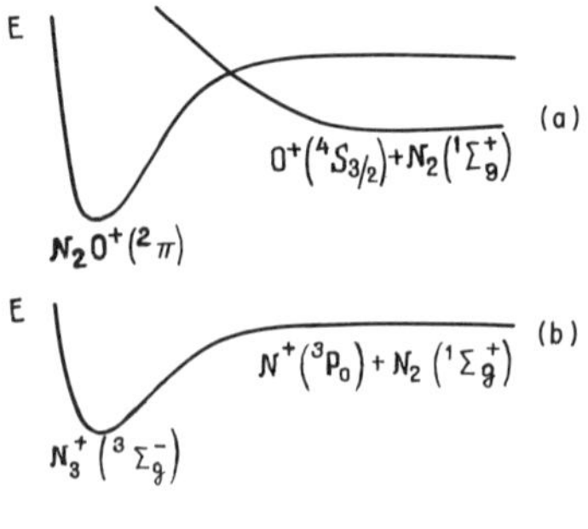

FIG. 2 Shematic reaction coordinate versus energy diagrams for (a) $O^+ + N_2 \rightarrow NO^+ + N$, (b) $^{14}N^+ + {}^{15}N_2 \rightarrow {}^{15}N^+ + {}^{14}N^{15}N$.

The examples given above show that even relatively simple qualitative considerations allow useful information on the dynamics of ion–molecule reactions.

To understand the dynamics of reactions, the method of finding correlations between separate orbitals of reactants and products of ion–molecule reactions proved to be very useful. Such an approach toward some ion–molecule reactions is considered in detail by Mahan (1971).

To explain the meaning of correlation diagrams for molecular orbitals, it is useful to recall the simple example of such diagrams plotted as early as the 1930s by Hund (1928) and Mulliken (1932) for the case of two atoms. The atom–atom collision process was considered from large distances, where their interaction was negligible, into physically sensible dimensions corresponding to molecular size and finally to the superposition of both nuclei and the formation of a united atom. The orbitals of the reactant atoms, as well as of the united one, are known for such a hypothetical reaction. The orbitals are symmetry classified and the levels of corresponding orbitals with identical symmetry are combined, taking into account the quantum-mechanical rule of nonintersection of orbitals with identical symmetry.

Correlation diagrams for more complicated multicentered systems with one hydrogen molecule were studied in Mahan (1971). By means of these diagrams one is able to show that reaction (5) must be slow. The correlation diagram for the lower orbitals is shown in Fig. 3. The 1s orbital of helium labeled α correlates with the lowest energy orbital α'' of the HeH^+ ion. The corresponding lowest orbital of $HeHH^+$, α', has no nodes (the orbitals and plots are given schematically in the diagram) and is heavily binding. The next lowest energy orbital, $\sigma_g(H_2)$ (β), should correlate with the 2σ antibonding orbital of HeH^+ (γ''). As the particles approach its energy must increase. Likewise, as the particles approach the antibonding orbital, H_2–$\sigma_u{}^*(H_2)$ (γ-orbital) should decrease its energy, correlating with 1s(H) (β'') orbital. These two upper orbitals form two orbitals of the $HeH_2{}^+$ molecule; the lower one with one node and the upper one with the two. In reaction (5) two electrons are originally in the σ_g orbital of H_2. If both of them adiabatically pass to the β''

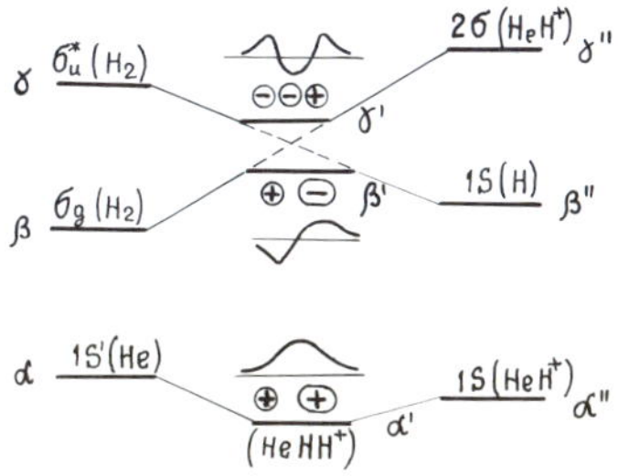

FIG. 3 The molecular orbital correlation diagram for the reactions $H_2{}^+ + He \rightarrow HeH^+ + H$ and $He^+ + H_2 \rightarrow HeH^+ + H$.

orbital, the products formed will be H^- and HeH^{2+}, which is energetically impossible. Another possibility is transition of one electron to the β'' orbital and the other to the γ'' orbital. γ'' is an antibonding orbital of the HeH^+ ion which must decompose into He^+ and H, i.e., the second possibility corresponds to dissociation of the H_2 molecule by the He^+ ion. The correlation diagram in Fig. 3 may also be used for the reaction

$$H_2{}^+ + He \rightarrow HeH^+ + H. \qquad (9)$$

In this case two electrons are originally in the 1s orbital of He and one is in the σ_g orbital of $H_2{}^+$. This latter electron can adiabatically pass to the β'' orbital so that reaction (9) may occur if the energy of the reacting particles exceeds the endothermicity of reaction (9), approximately equal to 0.8 eV. Analogous diagrams were plotted for more complicated systems as well in (Mahan, 1971).

Another useful approach for the estimation of the rates of ion–molecule reactions is the use of orbital symmetry conservation rules, or Woodward and Hoffman (1965a,b) rules. These rules suggest that a reaction occurs easily when there is a correspondence between the orbital symmetry of reagents and reaction products. If there is no such correspondence, reaction goes with difficulty (Woodward and Hoffman, 1971). Such a qualitative approach, which does not require complicated calculations, proved to be quite successful for the prediction of the "possibility" or "impossibility" of some classes of organic reactions (Woodward and Hoffman, 1971). To use these rules in chemical kinetics, one should determine the range of values the rate constants of symmetry allowed and symmetry forbidden reactions must correspond to. For example, the orbital symmetry allowed Diels-Alder [2 + 4] cycloaddition of ethelene to butadiene is considered in organic chemistry to be fast (Woodward and Hoffman, 1971), although its activation energy is 20 kcal/mole. On the other hand, a number of symmetry forbidden reactions of atoms and radicals with molecules may have activation energy only of 6–7 kcal/mole (Kondratiyev, 1970). (For example, $N + O_2 \rightarrow NO + O, E_a \approx$ 7 kcal/mole.)

Below we shall examine an approach based on orbital symmetry conservation rules for ion–molecule reactions. When applying these rules, the key is to determine orbital symmetry allowedness or forbiddenness of the reactions. For example, consider the reaction

$$H_2{}^+ + D_2 \rightarrow HD^+ + HD. \qquad (10)$$

We would note here that H_2D^+ and D_2H^+ ions will be formed by $H_2{}^+$ and D_2 collisions with high rate constant, but the correlation diagram for this reaction will differ from that for reaction (10). The diagram for reaction (10) will hold for the reaction $H_2 + D_2 \rightarrow 2HD$ as well, although the number

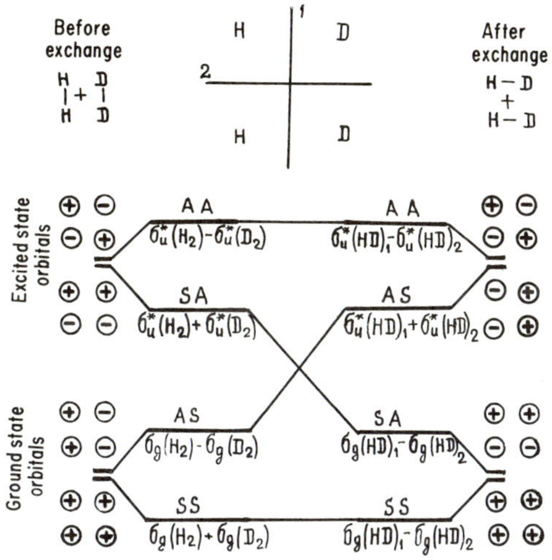

FIG. 4 Correlation diagram for the isotope exchange reactions $H_2^+ + D_2 \rightarrow HD^+ + HD$ and $H_2 + D_2 \rightarrow 2HD$.

of electrons will differ. Let us suppose that the molecules approach in such a way that the H and D atoms lie in the same plane and are situated at the corners of a square. We shall consider the delocalized orbitals of such a quadratic complex both before and after the exchange (in Fig. 4 these orbitals are to the left and to the right, accordingly).

We shall construct orbitals of the H_2D_2 complex as a combination of molecular orbitals of H_2 (D_2, HD) which, in turn, have been constructed as a combination of hydrogen atomic orbitals. Before the exchange, the lowest orbital of the complex is $\sigma_g(H_2) + \sigma_g(D_2)$. The next orbital in energy before the exchange will be the $\sigma_g(H_2) - \sigma_g(D_2)$ combination. This orbital is bonding for each of the H_2 and D_2 molecules, but corresponds to an antibonding orbital in the complex. These two orbitals represent the ground state of the complex before the exchange.

Excited state orbitals will be $\sigma_u^*(H_2) + \sigma_u^*(D_2)$ (antibonding in the separated molecules and bonding in the complex) and a higher one $\sigma_u^*(H_2) - \sigma_u^*(D_2)$ (antibonding in the separated molecules and in the complex). After the exchange, the orbitals of the complex may be constructed analogously. They will be for the lowest energy states, $\sigma_g(HD)_1 + \sigma_g(HD)_2$ and $\sigma_g(HD)_1 - \sigma_g(HD)_2$ and for the excited states, $\sigma_u^*(HD)_1 + \sigma_u^*(HD)_2$ and $\sigma_u^*(HD)_1 - \sigma_u^*(HD)_2$.

In order to label the orbitals, the following symmetry operations are useful:

(1) Reflection with respect to plane 1,
(2) Reflection with respect to plane 2.

If after reflection the sign does not change, the orbital will be symmetric with respect to this symmetry operation. If the sign changes, the orbital will be antisymmetric. Before the exchange and after it, the lowest orbital of the complex will be symmetric with respect to both symmetry operations and thus will have appropriate symmetry index SS. Orbital $\sigma_g(H_2) - \sigma_g(D_2)$ will be antisymmetric (A) with respect to reflection at plane 1 and symmetric (S) with respect to reflection at plane 2 (symmetry index AS). Similarly, $\sigma_g(HD)_1 - \sigma_g(HD)_2$ is labeled SA.

There are three electrons in reaction (10); two will be in the lowest energy orbital and one in the next lowest energy orbital. This orbital, $\sigma_g(H_2) - \sigma_g(D_2)$, has symmetry AS and does not correlate with the orbital of corresponding energy of the products. According to orbital symmetry conservation rules such a reaction is orbital symmetry forbidden. In fact, it will be shown below that radiolytic H–D exchange in H_2 and D_2 mixtures occurs only via the reaction of the ions, $H^+(H_2)_n$, but not in the reactions of type (10).

Reaction (11) provides a second example:

$$H_2^+ + H_2 \rightarrow H_3^+ + H. \tag{11}$$

In this case, it is assumed the geometry of the complex is that given in Fig. 5. This figure also outlines the correlation diagram for reaction (11). Before the reaction, the two lowest energy orbitals of the system [they are combinations of $\sigma_g(H_2^+)$ orbitals] will be symmetric with respect to rotation around the C_2 axis. After the reaction, the two lowest energy orbitals of the system [combinations of (H_3^+) and 1s(H) orbitals] will also be symmetric with respect to rotation around C_2 axis. Hence, one may conclude that reaction (11) is orbital symmetry allowed. Indeed reaction (11) is a well known example of a fast ion–molecule reaction (Eyring *et al.*, 1936).

There are other criteria to be considered as well. For instance, Pearson (1970) suggested that to determine the symmetry allowedness of a reaction, one should find out if the symmetry of the upper occupied orbitals coincides with that of the lower unoccupied ones or not. If these orbitals have different

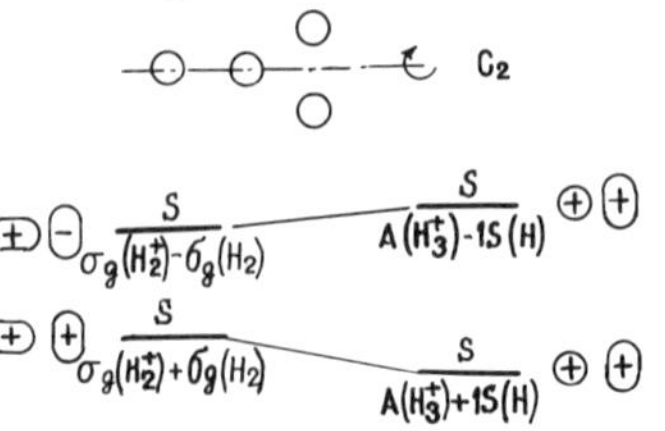

FIG. 5 Correlation diagram for the $H_2^+ + H_2 \rightarrow H_3^+ + H$ reaction.

symmetry, it must result in an activation barrier. For example, consider the reaction

$$N_2^+ + O_2 \rightarrow NO^+ + NO.$$

The upper bonding orbital of N_2^+ ion is $\sigma_g(2P_x)$. There are partially filled and unoccupied antibonding Π_g^* orbitals in the oxygen molecule. These two orbitals have different symmetry with respect to the reflection plane which is perpendicular to the molecular axis (the σ_g orbital is symmetric and the Π_g^* orbital is antisymmetric with respect to this symmetry operation). It is interesting to note that as an electron transits from the upper filled orbital of O_2 to the partially unoccupied orbital of N_2^+, the O—O bond is weakened, but the N_2 bond is strengthened with the result the reaction is impeded.

IV. ISOTOPE EXCHANGE REACTIONS OF ION–MOLECULE COLLISIONS

To check further to what extent orbital symmetry conservation rules may be used for ion–molecule reactions, we consider here a specific class of ion–molecule reactions, i.e., reactions of isotope exchange. These reactions are practically thermoneutral; the difference in energy between the initial and final states is determined strictly by zero point effects and these are usually negligible. This allows us to exclude the effect of energy on the rate constant.

Another feature is the chemical symmetry of such reactions, i.e., the chemical composition of the reagents and products is identical. This means that the reagents and products have the same orbitals in these reactions. But despite the similarity of reagents and products, orbital symmetry is not always preserved in these reactions. This may be illustrated in the following way. To determine conservation or nonconservation of orbital symmetry in the reactions, we first of all set the approach geometry of the reacting particles. This approach geometry in turn sets the symmetry elements of the system (for example, symmetry axis, symmetry planes). Although after the reaction, the types of orbitals are analogous to the initial ones, the formation of new, and termination of old, bonds results in a different (new) orientation of the orbitals with respect to the symmetry elements of the system. This may lead to orbital symmetry nonconservation, for example as it has been shown for reaction (10). Below we shall give a survey on the investigated reactions of isotope exchange. Wiltsbach (1957) demonstrated the relatively simple and now widespread method of labeling organic compounds with radioactive tritium. The appearance of tritium labels in organic molecules, which were mixed with tritium, was believed to be caused by radiolytic exchange via ion–molecule reactions. Since information on the radiolytic mechanism of tritium label introduction was of great practical interest in chemistry, biology, and medicine, a number of authors attempted to determine ion–molecule reaction mechanisms using mass spectral techniques. For example, Munson *et al.* (1963), in a single-chamber ion source of a mass spectrometer, estimated the rate constant for reactions (12)–(16):

$$CH_5^+ + D_2 \xrightarrow{k} CH_4D^+ + HD, \quad k \leq 10^{-13}\ cm^3/sec, \tag{12}$$

$$CH_4^+ + D_2 \rightarrow CH_3D^+ + HD, \quad k \approx 9.5 \times 10^{-12}\ cm^3/sec, \tag{13}$$

$$CH_3^+ + D_2 \begin{cases} \nearrow CDH_2^+ + HD \\ \searrow CD_2H^+ + H_2, \end{cases} \quad k \approx 1.7 \times 10^{-10}\ cm^3/sec, \tag{14}$$

$$CD_3^+ + H_2 \begin{cases} \nearrow CDH_2^+ + D_2 \\ \searrow CD_2H^+ + HD \end{cases} \quad k \approx 1.6 \times 10^{-10}\ cm^3/sec, \tag{15}$$

$$CH^+ + D_2 \rightarrow \text{products}, \quad k \approx 4.8 \times 10^{-10}\ cm^3/sec. \tag{16}$$

The values obtained, as the authors state, are of approximate character, since the peaks of primary and secondary ions overlap and the presence of a 12.5 V/cm repeller field makes the reactant ion energy uncertain (0–2.5 eV).

Reactions in the mixture of $CH_4 + D_2$ were studied using the thermal energy ion cyclotron resonance technique as well (Inoque and Wexler, 1969; Bowers and Elleman, 1970). Reactions of isotope exchange of the CH_3^+ ion, formed from CH_3Cl by electron impact

$$CH_3^+ + D_2 \rightarrow CH_2D^+ + HD, \quad k = 6 \times 10^{-10}\ cm^3/sec, \tag{17}$$

$$CH_3^+ + D_2 \rightarrow CHD_2^+ + H_2, \quad k = 3 \times 10^{-10}\ cm^3/sec, \tag{18}$$

were studied by Harrison and Heyer (1973) in a single-chamber ion source mass spectrometer at pressures of 10^{-2} torr. The relative rates of the reactions (17) and (18) (2 ± 0.4) suggest the equivalence of all H and D atoms in the ionic complex $CH_3D_2^+$. The large rate constants indicate that the sum of reaction (17) and (18) occur on every collision.

The exchange reactions (19) and (20),

$$^{15}N^{16}O^+ + {}^{14}N^{18}O_2 \rightarrow {}^{15}N^{18}O^+ + {}^{14}N^{16}O^{18}O, \tag{19}$$

$$^{15}N^{18}O^+ + {}^{14}N^{16}O \rightarrow {}^{15}N^{16}O^+ + {}^{14}N^{18}O, \tag{20}$$

were found (Ynon and Klein, 1971) to have rate constants of the order of 10^{-10} cm^3/sec.

Reactions in the mixtures $C_2H_2 + D_2$ and $C_2H_2 + H_2$ were studied using both ion cyclotron resonance and photoionization techniques (Buttrill *et al.*, 1974). Cross section dependence of the endothermal reaction

$$C_2H_2^+ + H_2 \rightarrow C_2H_3^+ + H \tag{21}$$

on photon energy was reported. This dependence was determined to have two abrupt steps which appear at wavelengths, $\lambda = 1067$ and 1046 Å. corresponding to the vibronic modes ($v = 1, v = 2$) of $C_2H_2^+$. The rate constant for (21) was measured at electron impact energy of 15 eV in the ICR; $(6.3 \pm 1.8) \times 10^{-11}$ cm^3/sec. As deuterium was introduced, ion peaks with $m/e = 27$ and 28 were observed and ascribed to the reactions

$$C_2H_2^+ + D_2 \rightarrow C_2HD^+ + HD, \tag{22}$$

$$C_2H_2 + D_2 \rightarrow C_2H_2D^+ + D. \tag{23}$$

The relative rate of reaction (22) to (23) was 1.6.

Rate constants of the following ion–molecule reactions were measured by ion cyclotron resonance (Kim *et al.*, 1975a).

$$CD_5^+ + H_2 \rightarrow CD_4H^+ + HD, \quad k \leq 5 \times 10^{-13}\ \text{cm}^3/\text{sec}, \tag{24}$$

$$CD_4^+ + H_2 \rightarrow CD_3H^+ + HD, \quad k \leq 10^{-12}\ \text{cm}^3/\text{sec}, \tag{25}$$

$$CD_3^+ + H_2 \rightarrow CD_2H^+ + HD, \quad k = (0.51 \pm 0.05) \times 10^{-9}\ \text{cm}^3/\text{sec}, \tag{26}$$

$$NH_4^+ + D_2 \rightarrow NH_3D^+ + HD, \quad k \leq 10^{-13}\ \text{cm}^3/\text{sec}, \tag{27}$$

$$NH_3^+ + D_2 \rightarrow NH_2D^+ + HD, \quad k \leq 2 \times 10^{-12}\ \text{cm}^3/\text{sec}, \tag{28}$$

$$H_3O^+ + D_2 \rightarrow H_2DO^+ + HD, \quad k \leq 10^{-12}\ \text{cm}^3/\text{sec}, \tag{29}$$

$$H_2S^+ + D_2 \rightarrow HDS^+ + HD, \quad k \leq 5 \times 10^{-13}\ \text{cm}^3/\text{sec}, \tag{30}$$

$$H_3S^+ + D_2 \rightarrow H_2DS^+ + HD, \quad k \leq 10^{-13}\ \text{cm}^3/\text{sec}. \tag{31}$$

Reactions (32) and (33) have been studied in a tandem mass spectrometer (Malkhasyan *et al.*, 1974) with relative

$$D_2H^+ + H_2 \rightarrow H_3^+ + D_2 \tag{32}$$

$$D_2H^+ + H_2 \rightarrow H_2D^+ + HD \tag{33}$$

abundances of channels (32) and (33) equal to 1.2. Over a kinetic energy range of 1.5–5 eV, the cross section of the reaction (32) changed from 1.5×10^{-15} to 1.5×10^{-16} cm^2.

In an experiment with crossed beams reaction (34)

$$C_2H^+ + D_2 \rightarrow C_2D^+ + HD \tag{34}$$

was shown to have an energy threshold of about 1 eV (Smith *et al.*, 1975). The reaction

$${}^{14}N^{16}O^+ + {}^{15}N^{18}O \rightarrow {}^{14}N^{18}O^+ + {}^{15}N^{16}O \tag{35}$$

similar to reaction (19), was studied by a double mass spectrometer over a kinetic energy range from a few tenths to 17 eV. Its cross section was shown not to exceed 3×10^{-18} cm^2 (Futrell and Tiernen, 1972). The results should be considered more reliable than those reported for reaction (19) because they were obtained by a more precise technique.

Isotope exchange reactions in three-centered systems have been studied by the flowing afterglow technique as well (Fehsenfeld *et al.*, 1974). For the exothermic ($\Delta H = -0.039$ eV)

$$D^+ + H_2 \rightarrow H^+ + HD \tag{36}$$

and endothermic ($\Delta H = 0.046$ eV)

$$H^+ + D_2 \rightarrow D^+ + HD \tag{37}$$

reactions, a temperature dependence of the rate constants was studied. Rate constants of reaction (36) and (37) are reported as $k_{36} = 10^{+5}_{-2.5} \times 10^{-10}$ cm^3/sec; $k_{37} = (3.7 \pm 1.8) \times 10^{-10}$ cm^3/sec at $T = 278$ K. Rate constants of isotope exchange reactions with heavier particles were also studied in this work (Fehsenfeld *et al.*, 1974);

$$^{14(15)}N^+ + {}^{15(14)}N_2 \rightarrow {}^{15(14)}N^+ + {}^{14}N^{15}N \qquad k = (1.8 \pm 0.7) \times 10^{10}\ \text{cm}^3/\text{sec}, \tag{38}$$

$$^{18(16)}O^+ + {}^{16(18)}O_2 \rightarrow {}^{16(18)}O^+ + {}^{16}O^{18}O \qquad k \leq 10^{-12}\ \text{cm}^3/\text{sec}, \tag{39}$$

$$^{18}O^+ + {}^{12}C^{16}O \rightarrow {}^{16}O^+ + {}^{12}C^{18}O \qquad k = (4.4 \pm 1.3) \times 10^{-10}\ \text{cm}^3/\text{sec}, \tag{40}$$

$$^{18}O^+ + {}^{14}N^{16}O \rightarrow {}^{16}O^+ + {}^{14}N^{18}O \qquad k \leq 5 \times 10^{-12}\ \text{cm}^3/\text{sec}. \tag{41}$$

A number of ion–molecule reactions of isotope exchange has been studied at our laboratory, in collaboration with Karachevtsev, for pressures of a few tenths of a torr in the ionization chamber of a mass spectrometer. The ions were formed by electron impact. The electron energy could be varied within wide limits. The use of an ion source with very small slits reduced the penetration of electric fields to such an extent that the ions could be considered to move with velocities corresponding to the primary ion energy within the largest part of the chamber. Velocity distribution of ions was controlled by a deflection voltage technique (Berry, 1950; Talrose and Frankevitch, 1958). Estimation of the number of ion collisions was made from the yield of secondary ions in ion–molecule reactions with known rate constants. A correction was introduced to account for the different residence times of the ions. In this work, we estimated the rate constants of reactions (42) and (43)

$$^{14}N_2{}^+ + {}^{15}N_2 \rightarrow {}^{14}N^{15}N^+ + {}^{14}N^{15}N, \tag{42}$$

$$^{15}N_2{}^+ + {}^{14}N_2 \rightarrow {}^{14}N^{15}N^+ + {}^{14}N^{15}N. \tag{43}$$

The rate constant of these reactions is less than 6×10^{-12} cm^3/sec at translational energies of some hundredths of an electron volt (Vinogradov and Karatchevtsev, 1974). Isotope exchange reactions were investigated in the mixture of ^{14}NO and $^{15}N_2$. In the reaction

$$^{14}N^+ + {}^{15}N_2 \rightarrow {}^{15}N^+ + {}^{14}N^{15}N, \tag{44}$$

the $^{14}N^+$ primary ions were formed by electrons with energy of 23 ± 0.5 eV, which is higher than the appearance potential of $^{14}N^+$ ion from NO (22.1 eV) but less than the appearance potential of $^{15}N^+$ ion from N_2 (24.6 eV) (Field and Franklin, 1957). The average kinetic energy of the N^+ fragment ion, determined in this experiment, was approximately 0.12 eV. Estimation of the rate constant of reaction (44) was $(1.2 \pm 0.4) \times 10^{-10}$ cm^3/sec (Vinogradov *et al.*, 1977), and is in a good agreement with the flowing afterglow result (Fehsenfeld *et al.*, 1974).

Another reaction studied in the mixture of ^{14}NO and $^{15}N_2$ was

$$^{15}N_2 + {}^{14}NO^+ \rightarrow {}^{15}NO^+ + {}^{14}N^{15}N, \tag{45}$$

for which $k \leq 1.5 \times 10^{-12}$ cm^3/sec (Vinogradov *et al.*, 1977) at a kinetic energy of the NO^+ molecular ion of about 0.03 eV. The energy of ionizing electrons was 14 eV which is lower than the N_2 ionization potential, and thus avoids complications due to the charge transfer reaction (46)

$$^{15}N_2{}^+ + {}^{14}NO \rightarrow {}^{15}NO^+ + {}^{14}N^{15}N. \tag{46}$$

The rate constants of the following reactions were estimated in a similar way:

$$^{14}N_2O^+ + {}^{15}N_2 \nearrow {}^{15}N^{14}NO^+ + {}^{14}N^{15}N \tag{47}$$

$$^{14}N_2O^+ + {}^{15}N_2 \searrow {}^{15}N_2O^+ + {}^{14}N_2 \tag{48}$$

These reactions proved to be slow processes as well: $k_{47}, k_{48} \leq 10^{-12}$ cm^3/sec. In the following reaction:

$$^{14}NF^+ + {}^{15}N_2 \rightarrow {}^{15}NF^+ + {}^{14}N^{15}N, \tag{49}$$

$^{14}NF^+$ primary ions were obtained by electron impact of $^{14}NF_3$ molecules. The ion peak at $m/e = 34(^{15}NF^+)$ was present at electron energies of 60 eV. But as the electron energy decreased, the $^{15}NF^+$ ion disappeared in proportion to the loss of the $^{15}N^+$ ion. This suggests that the $^{15}NF^+$ ion is formed by reaction (50)

$$^{15}N^+ + NF_3 \rightarrow {}^{15}NF^+ + NF_2. \tag{50}$$

The rate constant of the reaction (50) is in the range 10^{-9}–10^{-10} cm^3/sec.

At low electron energy (23 eV), the rate constant of the reaction (49) was estimated to be $k_{49} \leq 1.5 \times 10^{-12}$ cm^3/sec. The energy of $^{15}N^+$ ions in these experiments was 0.1 eV.

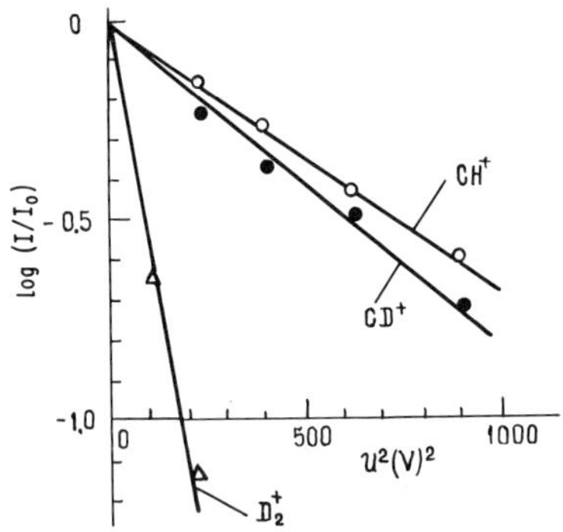

FIG. 6 Ion current logarithm versus the square of the deflecting voltage for D_2^+, △; CH^+, ○; CD^+, □.

Exchange processes with lighter particles, H and D atoms, were studied by the above described technique with increased pressure in the ion source. In the reaction

$$CH^+ + D_2 \rightarrow CD^+ + HD, \tag{51}$$

CH^+ primary ions were obtained by electron impact from $CHCl_3$ molecules. The CH^+ fragment ions had initial kinetic energies tens of times higher than D_2^+ molecular ions. Figure 6 represents the ion current dependence of the deflection voltage for D_2^+, CD^+, and CH^+ ions at low pressures of D_2. The ion distributions are considered to be quasi-Maxwell. The average energy of the CH^+ primary ion is only 16% higher than the CD^+ secondary ion and equal to approximately 0.3 eV. The rate constant of reaction (51), under our conditions, was found to be $(2.1 \pm 0.4) \times 10^{-10}$ cm³/sec, determined from the dependence of CD^+ ion peak intensity on D_2 pressure. The rate constant of the reaction (16) was estimated to be $(7 \pm 1.2) \times 10^{-10}$ cm³/sec, in good agreement with the data of Munson *et al.* (1963).

Rate constants of isotope exchange were determined in the same way for the following reactions:

$$C_2H_2^+ + D_2 \rightarrow C_2HD^+ + HD, \tag{52}$$

$$C_2H_2^+ + D_2 \rightarrow C_2D_2^+ + H_2. \tag{53}$$

To exclude complications due to D_2^+ ion reactions, the energy of the electron beam was reduced to 13 eV. Also, at such low electron energies, the contribution to $m/e = 28$, due to the endothermic reaction

$$C_2H_2^+ + D_2 \rightarrow C_2H_2D^+ + D, \tag{54}$$

is negligible. Rate constants obtained for the reactions (52) and (53) are $k_{52} = (8 \pm 2) \times 10^{-12}$ cm³/sec and $k_{53} = (3 \pm 1) \times 10^{-12}$ cm³/sec. For the processes of deuterium–hydrogen exchange in the reactions of molecular ions of ethylene at 12 eV,

$$C_2H_4^+ + D_2 \begin{array}{l} \nearrow C_2H_3D^+ + HD \\ \searrow C_2H_2D_2^+ + H_2. \end{array} \quad (55)$$

The rate constants of the reactions proved to be less than 10^{-11} cm^3/sec.

Reaction (56) was investigated in a mixture of H_2O and D_2 at 60 eV electron energy

$$H_2DO^+ + D_2 \rightarrow HD_2O^+ + HD, \quad (56)$$

where the H_2DO^+ reactant ions were formed through the processes

$$D_2^+ + H_2O \rightarrow H_2DO^+ + D, \quad (57)$$

$$H_2O^+ + D_2 \rightarrow H_2DO^+ + D. \quad (58)$$

The absence of an appreciable contribution (less than 3%) to the formation of HD_2O^+ ions by reactions (59) and (60),

$$D_2^+ + H_2O \rightarrow D_2HO^+ + H \quad (59)$$

$$H_2O^+ + D_2 \rightarrow D_2HO^+ + H, \quad (60)$$

was confirmed by the character of the dependence of peak growth of $m/e = 21$ on pressure. The rate constant of reaction (56) was found to be $k_5 = (1.1 \pm 0.2) \times 10^{-10}$ cm^3/sec (Vinogradov *et al.*, 1977) for an average translational energy of the H_2DO^+ ion of 0.13 eV. The value of the rate constant obtained for reaction (56) differs substantially from the one determined by ICR (Kim *et al.*, 1975a) for the similar reaction (29). This difference may be due to the fact that a large percentage of the reacting H_2DO^+ ions in (56) were formed in the strongly exothermic reaction (57) ($\Delta H = -4.4$ eV), under the conditions of our experiment. As a result, the H_2DO^+ ions most likely had a much greater internal excitation than the H_3O^+ ions in reaction (29) that formed in the less exothermic reaction

$$H_2O^+ + H_2O \rightarrow H_3O^+ + OH. \quad (61)$$

Now we would like to try to interpret the described data by means of orbital symmetry conservation rules and correlation diagrams. To begin with, consider the correlation diagram for the isotope exchange reactions (42) and (43) given in Fig. 7. The principles used in constructing this diagram are similar to those used in constructing Fig. 4. But in this case the combination of σ(2p) and π(2p) orbitals of the N_2 molecule should be considered both before and after the exchange. The same correlation diagram will hold for reactions of the neutral N_2 molecules and the N_2^+ ion. It is seen from Fig. 7 that the upper filled orbital $(\sigma_g 2p_z)_a - (\sigma_g 2p_z)_b$ before the exchange correlates

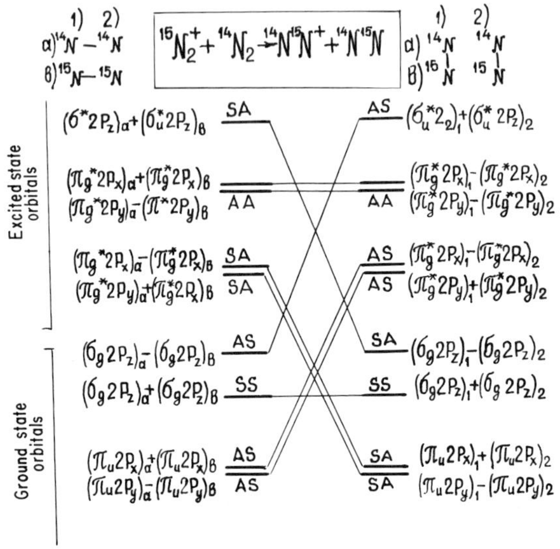

FIG. 7 Correlation diagram for the $^{14}N_2^+ + {}^{15}N_2 \rightarrow {}^{14}N^{15}N^+ + {}^{14}N^{15}N$ reaction.

in orbital symmetry with a higher energy antibonding orbital (σ_u*2p$_z$) + (σ_u*2p$_z$) after the exchange reaction. Hence, it follows that reactions of this type must be orbital symmetry forbidden, explaining the fact that the rate constants of reactions (42) and (43) are much lower than the collision limit.

It is interesting to note that, from analogous correlation diagrams, it follows that all reactions of the type $AA^{+(-)} + A'A' \rightarrow A'A^{+(-)} + A'A$, (A and A′ are isotopes of the elements of the first and second row of the Mendeliev periodic table) are orbital symmetry forbidden and therefore should go relatively slowly.

The NO^+ ion is an isoelectronic analog of the N_2 molecule, and the correlation diagram for reaction (46) will be similar with the correlation diagram shown in Fig. 7. Thus, reaction (46) is also orbital symmetry forbidden, explaining why it is slow. Further, isotope exchange of the molecular NO^+ ion with the NO molecule can also be shown to be an orbital symmetry forbidden process.

Figure 8 represents a simplified (plotted only for some of the upper levels) correlation diagram for the reaction of $^{14}NF^+$ and $^{15}N_2$. The upper orbitals of the system are constructed as a combination of the σ_g(p) orbitals of N_2 and σ_g(p) orbitals of NF^+. The diagram shows that the upper filled orbital $\sigma_g p(^{15}N_2) - \sigma_g p(^{14}NF^+)$ does not correlate in symmetry with the orbital $\sigma_g p(^{14}N^{15}N) - \sigma_g p(^{15}NF^+)$, and the reaction must be symmetry forbidden, a result consistent with the low experimental value of k_{49}.

To explain experimental results on exchange reactions in three-center systems one can make use of the Pearson approach, which was mentioned

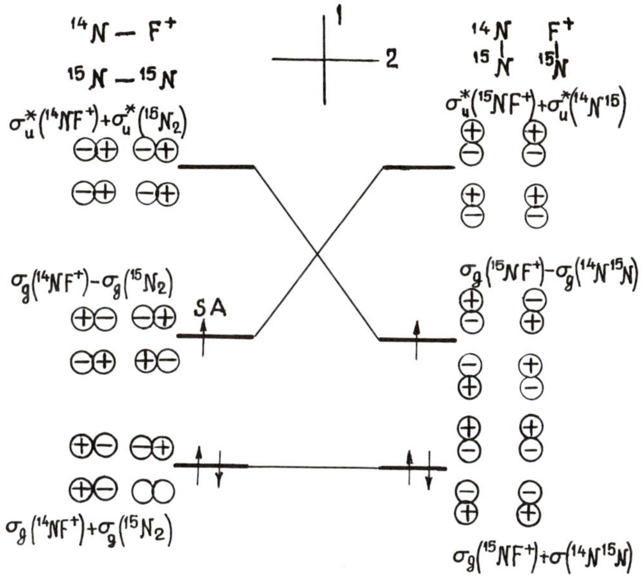

FIG. 8 Correlation diagram for the $^{14}NF^+ + {}^{15}N_2 \rightarrow {}^{15}NF^+ + {}^{14}N^{15}N$ reaction.

above, to determine if the reaction is symmetry allowed. This approach is based on symmetry correlation of upper filled and lower unfilled orbitals.

In the reaction of $H^+ + D_2$, the electrons pass from a σ_g orbital of D_2 to the 1s orbital of the H atom. These orbitals correlate in orbital symmetry.

In the reaction $O^+ + O_2$, when electron passes from the $\pi_u{}^*$ orbital of O_2 to the p orbital of the O atom, a "chemical" forbiddenness of the reaction is arising, i.e., the removal of an electron from the antibounding O—O bond orbital strengthens this bond. This effect might reduce the rate of reaction (39). It is interesting to note that the isotope exchange reaction of neutral O atoms with O_2 has a rather high rate constant. In this case the O atom is an electron donor. In addition, the fact that a combination of the $O(^3P_g)$ atom and $O_2(^3\Sigma_g{}^-)$ molecule forming a triangular O_3 complex correlates with the ground state of ozone $O_3(^1A_1)$ suggests the possibility of a fast isotope exchange reaction.

Consideration of orbital symmetry conservation in other three-centered reactions studied by Fehsenfeld *et al.* (1974) indicates that symmetry allowed three-centered ion–molecule reactions have rate constants of 10^{-9}–10^{-10} cm^3/sec, while symmetry forbidden reactions have rate constants much lower, at least for small values of translational energy.

For interpretation of the experimental results for the reaction $CH^+ + D_2$, we can use correlation diagrams plotted by Mahan (1971). The ground electron state of CH^+ is $^1\Sigma^+$ and its electron configuration is $K(2s\sigma)^2(2p\sigma)^2$. The corresponding configuration of the first excited state is $K(2s\sigma)^2(2p\sigma)(2p\pi)$, $(^1\Pi, ^3\Pi)$ (Herzberg, 1971). To plot the correlation diagram for reaction (51), we may use the diagrams given by Herzberg (1966) for molecular orbitals at large and small internuclear distances in planar molecules of the type XH_3 (the particles presumably approach on a surface of D_{3h} symmetry). Such a diagram is given in Fig. 9 for CH^+ and D_2. [It is analogous to the diagram for the reaction $OH^+ + H_2$ (Mahan, 1971).] The diagram shows that one of the upper filled $a_1(\sigma_g(H_2))$ orbitals correlates with a higher $a_1{}^*$ orbital of the complex $CH_3{}^+$, suggesting the isotope exchange reaction is forbidden. Nonetheless, reaction (51) has a considerable rate constant. A question arises whether the orbital symmetry forbiddenness estimated by these diagrams is absolute. Mahan (1971) suggested that orbital mixing due to vibrations might partially reduce the forbiddenness (in our case a degenerated $p\pi$ orbital, which is partially filled, with a σ_g orbital of D_2). At present, the qualitative character of this analysis does not yield theoretical criteria for predicting when the barrier reduction due to this mixing is considerable. In fact, the reaction $OH^+ + H_2$ is slow, not fast, although the qualitative correlation diagram is identical with reaction (51). A second possible reason for the high rate constant of reaction (51) is CH^+ may be internally excited since this ion is a fragment (a fact that might explain the effectiveness of orbital mixing due to vibrations). The data given in Fig. 6 testify to the presence of vibronic excitation of CH^+. If translational energy were statistically distributed between the reaction products, the dependence of log (I/I_0) on U^2 for CD^+ ions would be much lower than the experimentally observed curve. The ratio of the cotangents of the angle of the slope for CD^+ and CH^+ should be approximately 0.64 but not 0.83 as observed in the experiment. The latter is explained by the flow of vibrational energy of the CH^+ ion to the translational degrees of freedom of the reaction products. This in itself testifies to the existence of effective energy exchange between vibrational and translational degrees of freedom. Said another way, it appears an activation barrier in the reaction, arising due to its orbital symmetry, is overcome by ion vibronic excitation.

The correlation diagram for the collision of $C_2H_2{}^+ + H_2$ given by Mahan (1971) suggests the isotope exchange reaction $C_2H_2{}^+$ should be slow. This can be understood by looking at the more simple correlation diagram for the reaction of $C_2H_2{}^+ + D_2$ (Fig. 10). Such a diagram explains the symmetry forbiddenness of the reactions in which a π_u orbital of the C—C bond and a σ_g orbital of D_2 transform into two σ_{CH} orbitals. In our case, we are considering the symmetry operation which corresponds to the reflection with respect to the plane perpendicular to the C—C and D—D bonds.

Figure 10 shows that the upper filled orbital of the system before the formation of new CH bonds has a different symmetry than the orbital corresponding to the formation of CH bonds, which should result in low rates of

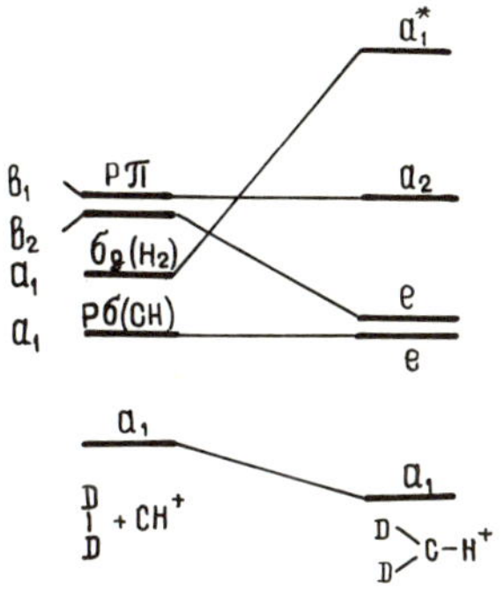

FIG. 9

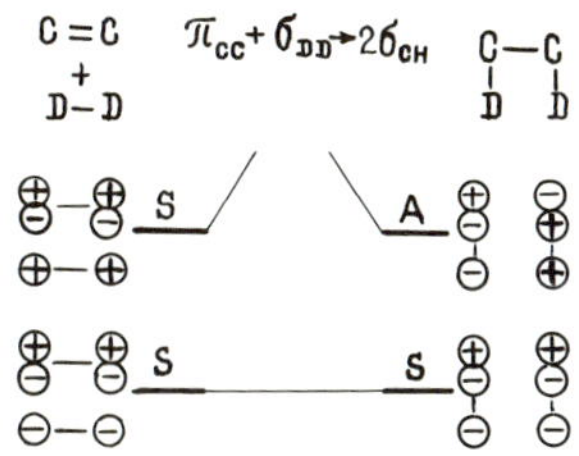

FIG. 10

FIG. 9 Correlation diagram for the $CH^+ + D_2 \rightarrow CD^+ + HD$ reaction.

FIG. 10 Correlation diagram indicating orbital symmetry forbiddenness for formation of two σ(CD) orbitals from πu(C—C) and $\sigma_g(D_2)$ orbitals.

the processes (52)–(54). The fact reaction (52) has a rate constant only 1–2 orders less than Langevin, indicating that perhaps in this case symmetry forbiddenness is reduced by effective mixing of bonding b_{3g} and antibonding b_{1u} CH orbitals of ethylene, due to two vibrations of ethylene of B_{2u} type or the motions of B_2 type particles, the possibility of which was pointed out by Mahan (1971).

The low rate of isotope exchange in reaction (55), ${C_2H_4}^+ + D_2$, may be explained by using the diagram given in Fig. 10. In this case, it is necessary to transform π_u and σ_g orbitals into two σ_{CH} orbitals for the complex ${C_2H_6}^+$ to be formed.

Next, we would like to consider isotope exchange reactions of the type $(XH_n)^+$ for which all the valence bonds of the X atom (or X radical) are saturated during collisions with D_2 molecules. Since the exchange reaction affects directly the X atom only or only one of the hyrdrogen atoms in the ion, it is worthwhile limiting consideration to correlation diagrams for four-centered systems, i.e., consideration of only one of the XH orbitals and the D_2 orbitals. The rest of the orbitals are filled and do not change in the course of the reaction. Another simplifying assumption is that since an orbital is usually a bonding XH orbital, it can be substituted for an equivalent orbital in such a way that the system would have symmetry elements crossing the changing bonds. [The $\sigma_g H_2$ orbital can be such an equivalent orbital (Fig. 11).] Using these approximations, correlation diagrams represented in Fig. 4 can be utilized. Keeping these approximations in mind, the exchange isotope reactions of the type

$$XH^+ + D_2 \rightarrow XD^+ + HD, \quad (62)$$

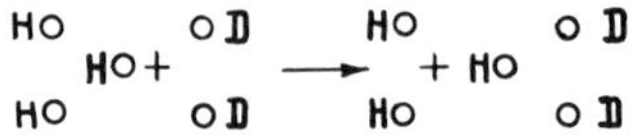

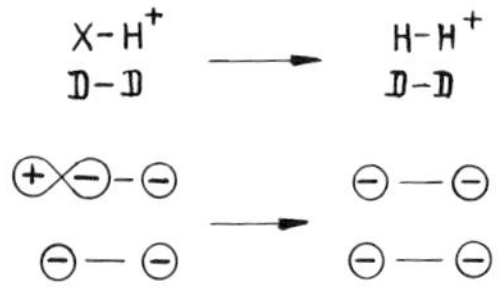

FIG. 11

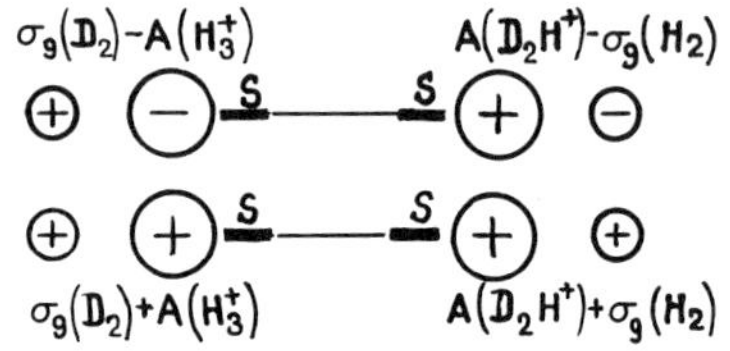

FIG. 12

FIG. 11 The scheme for the change of a σ_{XH} orbital for an equivalent σ_g orbital.

FIG. 12 Correlation diagram for the $H_3^+ + D_2 \rightarrow HD_2^+ + H_2$ reaction.

where X is an atom with a filled valence electron shell or a monovalent radical, should be orbital symmetry forbidden. Hence, the low rate constants of reactions (24), (25), (27),–(31) can be rationalized.

To conclude, we would note that isotope exchange in collisions of H_3^+ with D_2 is a symmetry allowed process as shown in Fig. 12. The upper filled orbitals of the system correlate in symmetry (a C_2 second-order axis was chosen as the symmetry element, which is perpendicular to the axis of D_2 molecule).

V. EXCHANGE REACTIONS OF ION CLUSTERS

Reactions of ion clusters are important and extremely widespread processes of charged particles. The usual mechanism for forming clusters is termolecular. Hence, the lower the temperature and the higher the neutral particle concentration and the greater the Gibbs free energy change, $\Delta G^0_{A^+ + B \rightarrow A^+ \cdot B}$, the larger the rate of formation of ion–molecule clusters by the process

$$A^+ + B + M \rightarrow A^+ \cdot B + M. \qquad (63)$$

The value of $\Delta G^0_{A^+ + B \rightarrow A^+ \cdot B}$ is determined by the relation

$$\Delta G^0_{A^+ + B \rightarrow A \cdot B^+} = \Delta H^0_{A^+ + B \rightarrow A \cdot B^+} - T\,\Delta S^0_{A^+ + B \rightarrow A \cdot B^+}, \qquad (64)$$

where $\Delta H^0_{A^+ + B \rightarrow A \cdot B^+}$ and $\Delta S^0_{A^+ + B \rightarrow A \cdot B^+}$ are the changes in enthalpy and entropy, respectively, in reaction (63). For polar molecules, the value of the A^+—B bond energy (i.e., $\Delta H^0_{A^+ + B \rightarrow A \cdot B^+}$) is tens of kilocalories per mole and decreases as the number of neutral molecules in cluster grows. For example, $\Delta H^0_{H_3O^+ + H_2O \rightarrow H_3O^+ \cdot H_2O} = -36$ kcal/mole, $\Delta H^0_{H_3O^+ \cdot H_2O + H_2O \rightarrow H_3O^+ \cdot (H_2O)_2} = -22.3$ kcal/mole, and $\Delta H^0_{H_3O^+ \cdot (H_2O)_2 + H_2O \rightarrow H_3O^+ \cdot (H_2O)^3} = -17$ kcal/mole

(Good *et al.*, 1970). Nonpolar molecules have lower binding energies to ions. Similar to polar molecules, this bond energy decreases with cluster growth. For instance, $\Delta H^0_{H_3^+ + H_2 \to H_3^+ \cdot H_2} = -9.6$ kcal/mole, $\Delta H^0_{H_3^+ \cdot H_2 + H_2 \to H_3^+ \cdot (H_2)_2} = -4.1$ kcal/mole, and $\Delta H^0_{H_3^+ \cdot (H_2)_2 + H_2 \to H_3^+ (H_2)_3} = -3.8$ kcal/mole (Hiraoka and Kebarle, 1975). The values of $\Delta S^0_{A^+ + B \to A \cdot B^+}$ are negative, of course, and make up 10–20 or more entropy units. These data give an idea of the equilibrium constants of the process (63). Even in the case of weakly polarized hydrogen molecules, H_5^+ ion clusters predominate at room temperature and a pressure of hundreds of torr (Kulagenkova *et al.*, 1972).

The concentration of ions formed thermally in gaseous hydrogen at 3000 K and atmospheric pressure is 5×10^7 cm^{-3} (Larin and Talrose, 1962). Under these conditions, $\Delta G_{H_3^+ + H_2 \to H_5^+} > 0$, suggesting a low concentration of H_5^+ ions in thermally ionized hydrogen at 3000 K (at higher temperatures contribution of H_5^+ will be still less). An analogous conclusion can be reached on the role of $H_3O^+ \cdot H_2O$ clusters in the thermal ionization of water vapor at $T = 3000$ K and $p = 1$ atm, yielding $H_2O^+ = 10^9$ cm^{-3} (Ratnasiri, 1972), $\Delta G_{H_3O^+ + H_2O \to H_3O^+ \cdot H_2O (3000\ K)} > 0$.

Quite a different situation arises when $\Delta H_{A^+ + B \to A^+ \cdot B}$ becomes comparable with the ionization potential of the gas, e.g., the case of

$$H^+ \cdot H_2O(\Delta H_{H^+ + H_2O \to H_3O^+} = 169 \text{ kcal/mole})$$

(Burt, 1970). Accounting for H_3^+ and H_3O^+ cluster formation in thermal hydrogen and water vapor ionization was shown to result in an appreciable degree of ionization at much lower temperatures than when only H_2^+ and H_2O^+ ions were considered. (The clusters, H_3^+ and H_3O^+, are the chief ions at low temperature ionization.) Thus, ion clusters can change not only ion composition but (under certain conditions) ionization degree as well, in the case of thermal ionization of a gas.

Rate constants of the processes like (63) are great and as a rule they range from 10^{-27} to 10^{-29} cm^6/sec, while rate constants of reverse processes, i.e., for destruction of ion clusters in bimolecular collisions, are less than Langevin values by at least the factor $\exp(-\Delta G^0_{A^+ + B \to A^+ \cdot B}/RT)$. On the whole, primary thermal ions, which are formed in a gas at pressures of 1 atm, evolve into stable long lived ion clusters after only a few termolecular collisions and then participate in further reactions with neutral molecules and ultimately vanish in recombination processes with electrons or with opposite charged ions or ion clusters.

In this connection, the problem of ion cluster reactivity assumes great importance. Strictly speaking, the matter in this case is to what extent charge screening by neutral molecules can affect the reactivity and mechanism of reaction of charged particles, and to what extent the effects may be connected with the number of neutral molecules surrounding the charged nucleus of the cluster. Understanding of these problems will bridge the gap currently existing between our understanding of ion processes in gaseous and condensed media.

Although the number of cluster/molecule reactions studied to the present time is great (mainly due to the growing interest in their role in the chemistry of the atmosphere) (Ferguson, 1975), the problem of ion cluster reactivity has not been well studied and the answers to the questions put above are formulated mainly as hypothesis sometimes supported by experimental data.

The enthalpy of recombination process, in which the cluster is destroyed, may be assumed to be higher than that of a corresponding ion–molecule reaction by a value of $-\sum_{i=1}^{n} \Delta H_i^0$, where i is the number of neutral molecules in cluster, and ΔH_i^0 is the bond energy of an i molecule with the cluster nucleus. This enthalpy effect may change the recombination process since it will proceed under more "soft" conditions and in some cases become endothermic. Thus, for instance, the destruction processes of oxygen negative ions (65) and (66)

$$O_2^- + O \rightarrow O_3 + e \tag{65}$$

$$O_2^- + O_2(^1\Delta g) \rightarrow 2O_2 + e \tag{66}$$

which are important in ionospheric chemistry, do not take place after attachment of one water molecule to the ion (i.e., $O_2^- \cdot H_2O$) (Ferguson, 1975). In this connection one may imagine a situation when the ion practically will not recombine at all in the gas phase since the process

$$A^+ \cdot (M)_n + B^-(M)_m \rightarrow A + B + (m+n) \cdot M$$

will be endothermic. This occurs when the following relation holds:

$$-I_{A^+} + S_{B^-} + \sum_{i=1}^{n} \Delta H_i^0(A^+ \cdot M) + \sum_{i=1}^{m} \Delta H_i^0(B^- \cdot M) > 0. \tag{67}$$

Relation (67) is possible for easily ionized alkali metal atoms (here I_{A^+} is the ionization potential of A and S_{B^-} the electron affinity of B).

The recombination rate of ion clusters may differ from that of simple ions. In fact at high enough pressure, when clusters dominate the ion abundance, the recombination coefficient $\bar{\alpha}$ is defined by (68)

$$\bar{\alpha} = 4\pi e(\mu^+ + \mu^-)^{1/2}, \tag{68}$$

where e is the elementary charge and μ^+, μ^- are mobilities of positive and negative charged particles which depend on their masses (Langevin, 1903). This dependence for a gas density, 2.69×10^{-19} cm^{-3} (1 atm), is given by (Dalgarno, 1958)

$$\mu = 35.9/(\alpha M_r)^{1/2} \text{ cm}^2/\text{V sec},$$

where α is the polarizability of the medium expressed in the unit a_0^3 (a_0 is the radius of the first Bohr orbit) and M_r the reduced mass in amu.

In the case of cluster molecule reactions, in which the cluster shell is preserved (the ion cluster is not destroyed), the reactivity might depend on a number of factors.

For this purpose we shall deliberately simplify the real situation and assume that the cluster has a completely filled shell, the dimensions of which correspond to the Langevin interaction sphere. It is further assumed that the shell does not screen to any essential extent the nucleus field. Finally, we shall consider ligand switching reactions

$$A^+ \cdot (B)_n + C \rightarrow A^+(B)_{n-1}C + B \qquad (69)$$

separately from the reactions in which both the cluster nucleus and its shell changes

$$A^+(B)_n + M \rightarrow N^+(B)_m + M + (n - m)B. \qquad (70)$$

For both types of reactions the question of intermediate complex formation is crucial. There are two cases possible, depending on the polarizabilities (or dipole moments) of shell molecules (α_c) and reagent molecule (α_a). If α_a is much greater than α_c, the Langevin radius will be greater than the cluster dimensions and an intermediate complex may be formed as if there were no shell at all. If $\alpha_a \leq \alpha_c$, the formation of a long lived intermediate complex is impossible since the Langevin sphere is occupied by cluster shell molecules. As a result, the process will differ little from the interaction of neutral particles.

It follows that in the case of $\alpha_a \gg \alpha_c$ reactions, (69) should not differ substantially from typical ion–molecule reactions without activation energy. If $\alpha_a \leq \alpha_c$, these reactions may have an activation energy which, similar to neutral systems, will be less than the bonding energy of shell molecules with the nucleus.

The following reactions are the examples of the above processes:

$$O_2^- \cdot H_2O + CO_2 \rightarrow CO_4^- + H_2O, \quad k = 5.8 \times 10^{-10}\ \text{cm}^3/\text{sec} \qquad (71)$$
(Adams *et al.*, 1970)

$$O_2^- \cdot H_2O + NO \rightarrow NO_3^- + H_2O, \quad k = 3.1 \times 10^{-10}\ \text{cm}^3/\text{sec} \qquad (72)$$
(Adams *et al.*, 1970)

$$NO^+ \cdot CO_2 + H_2O \rightarrow NO^+ \cdot H_2O + CO_2, \quad k = 10^{-9}\ \text{cm}^3/\text{sec} \qquad (73)$$
(Fehsenfeld and Ferguson, 1972)

$$NO^+ \cdot N_2 + H_2O \rightarrow NO^+ \cdot H_2O + N_2, \quad k = 10^{-9}\ \text{cm}^3/\text{sec} \qquad (74)$$
(Ratnasini, 1972)

Let us note here that high reaction rate constants correspond to the case when dipole moments or polarizabilities of the reacting particle are greater than those of the molecule shell.

All the above applies equally to the reactions with the changes in both nucleus and cluster shell. But for the process to proceed, it is necessary that the bonding energy of the nucleus to the shell be compensated for by the interactions of the nucleus with the incoming particle. If $\alpha_a \gg \alpha_c$, these energy expenditures might be compensated by the energy released by the ion-induced dipole interaction; if $\alpha_a \leq \alpha_c$, it is possible for an activation energy to appear and its magnitude to be about $\sum_{i=1}^{n} \Delta H_i^0$.

The real mechanism of the cluster molecule process is undoubtedly more complicated than the above scheme. Specificity of a cluster molecule process is likely when the number of molecules in the cluster is small. Thus, the reaction

$$O_2^-(H_2O)_n + O_3 \rightarrow O_3^-(H_2O)_m + O_2 + (n - m)H_2O \tag{75}$$

has a rate constant of 3×10^{-10} cm^3/sec which does not change with n changing from 0 to 2 (Ferguson, 1975). One of the key ionospheric reactions has a high rate constant as well:

$$O_2^+ \cdot H_2O + H_2O \rightarrow H_3O^+ + OH + O_2, \qquad k = 2.0 \times 10^{-10}\ \text{cm/sec} \tag{76}$$

(Fehsenfeld *et al.*, 1971).

Rate constants of interaction processes of the hydrogen ion cluster H_5^+ with xenon (2.7×10^{-9} cm^3/sec), nitrogen (2.1×10^{-9} cm^3/sec), krypton (1.6×10^{-10} cm^3/sec), and deuterium (3.0×10^{-9} cm^3/sec) are also high at room temperature and that is typical for fast ion–molecule reactions (Kulagenkova *et al.*, 1978).

In this connection, the data obtained on radiolytic deuterium–hydrogen exchange in the gas phase, which was shown (Kulagenkova *et al.*, 1978) to proceed by a chain mechanism with participation of the ion clusters $H_3^+ \cdot H_2$, $H_3^+(H_2)_2$ and heavier ones, are worth attention. The important result of this work is that the exchange rate decreases by four orders of magnitude as temperature drops from 300 to 77 K.

Isotope exchange in hydrogen influenced by ionizing radiation [in our case ^{60}Co γ-rays] provides a convenient probe to study the reactivity of ion clusters since the gross process

$$H_2 + D_2 \rightarrow 2HD \tag{77}$$

and the elementary act of chain propagation

$$H_3^+(H_2)_n + D_2 \xrightarrow{k_i} H_2D^+(H_2)_n + HD \tag{78}$$

are both thermoneutral processes. In addition, reaction (78) is allowed by orbital symmetry. Further, as the temperature decreases from 300 to 77 K, the dominant ion cluster changes from $H_3^+ \cdot H_2$ to $H_3^+ \cdot (H_2)_4$ (Kulagenkova *et al.*, 1978). (In an H_2, D_2 mixture these clusters will naturally have different isotope composition.) Hence, one can study the reactivity of various hydrogen ion clusters by measuring the rate of HD formation so long as the change in the number of molecules of H_2 and D_2 in the ion clusters depends only on the temperature of the mixture.

The experiment consists of irradiation with γ-rays from a ^{60}Co source, a specially prepared superpure equimolar mixture of H_2 and D_2 at pressures between 200–400 torr, with temperatures in the range 300–77 K, and in measurement of the radiation yield of HD, G(HD). The value of G(HD) (number of HD molecules formed per 100 eV of energy absorbed) characterizes the chain length and therefore reactivity of ion clusters participating in chain propagation since, under the conditions of chain termination by impurity molecules,

$$G(\mathrm{HD}) \cong \frac{k_i\{[\mathrm{H_2}] + [\mathrm{D_2}]\}}{K_d[\mathrm{M}]}, \tag{79}$$

where k_i is a rate constant for chain propagation (78), $[H_2]$, $[D_2]$, $[M]$ are concentrations of hydrogen, deuterium, and impurities terminating the chain, and K_d is a rate constant of chain termination which proceeds through the reaction

$$H_3^+(H_2)_n + M \rightarrow MH^+ + (n+1)H_2. \tag{80}$$

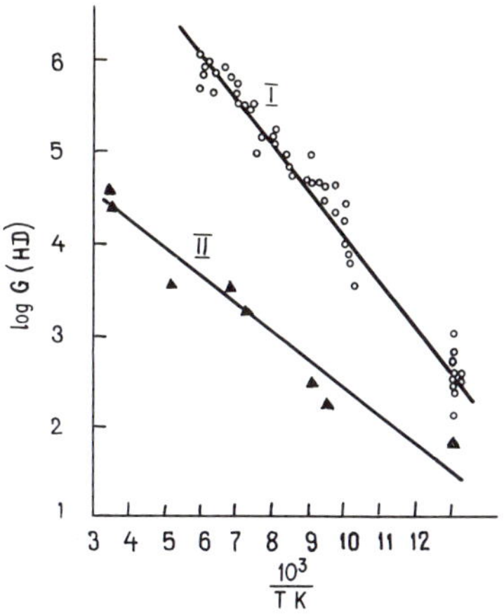

FIG. 13 The log G(HD) versus T^{-1} for a "pure" $H_2 + D_2$ mixture (curve I), for the mixture containing $6.9 \times 10^{-2}\%$ (volume percentage) of Kr (curve II). In the first case the effective activation energy is 2.3 ± 0.3 kcal/mole, in the second case it is 1.4 ± 0.25 kcal/mole.

The results obtained are shown in Fig. 13 which plots log G(HD) versus T^{-1} in the range 290–77 K. Curve 1 represents data of a "pure" mixture, in which chain termination is most likely, due to residual molecules of air which are present in concentrations of $10^{-4}\%$.* Curve II corresponds to the mixture of

* The authors paid exceptional attention to the purity of hydrogen and deuterium, and yet we could not purify them to better than 99.9999%.

H_2 and D_2 with $6.9 \times 10^{-2}\%$ krypton as a deliberately introduced isotope exchange inhibitor. In both cases, the exchange rate is seen to follow an Arrenius plot with an effective activation energy of 2.3 ± 0.3 kcal/mole for the "pure" mixture and 1.4 ± 0.25 kcal/mole for the mixture with krypton added. Similarly, an activation energy equal to 0.5 ± 0.1 kcal/mole was obtained when microadmixtures of N_2 or Xe were introduced into the mixture of H_2 and D_2 (Kulagenkova *et al.*, 1978). From (79) it follows that the measured effective activation energy of the process is equal to

$$E_{\text{eff}} = E_i - E_d \tag{81}$$

where E_i is the propagation activation energy and E_d is the activation energy of a chain termination process.

The results obtained can be explained by the previously given theory of cluster molecule reactions. In fact, since the same molecules are participating in cluster shell formation and chain propagation (when $\alpha_a' = \alpha_c$), one should expect that the activation energy E_i would be comparable with the bonding energy of the molecules to the cluster. On the other hand, molecules with polarizability α_a'' higher than that of the shell molecules (when $\alpha_a'' > \alpha_c'$) participate in chain termination, the activation energy of termination E_d should be either equal to zero or at least less than E_i. As a matter of fact, the experimental value of the effective activation energy (equal to $E_i - E_d$) was in all the cases positive and comparable to the binding energy of the H_2 molecule to the cluster shell.

The constancy of $E_i - E_d$ within a wide temperature range with various clusters participating in the exchange processes testifies to the near identical reactivity of the various clusters in both propagation and termination reactions. Both the chain propagation and termination steps are connected with the reactivity of nucleus of the cluster—the H_3^+ ion. The transfer of H^+ or D^+ from a nucleus to a molecule in the shell during a collision is likely to be the act of chain propagation, after which the cluster is rearranged around a new nucleus and a molecule of HD appears in a shell. This HD molecule is then transported to the bath gas either in an exchange reaction or in a chain termination step. In chain termination, the proton is transferred from H_3^+ to the impurity molecule.

VI. VIBRATION–TRANSLATION RELAXATION OF ION–ATOM COLLISIONS

In chemical kinetics, information on the rate of energy exchange is as fundamental as information on the rate constants of chemical transformations. Vibrationally excited ions are usually formed in ionization processes with electrons and photons and often as the products of exothermic ion–molecule reactions. Hence, the problem of deactivation of vibrational excitation of ions is of principal importance for the kinetics of nonequilibrium systems, particularly since the rate constants of a number of ion–molecule reactions depend heavily on the amount of vibrational excitation of reacting ions.

The problem of vibration–translation relaxation is of particular significance for relatively low values of reactant ion translational energies since such a situation is common in plasmachemistry, in radiation chemistry, and in the chemistry of the atmosphere and intersteller space. In addition, the problem of vibration–translational relaxation is of fundamental importance for the theory of elementary ion–molecule collisions.

Vibrational relaxation in collisions of diatomic neutral molecules is a relatively slow process with rare exception (for example, HF). Typical values for the probability of vibration–translation relaxation are in the range 10^{-3}–10^{-6} per collision. The theory of Landau and Teller (1938), as well as other theories of vibration–translation relaxation (Cottrel and McCourbey, 1959; Herzfeld and Litovitz, 1959; Chin, 1966), explain these low probabilities using a number of assumptions. One of the basic assumptions is that the vibrational amplitude A, of the excited molecule, is much smaller than a characteristic atomic dimension R. Attraction between the colliding partners at large distances, as well as the change in the frequency of the molecular vibration during the collision, are considered as small perturbations. We think that this approach is inapplicable for calculation of vibrational relaxation rates of ions colliding with atoms and molecules. As the ion and molecule approach, the polarization interaction distorts the potential energy surface of the free ion. Hence, the vibrational amplitude of the molecular ion increases considerably and the criterion $A \ll R_0$ might not be fulfilled.

The idea that one could expect much greater probabilities of vibration–translational relaxation for ion–molecule collisions than for neutral systems was first suggested at our laboratory by Karatchevtsev (1970). The statistical theory of ion–molecule reactions was used to calculate the energy states of the products of deactivation of the NO^+ ion in (82)

$$(NO^+)^* + Xe \rightarrow NO^+ + Xe. \tag{82}$$

For instance, it was shown that after the collision

$$NO^+(v = 4) + Xe \quad NO^+(v_1) + Xe, \tag{83}$$

the probability of NO^+ in vibration states $v = 0, 1, 2, 3, 4$ was accordingly 0.46, 0.3, 0.17, 0.07, 0.04. Using the data of Spohr and von Puttkamer (1967), it was shown that if the vibrational distribution of NO^+ ions formed with fast electrons corresponded to Frank–Condon factors (given by photoelectron spectroscopy), the average kinetic energy of the NO^+ ions should be 0.23 eV after deactivation via (82).

A number of authors noticed that the deactivation of vibrational excitation of H_3^+ and D_3^+ ions (formed in the exothermic reaction $H_2^+ + H_2 \rightarrow H_3^+ + H$) by collisions with hydrogen molecules was a rather fast process (Leventhal and Friedman, 1968, 1969; Ryan, 1974; Huntress and Bowers, 1973; Kim *et al.*, 1975b). In these studies they used the fact that the cross sections of endothermic ion–molecule reactions of H_3^+ ions depend on the internal energy of the ions. The more collisions of H_3^+ with H_2 before reaction, the less will be the yield of secondary ions in the endothermic reactions. The number of H_3^+ ions collisions was increased either by increasing the H_2 pressure in the ion source or by increasing the reaction time using an ion cyclotren resonance technique. The estimations of probabilities of deactivation of H_3^+ ion vibrational excitation obtained in Leventhal and Friedman (1968, 1969), Ryan (1974), Huntress and Bowers (1973), and Kim *et al.* (1975b) are the combination of probabilities of vibration–vibrational, vibration–rotational, rotation–translational, and vibration–translational relaxations.

The process of rotation–translational exchange of NO^+ ions with Xe atoms was studied by Vinogradov *et al.* (1972). In these experiments we used mass spectral techniques with pressures up to 10^{-1} torr in the ionization chamber of a mass spectrometer. A small exit slit in the source reduced the penetration of electric fields and allowed studies to be carried out at low translational energies. The ion kinetic energy distribution was measured by a deflection technique. The NO^+ ions were obtained by electron impact at 60 eV on nitromethane. The latter was introduced into the chamber in a small amount which remained invariable, while the Xe pressure ranged from 0 to greater than 10^{-2} torr. The dependence of the average kinetic energy of Cl^+ ions, formed by electron bombardment of CH_3Cl molecules, on pressure was investigated in independent experiments which served as a control.

In all the experiments, at any pressure the velocity distribution of Ar^+ ions was used to determine the temperature of the chamber walls. Argon was added in small quantities to methylene chloride. We used these experiments to take into account instrumental effects connected with the distortion of the velocity distribution due to the added Xe (such distortions, although small, were observed at high pressures) that resulted from the changes in space charge and ion–neutral scattering after the ions exited the ionization chamber. The experimentally measured distributions of Ar^+ were Maxwell–Boltzmann. Velocity distributions for Cl^+ and NO^+ ions appeared to be quasi-Maxwellian as determined from the dependence of the log of the ion current on the square of the deflection voltage.

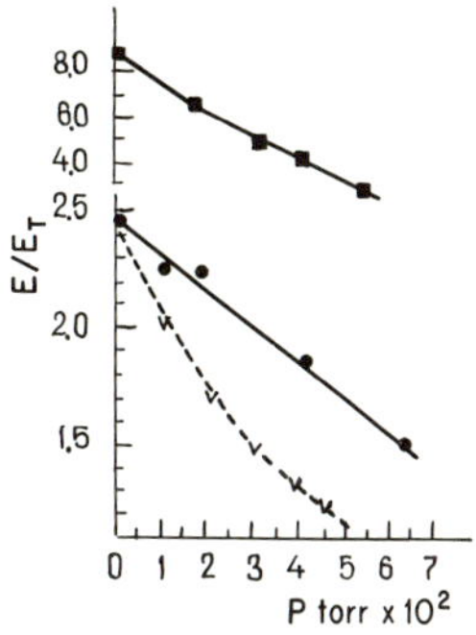

FIG. 14 The dependences of the ratio of average kinetic energies versus Xe pressure, ■, experimental dependence of the ratio for $^{35}Cl^+$ and Ar^+ ions; ●, experimental dependence for NO^+ and Ar^+ ions. The dashed line corresponds to the theoretical dependence of NO^+ ion kinetic energy loss calculated using formulas for ellastic collisions, i.e., when rotational translational energy exchange was not taken into account.

Figure 14 presents the dependence of the average kinetic energies of Cl^+ and NO^+ ions (in relative units, i.e., E/E_{Ar^+}) on Xe pressure. For example, Cl^+ atomic ions having excess initial translational energy lose it via collisions with Xe. The average kinetic energy, E_n, of these ions after n internally elastic collisions is

$$(E_n - E_T) = (E - E_T)\left(\frac{m_1{}^2 + m_2{}^2}{(m_1 + m_2)^2}\right)^n, \tag{84}$$

where E is the initial kinetic energy of ion and m_1 and m_2 are the masses of the colliding particles.

Using (84) and the experimental dependence of E_{Cl^+} versus Xe pressure, one is able to determine average number of Cl^+ collisions with Xe atoms as a function of Xe pressure (which was measured not in the chamber itself but at the point where the gases were introduced into the ion source). When calculating the average number of NO^+ ion collisions at different pressures of Xe, a correction was introduced to compensate for the different residence times of Cl^+ and NO^+ ions in the ion source. Figure 14 shows as well a calculated curve for the case of NO^+ ions retarded by elastic collisions with xenon. From Fig. 14 it is seen that the experimentally observed dependence differs from the calculated one. Using the data given in Fig. 14, we can estimate the average rotational energy of fragment NO^+ ions. For the first several collisions, vibration–translational exchange need not be taken into account (the rate of rotation–translational exchange is much greater than vibration–translational exchange for collisions of NO^+ and Xe, as shown in independent experiments which will be discussed). Thus, the difference between the calculated curve and the experiment is connected with the fact that rotation–translational energy exchange is occurring.

A theoretical model for rotation–translational exchange in ion–atom collisions has been suggested. According to this model, the relative translational energy of the particles and the rotational energy of the NO^+ ion will be evenly distributed in the system between the three translational degrees of freedom corresponding to the relative motion of NO^+ and Xe and the two rotational degrees of freedom of NO^+. Formulas for calculating the translational and rotational energy of NO^+ after the collision with Xe have been given (Vinogradov *et al.*, 1972). Using these formulas, the average initial rotational energy of NO^+ formed by electron impact at 60 eV from nitromethane was estimated to be approximately 0.17 eV (the initial translational energy of NO^+ was 0.2 eV).

Vibration–translational relaxation of diatomic NO^+ ions was also studied by similar technique (Vinogradov *et al.*, 1972). In these experiments the NO^+ ions were obtained by electron impact at 60 eV on NO molecules. If it is assumed that ionization with fast electrons yields vibrational distributions in NO^+, that correspond to Frank-Condon factors obtained from photoelectron spectroscopy, and that this energy will be statistically partitioned via collision with Xe, then it is possible to have an NO^+ average kinetic energy of 0.23 eV after vibration–translational relaxation with Xe.

The basic idea of the experiment is the following. The collisional deactivation of the vibrational excitation of the NO^+ ions should result in a portion of the energy ending up in the translational degrees of freedom of the ion. If the deactivation process is fast, one should notice a "heating" of the translational degrees of freedom of the ion as the gas pressure increases in the ion source.

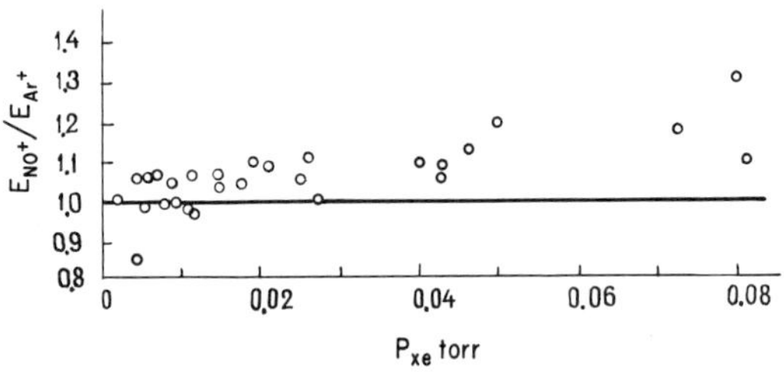

FIG. 15 A plot of E_{NO^+}/E_{Ar^+} versus Xe pressure.

Figure 15 represents a plot of E_{NO^+}/E_{Ar^+} versus P_{Xe} in torr (the pressure was measured at the entry to gas inlet system). For each pressure an average number of collisions of NO^+ ions with Xe atoms was calculated. Some increase in the average kinetic energy of the NO^+ ions can be seen from Fig. 15 as pressure increases. At the lowest pressures there is least chance for translationally hot NO^+ to be deactivated by subsequent collisions with Xe. But owing to a considerable dispersion of experimental values of E_{NO^+}/E_{Ar^+} at low pressure, which is comparable to the "heating" effect of

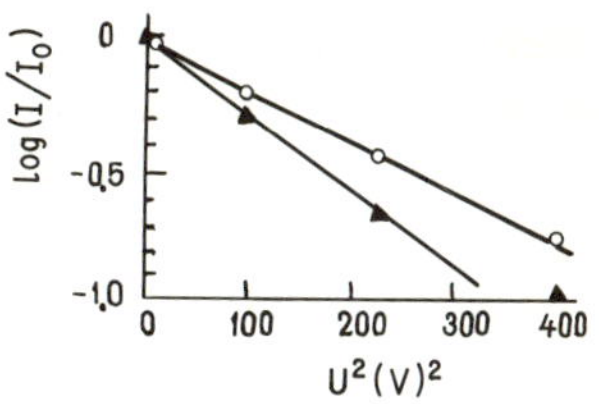

FIG. 16 An example of the ion current logarithm dependence versus the square of the deflecting voltage for H_2^+, ○ ; and Ar^+, ▲ ions at Ne pressure of 1.5×10^{-1} torr.

the translational degrees of freedom, it is not possible to extract meaningful results from these data. Rather a complicated procedure of experimental data processing (see Fig. 15) allows an estimate of the probability W of vibronic–translational relaxation of NO^+ ions colliding with Xe atoms within the limits of

$$\tfrac{1}{20} \geq W \geq \tfrac{1}{200}. \tag{85}$$

Strictly speaking, the estimate given above depends on the theoretical models used for the processing of the experimental data. These models are based on some approximations. So some remarks on accuracy range of the estimation must be given, for this accuracy is determined, in turn, by the accuracy of the model taken as the basis for the calculations.

One of the basic assumptions was the average translational energy of NO^+ ions must be equal to 0.23 eV after vibrational energy deactivation. For calculation of this energy it was assumed, following from statistical theory arguments, that the probability of finding an NO^+ ion in the ground vibrational state after deactivation, even in the case of high initial vibrational states, is very high. In principle, the average translational energy of NO^+ after deactivation can be less than 0.23 eV. Such would be the case if the vibrational quantum number change is restricted to unity in the deactivation process.

A further approximation ignores the fact there are really two ion distributions, thermal and "hot." A more detailed analysis shows that this approximation does not dramatically influence the limits of estimation of (85), and the upper limit of estimation (85) is valid within a factor of two. The values of probabilities of vibration–translational relaxation of NO^+ greatly exceed the corresponding probabilities for collisional deactivation of the isoelectron analog of NO^+, the N_2 molecule.

Vibration–translational relaxation of the complex ions $C_{12}H_{12}^+$ has been observed (Sergeev *et al.*, 1974). These ions were formed through the reaction $C_6H_6^+ + C_6H_6 \rightarrow C_{12}H_{12}^{*+}$ and were subsequently deactivated by collisions with helium atoms.

Deactivation of vibrational excitation in H_2^{+*} ions, formed by electron impact of H_2, by collisions with He atoms has been observed (Theard and Huntress, 1974). The authors used an ion cyclotron resonance technique. The idea used for estimation of the probability of deactivation of H_2^+ ions by He is as follows. The reaction

$$H_2^{+*} + He \xrightarrow{k_r} HeH^+ + H \tag{86}$$

is known to occur only if the H_2^+ ion is vibrationally excited ($v \geq 2$). Deactivation of vibronic excitation of H_2^+ will take place in competition with (86).

$$H_2^+(v \geq 2) + He \xrightarrow{k_d} H_2^+(v \leq 2) + He. \tag{87}$$

At low H_2 pressures, the reaction $H_2^+ + H_2 \rightarrow H_3^+ + H$ need not be taken into account, nor the processes of vibration–vibrational relaxation of H_2^{+*} by collision with H_2. The values of k_r and k_d can be found from experimental dependences of ion concentrations versus τ and solving the kinetics of (86) and (87). If it is assumed that k_d does not depend on vibrational excitation for $v \geq 2$, and that k_r is equal for all levels above the second, and that vibrational quantum number changes only by one in a collision, then the probability of vibration–translational relaxation of H_2^{+*} by collision with He is equal approximately to 0.7.

An estimate of the probability of vibration–translational relaxation of H_2^+ and D_2^+ ions in the collisions

$$H_2^{+*} + Ne \xrightarrow{\omega_1} H_2^+ + Ne, \tag{88}$$

$$H_2^{+*} + He \xrightarrow{\omega_2} H_2^+ + He, \tag{89}$$

$$D_2^{+*} + Ne \xrightarrow{\omega_3} D_2^+ + Ne, \tag{90}$$

was obtained using experimental technique similar to those described above for NO^+ relaxation studies. As in that work, vibrational deactivation events were detected by measuring the energy released in translational degrees of freedom. But unlike the NO^+ case, the ions were formed by photons from a helium resonance light source (photon energy equal to 21.4 eV). The number of ion collisions as a function of gas pressure was estimated from the pressure dependence of secondary ion production in reactions with well known rate constants. H_2 and D_2 were introduced into the ionization chamber at low pressures, which did not vary throughout the experiment. He or Ne pressures were variable and were changed within limits such that the average number of collisions of the ions with the atoms could vary from 0 to 2. The ion velocity distribution was measured as before. The Ar^+ velocity distribution was used as thermal distribution standard. The experiments showed that the increase in average kinetic energy with pressure of H_2^+ and D_2^+ ions was larger than with NO^+ ions. When processing the experimental data, cotangents of curves similar to the ones shown in Fig. 16 were found for each pair of velocity distributions at different pressures of inert gases and the relation $\xi = \text{ctg}\,\beta_{H_2^+(D_2^+)}/\text{ctg}\,\beta_{Ar^+}$ was calculated. Figure

17 plots ξ versus Ne pressure for D_2^+ and Ar^+ ions. From Fig. 17 it follows that this relation noticeably grows with the increase in Ne pressure. Hence, vibration–translational relaxation of D_2^+ by Ne is an efficient process.

We managed to process more accurately the results obtained for the collisions of the type (88)–(90) than for the collisions of NO^+ with Xe. This was possible because the increase in kinetic energy of H_2^+ and D_2^+ ions was appreciable, even at low pressures in the ionization chamber when the average number of collisions was less than 1.

While calculating probabilities of vibration–translational relaxation, the following approximations were used. First, vibrational energy was presumed to be changed in one quantum jumps. Second, the energy released in the vibrational deactivation was assumed to be statistically distributed in the center of mass system among the three translational degrees of freedom of relative motion of the colliding particles and the two rotational degrees of freedom of the molecule. These two assumptions made it possible to find E^*, the average energy of fast ions in the laboratory system. In addition, the deactivation probability was assumed to be independent of the vibrational level v. Finally, the fact that the velocity distributions of H_2^+ and D_2^+ are the sum of two distributions was taken into account; the first corresponds to a thermal distribution, the second to the distribution of fast ions with average translational energy E^*. This data processing gave the following values of vibronic–translational relaxation probabilities: $W_1 = 0.63 \pm 0.2$; $W_2 = 0.55 \pm 0.2$; $W_3 = 0.65 \pm 0.2$. Our value for W_2 is in a good agreement with the results obtained on the ICR (Theard and Huntress, 1974).

The results on vibronic–translational relaxation of H_2^+ and D_2^+ ions may be interpreted by using the properties of the potential energy surfaces of the system. The calculation of such a surface has been given (Cooper *et al.*, 1974). The map of the potential energy surface is shown schematically in Fig. 18 for the system $H_2^+(D_2^+) + He(Ne)$. The atoms are supposed to be on the same line at approach (though as it has been shown, (Cooper *et al.*, 1974), the surface is not greatly distorted if the approach is nonlinear). One of the basic properties of such a surface is that at relatively small distances there is an attractive well. In the valley corresponding to the formation of $XH_2^+(XD_2^+)$ molecules, the energy slowly increases to the value corresponding to the endothermicity of the formation of HeH^+ ions. In the case when the reactant ion is vibrationally excited, a point moving along the $H_2^+(D_2^+)$ valley will also oscillate along the H–H line (or between the walls of the valley). Upon reaching the region of energetically closest approach of the three particles, one of the walls evolves into the valley corresponding to the formation of the products of reaction. Thus, the vibrational amplitude increases; the higher the vibrational excitation is the greater the increase in amplitude.

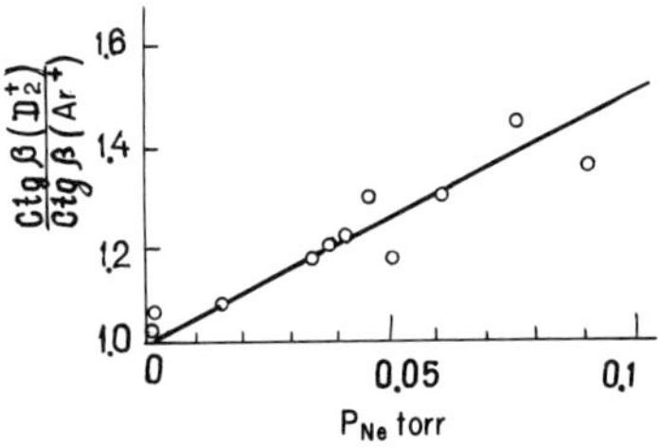

FIG. 17

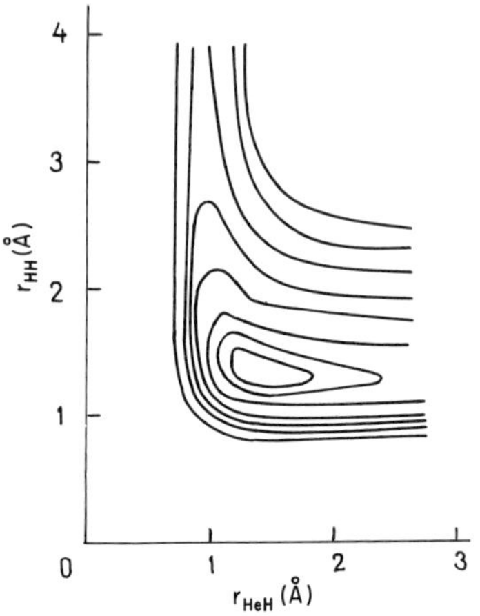

FIG. 18

FIG. 17 The dependence of the cotangent ratio for D_2^+ and Ar^+ ions of the velocity distribution curves slopes versus Ne pressure.

FIG. 18 Schematic map of potential energy surface for collisions: D^+ + Ne, H_2^+ + Ne, H_2^+ + He, and D_2^+ + He.

It is interesting to note that on the basis of the correlation diagram for a collision of H_2^* + He^+, one may conclude that at short distances between the colliding particles a noticeable energy barrier may arise. Hence, one should expect a reduced effectiveness of deactivation of H_2^* vibrational excitation by He^+ than for H_2^+ ions by He.

The results obtained point to a rather high efficiency of vibration–translational relaxation of diatomic ions by atoms, but this effect cannot yet be considered to be a universal rule. Further studies on these important systems should be made.

VII. SOME RULES FOR QUALITATIVE ESTIMATION OF RATE CONSTANTS OF EXOTHERMIC AND THERMONEUTRAL ION–MOLECULE REACTIONS

At present the number of known ion–molecule reactions is of the order of 10^4. This number is expected to rapidly grow in the future. That is why, along with precise experimental quantitative estimations of rate constants of ion–molecule reactions, a purely qualitative approach for *a priori* estimation of the rate constants of unknown reactions is of great importance. Understanding the dynamics of ion–molecule reactions is important for these estimations.

The major feature of ion–molecule reactions, as it has been already mentioned, is the presence of an electric charge in the system of colliding particles. It is this fact that allowed one of the authors of this chapter (Talrose and Ljubimova, 1952; Talrose, 1967) to establish the empirical rule that the *rapidity of ion/molecule reactions is determined by the sign of the enthalpy of the ion/molecule reaction.* Typical values of rate constants of exothermic and thermoneutral ion/molecule reactions are 10^{-9}–10^{-10} cm^3/sec.

The features of the reaction dynamics associated with the presence of the electric charge in the system of colliding particles were given in a qualitative manner in the first section of the chapter. It was suggested that these features should promote rearrangements of chemical bonds or electron shells of reacting particles. It was found out later there were exceptions to this rule for ion–molecule reactions. These exceptions proved to be connected with rearrangement of electron shells in the course of the reaction. Hence, the subsidiary criteria of orbital symmetry conservation and the presence or absence of reaction spin conservation were suggested. A practical use of this rule may be illustrated in the following way. We have analyzed the data on cross sections and rate constants of approximately 3000 ion–molecule reactions that require the transfer of heavy particles ($m \geq 1$ amu). Most of the slow processes* (62 reactions) were found to be orbital symmetry forbidden. Only in four cases (out of 62) were we not able to construct correlation diagrams due to the complexity of the processes and the absence of necessary information on the reagents and products. Hence, we state: *exothermic and thermoneutral ion–molecule reactions are fast processes if they are not orbital symmetry or spin forbidden.* When there is any forbiddenness, these reactions may be slow processes. It is usually thought that orbital symmetry forbiddenness results in an activation barrier. However, such forbiddenness might result in small values of the preexponential factor in the Arrhenius expression for reaction rate constants. For example, the symmetry forbidden reaction (35) was shown not to go even at kinetic energies up to 17 eV (Futrell and Tiernen, 1972). When symmetry forbiddenness of a reaction will result in an activation energy and when it will result in small values of the preexponential factor is a matter of fundamental importance in kinetic theory. Hence, experiments on the temperature dependence of the rate constants of symmetry forbidden reactions are of particular importance.

* By slow processes we mean reactions with rate constants less than 10^{-10} cm^3/sec or cross sections which are more than an order of magnitude less than Langevin.

The rate constant value which divides ion–molecule reactions into fast and slow ones may be assumed to be of the order of 10^{-11} cm^3/sec. Experimental data indicate that the majority of symmetry allowed reactions have rate constants which are higher than the above value and in a number of cases symmetry forbidden reactions are slower. Possible exceptions to the above mentioned rule may exist due to other features of the dynamics of these reactions which were not taken into account in this simple analysis. For example, vibronic excitation may be such a feature which, if taken into account, will allow further refinement of the rule.

The determination of the influence of translational, rotational, and vibrational energy on the rate of ion/molecule reactions is very important for the theory of ion/molecule reactions. Recently this problem has been studied by a number of authors (Honma *et al.*, 1976). The results yield information on active degrees of freedom and on the details of potential energy surfaces of ion–molecule reactions.

In 1957 one of the authors, in collaboration with Frankevich, formulated the rule for systems subjected to radiation dose rates up to 10^{25} eV cm^{-3}/sec that ion–molecule reactions go to completion long before the recombination of ions with electrons or with the ions of opposite sign starts. Stevenson independently formulated the same rule. Hence, recombination occurs with ions which as a rule are incapable of further ion–molecule reactions. Would existence of slow (due to orbital symmetry forbiddenness or to clusterization phenomenon) reactions introduce something new to this general rule? In exceptional cases of extremely slow ion–molecule reactions, one can suppose that primary ions will participate in recombination processes. That means that an energy equal to the ionization potential of the molecule may be released in the recombination reaction. A dissociative recombination explosion will give the largest number of chemically active neutral fragments. Such a situation occurs if the ion–molecule reaction rate constant

$$k_i \leq \frac{\sqrt{10^{-2} I G_i \bar{\alpha}}}{[\mathrm{M}]},$$

where I is the dose rate in electron volt per cubic centimeter second; G_i the radiation yield of ions per 100 eV of the energy absorbed, $\bar{\alpha}$ a recombination coefficient; and [M] a concentration of neutral molecules reacting with the ion.

It is interesting to note that the recombination of electrons with an ion cluster will occur with a smaller energy release due to solvatation stabilization. In this case a smaller number of chemically active particles will exist compared with a "cluster-free" situation.

ACKNOWLEDGMENT

The authors would like to thank O. Polukarova for technical assistance.

REFERENCES

Adams, N. G., Bohme, D. K., Dunkin, D. B., Fehsenfeld, F. C., and Ferguson, E. E. (1970). *J. Chem. Phys.* **52**, 3133.

Berry, C. E. (1950). *Phys. Rev.* **78**, 597.

Bowers, M. T., and Elleman, D. (1970). *J. Am. Chem. Soc.* **92**, 1847.

Brown, P. R., and Hayer, E. F. (1971). *J. Chem. Phys.* **55**, 922.

Burt, J. A. (1970). *J. Chem. Phys.* **52**, 6062.

Buttrill, S. E., Jr., Kim, J. K., Huntress, W. T., Jr., LeBreton, and Williamson, A., Jr. (1974). *J. Chem. Phys.* **61**, 2122.

Chapman, S., and Preston, R. K. (1974). *J. Chem. Phys.* **60**, 650.

Chin, H. (1966). "Ion Molecule Reactions in Gas Phase," p. 44. Am. Chem. Soc., Washington, D.C.

Cooper, W. G., Evers, N. S., and Kouri, D. J. (1974). *Mol. Phys.* **27**, 707.

Cottrel, T. L., and McCourbey, J. C. (1959). "Molecular Energy Transfer in Gases," p. 27. Butterworth, London.

Dalgarno, A. (1958). *Philos. Trans. R. Soc. London, Ser. A* **250**, 426.

Eyring, H., Hirschfelder, J. O., Taylor, H. S. (1936). *J. Chem. Phys.* **6**, 479.

Fehsenfeld, F. C., and Ferguson, E. E. (1972). Aeronom. Rep. N48, p. 241. Univ. of Illinois, Urbana.

Fehsenfeld, F. C., Mosesman, M., and Ferguson, E. E. (1971). *J. Chem. Phys.* **55**, 2115.

Fehsenfeld, F. C., Albriton, D. L., Bush, Y. A., Fournier, P. G., Govers, T. R., and Fournier, J. (1974). *J. Chem. Phys.* **61**, 2150.

Ferguson, E. E. (1975). *Atmos. Earth Planets, Proc. Summer Adv. Study Inst., Univ. Liege, 1974* p. 197.

Ferguson, E. E., Fehsenfeld, F. C., Golden, P. D., and Schmeltekopf, A. L. (1965). *J. Geophys. Res.* **70**, 4323.

Field, F. H., and Franklin, J. L. (1957). "Electron Impact Phenomena and the Properties of Gaseous Ions." Academic Press, New York.

Futrell, J. H., and Tiernen, T. O. (1972). *In* "Ion–Molecule Reactions" (J. L. Franklin, ed.), Vol. 2, p. 533. Butterworth, London.

Giese, C. F., and Maier, W. B. (1961). *J. Chem. Phys.* **35**, 1913.

Good, A., Durden, D. A., and Kebarle, P. (1970). *J. Chem. Phys.* **52**, 212.

Harrison, A. G., and Heyer, B. G. (1973). *Can. J. Chem.* **51**, 1265.

Herzfeld, K. F., and Litovitz, T. A. (1959). "Absorption and Dispersion of Ultrasonic Waves." Academic Press, New York.

Herzberg, G. (1966). "Molecular Spectra and Molecular Structure. III. Electronic Spectra and Electronic Structure of Poliatomic Molecules." Van Nostrand Reinhold, Princeton, NJ.

Herzberg, G. (1971). "The Spectra and Structure of Simple Free Radicals." Cornell Univ. Press, Ithaca, New York.

Hiraoka, K., and Kebarle, P. (1975). *J. Chem. Phys.* **62**, 2267.

Hoffman, R., and Woodward, R. (1965). *J. Am. Chem. Soc.* **87**, 2046.

Holliday, M., Muckerman, J., and Friedman, L. (1971). *J. Chem. Phys.* **54**, 1058.

Honma, K., Tanaka, K., Koyono, I., and Tanaka, I. (1976). *J. Chem. Phys.* **65**, 678.

Hund, F. (1928). *Z. Phys.* **51**, 759.

Huntress, W. T., and Bowers, M. T. (1973). *Int. J. Mass Spectrom. Ion Phys.* **12**, 1.

Inoque, M., and Wexler, S. (1969). *J. Am. Chem. Soc.* **91**, 5730.

Karatchevtsev, G. V. (1970). *Khim. Vys. Energ.* **4**, 397.
Kaufman, J. J., and Koski, W. S. (1969). *J. Chem. Phys.* **50**, 1942.
Kim, J. K., Theard, L. P., and Huntress, W. T., Jr. (1975a). *J. Chem. Phys.* **62**, 45.
Kim, J. K., Theard, L. P., and Huntress, W. T. (1975b). *Int. J. Mass Spectrom. Ion Phys.* **15**, 223.
Kondratiyev, V. N., and Nikitin, E. E. (1974). "Kinetika i Mekhanizm Gazofaznykh Reaktsii," p. 270. Nauka, Moscow.
Kondratiyev, V. N., Nikitin, E. E., Reznikov, A. I., and Umanskii, S. Y. (1976). "Termicheskiye Bimoleculyarniye Reaktsii v Gazakh." Nauka, Moscow.
Krenos, J. R., Preston, R. K., and Wolfgang, R. (1974). *J. Chem. Phys.* **60**, 1634.
Kulagenkova, N. A., Larin, I. K., and Talrose, V. L. (1972). *Dokl. Akad. Nauk SSSR* **207**, 129.
Kulagenkova, N. A., Larin, I. K., Talrose, V. L., and Chernovskii, S. M. (1978). *Khim. Vys. Energ.* **12**(1), 14.
Landau, L. D., and Teller, E. (1938) *Phys. Z. Sowjetunion* **10**. 34.
Langevin, P. (1903). *Ann. Chim. Phys.* **28**, 289, 433.
Larin, I. K., and Talrose, V. L. (1962). *Kinet. Katal.* **3**(3), 305.
Levental, J. J., and Friedman, L. (1968). *J. Chem. Phys.* **49**, 1974.
Levental, J. J., and Friedman, L. (1969). *J. Chem. Phys.* **50**, 2968.
Light, J. C. (1962). *J. Chem. Phys.* **36**, 1016.
Light, J. C. (1964). *J. Chem. Phys.* **40**, 3221.
McLaughlin, D. R., and Tompson, D. P. (1973). *J. Chem. Phys.* **59**, 4393.
Mahan, B. H. (1971). *J. Chem. Phys.* **55**, 1436.
Malkhasyan, R. G., Tikhomirov, M. V., Zhurkin, E. S., and Tunitskii, N. N. (1974). *Khim. Vys. Energ.* **8**, 187.
Meot-Ner, M., and Field, F. H. (1976). *J. Chem. Phys.* **64**, 277.
Moran, T. F., Hamill, W. H. (1963). *J. Chem. Phys.* **39**, 1413.
Mulliken, R. S. (1932). *Rev. Mod. Phys.* **4**, 1.
Munson, M. S., Field, F. H., and Franklin, J. L. (1963). *J. Am. Chem. Soc.* **85**, 3584.
Nikitin, E. E. (1970). "Teoriya Elementarnikh Atomno–Molekuyarnikh Protsessov v Gazakh." Khimiya, Moscow.
Pearson, R. G. (1970). *Chem. Eng. News* **48**, 66.
Preston, R. K., and Kross, R. J., Jr. (1973). *J. Chem. Phys.* **59**, 3616.
Ratnasiri, P. A. J. (1972). Aeronom. Rep. N48, p. 72. Univ. of Illinois, Urbana.
Ryan, K. R. (1974). *J. Chem. Phys.* **61**, 1559.
Sergeev, Y. L., Akopyan, M. E., and Golovin, A. V. (1974) *Vtoraya Vsesoyuznaya Konf. Mass Spektrom., Tezisy Dokl., Leningrad* p. 132.
Schissler, D. O., Stevenson, D. P. (1956). *J. Chem. Phys.* **24**, 926.
Smith, G. P. K., Karlson, R. D., and Cross, R. D., Jr. (1975). *J. Chem. Phys.* **62**, 199.
Spohr, R., and von Puttkamer, E. (1967). *Z. Naturforsch., Teil A* **22**, 705.
Su, T., and Bowers, M. T. (1973). *Int. J. Mass Spectrom. Ion Phys.* **12**, 347.
Talrose, V. L. (1967). *Pure Appl. Chem.* **5**, 455.
Talrose, V. L., and Frankevitch, E. L. (1956). *Dokl. Akad. Nauk SSSR* **111**(2), 376.
Talrose, V. L., and Frankevitch, E. L. (1958). *Tr. Vsesoyusnogo Soveschaniya Radiat. Khim.*, Moscow p. 13.
Talrose, V. L., and Karatchevtsev, E. V. (1971). *In* "Reactions under Plasma Conditions" (M. Venugopalan, ed.), Vol. 2, Ch. 12. Wiley, New York.
Talrose, V. L., and Ljubimova, A. K. (1951). *Dokl. Akad. Nauk SSSR* **86**(5), 209.
Talrose, V. L. (1952). PhD. Thesis, Institute of Chemical Physics, Moscow, Russia.
Talrose, V. L., Markin, M. I., Larin, I. K. (1962). *Discus.* Farad. Soc. **N 33**, 257.
Theard, L. P., and Huntress, W. T., Jr. (1974). *J. Chem. Phys.* **60**, 2840.
Tully, J. C. (1971). *J. Chem. Phys.* **54**, 1790.

Vinogradov, P. S., and Karatchevtsev, G. V. (1974). *Khim. Vys. Energ.* **8**, 379.
Vinogradov, P. S., Karatchevtsev, G. V., Marutkin, A. Z., Talrose, V. L., and Howansky, Y. K. (1972). *Ann. Geophys.* **28**, 859.
Vinogradov, P. S., Karatchevtsev, G. V., and Talrose, V. L. (1976). *Khim. Vys. Energ.* **10**(6), 475.
Vinogradov, P. S., Karatchevtsev, G. V., Travina, O. A., and Talrose, V. L. (1977). *Khim. Vys. Energ.* **11**, 6.
Woodward, R., and Hoffman, R. (1965a). *J. Am. Chem. Soc.* **87**, 395.
Woodward, R., and Hoffman, R. (1965b). *J. Am. Chem. Soc.* **87**, 2511.
Woodward, R., and Hoffman, R. (1971). "Sokhraneniye Orbitalnoi Simmetrii." Isdat. Mir., 30.
Wiltsbach, K. E. (1957). *J. Am. Chem. Soc.* **79**, 1013.
Ynon, Y., and Klein, F. (1971). *Adv. Mass Spectrum.* **5**, 241.

INDEX

Gas Phase Ion Chemistry

Volume 2

CONTRIBUTORS

DONALD H. AUE
JOHN E. BARTMESS
J. H. BEYNON
MICHAEL T. BOWERS
JOHN I. BRAUMAN
ROBERT C. DUNBAR
W. RONALD GENTRY
J. R. GILBERT
BRUCE K. JANOUSEK
K. R. JENNINGS
ROBERT T. McIVER, JR.
DONALD RAPP
ARTHUR B. WEGLEIN

Gas Phase Ion Chemistry

Edited by MICHAEL T. BOWERS

DEPARTMENT OF CHEMISTRY
UNIVERSITY OF CALIFORNIA
SANTA BARBARA, CALIFORNIA

Volume 2

ACADEMIC PRESS New York San Francisco London 1979
A Subsidiary of Harcourt Brace Jovanovich, Publishers

ACADEMIC PRESS, INC.
111 Fifth Avenue, New York, New York 10003

United Kingdom Edition published by
ACADEMIC PRESS, INC. (LONDON) LTD.
24/28 Oval Road, London NW1 7DX

Library of Congress Cataloging in Publication Data

Main entry under title:

Gas phase ion chemistry.

Includes bibliographies.
1. Chemical reaction, Conditions and laws of—Addresses, essays, lectures. 2. Gases, Ionized—Addresses, essays, lectures. I. Bowers, Michael T.
QD501.G324 541'.39 78–19666
ISBN 0–12–120802–8

PRINTED IN THE UNITED STATES OF AMERICA

79 80 81 82 9 8 7 6 5 4 3 2 1

CONTENTS

Chapter 11 The Gas-Phase Acidity Scale

JOHN E. BARTMESS AND ROBERT T. McIVER, JR.

Chapter 12 Chemical Ionization Mass Spectrometry

K. R. JENNINGS

Chapter 13 Energetics and Mechanisms of Unimolecular Reactions of Positive Ions: Mass Spectrometric Methods

J. H. BEYNON AND J. R. GILBERT

LIST OF CONTRIBUTORS

Numbers in parentheses indicate the pages on which the authors' contributions begin.

DONALD H. AUE (1), Department of Chemistry, University of California, Santa Barbara, California 93106

JOHN E. BARTMESS (87), Department of Chemistry, Indiana University, Bloomington, Indiana 47401

J. H. BEYNON (153), Royal Society Research Unit, University College of Swansea, Swansea, United Kingdom

MICHAEL T. BOWERS (1), Department of Chemistry, University of California, Santa Barbara, California 93106

JOHN I. BRAUMAN (53), Department of Chemistry, Stanford University, Stanford, California 94305

ROBERT C. DUNBAR (181), Department of Chemistry, Case Western Reserve University, Cleveland, Ohio 44106

W. RONALD GENTRY (221), Chemical Dynamics Laboratory, Department of Chemistry, University of Minnesota, Minneapolis, Minnesota 55455

J. R. GILBERT (153), Department of Chemistry, University of Essex, Colchester, United Kingdom

BRUCE K. JANOUSEK (53), Department of Chemistry, Stanford University, Stanford, California 94305

K. R. JENNINGS (123), Department of Chemistry and Molecular Sciences, University of Warwick, Coventry, United Kingdom

ROBERT T. McIVER, JR. (87), Department of Chemistry, University of California, Irvine, California 92717

DONALD RAPP (299), The University of Texas at Dallas, Richardson, Texas 75080

ARTHUR B. WEGLEIN* (299), The University of Texas at Dallas, Richardson, Texas 75080

*Present address: Cities Service Oil Company, Exploration and Production Research, Tulsa, Oklahoma 74150.

PREFACE

Fifteen years ago there was a handful of research groups involved in research on the chemistry of ions in the gas phase. Today there are hundreds of groups spanning a discipline that stretches from biochemistry to theoretical chemical physics. The growth has been dramatic, almost explosive, in the past decade, and the field is currently at the point where no one can even superficially keep up with all of it. There are three journals totally devoted to ion chemistry of one sort or another, and ion chemistry papers are making major contributions to most of the main line chemical journals.

The roots of gas phase ion chemistry, perhaps more than those of some disciplines, are found in technique development. The rapid technological advances in the post-Sputnik era provided the materials and vacuum and electronic capabilities to develop the sophisticated instrumentation we take for granted today. The chapter by Gentry on molecular beams in Volume 2 is illustrative of the inventive methods used to probe microscopic reaction mechanisms. Previous editors recognized this relationship and built their books around chapters that discussed the various techniques in use at that time.

The question arose, in attempting to put the present volumes together, of what topics should be emphasized and how they should be organized. In the end, the decision was to emphasize the topics I found most interesting and relevant, and to assign the organization of these topics to the various authors who graciously agreed to write the chapters. Each of the authors is a world leader in his respective area, and each area presents different requirements of both length and style of presentation. The areas I chose to emphasize are theory (3), kinetics and mechanism (6), thermochemistry (3), and technique and applications (4). The numbers in parentheses give the number of chapters in each section. The dominant theme of the treatise is the chemical physics aspect of gas phase ion chemistry; but a broad range of chemists should find most of the chapters of interest, particularly those on thermochemistry, where comprehensive compendiums of gas phase basicities, acidities, and electron affinities are given.

Finally, I would like to thank the many people who made these volumes possible. To the authors go the primary thanks for preparing high quality chapters and forwarding them promptly. Members of my research group all deserve thanks for carrying on during the editing process and for helping to proofread the galleys. The helpful, competent, and professional attitude of everyone at Academic Press minimized all of the difficulties encountered, and my thanks go to them. Last, my family, and especially my wife Diane, gave support and encouragement throughout, and I am deeply grateful and thankful for their presence in my life.

CONTENTS OF VOLUME 1

Gas Phase Ion Chemistry

Volume 2

Chapter 9

Stabilities of positive ions from equilibrium gas-phase basicity measurements

DONALD H. AUE

MICHAEL T. BOWERS

DEPARTMENT OF CHEMISTRY
UNIVERSITY OF CALIFORNIA
SANTA BARBARA, CALIFORNIA

ISBN 0-12-120802-8

I. INTRODUCTION

Proton-transfer reactions are among the most important classes of chemical reactions and have been studied extensively in solution over a long period of time. We report here the results of recent gas-phase basicity measurements that derive from a study of equilibrium constants for proton-transfer reactions among positive ions. Negative ion stability data from electron affinity measurements are presented in Chapter 10 by Janousek and Brauman. Results of studies of proton-transfer equilibria among negative ions are found in Chapter 11 by Bartmess and McIver.

In summarizing the results of work on the stabilities of positive ions, we have tried to include all of the published data from accurate equilibrium studies through early 1978 along with some unpublished data from our laboratories. The data are presented with only a minimal interpretive discussion aimed at pointing out some of the major structural effects observed.

High-pressure mass spectrometer sources were first used to study proton-transfer reactions in the gas phase. By assuming that exothermic proton transfers should occur rapidly in the gas phase, these studies were used to qualitatively and semiquantitatively determine relative gas-phase basicities and proton affinities of various molecules by the reaction bracketing technique (Beauchamp, 1971). The gas-phase basicity (GB) is defined as the negative of the free-energy change for the protonation reaction (1), while the proton affinity (PA) is the negative of the enthalpy change for reaction (1).

$$A + H^+ \rightleftarrows AH^+, \qquad \begin{aligned} -\Delta G^0 &= GB(A) \\ -\Delta H^0 &= PA(A) \end{aligned} \tag{1}$$

$$AH^+ + B \rightleftarrows A + BH^+, \qquad \begin{aligned} \Delta G^0 &= GB(B) - GB(A) \\ \Delta H^0 &= PA(B) - PA(A). \end{aligned} \tag{2}$$

Thus, the direction of the proton-transfer reaction (2) gives the relative gas-phase basicities of A and B. The proton affinity of water was deduced to be between 163 and 172 kcal/mole by Tal'roze and Frankevich (1956) on the basis of the occurrence and nonoccurrence, respectively, of reactions (3) and (4) and known heats of formation

$$H_2S^{+\cdot} + H_2O \rightarrow H_3O^+ + HS, \tag{3}$$

$$H_2C_2^{+\cdot} + H_2O \nrightarrow H_3O^+ + C_2H. \tag{4}$$

Munson (1965) determined the relative basicities of ammonia and the methylamines in the gas phase by such qualitative high-pressure mass spectrometric techniques. Later, using ion cyclotron double resonance techniques, Beauchamp and Buttrill (1968) and Brauman and Blair (1968, 1971) established relative basicities for some organic bases in the gas phase. Munson (1965) and Brauman and Blair (1968; Brauman *et al.*, 1971) observed that the order of gas-phase basicities of the methylamines was $NH_3 <$ $MeNH_2 < Me_2NH < Me_3N$. This order differs from that observed in aqueous solution ($NH_3 < Me_3N < MeNH_2 < Me_2NH$), which indicated immediately that such gas-phase data could supply a set of *intrinsic* basicities that could be used to clarify our understanding of the origin of solution-phase basicity orders.

Ion–molecule reactions in the gas phase were first studied under equilibrium conditions in a high-pressure mass spectrometer by Kebarle and Godbole (1963) for an ion clustering reaction (5) involving the gas-phase solvation of H_3O^+,

$$H_3O^+ + nH_2O \rightleftarrows H_3O^+ \cdot nH_2O. \tag{5}$$

Since then, Kebarle's group has studied a wide variety of such clustering reactions for both positive and negative ions (Kebarle, 1972, 1977). The measurement of accurate equilibrium constants for these reactions provides valuable insight into the energetics of ionic solvation.

Equilibrium constants for proton-transfer reactions were first measured by Bowers, Webb, and Aue using an ion cyclotron resonance drift cell at high pressures (10^{-4}–10^{-3} torr). These reactions were also found, in collaboration with McIver, to reach equilibrium under the low-pressure (10^{-7}–10^{-6} torr) conditions of a trapped ion cell (Bowers *et al.*, 1971). These observations made possible, for the first time, the measurement of quantitative relative acidities or basicities from acid–base reactions of ions in the gas phase. Since 1971, ion cyclotron resonance studies by groups of Aue and Bowers at Santa Barbara, Taft and McIver at Irvine, and Beauchamp at California Institute of Technology have led to an extensive set of such basicity data for negative and positive ions. In addition, proton-transfer reactions have been investigated under equilibrium conditions by Kebarle's group at Alberta using high-pressure mass spectrometry (Briggs *et al.*, 1972) and by Bohme's group at York using flowing afterglow techniques (Bohme *et al.*, 1971, 1973a,b). This chapter will present the results of these studies on proton-transfer reactions of positive ions. Reactions of ions which involve ligand transfer in clusters, halide transfer, hydride transfer, and electron transfer have also been studied in recent years and will be discussed, but in less detail.

II. TECHNIQUES FOR MEASUREMENT OF PROTON AFFINITIES

Measurements of equilibrium constants for proton-transfer reactions (2) give free energies equal to the difference in the gas-phase basicities of the molecules A and B. It has usually been assumed that these free energies differ from enthalpies for proton transfer by a $T\,\Delta S^0$ term approximately equal to that calculable from the symmetry changes in the reaction. Thus, the GBs and PAs differ by the entropy term for a free proton from the Sackur–Tetrode equation (7.75 kcal/mole) and the entropy term due to symmetry changes expressed in Eq. (6).

$$T\,\Delta S^0_{\text{sym}} = RT\ln(\sigma_{\text{AH}^+}/\sigma_A). \tag{6}$$

In a few cases these entropy terms have been determined by measuring equilibrium constants over a wide range of temperatures using a variable-temperature high-pressure mass spectrometer (Yamdagni and Kebarle, 1973) and variable-temperature ion cyclotron resonance drift and trapped ion cells (Wren *et al.*, 1978; Wren, 1976). In such measurements, the entropy terms for proton-transfer reactions have been very small and about equal to those calculated except in cases where intramolecular hydrogen bonds are formed. Generally, gas-phase basicity measurements have been made at only one temperature and the entropy term has been calculated to give reported proton affinity values.

A. *Absolute PAs*

While the measurement of proton-transfer equilibrium constants can give *relative* GBs and PAs, only a few experimentally determined *absolute* PAs are required in order to set up a scale of absolute proton affinity values. Where absolute PAs have been measured for several compounds and these compounds related by a series of proton-transfer equilibrium measurements, the absolute and relative PA data may be compared to give a valuable check on the consistency of these independent measurements. Absolute proton affinity measurements are also very important for molecules not interconnected with others on the relative PA scale by equilibrium proton-transfer reactions.

Absolute PAs have been most often measured by appearance potential measurements to give heats of formation of positive ions derived from photon- or electron-impact-induced molecular fragmentation or radical ionization. On the basis of such measurements for the heats of formation of *t*-butyl cation (Table XII) and ammonium ion (Ceyer *et al.*, 1979), we have assigned the proton affinity scale such that the PA of isobutene is 196.9 kcal/mole and that of ammonia is 205.0 kcal/mole. Where PAs from other laboratories have been reported, we have adjusted them to be consistent with this absolute PA scale. These absolute PA assignments are somewhat arbitrary and agree only within error limits of several kilocalories per mole with absolute proton affinity measurements of formaldehyde and acetaldehyde (Refaey and Chupka, 1968), water (Ng *et al.*, 1977), propene, 1,3-butadiene, and cyclopentadiene (see Table XII). Absolute PAs from appearance potential measurements may suffer from errors in heats of formation of the molecules and radicals involved and lack of true adiabaticity in the threshold measurements. The heats of formation must also be corrected from 0 K to 298 K by estimation of heat capacity changes on ionization and fragmentation. As a result, the absolute assignment of the PA scale is still in doubt and errors up to several kilocalories per mole are possible. Usually this causes little problem in the interpretation of PA data, however, since comparisons of relative PAs are generally made.

B. *Relative PAs from Drift-Cell Ion Cyclotron Resonance Spectrometry*

In ion cyclotron resonance spectrometry, a signal results when the cyclotron frequency of an ion, $\omega = qH/mc$, equals the frequency ω_0 of the marginal oscillator detector. At this point the ions of a particular mass m and charge q are in resonance, and they absorb power from the marginal oscillator when the magnetic field is fixed at H. The power absorbed is proportional to the number of ions and is thus a measure of ion intensity. The relative intensities of ions in an ICR cell can be measured in two ways. The first involves operating the marginal oscillator at constant frequency and varying the magnetic field. The cyclotron frequency of an ion is directly proportional to the magnetic field. As the magnetic field is varied, ions of different masses come into resonance. In the second method, the instrument is operated at a constant magnetic field and the frequency of the marginal oscillator is varied until it matches the cyclotron frequency of an ion.

In the course of carrying out double resonance experiments to determine qualitatively the direction of proton-transfer reactions of some heterocyclic amines, we discovered that equilibrium is reached in these proton-transfer reactions at pressures above (1–2) × 10^{-4} torr (Bowers *et al.*, 1971). When the pressure ratio of A and B is kept fixed but the total pressure increased through the 10^{-5} torr range, the relative intensities of the AH^+ and BH^+ ions are found to approach a constant value as the pressure increases above 1×10^{-4} torr. This behavior was interpreted as an indication that the proton-transfer reaction between A and B had reached equilibrium with an equilibrium constant $K = [BH^+][A]/[AH^+][B]$ for the reaction $AH^+ + B \rightleftharpoons BH^+ + A$. The ratio of the pressures of A and B can be measured directly using a Baratron capacitance manometer. The relative ion intensities can be measured by the cyclotron resonant power absorption at the appropriate marginal oscillator frequencies, as indicated above. A typical experimental trace of the ion intensities is shown for reaction (7) between azetidine (A) and pyrrolidine (B) in Fig. 1.

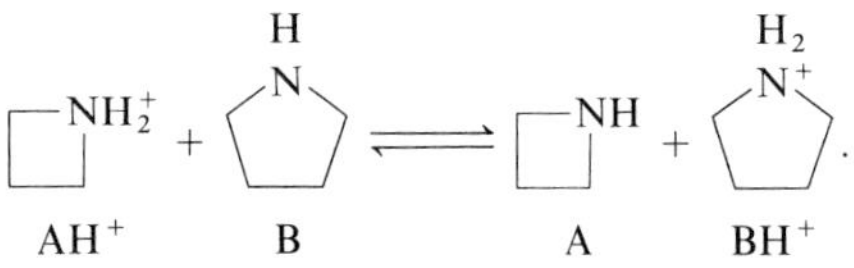

(7)

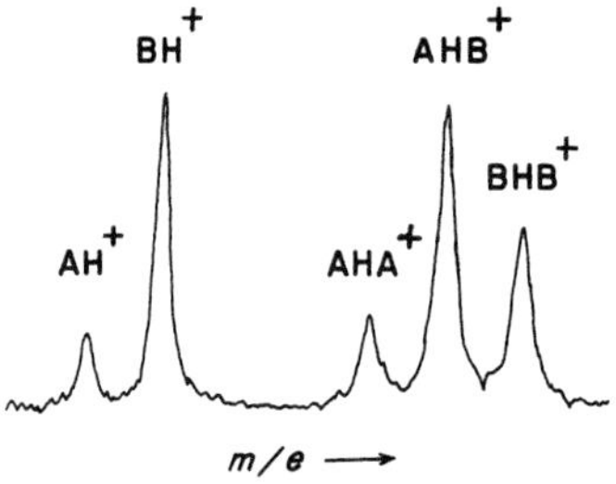

FIG. 1 Ion cyclotron resonance mass spectrum of azetidinium ion (AH^+, $m/e = 58$) and pyrrolidinium ion (BH^+, $m/e = 72$).

The equilibrium constant for this reaction can thus be calculated simply from the peak area ratio and the pressure ratio. The results of a number of such determinations over a wide pressure range are presented in Fig. 2. This figure shows the formation of AH^+ and BH^+ at low pressures by irreversible proton transfer from the radical cations $A^{+\cdot}$ and $B^{+\cdot}$ and the equilibration of AH^+ and BH^+ at high pressures. The drift time in a drift-cell ion cyclotron resonance spectrometer is usually ~2 msec, which permits hundreds of ion–molecule collisions to occur at the highest pressures shown in Fig. 2. The ratio of AH^+ to BH^+ remains essentially constant above 2×10^{-4} torr, although both ions are decreasing in intensity rapidly in this pressure range. At high pressures, the proton-bound dimers at *m/e* 115, 129, and 143 form corresponding to the structures AHA^+, AHB^+, and BHB^+, respectively.

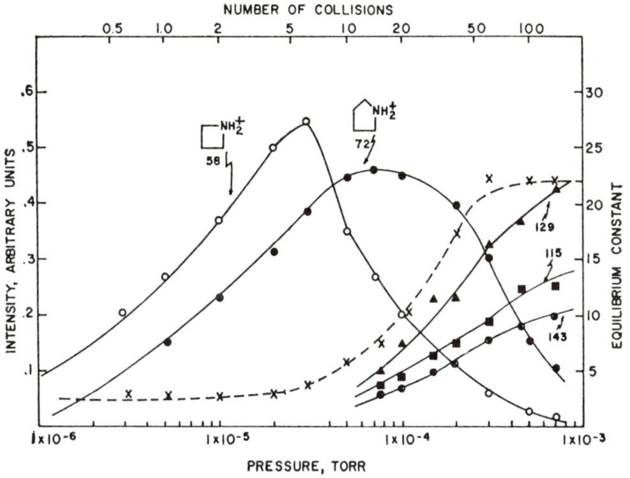

FIG. 2 Relative ion intensities versus pressure of the product ions in a 5.1:1.0 mixture of $(CH_2)_3NH$ and $(CH_2)_4NH$ at 9 eV. The protonated species (m/e = 58, 72) were formed by proton transfer from the parent ions (m/e = 57, 71) to the two neutral molecules. The dashed line denotes $K = [(CH_2)_4NH_2{}^+][(CH_2)_3NH]/[(CH_2)_3NH_2{}^+][(CH_2)_4NH]$. The number of collisions was approximated from the orbiting limit. K was determined directly from the ionic intensities without mass correction. [Reprinted with permission from Bowers *et al.* (1971). Copyright by the American Chemical Society.]

They form according to the detailed kinetic scheme I in a mechanism

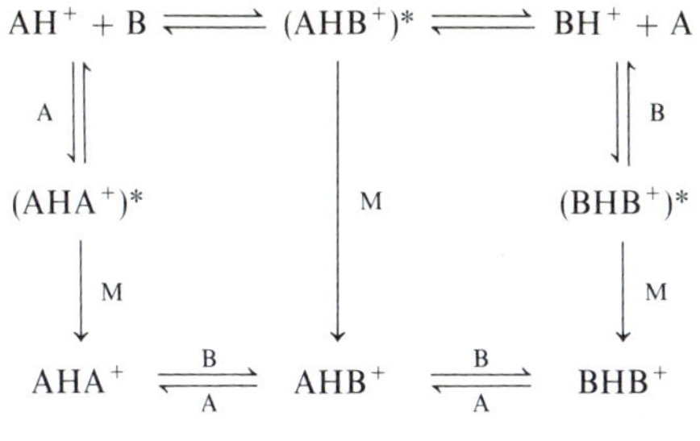

scheme I

involving collisional stabilization of the vibrationally excited species $[AHA^+]^*$, $[AHB^+]^*$, and $[BHB^+]^*$. This kinetic scheme for formation of proton-bound dimers has been confirmed in these laboratories by extensive kinetic studies on such dimerization reactions of amines (Neilson *et al.*, 1978). The intensity ratios of the ions AHA^+, AHB^+, and BHB^+ remain constant over the range of pressures shown, indicating that they too are in equilibrium as a result of base-transfer reactions among them. These reactions will be discussed in detail later.

The marked disappearance of AH^+ and BH^+ to proton-bound dimers presents some potential experimental and kinetic problems to the measurement of accurate equilibrium constants. We have analyzed these problems and discussed them in detail elsewhere (Davidson *et al.*, 1977). Problems caused by measurements at constant oscillator frequency are due to the variation in magnetic field which causes a change in drift times of the ions in the cell. This problem has now been solved by the development of techniques for the calibration of the marginal oscillator sensitivity at different frequencies, making it possible to operate at fixed magnetic field and variable frequency (Kemper and Bowers, 1977a).

A more serious problem is caused by the possibility that proton-bound dimer formation from AH^+ and BH^+ may occur at rates competitive with proton transfer and disrupt the attainment of proton-transfer equilibrium in the kinetic scheme II. This kinetic scheme has been analyzed and the

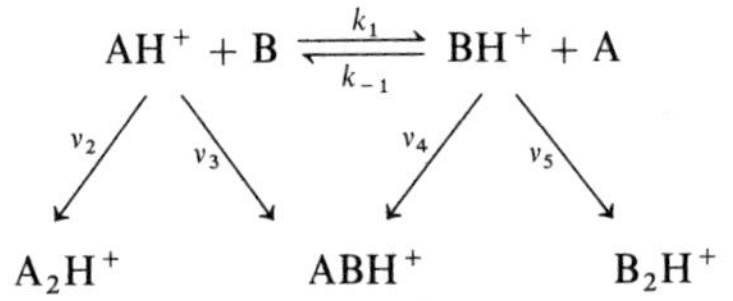

scheme II

differential equations solved exactly. The solution can also be conveniently approximated using a digital numerical integration technique developed at Santa Barbara (Shindell *et al.*, 1978). From the measurement of the dimerization rates v_2 and v_5 in scheme II and from experimental rate constants k_1 and k_{-1} for proton-transfer reactions, we are able to use this kinetic scheme to demonstrate how closely our experimentally measured equilibrium constants approximate the true equilibrium constant k_1/k_{-1}. The errors in the equilibrium measurements are likely to be largest when one of the two ions AH^+ or BH^+ dimerizes much more rapidly than the other and when this dimerization rate becomes competitive with proton transfer. We have shown experimentally that these dimerization reactions do not generally cause errors in ΔG^0 of more than 0.2 kcal/mole for equilibria involving amines. For hydrocarbons and oxygenated compounds, the dimerization rates are sometimes slow enough for rapid proton transfer to permit the measurement of equilibria using drift-cell techniques (Davidson *et al.*, 1977). In other cases, however, such measurements are not possible without substantial errors. In these cases, low-pressure trapped-ion-cell techniques may be often applied, since at lower pressures the proton-bound dimer formation is often slower because of the higher order of pressure dependence frequently seen in the dimerization rates versus proton-transfer rates. When applicable, the drift-cell technique for measurement of equilibrium constants is rapid, convenient, and less susceptible to experimental errors due to noise and pressure measurement than the trapped-ion-cell technique. Where errors from the dimerization kinetics are a potential problem, however, we use the trapped-ion-cell technique to measure equilibrium constants. This cell has an added advantage for molecules where the parent ions $A^{+\cdot}$ or $B^{+\cdot}$ or P-1 ions are unreactive and contribute to the intensity of the broad AH^+ or BH^+ peaks at high pressures in the drift cell or where AH^+ and BH^+ have nearly the same mass. The low-pressure trapped-ion spectra suffer from little collisional line broadening so that unit mass resolution is readily achievable.

C. *Relative PAs from Trapped-Ion-Cell Ion Cyclotron Resonance Spectrometry*

Proton-transfer equilibrium techniques on the McIver trapped ion cell (McIver, 1970) were developed in collaboration with McIver for the same reactions that had been studied by drift-cell techniques, in order to confirm the contention that equilibrium was being achieved in these systems (Bowers *et al.*, 1971). A somewhat different type of trapped ion cell was later developed by McMahon and Beauchamp (1972). In these cells, the ions do not drift out of the cell within a few milliseconds as in the drift cell. Instead, they are trapped within the cell for periods in excess of 1000 msec. Such long ion trapping times permit hundreds of collisions to occur and result in proton-transfer equilibrium at lower pressures ($\sim(1-4) \times 10^{-6}$ torr) than in the drift cell. At higher pressures, however, ion loss from the trapped ion becomes a major problem, so that the total number of ion–molecule collisions attainable is comparable to that in the drift cell. Absolute pressure measurement in the trapped ion cell must be accomplished indirectly using an ion gauge calibrated against a capacitance manometer at higher pressures. An example of an experimental equilibrium constant determination by this technique for the azetidine and pyrrolidine equilibrium (7) studied in the drift cell is shown in Fig. 3. Here the intensities of the azetidinium and pyrrolidinium ions reach a constant ratio after about 30 collisions to give an equilibrium constant in good agreement with the value obtained from drift-cell measurements. Because of the limitations on the total number of collisions attainable by either ICR technique, using current technology, the largest equilibrium constants that can be reliably measured are less than 50, corresponding to ΔG^0 values less than 2.0 kcal/mole at 25°C. Occasionally larger values of K can be measured in very well behaved systems, but these cases are not usual.

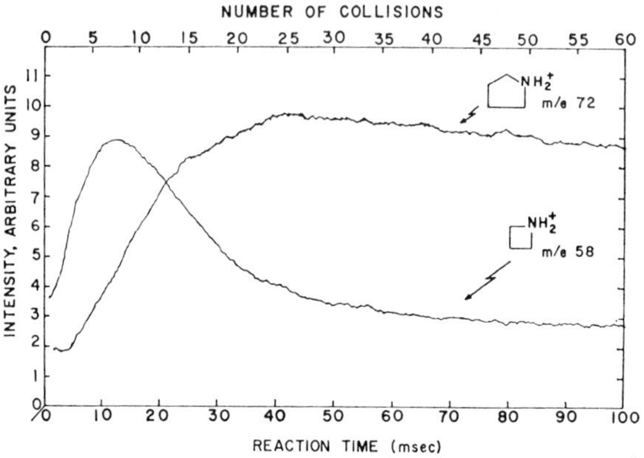

FIG. 3 Pulsed ICR data for the $(CH_2)_3NH_2^+$ and $(CH_2)_4NH_2^+$ peaks in a 6.4:1.0 mixture of $(CH_2)_3NH$ and $(CH_2)_4NH$ at 10 eV and approximately 1×10^{-5} torr. The protonated species ($m/e = 58, 72$) were formed by proton transfer from the parent ions ($m/e = 57, 71$) to the two neutral molecules. Each measured ionic intensity was multiplied by the mass of the ion to obtain K. [Reprinted with permission from Bowers *et al.* (1971). Copyright by the American Chemical Society.]

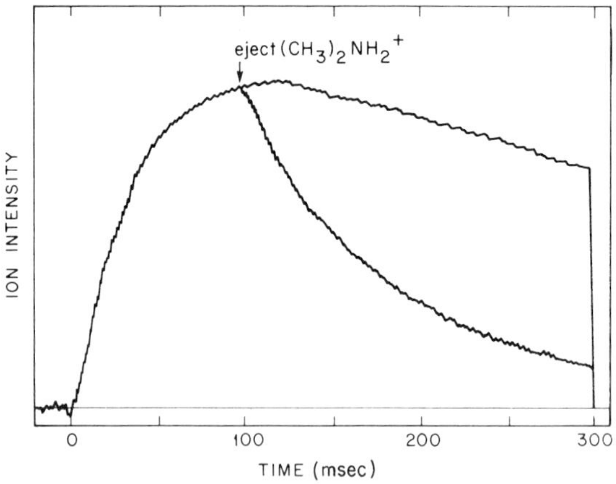

FIG. 4 Plot of ion intensity versus time of protonated pyridine, $C_5H_5NH^+$, in a mixture of 4×10^{-7} torr pyridine and 9×10^{-7} torr $(CH_3)_2NH$. The rate constant was determined from the decrease in pyridinium ion signal when protonated dimethylamine is ejected from the cell. [From Davidson *et al.* (1977).]

An important feature of the trapped ion cell is the possibility of measuring reaction rates for proton transfer by ejection of the ions and observation of the decay of the other ion (McIver and Dunbar, 1971; McMahon *et al.*, 1976). Such an experiment is illustrated in Fig. 4 for the disappearance of the pyridinium ion after ejection of dimethylammonium ion for equilibrium (8) (Davidson *et al.*, 1977).

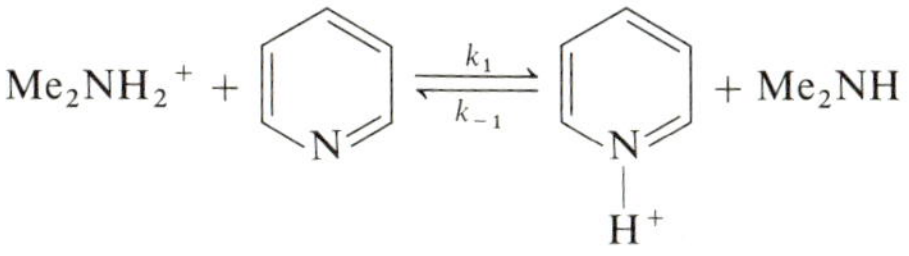

(8)

$$k_1 = 5.3 \times 10^{-10}\ \text{cm}^3\ \text{molecule}^{-1}\ \text{sec}^{-1}, \qquad K = 1.6$$
$$k_{-1} = 3.3 \times 10^{-10}\ \text{cm}^3\ \text{molecule}^{-1}\ \text{sec}^{-1}$$

Assuming that ion loss before ejection is negligible relative to that after ejection, the exponential decay of the pyridinium ion is equal to the pseudo-first-order rate constant k_{-1}. Similarly, the rate constant k_1 can be measured by ejection of the pyridinium ion. The equilibrium constant obtained from the ratio k_1/k_{-1} agrees well with that obtained from the ion intensity ratios of the drift- and trapped-ion-cell experiments.

For many reactions involving oxygenated species and hydrocarbons, the ion loss rates are rapid and competitive with proton-transfer rates. At low pressures in the trapped ion cell, we have found that ion loss processes by various condensation reactions have the same pressure dependence as proton transfer, so that operation at even the lowest pressures does not serve to suppress these reactions relative to proton transfer. In these cases we have attempted to assess the effect of these interfering reactions on the experimental determination of equilibrium constants by a kinetic analysis employing experimental rate constants. Rate constants for ion loss can be approximated from ion loss curves for each pure compound and for the mixture at long reaction times. The ion ejection technique in such cases gives traces, as illustrated in Fig. 5, that do not fit the assumption of slow ion loss relative to

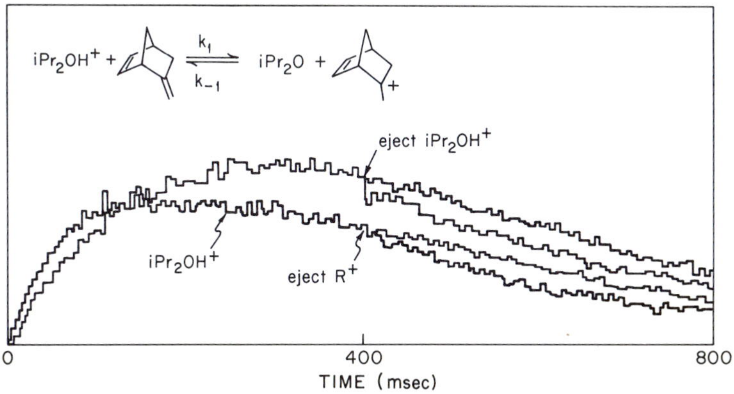

FIG. 5 Trapped ion ICR data showing approximate proton-transfer rate constants (in units of $\text{cm}^3\ \text{molecule}^{-1}\ \text{sec}^{-1}$) derived from continuous ion ejection at 400 msec reaction time. $k_1 = 1.4 \times 10^{-10}$, $k_1 = 1.0 \times 10^{-10}$, $\Delta G^0 = -0.3$, error $= -0.2$ kcal/mole.

the loss after ejection. To get proton-transfer rates from such data, these traces have been fit by an interactive digital kinetic simulator developed by Shindell *et al.* (1978) at Santa Barbara. Figure 5 shows an extreme case where the proton-transfer rates k_1 and k_{-1} are unusually small as indicated by the small effect observed on ejection. A computer fit of these ionic concentration curves can be obtained with the k_1 and k_{-1} values of about 10^{-10} cm^3 molecule^{-1} sec^{-1} and pseudo-first-order ion loss rates of 2–3 sec^{-1}. This fit is not unique and there can be significant errors in the k's derived, but it gives a reasonable estimate of the degree to which such a reaction can approach equilibrium. In this case, the equilibrium constant is nearly one, and the error in ΔG^0 is $\sim$0.2 kcal/mole. With proton-transfer rates this small, however, the measurement of larger equilibrium constants would have large errors. In these cases, double resonance data are sometimes required to ascertain the PAs within $\sim$2 kcal/mole error limits.

Thus, trapped-ion-cell ion cyclotron resonance spectrometer and drift-cell instruments are complementary in their application to thermochemical measurements. The drift-cell instrument is best for rapid determination of equilibrium constants for rapid proton-transfer reactions with moderately slow ion loss pathways. It is ideal for the study of base exchange reactions between proton-bound dimers, which occur at high pressures, and it is useful for semiquantitative double resonance experiments for the determination of relative gas-phase basicities. The drift-cell instrument is also well suited for the determination of reaction rates for many types of reactions (see, e.g., Anicich and Bowers, 1974; Kemper and Bowers, 1977b; Neilson *et al.*, 1978).

The trapped ion cell is very valuable for studying equilibria where the low pressures can slow down rapid third-order dimerization or condensation reactions relative to second-order proton-transfer reactions and where the higher resolution obtainable at low pressures is important. Resonant ion ejection experiments can be used to determine proton-transfer rates most conveniently in many cases and can be used in semiquantitative determinations of GBs when proton-transfer rates become small.

Extensive comparisons of data for the same sets of experiments and compounds demonstrate that the two techniques give closely comparable gas-phase basicities. Agreement is within a few tenths of a kilocalorie per mole for comparisons between drift- and trapped-cell results obtained at Santa Barbara and between results obtained at Santa Barbara by both techniques and those trapped-cell measurements at Irvine and California Institute of Technology (Taft, 1975; Wolf *et al.*, 1977a). Furthermore, all of these results agree well with those obtained by Kebarle by high-pressure mass spectrometric techniques (Briggs *et al.*, 1972; Yamdagni and Kebarle, 1973, 1976; Lau and Kebarle, 1977) and by the flowing afterglow technique (see, e.g., Staley and Beauchamp, 1975).

In the measurement of gas-phase basicities by all of these techniques, the internal consistency of the data can be checked by observing proton-transfer equilibria of each new compound with at least two or three other compounds. Such measurements lead to an extensive set of overlapping thermodynamic measurements which are consistent with each other to better than ±0.2 kcal/mole (Fig. 6). This internal consistency and the consistency of measurements by different techniques and in different laboratories are strong indications that the data obtained are true equilibrium data. That equilibrium has been established in these reactions is most convincingly demonstrated by rate measurements for the the proton-transfer reactions and analysis of the kinetics for equilibration (Davidson *et al.*, 1977).

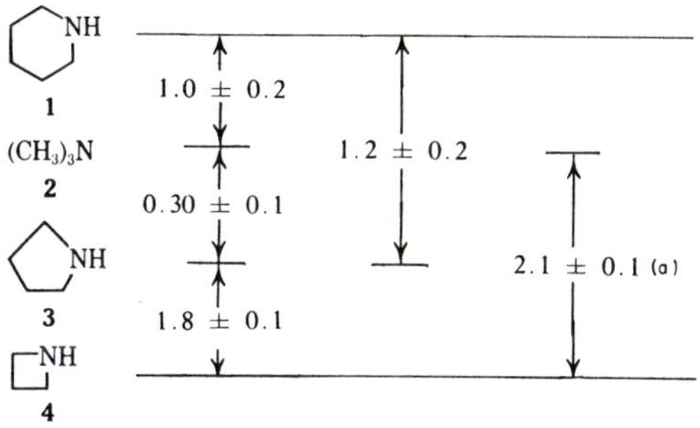

FIG. 6 Changes is free energy, ΔG^{25} (kcal/mole), for proton transfer reactions such as reaction (2). Piperidine (1) has the highest proton affinity in this series. (a) Determined by pulsed ICR only because at high pressure the ions at $m/e = 58$ and 60 were not resolved. [Reprinted with permission from Bowers *et al.* (1971). Copyright by the American Chemical Society.]

D. *Proton Affinity Measurements by High-Pressure Mass Spectrometry*

Clustering equilibria have been studied extensively by high-pressure mass spectrometric techniques by Kebarle (1972, 1977). This technique has more recently been applied to proton-transfer equilibria by Kebarle's group (Briggs *et al.*, 1972) and its application to proton-transfer reactions has been reviewed recently by Kebarle (1977). The method involves generation and reaction of ions in a field-free high-pressure region and diffusion of the ions through a small slit into a low-pressure region where they are accelerated and mass analyzed. A potential problem with this technique is that the need to sample the thermal reaction mixture for mass spectral analysis could lead to artifacts such as collisional decomposition of proton-bound dimers. Such problems can be largely avoided, however, and the high-pressure mass spectrometric technique has the advantage of a large sensitivity range, which makes possible the measurement of larger *K*s than by ion cyclotron resonance techniques. In addition to Kebarle's group, others have also applied high-pressure mass spectrometry to proton-transfer equilibria and other ionic reactions (see, e.g., Chang and Franklin, 1972a,b; Beggs and Field, 1971a,b; Solomon and Field, 1976a,b; Long and Franklin, 1974; Wei and Bone, 1974; Wincel, 1972; Arshadi and Futrell, 1974).

E. Proton Affinity Measurements by Flowing Afterglow Techniques

The flowing afterglow technique involves a fast helium flow system at 0.5 torr in a reaction tube. Ions are generated by electron impact from gases introduced into the helium flow upstream from the filament. The ions generated can react with other neutral molecules introduced through nozzles downstream. Further downstream the ions can be sampled through an orifice which leads to a quadrupole mass filter. This apparatus has been used extensively at Boulder by Ferguson *et al.* (1969), Fehsenfeld and Ferguson (1973), and Fehsenfeld (1975) and at York by Bohme (1975) to measure rate constants. From the rate constants of the forward and reverse reactions of an equilibrium, the equilibrium constant may be obtained, or the equilibrium constant may be measured directly at sufficiently high pressures of reactant gases. Such measurements have been the source of relative proton affinities for molecules of low basicity such as N_2, H_2, CO, CO_2, CH_4, and C_2H_6 (Bohme *et al.*, 1973a,b; Hemsworth *et al.*, 1973). Comparison of the methane–carbon dioxide proton-transfer equilibrium constant by this technique with ion cyclotron resonance measurements shows good agreement (Staley and Beauchamp, 1975).

F. Variable-Temperature Equilibrium Measurements

The temperature dependence of proton-transfer equilibrium constants has been measured by Yamdagni and Kebarle (1973) using a variable-temperature high-pressure mass spectrometer. From such measurements, the entropy change for proton-transfer between organic amines are found to be generally very small. In the case of α,ω-diaminoalkanes, however, the entropy changes are large because of the entropy loss on forming cyclic intramolecularly hydrogen bonded structures for the ammonium ions. Entropy terms for many ionic clustering reactions have been determined by high-pressure mass spectrometry as well (Kebarle, 1972, 1977). Variable-temperature ion cyclotron resonance drift- and trapped-ion-cell instruments have been recently developed (Wren, 1976; Wren *et al.*, 1978) for the measurement of rate and equilibrium constants. Results on α-ω-diaminoalkanes confirm Yamdagni and Kebarle's (1973) conclusions on the entropy terms for these bases.

III. STRUCTURAL EFFECTS ON ACID–BASE EQUILIBRIA

A. *Alkyl Group Effects*

The effect of alkyl groups on the basicity of alkylamines has been investigated extensively (Bowers *et al.*, 1971; Aue *et al.*, 1972, 1976a; Henderson *et al.*, 1972; Briggs *et al.*, 1972; Yamdagni and Kebarle, 1973). These results are summarized in Table I and show that alkyl groups stabilize positively charged ammonium ion sites. This alkyl group effect is also seen in the stabilization of negatively charged sites in alkoxide ions (McIver and Miller, 1974), suggesting that the mechanism for the stabilization involves a charge-induced dipole interaction between the charged site and the polarizable alkyl group (Brauman *et al.*, 1971). The magnitude of the effect decreases markedly as alkyl groups are appended at larger distances, as expected from Eq. (9) for the interaction energy E, where r is the distance of separation, ϵ the dielectric constant, α the group polarizability, and q the charge.

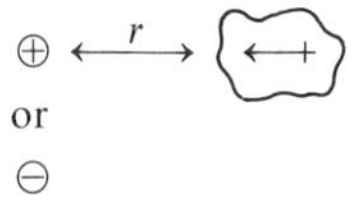

$$E = \frac{-\alpha q^2}{2\epsilon r^4}. \tag{9}$$

The PA data for the amines, phosphines, and arsines in Table I and the alcohols, ethers, and sulfides in Table II can be understood reasonably well in terms of such a model. Curiously, the long-chain amines, however, do not show the abrupt leveling of PA values with chain length that would be expected for a $1/r^4$ dependence of the interaction energy. Instead, there seems to be a small but continuous increase in PA from *n*-propyl- to *n*-decylamine. This appears to result from a coiling of the chain to bring the polarizable alkyl chain into close proximity to the charged site (Aue *et al.*, to be published). From vertical and adiabatic ionization potentials measured by photoelectron spectroscopy and from PAs, one can determine hydrogen atom affinities of these bases from Eq. (10) as shown in Table I and the following tables of data. Usually proton affinity changes and ionization potential changes approximately parallel each other within a related series of compounds,

$$\mathrm{HA(B^{+}\cdot) = IP(B:) + PA(B:) - IP(H\cdot)}, \tag{10}$$

(Aue *et al.*, 1976a) with the result that the HA values remain roughly constant within the series.

TABLE I

Proton Affinities, Adiabatic and Vertical Ionization Potentials, and Hydrogen Atom Affinities of Aliphatic Amines and Phosphines at 25°C[a]

Amine	GB[b]	PA[c]	aIP	aHA[d]	vIP	vHA[d]
NH_3	196.4	205.0	234.3[e]	124.5	250.2	140.1
$MeNH_2$	205.7	214.1	205	104	222.8	121.8
$EtNH_2$	208.7	217.1	202	104	217.4	119.4
n-$PrNH_2$	210.1	218.5	197	100	215.6	119.0
i-$PrNH_2$	211.0	219.4	199	103	214.9	119.2
n-$BuNH_2$	210.6	219.0	201[f]	105	214.9	118.8
i-$BuNH_2$	211.1	219.5	196	100	214.5	118.9
s-$BuNH_2$	212.1	220.5	195	100	214.5	119.9
t-$BuNH_2$	212.9	221.3	195	101	213.4	119.6
n-amyl NH_2	211.2[f]	219.6	200[f]	105	214.5[f]	119.0
Neopent NH_2	211.8	220.2	197	102	213.3	118.4
t-amyl NH_2	213.9	222.3	195	102	212.2	119.4
n-hexyl NH_2	211.7[f]	220.1	199[f]	104	214.1[f]	119.1
Cyclohexyl NH_2	212.9[f]	221.3	193[f]	99	210.5[f]	116.7
n-octyl NH_2	212.0[f]	220.4	197[f]	102	213.1[f]	118.4
n-decyl NH_2	212.3[f]	220.7	197[f]	103	212.8[f]	118.4
exo-2-aminonorbornane	213.3[f]	221.7	194[f]	101	209.5[f]	116.1
endo-2-aminonorbornane	213.3[f]	221.7	192[f]	99	209.9[f]	116.5
Me_2NH	212.3	220.5	188	93	205.9	111.3
MeEtNH	214.8[f]	222.6	188[f]	96	201.5[f]	109.0
Et_2NH	216.9	225.1	181	91	199.0	109.0
n-Pr_2NH	219.2	227.4	179	91	196.9	109.2
i-Pr_2NH	220.7	228.9	175	89	193.7	107.5
n-Bu_2NH	220.2	228.4	179	92	196.2	109.5
i-Bu_2NH	220.4	228.6	180	94	195.1	108.6
s-Bu_2NH	222.7	230.9	176	92	191.2	107.0
t-amyl-*t*-BuNH	224.7	232.5	180	97	185.6	103.0
Me_3N	216.5	224.3	179	88	196.7	105.9
Et_3N	223.4	231.2	169	85	185.2	101.3
n-Pr_3N	225.6	233.4	162	80	182.6	100.9
n-Bu_3N	227.0	234.8	161	81	182.2	101.9
Me_2NEt	218.8	226.6	176	88	193.4	104.9
$MeNEt_2$	221.1	228.9	173	87	189.3	103.1
Et_2N-*n*-Pr	224.2[f]	232.0	177[f]	94	192.1[f]	109.0
Pyrrolidine	216.1	224.3	185	94	201.8	111.0
Piperidine	217.2	225.4	181	91	199.7	110.0
N-methylpyrrolidine	220.0	227.8	—	—	193.9[g]	106.6

TABLE I (*Continued*)

Amine	GB[b]	PA[c]	aIP	aHA[d]	vIP	vHA[d]
N-methylpiperidine	221.0	228.8	—	—	191.2[g]	104.9
2,2,6,6-tetramethylpiperidine	223.9[f]	232.1	175	92	187.7[f]	104.7
Quinuclidine	223.6	231.4	173	89	185.6	101.9
$(CH_3)_3Si(CH_2)_3N(CH_3)_2$	223.3[h]	231.1				
$(CH_3)_3Si(CH_2)_2N(CH_3)_2$	223.3[h]	231.1				
$(CH_3)_3C(CH_2)_2N(CH_3)_2$	222.0[h]	229.8				
$(CH_3)_3SiCH_2N(CH_3)_2$	222.8[h]	230.6				
$(CH_3)_3CCH_2N(CH_3)_2$	221.5[h]	229.3				
$(CH_3)_3CSi(CH_3)_2N(CH_3)_2$	221.3[h]	229.1				
$(CH_3)_3CC(CH_3)_2N(CH_3)_2$	226.4[h]	234.2				
PH_3	182.5[i]	191.1	229.6	105.6	244.4[j]	120.4
$MePH_2$	195.4[k]	204.8	210.2[k]	99.9	—	—
Me_2PH	208.9[k]	217.1	195.3[k]	97.3	—	—
Me_3P	217.9[k]	225.7	184.7[k]	95.3	—	—
Et_3P	223.9[i]	231.7	175.5[i]	92.1	190.2[i]	106.8
AsH_3	175.0[l]	183.6	228.1	96.6	242.4[j]	110.9
$AsMe_3$	205.9[m]	213.7	182[m]	81	—	—

[a] All values in kilocalories per mole from revision of Aue *et al.* (1976a,b), unless otherwise indicated.

[b] GBs measured relative to ammonia (196.4 kcal/mole) and methylamine (205.7 kcal/mole). Numerous other GBs in this table generally agree to 0.6 kcal/mole with values from Taft (1975) and Yamdagni and Kebarle (1973).

[c] PAs calculated from GBs by correction for symmetry entropy changes on protonation (see text).

[d] Calculated for $HA(B^{+\cdot}) = IP(B:) + PA(B:) - IP(H\cdot)$ by using $IP(H\cdot)$ 315.1 kcal/mole from JANAF Thermochemical Tables (1971).

[e] See Aue *et al.* (1975a).

[f] Values from Aue, D. H., Betowski, L. D., and Bowers, M. T. (to be published).

[g] See Yoshukawa *et al.* (1974).

[h] Shea *et al.* (1978).

[i] Aue, D. H., Vidal, M. C., Davidson, W. R., and Bowers, M. T. (to be published).

[j] Potts and Price (1972).

[k] Staley and Beauchamp (1974).

[l] Wolf *et al.* (1977a).

[m] Hodges and Beauchamp (1975).

B. *Diamines and Intramolecular Hydrogen Bonding*

The diamines in Table III show much higher gas-phase basicities than would be expected from the GBs of alkylamines of comparable polarizability. These high GB values are the result of unusual stabilization of the ammonium

TABLE II

Gas-Phase Basicities, Proton Affinities, Vertical Ionization Potentials, and Vertical Hydrogen Atom Affinities of Oxides and Sulfides at 25°C[a]

Base	GB[b]	PA[c]	vIP[d]	vHA
H_2O	165.0[e]	173.0	291.0[f]	148.9
MeOH	177.1[e]	184.9	252.7[f]	122.5
EtOH	182.5	190.3	244.9[f]	120.1
n-PrOH	183.6	191.4	242.1[f]	118.4
t-BuOH	187	195	236.4[f]	116
Me_2O	185.8	193.1	232.9	110.9
MeOEt	189.2	202.0	226.2	108.1
Et_2O	193.1	200.4	222.8	108.1
i-PrOEt	195.6[e]	203.4	—	—
n-Pr_2O	195.6	202.9	219.8	107.6
i-Pr_2O	198.7	206.0	215.0	105.9
t-BuOMe	196.0	203.8	218.7	107.4
n-Bu_2O	196.6	203.9	217.8	106.6
i-PrO-*t*-Bu	201.0	208.8	212.1	105.8
n-$pentyl_2O$	197.9	205.2	—	—
Tetrahydrofuran	192.3	199.6	220.7[g]	105.2
Tetrahydropyran	193.4[e]	200.7	219.1[h]	104.7
H_2S	168.6[e]	176.6	241.7[i]	103.2
MeSH	180.8[e]	188.6	217.7[j]	91.2
Me_2S	193.4	200.7	200.2	85.8
MeSEt	195.8	203.6	196.9	85.4
Et_2S	198.3	205.6	194.2	84.7
i-Pr_2S	202.0	209.3	190.5[k]	84.7
H_2Se	169.4	177.4	227.8[j]	90.1

[a] All values in kilocalories per mole and from Aue *et al.* (1975a) and Aue, D. H., Davidson, W. R., Betowski, L. D., Vidal, M. C., and Bowers, M. T. (to be published). Unless otherwise indicated.

[b] GBs relative to NH_3 (196.4 kcal/mole).

[c] PAs calculated from GBs by correction for symmetry entropy changes in protonation (see text).

[d] Vertical IPs are measured at the top of the first sharp band; adiabatic values are 6–7 kcal/mole smaller for ethers and 2–4 kcal/mole for sulfides.

[e] GBs from Wolf *et al.* (1977a). Numerous GBs here have also been reported by Wolf *et al.* (1977a) and by Yamdagni and Kebarle (1976), with agreement generally to ±1.0 kcal/mole or better.

[f] Robin and Kuebler (1972).

[g] Bünzli *et al.* (1973).

[h] Sweigart and Turner (1972).

[i] Turner *et al.* (1970).

[j] Watanabe *et al.* (1962).

[k] Bock and Wagner (1972).

TABLE III

Gas-Phase Basicities and Proton Affinities of α,ω-Disubstituted Alkanes at 25°C[a]

	GB[b]	$-T\Delta S^0_{prot,g}$[c]	PA	ΔPA[d]
$NH_2CH_2CH_2NH_2$	215.5	11.0	226.5	8.0
$NH_2CH_2CH_2CH_2NH_2$	220.9	11.8	232.7	13.7
$NH_2CH_2CH_2CH_2CH_2NH_2$	223.9	14.2	238.1	18.5
$NH_2(CH_2)_5NH_2$	221.9	13.9	235.8	15.7
$NH_2(CH_2)_6NH_2$	221.9	14.7	236.6	16.3
$Me_2NCH_2CH_2NMe_2$	227.2	11.0	238.2	9
$Me_2NCH_2CH_2CH_2NH_2$	228.0	11.8	239.8	11
$Me_2NCH_2CH_2CH_2NMe_2$	230.2	11.8	242.0	13
$Me_2N(CH_2)_4NMe_2$	231.7	14.2	245.9	17
$Me_2N(CH_2)_6NMe_2$	229.3	19.7	249.0	15
$NH_2CH_2CH_2OMe$	212.3	11.0	223.3	4.3
$NH_2CH_2CH_2CH_2OH$	216.0	11.8	227.8	8.8
$NH_2(CH_2)_6OH$	216	14.7	231	10
$NH_2CH_2CH_2CH_2F$	210.3	11.8	222.1	3.6
$NH_2CH_2CH_2CH_2Cl$	210	11.8	222	4
$HOCH_2CH_2CH_2CH_2OH$	198	14.2	212	19.2
2,4-pentanedione	200.1	11.8	211.9	11.9

[a] All values in kilocalories per mole.

[b] GBs revised from Aue *et al.* (1973) and from Aue, D. H., Davidson, W. R., and Bowers, M. T. (to be published). See Yamdagni and Kebarle (1975).

[c] $T\Delta S^0_{prot,g}$ terms are estimated from corresponding alkane ⇄ cycloalkane equilibria from Stull *et al.* (1969). See Yamdagni and Kebarle (1973) for experimental values of $\Delta S^0_{prot,g}$.

[d] ΔPA values are the difference in PA between the α,ω-disubstituted alkane and monosubstituted alkanes of comparable polarizability.

ions by the formation of strong intramolecular hydrogen bonds (Aue *et al.*, 1973). Measurements of equilibrium constants for diamines at various temperatures confirm that ring formation occurs, since the large negative entropies expected for proton transfer with ring formation are, in fact, observed (Yamdagni and Kebarle, 1973). Comparisons of PAs of diamines with appropriate models give ΔPA values in Table III that serve as an estimate of the intramolecular hydrogen bond strengths in the molecules. Tertiary diamines show hydrogen bond strengths comparable to primary diamines. Amino alcohols show somewhat smaller hydrogen bond strengths and diols and diketones appear capable of strong intramolecular hydrogen bonding. Even weakly basic chloro and fluoro substituents seem to be capable of hydrogen bonding to some extent.

C. "*Inductive*" *Effects*

Halogenated amines and alcohols show lower proton affinities than would be expected for compounds of comparable polarizability (Staley *et al.*, 1977; Aue *et al.*, 1975a). The differences in PA and ΔPA, shown in Table IV, for these compounds appear to reflect electron-withdrawing inductive effects of the halogen substituent. Such effects must be interpreted with care, however, because of the possibility of intramolecular hydrogen

TABLE IV

Gas-Phase Basicities, Proton Affinities, and Vertical Adiabatic Ionization Potentials and Hydrogen Atom Affinities of Substituted Alkylamines and Alcohols at 25°C[a]

	GB	PA[b]	aIP	aHA	vIP	vHA	ΔPA[c]
$CF_2FCH_2NH_2$	204.6[d]	213.0	210[d]	108	227.4[d]	125.3	3.8
$CHF_2CH_2NH_2$	200.0[d,e]	208.4	216[d,e]	109	234.1[d,e]	127.4	8.4
$CF_3CH_2NH_2$	194.9[d,e]	203.3	224[d,e]	112	241.2[d,e]	129.4	13.5
$(CF_3)_3CNH_2$	183.1[e]	191.5	240	116	256.2[d,e]	132.6	29.9
$CF_3CH_2NHCH_3$	202.4[d,e]	210.6	—	—	—	—	12.0
$CF_3CH_2CH_2NH_2$	203.4[d,e]	211.8	215	112	223.7[d]	120.4	6.7
$CF_3CH_2CH_2CH_2NH_2$	206.9[d,e]	215.3	209	109	226.0[d]	126.2	3.7
$FCH_2CH_2CH_2NH_2$	210.3[e]	218.7	—	—	—	—	−0.2
$CF_3CH_2NMe_2$	207.1[d]	214.9	190	90	207.1[d]	106.9	11.7
CF_3NMe_2	187[c]	195	213	93	230.4[d]	110	29.3
CHF_2CH_2OH	172.8[f]	180.6	—	—	—	—	9.7
CF_3CH_2OH	172.1[f]	174.9	—	—	270.7[g]	130.5	15.4
CCl_3CH_2OH	173.9[f]	181.7	—	—	—	—	8.6
Piperazine	216[e]	224					1
1,4-dioxane	187.2[d,e]	194.1	—	—	217.5[h]	96.5	6.6
					245.6[h]	124.6	
Morpholine	211[e]	219	—	—	—	—	5
1,4-diazabicyclo-[2.2.2]octane	220[e,i]	228	—	—	173.4	86	3
					222.5	135	

[a] All values in kilocalories per mole.

[b] PAs calculated from GBs by using symmetry entropy corrections (see text).

[c] Difference in PA relative to an unsubstituted alkyl derivative.

[d] Staley *et al.* (1977).

[e] Aue *et al.* (1975a) and Aue, D. H., Davidson, W. R., Betowski, R. D., and Scherer, K. V. to be published).

[f] Wolf *et al.* (1977a).

[g] Robin and Kuebler (1972).

[h] Sweigart and D. W. Turner (1972).

[i] Staley and Beauchamp (1974).

bonding to the halogen in such systems as seen in the preceding section. The observed electron-withdrawing effects may be interpreted as the result of either a through-bond inductive effect or a through-space field effect. An unambiguous distinction between these two interaction mechanisms is not possible yet.

In eliminating the possibility of intramolecular hydrogen bonding and in beginning to probe the question of inductive versus field effects, a series of geometrically rigid substituted quinuclidines and other cyclic amines show considerable promise. The ΔPA value in Table V for 3,3-difluoroquinuclidine, for example, shows a larger electron-withdrawing effect (9.6 kcal/mole) than for 2,2-difluoroethylamine (8.4 kcal/mole, Table IV). This suggests that the hydrogen bonding or through-space field effect operates to produce a smaller destabilization energy for the freely rotating 2,2-difluoroethylammonium ion **1** than for the quinuclidinium ion **2**.

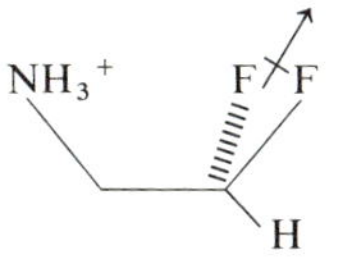

1

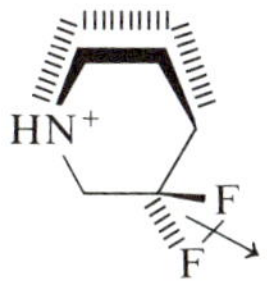

2

Interestingly, the amino- and methoxyquinuclidines show no electron-withdrawing effect, apparently as a result of the availability of conformations, **3** and **4**, of the substituent dipoles that lead to little or no net through-space destabilization of the ion

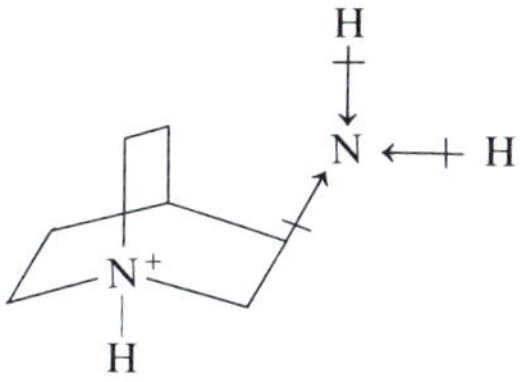

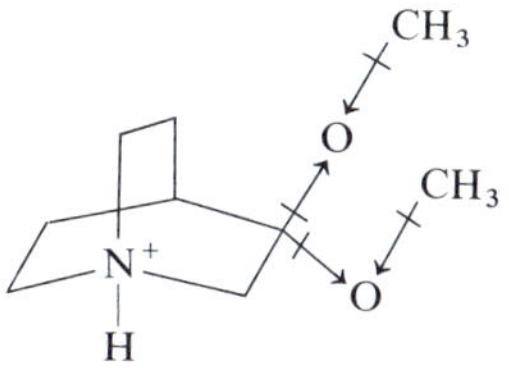

4

These effects and others in Table V are readily modeled by simple calculation of through-space field effects from

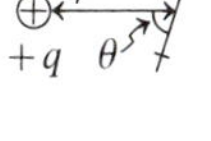

$$E = \frac{q\mu \cos\theta}{\epsilon r^2}, \tag{11}$$

TABLE V

Gas-Phase Basicities, Proton Affinities, and Vertical Ionization Potentials and Hydrogen Atom Affinities of Substituted Quinuclidines at 25° C[a]

	GB	PA[b]	vIP	vHA	ΔPA[c]
Quinuclidine	223.6	231.4	185.6	101.9	0
[x]3-methyl	223.9	231.7	—	—	+0.3
3-amino	224.0	231.8	183.6	100.3	+0.1
3-hydroxy	—		187.5		
3-chloro	217.8	225.6	202.9	113.4	−6.1
3-bromo	219.3	227.1	196.0	108.0	−4.6
3,3-difluoro	214.0	221.8	—	—	−9.6
3-keto	214.1	221.9	189.1	95.9	−9.8
3,3-dimethoxy	224	232	182.9	100	0
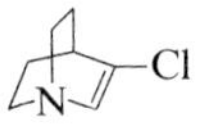	216.2	224.0	—	—	−7.7
	220.7	228.5	—	—	−2.9
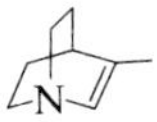	223.2	231.0	—	—	−0.7
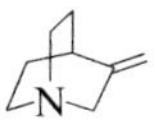	222.3	230.1	—	—	−1.6
	220.3	228.1	—	—	−3.3
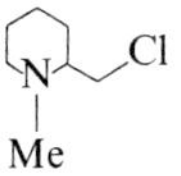	219.8	227.6	—	—	−3.4

[a] All values in kilocalories per mole. Data from Aue, D. H., Webb, H. M., Davidson, W. R., Pearlman, L., Betowski, L. D., and Bowers, M. T. (to be published).

[b] PAs calculated from GBs by using symmetry entropy correction (see text).

[c] ΔPA values are the difference in PA between the quinuclidine and an unsubstituted derivative of comparable polarizability.

TABLE VI

Gas-Phase Basicities, Proton Affinities, and Vertical and Adiabatic Ionization Potentials and Hydrogen Atom Affinities of Unsaturated Amines and Anilines at 25°C[a]

	GB	PA[b]	aIP	aHA	vIP	vHA
$CH_2{=}CH{-}CH_2NH_2$	208.1[c]	216.5	204	105	220.9	122.3
Cyclo-$C_3H_5NH_2$	206.6[c]	215.0	204	104	217.7	117.6
$CH_2{=}C(CH_3)CH_2NH_2$	209.6[c]	218.2	202	105	218.6	121.7
$HC{\equiv}C{-}CH_2NH_2$	203.1[d]	211.5	—	—	—	—
$(CH_2{=}CH{-}CH_2)_3N$	221.5[c]	229.3	176	90	192.6	106.8
$(CH_2{=}CCH_3{-}CH_2)_3N$	222.9[c]	230.7	179	95	193.2	108.8
$(HC{\equiv}C{-}CH_2)_3N$	211.4[d]	219.2	—	—	—	—
$C_6H_5NH_2$	203.1[e]	211.5	—	—	—	—
m-$CH_3C_6H_4NH_2$	205.9[f]	214.3	—	—	—	—
p-$CH_3C_6H_4NH_2$	205.9[f]	214.3	—	—	—	—
m-$CH_3OC_6H_4NH_2$	208.3[f]	216.7	—	—	—	—
p-$CH_3OC_6H_4NH_2$	216.6[f]	215.0	—	—	—	—
p-$ClC_6H_4NH_2$	201.2[f]	209.6	—	—	—	—
m-$ClC_6H_4NH_2$	200.2[f]	208.6	—	—	—	—
m-$FC_6H_4NH_2$	199.8[f]	208.2	—	—	—	—
C_6H_5NHMe	210.0[f]	218.2	—	—	—	—
$C_6H_5NMe_2$	216.0[f]	223.8	—	—	—	—
$C_6H_5CH_2NH_2$	211.3[c]	219.7	204	109	218.1	122.7
$C_6H_5CH_2NMe_2$	220.3[c]	228.1	181	94	195.3	108.3
	230.9[c]	238.7	163	87	152.2	75.8
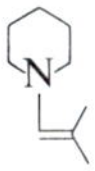	222.9[c]	230.7				
	220.7[c]	228.5				
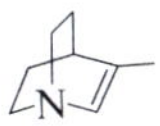	223.2[c]	231.0				
	201.6[c]	208.9	—	—	189.6[g]	83.4

[a] All values in kilocalories per mole.
[b] PAs calculated from GBs by using symmetry entropy corrections (see text).
[c] Aue, D. H., Vidal, M. C., Davidson, W. R., and Bowers, M. T. (to be published).
[d] Taft (1975).
[e] Briggs *et al.* (1972), Taft *et al.* (1973).
[f] Summerhays *et al.* (1977).
[g] Baker *et al.* (1970); Derrick *et al.* (1971); Baker and Turner (1970).

where q is the charge, μ the dipole moment, r the distance of separation, θ the angle of the dipole, and ϵ the dielectric constant. The data in Tables IV and V are consistent with such a model (Aue *et al.*, to be published).

D. *Unsaturated Amines and Ionic Structures*

Allyl and benzylamines (Table VI) show lower PAs than saturated analogs as a result of small electron-withdrawing effects of the electronegative sp^2 hybridized carbon. The unsaturated quinuclidinium ion **5** shows a destabilizing effect of the sp^2 carbon (2.9 kcal/mole) when that carbon is directly attached to the charged nitrogen. This is a larger effect than when the sp^2 carbon is further removed from the nitrogen in allylamine (2.0 kcal/mole lower PA than n-$PrNH_2$), but propargylamine shows a much larger electron-withdrawing effect for an sp hybridized carbon. Its PA is fully 7 kcal/mole lower than that of n-propylamine.

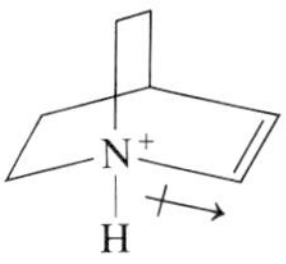

5

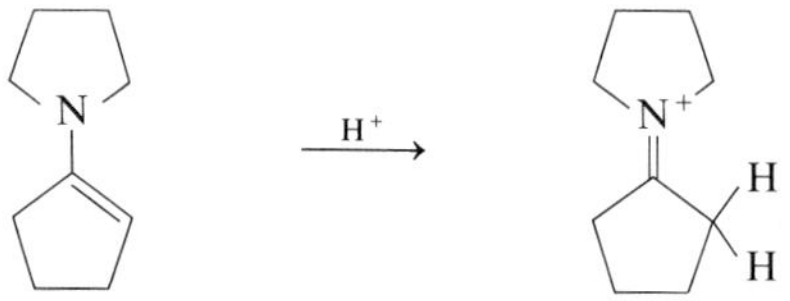

6 **7**

The enamine **6** has a larger PA than would be expected for protonation on nitrogen from the model **5**. Thus, **6** appears to protonate on carbon. Aniline might also protonate on carbon, but correlations with theory and substituent effects indicate that it prefers protonation on nitrogen by 1–3 kcal/mole (Pollack *et al.*, 1977b). Some substituted anilines seem, however, to protonate on carbon (Summerhays *et al.*, 1977). Labeling studies indicate that phenol prefers carbon protonation (PA = 194 kcal/mole) over oxygen protonation by about 15 kcal/mole (DeFrees *et al.*, 1977). Ionic structures in equilibrium studies must generally be assigned from rather indirect evidence at this stage. Further development of such spectroscopic techniques as photodissociation spectroscopy (Dunbar *et al.*, 1977; Freiser and Beauchamp, 1976) and collisional activation mass spectrometry (Dymerski and McLafferty, 1976) should provide much needed structural information for ions in the future.

E. *Hybridization Effects*

Proton affinities and ionization potentials of nitrogen and oxygen bases show large changes as a result of hybridization changes at N and O (Aue *et al.*, 1975a). An increase in the lone pair s character results in substantial decreases in PA and larger increases in the lone pair ionization potentials (Table VII). These effects at sulfur are similar to those at oxygen, but phosphirane's proton affinity is anomalously much lower (by 22 kcal/mole) than that of dimethylphosphine.

TABLE VII

Gas-Phase Basicities, Proton Affinities, and Adiabatic and Vertical Ionization Potentials and Hydrogen Atom Affinities of Compounds of N, O, P, and S with Various Hybridizations at the Heteroatom at 25° C[a]

	GB[b]	PA[c]	ΔPA[d]	aIP	aHA	vIP	ΔIP[d]	vHA	%s[e]
Aziridine	207.5	215.7	4.8	214	115	226.7	20.8	127.3	30
Me_2NH	212.3	220.5	—	188	93	205.9	—	111.3	25
N—methylaziridine	213.7	221.5	2.8	200	106	213.5	16.8	119.9	30
Me_3N	216.5	224.3	—	179	88	196.7	—	105.9	25
1-azabicyclo—[1.1.0]butane	204	212	12	—	—	225	18	122	34
Azetidine	214.5	222.7	−0.1	191	99	206.6	5.1	114.2	26
MeNHEt	214.8	222.6	—	188	96	201.5	—	109.0	25
2-methylaziridine	211.4	219.2	3.4	208	112	221.8	20.3	125.9	30
Pyridine	212.6	220.4	5.0	—	—	223.7[f]	24.0	129.0	33
Piperidine	217.2	225.4	—	181	91	199.7	—	110.0	25
MeCH=NEt	215.0	222.8	2.3	204	112	217.7	18.7	125.4	33
n-PrCH=NEt	217.5	225.3	0	201	111	214.5	15	124.7	33
Me_2C=NEt	221.7	229.5	−10	192	106	208.2	12	122.6	33
Manxine	223[g]	231	2	160[g]	76	161.7[g]	−29.9	—	0
HCN	171.1[h]	178.9	26.8	—	—	322.8(n)[i]	100.0	186.6	50
CH_3NH_2	205.7	214.1	—	205	104	222.8	—	121.8	25
MeCN	183.1	190.9	25.9	—	—	303.0(n)[j,g,k]	85.6	178.8	50
EtCN	185.0	192.8	25.6	—	—	296.3(n)[j,g]	80.7	174.0	50
n-PrCN	186.0	193.8	25.2	—	—	—	—	—	50
CH_2=CHCN	184.0[h]	191.8	24.6	—	—	300.7(n)[j]	—	177.4	50
ClCN	173.7[h]	181.5	—	—	—	318.2(n)[j,l]	—	184.6	50
BrCN	176.4[h]	184.2	—	—	—	312.7(n)[j,l]	—	181.8	50
CCl_3CN	173.9[h]	181.7	—	—	—	320.1(n)[j]	—	186.7	50
CH_2ClCN	177.1[h]	184.9	—	—	—	313.4(n)[j,h]	—	183.2	50
CH_2ClCH_2CN	182.3[h]	190.1	—	—	—	305.3(n)[h]	—	180.3	50
$NCCH_2CN$	173.2[h]	181.0	—	—	—	309.2(n_1)[k]	—	175.1	50
						313.4(n_2)[k]	—	179.3	—
i-PrCN	188.3[h]	196.1	23.4	—	—	—	—	—	50
n-BuCN	188.0[h]	195.8	23.8	—	—	—	—	—	50
c-C_3H_6CN	189.3[h]	197.1	17.9	—	—	—	—	—	50

TABLE VII (*Continued*)

	GB[b]	PA[c]	ΔPA[d]	aIP	aHA	vIP	ΔIP[d]	vHA	%s[e]
Oxirane	182.3	189.6	3.5	240	115	243.5	10.6	118.0	30
Me_2O	185.8	193.1	—	226	104	232.9	—	110.9	25
Oxetane	189.6	196.9	0.1	221	103	223.0	3.2	104.8	26
MeOEt	189.2	197.0	—	220	102	226.2	—	108.1	25
2-methyloxirane	186.9	194.7	2.3	232	112	235.2	9.0	114.8	30
CH_2O	169.9[m]	177.2	7.7	—	—	250.9[i]	−1.8	113.0	33
MeCHO	181.1	188.9	1.4	—	—	235.4[n]	−9.5	109.2	33
$Me_2C{=}O$	189.9	197.2	—	—	—	223.9[n]	—	106.0	33
Thiirane	188.5	195.8	4.9	206	87	208.2	8.0	88.9	30
Me_2S	193.4	200.7	—	198	84	200.2	—	85.8	29
Thietane	194.0	201.3	2.3	196	82	198.8	1.9	85.0	26
MeSEt	195.8	203.6	—	195	84	196.9	—	85.4	29
2-methylthiirane	192.8	200.6	3.0	201	87	203.9	7.0	89.4	30
Phosphirane	187.6	194.8	22.3	216	96	224.8	21	104.5	30
Me_2PH	208.9[o]	217.1	—	195[o]	97	—		—	

[a] All values in kilocalories per mole.
[b] GBs are on revised scale from Aue *et al.* (1975a) or Aue, D. H., Webb, H. M., Davidson, W. R., Betowski, L. D., Vidal, M. C., and Bowers, M. T. (to be published), unless otherwise noted.
[c] PAs calculated from GBs by using symmetry entropy corrections (see text).
[d] Difference relative to a model saturated derivative of comparable polarizability.
[e] Percent s character in lone pair hybrid, see Aue *et al.* (1975a).
[f] Batich *et al.* (1973).
[g] Aue *et al.* (1975b).
[h] Staley *et al.* (1976).
[i] Turner *et al.* (1970).
[j] Lake and Thompson (1970).
[k] Stafast and Bock (1973).
[l] Heilbronner *et al.* (1970).
[m] Wolf *et al.* (1977a).
[n] Tam *et al.* (1974).
[o] Staley and Beauchamp (1974).

Proton affinities of carbonyl compounds are summarized in Table VIII along with those of some imino ethers and hydrazines. Aldehydes, ketones, and imines appear to have low proton affinities as a result of their sp^2 hybridization, but resonance stabilization effects partially obscure these hybridization effects. Acids, esters, and amides show PAs that result from competing resonance stabilization effects and inductive and hybridization effects.

F. *Pyridines*

A large number of substituted pyridines have been studied in gas-phase proton-transfer equilibria. After correction for polarization effects, the inductive and resonance effects ($I + R$ in Table IX) of a variety of substituents can be evaluated. Proton affinities and ionization potentials for some substituted pyridines and fused bicyclic pyridines are summarized in Table X.

G. *Proton Affinity Scale below Water*

The proton affinities of compounds below water on the proton affinity scale have not been linked together in a contiguous set of equilibrium

TABLE VIII

Proton Affinities and Vertical Ionization Potentials and Hydrogen Atom Affinities of Carbonyl Compounds, Imino Ethers, and Hydrazines at 25° C[a]

	GB	PA[b]	aIP	aHA	vIP	vHA
Aldehydes						
CH_2O	169.9[c]	177.2			250.9[d]	113.0
MeCHO	181.1[c,e]	188.9			235.4[f]	109.2
EtCHO	183.6[c,e]	191.4			230.4[f]	106.7
n-PrCHO	185.8[c,e]	193.6			—	
n-BuCHO	185.5[c]	193.3			—	
i-PrCHO	185.5[c]	193.3			226.4[f]	104.6
Ketones						
$Me_2C{=}O$	189.9[c,g]	197.2			223.9[f]	106.0
cyclopentanone	192.5[e]	199.8			—	—
Esters						
HCO_2Me	182.6[c]	190.4	250.2[h]	125.5	254.1[h]	129.4
HCO_2Et	186.4[c,e,i]	194.2	244.9[h]	124.0	248.8[h]	127.9
HCO_2-*n*-Pr	187.4[c,e,i]	195.2	244.9[h]	125.0	248.8[h]	128.9
HCO_2-*n*-Bu	188.0[c,e]	195.8	243.0[h]	123.7	246.0[h]	127.7
$MeCO_2Me$	190.5[c,e,i]	198.3	238.2[h]	121.4	241.7[h]	124.9
$MeCO_2Et$	193.5[c,e,i]	201.3	236.1[h]	122.3	239.6[h]	125.8
$MeCO_2$-*n*-Pr	194.2[i]	202.0	—		—	
CF_3CO_2Me	175.2[c]	183.0	—		—	
CF_3CO_2Et	178.1[c]	185.9	—		267[h]	138
CF_3CO_2-*n*-Pr	179.1[c]	186.9	—		—	
CF_3CO_2-*n*-Bu	179.3[c]	187.1	—		—	
$HCO_2CH_2CF_3$	175.7[c]	183.5	260.8[h]	129.2	265.7[h]	133.1
$NCCO_2Et$	175.8[c]	183.6				
$CF_3CO_2CH_2CH_2F$	175.0[c]	182.8				
$(MeO)_2C{=}O$	192.5[c]	200.3				

TABLE VIII (*Continued*)

	GB	PA[b]	aIP	aHA	vIP	vHA
Carboxylic acids						
HCO_2H	174.6[c,i]	182.8	261.3[h]	129.0	256.4[h]	106.1
$MeCO_2H$	182.5[g,i]	190.7	245.9[j]	121.5	250.2[j]	125.8
$EtCO_2H$	185.2[i]	193.4	242.7[j]	121.0	246.0[j]	125.3
FCH_2CO_2H	178.6[i]	186.8				
$ClCH_2CO_2H$	178.1[i]	186.3				
CF_3CO_2H	167.8[c,i]	176.0				
Amides						
n-PrNHCHO	202.2[e]	210.0				
Me_2NCHO	204.6[e]	212.4			217.5[h]	114.8
N-Me-2-pyrrolidone	209.1[k]	216.9				
N-Me-2-piperidone	211.5[k]	219.3				
N-Me-2-pyridone	212.2[k]	220.0				
6-Cl-N-Me-2-pyridone	210.0[k]	217.8				
Imino ethers						
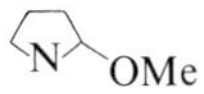	218.0[k]	225.8				
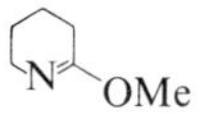	220.3[k]	228.1				
Hydrazines						
$MeNHNH_2$	206.3[e]	214.1				
	207.1[e]	214.9				

[a] All values in kilocalories per mole.
[b] PAs calculated from GBs by using symmetry entropy corrections (see text).
[c] Wolf *et al.* (1977a).
[d] Turner, *et al.*
[e] Aue, D. H., Davidson, W. R., Betowski, L. D., Vidal, M. C., and Bowers, M. T. (to be published).
[f] Tam *et al.* (1974).
[g] Aue *et al.* (1975a).
[h] Sweigart and Turner (1972).
[i] Yamdagni and Kebarle (1976).
[j] Watanabe (1974).
[k] Aue *et al.* (1979). See also Cook *et al.* (1976a).

measurements because there are too few compounds in this wide energy region of the scale. These proton affinity values (Table XI) depend upon absolute PA determinations and a few equilibrium proton-transfer experiments between compounds of comparable PA.

TABLE IX

Gas-Phase Basicities, Proton Affinities, and Derived Inductive and Resonance Effects for Substituted Pyridines at 25° C[a]

Pyridine	GB[b]	PA[c]	$I + R$[d]		GB[b]	PA[c]	$I + R$[d]
Pyridine	212.6	220.4	—	2-F	204.0	211.8	−8.6
2-Me	215.9	223.7	(3.3)[e]	3-F	207.0	214.8	−5.6
3-Me	215.0	222.8	(2.4)[e]	4-F	209.1	216.9	−3.5
4-Me	215.9	223.7	(3.3)[e]	2-Cl	207.0	214.8	−8.9
2-Et	217.1	224.9	(4.5)[e]	3-Cl	207.9	215.7	−7.1
3-Et	216.1	223.9	(3.5)[e]	4-Cl	210.0	217.8	−5.9
4-Et	216.8	224.6	(4.2)[e]	2-Br	207.8	215.6	−9.3
2-*t*-Bu	218.6	226.4	(6.0)	3-Br	208.5	216.3	−7.6
4-*t*-Bu	218.3	226.1	(5.3)[e]	4-Br	210.1	217.9	−6.7
2,3-diMe	218.4	226.2	(5.8)[e]	2-MeO	213.5	221.3	−3.6
2,4-diMe	219.1	226.9	(6.5)[e]	3-MeO	214.7	222.5	−1.4
2,5-diMe	218.4	226.2	(5.8)[e]	4-MeO	218.8	226.6	+2.0
2,6-diMe	219.3	227.1	(6.7)[e]	2-MeS	214.2	222.0	−4
3,4-diMe	218.4	226.2	(5.8)[e]	4-MeS	217.7	225.5	0
3,5-diMe	217.7	225.5	(5.1)[e]	2-NH_2	216.0	223.8	+0.1
2,4-di-*t*-Bu	223.6	231.4	(11.0)	3-NH_2	213.2	221.0	−1.8
2,6-di-*t*-Bu	223	231	(11)[e]	4-NH_2	222	229	+6
4-vinyl	215.4	223.2	−1.4	2-NMe_2	221.4	229.2	+4
2-CF_3	204.0	211.8	−11.9	3-NMe_2	222.1	229.9	+5
3-CF_3	205.0	212.8	−10.0	4-NMe_2	227.9	235.7	+10
4-CF_3	205.2	213.0	−10.7	2-Cl-4-Me	210.8	218.6	−8.3
2-CN	201.1	208.9	−16.0	2-Cl-6-Me	211	219	−8
3-CN	201.7	209.5	−14.4	2-Cl-6-MeO	208.1	215.9	−12.4
4-CN	202.8	210.6	−14.0	2-CH_2OCH_3	218.2	226.0	(1)[f]
4-CHO	207.4	215.2	−9.4	4-NO_2	201.7[g]	209.5	−16
4-$COCH_3$	209.6	217.4	−8				

[a] All values in kilocalories per mole.

[b] GB values from Aue *et al.* (1976c) and Aue, D. H., Webb, H. M., Davidson, W. R., Betowski, L. D., Vidal, M. C., Bowers, M. T., Thummel, R. P. (to be published). Some values have also been determined by Taft (1975) and Arnett *et al.* (1977).

[c] Calculated from GB using symmetry entropy correction (see text).

[d] $I + R$ is the calculated inductive and resonance effect after correction for the polarization effect of the substituent by using a polarization effect for an alkyl pyridine of corresponding polarizability.

[e] Polarization effect of alkyl group, which may include some inductive effect.

[f] The high GB here probably reflects an intramolecularly hydrogen-bonded pyridinium ion.

[g] See Taft (1975) and Arnett *et al.* (1977).

TABLE X

Gas-Phase Basicities, Proton Affinities, and Vertical Ionization Potentials and Hydrogen Atom Affinities for Substituted Pyridines at 25° C[a]

Pyridine	GB[b]	PA[c]	vIP(n)[d]	vHA
Pyridine	212.6	220.4	221.4	126.7
2-Me	215.9	223.7	219.1	127.7
4-Me	215.9	223.7	219.1	127.7
2-F	204.0	211.8	239.1	135.8
3-F	207.0	214.8	232.7	132.4
2-Cl	207.0	214.8	236.5	137.2
3-Cl	207.9	215.7	226.1	127.7
4-Cl	210.0	217.8	230.6	133.3
4-OMe	218.8	226.6	215	126
4-CN	202.8	209.6	236.5	133.0
4-NH_2	222	230	212	126
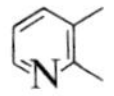	218.4	226.2	211[b]	122
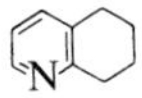	219.9	227.7	211[b]	124
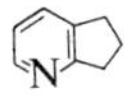	218.0	225.8	211[b]	122
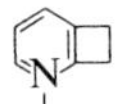	217.3	225.1	210[b]	120
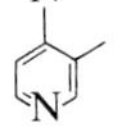	218.4	226.2	215[b]	126
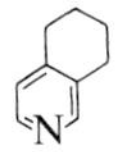	219.9	227.7	212[b]	124
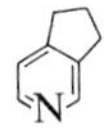	219.0	226.8	212[b]	124
	218.0	225.8	216[b]	127

[a] All values in kilocalories per mole.

[b] Values from Aue *et al.* (1976c) and Aue, D. H., Webb, H. M., Davidson, W. R., Betowski, L. D., Vidal, M. C., Bowers, M. T., Thummel, R. P. (to be published).

[c] Calculated from GBs by using symmetry entropy corrections (see text).

[d] IPs are for the nitrogen lone pairs as assigned by using correlations with semiempirical MO calculations; Ramsey and Walker (1974), Heilbronner *et al.* (1972), King *et al.* (1972), Turner *et al.* (1970).

TABLE XI

Proton Affinities of Substances Less Basic than Water at 25° C[a]

	PA
HF	112[b]
H_2	101[c,d]
O_2	101.1[e]
Kr	101.4[d]
N_2	113.7[e]
Xe	114.2[e]
CO_2	126.8[e]
CH_4	128.2[e]
N_2O	135.6[e]
CO	139.0[e]
C_2H_6	131.9[f]
	139.7[e,f]

[a] All values in kilocalories per mole.
[b] Foster and Beauchamp (1975), Ng *et al.* (1977).
[c] Duben and Lowe (1971).
[d] Payzant *et al.* (1975).
[e] Bohme (1975).
[f] Hiroaka and Kebarle, (1976). There appear to be two isomers of $C_2H_7^+$.

H. Carbocation Stabilities

Proton affinities of unsaturated hydrocarbons have been measured and heats of formation of the corresponding carbocations determined from these proton affinities and the heats of formation of the neutral hydrocarbons. The results of such measurements along with some heats of formation determined from appearance potentials and from ionization potentials of hydrocarbon radicals are summarized in Table XII. From the heat of formation of the carbocation $\Delta H_f^0(R^+)$, the hydride ion affinity HIA(R^+) can also be calculated. The HIA(R^+) value corresponds to the negative of ΔH^0 for reaction (12) for attachment of $H^\ominus$ to $R^\oplus$ in the gas phase.

$$R^+ + H^- \rightleftarrows R - H. \qquad \Delta H^\circ = -HIA(R^+). \tag{12}$$

Such a reaction is analogous to solvolysis reactions of R-X species in solution and provides a useful basis for comparison with solution results. The hydride

TABLE XII

Thermodynamic Data for Carbocations at 25° C[a]

R^+	$\Delta H_f^0(R^+)$	$D(R—H)$[b]	$IP(R\cdot)$	$HIA(R^+)$[c]	$\Delta H_f^0(R—H)$[d]	GB(olefin)	PA(olefin)	ΔH_f^0(olefin)[d]	Olefin
Alkyl cations:									
CH_3^+	261[e]	104[f]	226.9[e]	312	−17.9				
$CH_3CH_2^+$	219[e]	98.0[f,g]	193.3[e]	273	−20.2		163.5[f]	12.5	$CH_2{=}CH_2$
$CH_3CH_2CH_2^+$	208[e]	97.9[f,g]	186.8[e]	266	−24.8				
$CH_3CH^+CH_3$	192[e]	94.5[f,g]	174.1	247.4[e]	−24.8				
	187.2[h]			245.2		176.5[i]	184.9	4.9	$CH_3CH{=}CH_2$
$CH_3CH_2CH_2CH_2^+$	201[e]	100.5[f]	184.7[e]	265	−30.4				
$CH_3CH_2CH^+CH_3$	183[e]	95.9[f,g,i]	170.9[e]	247	−30.4				
$(CH_3)_2CHCH_2^+$	199[e]	99.5[f]	184.7[e]	265	−32.4				
$(CH_3)_3C^+$	164.0[e,k]	92.5[g,j]	154.5[e,k]	229.6	−32.4				
	166.0[h]			231.6[l]		188.3[m]	196.9	−4.3	$(CH_3)_2C{=}CH_2$
	165.1[n]			230.7					
$CH_3CH_2\overset{+}{C}(CH_3)_2$	159.8[h]	—	—	230.0	−36.0	189.5[m]	197.3	−10.1	$(CH_3)_2C{=}CHCH_3$
	159.0[c]			228.2[l]					
$CH_3CH_2CH_2\overset{+}{C}(CH_3)_2$	152.8[c]	—	—	227.9[l]	−41.8				
$CH_3CH_2\overset{+}{C}(CH_3)CH_2CH_3$	152.5[c]	—	—	226.8[l]	−41.1				
$(CH_3)_2CH\overset{+}{C}(CH_3)_2$	150.4[c]	—	—	226.2[l]	−42.6				
$C_2H_5C^+(C_2H_5)C_2H_5$	147.7[c]			226.2[o]	−45.3				
$(CH_3)_2C^+C(CH_3)_3$	141.7[c]			223.8[o]	−48.9				
$(CH_3)_2C^+CH_2CH(CH_3)_2$	145.5[c]			226.9[o]	−48.2				
$(CH_3)_2C^+CH_2CH_2CH_2CH_3$	147.8[c]			227.5[o]	−46.5				
Allyl cations									
	226[p,q]	89[r,s]	186[q]	256	4.9				
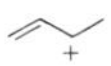	200[h]	—	—	236	−0.2	185[t]	193	26.1	
	204[p]	88.4[p]	174[p]	237					
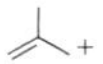	211[p]	85[p]	182[p]	248	−4.3	—	—	—	

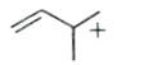	184.9[h]			224.7	−6.6	192.0[t]	200.4	18.1	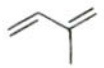
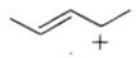	183.5[h]			224.6	−7.9	193.4[t]	201.8	18.1	
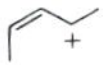	—			—	—	193.1[t]	—	—	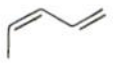
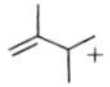	175.9[h]			224.3	−15.2	194.1[t]	202.1	10.8	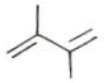
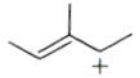	173[h]			221	−15.1	197.3[t]	205.7	11[f]	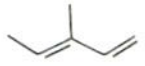
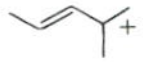	170[h]			219	−16.0	199.9[t]	207.9	11[f]	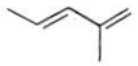
	158[h]			212	−21.2	204.3[t]	213.1	4[f]	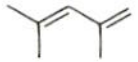
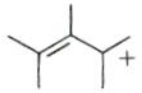	154[h]			211	−29[u]	201.8[t]	210.6	−3[f]	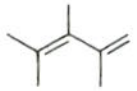
	199.1[h]	—	—	224.1	8.2	192.2[t]	200.0	31.9	
	200.7[v]	81.7[v]	161.4[v]	225.7					
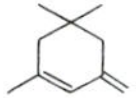	149[h]			207	−25[u]	207.7[t]	216.1	−2[u]	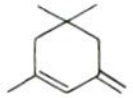
$HC{\equiv}CCH_2^+$	281[p]	88.4[p]	200.2[p]	270	44.4				
Substituted phenonium ions									
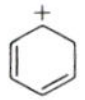	201.2	—	—	212.0	25.4	179.1[i]	185.8	19.8	
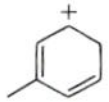	186.2					185.2[w]	193.0	12.0	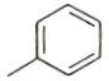
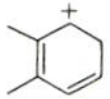	176.5					188.0[w]	195.3	4.6	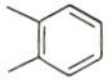

(continued)

TABLE XII (*Continued*)

R^+	$\Delta H_f^0(R^+)$	$D(R—H)^b$	$IP(R\cdot)$	$HIA(R^+)^c$	$\Delta H_f^0(R—H)^d$	GB(olefin)	PA(olefin)	$\Delta H_f^0(\text{olefin})^d$	Olefin
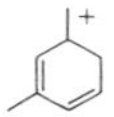	172.4					189.6^w	198.9	4.1	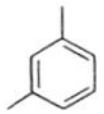
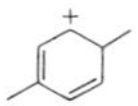	177.3					187.3^w	194.2	4.3	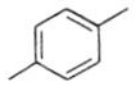
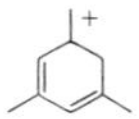	160.6					195.7^w	202.8	−3.8	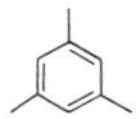
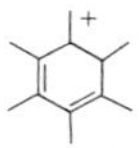	137.8					201.9^w	208.6	−20.8	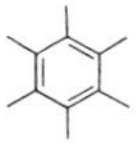
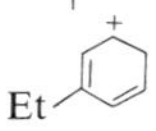	180.6					186.1^x	193.8	7.2	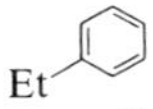
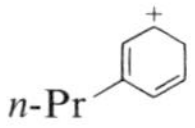	174.4					186.9^x	194.7	1.9	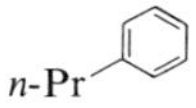
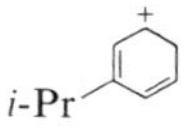	173.1					187.3^x	195.1	1.0	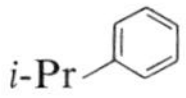
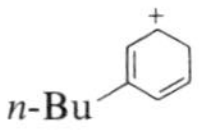	168.8					187.3^x	195.1	−3.3	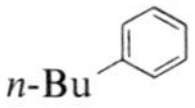
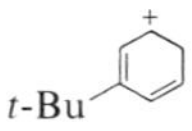	167					187^x	195	−5.4	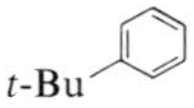
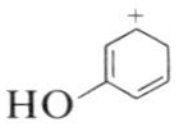	147					189^x	197	−23.0	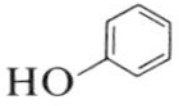

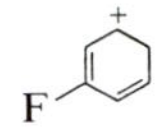	153.4					178.2[i]	186.0	−27.8	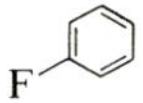
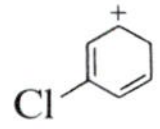	193.5					178.1[i]	185.9	12.2	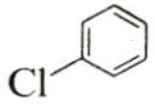
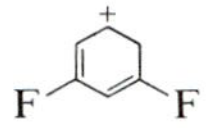	108.7					177.2[i]	184.5	−74.0	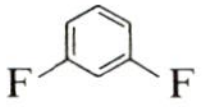
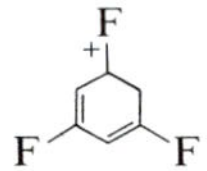	—					176.7[i]	183.8	—	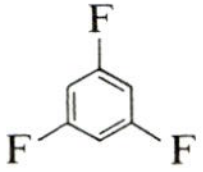
Vinyl cations									
	266[q]	99.2[q]	206.4[q]	287	12.5				
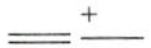	230[h]			258	4.9	174[y]	182	44.4	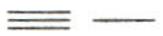
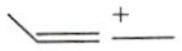	215[h]			251	−3.0	179[m]	187	34.7	
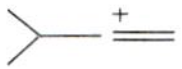	204[h]			244	−6.6	190[m]	198	35[u]	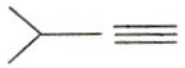
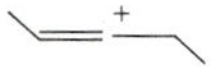	201[h]			242	−7.9	188[m]	196	30[u]	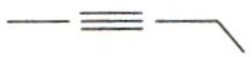
Halogenated cations									
CH_2F^+	200.3[z]	103.1	205.2	289.6	−55.9				
CHF_2^+	142.4[z]	101.0	201.5	283.9	−108.1				
CF_3^+	99.3[z]	106.2	211.5	299.0	−166.3				

(*continued*)

TABLE XII (*Continued*)

R^+	$\Delta H_f^0(R^+)$	D(R—H)[b]	IP(R·)	HIA(R^+)[c]	ΔH_f^0(R—H)[d]	GB(olefin)	PA(olefin)	ΔH_f^0(olefin)[d]	Olefin
Cycloalkyl cations									
 (?)	235[h]	100[r]	178	256	12.7	190[aa]	198	66.2	
	219[h]	97[u]	171	246	6[bb]	198[aa]	206	58.2	
	195[h]	97[u]	168	243	−15[u,bb]	206[aa]	214	42[u,bb]	
	214[h]	100[u]	170	248	−1[u,bb]	196[aa]	203	50[u,bb]	
	194[h]	—	—	231	−4[cc]	194[m]	202	28.3[cc]	
	185[h]	—	—	227	−9[u]	194[m]	201	19.8[cc]	
	198.2[v]	94.9[v,dd]	172.3[v]	249.8 241.5[l]	−18.4				
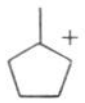	168.0[h]	—	—	226.5 225.2[l]	−25.3	190.4[m,ee]	198.6	−0.6	
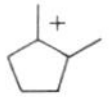	160			226	−32.7	191.7[ee]	198.6	−9[u]	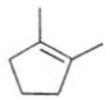
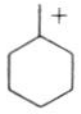	157.0			227.2	−37.0	191.7[ee]	199.9	−10.3	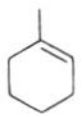
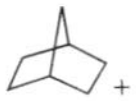	187[h]	—	—	232 230.1[l]	−12.4	193.8[m,ff]	201.6	21[cc]	
	172[h]			225 224.1[l]	−19.6[w]	199[m]	207	12[u]	

Conjugated cyclic cations									
	256[p]			223	66.2				
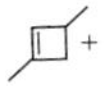	204			214	23[u]	204	212	48.3[cc]	
	257[v]	81.1[v]	193.9[v]	258	31.9				
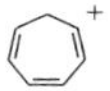	212[q,gg] 211[hh,ii]	76[q] 73[jj]	144[q] 144.8[hh]	201 200	43.9				
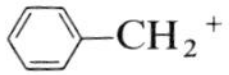	217.2[kk] 212[k] 213[q]	 85[q]	 167.6[q]	238.4 233 234	12.0				
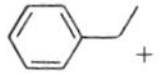	200.1			226.1	7.2	194.4[ll]	202.4	35.3	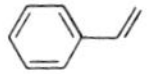
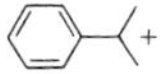	188			220	1.0	199.8[ll]	208.2	29[u]	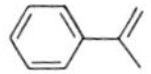
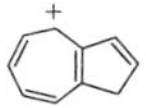	216.2			—	—	217.2[ll]	224.5	73.5	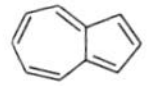
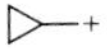 (?)	214[h,mm]			241	6[bb]	183[m]	191	37.5	
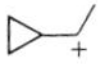	197[h]			230	1	190.8[nn]	198.8	29[u]	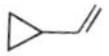
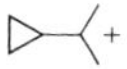	179[h]			218	−6[u]	200.9[m]	209.3	21[u]	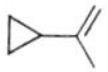
	182[h]			221	−6[u]	198	206	21[u]	

(*continued*)

TABLE XII (*Continued*)

R^+	$\Delta H_f{}^0(R^+)$	$D(R—H)^b$	$IP(R\cdot)$	$HIA(R^+)^c$	$\Delta H_f{}^0(R—H)^d$	GB(olefin)	PA(olefin)	$\Delta H_f{}^0(\text{olefin})^d$	Olefin
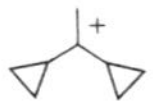	196			207	22^u	209.5^m	217.5	46^u	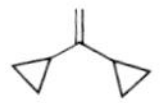
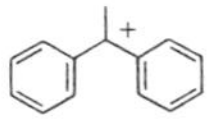	213.6			215	32^u	204.3^{ll}	212.3	58.7	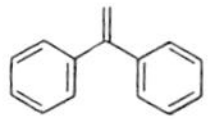

[a] All values in kilocalories per mole. Thermodynamic data are calculated using: $\Delta H_f{}^0(H^+) = 367.2$, $\Delta H_f{}^0(H\cdot) = 52.1$, $\Delta H_f{}^0(H^-) = 33.2$, from JANAF Thermochemical Tables (1971).

[b] Calculated from $D(R—H) = \Delta H_i{}^0(R\cdot) + \Delta H_f{}^0(H\cdot) - \Delta H_f{}^0(RH)$.

[c] Calculated from $HIA(R^+) = \Delta H_f{}^0(R^+) + \Delta H_f{}^0(H^-) - \Delta H_f{}^0(RH)$.

[d] Cox and Pilcher (1970).

[e] Lossing and Semeluk (1970); Lossing and Maccoll (1976). Values of $\Delta H_f{}^0(R^+)$ may be too high because of a nonadiabatic IP (See footnote *k*).

[f] Kerr (1966).

[g] Luria and Benson (1975); S. W. Benson (private communication).

[h] Calculated from PA using $\Delta H_f{}^0(R^+) = \Delta H_f{}^0(H^+) + \Delta H_f{}^0(\text{olef.}) - PA$.

[i] Yamdagni and Kebarle (1976).

[j] Tsang (1972).

[k] Beauchamp and Houle private communication (see footnote ii); Koenig *et al.* (1975). IP corrected by 1.5 kcal/mole for $\int_0^{298} \Delta Cp\, dT$ of e^- for the calculation of $\Delta H_f{}^0(R^+)$ 298°K.

[l] Solomon and Field (1975, 1976). Adjusted to $\Delta H_f{}^0(t\text{-}Bu^+) = 166.0$ kcal/mole.

[m] Aue *et al.* (to be published).

[n] Calculated from data in Chesnavich *et al.* (1978), correcting for heat capacity changes from 0°K to 298°K for e^-.

[o] Moet-Ner *et al.* (1976).

[p] Lossing (1972).

[q] Lossing (1971).

[r] Golden and Benson (1969).

[s] Golden *et al.* (1969, 1966).
[t] Aue *et al.* (1976d) and Aue *et al.*, to be published.
[u] Estimated from group equivalents.
[v] Lossing and Traeger (1975).
[w] Devlin *et al.* (1976).
[x] Hehre *et al.* (1974) and DeFrees *et al.* (1977).
[y] Revised value from Aue *et al.* (1976d).
[z] Blint *et al.* (1974); McMahon *et al.* (1972).
[aa] Aue *et al.* (1976d).
[bb] Good (1971).
[cc] Schleyer *et al.* (1970).
[dd] Ferguson and Whittle (1971).
[ee] Pollack *et al.* (1977a).
[ff] Staley *et al.* (1977).
[gg] Harrison *et al.* (1960).
[hh] Koenig and Chang (1978). IP corrected by 1.5 kcal/mole for $\int_0^{298} \Delta Cp\, dT$ of e^- for the calculation of $\Delta H_f{}^0(R^+)$ at 298°K.
[ii] Traeger and McLoughlin (1977).
[jj] Vincow *et al.* (1969).
[kk] Abboud *et al.* (1976).
[ll] Wolf *et al.* (1975b).
[mm] Pottie *et al.* (1961).
[nn] Wolf *et al.* (1975a).

TABLE XIII

Thermodynamic Properties of Amines at 25°C[a]

	GB[a]	$-\Delta G^0_{prot,s}$	$-\Delta G_s^0(B)$	$-\Delta G_s^0(BH^+)$[d]	PA[b,c]	$-\Delta H^0_{prot,s}$	$-\Delta H_s^0(B)$	$-\Delta H_s^0(BH^+)$[d]	$-\Delta H_s^0(BH^+)$[el]	$-T\Delta S^0_{prot,g}$[b]	$-T\Delta S^0_{prot,s}$	$-T\Delta S_s^0(B)$	$-T\Delta S_s^0(BH^+)$[d]
NH_3	196.4	12.61[e]	2.41[g]	79.1	205.0	12.49[e]	8.42[e]	85.7	77.3	8.57	−0.12	6.01	6.6
$MeNH_2$	205.7	14.53[e]	2.68[h]	72.0	214.1	13.18[e]	10.82[e]	79.7	68.9	8.40	−1.35	8.14	7.7
$EtNH_2$	208.7	14.56[e]	2.61[h]	69.0	217.1	13.71[i]	13.05[e]	79.5	66.4	8.40	−0.86	10.44	10.5
n-$PrNH_2$	210.1	14.42[e]	2.51[h]	67.3	218.5	13.84[i]	13.37[e]	78.5	65.1	8.40	−0.58	10.86	11.2
n-$BuNH_2$	210.6	14.51[e]	2.41[h]	66.8	219.0	13.98[i]	14.11[e]	78.9	64.8	8.40	−0.53	11.70	12.1
i-$BuNH_2$	211.1	14.30[j]			219.5	13.92[i]	13.66[e]	77.9	64.2	8.40	−0.38		
i-$PrNH_2$	211.0	14.56[j]			219.4	13.97[i]	13.37[e]	77.7	64.4	8.40	−0.59		
s-$BuNH_2$	212.1	14.41[e]			220.5	14.03[i]			63.3	8.40	−0.38		
t-$BuNH_2$	212.9	14.58[e]			221.3	14.35[e]	14.10[e]	77.0	62.9	8.40	−0.23		
Me_2NH	212.3	14.70[e]	2.41[h]	65.3	220.5	12.04[i]	12.69[e]	74.0	61.3	8.16	−2.66	10.28	8.8
Et_2NH	216.9	15.03[e]	2.19[h]	60.8	225.1	12.73[i]	15.31[e]	72.7	57.4	8.16	−2.30	13.12	12.0
n-Pr_2NH	219.2	15.01[k]	1.76[h]	58.1	227.4	13.17[i]	17.26[e]	72.8	55.6	8.16	−1.84	15.50	14.8
n-Bu_2NH	220.2	15.35[k]	1.34[h]	57.0	228.4	13.66[i]	18.81[e]	73.9	55.1	8.16	−1.69	17.55	17.0
i-Bu_2NH	220.4	14.32[i]			228.6	13.38[i]	15.56[e]	70.1	54.6	8.16	−0.74		
i-Pr_2NH	220.7	15.28[i]			228.9	13.55[i]	16.81[e]	71.3	54.5	8.16	−1.73		
s-Bu_2NH	222.7	14.88[i]			230.9	14.03[i]	17.81[e]	70.7	52.9	8.16	−0.85		
Me_3N	216.5	13.37[e]	1.34[g]	58.7	224.3	8.82[e]	13.23[e]	67.6	54.3	7.75	−4.55	11.89	8.9
Et_3N	223.4	14.62[e]	1.15[g]	52.8	231.2	10.32[i]	16.76[e]	65.7	48.9	7.75	−4.30	15.61	12.9
n-Pr_3N	225.6	14.54[i]	0.80[l]	50.2	233.4	10.50[i]			46.9	7.75	−4.04		
Pyrrolidine	216.1	15.42[k]	1.21	61.0	224.3	12.82[m]	15.19	73.5	58.3	8.16	−2.60	13.98	12.5
Piperidine	217.2	15.30[k]	0.83	59.4	225.4	12.71[m]	15.63	72.7	57.1	8.16	−2.59	14.80	13.4

N-Methyl-pyrrolidine	220.0	14.26[k]	−0.30	54.5	227.8	9.05[f]	15.16	66.2	51.1	7.75	−5.21	15.46	11.8
N-Methyl-piperidine	221.0	13.75[k]	−0.38	52.9	228.8	9.44[f]	15.22	65.7	50.4	7.75	−4.31	16.10	13.3
c-$PrNH_2$	206.6	12.41			215.0	11.72			66.5	8.40	−0.69		
Cyclohexylamine	212.9	14.56[k]			221.3	14.38[i]			62.9	8.40	−0.18		
Aniline	203.1	6.27[i]			211.5	7.24[i]			65.5	8.40	0.97		
Piperazine	216	13.41[k]			224	10.17[i]			56	8.16	−3.24		
Morpholine	211	11.58[k]			219	9.33[i]			60	8.16	−2.25		
1,4-Diazabicyclo-[2.2.2]octane	220	12.03[i]			228	7.30[i]			49	7.75	−4.73		
2,2,2-Trifluoro-ethylamine	194.9	7.65[n]			203.3	8.30[n]			74.8	8.40	+0.65		
Quinuclidine	223.6	15.21[k]			231.4	11.14			49.5	7.75			
Pyrrole	201.6	−4.18[k]			209.8					8.16	−2.6 (est)		
CH_3CN	183.1	−13.6[o]			190.9					7.75			

[a] All values in kilocalories per mole. GBs measured relative to methylamine (205.7 kcal/mole) and ammonia (196.4 kcal/mole).
[b] PAs calculated from GBs by correction for symmetry and the entropy of H^+ (Altschuller, 1955; see also Benson, 1958).
[c] Absolute PA of NH_3 taken as 205 = 3 kcal/mole (see text).
[d] Calculated using $\Delta G_s^0(H^+) = 260.5$ kcal/mole (Noyes, 1962) and $\Delta H_s^0(H^+) = -269.8$ kcal/mole (Friedman and Krishnan, 1973).
[e] Jones and Arnett (1974).
[f] Cabani *et al.* (1971).
[g] Bates and Pinching (1949).
[h] Christie and Crisp (1967).
[i] Christenson *et al.* (1969).
[j] Andon *et al.* (1954).
[k] Perrin (1965).
[l] Extrapolated from footnote *g* data.
[m] Cabani *et al.* (1968).
[n] Love *et al.* (1968).
[o] Kolthoff and Chantooni (1973).
[p] Sacconi *et al.* (1960a).
[q] Sacconi *et al.* (1960b).

TABLE XIV

Thermodynamic Data of Pyridines at 25°[a]

	GB[a]	$-\Delta G^0_{prot,s}$[b,e]	$-\Delta G_s{}^0(B)$	$-\Delta G_s{}^0(BH^+)$[c]	PA[d]	$-\Delta H^0_{prot,s}$	$-\Delta H_s{}^0(B)$	$-\Delta H_s{}^0(BH^+)$[c]	$-\Delta H_s{}^0(BH^+)$[el]	$-T\Delta S^0_{prot,s}$[d]	$-T\Delta S^0_{prot,s}$	$-T\Delta S_s{}^0(B)$	$-T\Delta S_s{}^0(BH^+)$
Pyridine	212.6	7.12	0.42[g]	55.4	220.4	4.80[e]	11.98[h]	66.2	54.2	7.75	−2.32	11.56	10.8
2-Me	215.9	8.13	0.36[g]	53.1	223.7	6.22[i]	13.28[h]	65.6	52.3	7.75	−1.91	12.92	12.6
3-Me	215.0	7.68	0.50[g]	53.7	222.8	5.87[e]	13.24[h]	66.1	52.9	7.75	−1.81	12.74	12.5
4-Me	215.8	8.15	0.66[g]	53.5	223.6	6.13[e]	13.27[h]	65.6	52.3	7.75	−2.02	12.61	12.1
2-Et	217.1	8.03	0.06[g]	51.5	224.9	6.58[j]	13.31[g]	64.8	51.5	7.75	−1.45	13.25	13.4
3-Et	216.1	7.58	0.33[g]	52.3	223.9		12.78[g]					12.45	
4-Et	216.8	8.00	0.46[g]	52.2	224.6		12.48[g]					12.02	
4-*t*-butyl	218.3	8.17			231.1								
2,4-diMe	219.1	9.04	0.59[g]	51.0	226.9	7.17[k]	14.51[h]	64.6	50.1	7.75	−1.87	13.92	13.6
2,5-diMe	218.4	8.73	0.44[g]	51.3	226.2	6.82[k]	14.54[h]	65.0	50.5	7.75	−1.91	14.10	13.7
2,6-diMe	219.3	9.16	0.33[g]	50.7	227.1	7.24[k]	14.82[h]	64.8	50.0	7.75	−1.92	14.49	14.1
3,5-diMe	217.7	8.39	0.57[g]	51.8	225.5	6.37[h]	14.46[h]	65.1	50.6	7.75	−2.02	13.89	13.4
2,6-di-*t*-Bu	223	6.8[f]			231								
2-NH_2	216.0	9.15			229.8	8.40[i]	10.3[j]		54.4		−0.75		
3-NH_2	213.2	8.24[j]			221.0	6.41[e]	17.0[j]	72.2	55.2		−1.83		
4-NH_2	222	12.44[j]			230	11.27[e]	17.3[j]	68.	51		−1.17		
2-NMe_2	221.4	9.32[j]			223.2	7.20[j]	15.24[j]	63.0	47.8		−2.13		
3-NMe_2	222.1	8.79[j]			229.9	7.16[j]	18.17[j]	65.2	47.1		−1.63		
4-NMe_2	227.9	13.37[j]			235.7	11.80[j]	16.3[j]	62.2	45.9		−1.57		
2-F	204.0	−0.6			211.8								
3-F	207.0	4.05			214.8								
4-F	209.4				217.2								
2-Cl	207.0	0.98			214.8	0.25[i,j]	8.3[j]	63.6	55.3		−0.73		
3-Cl	207.9	3.87			215.7	2.60[e]	13.0[l]	69.7	56.7		−1.27		
4-Cl	210.0	5.24			217.8	3.58[e]			55.6		−1.66		
2-Br	207.8	1.23			215.6	0.47[g,j]	12.1[j]	66.8	54.7		−0.76		
3-Br	208.5	3.87			216.3	2.78[e]	9.0[j]	65.3	56.3		−1.09		
4-Br	210.0	5.15			217.8	3.51[e]			55.5		−1.64		

2-CF_3	204.0				211.8								
3-CF_3	205.0				212.8	2.0[l]			59.0				
4-CF_3	205.2	3.59[m]			213.0	2.22[l]	11.66[l]	70.7	59.0		−1.37		
2-MeO	213.5	4.17			221.3	3.88[j]	10.4[n]	62.8	52.4		−0.29		
3-MeO	214.7	6.70			222.5								
4-MeO	218.8	8.82			226.6	6.85[e]	14.13[l]	64.2	50.1		−1.97		
2,4,6-triMe		10.13											
2-CHO		5.13	0.41[o]			3.84[o]	4.33[o]			7.75	−0.80	3.92	4.7
3-CHO		5.05	1.34[o]				3.48[o]			7.75	−1.57	2.14	2.1
4-CHO	207.4	6.18	−0.25[o]	59.0	215.2	4.92[o]	3.79[o]	63.3	59.5	7.75	−2.39	4.04	3.2
2-CN	201.1	−0.35			208.9	−0.91[j]	17.6[j]	77.6	60.0		−0.56		
3-CN	201.7	1.90			209.5	0.88	14.5[j]	75.7	61.2		−1.02		
4-CN	202.8	2.59			210.6	1.26[e]	9.8[j]	70.3	60.5		−1.33		
4-vinyl	215.3				223.1								
2-CH_3CO—	210.0				217.8								
2-CH_2OMe	218.2				226.0 H-bonded								

[a] All values in kcal/mole. GBs relative to methylamine (205.7 kcal/mol).
[b] Taken from D. D. Perrin (1965), except where noted.
[c] Calculated using $\Delta G_s^0(H^+) = -260.5$ kcal/mole (Noyes, 1962) and $\Delta H_s^0(H^+) = -269.8$ kcal/mole (Friedman and Krishnan, 1973).
[d] PAs calculated from GBs by using the entropy of H^+ (Altschuller, 1955). There is no symmetry correction for pyridines.
[e] Liotta *et al*., (1974).
[f] Estimated from differences in pK_as measured in H_2O and EtOH; see Brown and Kanner (1966), McDaniel and Ozcan (1969).
[g] Andon *et al*. (1954).
[h] Sacconi *et al*. (1960).
[i] Christenson *et al*. (1973).
[j] H. P. Hopkins (private communication).
[k] Sacconi *et al*. (1960a).
[l] Arnett *et al*. (1977).
[m] Taagepera *et al*. (1972).
[n] Calculated from $\Delta H^0_{soln}(B) = -0.65$ kcal/mole (Cook *et al*., 1976) and $\Delta H^0_{vap}(B) = +9.8$ kcal/mole (Beak *et al*., 1974).
[o] Cabani *et al*. (1969).

ion affinity scale is probably the most useful and convenient basis for the comparison of relative carbocation stabilities. Bromide and chloride affinities can also be derived, but the lack of reliable heats of formation for alkyl halides limits their use. In some cases, quantitative and semiquantitative halide affinities can be determined by observing halide-transfer reactions directly (Staley *et al.*, 1977; Weiting, *et al.* (1974); Abboud *et al.*, 1976). Hydride-ion-transfer reactions are generally slow, but equilibrium data is attainable in favorable cases (Solomon and Field, 1975, 1976a,b; Solomon *et al.*, 1974). Proton-transfer reactions with hydrocarbon ions can often be slow enough to limit the quantitative precision of proton affinities, but they are still the most generally useful means of determining accurate relative carbocation stabilities. Correlations of carbocation stability data with solution work and with theory are of considerable interest, and questions of cation structure pose important challenges. A discussion of these points, however, lies beyond the scope of this review.

I. Solution Basicity and Solvent Effects

Gas-phase basicity orders often differ dramatically from solution basicities. The origin of such changes must be either in solvation effects on the neutral base or its cationic conjugate acid, as shown in scheme III.

$$\begin{array}{ccccc}
B_g & + & H_g^+ & \xrightarrow[PA]{\Delta H^0_{prot,g}} & BH_g^+ \\
\downarrow{\scriptstyle \Delta H_s{}^0(B)} & & \downarrow{\scriptstyle \Delta H_s{}^0(H^+)} & & \downarrow{\scriptstyle \Delta H_s{}^0(BH^+)} \\
B_g & + & H_g^+ & \xrightarrow{\Delta H^0_{prot,s}} & BH_s^+
\end{array}$$

$$\Delta H^0_{prot,s} = -PA + \Delta H_s{}^0(BH^+) - \Delta H_s{}^0(B) - \Delta H_s{}^0(H^+)$$

scheme III. Thermodynamic cycle for analysis of solution effects on basicities.

Solvation data based on the most recent proton affinity data are presented in Tables XIII and XIV for ammonium ions and for pyridinium ions. Values for heats, free energies, and entropies of solvation for the cations can be calculated from proton affinity data and experimental data in the literature. Variation in the heats of solvation of the cationic conjugate acids of these bases generally play the largest role in altering gas-phase basicity orders in solution (Aue *et al.*, 1972, 1976b,c; Arnett *et al.*, 1972, 1977; Arnett, 1973; Arnett and Wolf, 1975; Taft *et al.*, 1978). From an analysis of the available data for ammonium ions and some carbocations (Wolf *et al.*, 1975b, 1977b), it appears that the major factor determining the relative heats of solvation is the size and charge density of the ion (Aue *et al.*, 1976b).

IV. BASE EXCHANGE REACTIONS AMONG PROTON-BOUND DIMER IONS. MOLECULE-PAIR GAS-PHASE BASICITIES

Ammonium ions and oxonium ions undergo efficient condensation reactions with neutral amines or oxides to form proton-bound dimer ions. The mechanism of these reactions has been investigated in detail to reveal that the proton-bound dimers are initially formed with excess vibrational energy which can be dissipated at high pressures to form stable dimer ions (see scheme I; see also Neilson *et al.*, 1978). Kebarle's (1972, 1977) group has directly measured equilibrium constants for reactions such as reaction (13) using high-pressure mass spectrometry,

$$BH^+ + B \rightleftarrows BHB^+. \tag{13}$$

Knowledge of the stabilities of these proton-bound dimer ions and of more highly condensed ions is basic to an understanding of the solvation of ions, since these species represent the first stages of solvation.

Drift-cell ion cyclotron resonance experiments are carried out in a convenient pressure range for the stabilization and study of proton-bound dimers. In such experiments on mixtures of amines A and B, the proton-bound dimers in Eq. (14) appear to be in equilibrium, since their ratios remain constant with increasing pressure (Bowers *et al.*, 1971)

$$AHA^+ \underset{A}{\overset{B}{\rightleftharpoons}} AHB^+ \underset{A}{\overset{B}{\rightleftharpoons}} BHB^+. \tag{14}$$

From these experiments and those of Yamdagni and Kebarle (1973), a scale of molecule-pair gas-phase basicities (MPGBs, Table XV) can be constructed, where the MPGB of a pair of molecules A and B is defined as the negative of ΔG^0 for

$$\begin{gathered} A + B + H^+ \rightleftarrows AHB^+, \\ -\Delta G^0 = \text{MPGB}; \qquad -\Delta H^0 = \text{MPPA}. \end{gathered} \tag{15}$$

Yamdagni and Kebarle's (1973) absolute MPGBs for dimethylamine and trimethylamine approximately agree with the relative values obtained in a series of equilibrium experiments for reaction type (14). This agreement and the self-consistency for overlapping thermodynamic cycles in the determination of relative MPGBs suggests that reaction (14) has indeed achieved equilibrium under drift-cell ICR conditions (Aue *et al.*, to be published; see also Taft, 1975).

The MPGBs in Table XV have values that generally parallel the GBs of the amines involved. In some cases, however, steric effects play a significant role in determining MPGBs. For example, the MPGB of diisopropylamine is ~2 kcal/mole lower than that of piperidine, even though its GB is 3.5 kcal/mole higher.

TABLE XV

Gas-Phase Basicities, Proton Affinities, Molecule-Pair Gas-Phase Basicities and Molecule-Pair Proton Affinities of Amines at 25°C[a]

Compound	GB	PA	MPGB	MPPA[b]	Mixture	Av. GB	Av. PA	MPGB	MPPA	Mixture	Av. GB	Av. PA	MPGB	MPPA
NH_3	196.4	205.0	213.5[c]	229.8[c]	allyl NH_2, $MeNH_2$	206.9	215.3	221.1	236.6	*s*-$BuNH_2$, Me_2NH	212.2	220.5	225.8	241.3
$MeNH_2$	205.7	214.1	220.4[c]	235.8[c]	$EtNH_2$, $MeNH_2$	207.2	215.6	221.4	236.9	*s*-$BuNH_2$, *c*-hexyl NH_2	212.5	220.9	226.1	241.6
Me_2NH	212.3	220.5	225.4[c]	241.3[c]	$EtNH_2$, allyl NH_2	208.4	216.8	222.3	237.8	Me_2NH, MeNHEt	213.6	221.6	226.8	242.3
Me_3N	216.5	224.3	229.5[c]	246.8[c]	*n*-$PrNH_2$, $EtNH_2$	209.4	217.8	223.1	238.6	Me_2NH, Azetidine	213.4	221.6	227.4	242.9
$MeNH_2$	205.7	214.1	219.8	235.7	*i*-$PrNH_2$, $EtNH_2$	209.9	218.3	223.5	239.0	Me_2NH, Me_3N	214.4	222.4	227.9	242.8
Allyl NH_2	208.1	216.5	221.5	237.4	*n*-$BuNH_2$, *n*-$PrNH_2$	210.4	218.8	224.1	239.6	*c*-hexyl NH_2, Me_3N	214.7	222.8	229.2	244.1
$EtNH_2$	208.7	217.1	222.1	238.0	*i*-$BuNH_2$, *n*-$PrNH_2$	210.6	219.0	224.3	239.8	*c*-hexyl NH_2, Et_2NH	214.9	223.2	229.3	244.8
n-$PrNH_2$	210.1	218.5	223.4	239.3	*n*-$BuNH_2$, *i*-$PrNH_2$	210.8	219.2	224.3	239.8	MeNHEt, Pyrrolidine	215.5	223.5	228.7	244.2
i-$PrNH_2$	211.0	219.4	223.7	239.6	*i*-$BuNH_2$, *i*-$PrNH_2$	211.1	219.5	224.4	239.9	MeNHEt, Et_2NH	215.9	223.9	228.5	244.0
n-$BuNH_2$	210.6	219.0	224.2	240.1	Me_2NH, *n*-$PrNH_2$	211.2	219.5	225.3	240.8	Azetidine, Et_2NH	215.7	223.9	229.5	245.0
i-$BuNH_2$	211.1	219.5	224.2	240.1	Me_2NH, *n*-$BuNH_2$	211.5	219.8	225.7	241.2	Azetidine, Me_2NEt	216.7		230.5	245.9
s-$BuNH_2$	212.1	220.5	224.9	240.8						Pyrrolidine, Me_3N	216.3		229.5	244.4
Me_2NH	212.3	220.5	225.4	241.3	*i*-$PrNH_2$, *s*-$BuNH_2$	211.6	220.0	224.7	240.2	Pyrrolidine, Piperidine	216.7		230.4	245.9
Pyridine	212.6	220.4	226.5	242.0	*i*-$PrNH_2$, Me_2NH	211.7	220.0	225.4	240.9	Pyrrolidine, *i*-Pr_2NH	218.4		230.8	246.3
c-hexyl NH_2	212.9	221.3	226.6	242.5	*i*-$PrNH_2$, *c*-hexyl NH_2	212.0	220.4	225.6	241.1	Pyrrolidine, *n*-Me Piperidine	218.6		231.9	247.4
Azetidine	214.5	222.7	228.7	244.6	*i*-$BuNH_2$, Pyridine	211.9	220.0	225.7	240.8	Me_3N, Et_2NH	216.7		229.1	244.0
MeNHEt	214.8	222.6	227.2	243.1	*i*-$BuNH_2$, Me_2NH	211.7	220.1	225.6	241.1	Me_3N, Piperidine	216.9	224.9	229.7	244.6
Pyrrolldine	216.1	224.3	229.9	245.8	Pyridine, *s*-$BuNH_2$	211.9	219.5	226.2	241.3	Et_2NH, Piperidine	217.1	225.3	230.0	245.5
Me_3N	216.5	224.3	228.3	243.6	Pyridine, Me_2NH	212.5	219.5	226.9	243.7	Et_2NH, *i*-Pr_2NH	217.8	227.0	229.7	245.2
Et_2NH	216.9	225.1	228.8	244.7	Pyridine, *c*-hexyl NH_2	212.8	220.9	227.1	242.2	Piperidine, Me_2NEt	218.0	226.0	230.6	246.1
Piperidine	217.2	225.4	230.2	246.1	Pyridine, MeNHEt	213.7	221.5	228.4	243.5	Piperidine, *i*-Pr_2NH	219.0	227.2	230.5	246.0
Me_2NEt	218.8	226.6	229.7	245.6	Pyridine, Azetidine	213.6	221.6	228.8	243.9	Piperidine, N-Me Piperidine	219.2	227.1	231.8	247.3
i-Pr_2NH	220.7	228.9	228.4	244.3	Pyridine, Me_3N	214.6	222.4	229.3	244.8	Azetidine, *i*-Pr_2NH	217.6	225.8	230.5	247.0
N-Me-Piperidine	221.0	228.8	231.1	247.0	Pyridine, Et_2NH	214.8	222.8	230.2	245.3					

[a] All values in kilocalories per mole taken from Aue *et al.* (1978d). See also Taft (1975). Relative values based on MPGB of Me_2NH from Yamdagni and Kebarle (1973).
[b] Calculated from MPGB using an entropy term based on that for Me_2NH ($-T\,\Delta S^0 = 15.9$ kcal/mole) from Yamdagni and Kebarle (1973) and corrected for symmetry entropy differences.
[c] Yamdagni and Kebarle (1973).

V. CONCLUSIONS

The gas-phase basicities of well over 300 compounds have now been measured from equilibrium proton-transfer reactions. These data cover most major organic structural types and provide a core of fundamental data on which to base theoretical interpretation. The influence of the common structural effects observed in solution is evident in the gas phase, e.g., electronegativity effects, inductive effects, resonance effects, hyperconjugation effects, aromaticity effects, polarization effects, hybridization effects, and steric effects. Polarization effects are especially strongly attenuated in solution, and chelation effects of intramolecular strong hydrogen bonding are unique to the gas phase. Other effects, however, are observable in much the same fashion as in solution and largely conform to established theory. The gas-phase data provide a means for determining heats of solvation of gaseous ions which are important in building theories of solvation.

From this fundamental basicity data we look forward to expansion of our insight into properties of ions through an increase in the scope of the structural types of ions investigated and an increase in the sophistication of the questions that can be asked in the future.

ACKNOWLEDGMENT

We wish to acknowledge the financial support of the National Science Foundation under Grant No. CHE-76-17573. We thank our co-workers, H. M. Webb, W. R. Davidson, L. D. Betowski, M. Vidal, and B. L. A. and D. B. A. for their contributions to this work.

REFERENCES

Abboud, J.-L. M., Hehre, W. J., and Taft, R. W. (1976). *J. Am. Chem. Soc.* **98**, 6072.

Altschuller, A. P. (1955). *J. Am. Chem. Soc.* **77**, 3480.

Andon, R. J. L., Cox, J. D., and Herington, E. F. G. (1954). *J. Chem. Soc.* p. 3188.

Anicich, V. G., and Bowers, M. T. (1974). *Int. J. Mass Spectrom. Ion Phys.* **11**, 329.

Arnett, E. M. (1973). *Acc. Chem. Res.* **6**, 404.

Arnett, E. M., and Wolf, J. F. (1975). *J. Am. Chem. Soc.* **97**, 3262.

Arnett, E. M., Jones, F. M., III, Taagepera, M., Henderson, W. G., Beauchamp, J. L., Holtz, D., and Taft, R. W. (1972). *J. Am. Chem. Soc.* **94**, 4724.

Arnett, E. M., Chawla, B., Bell, L., Taagepera, M., Hehre, W. J., and Taft, R. W. (1977). *J. Am. Chem. Soc.* **99**, 5729.

Arshadi, M. R., and Futrell, J. H. (1974). *J. Phys. Chem.* **78**, 1482.

Aue, D. H., Webb, H. M., and Bowers, M. T. (1972). *J. Am. Chem. Soc.* **94**, 4726.

Aue, D. H., Webb, H. M., and Bowers, M. T. (1973). *J. Am. Chem. Soc.* **95**, 2699.

Aue, D. H., Webb, H. M., and Bowers, M. T. (1975a). *J. Am. Chem. Soc.* **97**, 4136.

Aue, D. H., Webb, H. M., and Bowers, M. T. (1975b). *J. Am. Chem. Soc.* **97**, 4137.

Aue, D. H., Webb, H. M., and Bowers, M. T. (1976a). *J. Am. Chem. Soc.* **98**, 311.

Aue, D. H., Webb, H. M., and Bowers, M. T. (1976b). *J. Am. Chem. Soc.* **98**, 318.

Aue, D. H., Webb, H. M., Bowers, M. T., Liotta, C. L., Alexander, C. J., and Hopkins, H. P., Jr. (1976c). *J. Am. Chem. Soc.* **98**, 854.

Aue, D. H., Davidson, W. R., and Bowers, M. T. (1976d). *J. Am. Chem. Soc.* **98**, 6700
Aue, D. H., Betowski, L. D., Davidson, W. R., Bowers, M. T., Beak, P., and Lee, J. (1979). *J. Am. Chem. Soc.* **101**, 1361.
Baker, A. D., and Turner, D. W. (1970). *Philos. Trans. R. Soc. London, Ser. A*, 1931.
Baker, A. D., Betteridge, D., Kemp, N. R., and Kirby, R. E. (1970). *Anal. Chem.* **42**, 1064.
Bates, R. G., and Pinching, G. D. (1949). *J. Res. Natl. Bur. Stand.* **42**, 419.
Batich, C., Heilbronner, E., Hornung, V., Ashe, A. J., III, Clark, P. T., Cobley, V. T., Kilcast, D., and Scanlon, J. (1973). *J. Am. Chem. Soc.* **95**, 928.
Beak, P., Mueller, D. S., and Lee, J. (1974). *J. Am. Chem. Soc.* **96**, 3867.
Beauchamp, J. L. (1971). *Annu. Rev. Phys. Chem.* **22**, 527.
Beauchamp, J. L., and Butrill, S. E. (1968). *J. Chem. Phys.* **48**, 1783.
Beggs, D. P., and Field, F. H. (1971a). *J. Am. Chem. Soc.* **93**, 1567.
Beggs, D. P., and Field, F. H. (1971b). *J. Am. Chem. Soc.* **93**, 1576.
Benson, S. W. (1958). *J. Am. Chem. Soc.* **80**, 5151.
Blint, R. J., McMahon, T. B., and Beauchamp, J. L. (1974). *J. Am. Chem. Soc.* **96**, 1269.
Bock, H., and Wagner, G. (1972). *Angew. Chem., Int. Ed. Engl.* **11**, 150.
Bohme, D. K. (1975). *In* "Interactions Between Ions and Molecules" (P. Ausloos, ed.), p. 489. Plenum, New York.
Bohme, D. K., Lee-Rupp, E., and Brewster Young, L. (1971). *J. Am. Chem. Soc.* **93**, 4608.
Bohme, D. K., Fennelly, P., Hemsworth, R. S., and Schiff, H. I. (1973a). *J. Am. Chem. Soc.* **95**, 7512.
Bohme, D. K., Hemsworth, R. S., Rundle, H. W., and Schiff, H. I. (1973b). *J. Chem. Phys.* **58**, 3504.
Bowers, M. T., Aue, D. H., Webb, H. M., and McIver, R. T. (1971). *J. Am. Chem. Soc.* **93**, 4314.
Brauman, J. I., and Blair, L. K. (1968). *J. Am. Chem. Soc.* **90**, 5636.
Brauman, J. I., Riveros, J. M., and Blair, L. K. (1971). *J. Am. Chem. Soc.* **93**, 3914.
Briggs, J. P., Yamdagni, R., and Kebarle, P. (1972). *J. Am. Chem. Soc.* **94**, 5138.
Brown, H. C., and Kanner, B. (1966). *J. Am. Chem. Soc.* **88**, 9861.
Bünzli, J. C., Frost, D. C., and Weiler, L. (1973). *J. Am. Chem. Soc.* **95**, 7890.
Cabani, S., Conti, G., and Lepori, L. (1968). *Ric. Sci.* **38**, 1039.
Cabani, S., Conti, G., and Gianni, P. (1969). *J. Chem. Soc.* p. 1363.
Cabani, S., Conti, G., and Lepori, L. (1971). *Trans. Faraday Soc.* **67**, 1933.
Ceyer, S. T., Tiedemann, P. W., Mahan, B. H., and Lee, Y. T. (1979). *J. Chem. Phys.* **70**, 14.
Chang, S.-L., and Franklin, J. L. (1972a). *J. Am. Chem. Soc.* **94**, 6347.
Chang, S.-L., and Franklin, J. L. (1972b). *J. Am. Chem. Soc.* **94**, 6630.
Christenson, J. J., Izatt, R. M., Wrathall, D. P., and Hansen, L. D. (1969). *J. Chem. Soc. A* p. 1212.
Christenson, J. J., Smith, D. E., Slader, M. D., and Izatt, R. M. (1973). *Thermochim. Acta* **5**, 35.
Christie, A. O., and Crisp, D. J. (1967). *J. Appl. Chem.* **17**, 11.
Cook, M. J., Katritsky, A. R., Taagepera, M., Singh, T. D., and Taft, R. W. (1976a). *J. Am. Chem. Soc.* **98**, 6048.
Cook, M. J., Katritsky, A. R., Hepler, L. G., and Matsui, T. (1976b). *Tetrahedron Lett.* p. 2685.
Cox, J. C., and Pilcher, G. (1970). "Thermochemistry of Organic and Organometallic Compounds." Academic Press, New York.
Davidson, W. R., Bowers, M. T., Su, T., and Aue, D. H. (1977). *Int. J. Mass Spectrom. Ion Phys.* **24**, 83.
DeFrees, D. J., McIver, R. T., and Hehre, W. J. (1977). *J. Am. Chem. Soc.* **99**, 3853.
Derrick, P. J., Asbrink, L., Edguist, O., Jonsson, G. O., and Lindholm, E. (1971). *Int. J. Mass Spectrom. Ion Phys.* **6**, 191.
Devlin, J. L., III, Wolf, J. F., Taft, R. W., and Hehre, W. J. (1976). *J. Am. Chem. Soc.* **98**, 1990.

Duben, A. J., and Lowe, J. P. (1971). *J. Chem. Phys.* **55**, 4270.
Dunbar, R. C., Fu, E. W., and Olah, G. A. (1977). *J. Am. Chem. Soc.* **99**, 7502.
Dymerski, P. P., and McLafferty, F. W. (1976). *J. Am. Chem. Soc.* **98**, 6070.
Fehsenfeld, F. C. (1975). *Int. J. Mass Spectrom. Ion Phys.* **16**, 151
Fehsenfeld, F. C., and Ferguson, E. E. (1973). *J. Chem. Phys.* **59**, 6272.
Ferguson, E. E., Fehsenfeld, F. C., and Schmeltekopf, A. L. (1969). *Adv. At. Mol. Phys.* **5**, 1.
Ferguson, K. C., and Whittle, E. (1971). *Trans. Faraday Soc.* **67**, 2618.
Foster, M. S., and Beauchamp, J. L. (1975). *Inorg. Chem.* **14**, 1229.
Freiser, B. S., and Beauchamp, J. L. (1976). *J. Am. Chem. Soc.* **98**, 265.
Friedman, H., and Krishnan, C. V. (1973). *In* "Water, A Comprehensive Treatise" (F. F. Franks, ed.), Vol. 3, Ch. 1. Plenum, New York.
Golden, D. M., and Benson, S. W. (1969). *Chem. Rev.* **69**, 125.
Golden, D. M., Rodgers, A. S., and Benson, S. W. (1966). *J. Am. Chem. Soc.* **88**, 3196.
Golden, D. M., Gac, N. A., and Benson, S. W. (1969). *J. Am. Chem. Soc.* **91**, 2136.
Good, W. D. (1971). *J. Chem. Thermodyn.* **3**, 539.
Harrison, A. G., Honner, R. R., Dauben, H. J., Jr., and Lossing, F. P. (1960). *J. Am. Chem. Soc.* **82**, 5593.
Hehre, W. J., McIver, R. T., Pople, J. A., and Schleyer, P. von R. (1974). *J. Am. Chem. Soc.* **96**, 7162.
Heilbronner, E., Hornung, V., and Muszkat, K. A. (1970). *Helv. Chim. Acta* **53**, 347.
Heilbronner, E., Hornung, V., Pinkerton, F., and Thanes, S. F. (1972). *Helv. Chim. Acta* **55**, 289.
Hemsworth, R. S., Rundle, H. W., Bohme, D. K., Schiff, H. I., Dunkin, D. B., and Fehsenfeld, F. C. (1973). *J. Phys. Chem.* **59**, 61.
Henderson, W. G., Taagepera, D. Holtz, D., McIver, R. T., Beauchamp, J. L., and Taft, R. W. (1972). *J. Am. Chem. Soc.* **94**, 4728.
Hiroaka, K., and Kebarle, P. (1976). *J. Am. Chem. Soc.* **98**, 6119.
Hodges, R. V., and Beauchamp, J. R. (1975). *Inorg. Chem.* **14**, 2887.
"JANAF Thermochemical Tables" (1971). Natl. Stand. Ref. Data Ser., Natl. Bur. Stand., NSRDS-NBS 37 (2nd Ed.). U.S. Gov. Print. Off., Washington, D.C.
Jones, F. M., III, and Arnett, D. M. (1974). *Prog. Phys. Org. Chem.* **11**, 263.
Kebarle, P. (1972). *In* "Ions and Ion Pairs" (M. Szwarc, ed.), p. 27. Wiley, New York.
Kebarle, P. (1977). *Annu. Rev. Phys. Chem.* **28**, 445.
Kebarle, P., and Godbole, E. W. (1963). *J. Chem. Phys.* **39**, 1131.
Kemper, P. R., and Bowers, M. T. (1977a). *Rev. Sci. Instrum.* **48**, 1477.
Kemper, P. R., and Bowers, M. T. (1977b). *J. Chem. Phys.* **68**, 322.
Kerr, J. A. (1966). *Chem. Rev.* **66**, 465.
King, G. H., Murrel, J. N., and Suffolk, R. J. (1972). *J. Chem. Soc., Dalton Trans.* p. 564.
Koenig, T., Balle, T., and Snell, W. (1975). *J. Am. Chem. Soc.* **97**, 662.
Koenig, T., and Chang, J. C. (1978). *J. Am. Chem. Soc.*, **100**, 2240.
Kolthoff, I. M., and Chantooni, M. K., Jr. (1973). *J. Am. Chem. Soc.* **95**, 4768, 8539.
Lake, R. F., and Thompson, H. (1970). *Proc. R. Soc. London, Ser. A* 317, 187.
Lau, Y. K., and Kebarle, P. (1976). *J. Am. Chem. Soc.* **98**, 7452.
Liotta, C. L., Perdue, E. M., and Hopkins, H. P., Jr. (1974). *J. Am. Chem. Soc.* **96**, 7308.
Long, J. W., and Franklin, J. L. (1974). *J. Am. Chem. Soc.* **96**, 2320.
Lossing, F. P. (1971). *Can. J. Chem.* **49**, 356.
Lossing, F. P. (1972). *Can. J. Chem.* **50**, 3973.
Lossing, F. P., and Maccoll, A. (1976). *Can. J. Chem.* **54**, 990.
Lossing, F. P., and Semeluk, G. P. (1970). *Can. J. Chem.* **48**, 955.
Lossing, F. P., and Traeger, J. C. (1975). *J. Am. Chem. Soc.* **97**, 1579.
Love, P., Cohen, R., and Taft, R. W. (1968). *J. Am. Chem. Soc.* **90**, 2455.
Luria, M., and Benson, S. W. (1975). *J. Am. Chem. Soc.* **97**, 3342.

McDaniel, D. H., and Ozcan, M. (1969). *J. Org. Chem.* **33**, 1922.
McIver, R. T. (1970). *Rev. Sci. Instrum.* **41**, 555.
McIver, R. T., and Dunbar, R. C. (1971). *Int. J. Mass Spectrom. Ion Phys.* **7**, 471.
McIver, R. T., and Miller, J. S. (1974). *J. Am. Chem. Soc.* **96**, 4323.
McMahon, T. B., and Beauchamp, J. L. (1972). *Rev. Sci. Instrum.* **43**, 509.
McMahon, T. B., Blint, R. J., Ridge, D. P., and Beauchamp, J. L. (1972). *J. Am. Chem. Soc.* **94**, 8934.
McMahon, T. B., Miasek, P. G., and Beauchamp, J. L. (1976). *Int. J. Mass Spectrom. Ion Phys.* **21**, 63.
Munson, M. J. B. (1965). *J. Am. Chem. Soc.* **87**, 2332.
Neilson, P. V., Bowers, M. T., Chau, M., Davidson, W. R., and Aue, D. H. (1978). *J. Am. Chem. Soc.* **100**, 3649.
Ng, C. Y., Trevor, O. J., Tiedemann, P. N., Ceyer, S. T., Kronebusch, P. R., Mahan, B. H., and Lee, Y. T. (1977). *J. Chem. Phys.* **67**, 4235.
Noyes, R. M. (1962). *J. Am. Chem. Soc.* **84**, 513.
Payzant, J. D., Schiff, H. I., and Bohme, D. K. (1975). *J. Chem. Phys.* **63**, 149.
Perrin, D. D. (1965). "Dissociation Constants of Organic Bases in Aqueous Solution." Butterworth, London.
Pollack, S. K., Wolf, J. F., Levi, B. A., Taft, R. W., and Hehre, W. J. (1977a). *J. Am. Chem. Soc.* **99**, 1350.
Pollack, S. K., Devlin, J. L., Summerhays, K. Taft, R. W., and Hehre, W. J. (1977b). *J. Am. Chem. Soc.* **99**, 4583.
Pottie, R. F., Harrison, A. G., and Lossing, F. P. (1961). *J. Am. Chem. Soc.* **83**, 3204.
Potts, A. W., and Price, W. C. (1972). *Proc. R. Soc., Ser. A* **326**, 181.
Ramsey, B. G., and Walker, F. A. (1974). *J. Am. Chem. Soc.* **96**, 3314.
Refaey, K. M., and Chupka, W. A. (1968). *J. Chem. Phys.* **48**, 5205.
Robin, M. B., and Kuebler, N. A. (1972). *J. Electron Spectrosc. Relat. Phenom.* **1**, 13.
Sacconi, L., Paoletti, P., and Ciampolini, M. (1960a). *J. Am. Chem. Soc.* **82**, 3831.
Sacconi, L., Paoletti, P., and Ciampolini, M. (1960b). *J. Am. Chem. Soc.* **82**, 3828.
Schleyer, P. von R., Williams, J. E., and Blanchard, K. R. (1970). *J. Am. Chem. Soc.* **92**, 2377.
Shear, K. J., Gobeille, R., Bramblett, J., and Thompson, E. (1978). *J. Am. Chem. Soc.* **100**, 1611.
Shindell, D. M., Magagnosc, C., and Purich, D. L. (1978). *J. Chem. Educ.* **55**, 708.
Solka, B. H., and Russell, M. E. (1974). *J. Phys. Chem.* **78**, 1268.
Solomon, J. J., and Field, F. H. (1975). *J. Am. Chem. Soc.* **97**, 2625.
Solomon, J. J., and Field, F. H. (1976a). *J. Am. Chem. Soc.* **98**, 1025.
Solomon, J. J., and Field, F. H. (1976b). *J. Am. Chem. Soc.* **98**, 1567.
Solomon, J. J., Moet-Ner, M., and Field, F. H. (1974). *J. Am. Chem. Soc.* **96**, 3727.
Stafast, H., and Bock, H. (1973). *Z. Naturforsch., Teil B* **28**, 746.
Staley, R. H., and Beauchamp, J. L. (1974). *J. Am. Chem. Soc.* **96**, 1604.
Staley, R. H., and Beauchamp, J. L. (1975). *J. Chem. Phys.* **62**, 1998.
Staley, R. H., Kleickner, J. E., and Beauchamp, J. L. (1976). *J. Am. Chem. Soc.* **98**, 2081.
Staley, R. H., Wieting, R. D., and Beauchamp, J. L. (1977). *J. Am. Chem. Soc.* **98**, 5964.
Stull, D. R., Westrum, E. F., Jr., and Sinke, G. C. (1969). "The Chemical Thermodynamics of Organic Compounds." Wiley, New York.
Summerhays, K. D., Pollack, S. K., Taft, R. W., and Hehre, W. J. (1977). *J. Am. Chem. Soc.* **99**, 4585.
Sweigart, D. A., and Turner, D. W. (1972). *J. Am. Chem. Soc.* **94**, 5599.
Taagapera, M., Henderson, W. G., Brownlee, R. T. C., Beauchamp, J. L., Holtz, D., and Taft, R. W. (1972). *J. Am. Chem. Soc.* **94**, 1369.
Taft, R. W. (1975). *In* "Proton Transfer Reactions" (E. F. Caldin and V. Gold, eds.), p. 31. Chapman & Hall, London.

Taft, R. W., Taagepera, M., Summerhays, K. D., and Mitsky, J. (1973). *J. Am. Chem. Soc.* **95**, 3811.

Taft, R. W. Wolf, J. F., Beauchamp, J. L., Scorrano, G., and Arnett, E. M. (1978). *J. Am. Chem. Soc.* **100**, 1240.

Tal'roze, V. L., and Frankevich (1956). *Dokl. Akad. Nauk SSSR* **111**, 376.

Tam, W.-C., Yee, O., and Brian, C. E. (1974). *J. Electron. Spectrosc. Relat. Phenom.* **4**, 77.

Traeger, J. C., and McLoughlin, R. G. (1977). *J. Am. Chem. Soc.* **99**, 7351.

Tsang, W. (1972). *J. Phys. Chem.* **76**, 143.

Turner, D. W., Baker, C., Baker, A. D., and Brundle, C. R. (1970). "Molecular Photoelectron Spectroscopy." Wiley, New York.

Vincow, G., Dauben, H. J., Jr., Hunter, F. R., and Volland, W. V. (1969). *J. Am. Chem. Soc.* **91**, 2823.

Watanabe, J., Yakoyama, Y., and Ikeda, S. (1974). *Bull. Chem. Soc. Jpn.* **47**, 627.

Watanabe, K., Nakayama, T., and Mottl, J. (1962). *J. Quant. Spectrosc. Radiat. Transfer* **2**, 369.

Wei, L. Y., and Bone, L. I. (1974). *J. Phys. Chem.* **78**, 2527.

Wieting, R. D., Staley, R. H., and Beauchamp, J. L. (1974). *J. Am. Chem. Soc.* **96**, 7552.

Wincel, H. (1972). *Int. J. Mass Spectrom. Ion Phys.* **14**, 29.

Wolf, J. F., Harch, P. G., Taft, R. W., and Hehre, W. J. (1975a). *J. Am. Chem. Soc.* **97**, 2902.

Wolf, J. F., Harch, P. G., and Taft, R. W. (1975b). *J. Am. Chem. Soc.* **97**, 2904.

Wolf, J. F., Staley, R. H., Koppel, I., Taagepera, M., McIver, R. T., Beauchamp, J. L., and Taft, R. W. (1977a). *J. Am. Chem. Soc.* **99**, 5417.

Wolf, J. F., Abboud, J. L. M., and Taft, R. W. (1977b). *J. Org. Chem.* **42**, 3316.

Wren, A. (1976). M.S. Thesis, Univ. of California, Santa Barbara.

Wren, A., Gilbert, P., and Bowers, M. T. (1978). *Rev. Sci. Instrum.* **49**, 531.

Yamdagni, R., and Kebarle, P. (1973). *J. Am. Chem. Soc.* **95**, 3504.

Yamdagni, R., and Kebarle, P. (1976). *J. Am. Chem. Soc.* **98**, 1320.

Yoshikawa, K., Hashimoto, M., and Morishima, J. (1974). *J. Am. Chem. Soc.* **96**, 288.

Chapter 10
Electron affinities

BRUCE K. JANOUSEK AND JOHN I. BRAUMAN

DEPARTMENT OF CHEMISTRY
STANFORD UNIVERSITY
STANFORD, CALIFORNIA

I. INTRODUCTION AND SCOPE

One of the fundamental properties of a molecular negative ion is the lowest energy required to remove an electron. This energy is defined as the electron affinity (EA). For Eq. (1), $\Delta E_0 = \text{EA}$,

$$A^- \rightarrow A + e^-, \tag{1}$$

ISBN 0-12-120802-8

where A^- and A are in their ground rotational, vibrational, and electronic states and the electron has zero kinetic and potential energy.

This chapter is a review of some of the experimental methods used to determine molecular electron affinities—specifically, photoelectron spectroscopy, photodetachment spectroscopy, charge transfer, and collisional ionization. The values of electron affinities determined by these methods are presented at the end of the chapter. Also, a discussion of some of the interesting chemical and physical phenomena associated with magnitudes and trends in molecular electron affinities is included. While other methods such as electron swarm, surface ionization, dissociative electron attachment, and thermochemical cycles have been successful in attacking the general problem of obtaining accurate molecular electron affinities, we have restricted ourselves to the most direct, unambiguous determinations. The recent review by Rosenstock *et al.* (1977) covers electron affinities obtained by all types of methods. The subject of *atomic* electron affinities has been reviewed recently (Hotop and Lineberger, 1975) and will not be discussed in this chapter. In general, the references in the discussions are illustrative rather than comprehensive.

There are two general types of experiments that have been successful in determining molecular electron affinities. The first type consists of optical methods in which photons interact with negative ions to remove the least-tightly bound electron. The second type of experiment consists of chemical methods in which a chemical reaction (electron transfer) is utilized to determine the electron binding energy in the negative ion.

The optical methods that have been employed (see, e.g., Lineberger, 1974; Smyth and Brauman, 1972a) study the process

$$A^- + \hbar\omega \rightarrow A + e^- \tag{2}$$

by either using a fixed-frequency light source and measuring the energy of the ejected electron (photoelectron spectroscopy) or varying the frequency of the light source and following the wavelength dependence of (2) (photodetachment spectroscopy). The probability of photodetaching an electron from a negative ion is given in the thin-target approximation

$$P = (t)(f) \int \sigma(\lambda)\rho(\lambda)\, d\lambda, \tag{3}$$

where t is the time the ion spends in the light beam, f the geometrical overlap factor between the light beam and the ion beam, $\sigma(\lambda)$ the cross section for photodetachment, and $\rho(\lambda)$ the photon flux; the integration is over wavelength (λ). Photon–ion interactions are well suited to the determination of molecular electron affinities, since binding energies of most negative ions lie in the range 0.5–3.0 eV (~400–2500 nm). Thus, these methods should, in principle, give accurate electron affinities since the experimenter is only left to find the minimum photon energy needed to remove an electron from the ion. This is especially true considering the recent use of laser light sources in these experiments. Unfortunately, these methods must contend with vertical transitions between the negative ion and the resulting neutral, the intensities of the various transitions being governed by the Franck–Condon principle. Thus, if there is a substantial geometry change in going from the negative ion to the neutral, the 0–0 transition (i.e., the one corresponding to the adiabatic electron affinity) may be too weak to observe. However, in the case of small negative ions, the wealth of spectroscopic information available in these spectra can be unraveled to give not only highly accurate electron affinities, but also information on vibrational frequencies in the anion and neutral (Siegel *et al.*, 1972), spin–orbit splitting in the neutral (Smyth and Brauman, 1972b), and relative geometries of the anion and the neutral (Celotta *et al.*, 1972).

The chemical methods used to obtain molecular electron affinities also fall into two categories. The first type of experiment, exothermic charge transfer (Fehsenfeld *et al.*, 1966), determines if the electron transfer reaction between some primary negative ion A^- and a neutral molecule B,

$$A^- + B \rightarrow A + B^-, \tag{4}$$

proceeds efficiently to produce B^-. If the reaction does produce B^-, one can conclude that the electron affinity of B is greater than the electron affinity of A.

In a second type of experiment, the translational energy dependence for endothermic charge transfer reactions is investigated. Thus, for the reaction

$$A^- + B \rightarrow A + B^- \qquad \text{endothermic charge transfer} \tag{5}$$

or

$$A + B \rightarrow A^+ + B^- \qquad \text{collisional ionization} \tag{6}$$

the production of B^- is observed as a function of the translational energy of A^- in the first case (Lifshitz *et al.*, 1970) and A in the second case (Nalley and Compton, 1971). These chemical reactions are usually adiabatic, which would appear to be well suited to the determination of electron affinities. However, the most probable channel to the product ion B^- does not *necessarily* leave this ion in its ground state. This kinetic analog to the optical Franck–Condon principle makes the electron affinities determined by charge transfer reactions only lower limits to the adiabatic electron affinities.

II. EXPERIMENTAL METHODS

A. *Fixed-Frequency Laser Photoelectron Spectroscopy*

1. *Apparatus*

The method of studying negative ion detachment energies by photoelectron spectroscopy as developed by Hall and co-workers is described in detail by Siegel *et al.* (1972) and Celotta *et al.* (1972). Figure 1 is a diagram of the interaction region for this apparatus. In this particular experiment, a linearly polarized argon-ion laser (488.0 nm) intersects a 680 eV beam of negative ions perpendicularly. The electrons that are ejected from the negative ions are collected and energy analyzed at right angles to both the photon and ion beams. The data analysis requires kinematic corrections due to the coordinate transformation from the cm frame to the laboratory frame and corrections for drifts in the spatial overlap between the ion beam and the laser beam. Also, the contact potential energy difference between the interaction region and the electron energy analyzer is unknown and possibly

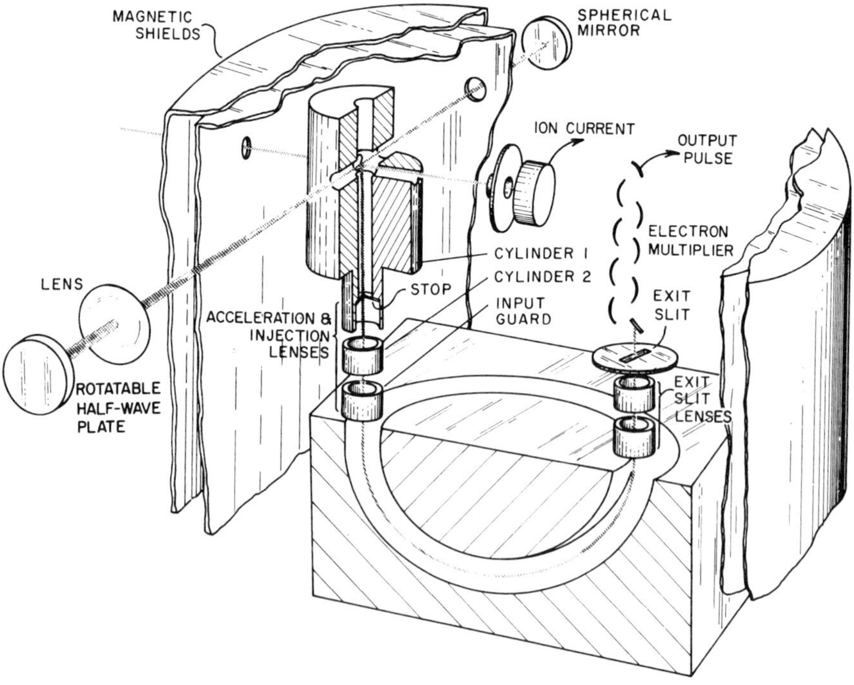

FIG. 1 Interaction region of the laser photoelectron spectroscopy apparatus employed by Lineberger and co-workers. [Siegel *et al.* (1972).]

varies with the chemical nature of the ion beam. Therefore, a reference ion whose electron affinity is known (usually O^-) is produced simultaneously, and the photoelectron energies obtained for the primary ion are calibrated against this known peak.

2. Discussion

The photoelectron spectrum of SO^- (Lineberger, 1974) is shown in Fig. 2. The peaks correspond to transitions from vibrational levels in SO^- to various vibrational levels in either $^3\Sigma$ or $^1\Delta$ neutral SO. To determine the adiabatic electron affinity, a vibrational assignment must be made to the peaks in the spectrum. Then the electron energy corresponding to the position of the 0–0 peak for the transition to the $^3\Sigma$ state of SO is subtracted from the photon energy (2.540 eV) to give the electron affinity of SO (1.126 ± 0.013 eV). In this case, the vibrational assignment was accomplished by measuring peak shifts upon isotopic substitution (i.e., comparing the spectra of $S^{16}O^-$ and $S^{18}O^-$). Alternatively, a Franck–Condon factor analysis can be made to determine the peak assignments (see, e.g., Siegel *et al.*, 1972). The resolution

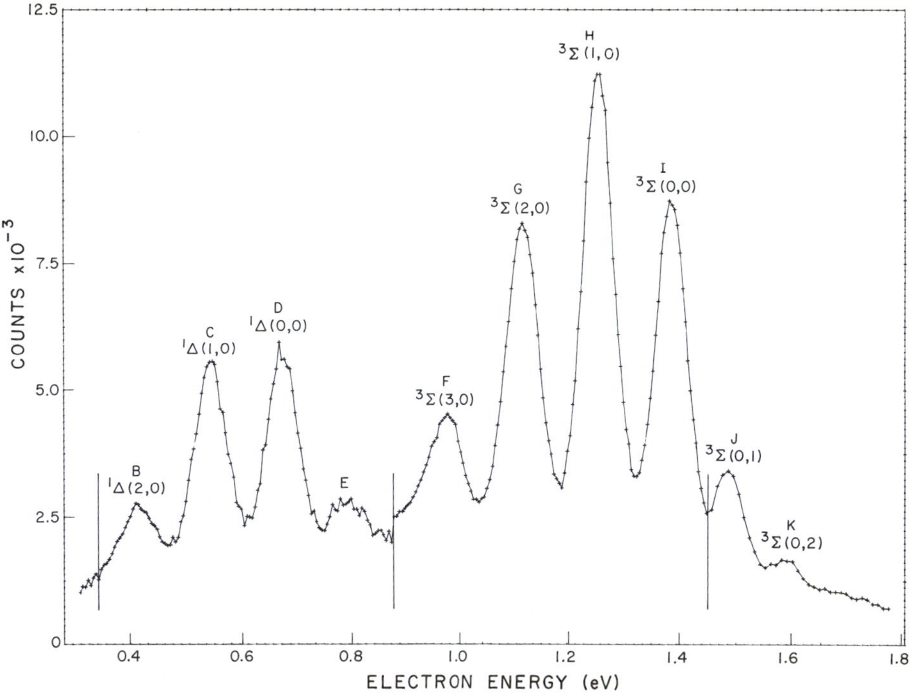

FIG. 2 Photoelectron energy spectrum of SO^- resulting from absorption of a 2.54 eV photon. [Lineberger (1974b).]

of the photoelectron spectra obtained by these experiments is determined by the resolution of the electron energy analyzer ~ 54 meV full-width at half-maximum (FWHM).

This technique has had two minor shortcomings, both of which will undoubtedly be eliminated in the future. First, molecules with electron affinities near or greater than the energy of the photon beam cannot be studied. This can be remedied by the use of a uv laser light source. Second, the negative ions have only been produced in a plasma source thus far. Consequently, the production of well-characterized ions has been difficult.

B. Photodetachment Spectroscopy

1. Apparatus

a. Crossed Photon–Molecular Beam Photodetachment. The pioneering work of Branscomb led to the first crossed-beam photodetachment apparatus (Smith and Branscomb, 1960). In these early experiments, ions were produced in a glow discharge at pressures $\sim 10^{-5}$ torr and accelerated to 300–500 eV. A carbon or xenon arc used in conjunction with interference filters or a monochromator served as the light source. The electrons produced via the process given in Eq. (2) were detected as a function of the photon flux and wavelength. This method was used to study the photodetachment of OH^-,

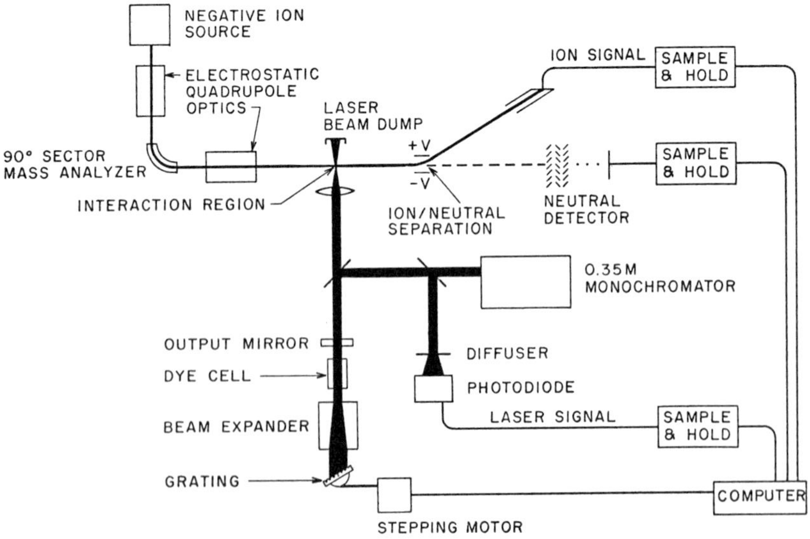

FIG. 3 Schematic diagram of the crossed-beam photodetachment apparatus employed by Lineberger and co-workers. [Hotop *et al.* (1973).]

OD^- (Smith and Branscomb, 1955; Branscomb, 1966), and O_2^- (Burch *et al.*, 1958, 1959). Feldmann has used a similar apparatus in which neutrals produced via Eq. (2) are detected to study several small negative ions (Feldmann, 1970, 1971).

These early experiments were followed by the crossed-beam method developed by Lineberger and co-workers. A schematic for this apparatus is shown in Fig. 3 (Hotop *et al.*, 1973). As in the photoelectron experiments, a mass-analyzed ion beam intersects the photon beam. In this method, the light source is a tunable laser and the production of neutral molecules via the process shown in Eq. (2) is followed as a function of wavelength. The ions are formed either in a hot cathode discharge source, a sputtering source, or an oven-type discharge source (Lineberger and Woodward, 1970; Lineberger, 1973; Hotop and Lineberger, 1973). Two electrostatic quadrupole doublets mass analyze and focus the beam. The neutrals formed via photodetachment are detected by a neutral particle multiplier. Pulsed operation with sample/hold circuitry is used to provide a neutral background correction. Similar circuitry is employed to measure the ion beam current and integrated laser power at a photodiode. Variations in the ion beam–laser beam spatial overlap are corrected for by measuring the neutral appearance at some wavelength after every three data points and normalizing the cross section to this wavelength.

b. Drift Tube Photodetachment. Drift tube mass spectrometers have been employed as a means of aging negative ions to thermal vibrational temperatures prior to the ion–photon interaction. This method has been successfully utilized by Woo *et al.* (1969), Sinnott and Beaty (1971), Burt (1972), and Huber *et al.* (1977). Figure 4 shows a diagram of the apparatus

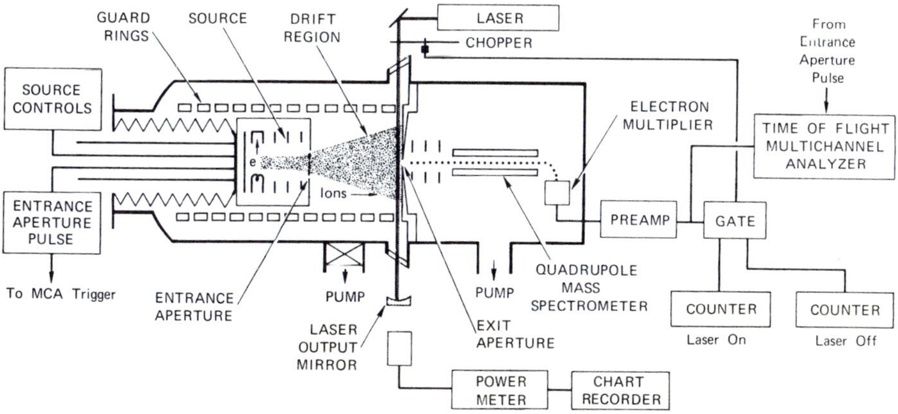

FIG. 4 Drift tube photodetachment apparatus employed by Moseley and co-workers. [Huber *et al.* (1977).]

used by Moseley, and co-workers. The ions are drifted under the influence of a weak electric field toward an extraction aperature. By varying the ratio of the electric field strength to the neutral gas density, the drift velocity of the ions is constrained to be much smaller than the mean thermal speed of the ions at room temperature. This assures that the ions will experience many thermalizing collisions following their production. The ions are intersected by a pulsed, intracavity photon beam before passing through an extraction aperture. The ions are then detected by an electron multiplier. Thus, the photodetachment cross section is determined by comparing the ion signal for alternate periods with the laser on and off.

c. Ion Cyclotron Resonance Photodetachment Spectroscopy. Brauman and co-workers have investigated ion–photon interaction by using an ion cyclotron resonance (ICR) spectrometer (Lehman and Bursey, 1976) as a negative ion trap. Figure 5 shows a schematic for the ICR cell. The cell is contained within a high vacuum can that enables experiments to be run in the pressure range from 10^{-7} to 10^{-4} torr. Ions are formed via dissociative electron capture or ion–molecule reaction and trapped for times on the order of a second by the electric and magnetic fields. The ions are detected by placing the upper and lower plates of the cell into the capacitance of a marginal oscillator circuit (Baldeschwieler, 1968) and bringing the ions into resonance by changing the magnetic field. In the photodetachment experiments, photons enter the ICR cell through a window at the end of the vacuum can. The marginal oscillator detects the change in negative ion concentration as the photon wavelength is varied. The photon flux is measured with a thermopile.

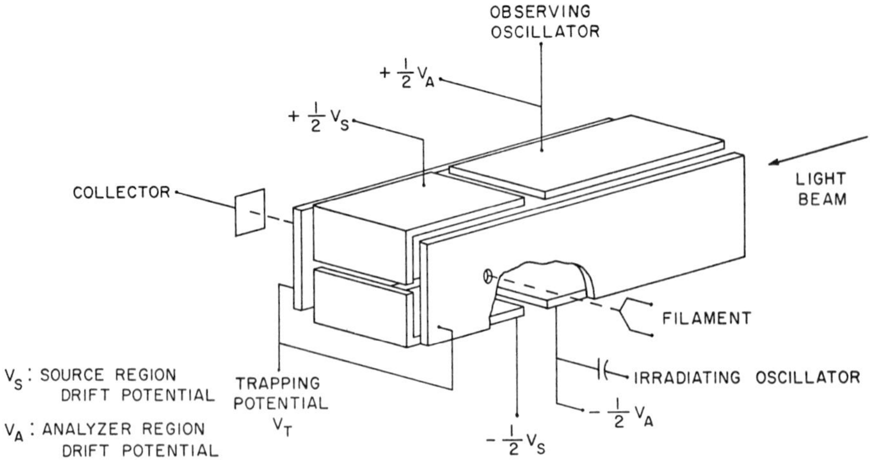

FIG. 5 Schematic of a drift ICR cell. (a) Ion source, (b) ion analyzer, and (c) ion collection.

Due to relatively long trapping times of the negative ions and the fact that the photon beam runs parallel to the ion motion, the ion–photon spatial overlap is large and cross sections can be determined with relatively low photon flux. A 1000 W arc lamp used in conjunction with a grating monochromator has proved to be successful in photodetaching several negative ions (see, e.g., Brauman and Smyth, 1969; Smyth and Brauman, 1972a; Reed and Brauman, 1974; Richardson *et al.*, 1975a). This photon source is capable of photodetaching approximately 5–10% of the ions in the cell at a resolution of 23.8 nm (FWHM). More recent work (Zimmerman and Brauman, 1977a) has employed a tunable dye laser light source. The high power and resolution (<1 Å) of this source has the potential of increasing the accuracy of electron affinities determined by ICR photodetachment.

To analyze the raw photodetachment data, a kinetic model of ion behavior in the ICR cell must be assumed. If signal decreases due to photodetachment exceed 10%, the thin-target approximation given in Eq. (3) does not provide a good description of ion behavior. In this regime, the model best suited to ICR photodetachment supposes steady-state behavior of the ions, i.e., the ions enter and leave the analyzer region of the cell at the same average rate, but the probability of a particular ion leaving the analyzer is independent of the time spent there (Dunbar, 1971, 1973).

2. *Discussion*

The photodetachment cross section onset for a ground state ion can be taken as an upper limit to the electron affinity of the resulting neutral molecule. If the cross section rises sharply at the onset, an accurate threshold wavelength can be assigned to the spectrum. This is the case for PH_2^- (Fig. 6). However, some ions, such as O_2^- (Fig. 7), have photodetachment cross sections that rise slowly, and it is difficult to make a good threshold assignment. Obviously, it would be desirable to have some *a priori* knowledge of the cross section shape as an aid in interpreting experimental cross sections.

Wigner has shown that the threshold energy dependence of a collision cross section for a process that has two product particles depends only on the long-range forces between the products (Wigner, 1948). For photodetachment of negative ions, the departing electron interacts with the neutral via a potential that falls off faster than r^{-2} if charge–dipole interactions are assumed to be negligible. For this case, the angular momentum $l\hbar$ of the two product particles with respect to their common center of mass determines the threshold energy dependence. Wigner showed that

$$\sigma(E) \propto k^{2l+1}, \tag{7}$$

where k is the magnitude of the reduced linear momentum of the photoelectron and $k = (2m\Delta E)^{1/2}/\hbar$. ΔE is the amount by which E, the energy

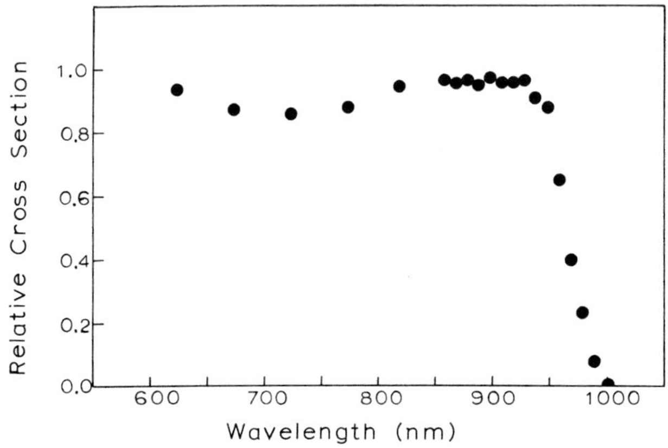

FIG. 6 Photodetachment spectrum of PH_2. [Smyth and Brauman (1972a).]

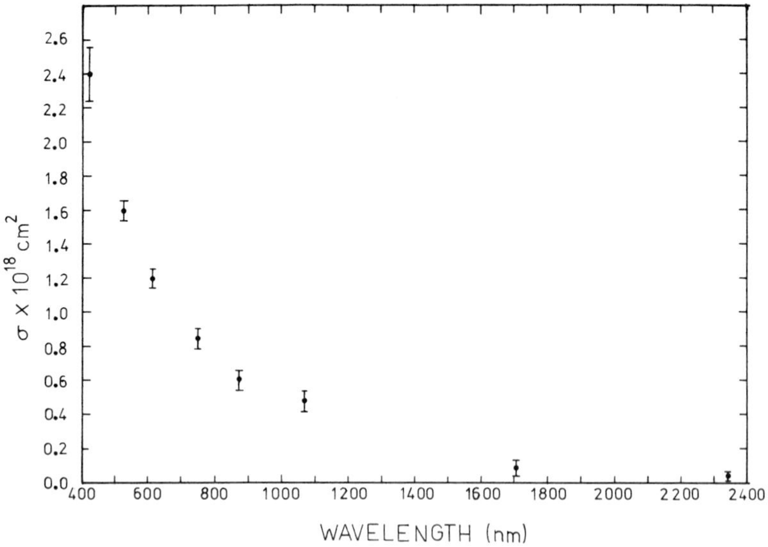

FIG. 7 Photodetachment spectrum of O_2^-. [Burch *et al.* (1958).]

of the reactants, exceeds the threshold energy. For atomic negative ions, $\Delta l = \pm 1$ and Eq. (7) allows the threshold shape to be easily correlated with the electronic angular momentum of the initial state.

Geltman has treated the threshold shape for *diatomic* negative ion photodetachment as a one-electron problem in which an electron moves in a cylindrically symmetric field (Geltman, 1958). This analysis leads to the expansion

$$\sigma \propto k^m(a_0 + a_1 k^2 + a_2 k^4 + \cdots), \tag{8}$$

where m is an odd positive integer related to the axial angular momentum λ_0 of an electron in the highest occupied molecular orbital (HOMO).

Reed *et al.* (1976) have extended Wigner's result to develop a theory for molecular photodetachment where the total cross section $\sigma(E)$ is expanded in terms of partial cross sections $\sigma_l(E)$. The partial cross section represents photodetachment into a final state in which the electron and the neutral have a relative orbital angular momentum of $l\hbar$. This expansion gives

$$\sigma(E) = A_{l*}k^{2l^*+1}[1 + O(k^2)], \tag{9}$$

where l^* is the smallest value of l for which $\sigma_l(E)$ is not identically zero for symmetry reasons. The value of l^* can be determined from group theory. For example, in O_2^-, the HOMO is π_g and is labeled by xy and yz in the point group $D_{\infty h}$. The effective angular momentum of the HOMO, therefore, is two and the selection rule $\Delta l = \pm 1$ gives $l^* = 1$. Thus, $\sigma(E) \propto k^3$ or $\sigma(E) \propto (\Delta E)^{3/2}$ and the slope of the cross section is zero at threshold, in agreement with experiment.

A one-electron formalism has also been developed by Reed *et al.* to calculate the cross section behavior for molecular negative ions. In general, the cross section is proportional to the square of the dipole matrix element

$$\sigma \propto |\langle \psi_i | \mathbf{r} | \psi_f \rangle|^2. \tag{10}$$

The initial state is taken to be the HOMO of the anion and is determined by an LCAO calculation, and the final state is approximated as a plane wave orthogonalized to this HOMO (partially orthogonalized plane wave). The Slater exponents in the HOMO expansion are adjusted to fit atomic photodetachment cross sections; this takes into account the diffuseness of anionic atomic orbitals. The calculations agree well with experimental data and can be used to predict photodetachment cross section behavior. However, high-resolution threshold measurements (Hotop *et al.*, 1974) show that cross sections can be sharper than predicted when the final-state neutral has a large dipole moment.

Even in cases where the cross section has a sharp onset and the threshold can be easily identified, poor Franck–Condon overlap (Section I) between the ion and the neutral can result in a threshold energy (the vertical detachment energy) which exceeds the adiabatic electron affinity. An example of this is CF_3^-, where the geometry change in going from the anion to the neutral is substantial and the threshold energy (2.96 ± 0.02 eV) has been estimated to exceed the adiabatic electron affinity by 0.8 eV (Richardson *et al.*, 1975b). Geometry optimizations of anions and neutrals have been attempted by both *ab initio* (Pearson *et al.*, 1974; Herbst *et al.*, 1974) and CNDO (Richardson *et al.*, 1975b) methods as a means of estimating Franck–Condon overlaps. These calculations have been limited to small molecules but have been helpful in relating experimental cross sections to electron affinities.

Most of the large molecular system photodetachment experiments have employed a lamp-monochromator photon source with resolution such that structure of $\sim 1000\ \text{cm}^{-1}$ would be barely discernable in the spectrum. In principle, a high-resolution light source could reveal vibronic fine structure in a photodetachment cross section, although this structure might be difficult to observe since the cross section measured in an integral spectrum contains a superposition of numerous vibronic transitions. However, if the cross section rises sharply at threshold, a laser photodetachment cross section could show a series of steps and could provide the same type of spectroscopic information derivable from photoelectron spectra (Section II.B).

If an ion is not in its ground vibrational state, it is possible for an electron to be detached at a photon energy less than the adiabatic electron affinity. In this case, the photodetachment threshold can no longer be taken as an upper limit to the electron affinity of the resulting neutral. This is a potential problem in experiments in which ions are formed at high temperatures and do not have time to relax. The ICR method allows ions to undergo ~ 10–100 collisions while trapped in the cell. These ions have the opportunity to lose most of any excess vibrational energy gained on formation. In these experiments, it is desirable to form the primary ion by methods differing in the degree of excess energy available to the ion and to measure the photodetachment cross section in each case. If the collisions in the cell are not effective in relaxing the ions, the photodetachment cross section of an ion with initial excess vibrational energy should exhibit hot band intensity. The drift tube experiments of Moseley and co-workers are ideally suited to the study of vibrationally excited ions since the vibrational temperature of the ions can be varied. Recent work by this method has determined the vibrational relaxation rate via collision for NO_2^- by observing the intensity of the hot band region of the NO_2^- photodetachment spectrum as a function of drift conditions (Huber *et al.*, 1977).

For large negative ions, hot bands are expected to play a role in the photodetachment cross section because low frequency vibrations in the ion will be excited (a $250\ \text{cm}^{-1}$ vibration will $\sim 25\%$ populated at 300 K). If these hot band transitions are allowed by their Franck–Condon overlap, the cross section will have a tail at longer wavelength than the adiabatic threshold. At high resolution it should be possible to observe the thermal hot bands as a series of steps, but with no *a priori* knowledge of the Franck–Condon overlaps it is difficult to determine which is the adiabatic transition. To determine how much of the cross section is due to thermal hot bands, one is left with the problem of finding a way to excite the ions while observing the effect on the cross section in the threshold region.

C. *Charge Transfer*

1. *Apparatus*

a. Exothermic Charge Transfer. Dunkin *et al.* (1972) have used a variable temperature flowing afterglow system to study exothermic charge transfer reactions. The afterglow was operated at a buffer gas (helium) pressure of ~0.4 torr and long residence times so that reactant ions were allowed to relax to their ground state. Lifshitz and Grajower (1969) have used a pulsed ion source mass spectrometer to study the charge transfer reactions of fluorocarbon negative ions. In these experiments, the ion source pressure was $(1-5) \times 10^{-4}$ torr and the electron energy ~0 eV.

b. Endothermic Charge Transfer. Lifshitz *et al.* (1970) have studied the translational energy dependence of charge transfer reactions with a tandem mass spectrometer which is capable of producing a mass and energy resolved ion beam. The translational energy of the beam can be controlled over a range from 0.3 to 170 eV with a resolution of ± 0.3 eV. The ion beam reacts with neutrals in a field-free collision chamber kept at pressures of $(5-30) \times 10^{-3}$ torr. Product ions emerging from the collision chamber are reaccelerated, mass analyzed, and detected with an electron multiplier in the pulse-counting mode. Due to the thermal velocity distribution of the neutral target molecules in the collision chamber, the data must be corrected for Doppler broadening of the effective reaction energy.

Chupka *et al.* (1971) have developed a double ionization chamber apparatus to study reactions in which halide ions were employed as the reactant ion. The halide ions are formed by uv photodissociation of halogens and then accelerated to the desired collision energy by a potential applied between two grids. The charge transfer reaction occurs in a field-free region. The product and reactant ions are focused into a mass spectrometer and detected.

c. Collisional Ionization. Helbing and Rothe (1969) have used a velocity selected Cs beam to study the translational energy dependence of the reaction

$$Cs + Br_2 \rightarrow Cs^+ + Br_2^-. \tag{11}$$

To produce a Cs beam, Cs^+ ions are formed by surface ionization on a heated filament and are accelerated through a nickel grid. The ions are converted to neutral atoms by charge exchange. The atom and ion beam is then stripped of the ions electrostatically and the atom beam passes through a velocity selector. The velocity selector has 3.4% velocity resolution but the beam divergence brings the resolution to 10% (FWHM). A Br_2 beam emerges from an array of sintered glass capillaries and intersects the primary beam perpendicularly. The Cs^+ formed by collision is accelerated into a thoriated tungsten wire which emits secondary electrons into an electron multiplier. After passing through the collision region, the primary Cs beam is ionized and the resulting Cs^+ accelerated to a multiplier for detection. The Br_2 beam is mechanically chopped and the outputs of the two multipliers are amplified and compared.

Baede *et al.* (1973) have studied the formation of molecular halogen negative ions by collision with alkali metal beams. A sputtering source is used to produce the alkali atom beam. Argon ions, produced in a discharge chamber, are accelerated onto an alkali metal atom surface. The sputtered atoms are collimated and velocity selected before crossing the target gas beam perpendicularly. The negative ions formed in the interaction region are accelerated toward a 30° sector-magnet mass spectrometer where they are detected.

Compton and co-workers (Nalley *et al.*, 1973; Compton and Cooper, 1973) have successfully utilized collisional ionization to determine the electron affinity of a number of molecules. A Cs atom beam, produced by the method of Helbing and Rothe (*vide ante*), reacts with the target gas in a reaction chamber near the entrance to a time-of-flight mass spectrometer. Detection of the Cs atom beam is accomplished by ionizing the Cs on a 1300 K filament and accelerating the ions onto a channel-electron multiplier. The mass spectrometer terminates at a magnetic electron multiplier chamber where the product negative ions are counted. In this method, corrections must be made for collision Doppler broadening due to the velocity distribution of the target gas.

2. *Discussion*

The early experiments of Henglein and Muccini (1959) showed that charge transfer reactions between negative ions and neutrals are fast in the gas phase. Therefore, the fact that a charge transfer reaction is observed can yield information about a thermodynamic quantity such as the electron affinity. For example, Dunkin *et al.* (1972) have observed the reactions

$$NO_2^- + Cl_2 \rightarrow NO_2 + Cl_2^-, \tag{12}$$

$$SH^- + NO_2 \rightarrow SH + NO_2^-, \tag{13}$$

which leads to the conclusion that

$$EA(SH) < EA(NO_2) < EA(Cl_2). \tag{14}$$

Only by relying on electron affinities determined by other methods can exothermic charge transfer data be used as a quantitative bracket. For NO_2,

Dunkin and co-workers have used the electron affinities of SH and Cl_2 determined by photodetachment and endothermic charge transfer reactions in conjunction with (12) and (13) to conclude that

$$2.32 \text{ eV} < EA(NO_2) < 2.43 \text{ eV}. \tag{15}$$

For endothermic charge transfer, one must determine the translational energy of A^- necessary to produce B^- via

$$A^- + B \rightarrow A + B^-. \tag{5}$$

The electron affinity of B is then determined by subtracting the threshold energy for Eq. (5) from the electron affinity of A. Thus, these experiments depend on a previously determined electron affinity value in order to be quantitative. Similarly, for collisional ionization, one must find the translational energy necessary to produce B^- via

$$A + B \rightarrow A^+ + B^-. \tag{6}$$

The electron affinity of B is determined by subtracting the threshold energy for (6) from the ionization potential of A. Since A is an alkali metal atom (usually Cs), its ionization potential is known accurately. Electron affinities can therefore be determined by this method without relying on previously determined electron affinities. As stated in Section I, both endothermic charge transfer and collisional ionization could leave the product ion B^- in an excited state. Although there is evidence that excited product ions were formed in reactions producing NO_2^- (Hughes *et al.*, 1973) and SF_6^- (Lifshitz *et al.*, 1975), this is not a general problem, and these methods will usually yield good absolute values for molecular electron affinities.

The threshold energy for both endothermic charge transfer and collisional ionization is difficult to determine accurately due to the Doppler broadening caused by the motion of the target neutrals and the slowness of the charge transfer reaction near threshold. These problems produce a slowly rising charge transfer cross section in the threshold region. To analyze a charge transfer cross section, the typical procedure (see, e.g., Hughes *et al.*, 1973) is to calculate the energy distribution of the target neutrals and then fold this distribution into some assumed threshold function. This calculated curve is then compared with the experimental data. The threshold law which gives the best fit with the experimental data is taken to be the "true" threshold and is used to find the threshold energy. The electron affinities determined by endothermic charge transfer and collisional ionization are accurate to no better than 0.1 eV.

III. TRENDS IN MOLECULAR ELECTRON AFFINITIES

With the increase in the number of experimentally determined electron affinities, it has now become possible to explore the relationship between structure and stability in negative ions. We attack this problem by considering the effects on molecular electron affinities of central atom electron-attracting power, conjugation, and alkyl substitution.

A. Central Atom Electron-Attracting Power

The electron-attracting power of atoms exhibits the well-known trend that elements to the upper right in the periodic table have an increasing tendency to bind electrons (increasing ionization potential). One might expect the electron affinities of the elements to follow a similar trend. In general, this is the case (Table I), but there are features that are unusual at first glance. For example, the group VA atoms (N, P, and As) have low electron affinities while the group IVA atoms (C, Si, and Ge) have relatively high electron affinities. This behavior can be understood by considering the electron configuration of the atoms. The neutral atoms N, P, and As have half-filled p shells, and by adding an electron to one of these atoms a p orbital will become doubly occupied, resulting in strong electron–electron repulsion. This repulsion raises the total energy of the group VA negative atomic ions relative to the neutrals and results in small electron affinities for atoms in this group. On the other hand, the group IVA atoms have an empty p orbital and an additional electron can be accomodated with a smaller increase in electron–electron repulsion. Electron–electron repulsion is also responsible for the ionization potential of N being larger than that of O (also P > S and As > Se). Here it is the electron–electron repulsion in neutral oxygen (and the group VIA atoms) which makes electron removal easier.

TABLE I

Selected Atomic Electron Affinities[a,b]

C	N	O	F
1.268	−0.07	1.462	3.399
Si	P	S	Cl
1.385	0.743	2.0772	3.615
Ge	As	Se	Br
1.2	0.8	2.0206	3.364

[a] See Hotop and Lineberger (1975) for references.

[b] All units are in electron volts.

TABLE II

Electron Affinities of Isoelectronic Species[a,b]

CH_3	NH_2	OH	F
0.08	0.779	1.829	3.399
SiH_3	PH_2	SH	Cl
<1.44	1.271	2.319	3.615
GeH_3	AsH_2	SeH	Br
<1.74	1.27	2.21	3.364

[a] See Table V for references.
[b] All units are in electron volts.

The instability produced by pairing electrons in an atomic orbital suggests that the electron configuration of atoms strongly affects their electron-attracting power. Therefore, it is important that we compare *isoelectronic* species when considering the effect of the central atom on the electron affinities of small molecules. Thus, in Table II we consider the electron affinities of the isoelectronic binary hydrides of the upper right corner of the periodic table. The electron-attracting power of the central atom is clearly responsible for the observed trend in this case (F > OH > NH_2; Cl > SH > PH_2).

We can also consider the effect of central atom electron-attracting power in larger molecules by replacing one of the CH_2 groups in the allyl anion with an oxygen atom to give the π-isoelectronic acetaldehyde enolate anion (Fig. 8). The enolate radical has an electron affinity about 1.25 eV greater than that of the allyl radical (Zimmerman and Brauman, 1977b; Zimmerman *et al.*, 1977). It appears that the stronger electron-attracting power of oxygen relative to carbon has manifested itself in stabilizing the acetaldehyde enolate anion relative to the allyl anion. Similarly, we can compare the π-isoelectronic species cyclopentadienyl anion and pyrrolate anion. The stronger electron-attracting power of the nitrogen atom gives the pyrrolate radical a greater electron affinity by 0.60 eV (Richardson *et al.*, 1975a; Engelking and Lineberger, 1977b).

FIG. 8 π-isoelectronic acetaldehyde enolate and allyl radicals.

$\dot{C}H_2{-}CH{=}O$ EA = 1.83 eV

$\dot{C}H_2{-}CH{=}CH_2$ EA = 0.550 eV

B. *Conjugation*

In general, conjugation helps to stabilize ions by charge dispersal. The most obvious kind of conjugation in negative ions is π delocalization, of which the phenoxide anion provides a good example. The electron affinity of the phenoxyl radical [~2.36 eV (Richardson *et al.*, 1975c)] is considerably greater than the electron affinities of alkoxyl radicals [~1.6–1.9 eV (Janousek *et al.*, 1978)],

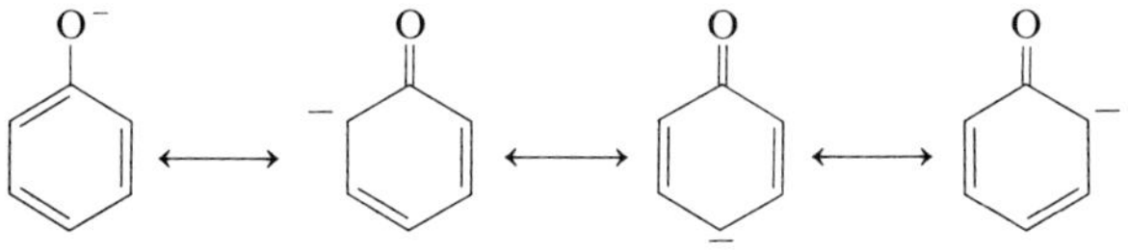

which have no π delocalization. This suggests that, as expected, the delocalized negative ion is highly stabilized. More quantitatively, as the size of a series of neutral π systems increases, the energy of the lowest unoccupied molecular orbital ($E_{\rm LUMO}$) should be lowered, and we should expect a correlation between EA and $E_{\rm LUMO}$. This is in accord with Koopmans' theorem (Koopmans, 1934), which suggests that $E_{\rm LUMO}$ of a closed shell neutral can be taken as a good approximation to -EA of the molecule. Equivalently, the energy of the highest occupied molecular orbital ($E_{\rm HOMO}$) of a closed shell anion can be taken as a good approximation to -EA of the neutral produced when an electron is removed from this orbital. Experimentally, this expectation is confirmed by the linear correlation between the polarographic half-wave potential of a molecule and $E_{\rm LUMO}$ of a molecule as determined by Hückel molecular orbital (HMO) theory (Hoijtink, 1955). The solution half-wave potential of a molecule is related to the gas-phase molecular electron affinity by (Streitwieser, 1961)

$$\epsilon_{1/2} = \mathrm{EA} + \Delta\Delta G^0_{\rm solv} + \mathrm{const}, \tag{16}$$

where $\Delta\Delta G^0_{\rm solv}$ is the difference in free energies of solvation of the molecule and its anion. For a series of hydrocarbons, $\Delta\Delta G^0_{\rm solv}$ changes in a predictable way, so that the intimate relationship between $\epsilon_{1/2}$ and EA is preserved. Thus, this work found a linear correlation between EA and the HMO $E_{\rm LUMO}$ for a series of delocalized hydrocarbons.

Hückel molecular orbital theory, however, neglects the effects of electron-electron repulsion. The importance of this effect in molecular anions can be illustrated by considering a series of polyene negative ions. Using HMO theory, one predicts that all odd alternant linear polyene anions should have identical electron affinities. However, a many-electron semiempirical calculation (PPP) on these anions (i.e., a calculation which includes some electron–electron repulsion) predicts

(Zimmerman *et al.*, 1978). This agrees with the observed experimental trend and is consistent with greater charge dispersal favoring a greater electron affinity by reducing electron–electron repulsion.

TABLE III

Electron Affinity and E_{LUMO} Values[a]

Molecule	EA (eV)	E_{LUMO} (eV)
Benzene	-1.14 ± 0.05	-0.87
Naphthalene	-0.20 ± 0.05	-1.94
Pyridine (1st resonance)	-0.62 ± 0.05	-1.26
Pyridine (2nd resonance)	-1.20 ± 0.05	-0.91
Pyrimidine (2nd resonance)	-0.77 ± 0.05	-1.09
Pyrazine (2nd resonance)	-0.87 ± 0.05	-1.01
Maleic anhydride	1.4 ± 0.2	-3.35
Para-benzoquinone	1.8 ± 0.2	-3.71
TCNQ	2.8 ± 0.2	-5.12

[a] See Younkin *et al.* (1976) for references.

Previously, Compton had employed PPP calculations to determine E_{LUMO} for a number of conjugated systems containing C, N, and O atoms (Younkin *et al.*, 1976) and found a linear correlation between experimentally determined gas-phase electron affinities and E_{LUMO} (Table III). A least-squares fit to these data gives

$$EA = -0.95E_{LUMO} - 1.9 \text{ eV}. \tag{17}$$

Koopmans' theorem predicts a slope of -1 and an intercept of zero. It is not surprising that Koopmans' theorem is not obeyed, since it neglects orbital reorganization in going from the neutral to the anion and electron correlation. The latter effect is expected to play a major role in negative ion stabilities, where much of the binding energy results solely from this effect. So, while Compton's empirical linear correlation can be used to predict electron affinities, Hartree–Fock calculations predict absolute values of molecular electron affinities rather poorly. In order to calculate molecular electron affinities properly, it is necessary to use techniques such as the equation-of-motion (EOM) method (Simons, 1977). The reliability of this method with respect to polyatomic systems shows considerable promise.

C. *Alkyl Substitution*

Alkyl substitution can affect negative ion stability in a variety of ways. The first of these can best be observed by considering a series of negative ions which differ only in the size of the alkyl group, e.g., a series of alkoxyl anions RO^-, $R'O^-$, etc. Alkoxyl anion stabilities were first studied in the gas phase when the relative acidities of a series of aliphatic alcohols were determined (Brauman and Blair, 1970). The acidity order neopentyl alcohol > *t*-butyl alcohol > ethanol > methanol was established. By the construction of an appropriate thermochemical cycle (Fig. 9), the acidity of A–H can be decomposed into the bond dissociation energy DH^0 (A–H), the electron affinity of A, and the ionization potential of H. For a series of aliphatic alcohols, the O–H bond dissociation energy remains constant as the size of the alkyl group is changed (Benson and Shaw, 1968). Therefore, the relative acidity order is governed by the differences in the electron affinities of the alkoxyl radicals. The results of ICR photodetachment experiments show the expected trend: EA(MeO) = 1.59 ± 0.04 eV, EA(Me_3CO) = 1.90 ± 0.01 eV or 1.87 ± 0.01 eV, and EA(Me_3CCH_2O) = 1.93 ± 0.05 eV (Janousek *et al.*, 1978).

These results suggest that larger alkyl groups are more effective at stabilizing a negatively charged center. Figure 10 shows the semiclassical model based on the ion-induced dipole interaction suggested by Brauman and Blair (1970) to explain the trends in alkoxyl anion stability, where $U(r)$ is the energy and α the polarizability of a nonpolar alkyl group which interacts with a unit charge e at a distance r. In this model, an alkoxyl anion is considered to be a polarizable alkyl group connected to a negatively charged oxygen atom; larger alkyl groups are more polarizable and therefore will be more effective at stabilizing the charge on oxygen. Although this model is simple and correctly predicts the electron affinity trend, it does not include explicitly the possibility for electron delocalization. It is also possible to use a molecular orbital perturbation model incorporating electron delocalization which considers the effect of increasing alkyl group size in alkoxyl anions.

Koopmans' theorem can be used in conjunction with a molecular orbital model to provide a simple explanation of alkyl substituent effects. We note that E_{HOMO} of a molecule is a function of the size of a molecule; larger neutral molecules generally have HOMOs at higher energies. Recalling that EA (or IP) $\approx -E_{\text{HOMO}}$, one predicts the trend in the ionization potentials of a

THERMOCHEMICAL CYCLE

$AH \longrightarrow A\cdot + H\cdot$	DH°
$A\cdot + e^- \longrightarrow A^-$	$-EA$
$H\cdot \longrightarrow H^+ + e^-$	IP
$AH \longrightarrow A^- + H^+$	ΔH° = Acidity

FIG. 9 Thermochemical cycle.

$$\overset{\leftrightarrow}{R}-\overset{-}{Z} \qquad U(r) = -\frac{1}{2}\frac{\alpha e^2}{r^4}$$

FIG. 10 Semiclassical ion-induced dipole model.

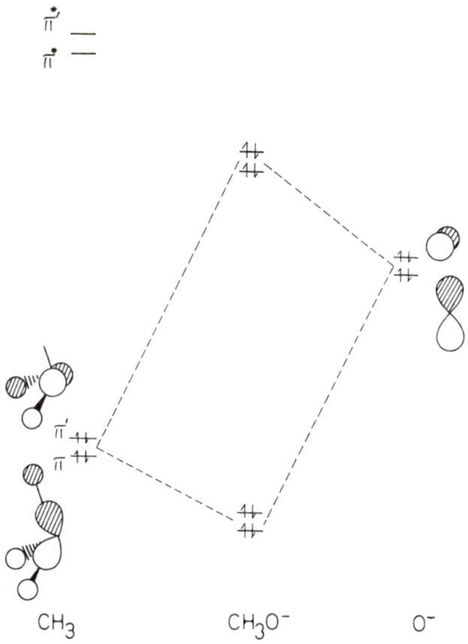

FIG. 11 Interaction between alkyl π orbital and oxygen p orbital.

series of alkyl chlorides: $CH_3Cl > CH_3CH_2Cl > (CH_3)_2CHCl$, which agrees with the observed trend (Watanabe, 1957; Watanabe *et al.*, 1962). If this model is used to explain the trend of alkyl substituent effects in the alkoxyl anions, the prediction is contrary to the experimental result. As noted above, larger molecules have E_{HOMO} at a higher energy. If this is also true for negative ions, then anions with larger alkyl groups presumably will have smaller electron binding energies. This does not agree with the observed electron affinity trend $Me_3CCH_2O > Me_3CO > MeO$ and suggests that this simple model does not adequately describe the alkyl substituent effects in negative ions.

The model best suited to explain the observed electron affinity trend in the alkoxyl radicals is an extension of that of Hudson *et al.* (1975) and, more recently, Hehre and co-workers (DeFrees *et al.*, 1977). In this perturbation picture, an alkoxyl anion RO^- is viewed as a combination of an alkyl radical R· and O^-. As shown for CH_3O^- in Fig. 11, the π orbitals on the alkyl group have the correct symmetry to mix with the filled oxygen p orbitals. The interaction between the π orbital on R and the oxygen p orbital is closed shell and destabilizing. However, the interaction between the π^* orbital on R and the oxygen p orbital is stabilizing as shown for CH_3O^- in Fig. 12. If the R group is increased in size, there are more π^* orbitals and the stabilizing interactions will be stronger. In anions, this appears to be the stronger interaction and results in a lowering of E_{HOMO}; therefore, the electron affinity trend $R'O > RO$ should be observed when R′ is larger than R.

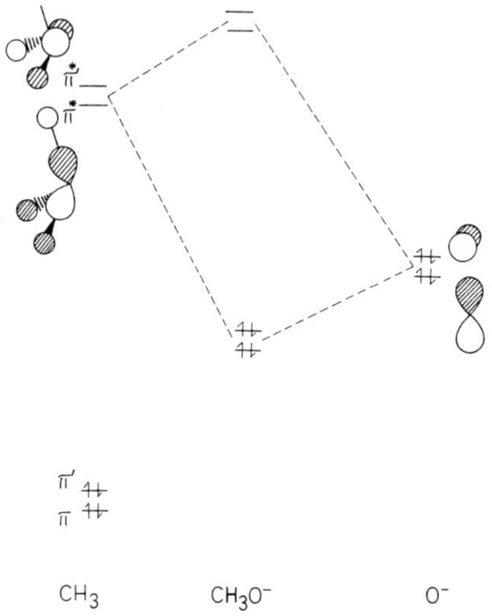

FIG. 12 Interaction between alkyl π^* orbital and oxygen p orbital.

The π^*–p orbital interaction which is important in the alkoxyl anions is not dominant in determining the relative ionization potentials of the neutral alkyl chlorides. In this case, the p orbitals of chlorine are lower in energy than the p orbitals of O^- and the π–p orbital repulsive interaction dominates. Alkyl chlorides with larger alkyl groups will have HOMOs at higher energies and have the lower Koopmans' ionization potential; this agrees with the observed trend in ionization potentials.

Theoretical work provides evidence for the importance of the π^*–p orbital interaction in the alkoxyl anions. Both CNDO/2 calculations (Janousek *et al.*, 1978) and *ab initio* calculations (Hudson *et al.*, 1975) have found when comparing CH_3O^- to CH_3O the C–H overlap populations are smaller in the anion than in the neutral. Thus, it appears that some anion stabilization occurs via hyperconjugative release of electrons onto the alkyl groups where partial occupation of the alkyl π^* orbitals is of prime importance (see also Lewis, 1969; Hehre and Pople, 1970; Owens *et al.*, 1970).

It is important to note that the original ion-induced dipole model is a classical consequence of the quantum mechanical model. The polarizability α of a molecule is defined as the ratio of the induced electric dipole moment to the strength of the applied electric field. However, this induced dipole moment can be related to the second-order Stark effect (Schiff, 1968), in which case

$$\alpha = -2\omega_2/E, \tag{18}$$

where

$$\omega_2 = S_n'(|\langle\psi_0|H'|\psi_n\rangle|^2)/(E_0 - E_n), \tag{19}$$

E is the electric field strength, E_0 the ground state energy of the molecule, E_n the energy of the nth excited state of the molecule, ψ_0 the wave function of the ground electronic state, ψ_n the wave function of the nth excited state, and S_n' a summation over the discrete set of wave functions together with an integration over the continuous set of eigenfunctions. (The term $n = m$ is omitted and the sum and integration are over n.) H' is the perturbation Hamiltonian arising from the influence of the external electric field.

From this equation it is clear that larger alkyl groups will give larger negative values for ω_2. This is due to the excited states being closer in energy to the ground state for larger groups (making the denominator smaller) and the number of excited states being larger, which makes the sum of numerators larger. Thus, the presence of low-lying unfilled orbitals will increase the polarizability of a molecule. Consequently, from the perturbation model description of alkoxyl anion stabilities, we predict that the polarizability of alkyl groups should correlate with the stabilities of negative ions.

The perturbation model used to explain the electron affinity trend in the alkoxyl radicals is valid only if Koopmans' theorem is a good approximation. In general, because it neglects the effects of orbital reorganization in going from the anion to the neutral (Section III.B), Koopmans' theorem does not work well in dealing with negative ions. It is not surprising to see large electron correlation effects in closed shell negative ions (see, e.g., Simons, 1977) and one would expect that the removal of an electron from a system in which electron correlation is so important should result in substantial orbital relaxation. The magnitude of this relaxation can be determined by comparing the Koopmans' electron affinity of an open shell molecule (E_{HOMO} of the anion) to the energy difference between an SCF calculation on the neutral molecule and the molecular anion (ΔSCF). The difference E_{HOMO} − ΔSCF is over 3 eV for OH^- (Simons, 1977) and is indicative of the problems in applying Koopmans' theorem to negative ions. However, for the series of aliphatic alcohols, we note that the RO—H bond dissociation energies are essentially constant. This implies that there are no stabilizing effects in the alkoxyl radicals that depend on the size of the alkyl groups and suggests that orbital reorganization will not be important in controlling the electron affinity trend in the alkoxyl radicals. Therefore, the trend in E_{HOMO} for the alkoxyl anions should agree with the observed electron affinity trend for the radicals. Although Koopmans' theorem is qualitatively valid in this case, this could not have been predicted without the RO—H bond strength data.

A comparison of the electron affinities of CH_3O (EA ≈ 1.59 eV) (Janousek *et al.*, 1978) and OH (EA = 1.829 eV) (Celotta *et al.*, 1974) provides a good example of the problems in applying Koopmans' theorem. As shown above, the π^*–p interaction should stabilize CH_3O^- relative to OH^- and make $EA(CH_3O) > EA(OH)$ by Koopmans' theorem (wrong prediction). However, if the perturbation model is applied to CH_3O we note that there is also a stabilizing interaction in this system. CH_3O is considered to be a mixing of CH_3 and an oxygen atom. The oxygen atom p orbitals are lower in energy than those on O^- and therefore the mixing of a methyl π orbital with a half-filled oxygen p orbital will result in a substantial stabilizing interaction in CH_3O which is not present in OH. Evidence that this effect is important lies in the fact that the O—H bond strength in CH_3OH is 15 kcal/mole lower than the O—H bond strength in H_2O (Kerr, 1966). Thus, there are stabilizing interactions present in both CH_3O^- and CH_3O not present in either OH^- or OH, and it is impossible to tell by the perturbation model which of the interactions in the methoxyl system will be stronger. Clearly, the stabilizing effect in the neutral causes Koopmans' theorem to be a poor approximation when comparing the electron affinities of CH_3O and OH due to the different extent of orbital relaxation occurring in CH_3O^-/CH_3O compared with OH^-/OH. Because of these difficulties, it appears that the Koopmans' approximation should be used with caution in dealing with negative ions.

Recent photodetachment experiments on substituted enolate anions provide further evidence of the stabilizing effects of larger alkyl groups. Some of the results of these experiments are shown in Table IV. By comparing

TABLE IV

Photodetachment Thresholds for Substituted Enolate Anions[a]

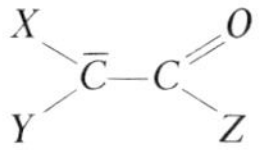

	X	Y	Z	Threshold (kcal/mole) = EA of resulting neutral
(1)	H	H	CH_3	41.2 ± 1.3
(2)	H	H	CD_2CH_3	40.4 ± 1.3
(3)	H	H	*t*-Bu	43.4 ± 1.5
(4)	D	CH_3	CH_3	38.6 ± 1.2
(5)	H	CH_3	CH_2CH_3	38.9 ± 1.2
(6)	H	CH_2CH_3	$CH_2CH_2CH_3$	39.7 ± 1.3
(7)	H	CH_3	H	38.9 ± 1.2
(8)	H	CH_2CH_3	H	38.5 ± 1.2
(9)	H	*t*-Bu	H	42.0 ± 1.4
(10)	H	H	H	42.3 ± 1.4
(11)	CH_3	CH_3	$CH(CH_3)_2$	33.8 ± 1.0

[a] Zimmerman *et al.* (1977).

the electron affinities of the first three entries in the table we see that, within experimental error, the larger, more polarizable alkyl groups increase the electron affinity. The same effect is observed by comparing entries 4, 5, and 6 and, likewise, by comparing 7, 8, and 9.

The enolate anion data allow us to investigate the effect of a given substituent at each carbon in the enolate system, i.e., the relative effects of methyl and *t*-butyl substituents at both the enolate α-carbon and the carbonyl carbon. At the α-carbon, the stabilities are methyl (38.9) $<$ *t*-butyl (42.0), while at the carbonyl carbon, methyl (41.2) $<$ *t*-butyl (43.4). Thus, substituents on the enolate α-carbon appear to be better at stabilizing the anion than those at the carbonyl carbon. This is expected since an enolate anion has considerable charge density on the α-carbon and considerable electron density on the α-carbon in the HOMO. The carbonyl carbon, on the other hand, carries little charge and little electron density in the HOMO. Therefore, one would expect substituents to act more effectively at the enolate α-carbon.

The lowering of the electron affinity upon replacing H with an alkyl group is observed where substitution has taken place at a trigonal carbon. For example, replacing H by CH_3 in cyclopentadienyl radical lowers the electron affinity by $\sim$0.17 eV:

$$EA(C_5H_5) \approx 1.84 \pm 0.03 \text{ eV} \quad \text{and} \quad EA(CH_3C_5H_4) \approx 1.67 \pm 0.04 \text{ eV}$$

(Richardson *et al.*, 1973). More examples are found for the enolate system (compare entries 1, 4, 7, 10, and 11 in Table IV). The effect is enhanced for substitution at the α-carbon as compared to the carbonyl carbon: -3.4 kcal/mole for acetaldehyde versus propionaldehyde and -1.1 kcal/mole for acetaldehyde versus acetone. Substitution of a methyl group for hydrogen appears to give rise to a dipolar destabilization (Brauman and Blair, 1971; Yamdagni and Kebarle, 1973) so the stronger effect at the α-carbon implies substantial negative charge on this carbon.

Finally, we note that if CH_3 is replaced by a strong electron-withdrawing group such as CF_3, there is a strong stabilization of the negative ion. For example,

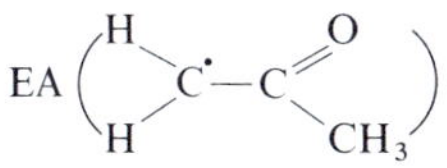

= 41.2 kcal/mole

and

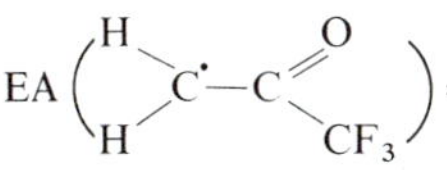

= 59.5 kcal/mole.

(Zimmerman *et al.*, 1977).

IV. TABLE OF EXPERIMENTALLY DETERMINED MOLECULAR ELECTRON AFFINITIES

In Table V we present a compilation of experimentally determined electron affinities.

TABLE V

Experimentally Determined Molecular Electron Affinities

Molecule	EA (eV)	Method[a]–Reference
	Diatomics	
As_2	0	PD Feldmann *et. al.* (1977)
AsH	1.0 ± 0.1	PD Rackwitz *et al.* (1977)
BeH	0.70 ± 0.1	PD Rackwitz *et al.* (1977)
Br_2	2.51 ± 0.10	ECT Chupka *et al.* (1971)
Br_2	2.55 ± 0.10	ECT Baede (1972)
Br_2	2.62 ± 0.2	ECT Hughes *et al.* (1973)
Br_2	2.6 ± 0.2	ECT Dispert and Lacmann (1977b)
C_2	$\leq 3.54 \pm 0.05$	PD Feldmann (1970)
CH	0.74 ± 0.05	PD Feldmann (1970)
CH	1.238 ± 0.008	PES Kasdan *et al.* (1975b)
CaH	0.93 ± 0.05	PD Rackwitz *et al.* (1977)
CN	3.82 ± 0.02	PD Berkowitz *et al.* (1969)
Cl_2	2.38 ± 0.10	ECT Chupka *et al.* (1971)
Cl_2	2.45 ± 0.15	ECT Baede (1972)
Cl_2	≥ 2.32	CT Dunkin *et al.* (1972)
Cl_2	2.32 ± 0.1	ECT Hughes *et al.* (1973)
Cl_2	2.4 ± 0.2	ECT Dispert and Lacmann (1977b)
F_2	3.08 ± 0.10	ECT Chupka *et al.* (1971)
FeO	1.492 ± 0.020	PES Engelking and Lineberger (1977a)
HS	2.319 ± 0.010	PD Steiner (1968)
HS	2.28 ± 0.15	PD Brauman and Smyth (1969)
IBr	$2.? \pm 0.2$	ECT Chopka *et al.* (1971)
IBr	2.55 ± 0.10	ECT Baede (1972)
I_2	2.58 ± 0.10	ECT Chopka *et al.* (1971)
I_2	2.55 ± 0.05	ECT Baede (1972)
I_2	2.42 ± 0.2	ECT Hughes *et al.* (1973)
LiCl	0.61 ± 0.02	PES Carlsten *et al.* (1976)
MgH	1.05 ± 0.06	PD Rackwitz *et al.* (1977)
NH	0.381 ± 0.014	PES Engelking and Lineberger (1976)
NH	0.38 ± 0.03	PES Celotta *et al.* (1974)
NO	$0.024 + 0.010 - 0.005$	PES Siegel *et al.* (1972)
NO	≥ 0.09	ECT Berkowitz *et al.* (1971)
NO	≤ 0.46	CT Fehsenfeld *et al.* (1966)

TABLE V (*Continued*)

Molecule	EA (eV)	Method[a]–Reference
NO	0.8 ± 0.2	ECT Refaey (1976)
NO	0.015 ± 0.1	ECT Hughes *et al.* (1973)
NO	$\geq 0.1 \pm 0.1$	CI Nalley *et al.* (1973)
OD	1.823 ± 0.002	LPD Hotop *et al.* (1974)
OH	1.83 ± 0.04	PD Branscomb (1966)
OH	1.825 ± 0.002	LPD Hotop *et al.* (1974)
OH	$1.829 + 0.010 - 0.014$	PES Celotta *et al.* (1974)
O_2	0.440 ± 0.008	PES Celotta *et al.* (1972)
O_2	0.45 ± 0.10	ECT Tiernan *et al.* (1971)
O_2	≥ 0.48	ECT Berkowitz *et al.* (1971)
O_2	0.5 ± 0.1	ECT Baede (1972)
O_2	0.15 ± 0.05	PD Burch *et al.* (1958)
O_2	$\geq 0.46 \pm 0.05$	CI Nalley and Compton (1971)
PH	1.028 ± 0.010	PES Zittel and Lineberger (1976)
PH	1.00 ± 0.06	PD Rackwitz *et al.* (1977)
P_2	≤ 0.65	PD Feldmann *et al.* (1977)
PO	1.092 ± 0.010	PES Zittel and Lineberger (1976)
SO	1.126 ± 0.013	PES Lineberger (1974)
SO	$\leq 1.09 \pm 0.05$	PD Feldmann (1970)
SO	≥ 1.0; ≤ 1.27	CT Kraus *et al.* (1961)
S_2	1.663 ± 0.040	PES Celotta *et al.* (1974)
SeH	2.21 ± 0.03	PD Smyth and Brauman (1972b)
SiH	1.277 ± 0.009	PES Kasdan *et al.* (1975a)
ZnH	≤ 0.95	PD Rackwitz *et al.* (1977)
	Triatomics	
AsH_2	1.27 ± 0.03	PD Smyth and Brauman (1972c)
C_2H	$\leq 3.73 \pm 0.05$	PD Feldmann (1970)
CH_2	0.210 ± 0.03	PES Zittel *et al.* (1976)
CH_2	≤ 0.6	PD Feldmann *et al.* (1977)
COS	0.46 ± 0.2	CI Compton *et al.* (1975)
CS_2	1.0 ± 0.2	CI Compton *et al.* (1975)
CS_2	0.50 ± 0.2	ECT Hughes *et al.* (1973)
CS_2	0.62 ± 0.2	CI Dispert and Lacmann (1975)
NH_2	0.744 ± 0.022	PD Smyth and Brauman (1972c)
NH_2	0.74	LPD Smyth *et al.* (1971)
NH_2	0.76 ± 0.04	PD Feldmann (1971)
NH_2	0.779 ± 0.037	PES Celotta *et al.* (1974)
N_2O	≤ 1.465	CT Paulson (1966)
NO_2	3.10 ± 0.05	PD Warneck (1969)
NO_2	≥ 2.32; ≤ 2.5	CT Dunkin *et al.* (1972)
NO_2	2.5 ± 0.1	ECT Baede (1972)
NO_2	≥ 2.04	ECT Berkowitz *et al.* (1971)
NO_2	≥ 3.6	CT Curran (1962)
NO_2	≥ 1.8	CT Ferguson *et al.* (1969)

TABLE V (*Continued*)

Molecule	EA (eV)	Method[a]–Reference
NO_2	2.1	ECT Refaey (1976)
NO_2	2.28 ± 0.1	ECT Hughes *et al.* (1973)
NO_2	$\geq 2.5 \pm 0.1$	CI Nalley *et al.* (1973)
NO_2	2.36 ± 0.10	LPD Herbst *et al.* (1974)
NO_2	2.50 ± 0.05	CI Leffert *et al.* (1973)
O_3	1.92 ± 0.13	PD Wong *et al.* (1971)
O_3	≥ 1.96	ECT Berkowitz *et al.* (1971)
O_3	2.14 ± 0.15	CI Rothe *et al.* (1975)
O_3	≤ 2.09	LPD Sinnott and Beatty (1971)
O_3	≥ 1.8234	CT Dotan *et al.* (1977)
O_3	$\geq 2.26 + 0.04 - 0.06$	ECT Lifshitz *et al.* (1978)
PH_2	1.25 ± 0.03	LPD Smyth and Brauman (1972a)
PH_2	1.26 ± 0.03	LPD Smyth *et al.* (1971)
PH_2	1.271 ± 0.010	PES Zittel and Lineberger (1976)
SO_2	$\leq 1.0 \pm 0.1$	PD Feldmann (1970)
SO_2	≥ 1.0; ≤ 1.27	CT Kraus *et al.* (1961)
SO_2	1.05	ECT Refaey and Franklin (1976)
SO_2	1.14 ± 0.15	CI Rothe *et al.* (1975)
SO_2	0.99 ± 0.1	ECT Hughes *et al.* (1973)
SO_2	1.097 ± 0.036	PES Celotta *et al.* (1974)
S_3	2.0 ± 0.15	PD Feldmann *et al.* (1977)
SiH_2	1.124 ± 0.020	PES Kasdan *et al.* (1975a)
	Tetra-atomics	
CCH_2	≤ 0.43	CT Lindinger *et al.* (1975)
CH_3	≤ 0.5	PD Feldmann *et al.* (1977)
CH_3	0.08 ± 0.03	PES Ellison *et al.* (1978)
CO_3	$2.69 + 0.07 - 0.14$	PD Hong *et al.* (1977)
GeH_3	$\leq 1.74 \pm 0.04$	PD Reed and Brauman (1974)
PBr_3	1.6	CI Mathur *et al.* (1976a)
PBr_2Cl	1.6	CI Mathur *et al.* (1976a)
PCl_2Br	1.5	CI Mathur *et al.* (1976a)
PCl_3	0.8	CI Mathur *et al.* (1976a)
$POCl_2$	3.8	CI Mathur *et al.* (1976a)
SO_3	$\geq 1.70 \pm 0.15$	CI Rothe *et al.* (1975)
SiH_3	$\leq 1.44 \pm 0.03$	PD Reed and Brauman (1974)
	Larger Polyatomics	
CD_3O	1.552 ± 0.022	PES Engelking *et al.* (1978)
CF_3Br	0.91 ± 0.2	CI Compton *et al.* (1978b)
CF_3I	1.4 ± 0.2	CI Tang *et al.* (1976)
CF_3I	1.57 ± 0.2	CI Compton *et al.* (1978b)
CH_3I	0.2 ± 0.1	ECT Moutinho *et al.* (1974)

TABLE V (*Continued*)

Molecule	EA (eV)	Method[a]–Reference
CH_3NO_2	0.44 + 0.1–0.2	CI Compton *et al.* (1978b)
CH_3O	≤1.59 ± 0.04	PD Reed and Brauman (1975)
CH_3O	1.570 ± 0.022	PES Engelking *et al.* (1978)
CH_3O	1.59 ± 0.04	LPD Janousek *et al.* (1978)
CH_3S	1.882 ± 0.024	PES Engelking *et al.* (1978)
$CO_3[H_2O]$	2.1 ± 0.2	PD Burt (1972)
C_2H_2FO (acetyl fluoride enolate)	2.22 ± 0.09	PD Zimmerman *et al.* (1977)
C_2H_2N (cyanomethyl)	1.507 ± 0.018	PD Zimmerman and Brauman (1977b)
C_2H_5N (ethyl nitrene)	≤1.87 ± 0.16	PD Richardson *et al.* (1974)
C_2H_3O (acetaldehyde enolate)	1.83 ± 0.06	PD Zimmerman *et al.* (1977)
C_3H_5 (allyl)	0.550 ± 0.054	PD Zimmerman and Brauman (1977b)
$C_3H_2F_3O$ (1,1,1-trifluoroacetone enolate)	2.58 ± 0.12	PD Zimmerman *et al.* (1977)
C_3H_5O (acetone enolate)	1.79 ± 0.06	PD Zimmerman *et al.* (1977)
C_3H_5O (propionaldehyde enolate)	1.69 ± 0.05	PD Zimmerman *et al.* (1977)
$C_3H_5O_2$ (pinacolone enolate)	1.80 ± 0.06	PD Zimmerman *et al.* (1977)
C_4H_6DO (2-butanone-3-*d*1 enolate)	1.67 ± 0.05	PD Zimmerman *et al.* (1977)
$C_4H_5D_2O$ (2-butanone-3,3-*d*2 enolate)	1.75 ± 0.06	PD Zimmerman *et al.* (1977)
$C_4F_4O_3$ (tetrafluorosuccinic anhydride)	0.5 ± 0.2	CI Cooper and Compton (1974)
C_4H_4N (pyrrolate)	2.39 ± 0.13	PD Richardson *et al.* (1975a)
$C_4H_2O_3$ (maleic anhydride)	1.4 ± 0.2	CI Compton *et al.* (1974)
C_4H_5O (cyclobutanone enolate)	1.84 ± 0.06	PD Zimmerman *et al.* (1978a)
C_4H_7O (butyraldehyde enolate)	1.67 ± 0.05	PD Zimmerman *et al.* (1977)
C_4H_9O (*t*-butoxyl)	<1.87 + 0.04	PD Reed and Brauman (1975)
C_4H_9O (*t*-butoxyl)	1.87 or 1.90 ± 0.01	LPD Janousek *et al.* (1978)
$C_5F_6O_3$ (hexafluoroglutaric anhydride)	1.5 ± 0.2	CI Cooper and Compton (1974)
C_5H_5 (cyclopentadienyl)	<1.84 ± 0.03	PD Richardson *et al.* (1973)
C_5H_5	1.786 ± 0.020	PES Engelking and Lineberger (1977b)
C_5H_7 (pentadienyl)	0.91 ± 0.03	PD Zimmerman *et al.* (1978b)
C_5H_7O (cyclopentanone enolate)	1.62 ± 0.06	PD Zimmerman *et al.* (1978a)
C_5H_9O (3-pentanone enolate)	1.69 ± 0.05	PD Zimmerman *et al.* (1977)
$C_5H_{11}O$ (neopentoxyl)	<1.93 ± 0.06	PD Reed and Brauman (1975)
$C_5H_{11}O$ (neopentoxyl)	1.93 ± 0.05	LPD Janousek *et al.* (1978)
$C_6Br_4O_2$ (*p*-bromanil)	2.44 ± 0.2	ECT Cooper *et al.* (1978)
$C_6Cl_4O_2$ (*p*-chloranil)	2.76 ± 0.2	ECT Cooper *et al.* (1978)
$C_6F_4O_2$ (*p*-fluoranil)	2.92 ± 0.2	ECT Cooper *et al.* (1978)
C_6F_6 (hexafluorobenzene)	>1.8 ± 0.3	ECT Lifshitz *et al.* (1973)
C_6F_{10} (perfluorocyclohexene)	>1.4 ± 0.3	ECT Lifshitz *et al.* (1973)
C_6H_7 (methylcyclopentadienyl)	<1.67 ± 0.04	PD Richardson *et al.* (1973)

TABLE V (*Continued*)

Molecule	EA (eV)	Method[a]–Reference
C_6N_4 (TCNE)	2.3 ± 0.3	PD Lyons and Palmer (1976)
C_6H_4ClO (*o*-chloro phenoxide)	$\leq 2.58 \pm 0.08$	PD Richardson *et al.* (1975e)
$C_6H_5NO_2$ (nitrobenzene)	$>0.7 \pm 0.2$	ECT Lifshitz *et al.* (1973)
$C_6H_4O_2$ (*p*-benzoquinone)	$1.89 + 0.2 - 0.3$	CI Cooper *et al.* (1975)
C_6H_5O (phenoxide)	$\leq 2.36 \pm 0.06$	PD Richardson *et al.* (1975c,e)
C_6H_5S (thiophenoxide)	$\leq 2.47 \pm 0.06$	PD Richardson *et al.* (1975e)
C_6H_9O (cyclohexanone enolate)	1.55 ± 0.05	PD Zimmerman *et al.* (1978a)
$C_6H_{11}O$ (methyl acetate enolate)	1.88 ± 0.06	PD Zimmerman *et al.* (1977)
$C_6H_{11}O$ (3,3-dimethylbutanal enolate	1.82 ± 0.06	PD Zimmerman *et al.* (1977)
C_7F_8 (octafluorotoluene)	$>1.7 \pm 0.3$	ECT Lifshitz *et al.* (1973)
$C_6H_5CH_2$ (benzyl)	$<0.88 \pm 0.06$	PD Richardson *et al.* (1975d)
C_7H_9 (heptatrienyl)	1.27 ± 0.03	PD Zimmerman *et al.* (1978b)
C_7H_7O (*o*-methyl phenoxide)	$\leq 2.36 \pm 0.06$	PD Richardson *et al.* (1975e)
C_7H_9O (2-norbornanone enolate)	1.61 ± 0.05	PD Zimmerman *et al.* (1978a)
$C_7H_{11}O$ (cycloheptanone enolate)	1.48 ± 0.04	PD Zimmerman *et al.* (1978a)
$C_7H_{11}O$ (2,5-dimethylcyclopentanone enolate)	1.49 ± 0.04	PD Zimmerman *et al.* (1978a)
$C_7H_{13}O$ (4-heptanone enolate)	1.72 ± 0.06	PD Zimmerman *et al.* (1977)
$C_7H_{13}O$ (di-isopropyl ketone enolate)	1.46 ± 0.04	PD Zimmerman *et al.* (1977)
C_8H_7O (acetophenone enolate)	2.06 ± 0.08	PD Zimmerman *et al.* (1977)
C_8H_7O (phenylacetaldehyde enolate)	$2.10 + 0.08$	PD Zimmerman *et al.* (1977)
$C_8H_{13}O$ (cyclooctanone enolate)	1.63 ± 0.06	PD Zimmerman *et al.* (1978a)
$C_8H_{13}O$ (cyclooctanone enolate)	1.63 ± 0.06	PD Zimmerman *et al.* (1978a)
$C_9H_{15}O$ (cyclononanone enolate)	1.69 ± 0.06	PD Zimmerman *et al.* (1978a)
$C_{10}H_{17}O$ (cyclodecanone enolate)	1.83 ± 0.06	PD Zimmerman *et al.* (1978a)
$C_{12}H_4N_4$ (TCNQ)	$2.8 + 0.1$	CI Klots *et al.* (1974)
$C_{12}H_4N_4$ (TCNQ)	$2.8 + 0.05 - 0.3$	CI Compton and Cooper (1977)
$C_{12}H_9$ (perinaphthenyl)	1.07	PD Gygax *et al.* (1979)
$C_{12}H_{21}O$ (cyclododecanone enolate)	1.90 ± 0.07	PD Zimmerman *et al.* (1978a)
HNO_3	0.57 ± 0.15	CI Mathur *et al.* (1976b)
H_3O_2 (hydrated hydroxyl radical)	$<2.95 \pm 0.15$	PD Golub and Steiner (1968)
$POCl_3$	1.4	CI Mathur *et al.* (1976a)
SF_6	0.75 ± 0.1	CI Leffert *et al.* (1974)
SF_6	$>0.6 \pm 0.1$	ECT Lifshitz *et al.* (1973)
SF_6	$0.54 + 0.1 - 0.17$	CI Compton and Cooper (1973)
SF_6	$>0.43; <1.465$	CT Fehsenfeld (1971)
SF_6	0.46 ± 0.2	CI Compton *et al.* (1978a)
SeF_6	2.9 ± 0.2	CI Compton *et al.* (1978a)
TeF_6	$3.34 + 0.1 - 0.17$	CI Compton and Cooper (1973)
TeF_6	3.3 ± 0.2	CI Compton *et al.* (1978a)
UF_6	$\geqslant 5.1$	CI Compton (1977)
WF_6	3.7 ± 0.3	CI Dispert and Lacmann (1977a)

[a] CI is collisional ionization, CT is charge transfer, ECT is endothermic charge transfer, PES is photoelectron spectroscopy, PD is photodetachment spectroscopy, and LPD is laser photodetachment spectroscopy.

V. CONCLUSION

In this chapter we have outlined some of the experimental procedures that are being employed to determine molecular electron affinities. We have also commented on the observed trends in electron affinities and compiled a table of experimentally determined electron affinities.

ACKNOWLEDGMENT

We gratefully acknowledge the financial assistance of the National Science Foundation for fellowship and research support.

REFERENCES

Baede, A. P. M. (1972). *Physica* (*Utrecht*) **59**, 541.
Baede, A. P. M., Auerbach, D. J., and Los, J. (1973). *Physica* (*Utrecht*) **64**, 134.
Baldeschwieler, J. D. (1968). *Science* **159**, 263.
Benson, S. W., and Shaw, R. (1968). *Adv. Chem. Ser.* No. 75, p. 288.
Berkowitz, J., Chupka, W. A., and Walter, T. A. (1969). *J. Chem. Phys.* **50**, 1497.
Berkowitz, J., Chupka, W. A., and Gutman, D. (1971). *J. Chem. Phys.* **55**, 2733.
Branscomb, L. M. (1966). *Phys. Rev.* **148**, 11.
Brauman, J. I., and Blair, L. K. (1970). *J. Am. Chem. Soc.* **92**, 5986.
Brauman, J. I., and Blair, L. K. (1971). *J. Am. Chem. Soc.* 93, 4315.
Brauman, J. I., and Smyth, K. C. (1969). *J. Am. Chem. Soc.* **91**, 7778.
Burch, D. S., Smith, S. J., and Branscomb, L. M. (1958). *Phys. Rev.* **112**, 171.
Burch, D. S., Smith, S. J., and Branscomb, L. M. (1959). *Phys. Rev.* **114**, 1652.
Burt, J. A. (1972). *J. Chem. Phys.* **57**, 4649.
Carlsten, J. L., Peterson, J. R., and Lineberger, W. C. (1976). *Chem. Phys. Lett.* **37**, 5.
Celotta, R. J., Bennett, R. A., Hall, J. L., Siegel, M. W., and Levine, J. (1972). *Phys. Rev.* **A 6**, 631.
Celotta, R. J., Bennett, R. A., and Hall, J. L. (1974). *J. Chem. Phys.* **60**, 1740.
Chupka, W. A., Berkowitz, J., and Gutman, D. (1971). *J. Chem. Phys.* **55**, 2724.
Compton, R. N. (1977). *J. Chem. Phys.* **66**, 4478.
Compton, R. N., and Cooper, C. D. (1973). *J. Chem. Phys.* **59**, 4140.
Compton, R. N., and Cooper, C. D. (1977). *J. Chem. Phys.* **66**, 4325.
Compton, R. N., Reinhardt, P. W., and Cooper, C. D. (1974). *J. Chem. Phys.* **60**, 2953.
Compton, R. N., Reinhardt, P. W., and Cooper, C. D. (1975). *J. Chem. Phys.* **63**, 3821.
Compton, R. N., Reinhardt, P. W., and Cooper, C. D. (1978a). *J. Chem. Phys.* **68**, 2023.
Compton, R. N., Reinhardt, P. W., and Cooper, C. D. (1978b). *J. Chem. Phys.* **68**, 4360.
Cooper, C. D., and Compton, R. N. (1974). *J. Chem. Phys.* **60**, 2424.
Cooper, C. D., Frey, W. F., and Compton, R. N. (1978). *J. Chem. Phys.* **69**, 2367.
Cooper, C. D., Naff, W. T., and Compton, R. N. (1975). *J. Chem. Phys.* **63**, 2752.
Curran, R. K. (1962). *Phys. Rev.* **125**, 910.
DeFrees, D. J., Bartmess, J. E., Kim, J. K., McIver, R. T., and Hehre, W. J. (1977). *J. Am. Chem. Soc.* **99**, 6451.
Dispert, H., and Lacmann, K. (1975). *Hahn-Meitner-Inst. Kernforsch., Berlin* **HMI-B 198**, 34.
Dispert, H., and Lacmann, K. (1977a). *Chem. Phys. Lett.* **45**, 311.
Dispert, H., and Lacmann, K. (1977b). *Chem. Phys. Lett.* **47**, 533.
Dotan, I., Davidson, J. A., Streit, G. E., Albritton, D. L., and Fehsenfeld, F. C. (1977). *J. Chem. Phys.* **68**, 247.

Dunbar, R. C. (1971). *J. Am. Chem. Soc.* **93**, 4354.
Dunbar, R. C. (1973). *J. Am. Chem. Soc.* **95**, 6191.
Dunkin, D. B., Fehsenfeld, F. C., and Ferguson, E. E. (1972). *Chem. Phys. Lett.* **15**, 257.
Ellison, G. B., Engelking, P. C., and Lineberger, W. C. (1978). *J. Am. Chem. Soc.* **100**, 2556.
Engelking, P. C., and Lineberger, W. C. (1976). *J. Chem. Phys.* **65**, 4323.
Engelking, P. C., and Lineberger, W. C. (1977a). *J. Chem. Phys.* **66**, 5054.
Engelking, P. C., and Lineberger, W. C. (1977b). *J. Chem. Phys.* **67**, 1412.
Engelking, P. C., Ellison, G. B., and Lineberger, W. C. (1978). *J. Chem. Phys.* **69**, 1826.
Fehsenfeld, F. C. (1971). *J. Chem. Phys.* **54**, 438.
Fehsenfeld, F. C., Ferguson, E. E., and Schmeltekopf, A. L. (1966). *J. Chem. Phys.* **45**, 1844.
Feldmann, D. (1970). *Z. Naturforsch., Teil A* **25**, 621.
Feldmann, D. (1971). *Z. Naturforsch., Teil A* **26**, 1100.
Feldmann, D., Rackwitz, R., Kaiser, H. J., and Heinicke, E. (1977). *Z. Naturforsch., Teil A* **32**, 600.
Ferguson, E. E., Fehsenfeld, F. C., and Schmeltekopf, A. L. (1969). *Adv. Chem. Ser.* **80**, 83.
Geltman, S. (1958). *Phys. Rev.* **112**, 176.
Golub, S., and Steiner, B. (1968). *J. Chem. Phys.* **49**, 5191.
Gygax, R., McPeters, H. L., and Brauman, J. I. (1979). *J. Am. Chem. Soc.* (in press).
Hehre, W. A., and Pople, J. P. (1970). *Tetrahedron Lett.* **34**, 2959.
Helbing, R. K. B., and Rothe, E. W. (1969). *J. Chem. Phys.* **51**, 1607.
Henglein, A., and Muccini, G. A. (1959). *J. Chem. Phys.* **31**, 1426.
Herbst, E., Patterson, T. A., and Lineberger, W. C. (1974). *J. Chem. Phys.* **61**, 1300.
Hoijtink, G. J. (1955). *Rec. Trav. Chim. Pays-Bas* **74**, 1525.
Hong, S. P., Woo, S. B., and Helmy, E. M. (1977). *Phys. Rev. A* **15**, 1563.
Hotop, H., and Lineberger, W. C. (1973). *J. Chem. Phys.* **58**, 2379.
Hotop, H., and Lineberger, W. C. (1975). *J. Phys. Chem. Ref. Data* **4**, 539.
Hotop, H., Patterson, T. A., and Lineberger, W. C. (1973). *Phys. Rev.* **A 8**, 762.
Hotop, H., Patterson, T. A., and Lineberger, W. C. (1974). *J. Chem. Phys.* **60**, 1806.
Huber, B. A., Crosby, P. C., Peterson, J. R., and Moseley, J. T. (1977). *J. Chem. Phys.* **66**, 4520.
Hudson, R. F., Eisenstein, O., and Anh, N. T. (1975). *Tetrahedron* **31**, 751.
Hughes, B. M., Lifshitz, C., and Tiernan, T. O. (1973). *J. Chem. Phys.* **59**, 3162.
Janousek, B. K., Zimmerman, A. H., Reed, K. J., and Brauman, J. I. (1978). *J. Am. Chem. Soc.* **100**, 6142.
Kasdan, A., Herbst, E., and Lineberger, W. C. (1975a). *J. Chem. Phys.* **62**, 541.
Kasdan, A., Herbst, E., and Lineberger, W. C. (1975b). *Chem. Phys. Lett.* **31**, 78.
Kerr, J. A. (1966). *Chem. Rev.* **66**, 465.
Klots, C. E., Compton, R. N., and Raaen, V. F. (1974). *J. Chem. Phys.* **60**, 1177.
Koopmans, T. (1934). *Physica (Utrecht)* **1**, 104.
Kraus, K., Muller-Duysing, W., and Neuert, H. (1961). *Z. Naturforsch., Teil A* **16**, 1385.
Leffert, C. B., Jackson, W. M., and Rothe, E. W. (1973). *J. Chem. Phys.* **58**, 5801.
Leffert, C. B., Tang, S. Y., Rothe, E. W., and Cheng, T. C. (1974). *J. Chem. Phys.* **61**, 4929.
Lehman, T. A., and Bursey, M. M. (1976). "Ion Cyclotron Resonance Spectrometry." Wiley, New York.
Lewis, T. P. (1969). *Tetrahedron* **25**, 4117.
Lifshitz, C., and Grajower, R. (1969). *Int. J. Mass Spectrom. Ion Phys.* **3**, Appendix 5.
Lifshitz, C., Hughes, B. M., and Tiernan, T. O. (1970). *Chem. Phys. Lett.* **7**, 469.
Lifshitz, C., Tiernan, T. O., and Hughes, B. M. (1973). *J. Chem. Phys.* **59**, 3182.
Lifshitz, C., Wu, R. L. C., Tiernan, T. O., and Terwilliger, D. T. (1978). *J. Chem. Phys.* **68**, 247.
Lindinger, W., Albritton, D. L., Fehsenfeld, F. C., and Ferguson, E. E. (1975). *J. Chem. Phys.* **63**, 3238.

Lineberger, W. C. (1973). *In* "Energy, Structure, and Reactivity" (D. W. Smith and W. B. McRae, eds.), pp. 333–355. Wiley, New York.
Lineberger, W. C. (1974a). *In* "Laser Spectroscopy" (R. Brewer and A. Mooradian, eds.) pp. 581–595. Plenum, New York.
Lineberger, W. C. (1974b). *In* "Chemical and Biochemical Applications of Lasers," Vol. 1, (C. B. Moore, ed.) pp. 71–101. Academic Press, New York.
Lineberger, W. C., and Woodward, B. W. (1970). *Phys. Rev. Lett.* **25**, 424.
Lyons, L. E., and Palmer, L. D. (1976). *Aust. J. Chem.* **29**, 1919.
Mathur, B. P., Rothe, E. W., Tang, S. Y., and Reck, G. P. (1976a). *J. Chem. Phys.* **65**, 565.
Mathur, B. P., Rothe, E. W., Tang, S. Y., Mahajan, K., and Reck, G. P. (1976b). *J. Chem. Phys.* **64**, 1247.
Moutinho, A. M. C., Aten, J. A., and Los, J. (1974). *Chem. Phys.* **5**, 84.
Nalley, S. J., and Compton, R. N. (1971). *Chem. Phys. Lett.* **9**, 529.
Nalley, S. J., Compton, R. N., Schweinler, H. C., and Anderson, V. E. (1973). *J. Chem. Phys.* **59**, 4125.
Owens, P. H., Wolf, R. A., and Streitwieser, A. (1970). *Tetrahedron Lett.* **38**, 3385.
Paulson, J. F. (1966). *Adv. Chem. Ser.* **58**, 28.
Pearson, P. K., Schaefer, H. F., Richardson, J. H., Stephenson, L. M., and Brauman, J. I. (1974). *J. Am. Chem. Soc.* **96**, 6778.
Rackwitz, R., Feldmann, D., Kaiser, H. J., and Heinicke, E. (1977). *Z. Naturforsch., Teil A* **32**, 594.
Reed, K. J., and Brauman, J. I. (1974). *J. Chem. Phys.* **61**, 4830.
Reed, K. J., and Brauman, J. I. (1975). *J. Am. Chem. Soc.* **97**, 1625.
Reed, K. J., Zimmerman, A. H., Andersen, H. C., and Brauman, J. I. (1976). *J. Chem. Phys.* **64**, 1368.
Refaey, K. M. A. (1976). *Int. J. Mass Spectrom. Ion Phys.* **21**, 21.
Refaey, K. M. A., and Franklin, J. L. (1976). *J. Chem. Phys.* **65**, 1994.
Richardson, J. H., Stephenson, L. M., and Brauman, J. I. (1973). *J. Chem. Phys.* **59**, 5068.
Richardson, J. H., Stephenson, L. M., and Brauman, J. I. (1974). *Chem. Phys. Lett.* **25**, 321.
Richardson, J. H., Stephenson, L. M., and Brauman, J. I. (1975a). *J. Am. Chem. Soc.* **97**, 1160.
Richardson, J. H., Stephenson, L. M., and Brauman, J. I. (1975b). *Chem. Phys. Lett.* **30**, 17.
Richardson, J. H., Stephenson, L. M., and Brauman, J. I. (1975c). *J. Chem. Phys.* **62**, 1580.
Richardson, J. H., Stephenson, L. M., and Brauman, J. I. (1975d). *J. Chem. Phys.* **63**, 74.
Richardson, J. H., Stephenson, L. M., and Brauman, J. I. (1975e). *J. Am. Chem. Soc.* **97**, 2967.
Rosenstock, H. M., Droxl, K., Steiner, B. W., and Herron, J. T. (1977). *J. Phys. Chem. Ref. Data* **6**, Suppl. No. 1.
Rothe, E. W., Tang, S. Y., and Reck, G. P. (1975). *J. Chem. Phys.* **62**, 3829.
Schiff, L. I. (1968). "Quantum Mechanics," p. 265. McGraw-Hill, New York.
Siegel, M. W., Celotta, R. J., Hall, J. L., and Bennett, R. A. (1972). *Phys. Rev.* **A 6**, 607.
Simons, J. (1977). *Annu. Rev. Phys. Chem.* **28**, 15.
Sinnott, G., and Beaty, E. C. (1971). *Proc. Int. Conf. Phys. Electron. At. Collisions, 7th, Amsterdam* Abstr. p. 176.
Smith, S. J., and Branscomb, L. M. (1955). *Phys. Rev.* **99**, 1657.
Smith, S. J., and Branscomb, L. M. (1960). *Rev. Sci. Instrum.* **31**, 733.
Smyth, K. C., and Brauman, J. I. (1972a). *J. Chem. Phys.* **56**, 1132.
Smyth, K. C., and Brauman, J. I. (1972b). *J. Chem. Phys.* **56**, 5993.
Smyth, K. C., and Brauman, J. I. (1972c). *J. Chem. Phys.* **56**, 4620.
Smyth, K. C., McIver, R. T., Brauman, J. I., and Wallace, R. W. (1971). *J. Chem. Phys.* **54**, 2758.
Steiner, B. (1968). *J. Chem. Phys.* **49**, 5097.

Streitwieser, A. (1961). "Molecular Orbital Theory for Organic Chemists," p. 175. Wiley, New York.

Tang, S. Y., Mathur, B. P., Rothe, E. W., and Reck, G. P. (1976). *J. Chem. Phys.* **64**, 1270.

Tiernan, T. O., Hughes, B. M., and Lifshitz, C. (1971). *J. Chem. Phys.* **55**, 5692.

Warneck, P. (1969). *Chem. Phys. Lett.* **3**, 532.

Watanabe, K. (1957). *J. Chem. Phys.* **26**, 542.

Watanabe, K., Nakayama, T., and Mottl, J. (1962). *J. Quant. Spectrosc. Radiat. Transfer* **2**, 369.

Wigner, E. P. (1948). *Phys. Rev.* **73**, 1002.

Wong, S. F., Vorburger, T. V., and Woo, S. B. (1971). *Bull. Am. Phys. Soc.* **16**, 213.

Woo, S. B., Branscomb, L. M., and Beaty, E. C. (1969). *J. Geophys. Res.* **74**, 2933.

Yamdagni, R., and Kebarle, P. (1973). *J. Am. Chem. Soc.* **95**, 4050.

Younkin, J. M., Smith, L. J., and Compton, R. N. (1976). *Theor. Chim. Acta* **41**, 157.

Zimmerman, A. H., and Brauman, J. I. (1977a). *J. Chem. Phys.* **66**, 5823.

Zimmerman, A. H., and Brauman, J. I. (1977b). *J. Am. Chem. Soc.* **99**, 3565.

Zimmerman A. H., Jackson, R. L., Janousek, B. K., and Brauman, J. I. (1978a). *J. Am. Chem. Soc.* **100**, 4674.

Zimmerman, A. H., Reed, K. J., and Brauman, J. I. (1977). *J. Am. Chem. Soc.* **99**, 7203.

Zimmerman, A. H., Gygax, R., and Brauman, J. I. (1978b). *J. Am. Chem. Soc.* **100**, 5595.

Zittel, P. F., and Lineberger, W. C. (1976). *J. Chem. Phys.* **65**, 1236.

Zittel, P. F., Ellison, G. B., O'Neil, S. V., Herbst, E., Lineberger, W. C., and Reinhardt, W. P. (1976). *J. Am. Chem. Soc.* **98**, 3731.

Chapter 11
The gas-phase acidity scale

JOHN E. BARTMESS
DEPARTMENT OF CHEMISTRY
INDIANA UNIVERSITY
BLOOMINGTON, INDIANA

ROBERT T. MCIVER, JR.
DEPARTMENT OF CHEMISTRY
UNIVERSITY OF CALIFORNIA
IRVINE, CALIFORNIA

ISBN-0-120802-8

I. INTRODUCTION

The heterolytic cleavage of a neutral molecule into a proton and a negative ion, as in reaction (1),

$$AH \rightleftharpoons H^+ + A^-, \tag{1}$$

is one of the most generally important of all chemical processes. Any molecule containing a hydrogen atom is potentially a proton donor, or Brønsted acid. A fundamental understanding of this process is of central importance to studies of the rates and mechanisms of wide variety of chemical and biochemical reactions, many of which often involve homogeneous or heterogeneous catalysis. Furthermore, a tremendous amount of structural variations is possible for the acid AH. This is ideal for fundamental studies of structure–reactivity and structure–energy relationships.

The classical approach to this subject involves dissolving the acid AH in a liquid solvent medium and measuring the degree of proton transfer to some reference base B, as in reaction (2),

$$AH + B \rightleftharpoons BH^+ + A^-, \tag{2}$$

where B can be either a neutral or an anionic species. An enormous amount of valuable data has been obtained in this way, but several problems limit its range of applicability and complicate interpretation of the data. First, the range of structural effects which can be investigated in a given solvent medium SH is limited by interfering reactions such as

$$AH + SH \rightleftharpoons SH_2{}^+ + A^- \tag{3}$$

$$A^- + SH \rightleftharpoons S^- + AH \tag{4}$$

involving proton transfer between the acid or its conjugate base and the solvent. If AH is a stronger Brønsted acid than SH, the acid $SH_2{}^+$ is the predominant species in solution, and if AH is a much weaker acid than SH, the base S^- predominates over A^-. This well-known "leveling effect" (King, 1965) limits the range of measurable acidities in aqueous solution to about 19 kcal/mole.* As a result, much more is known about structural effects on the acidity of carboxylic acids, phenols, and similar oxygen acids, whose acidity happens to fall in this range, than for the strong mineral acids or the weak carbon and nitrogen acids. Different solvent systems such as fluorosulfonic acid for strong acids (Arnett *et al.*, 1974), and dimethylsulfoxide (Matthews *et al.*, 1975) and cyclohexylamine (Streitwieser and Nebenzahl,

* 1 kcal = 4184 J.

1976) for weak acids have expanded considerably the accessible range for pK_a measurements, but interpretation of the data are often hampered by the formation of homoconjugate ion pairs ($A^- \cdots HA$) and counter ion pairs ($M^+ \cdots A^-$).

An obvious solution to all the above-mentioned difficulties is to eliminate the solvent medium entirely and measure the proton donating ability of molecules in the gas phase. During the last decade, in many laboratories throughout the world, specialized mass spectrometers have been developed which are capable of generating gaseous negative ions and studying ion–molecule reactions such as

$$OH^- + CH_3OH \rightarrow CH_3O^- + H_2O. \qquad (5)$$

Relative acidities measured in the gas phase truly reflect intrinsic structural effects, and comparison with solution pK_as provides a clear elucidation of the effects of solvation. Furthermore, substituent effects are usually larger in the gas phase than in solution, and this allows one to study effects that are compressed and masked by solvation.

In this chapter we focus on the results of experimental investigations of gas-phase acidities. Special attention has been given to developing a consistent scale of gas-phase acidities and using the scale to derive a wide variety of thermochemical data for heats of formation of negative ions and radicals, electron affinities of radicals, and bond dissociation energies. Our coverage includes published literature up to January 1978 and much previously unpublished data.

II. CALCULATIONS USING BOND STRENGTHS AND ELECTRON AFFINITIES

For most structures, heterolytic bond dissociation, as in reaction (1), is 300–400 kcal/mole endothermic in the gas phase. The only practical way of determining such a large energy is to use a thermochemical cycle such as

$$AH \rightarrow A + H, \qquad \Delta H^0_{298} = DH^0(A\text{—}H); \qquad (6)$$

$$\underline{A + H \rightarrow A^- + H^+}, \qquad \Delta H^0_{298} \cong \Delta H_0{}^0 = IP(H) - EA(A); \qquad (7)$$

$$AH \rightarrow A^- + H^+, \qquad \Delta H^0_{298} = \Delta H^0_{acid}; \qquad (8)$$

$$\Delta H^0_{acid} \cong DH^0(A\text{—}H) + IP(H\cdot) - EA(A). \qquad (9)$$

scheme I

Since the ionization potential of hydrogen IP(H) is 313.6 kcal/mole (Stull and Prophet, 1971) and a constant for all acids, the enthalpy change ΔH^0_{acid} is

TABLE I

Calculated Enthalpies for $AH = A^- + H^+$ *at 298 K*

Acid	DH^0_{298}(AH)	Ref.	EA(A·)	Ref.	ΔH^0_{acid}(AH)
CH_4	104.8 ± 0.3	*a*	1.84 ± 0.7	*b*	416.6 ± 1.0
NH_3	103.2 ± 3.1	*a*	17.2 ± 0.5	*c*	399.6 ± 3.6
H_2	104.2 ± 0.001	*a*	17.392 ± 0.5	*c*	400.4 ± 0.5
H_2O	119.3 ± 0.3	*a*	42.1 ± 0.05	*c*	390.8 ± 0.4
Propene	89 ± 2	*d*	12.7 ± 1.2	*e*	389.9 ± 3.2
HO·	102.3 ± 0.3	*a*	33.8 ± 0.1	*c*	382.1 ± 0.5
$PhCH_3$	85 ± 1	*d*	20.4 ± 1.5	*f*	378.2 ± 2.5
MeOH	104 ± 1	*g*	39.4 ± 1.4	*h*	378.2 ± 2.4
			36.2 ± 0.7	*i*	381.4 ± 1.7
EtOH	104 ± 1	*g*	39.8 ± 0.7	*i*	377.8 ± 1.7
n-PrOH	104 ± 1	*g*	41.2 ± 0.7	*i*	376.4 ± 1.7
i-PrOH	104.8 ± 1	*g*	42.4 ± 0.7	*i*	376.0 ± 1.7
t-BuOH	105.1 ± 1	*g*	44.8 ± 0.1	*h*	373.9 ± 1.2
		g	43.8 ± 0.7	*i*	374.9 ± 1.7
t-$BuCH_2OH$	104 ± 1	*g*	45.2 ± 0.2	*h*	372.4 ± 2.4
HCF_3	106 ± 1	*d*	42 ± 1.5	*j*	377.6 ± 2.5
SiH_4	94 ± 2	*k*	33.2 ± 0.7	*l*	374.4 ± 2.7
HF	136.1 ± 0.6	*a*	78.4 ± 0.05	*c*	371.3 ± 0.7
CH_3CN	92.9 ± 2.5	*e*	34.8 ± 0.4	*e*	371.7 ± 2.9
PH_3	83.9 ± 3	*m*	28.9 ± 0.07	*c*	368.6 ± 3.1
Me_2CO	98.0 ± 2.6	*n*	40.6 ± 1.3	*o*	371.0 ± 3.9
			40.5 ± 0.7	*i*	371.1 ± 3.3
			<44.0 ± 1.0	*p*	>367.6 ± 3.6
GeH_4	87.2 ± 2	*q*	40.1 ± 0.9	*l*	360.7 ± 2.9
MeSH	91.8 ± 2	*r*	44.3 ± 0.7	*h*	361.1 ± 2.7
	87 ± 2	*s*			356.3 ± 2.7
EtSH	87 ± 2	*s*	44.9 ± 0.2	*h*	355.7 ± 2.2
n-PrSH	87 ± 2	*s*	46.8 ± 0.7	*h*	353.8 ± 2.7

determined by the difference between the bond strength DH^0(A—H) and the electron affinity of the radical EA(A). A weak A—H bond and/or a high electron affinity of the radical A causes molecules to be strong acids. A major limitation of this approach in the past has been the paucity of reliable EA values, but great progress has been made in the last five years (see Janousek and Brauman, Chapter 10, in this volume) due largely to the development of new electron photodetachment and negative ion photoelectron spectroscopy experiments. Another problem encountered with calculations of this type is that bond dissociation energies DH^0(A—H) are usually reported at 298 K, whereas electron affinities and ionization potentials obtained from spectroscopic measurements refer to enthalpies at 0 K. We find, however, that ΔC_p for the electron transfer reaction (7) is generally quite small due to the

TABLE I (*Continued*)

Acid	DH^0_{298}(AH)	Ref.	EA(A·)	Ref.	ΔH^0_{acid}(AH)
i-PrSH	87 ± 2	*s*	47.4 ± 1.0	*h*	353.2 ± 1.2
t-BuSH	87 ± 2	*s*	48.0 ± 0.3	*h*	352.6 ± 2.3
Cyclopentadiene	81.2 ± 1.2	*t*	42.4 ± 0.7	*u*	352.4 ± 1.9
			41.2 ± 0.5	*v*	353.6 ± 1.7
H_2S	91.6 ± 4	*a*	53.5 ± 0.2	*c*	351.7 ± 4.2
HO_2·	47.1 ± 2.0	*a*	10.1 ± 0.2	*c*	350.6 ± 2.2
HCN	123.8 ± 2	*a*	88.1 ± 0.4	*c*	349.3 ± 2.4
PhOH	86.5 ± 2	*w*	54.4 ± 1.4	*x*	345.7 ± 3.4
H_2Se	76	*y*	50.9 ± 0.7	*c*	339 ± 5
PhSH	82.2 ± 2	*r*	56.9 ± 1.4	*x*	338.9 ± 3.4
HONO	77.7 ± 0.7	*z*	53 ± 7	*z*	338.3 ± 8
HCl	103.1 ± 2	*a*	83.4 ± 0.07	*c*	333.3 ± 0.3
HBr	87.6 ± 0.2	*a*	77.6 ± 0.07	*c*	323.6 ± 0.3
HI	71.3 ± 0.1	*a*	70.6 ± 0.07	*c*	314.3 ± 0.2

[a] Stull and Prophet (1971).
[b] Janousek and Brauman (this volume).
[c] Rosenstock *et al.* (1977).
[d] Golden and Benson (1969).
[e] Zimmerman and Brauman (1977b).
[f] Richardson *et al.* (1975c).
[g] From data in Batt *et al.* (1974) and Benson *et al.* (1969).
[h] Zimmerman and Brauman (1977a).
[i] Ellison (1977).
[j] Page and Goode (1969).
[k] Steele *et al.* (1962).
[l] Reed and Brauman (1974).
[m] McAllister and Lossing (1968).
[n] Solly *et al.* (1970).
[o] Zimmerman *et al.* (1977).
[p] Eyler (1978).
[q] Darwent (1970).
[r] Colussi and Benson (1977).
[s] Fine and Westmore (1970).
[t] Furuyama *et al.* (1971).
[u] Richardson *et al.* (1973).
[v] Engelking and Lineberger (1977).
[w] Colussi *et al.* (1977).
[x] Richardson *et al.* (1975).
[y] Dixon *et al.* (1972).
[z] Ferguson *et al.* (1972).

structural similarity of the reactants and products. Thus, for the reactions

$$F\cdot + H\cdot \rightarrow H^+ + F^- \quad (10a)$$

$$OH\cdot + H\cdot \rightarrow H^+ + OH^- \quad (10b)$$

available data (Stull and Prophet, 1971) indicate $\Delta H_0{}^0 - \Delta H^0_{298}$ is 0.12 and 0.16 kcal/mole, respectively.

Table I gives all of the ΔH^0_{acid} values that can currently be calculated in this way from reported bond strengths and electron affinities. It is apparent that there is considerable variation in the values of DH^0 and EA obtained by different workers, especially for EA of acetone and the DH^0 for the alkanethiols. Typically, the error in ΔH^0_{acid} is on the order of 1–2 kcal/mole, making it difficult to assign the relative acidity order for close pairs such as H_2O–propene, PH_3–HF, and PhSH–HCl.

III. MEASUREMENT OF RELATIVE ACIDITIES

During the last five years experimental methods have been developed in many laboratories for measuring relative gas-phase acidities. Gaseous proton-transfer reactions such as

$$AH + B^- \underset{k_2}{\overset{k_1}{\rightleftharpoons}} BH + A^- \tag{11}$$

are generally very rapid, thus making it possible to measure experimentally the forward and reverse rate constant k_1 and k_2 and the equilibrium abundance of the reactant and product species. In this way the equilibrium constant K for reaction (11) can be determined:

$$K = \frac{[BH][A^-]}{[AH][B^-]} = \frac{k_1}{k_2}. \tag{12}$$

The relative gas-phase acidity $\delta\,\Delta G^0_{acid}$ of the two acids AH and BH can be calculated from the equilibrium constant by knowing the temperature T of the system

$$\delta\,\Delta G^0_{acid} = -RT \ln K, \tag{13}$$

where R is the ideal gas constant (1.987 cal/(mole-deg). A scale of relative acidities can be constructed by measuring AH and BH relative to other compounds, in the same manner as has been used to construct the relative solution acidity scales using spectrophotometric techniques to determine the anion abundances.

It is apparent from reaction (11) that the measured relative acidity $\delta\,\Delta G^0_{acid}$ is equal to the difference in the absolute acidities of AH and BH. Thus,

$$\delta\,\Delta G^0_{acid} = \Delta G^0_{acid}(AH) - \Delta G^0_{acid}(BH), \tag{14}$$

where $\Delta G^0_{acid}(AH)$ is the Gibbs free energy for reaction (1) at 298 K. Absolute acidities can be calculated for certain of the compounds in Table I by knowing the standard entropies $S^0(AH)$, $S^0(H^+)$, and $S^0(A^-)$, and using the equation

$$\Delta G^0_{acid} = \Delta H^0_{acid} - T\,\Delta S^0_{acid}, \tag{15}$$

where $T = 298$ K and $\Delta S^0_{acid} = S^0(H^+) + S^0(A^-) - S^0(AH)$. Thus, the experimental scale of relative acidities can be converted to a scale of absolute acidities by including some of the Table I compounds as "anchor" points.

Most of the equilibrium measurements of relative acidities in the gas phase have been done by Kebarle and co-workers at the University of Alberta using high-pressure pulsed mass spectrometry (HPMS) and by McIver and co-workers at the University of Calfornia, Irvine, using pulsed ion cyclotron resonance spectroscopy (ICR). The two techniques use the same basic chemical reactions, as presented in scheme II:

$$XY + e^- \rightarrow X^- + Y, \tag{16}$$

$$X^- + AH \rightarrow A^- + HX, \tag{17}$$

$$X^- + BH \rightarrow B^- + HX, \tag{18}$$

$$A^- + BH \rightleftharpoons B^- + AH. \tag{19}$$

scheme II

Compound XY is a compound such as NF_3, SO_2F_2, CH_3ONO, NH_3, or H_2O, and has a large cross section for production of gaseous negative ions upon electron impact. XY is chosen so that X^- will rapidly and irreversibly deprotonate AH and BH. After a sufficient number of reactive collisions, equilibrium according to reaction (19) is established and the ratio of the signals due to A^- and B^- determined. This, combined with the ratio of the pressures of AH and BH introduced into the reaction chamber, yields the equilibrium constant and thus $\delta \Delta G^0_{\text{acid}}$.

In both the HPMS and ICR experiments, gaseous ions are stored for an extended period of time and caused to suffer hundreds of stabilizing ion–neutral collisions. This is essential to insure thermalization of excited ions produced by electron impact or exothermic ion–molecule reactions. Furthermore, reliable equilibrium data can only be obtained when the time constant τ for attainment of equilibrium in reaction (11)

$$\tau = (k_1[AH] + k_2[BH])^{-1} \tag{20}$$

is much greater than the characteristic lifetime of ions in the apparatus.

In HPMS (Cunningham *et al.*, 1972) ions are generated in a field-free chamber containing $\sim 10^{-2}$ torr of the neutral acids and several torr of methane to serve as a thermalizing collision gas. After an initial ionizing electron pulse, a small fraction of the ions continuously escape from the chamber through a exit port into a low-pressure region where they are accelerated and mass analyzed in a magnetic sector. The ions are observed for up to 3 msec and suffer several hundred collisions with the neutral acids before they are neutralized on the walls or diffuse out of the collision chamber. Equilibrium is assumed to be attained if the ratio of ion abundances $[A^-]/[B^-]$ is invariable with time. At the relatively high pressures used in the source region of the mass spectrometer, formation of "cluster" ions such as AHA^- and $A^-(HA)_2$ occurs rapidly. While the chemistry of such species as a stepwise model of solvation is of considerable interest, their formation tends to interfere with HPMS measurements of relative acidities. Thus, source temperatures of 600 K are typically used to suppress the formation of the cluster ions. In order to relate equilibria at 600 K to thermochemical data at 298 K, the experimental $\delta\,\Delta G^0_{\rm acid}$ values are adjusted using appropriate entropy corrections, as is discussed below.

In pulsed ICR studies of ion–molecule equilibrium reactions, much lower (10^{-6}–10^{-5} torr) pressures of neutral acids are used, but the ions are stored for a considerably longer time (up to several seconds) in a static magnetic ion trap. While the ions are reacting they are trapped by static magnetic and electric fields, and each ion undergoes several hundred collisions before it is lost from the trap. After a certain reaction period ions of a chosen mass-to-charge ratio are detected by accelerating them with an alternating electric field. The principles of pulsed ICR are thoroughly presented in the literature (McIver, 1978). An important feature of the ICR method is the use of double resonance ejection to measure the forward and reverse rate constants (Wolf *el al.*, 1977) for the proton-transfer equilibria, reaction (11). This provides evidence of whether the equilibrium is established in a time short compared to the lifetime of ions in the ICR cell. Several examples are available now of ion–molecule reactions which are so slow that correct equilibrium constants cannot be measured directly. Carbon acids are particularly likely to be in this class (Farneth and Brauman, 1976).

IV. ESTIMATION OF ENTROPIES

In order to place these acidity measurements done at various temperatures on a common $\Delta H^0_{\rm acid}$ scale and to interface them with the extensive existing body of enthalpic thermochemical data, it is necessary to determine the entropy change $\Delta S^0_{\rm acid}$ for reaction (11)

$$\Delta S^0_{\rm acid} = S^0({\rm H}^+) + S^0({\rm A}^-) - S^0({\rm AH}). \tag{21}$$

The entropy of the proton is well known (Stull and Prophet, 1971), and the entropies of the neutral acids either are obtainable from the literature or can be estimated by group additivity (Benson *et al.*, 1969; Benson, 1976). The entropies of the anions, however, are not generally available and must be estimated in various ways. We adopt the general approach of Cumming and Kebarle (1978) to obtain $\Delta S^0_{\rm acid}$.

The total entropy of a species is the sum of translational, vibrational, rotational, and electronic terms,

$$S^0_{\rm tot} = S^0_{\rm tr} + S^0_{\rm vib} + S^0_{\rm rot} + S^0_{\rm elec}. \tag{22}$$

Each of these can be evaluated by procedures from statistical mechanics. Since the quantity $S^0({\rm A}^-) - S^0({\rm AH})$ is primarily needed for Eq. (21), these calculations can be simplified by certain assumptions. For acids of molecular weight greater than 30 amu, $S^0_{\rm tr}({\rm HA})$ and $S^0_{\rm tr}({\rm A}^-)$ are less than 0.1 eu different, so they essentially cancel and may be neglected. The acid and its anion are isoelectronic and are closed-shell species; therefore, the difference in electronic entropies is likely to be small. The structures of AH and A^- are usually similar enough that those vibrations of $<1000\ cm^{-1}$ that contribute appreciably to S^0_{vib} at 300 K should nearly cancel (exceptions are noted below). It is thus rotational entropy, primarily from symmetry changes and freezing out or loss of internal rotations, that contributes to changes in ΔS^0_{acid} with structural variation. The entropy $S_f{}^0$ of a one-dimensional free rotor bearing the acidic hydrogen is calculable from structural parameters (Benson, 1976). This is only an upper limit to the rotational entropy lost upon proton transfer, however, since most rotors are hindered by some barrier, V_0 attributable to the neighboring structure. The reduction in rotational entropy due to this hindrance $S_h{}^0$ is likewise calculable using barriers available from molecular orbital calculations, microwave spectra, or ir torsional frequencies (Lowe, 1972). The entropy change is then

$$\Delta S^0_{ioniz}(AH) = -[S_f{}^0(AH) - S_h{}^0(AH)] + R \ln(\sigma_{AH} n_{AH}/\sigma_{A^-} n_{A^-}), \quad (23)$$

$$\Delta S^0_{acid} = \Delta S^0_{ioniz} + S^0(H^+), \quad (24)$$

where σ is the external rotational symmetry number and n the corresponding internal value. This assumes that the internal rotation lost upon proton transfer, such as about the C—O bond in alcohols, is not strongly coupled to other rotations in the acids, so that these others tend to cancel in the acid and anion. For the carbon acids CH_3—EWG (EWG is an electron withdrawing group, such as —NO_2, —CHO, etc.) there remains a possible internal rotation of $\bar{C}H_2$— after proton transfer. Cumming and Kebarle assume a barrier of half that of ethylene for their calculations at 600 K. Due to the uncertainty in the barrier height, for the entropy around 300 K we assume that this is a torsional vibration rather than a hindered rotation, resulting in a negligibly small correction (Benson, 1976). The acids $MeCH_2$—EWG and Me_2CH—EWG have a sufficiently lower torsion frequency in the anion to increase its entropy by 2–3 eu; however, the larger groups also have a larger $S_f{}^0$ in the acid by a similar amount. These two nearly cancel so that $\Delta S^0_{acid}(HA)$ is taken here as the same for the methyl, ethyl, and isopropyl acids of a given EWG, save for symmetry number differences. Since we assume that rotation about the double bond is merely a vibration, we need not include internal symmetry correction for it, as must Cumming and Kebarle.

For the alcohols, $S_f{}^0(R\text{—}OH) = 4.6$ eu, $V_0 = 1.1$ kcal/mole, and $S_h{}^0 = 0.6$ eu at 300 K to give $\Delta S^0_{ioniz} = -4.0$ eu, save for those with intramolecular hydrogen bonding (Craven and Bentley, 1972). For the alkanethiols, $\Delta S^0_{ioniz} = -5.0$. It may be argued that the larger, more branched members of these series should have larger $S_h{}^0$ values; however, corrections to the entropies of the analogous alkanes for 1,5 interactions are not necessary. For the localized carbanions on a symmetry axis in the substituted benzenes and acetylenes,

$S_f{}^0 - S_h{}^0$ is taken as zero. Based on the similarity of V_0 for $CH_2{=}NMe$ and $CH_2{=}CH{-}Me$ (Lowe, 1972) the barrier to rotation about N—OH in oximes is taken as that for vinyl alcohol (Sandal and Seip, 1975) for $S^0_{\text{ioniz}} = -2.8$ eu. Barriers to rotation in phenols (Fately *et al.*, 1975) and anilines (Carlson and Fately, 1977) are available from far-ir studies. The barrier to methyl rotation in acetophenone is taken as that in 3-butene-2-one (Lowe, 1972). For $CH_3{-}CO_2Me$ and $CH_3{-}CONMe_2$, the barrier to methyl rotation is assigned as that of $CH_3{-}CO_2H$. It is possible that the barrier to methoxy or dimethylamino rotation in these two is considerably reduced in the anion relative to the acid, since the resonance forms that contribute to this barrier require localization of the negative charge on carbon:

$$CH_2{=}\overset{\overset{\large O^-}{|}}{C}{-}NMe_2 \leftrightarrow \bar{C}H_2{-}\overset{\overset{\large O^-}{|}}{C}{=}\overset{+}{N}Me_2$$

This would increase the anion's entropy and thus $\Delta S^0_{\text{ioniz}}$. Lacking experimental verification, no correction is applied.

For cyclopentadiene and cycloheptatriene, proton removal results in formation of highly delocalized anions, though no rotations are frozen out. For the former, a value of $\Delta S^0_{\text{ioniz}} = -2$ eu plus symmetry terms is adopted. The C_7 anion is probably not planar due to the antiaromaticity and angle strain that occurs. It is likely, however, to exist as a number of rapidly interconverting fluxional forms to maximize charge delocalization. We thus assign $\Delta S^0_{\text{ioniz}} = -1\text{ eu} + R\ln(2/14)$, considering it *effectively* planar for symmetry purposes. Pyrrole is already a 6-electron π system in the acid, so there is little decrease in delocalization in the anion for $\Delta S^0_{\text{ioniz}} = 0$ eu.

The site of deprotonation of acetonylacetate is somewhat uncertain, considering the propensity of first row α-heteroatoms to reduce carbanion stability (Adolph and Kamlet, 1966; Hine *et al.*, 1967). Lacking labeling studies, the assignment of Kebarle and co-workers is accepted.

These derivations of $\Delta S^0_{\text{ioniz}}$ involve numerous assumptions. As support for their validity, the entropies of the anions may be obtained in another way for certain compounds. The entropy of an anion can be taken as that of its isoelectronic neutral, i.e., $S^0(CH_3O^-) = S^0(CH_3F)$. The similarity of electronic and nuclear structure for isoelectronic pairs lead to all the terms of Eq. (22) being nearly identical for the two species (Benson, 1976). This is the method of choice for obtaining entropies of atomic anions such as F^- and Cl^- (Stull and Prophet, 1971) and is used here for all inorganic acids.

Table II shows several comparisons of ΔS^0_{acid} calculated from both the isoelectronic method and the statistical mechanical method. Generally there is excellent agreement between the two methods. The deviation for benzyl anion is due to the lower barrier for $-NH_2$ rotation in aniline than for the much more delocalized $-\bar{C}H_2$ in the carbanion. In addition, the experi-

TABLE II

Comparison of ΔS^0_{acid} Values Derived from the Isoelectronic Approximation and from Statistical Mechanics Calculations[a]

Acid	ΔS^0_{acid}(iso)[b]	ΔS^0_{acid}(stat. mec.)[c]	$\delta\,\Delta S^0_{acid}$
CH_3OH	22.0	22.1	−0.1
CH_3CH_2OH	21.8	22.0	−0.2
PhOH	23.3	23.0	−0.3
CH_3SH	21.0	21.3	−0.3
CH_3CH_2SH	21.3	21.0	0.3
PhSH	20.7	20.9	−0.2
$PhCH_3$	25.8	22.4	3.2
$PhNH_2$	24.7	24.4	0.3
HCF_3	26.3	26.0	−0.3
HCCH	26.3	27.2	−0.9
CH_3CCH	25.4	26.0	−0.6
PhCCH	26.1	26.0	−0.1
CH_3NO_2	23.7	22.4	−1.3
CH_3CN	25.1	26.3[d]	−1.2
CH_3CHO	23.2	22.8	0.4

[a] Cal/(mole deg) for the process $AH = A^- + H^+$.

[b] Equation (21), using the entropy of the isoelectronic neutral as an approximation to $S^0(A^-)$.

[c] Equations (23) and (24).

[d] External rotations used, with ketene as the model for the anion.

mental entropies of some alkyl-amine anions (Mackay *et al.*, 1976) are correctly predicted by the isoelectronic method to 0.1 eu. We conclude that the various methods used here for estimating ΔS^0_{acid} are good to ± 2 eu in most cases, and note is made in the tables for those cases where less confidence in the results is warranted, due to uncertain ΔS^0_f and V_0.

V. CONSTRUCTION OF THE GAS-PHASE ACIDITY SCALE

Table III summarizes the thermochemical data derived from pulsed ICR studies of gas-phase acidities. Experimental equilibrium constants were first converted to $\delta\,\Delta G^0_{acid}$ values using Eq. (13). We have found that there is some uncertainty in the temperature of the ICR analyzer cell. The temperature depends on the current supplied to the filament which emits the ionizing electrons, and this causes a slight temperature gradient across the cell. Therefore, in computing $\delta\,\Delta G^0_{acid}$, the temperature used is the average of the

TABLE III

Thermochemical Data Derived from Pulsed ICR Experiments[a]

Acid	ΔS^0_{acid} [b]	ΔH^0_{acid} [c]	ΔG^0_{acid}(298 K)[d]	ΔH^0_{f298}(HA)[e]	$\Delta H^0_{f298}(A^-)$[f]
MeOH	22.0	379.2	372.6	−48.0	−36.0
EtOH	22.0	376.1	369.5	−56.2	−47.5
n-PrOH	22.0	374.7	368.1	−61.3	−53.8
i-PrOH	22.0	374.1	367.5	−65.2	−58.3
i-BuOH	22.0	373.4	366.8	−67.9	−61.7
t-BuOH	22.0	373.3	366.7	−74.7[g]	−68.6
$MeO(CH_2)_2OH$	23.2	372.5	365.6	−87.4	−82.1
t-$BuCH_2OH$	22.0	371.8	365.2	−76.1[g]	−71.5
t-BuCH(Me)OH	22.0	370.7	364.1	−83.8[g]	−80.3
t-BuCH(Et)OH	22.0	369.6	363.0	−88.7[g]	−86.3
$PhCH_2OH$	22.0	369.6	363.0	−22.5	−20.1
t-BuCH(*i*-Pr)OH	22.0	368.5	361.9	−93.6[g]	−92.3
$(t\text{-}Bu)_2CHOH$	22.0	367.3	360.7	−99.1[g]	−99.0
F_2CHCH_2OH	24.0	367.0	359.8	−148.3[g]	−148.5
F_3CCH_2OH	25.5	364.4	356.8	−207.0[g]	−209.8
$Me_2C{=}NOH$	23.2	366.7	359.8	−13.0[g]	−13.5
MeCH=NOH	23.2	366.2	359.3	−4.7	−5.7
t-BuCH=NOH	23.2	364.6	357.7	−23.2[g]	−25.8
PhCH=NOH	23.2	355.1	348.2	25.8	13.7
PhOH	23.0	351.4	344.5	−23.0	
p-MeC_6H_4OH	23.0	352.6	345.7	−30.0	
MeSH	21.0	359.0	352.7	−5.5	−13.7
EtSH	21.0	357.4	351.1	−11.0	−20.8
n-PrSH	21.0	356.4	350.1	−16.1	−26.9
i-PrSH	21.0	355.6	349.3	−18.2	−29.8
t-BuSH	21.0	354.7	348.4	−26.1	−38.6
HF	19.3	371.5	365.7	−64.8	−60.5
HCF_3	26.0	375.6	367.9	−165.7	−157.2
SiH_4	27.1	371.5	363.4	8.2	12.5
PH_3	24.9	370.4	363.0	1.3	4.5
H_2S	21.3	353.4	347.1	−4.8	−18.5
HCN	24.6	353.1	345.8	32.3	18.2
MeC≡CH	26.0	379.6	371.9	44.3	56.8
n-PrC≡CH	26.0	378.3	370.6	34.5	45.7
t-BuC≡CH	26.0	376.6	368.9	25.1[h]	34.6
HC≡CH	27.4	375.4	367.2	54.3	62.5
PhC≡CH	26.0	370.3	362.6	78.3	81.5
Cyclopentadiene	20.8	356.1	349.9	32.4	21.3
Cycloheptatriene	20.1	373.9	367.9	44.5	51.2
$ArNH_2$					
p-MeO	24.0	367.9	360.7		
p-Me	24.1	368.2	361.0	12.9[g]	13.9
m-Me	24.4	367.6	360.3	12.9[g]	13.3
H	24.4	367.1	359.8	20.8	20.7
p-F	24.4	365.5	358.2		

TABLE III (*Continued*)

Acid	ΔS^0_{acid}[b]	ΔH^0_{acid}[c]	$\Delta G^0_{\text{acid}}(298\text{ K})$[d]	$\Delta H^0_{f298}(\text{HA})$[e]	$\Delta H^0_{f298}(\text{A}^-)$[f]
o-F	24.4	364.5	357.2		
m-MeS	24.4	364.0	356.7	24.8[g]	21.6
m-F	24.4	363.1	355.8	−25.3[g]	−29.4
2,4-F_2	24.4	362.5	355.2		
p-Cl	24.4	362.5	354.2		
m-Cl	24.4	361.1	353.8	13.7[g]	7.6
m-CF_3	24.4	359.1	351.8	−139.7[g]	−147.8
p-CF_3	24.5	355.6	348.3		
Pyrrole	26.0	360.7	353.0	25.9	19.5
CH_3Ph	22.4	379.0	372.3	12.0	23.8
$MeCH_2Ph$	21.6	378.3	371.9	7.1	18.3
Me_2CHPh	20.2	377.5	371.5	1.0	11.3
$CH_3C_6H_4$—*p*-Me	23.8	380.5	373.4	4.3	17.6
CH_3CN	26.3	372.2	364.4	19.0	24.1
$MeCH_2CN$	25.5	373.7	366.1	12.3	18.8
Me_2CHCN	24.2	373.8	366.6	5.6	12.2
c-PrCN	25.5	374.1	366.5	43.5[g]	50.4
$MeOCH_2CN$	25.5	371.4	363.8	−12.5[h]	−8.3[j]
$PhCH_2CN$	22[i]	353.3[j]	346.7[j]	44.5[h]	29.4[j]
CH_3CHO	22.8	366.4	359.6	−39.7	−40.5
$MeCH_2CHO$	21.9	365.9	359.4	−45.5	−46.8
$(CH_3)_2CO$	24.0	368.8	361.6	−51.7	−50.1
CH_3COAr					
H	22.8	363.2	356.4	−22.0	−26.0
m-MeO	22.8	362.8	356.0	−59.8[g]	−64.2
m-Cl	22.8	359.3	352.5	−29.1[g]	−37.0
$MeCH_2COPh$	22.0	362.4	355.8	−26.0	−30.8
$PhCH_2COMe$	19[i]	352.5[j]	346.8[j]	−24.9	−39.6[j]
CH_3CO_2Me	22.6[i]	371.0[j]	364.3[j]	−98.0	−94.2[j]
CH_3CONMe_2	22.6[i]	373.5[j]	366.8[j]	−63.0[h]	−56.7[j]
CH_3SOCH_3	23.7	372.7	365.6	−34.6	−29.1
$CH_3SO_2CH_3$	25.4	366.6	359.0	−89.1	−89.7
CH_3NO_2	22.4	358.7	352.0	−17.9	−26.4
$MeCH_2NO_2$	21.6	358.1	351.7	−23.8	−32.8
Me_2CHNO_2	20.2	358.2	352.2	−33.9	−42.9
t-$BuCH_2NO_2$	21.6	357.0	350.6	−45.2[g]	−55.3
Ph_2CH_2	18[i]	364.5[j]	359.1[j]	39.7[h]	37.0[j]
$CH_3CH{=}CH_2$	22.3	390.8	384.2	4.9	28.5

[a] From Miller (1975) and Bartmess *et al.* (1979).

[b] Entropy change for the process $AH = A^- + H^+$, ±2 cal/(mole deg) unless otherwise stated.

[c] Enthalpy change for the process $AH = A^- + H^+$, ±2 kcal/mole.

[d] $\Delta H^0_{\text{acid}} - (298\text{ K})\,\Delta S^0_{\text{acid}}$, kcal/mole.

[e] From Benson *et al.* (1969) or Benson (1976).

[f] Equation (25), ±3 kcal/mole except as noted.

[g] Estimated by group additivity methods (Benson, 1976).

[h] From group equivalents estimated by the authors.

[i] ±5 eu, see text. [j] ±4 kcal/mole.

temperatures of the two side plates at the particular filament current used for the experiment. For most of the data, T is between 313 and 330 K, averaging about 320 K. Each relative acidity is converted to a relative enthalpy change $\delta\Delta H^0_{\text{acid}}$ using the entropy values derived above and given in Table III. Several experiments were run for each compound to obtain a consistent scale of $\delta\Delta H^0_{\text{acid}}$ values with multiple overlaps. The precision of the overall relative scale is estimated as ± 1 kcal/mole. This is half from the uncertainty in ΔS^0_{acid}, and half from the experimental uncertainties in the $\delta\Delta G^0_{\text{acid}}$ values obtained by multiple overlaps.

The relative scale has been anchored by a weighted least squares procedure to the data in Table I, the weighting being the uncertainty of the absolute values. The ΔH^0_{acid} values obtained in this way are presented in Table III, along with the calculated ΔG^0_{acid}(298 K) acidities. It is assumed that ΔH^0_{acid} and ΔS^0_{acid} are independent of temperature over this small (20–30° K) range.

Cumming and Kebarle (1978) have constructed their ΔH^0_{acid} scale in a similar way, using data from HPMS at 500 or 600 K and HCl as the anchor point for assigning absolute values. Appropriate $\Delta C_p{}^0$ corrections were used to convert ΔH^0_{acid}(600 K) to ΔH^0_{acid}(300 K). These values, along with ΔG^0_{acid}(298 K) are given in Table IV. Table V presents those acids for which ΔH^0_{acid} is known from both methods (and temperatures). The agreement is quite good, with some deviation apparent at the low acidity end of the HPMS scale. The temperature range is too large, unfortunately, to determine ΔH^0_{acid} and ΔS^0_{acid} values by the van't Hoff method, since they are not constant over this range. The agreement indicates that the estimated ΔS^0_{acid} values are reasonably good. For internal consistency in discussing substituent effects later, we adopt the room-temperature ICR values for the acids in Table V, save for the phenols. Those ketones and sulfones in Table VI are assigned absolute ΔH^0_{acid} values based on the relative HPMS values at 500 K (Cumming and Kebarle, 1978) but anchored to the absolute ICR values, once again for internal consistency in substituent effects. The difference is no more than 1 kcal/mole.

Table VII contains the $\delta\Delta G^0_{\text{acid}}$(600 K) values for the substituted phenols and benzoic acids (McMahon and Kebarle, 1977). No ΔH^0_{acid}(300 K) values were assigned these by Cumming and Kebarle due to the uncertainty of the ΔS^0_{acid} values, especially for the *ortho* substituents. It is probable that ΔS^0_{acid} is the same for the *meta* and nonresonance interacting *para* substituents as for the parents in both series, so for those compounds $\delta\Delta G^0_{\text{acid}}(600\text{ K}) \approx \delta\Delta H^0_{\text{acid}}(600\text{ K}) \approx \delta\Delta H^0_{\text{acid}}(300\text{ K})$. Entropies for the *ortho* compounds are considerably more complicated (Fateley *et al.*, 1975). Table VIII contains two groups of relative ΔH^0_{acid} values anchored to other standard acids: the alkylamines to H_2 and nitric acid to HBr.

TABLE IV

Acidity Scale Based on HPMS Measurements (500 or 600 K)[a]

Acid	ΔH^0_{acid}[a]	ΔG^0_{acid}(298 K)[a]	ΔH^0_{f298}(AH)[b]	$\Delta H^0_{f298}(A^-)$[c]
HCl	(333.6)	328.1	−22.0	(−55.9)
$MeCO_2H$	348.5	341.5	−103.8	−122.5
$EtCO_2H$	347.3	340.3	−108.6[d]	−128.8
n-$PrCO_2H$	346.5	339.5	−113.5[d]	−134.6
$Cl(CH_2)_3CO_2H$	345.4	338.4	−119.8[d]	−141.6
HCO_2H	345.2	338.2	−90.5	−112.5
$ClCH(Me)CH_2CO_2H$	341.9	334.9	−123.4[d]	−148.7
$Cl(CH_2)_2CO_2H$	340.8	333.8	−114.9[d]	−141.3
$PhCO_2H$	338.8	331.7	−78.2[d]	−106.6
FCH_2CO_2H	337.6	331.0		
$ClCH(Et)CO_2H$	337.5	330.9		
$ClCH(Me)CO_2H$	336.4	329.8		
$ClCH_2CO_2H$	335.4	328.8		
$BrCH_2CO_2H$	334.0	327.4	−94.4[e]	−127.6
F_2CHCO_2H	330.0	323.5		
Cl_2CHCO_2H	328.4	321.9		
CF_3CO_2H	322.7	316.3	−245.6[f]	−290
p-$MeOC_6H_4OH$	351.7	344.9	−30.0	−45.5
PhOH	349.8	343.4	−23.1	−40.5
PhNHCOMe	352.8	345.8	−24.9	−39.3
$NH_2CONHCOMe$	348.5	341.1		
$(MeCO)_2NH$	346.8	339.8	−102.7	−123.1
Succinimide	345.4	338.0	−86.0	−107.8
p-$NO_2C_6H_4CH_3$	353.1	345.9	7.4[g]	−6.7
Fluorene	353.3	345.5		
$MeCO_2CH_2COMe$[h]	350.3	343.3		
CH_3COCF_3	350.3	343.2	−193.9[f]	−210.8
$CH_2(CO_2Et)_2$	348.3	342.3	−200.6[d]	−219.5
$CH_2(COMe)_2$	343.7	336.6	−90.6	−114.1
$MeCOCH_2COPh$	339.9	333.0	−59.8[d]	−87.1
Dimedone	338.9	331.1		
$CH_2(CN)_2$	336.0	328.2	63.9	
Barbituric acid	335.1	327.3		
$MeCOCH_2COCF_3$	328.5	322.1	−232.9[f]	−271.2

[a] Cumming and Kebarle (1978), ±2 kcal/mole.
[b] From Benson *et al.* (1969) or Benson (1976).
[c] Equation (25), ±3 kcal/mole.
[d] Estimated from group additivity contributions (Benson, 1976).
[e] C—$(Br)(H)_2(CO)$ from King *et al.* (1971).
[f] C—$(F)_3(CO)$ from Liu *et al.* (1977).
[g] Lenchitz *et al.* (1971).
[h] Site of deprotonation uncertain, see text.

TABLE V

Comparison of ΔH^0_{acid} Values from HPMS and ICR

Acid	ΔH^0_{300}(HPMS)[a]	ΔH^0_{300}(ICR)[b]	$\delta \Delta H^0$ [c]
CH_3SOCH_3	374.6	372.7	1.9
CH_3CN	373.5	372.2	1.3
CH_3COCH_3	370.0	368.8	1.2
CH_3CHO	367.0	366.4	0.6
$MeCH_2CHO$	366.4	365.9	0.5
$CH_3SO_2CH_3$	366.4	366.6	−0.2
Ph_2CH_2	362.7	364.5	−1.8
CH_3COPh	362.8	363.2	−0.4
$MeCH_2COPh$	361.2	362.4	−1.2
Pyrrole	359.2	360.7	−1.5
CH_3NO_2	357.6	358.7	−1.1
$MeCH_2NO_2$	357.3	358.1	−0.8
Me_2CHNO_2	356.6	358.2	−1.6
Cyclopentadiene	355.5	356.1	−0.6
H_2S	352.0	353.4	−1.4
$PhCH_2COMe$	351.8	352.5	−0.7
$PhCH_2CN$	351.9	353.3	−1.4
p-MeC_6H_4OH	351.7	352.6	−0.9
PhOH	349.8	351.4	−1.6

[a] Cumming and Kebarle (1978).
[b] Bartmess *et al.* (1979).
[c] $\delta \Delta H^0 = \Delta H^0_{acid}(HPMS) - \Delta H^0_{acid}(ICR)$.

TABLE VI

Anchoring of Some Relative ΔH^0_{acid}(HPMS) Values by Absolute ΔH^0_{acid}(ICR) Values

Acid	ΔH^0_{acid} [a]	ΔH^0_{f298}(HA)[b]	$\Delta H^0_{f298}(A^-)$[c]
$MeCOCH_2Me$	367.4	−57.0	−54.8
$(MeCH_2)_2CO$	367.4	−61.7	−61.5
$MeCOCHMe_2$	366.9	−62.8	−63.1
$MeCH_2SO_2Ph$	364.9	−71.4	−73.7
$EtCH_2SO_2Ph$	364.0	−76.3	−79.5
i-$PrCH_2SO_2Ph$	363.0	−83.5	−87.7
CH_3SO_2Ph	363.1	−63.6	−67.7
c-$PrSO_2Ph$	361.7	−38.4	−43.9

[a] ± 2 kcal/mole, see text.
[b] From Benson (1976) or Benson *et al.* (1969).
[c] Equation (25), ± 3 kcal/mole.

TABLE VII

Relative $\delta\Delta G^0_{acid}$(600 K) Values for Substituted Benzoic Acids and Phenols[a]

Substituent	$ArCO_2H$ *o*	*m*	*p*	ArOH *o*	*m*	*p*
H	(0)			(0)		
Me	0.8	−0.7	−1.1	0.7	−0.4	−1.3
OMe	0.5	0.5	−0.7	0.9	1.5	−0.8
OH		1.3		10.5	5.1	
NH_2	2.7	−1.5	−2.3	2.1	−0.9	−4.2
F	2.3	3.8	2.9	3.9	5.8	2.6
Cl	4.7	4.7	4.4	6.3	7.9	6.6
CN		10.2	10.9	16.2	14.3	17.7
NO_2	9.4	9.6	11.7	13.9	15.7	
CO_2H				10.5		4.1

[a] McMahon and Kebarle (1977), kcal/mole.

TABLE VIII

Other Quantitative Acidities

Acid	ΔS^0_{acid}[a]	ΔH^0_{acid}[b]	$\Delta G^0_{acid298}$	$\Delta H^0_{f298}(HA)$[c]	$\Delta H^0_{f298}(A^-)$[d]
HNO_3[e]	23.5	324.6	317.3	−32.1	−74.7
NH_3[f]	25.0	403.6	396.1	−11.0	25.4
$MeNH_2$[f]	25.2	403.2	395.7	−5.5	30.5
$EtNH_2$[f]	25.4	399.3	391.7	−11.0	21.1
Me_2NH[f]	23.3	396.4	389.2	−4.5	24.7

[a] ±1 cal/(mole deg), experimental.
[b] ±1 kcal/(mole deg).
[c] From Benson (1976).
[d] Equation (25), ±2 kcal/mole.
[e] Ferguson *et al.* (1972), relative to $\Delta H^0_{acid}(HBr) = 323.6$ kcal/mole (Table I).
[f] Mackay *et al.* (1976), relative to $\Delta H^0_{acid}(H_2) = 400.4$ kcal/mole (Table I).

There are several studies in which qualitative acidity orders in the gas phase have been assigned (Fig. 1). With the present scale as a framework, quantitative bounds may be placed on some of these data, as given in Table IX. The order $PH_3 < Me_3B < AsH_3 < C_2H_4O < H_2S$ has been determined by the kinetic method in an ICR spectrometer (Murphy and Beauchamp, 1976; Wyatt *et al.*, 1974). If $C_2H_3O^-$, generated by ethyl nitrite pyrolysis, is the enolate of acetaldehyde, then AsH_3 and Me_3B are in a 3 kcal/mole range between acetaldehyde and phosphine. Such a small range seems unlikely, since endoenergetic double resonance is not observed for any of these

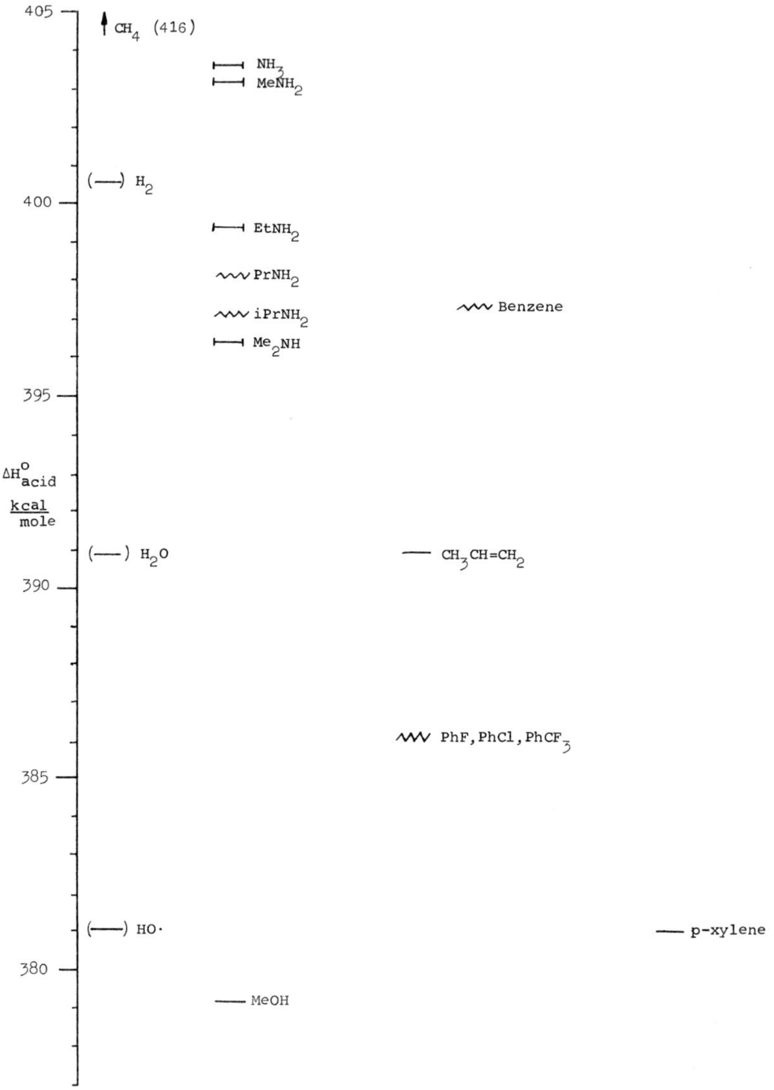

FIG. 1 The gas phase ΔH^0_{acid} (298 K) scale. The values of those acids marked (———) are absolute numbers from Table I and scheme I. A solid line indicates data from Table III (ICR, 320 K). A dotted line is used for data from Tables IV–VII (HPMS, 500 or 600 K). For meta and para substituted compounds, ΔS^0_{acid} is taken as that of the parent acid. The symbol ├──┤ denotes data from Table VIII. The symbol ∿∿∿ represents qualitative data from Table IX; error bounds, relative to the other acids, for these last values may be up to 7 kcal/mole.

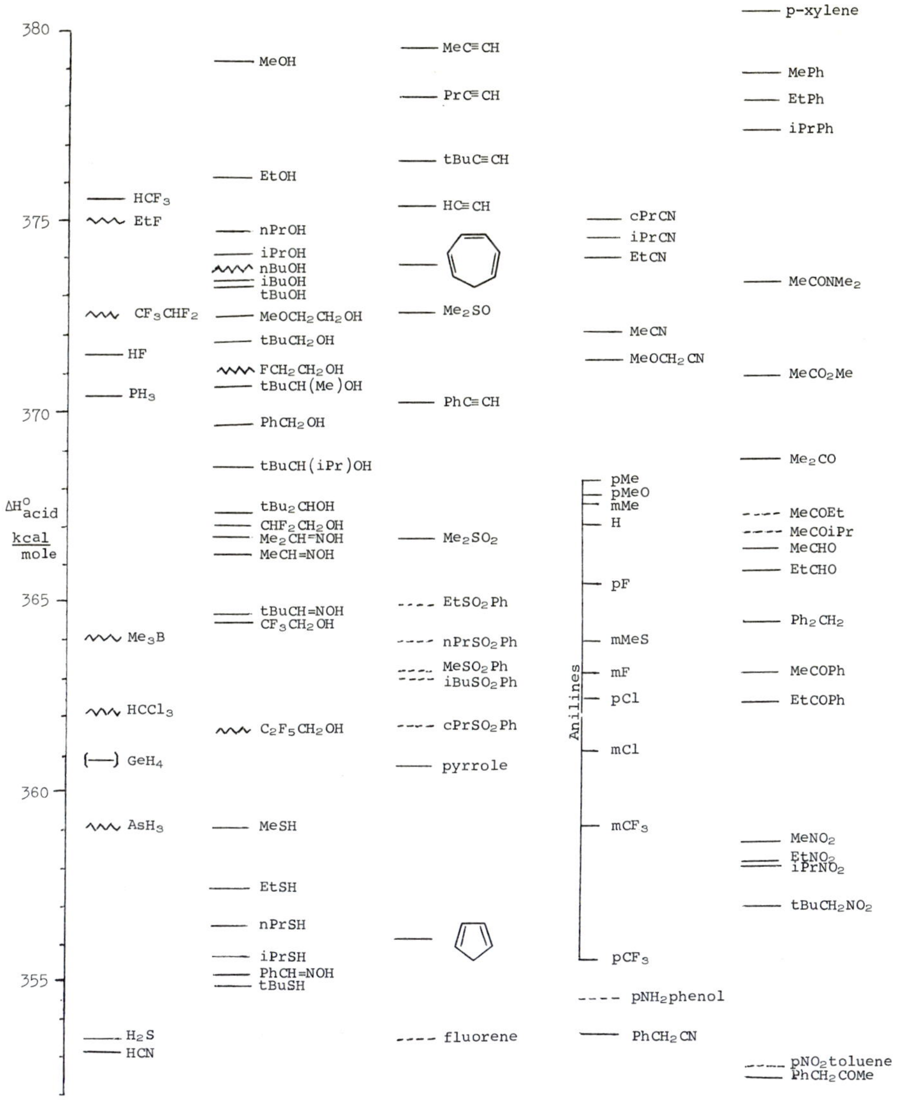

FIG. 1 (*Continued*)

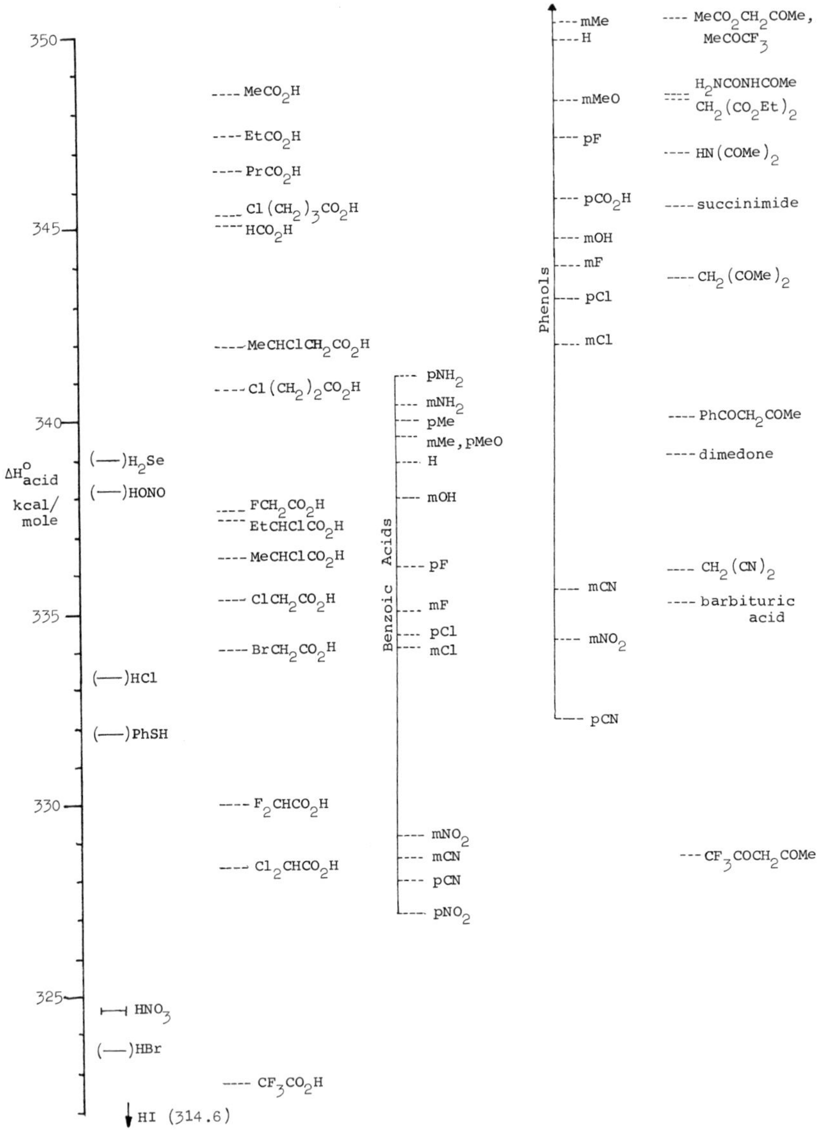

FIG. 1 (*Continued*)

TABLE IX

Assignment of Absolute Acidities to Compounds Qualitatively Bracketed by Known Acids

Acid	ΔH^0_{acid}	Bracketing acids	Ref.
$CH_2{=}CH_2$	>404	NH_3	*a*
Cyclopropane	>404	NH_3	*a*
Me_3N	>404	NH_3	*b*
n-$PrNH_2$	398 ± 1	$EtNH_2$, Me_2NH	*c*
i-$PrNH_2$	397 ± 1	$EtNH_2$, Me_2NH	*c*
C_6H_6	398 ± 7	NH_3, H_2O	*a*
C_6H_5Cl	386 ± 7	H_2O, MeOH	*d*
C_6H_5F	386 ± 7	H_2O, MeOH	*d*
$C_6H_5CF_3$	386 ± 7	H_2O, MeOH	*d*
n-BuOH	374.0 ± 3	*n*-PrOH, *t*-BuOH	*e*
n-$C_5H_{11}OH$	374.0 ± 3	*n*-PrOH, *t*-BuOH	*e*
FCH_2CH_3	375 ± 3	EtOH, *i*-PrOH	*f*
F_2CHCH_3	375 ± 3	EtOH, *i*-PrOH	*f*
F_3CCH_3	375 ± 3	EtOH, *i*-PrOH	*f*
F_2CHCH_2F	372.5 ± 3	*t*-BuOH, HF	*f*
F_2CHCHF_2	372.5 ± 3	*t*-BuOH, HF	*f*
F_3CCH_2F	372.5 ± 3	*t*-BuOH, HF	*f*
F_3CCHF_2	372.5 ± 3	*t*-BuOH, HF	*f*
Me_3B	364 ± 5	PH_3, AsH_3	*g*
AsH_3	359 ± 7	Me_3B, H_2S	*g*
FCH_2CH_2OH	371.0 ± 5	*t*-BuOH, Me_2CO	*h*
$CF_3CH(Me)OH$	364 ± 4	$=CF_3CH_2OH$	*h*
$C_2F_5CH_2OH$	361.5 ± 4	CF_3CH_2OH, CH_3NO_2	*h*
$(CF_3)_2CHOH$	<351	>HCN	*h*
CH_3COCN	349 ± 5	HCN, $CH_2(COMe)_2$	*i*
$HCCl_3$	362 ± 6	Me_2CO, cyclopentadiene	*a*

[a] Bohme *et al.* (1972).
[b] Mackay *et al.* (1976).
[c] Brauman and Blair (1971).
[d] Miller (1975).
[e] Brauman and Blair (1970).
[f] Sullivan and Beauchamp (1976).
[g] Murphy and Beauchamp (1976); Wyatt *et al.* (1974).
[h] Dawson and Jennings (1977).
[i] Brauman and Blair (1968).

pairs and that usually requires 2–3 kcal/mole difference, unless appreciable energies of activation are present. The placement of $C_2H_3O^-$ in the ordering by observation of the reaction.

$$AsH_2^- + X \rightarrow C_2H_3O^- + AsH_3$$

may be due to X not being acetaldehyde, or due to spurious results arising from the detuning effect (DeFrees *et al.*, 1979). If $C_2H_3O^-$ is omitted, the possible acidity values are more consistent with the observed double resonance results and the acidity range of 18 kcal/mole from PH_3 to H_2S. The value for Me_3B has been according assigned $\Delta H^0_{acid} = 364 \pm 5$. The report that diethyl amine is more acidic than H_2O (Brauman and Blair, 1971) could not be verified (Miller, 1975) so its acidity is still uncertain. Assignment of the acidity of acetyl cyanide, bracketed by HCN and acetylacetone (Brauman and Blair, 1968) depends on the acidity of the enol form of the latter compound. If ΔH_f^0 (enol) is 6 kcal/mole lower than that of the keto form (Andreassen and Bauer, 1972), whose acidity is known, this would put $\Delta H_{\text{acid}}^0(\text{enol}) = 346$, or 6 kcal/mole more acidic than HCN. This gives acetyl cyanide as 349 ± 5 kcal/mole.

By the kinetic method in the flowing afterglow, allene is 0.1 kcal/mole more acidic than methanol (Bohme *et al.*, 1972). The reverse rate (allene anion plus methanol) is not well defined, due to the ambident nature of allene anion, giving either allene or the more thermodynamically stable (1.5 kcal/mole) propyne on reprotonation. Deuterium labeling reveals that propyne is less acidic than methanol at the $C\underline{H}_3$-site, but more acidic at the localized acetylenic $C{\equiv}C{-}\underline{H}$ site. The latter is the predominant ion at long times in the ICR, enabling its acidity to be measured.* The acidity of allene is thus given large error bounds, with $\Delta H_{\text{acid}}^0 = 380 \pm 5$ kcal/mole.

There are a number of ketones for which the electron affinity of the corresponding radical has been determined (Zimmerman *et al.*, 1977), but the bond energies are not available. These may be estimated from the DH^0 values derived from similar ketones in Section VI.C, since it appears that the C—H bond energy in RR′CHCOX is sensitive to the structure of R, but much less so for variable X. For methyl ketones (R, R′ = H), $DH^0(C{-}H) = 97 \pm 2$ kcal/mole is adopted; for secondary systems, $DH^0 = 92 \pm 2$, and for tertiary C—H(R = R′ = Me), $DH^0 = 87$ is estimated. The calculated ΔH_{acid}^0 values, with appropriate large uncertainties, are given in Table X.

TABLE X

Absolute Acidities of Some Carbon Acids from EA and Estimated DH^0 [a]

Acid	EA[a]	DH^0 [b]	ΔH_{acid}^0 [c]
CH_3CO-*t*-Bu	42.5 ± 1.5	97 ± 2	368 ± 4
$(Me_2CH)_2CO$	33.8 ± 1.0	87 ± 3	367 ± 4
$EtCH_2CHO$	38.5 ± 1.2	91 ± 3	366 ± 4
$(EtCH_2)_2CO$	39.7 ± 1.3	92 ± 3	366 ± 4
t-$BuCH_2CHO$	42.0 ± 1.4	92 ± 3	364 ± 4
CH_3COF	51.2 ± 2.1	97 ± 2	359 ± 6

[a] Zimmerman *et al.* (1977), kcal/mole.
[b] Estimated, see text, kcal/mole.
[c] In kcal/mole.

* D. J. DeFrees and R. T. McIver, Jr., University of California, Irvine, unpublished results.

VI. THERMOCHEMICAL DATA DERIVED FROM GAS-PHASE ACIDITIES

A. *Heats of Formation of Anions*

With $\Delta H^0_{acid}(AH)$ now available, it can be related to the considerable body of thermochemical data in the literature to derive new quantities. From Eq. (25), $\Delta H^0_{f298}(A^-)$ can be derived if $\Delta H^0_{f298}(AH)$ is known. For reaction (8),

$$\Delta H^0_{f298}(A^-) = \Delta H^0_{acid}(AH) - \Delta H^0_{f298}(H^+) + \Delta H^0_{f298}(AH). \qquad (25)$$

Using $\Delta H^0_{f298}(H^+) = 367.2$ kcal/mole (Stull and Prophet, 1971), the anion heats of formation shown in Tables III, IV, VI, and VIII are derived. These considerably expand the number of $\Delta H_f{}^0(A^-)$ values available (Rosenstock *et al.*, 1977). The uncertainty in ΔH^0_{acid} for the compounds in Tables IX and X are such that it is felt that further thermochemical derivations are not warranted. This body of $\Delta H_f{}^0(A^-)$ values should be a valuable asset for further investigations into negative ion–molecule chemistry. The limiting factor now appears to be the lack of $\Delta H_f{}^0(AH)$ values for many of these acids.

B. *Electron Affinities*

Through an inversion of scheme I, the electron affinity of A· is obtainable if the bond dissociation energy of A—H is known, via Eq. (26),

$$EA(A\cdot) = IP(H\cdot) - \Delta H^0_{acid}(AH) + DH^0(A\text{—}H). \qquad (26)$$

A number of values derived this way are given in Table XI. They are in accord with expected trends: $EA(PhCH_2\cdot) < EA(PhNH\cdot) < EA(PhO\cdot)$, $EA(NO_2) < EA(NO_3)$, and EA(propene) $\approx$ EA(cycloheptatriene) $\ll$ EA(cyclopentadiene). The EA of acetoxy is about twice that of alkoxy, consistent with it having twice as many oxygens capable of supporting the charge. The uncertainty in the acetylene C—H bond strength yields considerable uncertainty in $EA(HC_2)$; the values obtained are consistent with the upper limit to EA of ≤ 86 kcal/mole (see Janousek and Brauman, Chapter 10, in this volume). Reexamination of Table I also reveals some interesting trends. For all the alcohols, either the EA is too low by $\sim$1–2 kcal/mole or the bond energy too high. A similar, though smaller, deficit is seen for the alkanethiols. For acetone, the EA of Eyler (1978), where special care was taken to thermalize the ions before laser photodetachment in the ICR, agrees with equilibrium experiments better than the values of Ellison or Zimmerman and Brauman. The absolute ΔH^0_{acid} for phenol from Table I is lower than that derived from these equilibrium measurements, consistent with the reported EA being only an upper limit.

TABLE XI

Derivation of Electron Affinities of A· from Bond Energies and $\Delta H^0_{acid}(AH)$

Acid	$DH^0(A—H)^a$	Ref.	$\Delta H^0_{acid}(AH)^b$	$EA(A)^c$
$MeNH_2$	101	*d*	403.2	11.4 ± 3
$EtNH_2$	101	*d*	399.3	15.3 ± 3
Me_2NH	96	*d*	396.4	13.2 ± 3
p-$MeC_6H_4CH_3$	84.8	*e*	380.5	7.9 ± 4
$MeCH_2Ph$	78	*f*	378.3	13.3 ± 4
Cycloheptatriene	73.2	*e*	373.9	13.1 ± 4
C_6H_6	110	*e*	398	26 ± 9
HC≡CH	115 ± 7	*f*	375.4	53 ± 11
	128	*d*		66 ± 10
$PhNH_2$	80 ± 3	*f*	367.1	26.5 ± 7
HNO_3	101 ± 5	*g*	324.6	90 ± 6
$MeCO_2H$	110	*e*	348.5	75.1 ± 4

[a] ±2 kcal/mole, unless otherwise noted.
[b] Tables III, IV, VI, and VIII, kcal/mole.
[c] Eq. 26, kcal/mole.
[d] Sen Sharma and Franklin (1973).
[e] Egger and Cocks (1973).
[f] Kerr (1966).
[g] Stull and Prophet (1971).

C. *Bond Dissociation Energies and Heats of Formation of Radicals*

The bond dissociation energy $DH^0(A—H)$ is likewise obtainable from EA(A·) and scheme I. Rearranging Eq. (26) gives

$$DH^0(A—H) = \Delta H^0_{acid}(AH) - IP(H\cdot) + EA(A\cdot). \tag{27}$$

Furthermore, the definition of DH^0 in terms of heats of formation yields

$$\Delta H^0_{f298}(A\cdot) = DH^0_{298}(A—H) - \Delta H^0_{f298}(H\cdot) + \Delta H^0_{f298}(AH). \tag{28}$$

Values calculated in these ways are presented in Table XII. It is seen that $DH^0(H—CH_2COX) = 97 \pm 3$ for X = —OMe, —Ph, and —CF_3, but is lower for X = H. The value for phenol is consistent with the literature (Table I). For pyrrole, $DH^0(N—H)$ is rather high for a secondary amine (see Table XII), but this is consistent with little delocalization of the radical. Pyrrole is already a six-electron π system, and has nothing to gain by becoming a five-electron system. Finally it should be noted that EA(A·) is not the same as $\Delta H^0_{f298}(A\cdot) - \Delta H^0_{f298}(A^-)$, since the electron affinity is defined at 0 K, and requires IP(H·), also at 0 K with similar $\Delta C_p{}^0$, to cancel it in scheme I.

TABLE XII

Derivation of Bond Energies of A—H *and Radical* ΔH_f^0 *from Electron Affinities and* ΔH^0_{acid}(AH)

Acid	EA(A·)[a]	Ref.	ΔH^0_{acid}(AH)[b]	DH^0(A—H)[c]	ΔH_f(A·)[d]
CH_3CO_2Me	41.5 ± 1.4	*e*	371.0	98.9 ± 3.4	−51.2
CH_3CHO	41.7 ± 1.4	*e*	366.4	94.5 ± 3.4	2.7
$MeCH_2CHO$	38.9 ± 1.2	*e*	365.9	91.2 ± 3.2	−6.4
CH_3COPh	47.6 ± 1.2	*e*	363.2	97.2 ± 3.2	23.1
Pyrrole	55.1 ± 0.3	*f*	360.7	102.2 ± 2.3	76.0
PhOH	<54.4 ± 1.4	*g*	349.8	<90.6 ± 3.4	<15.5
CH_3COCF_3	59.5 ± 2.9	*e*	350.3	96.2 ± 4.9	−149.8

[a] In kcal/mole.
[b] Tables III and IV.
[c] Equation (27), kcal/mole.
[d] Equation (28), ±5 kcal/mole.
[e] Zimmerman *et al.* (1977).
[f] Richardson *et al.* (1975b).
[g] Richardson *et al.* (1975a).

D. *Methyl Cation Affinities of Anions*

With $\Delta H_f^0(A^-)$ available, numerous thermochemical cycles can be evaluated. For the reaction

$$\begin{gathered} ACH_3 \rightarrow A^- + CH_3^+ \\ \Delta H^0_{\text{MCA}}(A^-) = \Delta H_f^0(A^-) + \Delta H_f^0(CH_3^+) - \Delta H_f(ACH_3), \end{gathered} \tag{29}$$

the enthalpy change $\Delta H^0_{\text{MCA}}(A^-)$ is the methyl cation affinity of the anion. A scale of these values is a measure of nucleophilicities of the anions toward carbon, just as ΔH^0_{acid} is a measure of nucleophilicity toward hydrogen, or the anion's proton affinity. The quantity ΔH^0_{MCA} can also be arrived at through knowledge of the relative A—H and A—CH_3 homolytic bond energies. These are not available in many cases, however, so the formulation here is a useful source. A few values are presented in Table XIII. It is evident that carbanions are better nucleophiles toward CH_3^+ than toward the proton, relative to alkoxides, and fluoride relatively weak, though a linear correlation is seen for a given type of anion, such as alkoxides.

E. *Hydride Affinities of Unsaturated Compounds*

The propensity of hydride ion to react with some unsaturated compound to give another anion can be calculated using Eq. (30)

$$\begin{gathered} HXY^- \rightarrow H^- + XY \\ \Delta H^0_{\text{redn}}(XY) = \Delta H_f^0(H^-) + \Delta H_f^0(XY) - \Delta H_f^0(HX{-}Y^-), \end{gathered} \tag{30}$$

TABLE XIII

Comparison of Methyl Cation Affinities and Proton Affinities for Some Anions

Anion	ΔH^0_{MCA} [a]	ΔH^0_{acid} [b]
$MeC{\equiv}C^-$	283.8	379.6
$PhCH_2^-$	278.7	379.0
HO^-	275.8	390.8
CF_3^-	276.0	375.6
MeO^-	270.0	379.2
t-BuO^-	263.4	373.3
$PhNH^-$	261.1	367.1
CN^-	261.2	353.1
F^-	257.5	371.5
PhO^-	239.5	349.8
$MeCO_2^-$	237.5	348.5

[a] ±4 kcal/mole (relative), Eq. (29), with $\Delta H_f^0(CH_3^+) = 262$ kcal/mole (Rosenstock *et al.*, 1977).

[b] Tables I, III, and IV.

where $\Delta H^0_{redn}(XY)$ is the hydride affinity of the unsaturated compound XY. For example, H^- plus acetone yields isopropoxide anion. Values for some of these gas phase reductions are given in Table XIV. For carbonyl compounds, ΔH^0_{redn} increases with alkyl group elaboration (product anion

TABLE XIV

Hydride Affinities of Some Unsaturated Compounds

$HXY^- = XY + H^-$

XY	HXY^-	ΔH^0_{redn} [a]
$CH_2{=}O$	CH_3O^-	43.2
$MeCH{=}O$	EtO^-	41.0
$Me_2C{=}O$	i-PrO^-	39.8
$EtCH{=}O$	n-PrO^-	41.4
i-$PrCH{=}O$	i-BuO^-	43.7
t-$BuCH{=}O$	t-$BuCH_2O^-$	46.4
$CH_2{=}CHCOCH_3$	$MeCH{=}C(Me)O^-$	59.2
$PhCH{=}CH_2$	$Ph\bar{C}HMe$	50.1
CO_2	HCO_2^-	51.7
$CH_2{=}C{=}O$	$CH_2{=}CHO^-$	61.3

[a] ΔH^0_{redn} is the hydride affinity of the unsaturated compound; ±4 kcal/mole (absolute value), with $\Delta H_f^0(H^-) = 33.2$ kcal/mole (Rosenstock *et al.*, 1977).

stability controlled) but decreases with R for H substitution at carbonyl (reactant stability controlled). Reduction reactions of this type have been observed in the gas phase (Bierbaum *et al.*, 1978).

F. *Acidity of Tautomers*

Some of the carbon acids mentioned herein can exist in two tautomeric forms. If the difference in $\Delta H_f{}^0$ for two forms is known, then the less stable form is that much greater in acidity than the more stable tautomer. As mentioned above, the enol form of acetylacetone is 6 kcal/mole less acidic than the keto form because the enol form is more stable at 298 K. Also, acetone is 14 kcal/mole less acidic than the enol of acetone, 2-propenol (Pollack and Hehre, 1977). Other enols are similarly 10–15 kcal/mole more acidic than their keto forms. Nitroso compounds are 13–19 kcal/mole more acidic than the corresponding oximes (Benson *et al.*, 1969) and allene should be 1.5 kcal/mole more acidic than propyne (methyl protons).

VII. STRUCTURAL EFFECTS ON ACIDITY

The relationship between intrinsic structure and acidity has been discussed in the primary works mentioned above. A brief review will be given here, along with discussion of points pertaining to the acidity scale as a whole.

A. *Nature of the Acidic Site*

As shown in Fig. 2, the simple hydrides increase in acidity across each row and down each column (Brauman *et al.*, 1971), with the possible exception of germane and arsine which are within experimental uncertainty of

CH_4	$<$	NH_3	$<$	H_2O	$<$	HF
416.6 ± 1.0		403.6 ± 1.0		390.8 ± 0.4		371.5 ± 0.7
$\wedge$		$\wedge$		$\wedge$		$\wedge$
SiH_4	$<$	PH_3	$<$	H_2S	$<$	HCl
371.5 ± 2		370.4 ± 2		353.4 ± 2		333.3 ± 0.3
$\wedge$		$\wedge$		$\wedge$		
GeH_4	$\approx$	AsH_3	$<$	H_2Se	$<$	HBr
360.7 ± 2.9		359 ± 7		339 ± 5		323.6 ± 0.3
						$\wedge$
						HI
						314.3 ± 0.2

FIG. 2 Acidity order of the simple hydrides. Values are ΔH^0_{acid} from Tables I and III–V.

each other. While the horizontal trend can be explained by electronegativity increases down an isoelectronic sequence, the vertical order is opposite this. In terms of scheme I, the acidities in each row are controlled by the EA; the change in DH^0 opposes this but is smaller. For the columns, the changes in DH^0 are much larger than those for EA and thus control the acidity, save for the group IVA hydrides, where the situation is much more complex (Brauman *et al.*, 1971). This order may vary upon substitution of groups for the hydrogens, however, since the acidic atoms have varying sensitivities to the polar, resonance, etc., interactions which can occur. Thus HF < CF_3CH_2OH < pyrrole < HCN. For a given substituent, however, the acidities usually follow the order of the hydrides.

B. *Alkyl Group Effects*

1. *Methyl versus Hydrogen*

Substitution of a methyl group for hydrogen in the hydrides yields a confusing pattern of acidities: MeOH $\gg$ H_2O, $MeNH_2 \geq NH_3$, and MeSH < H_2S. Theoretical calculations (Streitwieser *et al.*, 1974; Kollman *et al.*, 1975) indicate methyl may be slightly ($\sim$1 kcal/mole) acid strengthening in ethane so at least the first row trend appears consistent. For the hydrides of elements lower in the periodic table such as Ge, As, and P, solution acidities (Issleib and Kümmel, 1965; Birchall and Jolly, 1966) show methyl as acid weakening, though extrapolation to the gas phase is very uncertain. From polarizability considerations (Brauman and Blair, 1968), methyl should be acid strengthening in all cases, though perhaps less so for sulfur due to the more diffuse anion and longer bond length there. The methyl effects plot linearly with the secondary isotope effects (CD_3 versus CH_3) for each acid (DeFrees *et al.*, 1977), indicating that anionic hyperconjugation is the controlling factor. The order MeSH < H_2S is attributed to an electron-donating polar effect by methyl, destabilizing the anion (Bartmess *et al.*, 1979). The *balance* of these three effects controls the net acidity.

Methyl substitution into more complex acids generally results in small (<3 kcal/mole) effects. Three categories can be defined by structure and function: (1) substitution at a sp^2 or sp carbon which does not rehybridize upon ionization, (2) substitution at sp^3 carbon which does not rehybridize, and (3) substitution at sp^3 carbon which rehybridizes to a sp^2 carbanion upon proton loss. To explain the observed methyl effects, it is assumed that methyl stabilizes a structure relative to hydrogen when sited on sp^2 carbon, due to hyperconjugation, and that methyl is electron donating, relative to hydrogen when sited on sp^2 carbon, due to electronegativity and polarizability differences (Brauman and Blair, 1970) but has negligible polar effect when on sp^3 carbon. Methyl is acid weakening in the first group (acetylenes, oximes, RCO_2H, CH_3COR, and meta- or paraphenyl ring substitution) due to the polar effect; the hyperconjugation effect tends to cancel out between the acid and anion. Substitution at sp^3 carbon, group 2, is dealt with in Section VII.B.2. Substitution at the rehybridizing (acidic) carbon occurs in the acids R_2CH—EWG, with EWG (electron withdrawing group) as —Ph, —CN, —SO_2Ph, —COMe, —CHO, and —COPh. Methyl is acid strengthening by small amounts for all these save for —CN and —SO_2Ph. This appears to be due to a balance between a polar electron donating (and therefore acid-weakening) effect and a hyperconjugative acid-strengthening effect, both operating in the anion (Bordwell *et al.*, 1978). The polar effect depends on the proximity of negative charge, so is greatest for —CN and —SO_2Ph where delocalization of change is slight relative to polar effects of EWG, since $\sigma_R^- < \sigma_I$ (Ehrenson *et al.*, 1973). Thus the polar effect of methyl controls the acidities here and is determined by the amount of residual negative charge on carbon. Polarizability probably also is important in these effects, but does not appear to be a *controlling* factor.

2. Different Alkyl Groups

In all observed cases save one, acidity increases with methyl substitution on sp^3 carbon, i.e., with larger alkyl groups, primarily due to polarizability (Brauman and Blair, 1970). The effect falls off with distance of the alkyl group from the anionic site, but not as rapidly as the $1/r^4$ theoretical prediction. The difference is due to the anisotropic polarizability of alkyl groups (LeFevré, 1965) and to the presence, on the angstrom distance scale, of hyperpolarizability interactions not present in the macroscopic model (Coulson *et al.*, 1952). Alkyl groups may also adopt "bunched-up" conformations, as in

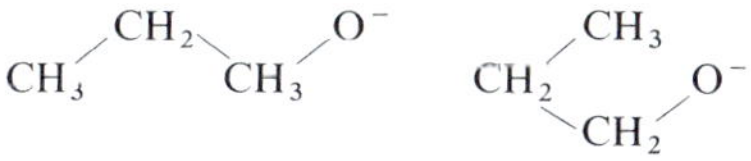

to maximize polarizability interactions. Thus by through-space interactions, β-methyl substitution into ethanol or isopropyl alcohol can have a larger effect on acidity than α-methyl substitution into these compounds, even though it is farther away in bonding terms. Polarizability also explains the gas-phase acidity orders fluorene > cyclopentadiene and toluene > propene, both of which are the opposite of that predicted by molecular orbital calculations (Streitwieser, 1961) and by solution acidities. Similarly, benzene is more acidic than ethylene, even though both give localized sp^2 carbanions, and *p*-nitrotoluene is more acidic than nitromethane in spite of the quinoid-type structure and consequent loss of the benzene resonance energy required in the anion for the aromatic case.

C. *Polar Effects*

In contrast to the polarizability effects discussed above which often invert acidity orders between the gas phase and solution, polar effects in the gas phase largely parallel solution values. Reasonably good linear free energy relationships are seen for the alcohols, anilines (Bartmess *et al*., 1979), benzoic acids, phenols, and carboxylic acids (McMahon and Kebarle, 1977) versus solution phase substituent constants such as σ_I or σ. In general, polar effects are larger in the gas phase, compared to solution, due to the *compression effect*: the substituent, which acts alone in the gas phase to stabilize the anion, in solution must share this task with the solvent, so that its apparent effect in the condensed phase is only a fraction of the total. This is seen in the two- to tenfold increase in ρ or ρ_I for the various series cited upon going from solution to the gas phase. The effect of a polar group falls off with distance, much as in solution (Yamdagni and Kebarle, 1974).

There are some inversions of polar effects seen upon solvation. For the haloacetic acids in the gas phase Br > Cl > F, opposite to the aqueous acidities (Hiraoka *et al*., 1973). This can be attributed to the greater polarizability of the larger halo substituents, even though they have lower electronegativity. Solvation effects, primarily on the anion, invert this order in solution (Haberfield and Rakshit, 1976) and cause $\sigma_I(F) > \sigma_I(Cl) > \sigma_I(Br)$. A similar effect is seen with $HCF_3 < HCCl_3$ in the gas phase.

The decrease in acidity with alkyl substitution into chloroacetic and 2-chloropropanoic acid (Yamdagni and Kebarle, 1974) is in contrast to every other alkyl group effect (Section VII.B.2.). Similarly, it appears $CF_3CH_2OH \approx CF_3CH(Me)OH$ (Dawson and Jennings, 1977). Possible causes include: (1) intramolecular hydrogen bonding in the acid, creating a partial positive charge on chlorine which the alkyl groups can stabilize via their polarizability, (2) steric hindrance of this hydrogen bonding by the alkyl group, thus increasing the acid's entropy and reducing acidity, or (3) steric hindrance by the alkyl group to maximum separation of the electronegative substituent and the negatively charged atom in the anion. It is evident that further experimental work is in order.

The comparisons in Table XV reveal other interesting trends. Replacement of hydrogen by acetyl in the CH_3COX acids reveals a resonance saturation effect (Bordwell and McCollum, 1976) of decreasing acetyl effect with increasing acidity. In group b, CF_3 for CH_3 substitution increases acidity by amounts depending on the proximity of charge to the substituent as expected for polar effects. The nitrogen acids in group c should be inherently more acidic than carbon acids for electronegativity reasons (Section VII.A); however, when the nitrogen lone pair is capable of delocalization to oxygen, as in acetimide, the stabilization of the acid results in lowered net acidity (McMahon and Kebarle, 1976). The resonance forms in aniline involving

TABLE XV

Substituent Effects on ΔH^0_{acid}

	AH[a]	BH[a]	$\delta\Delta H^0_{acid}$ [b]
(a)	$C\underline{H}_3COCH_3$	$AcC\underline{H}_2COCH_3$	25.1
	$C\underline{H}_3COPh$	$AcC\underline{H}_2COPh$	23.3
	$C\underline{H}_3COCF_3$	$AcC\underline{H}_2COCF_3$	21.8
(b)	$CH_3CO_2\underline{H}$	$CF_3CO_2\underline{H}$	25.8
	$CH_3COC\underline{H}_3$	$CF_3COC\underline{H}_3$	18.5
	$CH_3COC\underline{H}_2COCH_3$	$CF_3COC\underline{H}_2COCH_3$	15.2
	$CH_3C_6H_4N\underline{H}_2$	$pCF_3C_6H_4N\underline{H}_2$	12.6
(c)	cyclopentadiene	Pyrrole	−4.6
	$CH_3COC\underline{H}_2COCH_3$	$CH_3CON\underline{H}COCH_3$	−3.1
	$PhC\underline{H}_2COCH_3$	$PhN\underline{H}COCH_3$	−0.3
	$PhC\underline{H}_3$	$PhN\underline{H}_2$	11.9
	$C\underline{H}_4$	$N\underline{H}_3$	13.0
(d)	$C\underline{H}_4$	$C\underline{H}_3Ph$	37.6
	$C\underline{H}_3Ph$	$C\underline{H}_2Ph_2$	14.5
	$C\underline{H}_3Ph$	Fluorene	25.7
	$C\underline{H}_4$	$C\underline{H}_3CN$	44.4
	$C\underline{H}_3CN$	$C\underline{H}_2(CN)_2$	36.2
	$C\underline{H}_4$	$C\underline{H}_3CO_2Me$	45.6
	$C\underline{H}_3CO_2Me$	$C\underline{H}_2(CO_2Et)_2$	22.7
	$C\underline{H}_4$	$C\underline{H}_3COCH_3$	47.8
	$C\underline{H}_3COCH_3$	$C\underline{H}_2(COCH_3)_2$	25.1

[a] Acidic proton underlined.
[b] $\delta\Delta H^0_{acid} = \Delta H^0_{acid}(AH) - \Delta H^0_{acid}(BH)$, kcal/mole.

negative charge on carbon are not of sufficient importance, however, to lower its acidity below that of toluene. For acetanilide, with one group of each type, the effect is nearly the average of these two cases, with the electronegative slightly winning out. The greater acidity of cyclopentadiene versus pyrrole is due to the nitrogen acid being aromatic in both acid and anion, while the carbon acid has this extra stabilization only in the anion.

The effect of the second phenyl group on the acidity of methane is considerably less than the first, 14.5 versus 37.6 kcal/mole. Molecular models indicate that both rings cannot become coplanar in the anion, due to steric hindrance of the orthohydrogens. Forcing the rings coplanar in fluorene increases the "second" phenyl effect but still to only 68% of the first. A similar effect is seen with acetyl, carbethoxy, and cyano groups. In the cyano case, the reduced effect of the second —CN, 82% of the first, cannot be due to steric hindrance to resonance due to the low steric demand of CN. It is concluded that the falloff in effect is due to a resonance saturation effect (Bordwell and McCollum, 1976).

For EWG equal to —BMe_2 and —NMe_2, the observed carbon acidities are predicted reasonably well by calculations on the anions $CH_2BH_2^-$ and $CH_2NH_2^-$ (Kollman *et al.*, 1975). Since —BR_2 is conceptually the simplest EWG involving both resonance and polar effects, it is unfortunate that σ_I and σ_R^- values do not exist for this group. The small effect seen for —NR_2, presumably due to destabilization of the carbanion by *p*–*p* repulsion, is similar to that seen for other first-row electronegative elements when adjacent to solution phase carbanion (Adolph and Kamlet, 1966; Hine *et al.*, 1967; Streitwieser and Mares, 1968).

D. *Resonance Effects in Hydrocarbons*

Delocalization of negative charge away from the primary site into a π system should stabilize an anion and increase the acidity of the corresponding

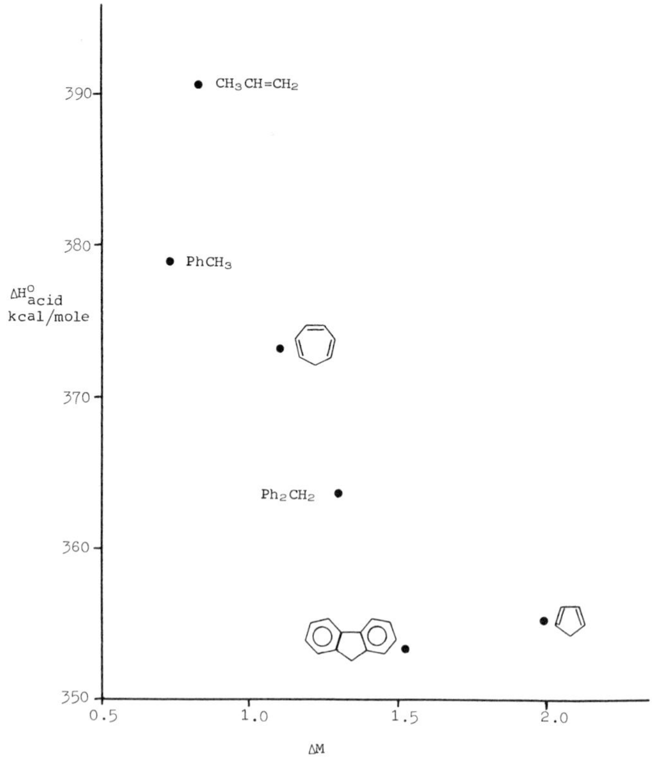

FIG. 3 The ΔH^0_{acid} value of some hydrocarbons plotted against the HMO parameter ΔM.

conjugate acid. The CH_3—EWG acids show this trend (Bartmess *et al.*, 1979); however, appreciable polar effects operating in these compounds hinder the exact analysis of the resonance component. Polar effects are much smaller in hydrocarbons, so a better correlation with some parameter of delocalization might be expected. The parameter ΔM from Hückel molecular orbital theory has been shown to correlate reasonably well with solution acidities of hydrocarbons, such as the MSAD scale of ion-pair acidities (Streitwieser, 1961). A similar correlation is seen in Fig. 3: a general trend is evident with acids of low polarizability near the top of the trend and high polarizability on the bottom. For Eq. (31), a dual substituent parameter analysis

$$\Delta H^0_{\text{acid}} = -\beta \Delta M - A\alpha + \Delta H_o{}^0 \tag{31}$$

of ΔH^0_{acid} for the planar acids versus ΔM and α, the bulk polarizability of the acids in cubic angstroms, gives $\beta = 23.7$ and $A = 1.2$, with the intercept $\Delta H_o{}^0$ as 414.1 kcal/mole, quite close to methane at 416. The value for β is the expected magnitude (Streitwieser, 1961), but should not be taken as definitive, since the bulk polarizability α is probably only a qualitative measure of polarizability when applied at the molecular level.

REFERENCES

Adolph, H. G., and Kamlet, M. J. (1966). *J. Am. Chem. Soc.* **88**, 4761.

Andreassen, A. L., and Bauer, S. H. (1972). *J. Mol. Struct.* **12**, 381.

Arnett, E. M., Mitchell, E. J., and Murty, T. S. S. R. (1974). *J. Am. Chem. Soc.* **96**, 3875.

Bartmess, J. E., McIver, R. T., Jr., and Miller, J. S. (1979). *J. Am. Chem. Soc.* (submitted).

Batt, L., Christie, K., Milne, R. T., and Summers, A. J. (1974). *Int. J. Chem. Kinet.* **6**, 877.

Benson, S. W. (1976). "Thermochemical Kinetics," 2nd ed. Wiley (Interscience), New York.

Benson, S. W., Cruickshank, F. R., Golden, D. M., Haugen, G. R., O'Neal, H. E., Rodgers, A. S., Shaw, R., and Walsh, R. (1969). *Chem. Rev.* **69**, 279.

Bierbaum, V. M., DePuy, C. H., Schmitt, R. J., and Shapiro, R. H. (1978). Personal *J. Am. Chem. Soc.* **100**, 2920.

Birchall, T., and Jolly, W. L. (1966). *Inorg. Chem.* **5**, 2177.

Bohme, D. K., Lee-Ruff, E., and Young, L. B. (1972). *J. Am. Chem. Soc.* **94**, 5152.

Bordwell, F. G., and McCollum, C. J. (1976). *J. Org. Chem.* **41**, 2391.

Bordwell, F. G., Bartmess, J. E., and Hautala, J. A. (1978). *J. Org. Chem.* **43**, 3095.

Brauman, J. I., and Blair, L. K. (1968). *J. Am. Chem. Soc.* **90**, 5636.

Brauman, J. I., and Blair, L. K. (1970). *J. Am. Chem. Soc.* **92**, 5986.

Brauman, J. I., and Blair, L. K. (1971). *J. Am. Chem. Soc.* **93**, 3911.

Brauman, J. I., Eyler, J. R., Blair, L. K., White, M. J., Comisarow, M. B., and Smyth, K. C. (1971). *J. Am. Chem. Soc.* **93**, 6366.

Carlson, G. L., and Fateley, W. G. (1977). *J. Phys. Chem.* **81**, 2308.

Colussi, A. J., and Benson, S. W. (1977). *Int. J. Chem. Kinet.* **9**, 295.

Colussi, A. J., Zabel, F., and Benson, S. W. (1977). *Int. J. Chem. Kinet.* **9**, 161.

Coulson, C. A., MacColl, A., and Sutton, L. E. (1952). *Trans. Faraday Soc.* **48**, 106.

Craven, S. M., and Bentley, F. F. (1972). *Appl. Spectrosc.* **26**, 242.

Cumming, J. B., and Kebarle, P. (1978). *Can. J. Chem.* **56**, 1.

Cunningham, A. J., Payzant, J. D., and Kebarle, P. (1972). *J. Am. Chem. Soc.* **94**, 7627.

Darwent, B. D. (1970). "Bond Dissociation Energies in Simple Molecules," Natl. Stand. Ref. Data Ser., Natl. Bur. Stand., NSRDS-NBS 31. U.S. Govt. Print. Off., Washington, D.C.
Dawson, J. H. J., and Jennings, K. R. (1977). *Int. J. Mass Spectrom. Ion Phys.* **25**, 47.
DeFrees, D. J., Bartmess, J. E., Kim, J. K., McIver, R. T., Jr., and Hehre, W. J. (1977). *J. Am. Chem. Soc.* **99**, 6451.
DeFrees, D. J., Hehre, W. J., McIver, R. T., Jr., and McDaniel, D. H. (1979). *J. Chem. Phys.* (submitted).
Dixon, D. A., Holtz, D., and Beauchamp, J. L. (1972). *Inorg. Chem.* **11**, 960.
Egger, K. W., and Cocks, A. T. (1973). *Helv. Chim. Acta* **56**, 1517.
Ehrenson, S., Brownlee, R. T. C., and Taft, R. W. (1973). *Prog. Phys. Org. Chem.* **10**, 1.
Ellison, B. (1977). Personal communication.
Engelking, P. C., and Lineberger, W. C. (1977). *J. Chem. Phys.* **67**, 1412.
Eyler, J. R. (1978). Personal communication.
Farneth, W. E., and Brauman, J. I. (1976). *J. Am. Chem. Soc.* **98**, 7891.
Fateley, W. G., Carlson, G. L., and Bentley, F. F. (1975). *J. Phys. Chem.* **79**, 199.
Ferguson, E. E., Dunkin, D. B., and Fehsenfeld, F. C. (1972). *J. Chem. Phys.* **57**, 1459.
Fine, D. H., and Westmore, J. B. (1970). *Can. J. Chem.* **48**, 395.
Furuyama, S., Golden, D. M., and Benson, S. W. (1971). *Int. J. Chem. Kinet.* **3**, 1237.
Golden, D. M., and Benson, S. W. (1969). *Chem. Rev.* **69**, 125.
Haberfield, P., and Rakshit, A. K. (1976). *J. Am. Chem. Soc.* **98**, 4393.
Hine, J., Malone, L. G., and Liotta, C. L. (1967). *J. Am. Chem. Soc.* **89**, 5911.
Hiraoka, K., Yamdagni, R., and Kebarle, P. (1973). *J. Am. Chem. Soc.* **95**, 6833.
Issleib, K., and Kümmel, R. (1965). *J. Organomet. Chem.* **3**, 84.
Kerr, J. A. (1966). *Chem. Rev.* **66**, 465.
King, E. J. (1965). "Acid-Base Equilibria," pp. 299–301. Pergamon, New York.
King, K. D., Golden, D. M., and Benson, S. W. (1971). *J. Chem. Thermodyn.* **3**, 129.
Kollman, P., McKelvey, J., and Grund, P. (1975). *J. Am. Chem. Soc.* **97**, 1640.
LeFevré, R. J. W. (1965). *Adv. Phys. Org. Chem.* **3**, 1.
Lenchitz, C., Velicky, R. W., Silvestro, G., and Schlosberg, L. P. (1971). *J. Chem. Thermodyn.* **3**, 689.
Liu, M. T. H., Loucks, L. F., and Hooper, D. G. (1977). *Int. J. Chem. Kinet.* **9**, 589.
Lowe, J. P. (1972). *Prog. Phys. Org. Chem.* **6**, 1.
McAllister, T., and Lossing, F. P. (1968). *J. Phys. Chem.* **73**, 2996.
McIver, R. T., Jr. (1978). *Rev. Sci. Instrum.* **49**, 111.
Mackay, G. I., Hemsworth, R. S., and Bohme, D. K. (1976). *Can. J. Chem.* **54**, 1624.
McMahon, T. B., and Kebarle, P. (1976). *J. Am. Chem. Soc.* **98**, 3399.
McMahon, T. B., and Kebarle, P. (1977). *J. Am. Chem. Soc.* **99**, 2272.
Matthews, W. S., Bares, J. E., Bartmess, J. E., Bordwell, F. G., Cornforth, F. J., Drucker, G. E., Margolin, Z., McCallum, R. J., McCollum, G. J., and Vanier, N. R. (1975). *J. Am. Chem. Soc.* **97**, 7006.
Miller, J. S. (1975). Ph.D. Thesis, Univ. of California, Irvine.
Murphy, M. K., and Beauchamp, J. L. (1976). *J. Am. Chem. Soc.* **98**, 1433.
Page, F. M., and Goode, G. C. (1969). "Negative Ions and the Magnetron." Wiley, New York.
Pollack, S. K., and Hehre, W. J. (1977). *J. Am. Chem. Soc.* **99**, 4845.
Reed, K. J., and Brauman, J. I. (1974). *J. Chem. Phys.* **61**, 4830.
Richardson, J. H., Stephenson, L. M., and Brauman, J. I. (1973). *J. Chem. Phys.* **59**, 5068.
Richardson, J. H., Stephenson, L. M., and Brauman, J. I. (1975a). *J. Am. Chem. Soc.* **97**, 2967.
Richardson, J. H., Stephenson, L. M., and Brauman, J. I. (1975b). *J. Am. Chem. Soc.* **97**, 1160.
Richardson, J. H., Stephenson, L. M., and Brauman, J. I. (1975c). *J. Chem. Phys.* **63**, 74.
Rosenstock, H. M., Draxl, K., Steiner, B. W., and Herron, J. T. (1977). *J. Phys. Chem. Ref. Data* **6**, *Suppl.* No. 1, p. 736.

Sandal, S., and Seip, H. M. (1975). *J. Mol. Struct.* **28**, 193.

Sen Sharma, D. K., and Franklin, J. L. (1973). *J. Am. Chem. Soc.* **95**, 6562.

Solly, R. K., Golden, D. M., and Benson, S. W. (1970). *Int. J. Chem. Kinet.* **2**, 11.

Steele, W. C., Nichols, L. D., and Stone, F. G. A. (1962). *J. Am. Chem. Soc.* **84**, 4441.

Streitwieser, A., Jr. (1961). "Molecular Orbital Calculations for Organic Chemists." Wiley, New York.

Streitwieser, A., Jr., and Mares, F. (1968). *J. Am. Chem. Soc.* **90**, 2444.

Streitwieser, A., Jr., and Nebenzahl, L. L. (1976). *J. Am. Chem. Soc.* **98**, 2188, and references therein.

Streitwieser, A., Jr., Owens, P. H., Wolf, R. A., and Williams, J. E. (1974). *J. Am. Chem. Soc.* **96**, 5448.

Stull, D. R., and Prophet, H., eds. (1971). "JANAF Thermochemical Tables," Natl. Stand. Ref. Data Ser., Natl. Bur. Stand., NSRDS-NBS 37. U.S. Govt. Print. Off., Washington, D.C.

Sullivan, S. A., and Beauchamp, J. L. (1976). *J. Am. Chem. Soc.* **98**, 116.

Wolf, J. F., Staley, R. H., Koppel, T., Taagepera, M., McIver, R. T., Jr., Beauchamp, J. L., and Taft, R. W. (1977). *J. Am. Chem. Soc.* **99**, 5417.

Wyatt, R. H., Holtz, D., McMahon, T. B., and Beauchamp, J. L. (1974). *Inorg. Chem.* **13**, 1511.

Yamdagni, R., and Kebarle, P. (1974). *Can. J. Chem.* **52**, 861.

Zimmerman, A. H. and Brauman, J. I. (1977a). *Struct.–Energy Relationships Conf., 5th Santa Barbara, Calif.*

Zimmerman, A. H., and Brauman, J. I. (1977b). *J. Am. Chem. Soc.* **99**, 3565.

Zimmerman, A. H., Reed, K. J., and Brauman, J. I. (1977). *J. Am. Chem. Soc.* **99**, 7203.

Chapter 12

Chemical ionization mass spectrometry

K. R. JENNINGS

DEPARTMENT OF CHEMISTRY
AND
MOLECULAR SCIENCES
UNIVERSITY OF WARWICK
COVENTRY, UNITED KINGDOM

ISBN 0-12-120802-8

I. INTRODUCTION

The technique of chemical ionization mass spectrometry (CIMS) was first described by Munson and Field (1966) and is now well established as an analytical technique of considerable versatility, finding applications in many branches of chemistry and biochemistry, and in medical and environmental studies. The increased use of CIMS, particularly over the last five years, stems partly from developments in the technique itself and partly from the increased availability of commercial ion sources which can operate at pressures of 0.5–1.0 torr and which are necessary for the obtaining of CI mass spectra. Early work in this field has been reviewed (Field, 1968, 1972a,b; Wilson, 1971, 1973), and more recently attempts have been made to assess the relative merits of CIMS, electron impact (EI), field ionization (FI), and field desorption (FD) mass spectrometry in solving various analytical problems (Milne and Lacey, 1974). Recent, more general reviews of mass spectrometry also contain many references to applications of CIMS (Wilson, 1975, 1977; Jennings, 1977; Games, 1977; Burlingame *et al.*, 1976). This article will therefore concentrate on surveying the basis of the technique together with recent developments.

II. FUNDAMENTAL PRINCIPLES

A typical electron impact mass spectrum is obtained by transferring a range of energies from the electron beam to the sample molecules at pressures of 10^{-5}–10^{-6} torr so as to produce molecular radical cations $M^{+\cdot}$ with a range of internal energies. For many compounds, the structure of $M^{+\cdot}$ closely resembles that of M so that the structure of the neutral may be inferred from the observed fragment ions which result from competing and consecutive unimolecular decompositions of $M^{+\cdot}$ ions and fragment ions. However, there are two types of behavior which reduce the utility of EIMS for structure determination, both of which derive from the fact that EIMS is essentially a high energy process. For some classes of compounds, $M^{+\cdot}$ is intrinsically unstable with respect to dissociation, irrespective of its internal energy, so that almost all $M^{+\cdot}$ ions fragment within the ion source and the intensity of the peak given by these ions in the mass spectrum is negligible. In other cases, isomeric samples yield identical EI mass spectra, indicating that structural information is lost before fragmentation occurs, owing to the fact that the molecular ions formed from the different isomers rearrange to a common structure or mixture of structures on a time scale that is short compared with that required for fragmentation.

A quite different approach is used in chemical ionization mass spectrometry in which ions characteristic of the sample molecules are formed in ion–molecule reactions between reagent ions and the sample. A reagent gas or mixture of gases is bombarded with a beam of high energy electrons (typically ~400 eV) at a pressure of ~1 torr to produce a high yield of a reagent ion which may be positively or negatively charged. The reagent ion may be a molecular ion, fragment ion, or product of an ion–molecule reaction between a primary ion and a reagent gas molecule (e.g., $N_2^{+\cdot}$ from N_2, $O^{-\cdot}$ from N_2O or CH_5^+ from CH_4) which is stable with respect to further reaction with the reagent gas. If a low concentration (0.1%) of sample is introduced into the ion source, very few ions will be formed from it through direct ionization by electron impact. However, reagent ions may undergo ion–molecule reactions with the sample, the product ions of which constitute the chemical ionization mass spectrum of the compound. It is clear that, for a given sample, the CI mass spectrum that is observed depends very much on the type of reagent ion that is used to produce the spectrum, and an important feature of CIMS is that by making an appropriate choice of reagent ion, different types of ion–molecule reactions can be used to produce a variety of CI mass spectra. It is therefore possible to "tailor" a CI mass spectrum to suit a particular application under investigation. For example, a gas phase Brønsted acid XH^+ will donate a proton to a sample molecule M, providing that the proton affinity (PA) of M is greater than that of X:

$$XH^+ + M \rightarrow MH^+ + X \qquad (\Delta H_1), \tag{1}$$

where $\Delta H_1 = PA(X) - PA(M)$. The extent of fragmentation of MH^+ depends upon the internal energy with which it is formed, which, in turn, is dependent upon ΔH_1. The magnitude of ΔH_1 can be controlled by suitable choice of X and the use of a strong gas phase acid such as H_3^+ leads to considerable fragmentation of MH^+, whereas the much weaker acid $C_4H_9^+$ frequently gives MH^+ as the most abundant ion. Factors affecting the choice of reagent gases are discussed in a later section.

The above brief description of the essentials of CIMS is sufficient to illustrate several important differences between this technique and that of EIMS. As has already been pointed out, the latter is a high energy technique which is not dissimilar to pyrolysis. In CIMS, on the other hand, the sample molecules have an internal energy characteristic of the ion source temperature (typically 100–200°C) and structural information is inferred from the reactions of a variety of reagent ions with these molecules. If a reagent ion is chosen which undergoes a very exothermic ion–molecule reaction with the sample molecules, a large number of fragment ions may result and the CI mass spectrum may resemble the EI mass spectrum quite closely. Under less energetic conditions, however, the gas phase ion–molecule reactions may closely resemble reactions which occur in solution and it is not unreasonable to search for reactant ions that will give information concerning the presence of particular functional groups or molecular geometries. If the sample molecule in its ground electronic state participates in an ion–molecule reaction of low exothermitity, it is to be expected that isomeric compounds, which frequently yield identical EI mass spectra, will give CI mass spectra that will differ sufficiently for the isomers to be distinguished. A further factor that contributes to the differences between CIMS and EIMS is that most of the commonly used reagent gases generate even electron reagent ions which react by the transfer of an even electron species to or from the sample molecule to give even electron ions such as MH^+, $(M - H)^+$, or $(M - H)^-$, the fragmentations of which are usually simpler than those of radical ions produced by electron impact or electron attachment. Furthermore, it is often possible to rationlize a CI mass spectrum by assuming that the reagent ion attacks the sample molecule at one or more specific positions, and, as will be seen in a later section, this can be demonstrated in favorable cases by the use of isotopically labeled reagent ions or sample molecules. It is this specific site attack by certain reagent ions that makes possible the obtaining of information on the locations of particular functional groups within a sample molecule in certain cases, information which may be difficult to obtain directly from an EI spectrum.

III. INSTRUMENTAL ASPECTS OF CIMS

Almost all chemical ionization studies have been carried out in ion sources capable of operating within the pressure range 0.5–2.0 torr and with the reagent gas present in at least one-thousandfold excess. In order to achieve maximum sensitivity, the extraneous flow of gas out of the source must be minimized and the pumping speed available in the source housing must be as high as possible. A pressure of $< 10^{-4}$ torr can usually be maintained in the source housing by the use of high speed diffusion pumps or turbomolecular pumps, and differential pumping of the source and analyzer regions produces an analyzer tube pressure in the region of 10^{-6} torr in a magnetic deflection instrument and a somewhat higher pressure in a quadrupole instrument.

Careful machining of all components of the source block is necessary in order to ensure a gas-tight fit. The aperture for the entrance of the electron beam is typically 0.10–0.15 mm^2 in area, and a grid plate between the filament and the source block facilitates focusing of the electron beam on the aperture and reduces the gas flow to the filament. An electron energy of ~ 400 eV is employed in order that the electrons penetrate into the source, but at typical working pressures they will not reach the far wall of the ion source. In a source designed exclusively for CIMS, there is therefore no electron exit aperture to an electron trap, since the electron beam is filament- or emission-current regulated, and no source collimating magnet is required. The ion exit slit must be large enough to provide adequate sensitivity but at the same time be of low enough conductance both to maintain the source pressure and to minimize the pressure within the accelerating region in a magnetic deflection instrument. The width of the slit is typically in the region of 3×10^{-2} mm and the low conductance can be maintained either by the use of a short slit made from blade edges or of a longer slit made in a thicker plate, the area of the slit being in the region of 0.05–0.15 mm^2. Under these conditions, the collision rate in the source is $\sim 3 \times 10^7$ s^{-1} and if the electron beam enters at a point 5 mm behind the ion exit slit, the average source residence time in the absence of a repeller field is of the order of a few milliseconds. For analytical work, it is normal to work with a repeller field in order to maximize sensitivity, thereby reducing the residence time to 10^{-4}–10^{-5} s, depending on the field strength and the geometry of the source. Many commercial sources are designed to serve as EI sources at low pressures and CI sources at high pressures, and their designs are necessarily compromises. Their sensitivities in either mode of operation are therefore below what could be attained with a specially designed source although on some instruments, adjustments can be made external to the vacuum system which make the source more gas tight when operated as a CI source. Although most manufacturers now offer a CI source as a standard accessory, various accounts of methods of modifying EI sources for CI operation have appeared in the literature (Beggs *et al.*, 1971; Yinon and Boettger, 1972; Garland *et al.*, 1974; Hogg, 1972; Arsenault *et al.*, 1971), allowing the sources to be used in either mode. Although most commonly used reagent gases do not lead to an unacceptably short filament life, the use of strongly oxidizing gases is precluded. This limitation can be overcome by replacing the filament assembly by a small Townsend discharge assembly which may be used to supply energetic electrons and/or ions to effect ionization in both positive and negative CIMS (Hunt *et al.*, 1975a). This may be fitted in the hole in the ion chamber wall which is normally made use of for the electron trap assembly. The discharge assembly consists of a nickel screen over the hole in the ion chamber, a short length of ceramic tubing, and a stainless steel cathode. A potential of approximately 1 kV between the cathode and the source block gives 10–30 μA of electrons when used in the positive ion mode.

Since the ion sources of magnetic deflection instruments are normally operated at a potential of several kilovolts with respect to the source housing and inlet lines which are at earth potential, care must be taken to prevent the striking of an arc when the instrument is operated under CI conditions. The inlet line for the admission of gases and volatile liquids should be entirely of glass on the low pressure side of the leak. For the introduction of samples from a gas chromatograph through a heated inlet line, a short length of capillary tubing made from quartz or other electrically insulating glass is usually inserted between the ion source and the jet separator or valve of a direct inlet system since pyrex is a good conductor of electricity at 250–300°C. A direct flow interface giving satisfactory results up to 250°C at 6 kV and 300°C at 4 kV has recently been described (Hatch and Monson, 1977a). On the Warwick MS50 instrument, the source pressure is measured by means of an MKS "Baratron" capacitance manometer, the output of which is fed to a Granville–Philips automatic pressure regulator and valve. All of this equipment is floated at source potential behind a protective screen and the control units are supplied *via* a mains isolation transformer. CIMS studies carried out on quadrupole instruments are free from these problems since very much lower potentials are applied to the ion source and focusing electrodes.

When a magnetic deflection instrument is operated under negative chemical ionization (NCI) conditions, no difficulty is experienced in detecting negative ions by means of an electron multiplier if the potential of the ion source is appreciably more negative with respect to earth than the potential of the first dynode of the multiplier. If the two potentials are similar, e.g., −2 kV, no signal is detected unless the detection system is modified. If the first dynode of the multiplier is held at earth potential, the output signal is typically at 2 kV and must be passed through a suitable electrical feedthrough to the normal preamplifier *via* a triaxial cable and isolation amplifier (Smit *et al.*, 1976). Alternatively, an extra dynode at a suitably high positive potential with respect to earth may be placed in the path of the ions immediately in front of the electron multiplier, the first dynode of which is negative with respect to earth (Stafford, 1977). Negative ions impacting on to the extra dynode give rise to positive ions which strike the first dynode of the multiplier and give a signal in the normal way. Similarly, a pair of dynodes can be placed in front of the scintillation plate of a Daly metastable detector (Daly *et al.*, 1973) so that the electrons produced by the second dynode strike the plate and give a signal. Since the ion source of a quadrupole instrument is close to earth potential, the detector must be modified as described above when the instrument is operated under NCI conditions.

Since the quadrupole mass filter will transmit ions of a given mass-to-charge ratio, irrespective of the sign of the charge, a source has recently been developed (Hunt *et al.*, 1976) that allows the recording of both positive and negative CI mass spectra simultaneously by pulsing the polarity of ion source and focusing lens potentials at 10 kHz. Packets of positive and negative ions are mass analyzed by the mass filter and ions are detected by two electron multipliers placed side by side at the exit of the mass filter. Ions of opposite charge are separated by applying potentials of opposite polarity to the first dynode of each multiplier. Different types of positive and negative ion mass spectra can be produced by appropriate reagent gas mixtures, details of which are discussed in a later section.

Most samples are introduced into the ion source either through a standard gas inlet or GC effluent line or on a probe, but these methods cannot be used for thermally unstable samples. A recent development of considerable promise is to make use of a field desorption emitter on to which a solution of the sample is placed (Hunt *et al.*, 1977). Rapid desorption of the sample with little thermal decomposition occurs from the microneedle points when a predetermined current is passed through the emitter. The emitter replaces the repeller plate in a quadrupole CI source and allows the obtaining of methane CI mass spectra of underivatised compounds such as guanosine, cyclic adenosine monophosphate, and tetrapeptides, none of which give satisfactory spectra when introduced by conventional methods.

Although almost all CIMS studies have been carried out at pressures in the region of 1 torr, similar studies can be carried out at much lower pressures making use of ion cyclotron resonance (ICR) techniques (Lehman and Bursey, 1976) or quadrupole ion storage (Quistor) devices (Todd *et al.*, 1976). The use of double resonance and ion ejection techniques in ICR studies provides valuable information on the mechanisms of chemical ionization processes, but since the technique is the subject of another chapter in this volume, it will not be discussed further here. The quadrupole ion storage trap allows ions to be stored for up to 4 ms before they pass into a quadrupole mass filter and this allows the observation of CI mass spectra at pressures of 10^{-4} torr (Lawson *et al.*, 1973). Applications of this device have recently been reviewed (Lawson and Todd, 1977). At the other end of the pressure range, CI mass spectra can be obtained in an atmospheric pressure ionization (API) source in which ions are produced either by radiation from a ^{63}Ni source (Horning *et al.*, 1973), or more recently by a corona discharge (Carroll *et al.*, 1975), from a carrier gas which contains a trace of reagent gas. The carrier gas is typically nitrogen and the reagent gas is frequently the solvent in which the sample is introduced. Reagent ions are formed from N_2^+ ions by charge exchange and these together with ions derived from reactions with the sample molecules are analyzed in a quadrupole mass filter. When the API technique is used for the production of NCI mass spectra, it is almost impossible to exclude traces of oxygen from the system. The formation of $O_2^{-}\cdot$ reagent ions leads to irreproducible NCI mass spectra unless a trace of oxygen is added to the system so that this ion is the principal reactant ion (Dzidic *et al.*, 1978).

For many applications, an important parameter in assessing the utility of CIMS is its sensitivity, particularly relative to that of EIMS. A quantitative definition of the sensitivity or detection limit of a technique or instrument is difficult to establish in view of its dependence on operating conditions, sample purity, etc., and in considering the sensitivity obtainable using positive or negative CIMS, an additional variable is the choice of a suitable reagent gas for a particular type of sample. In terms of total ionization current derived from sample molecules, there is often little to choose between EIMS and positive CIMS and in a combined EI/CI source, the relative sensitivities will also be affected by the compromises inherent in the design of a dual-purpose source. If the intensities of base peaks are compared, a reagent gas that produces a single ion from sample molecules should, in principle, lead to a higher sensitivity for CIMS. Under negative CIMS conditions, sample molecules that have a positive electron affinity will yield electron capture spectra and in favorable circumstances, the sensitivity obtainable by this method is two to three orders of magnitude greater than that obtainable from positive CIMS and affords quantitation at the 100 fg level (Hunt *et al.*, 1976). The main reason for the high sensitivity observed for electron capture spectra is believed to be the high rate of diffusion of electrons together with high rate constants for electron capture processes ($\sim 10^{-7}$ cm^3 molecule^{-1} s^{-1}) compared with those for ion molecule reactions ($\sim 10^{-9}$ cm^3 molecule^{-1} s^{-1}) since the rate constants for ion loss processes for positive and negative ions are similar. True negative CIMS, in which a reagent ion reacts with a sample molecule, will in general exhibit a sensitivity similar to its positive ion counterpart and, in favorable cases, each method will allow quantitation down to picogram levels. Most instruments commercially available are capable of giving sensitivities in the region of 10^{-9}–10^{-10} C μg^{-1} for appropriate samples using any of these techniques, and the advantage of chemical ionization and electron capture spectra is that the ion current is frequently concentrated in one or two intense peaks.

IV. REAGENT GASES USED IN POSITIVE CHEMICAL IONIZATION MASS SPECTROMETRY

The CI mass spectrum of a compound is clearly very largely dependent on the reagent gas or gas mixture used to produce the reagent ions which react with the sample. A general requirement of any CI reagent gas is that it should produce one major reactant ion, preferably of low mass-to-charge ratio, with a minimum number of other ions. The relative abundances of the ions should be insensitive to small changes of reagent gas pressure in order that spectra should be reproducible. Reagent gases should also be obtainable in a high state of purity since the impurities will compete with the sample molecules for the reactant ions and may modify the spectrum significantly. The ions produced by reaction with the sample molecules should not react further with the reagent gas, e.g., in the formation of proton-bound dimers or cluster ions, since these will complicate interpretation of the spectrum and the relative abundances of such ions are very dependent on source pressure and temperature, thereby leading to irreproducibility. In order to establish criteria by which the behavior of different reagent gases may be judged, it is essential to be clear what type of information is being sought from the mass spectrum, as may be seen from the following examples.

A. *Quantitation of a Known, Pure Compound*

In this case, the ideal reagent gas will produce a CI mass spectrum consisting of a single intense peak, in order to maximize the sensitivity. Provided that this peak is not present in the spectrum given by the reagent gas alone, the mass-to-charge ratio of the ions giving rise to the peak is unimportant.

B. *Quantitation of a Known Compound in a Mixture*

For this application, the requirements are a little more demanding than in the previous example. It is still desirable that the known compound has a CI mass spectrum consisting of a single intense peak but the mass-to-charge ratio of the ions giving rise to this peak must be characteristic of the known compound, and these ions must not be given by any other components in the mixture. Alternatively, if in a mixture of chemically similar compounds (e.g., homologs), the known compound is an impurity that belongs to a different class of compounds, it may be possible to simplify the spectrum by selecting a reagent ion which reacts only with the impurity.

C. *Single-Ion Monitoring in GC/CIMS*

The requirements for this application are similar to those set out in Section IV.B if the quantitation of a single compound is required. If, on the other hand, the requirement is to identify which peaks in the gas chromatogram are due to a related class of compounds (e.g., unsaturated esters in a mixture of unsaturated and saturated esters), the aim would be to produce a reagent ion capable of reacting only with the unsaturated esters. The GC peaks due to these esters could then be identified by monitoring the intensity of the peak given by this ion (reactant ion monitoring) (Hatch and Munson, 1977b). Alternatively, a reactant ion may be chosen which gives a particular ion with each of a series of related compounds but gives different ions with other components of the mixture.

D. *Molecular Weight Information*

Since many classes of compound give molecular ions of low abundance under electron impact conditions, CIMS is frequently employed as a method of determining the molecular weight of a compound. In this case, it is necessary to generate a reagent ion that reacts in a known way with the sample to produce an ion, the mass-to-charge ratio of which can readily be related to the molecular weight of the sample. The ions which are most commonly used for molecular weight determination under CIMS conditions are $(M + H)^+$, $(M - H)^+$, $M^{-}\cdot$, and $(M - H)^-$, although in specific instances, others such as $(M + NH_4)^+$ and $(M + O_2)^{-}\cdot$ may be of use. If the reagent ion reacts to produce only one ion from each component in a sample, the molecular weight of each component can be established in this way, provided that the reaction is of a known type in each case.

E. *Structural Information*

Whereas the above examples all required that the reagent gas should give rise only to one peak in the CI mass spectrum, applications in which structural information is the main objective ideally require the production of a mass spectrum containing a number of peaks. A reagent gas may therefore be chosen that produces a reagent ion capable of reacting with the sample molecule to yield an ion that undergoes limited fragmentation. The extent of fragmentation can to some extent be varied by controlling the exothermicity of the reaction between the reactant ion and the sample molecule. An alternative approach would be to seek a reagent ion that reacts in a known manner with a particular functional group or which attacks at a specific site in a molecule so that with the help of deuterium labeling, structural information may be derived from the products of the reaction. It is in this type of application that CIMS comes closest to classical organic chemistry in which the structure of a molecule is inferred from the products of its reactions with a variety of reagents. The characteristics of the more commonly used reagent gases will now be considered in the light of these requirements.

V. GASES GIVING CHARGE TRANSFER REAGENT IONS

The rare gases (Field *et al.*, 1967) and nitrogen (Whitney *et al.*, 1971) give reactant ions $X^{+}\cdot$ which react with sample molecules M to produce excited molecular ions $M^{+\cdot}$, the internal energy of which is determined by the difference in ionization energies of X and M. The most commonly used reagent gases of this type are nitrogen and argon, for which the ionization energies are, respectively, 15.6 and 15.8 eV, which compare with values of 7–10 eV for many organic molecules. The internal energy of $M^{+\cdot}$ will therefore be equal to the difference in ionization energies of the reagent gas and the sample. This is similar to the average energy of the ion produced by electron impact so that CI mass spectra given by nitrogen or argon will resemble EI mass spectra except that the intensity of the molecular ion peak will usually be much smaller since no molecular ions with low internal energies are formed. Nitrogen and argon are therefore rarely used alone as reagent gases but are useful when mixed with a second gas, particularly if the nitrogen or argon is a GC carrier gas. The second component is usually added after the sample has passed through the column and is chosen to react with the sample to give an ion in the region of the molecular ion. If water is added, the ion H_3O^+ acts as a Brønsted acid (see below) and will protonate many compounds giving abundant $(M + H)^+$ ions (Hunt and Ryan, 1972a). The relative abundances of these ions and fragment ions can be controlled by varying the partial pressures of argon and water in the reagent gas mixture. Similarly, the abundance of $M^{+\cdot}$ ions can be enhanced by using a N_2/NO reagent gas mixture (Jelus *et al.*, 1974a; Hunt and Harvey, 1975) which gives $M^{+\cdot}$ ions of low internal energy by charge transfer from NO^+ ions (ionization energy of NO is 9.3 eV). However, care must be taken in the use of this mixture since NO^+ ions undergo other reactions with certain classes of compounds, as discussed below. In certain applications, a charge transfer reagent ion may be used to demonstrate the presence of a particular class of compound in a mixture containing other classes of compounds. For example, the presence of polynuclear aromatic hydrocarbons in the presence of other organic material can be shown using $Ar^{+\cdot}$ as the reagent ion (Ryan, 1977) since they give intense peaks due to the great stability of the molecular ions of the aromatic compounds, whereas other components will give mainly low-mass fragment ions. In a photoionization study (Sieck, 1979), in which krypton resonance radiation was used to produce high yields of the molecular ion of cyclohexane (ionization energy of 9.83 eV), charge transfer spectra of gasolines and fuel oils consist largely of ions derived from aromatic components of the mixtures, and the spectra afford a rapid method of distinguishing samples from different sources. Aromatic compounds may also be used as reagent gases, and benzene (Subba Rao and Fenselau, 1977) (ionization energy of 9.2 eV) gives a high yield of $C_6H_6{}^{+\cdot}$ ions under CI conditions which react by charge exchange with esters of unsaturated fatty acids, whereas no ionization of esters of saturated fatty acids occurs since their ionization energies are greater than 9.2 eV. The $C_6H_6{}^{+\cdot}$ ion is preferable to certain other charge transfer reagent ions in that other reactions such as proton or hydride transfer or ion attachment do not occur, but under most operating conditions, dimer ions $(C_6H_6)_2^{+\cdot}$ are readily formed.

VI. GASES GIVING BRØNSTED ACID REAGENT IONS

By far, the most widely used reagent gases are those that give rise to ions which react as Brønsted acids (XH^+) with samples that have a higher proton affinity than X, as briefly discussed in an earlier section. The most commonly used reagent gases of this type (Field, 1972a,b; Milne and Lacey, 1974) are methane, water, isobutane, and ammonia, and the main ions formed from these gases under CI conditions are, respectively, CH_5^+, H_3O^+, $C_4H_9^+$, and NH_4^+. The proton affinities of CH_4, H_2O *i*-C_4H_8, and NH_3 are, respectively (Yamdagni and Kebarle, 1976), 535, 706, 806, and 840 kJ mole^{-1}. The range of compounds which these ions can protonate falls in the series $CH_5^+ > H_3O^+ > C_4H_9^+ > NH_4^+$. The proton affinities of many organic compounds that contain oxygen as the only heteroatom fall within the range (Yamdagni and Kebarle, 1976) 710–830 kJ mole^{-1} so that whereas CH_5^+ and H_3O^+ ions protonate these compounds, very few are protonated by the NH_4^+ ion. In general, the NH_4^+ ion will protonate organic compounds which contain nitrogen, such as amines and heterocyclic bases (Dzidic and McCloskey, 1972), but with oxygen containing compounds such as sugars, ion attachment to form $(M + NH_4)^+$ ions occurs (Hogg *et al.* 1972; Horton *et al.*, 1974).

For a given sample, the exothermicity of reaction (1) can be controlled to some extent by using different reagent gases and hence varying PA(X). If PA(X) is low, as in the case of methane, the reaction is highly exothermic and some of this exothermicity appears as vibrational energy in the MH^+ ions, leading to considerable fragmentation. Since the proton affinity of isobutene is 271 kJ mole^{-1} higher than that of methane, the use of isobutane as a reagent gas leads to a much less exothermic reaction, and frequently the peak due to MH^+ ions is the base peak. It is, therefore, not uncommon to obtain both methane and isobutane CI mass spectra since the former will frequently exhibit considerable fragmentation giving structural information, whereas the latter will give an intense MH^+ peak and hence give molecular weight information.

Whereas CI mass spectra given by charge transfer reagent ions resemble EI mass spectra in many ways, those produced by Brønsted acid reagent ions are often very different since they arise from the fragmentation of the even-electron species MH^+, rather than the odd-electron species $M^{+\cdot}$. If the sample molecule contains a heteroatom, this can be regarded formally as the site of protonation leading to the general reaction type

$$XH^+ + AB \rightarrow X + (A - B - H)^{+*} \rightarrow A^+ + HB.$$

The leaving tendency of the group B is in general inversely related to the proton affinity of HB so that the leaving tendencies of common functional groups is in the order (Field, 1972b)

$$NH_2 < CH_3S, C_6H_5, CH_3O, COOH < CN, SH < OH < I, Br, Cl.$$

Hydrogen atom rearrangement reactions are much less common than in EI spectra and this can probably be ascribed to the protonation of heteroatoms, thereby reducing their ability to participate in such reactions. However, they may occur if a molecule possesses more than one nucleophilic center. The mass spectra of such compounds can often be rationalized by assuming that each heteroatom or nucleophilic center competes for the proton donated by the reagent ion so that formally one might envisage MH^+ ions of different structures, each of which eliminates a different functional group in reactions of the type discussed above.

Although the formation of MH^+ is the major reaction with molecules of higher proton affinities in methane, isobutane, and ammonia CI mass spectra, peaks due to the products of other reactions are also observed (Field, 1972a,b). In the case of methane, $C_2H_5^+$ and $C_3H_5^+$ are also present as reagent ions, and these frequently give rise to the low intensity peaks $(M + C_2H_5)^+$ and $(M + C_3H_5)^+$ by addition reactions. In cases for which the identification of the MH^+ peak is in doubt, the observation of peaks at 28 and 40 mass numbers higher on the mass scale serves to confirm the identification. With compounds of low proton affinity, such as aliphatic hydrocarbons, the major ion formed is $(M - H)^+$, partly by dissociative protonation but primarily by hydride abstraction (Field, 1968; Hunt and McEwen, 1973; Clow and Futrell, 1972; Houriet *et al.*, 1977). For most other types of molecule, protonation predominates. The use of isobutane as the CI reagent gas usually leads to a very simple CI mass spectrum, and for many amino acids and drugs the MH^+ ion gives the only peak of any significant intensity (Milne and Lacey, 1974). At source temperatures in the region of 150°C, low intensity peaks due to the formation of $(M + C_3H_3)^+$ and $(M + C_4H_9)^+$ in addition reactions are observed in the isobutane CI mass spectra of some compounds, but their intensities decrease at higher source temperatures (Field, 1968, 1972a). In contrast, ammonia CI mass spectra of various sugars (Horton *et al.*, 1974; Hogg and Nagabhushan, 1972) and their derivatives (Dougherty *et al.*, 1974) contain intense peaks due to $(M + NH_4)^+$ ions and the formation of ions of this type have recently been shown to be useful in the analysis of triglycerides (Murata and Takahasi, 1977). The simple spectra given by isobutane and ammonia make them ideal reagent gases in many cases for molecular weight determination and quantitation studies. Since $PA(i\text{-}C_4H_8)$ and $PA(NH_3)$ are much higher than $PA(H_2O)$, $C_4H_9^+$ and NH_4^+ cannot protonate water, and the presence of moisture in the source does not affect the spectrum appreciably. The situation is different in the case of methane since $PA(CH_4) < PA(H_2O)$, which results in the formation of H_3O^+ under methane CIMS conditions if any moisture is present in the ion source (Michnowicz and Munson, 1972a). If the concentration of water vapor in methane is gradually increased, the major reactant ions change from CH_5^+ and $C_2H_5^+$ to H_3O^+ and this leads to significant decreases in the amount of fragmentation. The H_3O^+ ion competes with the CH_5^+ ion for the sample molecules and irreproducible results are obtained if the water vapor concentration is not controlled.

Water vapor alone has not found wide use as a CI reagent gas, owing to its tendency to form cluster ions of the type $H^+(H_2O)_n$ and $MH^+(H_2O)_m$. Although the former ions complicate the spectrum, they are readily recognized. The latter are more troublesome since ions for which $m = 0-3$ are frequently formed, the relative intensities of which are very dependent on the structure of the sample molecule and the pressure and temperature at which the ion source is operated (Price *et al.*, 1976). The intensities of peaks given by hydrated ions are reduced by raising the source temperature or by passing the ions through a buffer gas in a drift tube just outside the source when the hydrated ions are preferentially removed by collision-induced decomposition (Price *et al.*, 1977; Kambara and Kanomata, 1977). The dependence of the relative intensities of $MH^+(H_2O)_m$ ions on the structure of M has been used to distinguish isomeric dicyanobenzenes and it has recently been shown (Buttrill, 1977) that for a series of substituted phenols, the intensity ratios $I(M + 19)^+/I(M + 1)^+$ measured under carefully controlled conditions correlate very well with equilibrium free energy changes calculated from equilibrium constants measured for the reactions

$$MH^+ + H_2O \rightleftharpoons MH^+{\cdot}H_2O.$$

The use of D_2O as a CI reagent gas (Hunt *et al.*, 1972; Martinsen and Buttrill, 1976; Frieser *et al.*, 1975) allows one to determine the site of protonation in favorable cases and additional H/D interchange provides interesting information, as discussed in a later section.

VII. GASES GIVING OTHER POSITIVE REAGENT IONS

The vast majority of CIMS studies have been carried out making use of the reagent ions described in the Sections V and VI, but in the search for selective reagent ions, a number of other reagent gases have been used in specific applications. Nitric oxide, either pure or in excess nitrogen, has been used in a number of studies. Its use as a charge transfer reagent ion to give molecular ions of compounds having low ionization energies has already been mentioned (Jelus *et al.*, 1974a; Hunt and Harvey, 1975). In the case of compounds for which charge transfer is an endothermic process, the $(M + NO)^+$ ion is frequently the most abundant, together with ions arising from hydride, hydroxide, or halide abstraction (Munson, 1975; Jardine and Fenselau, 1975a,b; Hunt and Ryan, 1972b), the relative abundances of which vary with structure and can for example be used to distinguish primary, secondary, and tertiary alcohols (Hunt and Ryan, 1972b).

Tetramethylsilane gives predominantly $Si(CH_3)_3{}^+$ ions which react to give $(M + Si(CH_3)_3)^+$ ions, $(M + 73)^+$, with compounds containing *n* or π electrons (Odiorn *et al.*, 1973). Dimethylamine (Vouros and Carpino, 1974), trimethylamine (Wilson and McCloskey, 1975), and isobutane admixed with ethanolamine or ethylenediamine (Bowen and Field, 1974) have all been used to form addition ions $(MHX)^+$, where X is the amine, with a sulfone (Vouros and Carpino, 1974), nucleotides (Wilson and McCloskey, 1975), and a bifunctional oxygen compound, respectively (Bowen and Field, 1974).

A method of locating olefinic double bonds without derivatization (Ferrer-Correia *et al.*, 1976) makes use of vinyl methyl ether, either pure or diluted with CO_2, to produce $(CH_2{=}CHOCH_3)^{+\cdot}$ ions which add across the olefinic bond of a compound $R^1CH{=}CHR^2$ and eliminate an olefin to form an even mass ion $R^2CH{=}CHOCH_3{}^{+\cdot}$. With mono-olefins, the lower molecular weight olefin $R^1CH{=}CH_2$ is eliminated and it has been possible to distinguish isomeric *n*-octenes in this way. With unsaturated esters, the ester group is retained in the ion, and this method has been used to determine the positions of olefinic bonds in a series of C_{14} unsaturated acetates which are insect sex attractants (Burke *et al.*, 1978).

Biacetyl diluted with helium gives rise to the reagent ion CH_3CO^+ which reacts with certain ketones, esters, nitriles, and nitrocompounds to give both MH^+ and $(M + CH_3CO)^+$ ions, the relative abundances of which show some structural dependence (Hass *et al.*, 1977). Differences in CI mass spectra given by stereoisomers (discussed more fully in a later section) are much less marked than spectra obtained under ICR conditions but in each case studied, acetylation is more selective than protonation. The Li^+ ion has been observed to undergo a number of useful reactions with organic substrates (Wieting *et al.*, 1975), and with certain halides and aromatic compounds, addition to form LiM^+ ions is the major process (Hodges and Beauchamp, 1976).

VIII. NEGATIVE CHEMICAL IONIZATION MASS SPECTROMETRY

Under the conditions normally used for positive ion electron impact mass spectrometry, many compounds give relatively few negative ions and frequently no molecular anions, or ions from which the molecular weight may be readily determined. This is because 70 eV electrons can form negative ions directly only in ion pair production processes, and the more useful negative ions resulting from resonance capture, three-body attachment or dissociative resonance capture processes, arise only from secondary electrons formed either at metal surfaces or in the gas phase. At the much higher pressures at which a CI source operates, many low energy electrons are present which can lead to the formation of a stable molecular anion if the substance has a positive electron affinity or to a negative CI mass spectrum in the presence of a suitable reagent gas. This is now a rapidly growing branch of CIMS and work published before mid-1976 has recently been reviewed (Jennings, 1977).

The simplest type of process which leads to a negative ion mass spectrum under CI operating conditions is that in which the reagent gas acts simply as a buffer gas in producing a high yield of thermal electrons. These give rise to an electron capture mass spectrum of substances that possess a positive electron affinity, and this is the counterpart to the charge transfer mass spectrum in positive CIMS. No reagent anion is employed under these conditions, and although nitrogen was used as the buffer gas at low pressures in early experiments (Dougherty and Weisenberger, 1968), methane or isobutane is now commonly used (Hunt *et al.*, 1976; Dougherty *et al.*, 1972, 1975). In pulsed positive and negative-ion chemical ionization mass spectrometry (PPNICI) (Hunt *et al.*, 1976), these yield Brønsted acids in the positive ion mode, leading to the formation of MH^+ ions, and act as buffer gases in the negative ion mode, leading to $M^{-}\cdot$ ions in many cases. If a substance does not form a stable molecular anion, derivatives containing electronegative groups give excellent results; pentafluorobenzaldehyde is used to form Schiff bases with aromatic amines, pentafluorobenzoyl chloride forms derivatives with phenols, and perfluoro acid anhydrides are also used for this purpose. The high sensitivity obtainable in the electron capture mode has already been commented upon and is typically 100–1000 times that given by positive CIMS. Sensitivity ratios of 156 and ~1000 were measured for Aflatoxin B and trinitrotoluene, respectively (Hunt *et al.*, 1976), and in favorable cases, the detection limit in this mode is 100 fg–1 pg. Usually, simple spectra are observed consisting of very few peaks of which $M^{-}\cdot$ is often the most intense. Any oxygen-containing impurities in the reagent gas may produce low concentrations of H^-, OH^-, or $O^{-}\cdot$ ions which may complicate the interpretation of the spectrum.

In NCI mass spectrometry, Brønsted bases play a role analogous to that played by Brønsted acids in positive CIMS:

$$X^- + M \rightarrow (M - H)^- + HX \qquad \Delta H_4 = PA(M - H)^- - PA(X)^-.$$

Reagent gas mixtures yielding Brønsted bases include $CH_4 + \sim 1\%$ CH_3ONO, which gives a high yield of CH_3O^- ions (Hunt *et al.*, 1976), and $N_2O/He/H_2$ or $N_2O/He/CH_4$ in 1:1:1 ratios, which give abundant OH^- ions (Smit and Field, 1977). Other bases X^- can be produced by adding HX to the CH_4/CH_3ONO mixture until the CH_3O^- ion at $m/e = 31$ just disappears (HX = $(CH_3)_2CO$, C_2H_2, CH_3NO_2, RSH, etc.) (Hunt *et al.*, 1976). The utility of these as specific reagent ions has yet to be investigated. Since much of the exothermicity of reaction (4) is likely to appear as vibrational energy of the neutral HX, little fragmentation of $(M - H)^-$ ions occurs in many cases and these ions frequently constitute the base peak. For example, the use of CH_4/CH_3ONO as the reagent gas gives only the $(M - H)^-$ peak when *trans*-4-phenyl-3-buten-2-one is the sample (Hunt *et al.*, 1976), and in the N_2O/CH_4 NCI spectrum of cholesterol (Smit and Field, 1977), the $(M - H)^-$ ion constitutes 76% of the total ion current, the other ions being $(M - 3)^-$, $(M - 5)^-$, $(M - 19)^-$, and $(M - 21)^-$.

Nitrous oxide, usually diluted with nitrogen, is a convenient source of $O^{-}\cdot$ ions which undergo a range of reactions with different types of compounds (Mitchum *et al.*, 1976). The $(M - H)^-$ ion is frequently the most abundant ion, but an unusual reaction is the abstraction of $H_2{}^+$ from suitably activated CH_2 groups in terminal olefins (Goode and Jennings, 1974; Dawson and Jennings, 1976), nitriles, and carbonyl compounds (Harrison and Jennings, 1976) to produce $(M - 2H)^{-}\cdot$ radical anions. Almost all aromatic compounds give a peak at $(M + 15)^-$ due to the displacement of a hydrogen atom from the ring (Bruins *et al.*, 1978a) and the relative abundances of the $(M - 2)^{-}\cdot$, $(M - 1)^-$, and $(M + 15)^-$ ions have been used to distinguish between the isomeric xylenes and ethylbenzene. A mixture of hydrogen and oxygen gives $O_2{}^{-}\cdot$ as the reagent ion, $O^{-}\cdot$ ions being removed by the hydrogen (Hunt *et al.*, 1976). With polynuclear aromatic hydrocarbons, the principal ions formed are $(M + 30)^{-}\cdot$, $(M + 15)^-$, and $M^{-}\cdot$, the relative intensities of which vary considerably with the structure of the hydrocarbon. If the molecule contains a five-membered ring in which a CH_2 group is present, an $(M + 14)^{-}\cdot$ peak is observed and other peaks serve to characterize compounds containing sulfur or nitrogen. If oxygen is used as the reagent gas in conjunction with a Townsend discharge operated in the negative ion mode, high yields of electrons and $O_2{}^{-}\cdot$ ions are given. These have been used (Hunt *et al.*, 1975b) in the analysis of 2,3,7,8-tetrachlorodibenzo-*p*-dioxin (TCDD), molecular weight of 320. Ions arising from both Cl and H atom displacement by $O_2{}^{-}\cdot$ ions to give phenoxide ions, and ·OCl and ·OH radicals are observed at $m/e = 301$ and 335, respectively, and the molecular anion is also observed at $m/e = 320$. By far the most intense peak is given at $m/e = 176$ due to ions formed in the reaction sequence initiated by electron capture:

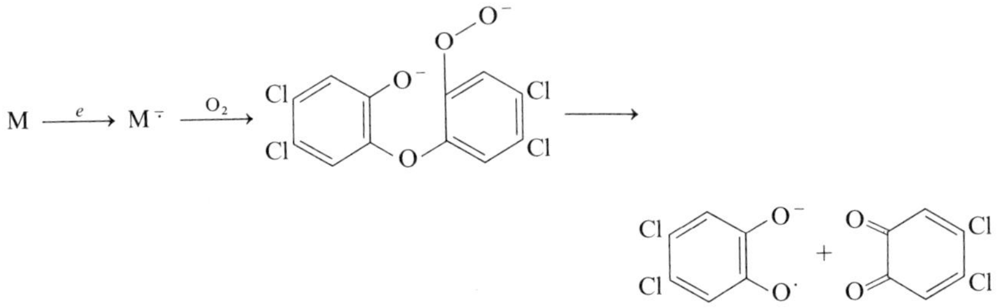

The peaks are well removed from those given by other contaminating chlorocarbons, and with the reduction of the number of isotopic peaks, it is possible to detect TCDD in picogram quantities. Methylene chloride, CH_2Cl_2, has been used as a reagent gas giving mainly Cl^- as reagent ion together with electrons (Tannenbaum *et al.*, 1975). This therefore leads to an electron capture spectrum together with the addition product $(M + Cl)^-$.

IX. SELECTED TOPICS IN CHEMICAL IONIZATION MASS SPECTROMETRY

A. *Effects of Functional Group Interactions in CIMS*

The rate of fragmentation of ions formed from monofunctional compounds by charge or proton exchange is largely controlled by the exothermicity of the ionization processes, and in most cases the reagent gas plays no further role in determining the types of fragmentation reactions that occur. On the other hand, if ionization occurs primarily by addition to form $(M + X)^+$ ions, subsequent fragmentation reactions will depend on the chemical nature of X^+ (Jelus *et al.*, 1974b). For certain bifunctional compounds, other types of reactions can occur such that the resulting CI mass spectrum is not simply related to those of related monofunctional compounds.

There are now many examples in the literature of methane and isobutane CI mass spectra of α,ω-substituted alkenes which are interpreted in terms of the formation of an MH^+ ion in which intramolecular hydrogen-bond formation leads to the proton being coordinated between the two functional groups (Dzidic and McCloskey, 1971; Morton and Beauchamp, 1972; Aue *et al.*, 1973; Yamdagni and Kebarle, 1973; Weinkam and Gal, 1976a). This may enhance the stability of the MH^+ ions, as in the methane CI mass spectra of aliphatic diols (Dzidic and McCloskey, 1971), or the energy released in hydrogen-bond formation may promote the elimination of water to form a stabilized cyclic ion as in the methane CI mass spectra of bifunctional alkyl acids (Weinkam and Gal, 1976a). Methyl esters of monoepoxyfatty acids (Weinkam, 1974) and bifunctional alkyl acids (Weinkam and Gal, 1976a) give abundant $(MH—CH_3OH)^+$ ions by a similar mechanism. Intramolecular hydrogen-bond formation increases the proton affinities of bifunctional compounds above those of the individual functional groups so that, for example, MH^+ ions are observed in the ammonia CI mass spectra of methyl dicarboxylates, $CH_3O_2C(CH_2)_nCO_2CH_3$, and other disubstituted alkanes even though the proton affinity of the monofunctional compound is over 70 kJ $mole^{-1}$ less than that of ammonia (Weinkam and Gal, 1976a).

Reagent ions such as H_3O^+, $CH_3OH_2{}^+$, and $NH_4{}^+$ all react with certain compounds to form solvated MH^+ ions in which the proton forms a hydrogen bond between the sample and solvent molecules. Hydrogen/deuterium exchange reactions occur between the two molecules (Hunt *et al.*, 1972), but with bifunctional sample molecules, more extensive exchange reactions are observed (Weinkam, 1974; Haegele and Desiderio, 1974; Weinkam and Gal, 1976b). The ions H_3O^+ and $CD_3OD_2{}^+$ react with monoexpoxyfatty acid methyl esters to give ions having the molecular formulas of the hydrolysis and transesterification products (Weinkam, 1974). In the ammonia CI mass spectra of *N*-acetyl prolyl dipeptides and antimycin A, ions corresponding to the products of ammonolysis of lactone, ester, and amide bonds are observed (Haegele and Desiderio, 1974). Hydrolysis, methanolysis, and ammonolysis of a series of dicarboxylic acids and their methyl esters has also been reported (Weinkam and Gal, 1976b). For example, in the water CI mass spectra of the esters $CH_3O_2C(CH_2)_nCO_2CH_3$, the main peaks are given by the ions MH^+, $(MH - CH_3OH)^+$, and $(CH_3O_2C(CH_2)_nCO_2H_2)^+$, the relative intensities of which vary with n. Similarly, in the methanol CI mass spectrum of dodecanedioic acid, $HO_2C(CH_2)_{10}CO_2H$, the four most intense peaks are observed at $m/e = 259, 245, 231$, and 227, which correspond to $(MH + 2CH_3OH—2H_2O)^+$, $(MH + CH_3OH—H_2O)^+$, MH^+, and $(MH + CH_3OH—2H_2O)^+$ ions, respectively. In the isobutane CI mass spectra of dicarboxylic acids, the abundance ratio $(M + H—H_2O)^+/[(M + H)^+ + (M + H—H_2O)^+]$ varies both with the chain length and the sample pressure, the latter effect being ascribed to the stabilization of MH^+ by the reversible formation of the protonated dimer M_2H^+ (Grønneberg, 1977).

Functional group interactions can also lead to distinctive differences in the CI mass spectra of geometrical and stereochemical isomers since isomerization to a common structure prior to fragmentation is less likely to occur under CI conditions than under EI conditions. The diethyl esters of maleic and fumaric acids can be readily distinguished (Fales *et al.*, 1971) by the fact that the successive loss of two ethylene molecules by the MH^+ ion is much more important in the CI mass spectrum of the fumarate, whereas the loss of ethanol, which is formulated as proceeding through a cyclic intermediate, is a major process only in the fragmentation of the MH^+ ion given by the maleate. The isobutane CI mass spectra of a number of 1,2 and 1,3 steroidal amino alcohols show striking differences in the intensity ratio $(MH^+)/(MH - H_2O)^+$ and H_2O loss occurs only when the distance between the oxygen and nitrogen atoms is too great to allow the formation of a hydrogen bond (Longevialle *et al.*, 1973), as indicated by the O—H stretching frequency in the infrared spectrum. The lack of $(M + H)^+$ ions in the methane CI mass spectra of andostane-3,17-diols is similarly attributed to the inability of the rigidly separated hydroxyl groups to participate in intramolecular hydrogen bonding (Michnowicz and Munson, 1972b).

The effects of functional group interactions have also been demonstrated in the CI mass spectra of a number of bifunctional cyclopentane and cyclohexane derivatives (Munson, 1975; Winkler and McLafferty, 1974; Van Gaever *et al.*, 1977; Claeys and Van Haver, 1977; Van de Sande *et al.*, 1977a). The *cis*- and *trans*-isomers of 1,4-cyclohexanediol can readily be distinguished from their isobutane CI mass spectra in which the intensity ratio $(MH^+)/(MH - H_2O)^+$ is approximately eight for the *cis*-isomer, in which intramolecular hydrogen bonding can stabilize the MH^+ ion, but is only 0.6 for the *trans*-isomer. The 1,2-cyclohexanediols could not be similarly distinguished since the separation of the hydroxyl groups is almost the same in the two isomers (Munson, 1975), and in the case of the 1,3-diols, the stability of the proton bridge is affected by steric interactions with *cis*-5-substituents (Winkler and McLafferty, 1974). In the methane and isobutane CI mass spectra of 3-methoxycyclopentyl and 3-methoxycyclohexyl acetic acid esters, only the *cis*-isomers yield MH^+ ions, stabilised by intramolecular hydrogen bonding, and $(MH - CH_3OH)^+$ ions are more abundant for the *trans*-isomers (Van Gaever *et al.*, 1977). Similar stereochemical effects were observed in the methane and isobutane CI mass spectra of derivatives of 2,3-dimethyl-1,4-cyclopentanediols (Claeys and Van Haver, 1977), and 3,4-dimethyl-1,2-cyclopentanediols (Van de Sande *et al.*, 1977a). The *cis*- and *trans*-isomers of 4-*t*-butylcyclohexyl acetate have been investigated using biacetyl diluted with helium as the reagent gas and the less hindered equatorial (*trans*) isomer gives a larger $(M + CH_3CO)^+/(MH)^+$ intensity ratio than that given by the *cis* isomer and similar results were obtained for the propionate. Acetylation is more selective than protonation in each case, but conditions need to be optimized for stereospecific acetylation (Hass *et al.*, 1977). The use of CIMS in probing the stereochemistry of isomeric samples has recently been reviewed (Van de Sande *et al.*, 1977b).

The difference in vapor pressures of the optical isomers and racemic compound of dimethyl tartrate allow an optical isomer to be preferentially vaporized. When dimethyl-d_6-D-tartrate (10%) and dimethyl-d_0-L-tartrate (90%) are introduced on a direct insertion probe, the CI mass spectrum initially consists of MH^+ and M_2H^+ peaks derived entirely from the latter. At a later time, the peaks at $m/e = 179$ and 185, due to the MH^+ ions of the two isomers, become of equal intensity, but the intensities of peaks at $m/e =$ 357, 363, and 369, due to the various M_2H^+ ions, are not in the expected 1:2:1 ratio, that at $m/e = 363$ being only 78% of the expected intensity. In the case of the di-isopropyl esters, the corresponding peak is only 46% of the expected intensity. These observations suggest that *DD* and *LL* proton-bound dimers are in some way slightly more stable than *DL* proton-bound dimers in this type of compound, although similar effects were not observed in the corresponding spectra of camphor and *N*-acetylamphetamine isomers (Fales and Wright, 1977).

B. *Active Hydrogen Atoms and the Site of Protonation*

Implicit in the discussion in the previous section is the assumption that protonation of a molecule containing a functional group in which an oxygen or nitrogen atom is present may be considered as taking place on the heteroatom. If the molecule contains more than one functional group, Brønsted acid CI mass spectra can often be rationalized by assuming that protonation may occur on any of the functional groups, each of which competes with the others for the proton.

Some insight into this problem can be obtained by the use of D_2O as a reagent gas (Hunt *et al.*, 1972). Deuterium exchange occurs of the hydrogen atoms bonded to the heteroatom in compounds such as alcohols, thiols, phenols, carboxylic acids, amides, and amines, but not with esters, aldehydes, ketones, and unsaturated aliphatic hydrocarbons. Two more recent studies of the protonation of substituted benzenes under ICR (Frieser *et al.*, 1975) and CI (Martinsen and Buttrill, 1976) conditions, using D_2O as the reagent gas, indicate that two types of behavior may be observed. Compounds such as aniline, phenol, nitrobenzene, and benzaldehyde protonate at the substituent: hydrogen atoms attached to the heteroatom are exchanged and cluster ions containing one or two D_2O molecules are observed at 20 and 40 mass units above the deuterated analogue of MD^+. For example, aniline exchanges both hydrogen atoms attached to the nitrogen atom to give the $C_6H_5ND_3{}^+$ ion, $m/e = 97$, together with cluster ion peaks at $m/e = 117$ and 137. Hydrocarbons such as benzene itself, toluene, biphenyl, and halogenated benzenes such as bromobenzene and iodobenzene, protonated on the ring rather than on the substituent. Their CI mass spectra are characterized (Martinsen and Buttrill, 1976) by the absence of D_2O cluster ions and by peaks corresponding to $MD^{+\cdot}$ ions in which various ring hydrogen atoms have been replaced by deuterium atoms. In the case of benzene, all six hydrogen atoms are exchanged to give $C_6D_7^+$ as the main peak and up to five are exchanged in the other cases of monosubstituted benzenes which behave in this way. Differences were found in the behavior of the three xylenes, *o*- and *p*-xylenes undergoing rapid exchange of up to four ring hydrogen atoms, whereas *m*-xylene gives MD^+ as the base peak together with smaller peaks which indicate much slower exchange of up to three ring hydrogen atoms. This behavior is paralleled by mesitylene, 1,3,5-trimethylbenzene, which suggests that it is the hydrogen atom in the position *meta* to the two methyl groups, which is much more difficult to exchange.

This technique may be extended by the use of more acidic sources of deuterium such as CH_3OD or CF_3CH_2OD (Hunt, 1977), when the number and types of hydrogen atoms exchanged may be increased. Anions generated by proton abstraction by ions such as OH^- and NH_2^- may also exchange hydrogen atoms, and under both flowing afterglow (Stewart *et al.*, 1977) and ICR (Dawson *et al.*, 1979) conditions, the allyl anion has been shown to exchange only the four terminal hydrogen atoms in the presence of D_2O, and this method has been used (Stewart *et al.*, 1977) to distinguish isomeric anions with different numbers of potentially exchangeable hydrogen atoms such as cyclopropyl and allyl, and benzyl and cycloheptatrienyl. Less basic anions such as those formed from aldehydes, ketones, and dienes, do not exchange with D_2O but will do so with CH_3OD (De Puy *et al.*, 1978). The 1,3-pentadienyl anion derived from the *cis*-isomer exchanges less readily than that derived from the *trans*-isomer, indicating that the ions do not interconvert. From the few publications which have so far appeared on this topic, it seems that hydrogen/deuterium exchange reactions between acids and bases of different strengths is likely to continue to yield information about the structures of gas phase cations and anions and the mechanisms of ion molecule reactions which lead to the CI mass spectra of compounds.

C. The Collision-Induced Decomposition of Ions Formed by Chemical Ionization

One of the most useful features of CIMS is that by the choice of a suitable reagent gas, it is usually possible to arrange for a compound to give rise very largely, if not completely, to a single ion in its CI mass spectrum. This is useful for quantifying a known compound or for indicating its presence in a mixture if no other components give rise to ions of the same mass-to-charge ratio. If structural information is required, however, a reagent gas giving a CI mass spectrum containing peaks due to fragment ions must be chosen and although this is satisfactory for a pure compound, there are obvious disadvantages if this gas is used with a mixture without prior separation of the components, e.g., by gas chromatography. The separation of neutral species before ionization is often time consuming and an alternative procedure is to make use of collision-induced decomposition of ions and the characteristics of a double focusing mass spectrometer (Jennings, 1968; McLafferty *et al.*, 1973; Levsen, 1975). Ions from the source are passed through a collision chamber at a focal point in a double focusing instrument of Nier–Johnson or reverse geometry, and the resulting fragment ions are brought to focus by the use of one of several types of scans of the electric and/or magnetic sectors (Beynon and Cooks, 1976; Bruins *et al.*, 1978b; Beynon and Boyd, 1978).

The power of the technique is amply demonstrated in several recent publications in which methane, water, isobutane, or ammonia was used as reagent gas to produce MH^+ ions in a CI source of a reverse geometry instrument (Kruger *et al.*, 1976; Soltero-Rigav *et al.*, 1977; Kondrat and Cooks, 1978; McLafferty and Bockhoff, 1978). The magnet current was adjusted so as to select the protonated molecular ion of interest, thereby separating it from all other ions formed from other components in the mixture, after which collision-induced decomposition and energy analysis of the products gave a "collisional activation" (CA) spectrum. Methane chemical ionization was used to produce MH^+ and $(M + C_2H_5)^+$ ions from a number of isomeric barbiturates which give little or no molecular ion abundance under electron impact conditions (Soltero-Rigav *et al.*, 1977). An energy analysis of the unimolecular and collision-induced decompositions of these two ions was carried out for the seven most difficult samples and allowed each to be identified. The minimal sample preparation that is required for the use of this technique is illustrated by the identification of coniine in fresh stem and leaf material crushed under liquid nitrogen and introduced into the mass spectrometer together with ammonium acetate. The magnet was adjusted to transmit ions of $m/e = 128$ and the CA spectrum given by these ions was identified by comparison with that from an authentic sample. Similarly, the presence of cocaine in 1 mg of coca leaves and of caffeine in a cola beverage were established by this method (Kondrat and Cooks, 1978), which was also used to investigate DNA pyrolysis products. The sensitivity of the technique is in the 0.1 ng region at present, which compares with a sensitivity of 1–10 pg using GC/MS single ion monitoring, but its big advantages are the much shorter analysis time, due to the fact that sample separation is effected by the choice of a suitable reagent gas and magnet current rather than by a chromatographic method, and the reliability of the identification.

The use of an isotopically labeled internal standard may further shorten the analysis time. Similar information can be obtained from a double focusing instrument of conventional Nier–Johnson geometry by making use of a linked scan of the electric and magnetic sectors (Bruins *et al.*, 1978b; Stradling *et al.*, 1978). The full potential of the technique has yet to be exploited, but in view of its versatility, sensitivity, speed, and reliability, it seems very likely that it will find widespread use.

D. *Reactant Ion Monitoring*

For certain applications the simplicity of the mass spectrum often obtained under CI conditions may be a disadvantage in that samples that contain a specific functional group do not all give a characteristic ion of fixed mass, as is often found in EI mass spectra. One cannot, therefore, use the normal GC/MS single-ion monitoring technique to identify compounds of a particular type when they are present in a complex mixture. An alternative procedure which has recently been suggested (Hatch and Munson, 1977b) is to monitor the ion current due to the reagent ion itself during a GC/MS experiment and if the reagent ion has been carefully chosen, it will react with a certain class of compounds but not with others. The trace of the recorded reagent ion current is compared with that given by the GC detector, allowing one to identify those peaks in the gas chromatogram which are given by the class of compounds with which the reagent ion reacts. The use of the $C_6H_6^{+}\cdot$ ion to distinguish esters of saturated and unsaturated acids has already been mentioned (Subba Rao and Fenselau, 1977). Helium containing a trace of isobutane was used to generate $C_4H_9^{+}$ ions which were used to investigate a sample containing four alcohols and nine hydrocarbons which was passed through the column of a gas chromatograph. This ion reacts rapidly with the alcohols so that the depletion of its ion current gives rise to four large peaks together with minor peaks due to the much slower reaction with the hydrocarbons. Similarly, protonated acetone, $C_3H_7O^{+}$, was used to detect a series of ketones in a hydrocarbon/ketone mixture.

The area of the peak given by depletion of the reagent ion current is proportional both to the sample concentration and to the rate constant for the reaction of the reagent ion with the sample. The analytical use of the technique depends upon the differences in rate constants for reaction of the reagent ion with samples of similar concentration, but if the total amount of different samples injected into the system are known and if a constant fraction of the effluent of the column is admitted to the ion source by using a stream splitter rather than a separator, relative total rate constants for reaction of a reagent ion with different samples can be determined from the expression

$$\frac{k_B}{k_A} = \frac{W_B \ln(1 - \Delta I/I)_B \cdot \alpha}{W_A \ln(1 - \Delta I/I)_A \cdot \beta},$$

where I is the reagent ion current and ΔI the amount by which it is depleted, α and β are the amounts of A and B injected into the ion source, and W_A and W_B the widths of the peaks given by the two samples (Munson and Hatch, 1977). Relative total rate constants of reactions of CH_5^+ and $C_2H_5^+$ with a series of hydrocarbons have been obtained in this way and are in fair agreement with ICR data and the collision rate constants predicted by the Langevin expression.

This technique is equally applicable to negative ion studies although none have been reported to date. The sensitivity is unlikely to approach that obtainable by using single ion monitoring of peaks given by electron capture, but the method is of general applicability and should find use in the qualitative analysis of complex mixtures.

X. CONCLUSION

In this short review, emphasis has been placed on the fundamental principles underlying the analytical applications of CIMS. These principles are derived from the large body of data on the thermochemistry and kinetics of positive ions in the gas phase, and although similar data on negative ions are much less extensive, this deficiency is being remedied. Proton affinities and gas phase acidities are reviewed elsewhere in this volume and the thermochemical data of this type are now well established and provide a firm basis for the interpretation of many features of CI mass spectra produced by reagent gases which yield Brønsted acids or bases. Similarly, ionization energies provide the basis for the interpretation of CI mass spectra produced by charge transfer reagent ions, and electron affinities play a similar role in electron capture mass spectra.

The situation is less satisfactory in relation to the thermochemistry and kinetics of reactions of other ions used as reagent ions, and in several cases our understanding of their gas phase ion chemistry is still at the qualitative stage. This is particularly true of selective reagent ions, and it is likely that during the next few years new reagent ions will come into use as our knowledge of their gas phase ion chemistry increases. Although the presence of specific functional groups can often be inferred from the use of other spectroscopic techniques, the great sensitivity of CIMS makes it very desirable that these reagent ions should be developed. The judicious choice of reagent ions which undergo only slightly exothermic reactions with sample molecules has already indicated that interesting information on structural isomers and stereoisomers may be obtained, and it is to be hoped that this work will be extended to yield information on the conformation of ions in the gas phase.

In view of the comparitive ease with which EI mass spectra are normally obtained, CIMS should be seen as a complementary technique. For many workers in the field the attraction of CIMS lies in its sensitivity, selectivity, and simplicity of the mass spectrum and the ease with which the molecular weight of the sample may be determined. However, as the volatility of the sample decreases, there is an increasing possibility that the highest mass peak observed in a CI mass spectrum may not be due to MH^+ but rather to $(MH - H_2O)^+$, as in the case of cholest-5-enediol or $(MH—RCOOH)^+$ as in the isobutane CI mass spectra of higher triglycerides (Fales *et al.*, 1975). For such molecules, field desorption tends to give more reliable molecular weight information but it will be interesting to see if the use of a field desorption emitter in a CI ion source will extend the utility of CIMS in the study of large, biologically important molecules (Hunt *et al.*, 1977). Other rapid desorption techniques have also given promising results (Beuhler *et al.*, 1974; Buttrill and Findeis, 1976; Torgerson *et al.*, 1974), and further developments in sample introduction methods are to be expected in view of the increasing use of CIMS in the biomedical field. Increasing interest is being shown in the coupling of liquid chromatographs to mass spectrometers and the use of water, methanol, and acetonitrile as both solvents and reagent gases has been shown to give a sensitivity in the region of 10 ng (McLafferty *et al.*, 1975). As new methods of interfacing the two instruments are developed, the use of a wider range of reagent ions should become possible.

REFERENCES

Arsenault, G. P., Dolhun, J. J., and Biemann, K. (1971). *Anal. Chem.* **43**, 1722.

Aue, D. H., Webb, H. M., and Bowers, M. T. (1973). *J. Am. Chem. Soc.* **95**, 2699.

Beggs, D. P., Vestal, M. L., Fales, H. M., and Milne, G. W. A. (1971). *Rev. Sci Instrum.* **42**, 1578.

Beuhler, R. J., Flanigan, E., Greene, L. J., and Friedman, L. (1974). *J. Am. Chem. Soc.* **96**, 3990.

Beynon, J. H., and Boyd, R. K. (1978). *Adv. Mass. Spectrom.* **7**, 1115.

Beynon, J. H., and Cooks, R. G. (1976). *Int. J. Mass Spectrom. Ion Phys.* **19**, 107.

Bowen, D. V., and Field, F. H. (1974). *Org. Mass Spectrom.* **9**, 195.

Bruins, A. P., Ferrer-Correia, A. J. V., Harrison, A. G., Jennings, K. R., and Mitchum, R. K. (1978a). *Adv. Mass Spectrom.* **7**, 355.

Bruins, A. P., Jennings, K. R., and Evans, S. (1978b). *Int. J. Mass Spectroms. Ion Phys.* **26**, 395.

Burke, P. D., Friesen, M. L., Golding, B. T., Hall, D. R., Jennings, K. R., and Stradling, R. S. (1978). Unpublished observations.

Burlingame, A. L., Kimble, B. J., and Derrick, P. J. (1976). *Anal. Chem.* **48**, 368R.

Buttrill, S. E., Jr. (1977). *ASMS Conf., 25th, Washington, D.C.* Pap. MD-8, p. 132.

Buttrill, S. E., Jr., and Findeis, A. F. (1976). *Anal. Chem.* **48**, 626.
Carroll, D. I., Dzidic, I., Stillwell, R. N., Haegele, K. D., and Horning, E. C. (1975). *Anal. Chem.* **47**, 2369.
Claeys, M., and Van Haver, D. (1977). *Org. Mass Spectrom.* **12**, 531.
Clow, R. P., and Futrell, J. H. (1972). *J. Am. Chem. Soc.* **94**, 3784.
Daly, N. R., McCormick, A., Powell, R. E., and Hayes, R. (1973). *Int. J. Mass Spectrom. Ion Phys.* **11**, 225.
Dawson, J. H. J., and Jennings, K. R. (1976). *J. Chem. Soc., Faraday Trans. II* **72**, 700.
Dawson, J. H. J., Noest, A. J., and Nibbering, N. M. M. (1979). *Int. J. Mass Spectrom. Ion Phys.* (in press).
De Puy, C. H., Bierbaum, V. M., King, G. K., and Shapiro, R. H. (1978). *J. Am. Chem. Soc.* **100**, 2921.
Dougherty, R. C., and Weisenberger, C. R. (1968). *J. Am. Chem. Soc.* **90**, 6570.
Dougherty, R. C., Dalton, J., and Biros, F. J. (1972). *Org. Mass Spectrom.* **6**, 1171.
Dougherty, R. C., Roberts, J. D., Binkley, W. W., Chizhov, O. S., Kadentsev, V. I., and Solov'-yov, A. A. (1974). *J. Org. Chem.* **39**, 491.
Dougherty, R. C., Roberts, J. D., and Biros, F. J. (1975). *Anal. Chem.* **47**, 54.
Dzidic, I., and McCloskey, J. A. (1971). *J. Am. Chem. Soc.* **93**, 4955.
Dzidic, I., and McCloskey, J. A. (1972). *Org. Mass Spectrom.* **6**, 939.
Dzidic, I., Carroll, D. J., Stillwell, R. N., Horning, M. G., and Horning, E. C. (1978). *Adv. Mass Spectrom.* **7**, 359.
Fales, H. M., and Wright, G. J. (1977). *J. Am. Chem. Soc.* **99**, 2339.
Fales, H. M., Milne, G. W. A., and Nicholson, R. S. (1971). *Anal. Chem.* **43**, 1785.
Fales, H. M., Milne, G. W. A., Winkler, H. V., Beckey, H. D., Damico, J. N., and Barron, R. (1975). *Anal. Chem.* **47**, 207.
Ferrer-Correia, A. J. V., Jennings, K. R., and Sen Sharma, D. K. (1976). *Org. Mass Spectrom.* **11**, 867.
Field, F. H. (1968). *Acc. Chem. Res.* **1**, 42.
Field, F. H. (1972a). *In* "MTP International Review of Science" (A. Maccoll, ed.), Vol. 5, p. 133. Butterworth, London.
Field, F. H. (1972b). *In* "Ion Molecule Reactions" (J. L. Franklin, ed.), Vol. 1, p. 261. Plenum, New York.
Field, F. H., Hamlett, P., and Libby, W. F. (1967). *J. Am. Chem. Soc.* **89**, 6035.
Frieser, B. S., Woodin, R. L., and Beauchamp, J. L. (1975). *J. Am. Chem. Soc.* **97**, 6893.
Games, D. E. (1977). *Mass Spectrom.* **4**, 242.
Garland, W. A., Weinkam, R. J., and Trager, W. F. (1974). *Chem. Instrum.* **5**, 271.
Goode, G. C., and Jennings, K. R. (1974). *Adv. Mass Spectrom.* **6**, 797.
Grønneberg, T. (1977). *Org. Mass Spectrom.* **12**, 769.
Haegele, K. D., and Desiderio, D. M. (1974). *J. Org. Chem.* **39**, 1078.
Harrison, A. G., and Jennings, K. R. (1976). *J. Chem. Soc., Faraday Trans. I* **72**, 1601.
Hass, J. R., Nixon, W. B., and Bursey, M. M. (1977). *Anal. Chem.* **49**, 1071.
Hatch, F., and Munson, M. S. B. (1977a). *Anal. Chem.* **49**, 169.
Hatch, F., and Munson, M. S. B. (1977b). *Anal. Chem.* **49**, 731.
Hodges, R. V., and Beauchamp, J. L. (1976). *Anal. Chem.* **48**, 825.
Hogg, A. M. (1972). *Anal. Chem.* **44**, 227.
Hogg, A. M., and Nagabhushan, T. L. (1972). *Tetrahedron Lett.* p. 4827.
Horning, E. C., Horning, M. G., Carroll, D. I., Dzidic, I., and Stillwell, R. N. *Anal. Chem.* **45**, 936.
Horton, D., Wander, J. D., and Foltz, R. L. (1974). *Carbohydr. Res.* **36**, 75.
Houriet, R., Parisod, G., and Gäumann, T., (1977). *J. Am. Chem. Soc.* **99**, 3599.
Hunt, D. F. (1977). *Br. Mass Spectrom. Group Meet., Swansea.*

Hunt, D. F., and Harvey, T. M. (1975). *Anal. Chem.* **47**, 1965.
Hunt, D. F., and McEwen, C. N. (1973). *Org. Mass Spectrom.* **7**, 441.
Hunt, D. F., and Ryan, J. F. (1972a). *Anal. Chem.* **44**, 1306.
Hunt, D. F., and Ryan, J. F. (1972b). *J. Chem. Soc., Chem. Commun.* 620.
Hunt, D. F., McEwen, C. N., and Upham, R. A. (1972). *Anal. Chem.* **44**, 1292.
Hunt, D. F., McEwen, C. N., and Harvey, T. M. (1975a). *Anal. Chem.* **47**, 1730.
Hunt, D. F., Harvey, T. M., and Russell, J. W. (1975b). *J. Chem. Soc., Chem. Commun.* p. 151.
Hunt, D. F., Stafford, G. C., Crow, F. W., and Russell, J. W. (1976). *Anal. Chem.* **48**, 2098.
Hunt, D. F., Shabanowitz, J., Botz, F. K., and Brent, D. A. (1977). *Anal. Chem.* **49**, 1160.
Jardine, I., and Fenselau, C. (1975a). *Anal. Chem.* **47**, 730.
Jardine, I., and Fenselau, C. (1975b). *Org. Mass Spectrom.* **10**, 748.
Jelus, B., Munson, M. S. B., and Fenselau, C. (1974a). *Anal. Chem.* **46**, 729.
Jelus, B. L., Munson, M. S. B., and Fenselau, C. (1974b). *Biomed. Mass Spectrom.* **1**, 96.
Jennings, K. R. (1968). *Int. J. Mass Spectrom. Ion Phys.* **1**, 227.
Jennings, K. R. (1977). *Mass Spectrom.* **4**, 203.
Kambara, H., and Kanomata, I., (1977). *Anal. Chem.* **49**, 270.
Kondrat, R. W., and Cooks, R. G. (1978). *Anal. Chem.* **50**, 81A.
Kruger, T. L., Litton, J. F., Kondrat, R. W., and Cooks, R. G. (1976). *Anal. Chem.* **48**, 2113.
Lawson, G., and Todd, J. F. J. (1977). *Anal. Chem.* **49**, 1619.
Lawson, G., Bonner, R. F., and Todd, J. F. J. (1973). *J. Sci. Instrum.* **6**, 357.
Lehman, T. A., and Bursey, M. M. (1976). "Ion Cyclotron Resonance Spectrometry." Wiley (Interscience), New York.
Levsen, K. (1975). *Org. Mass Spectrom.* **10**, 55.
Longevialle, P., Milne, G. W. A., and Fales, H. M. (1973). *J. Am. Chem. Soc.* **95**, 6666.
McLafferty, F. W., and Bockhoff, F. M. (1978). *Anal. Chem.* **50**, 69.
McLafferty, F. W., Bente, P. F., Kornfield, R., Tsai, S.-C., Howe, I. (1973). *J. Am. Chem. Soc.*, 2120.
McLafferty, F. W., Knutti, R., Venkataraghavan, R., Arpino, P. J., and Dawkins, B. G. (1975). *Anal. Chem.* **47**, 1503.
Martinsen, D. P., and Buttrill, S. E., Jr. (1976). *Org. Mass Spectrom.* **11**, 762.
Michnowicz, J., and Munson, M. S. B. (1972a). *Org. Mass Spectrom.* **6**, 283.
Michnowicz, J., and Munson, M. S. B. (1972b). *Org. Mass Spectrom.* **6**, 765.
Milne, G. W. A., and Lacey, M. J. (1974). *Crit. Rev. Anal. Chem.* p. 45.
Mitchum, R. K., Harrison, A. G., Ferrer-Correia, A. J. V., and Jennings, K. R. (1976). Unpublished observations.
Morton, T. L., and Beauchamp, J. L. (1972). *J. Am. Chem. Soc.* **94**, 3671.
Munson, M. S. B. (1975). *In* "Ion Molecule Interactions" (P. Ausloos, ed.), p. 505. Plenum, New York.
Munson, M. S. B., and Field, F. H. (1966). *J. Am. Chem. Soc.* **88**, 2621.
Munson, M. S. B., and Hatch, F. (1977). *ASMS Meet., 25th, Washington, D.C.* Pap. MD-7, p. 29.
Murata, T., and Takahashi, S. (1977). *Anal. Chem.* **49**, 728, 2209.
Odiorne, T. J., Harvey, D. J., and Vouros, P. (1973). *J. Org. Chem.* **38**, 4274.
Price, P. C., Swofford, H. S., Jr., and Buttrill, S. E., Jr. (1976). *Anal. Chem.* **48**, 494.
Price, P. C., Swofford, H. S., Jr., and Buttrill, S. E., Jr. (1977). *Anal. Chem.* **49**, 1497.
Ryan, P. W. (1977). *ASMS Meet., 25th, Washington, D.C.* Pap. WP-10, p. 23.
Sieck, L. W. (1978). Personal communication.
Smit, A. L. C., and Field, F. H. (1977). *J. Am. Chem. Soc.* **99**, 6471.
Smit, A. L. C., Rossetto, M. A. J., and Field, F. H. (1976). *Anal. Chem.* **48**, 2042.
Soltero-Rigau, E., Kruger, T. L., and Cooks, R. C. (1977). *Anal. Chem.* **49**, 435.
Stafford, G. C. (1977). *Meet Negative Ions, N.C.*

Stewart, J. H., Shapiro, R. H., DePuy, C. H., and Bierbaum, V. M. (1977). *J. Am. Chem. Soc.* **99**, 7650.
Stradling, R. S., Jennings, K. R., and Evans, S. (1978). *Org. Mass Spectrom.* **13**, (429).
Subba Rao, S. C., and Fenselau, C. (1977). *ASMS Meet., 25th, Washington D.C.* Pap. WP-5, p. 613.
Tannenbaum, H. P., Roberts, J. D., and Dougherty, R. C. (1975). *Anal. Chem.* **47**, 49.
Todd, J. F. J., Lawson, G., and Bonner, R. F. (1976). *In* "Quadrupole Mass Spectrometry" (P. H. Dawson, ed.), p. 181. Elsevier, Amsterdam.
Torgenson, D. F., Skowronski, R. P., and Macfarlane, R. D. (1974). *Biochem. Biophys. Res. Commun.* **60**, 615.
Van de Sande, C. C., Van Gaever, F., Hanselaer, R., and Vandewalle, M. (1977a). *Z. Naturforsch., Teil B* **32**, 810.
Van de Sande, C. C., Van Gaever, F., Sandra, P., and Monstrey, J. (1977b). *Z. Naturforsch., Teil B* **32**, 573.
Van Gaever, F., Monstrey, J., and Van de Sande, C. C. (1977). *Org. Mass Spectrom.* **12**, 200.
Vouros, P., and Carpino, L. A. (1974). *J. Org. Chem.* **39**, 3777.
Weinkam, R. J. (1974). *J. Am. Chem. Soc.* **96**, 1032.
Weinkam, R. J., and Gal, J. (1976a). *Org. Mass Spectrom.* **11**, 188.
Weinkam, R. J., and Gal, J. (1976b). *Org. Mass Spectrom.* **11**, 197.
Whitney, T. A., Kleemann, L. P., and Field, F. H. (1971). *Anal. Chem.* **43**, 1084.
Wieting, R. D., Staley, R. H., and Beauchamp, J. L. (1975). *J. Am. Chem. Soc.* **97**, 924.
Wilson, J. M. (1971). *Mass Spectrom.* **1**, 2.
Wilson, J. M. (1973). *Mass Spectrom.* **2**, 6.
Wilson, J. M. (1975). *Mass Spectrom.* **3**, 86.
Wilson, J. M. (1977). *Mass Spectrom.* **4**, 102.
Wilson, M. S., and McCloskey, J. A. (1975). *J. Am. Chem. Soc.* **97**, 3436.
Winkler, J., and McLafferty, F. W. (1974). *Tetrahedron* **30**, 2971.
Yamdagni, R., and Kebarle, P. (1973). *J. Am. Chem. Soc.* **95**, 3504.
Yamdagni, R., and Kebarle, P. (1976). *J. Am. Chem. Soc.* **98**, 1320.
Yinon, J., and Boettger, H. G. (1972). *Chem. Instrum.* **4**, 103.

Chapter 13

Energetics and mechanisms of unimolecular reactions of positive ions: mass spectrometric methods

J. H. BEYNON

ROYAL SOCIETY RESEARCH UNIT
UNIVERSITY COLLEGE OF SWANSEA
SWANSEA, UNITED KINGDOM
AND

J. R. GILBERT

DEPARTMENT OF CHEMISTRY
UNIVERSITY OF ESSEX
COLCHESTER, UNITED KINGDOM

ISBN-0-12-120802-8

I. INTRODUCTION AND EXPERIMENTAL METHOD

The range of times over which unimolecular ion reactions can be observed, and thus the range of rate constants that may be measured, depend both upon the type of equipment available and the experimental method used. The relationship between rate constant and the excess energy above the minimum necessary for reaction is generally depicted with the aid of a diagram such as Fig. 1. The rate constant at a particular energy E_x depends both upon the activation energy for this reaction and also upon an entropy factor. A simple representation of the relationship between k and E which is often used (Rosenstock *et al.*, 1952) is

$$k = \nu\left(\frac{E - \epsilon_0}{E}\right)^{s-1}, \tag{1}$$

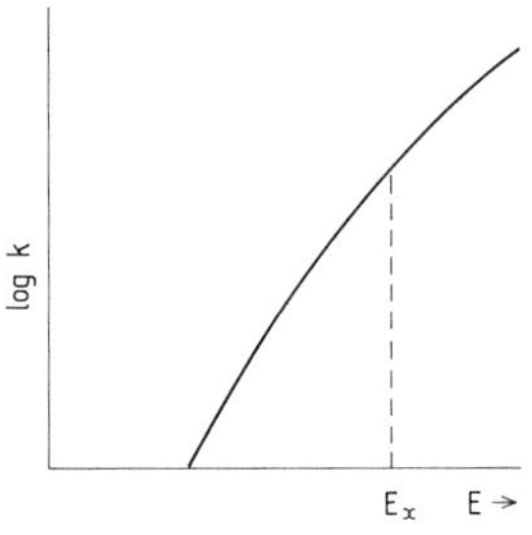

FIG. 1 A log k versus E plot.

where ϵ_0 is the activation energy for the reaction being considered, ν a frequency factor, and s the number, or effective number, of oscillators in the ion. A more detailed discussion, using the Rice, Rampsburger, Kassel, Marcus (RRKM) formulation (Forst, 1973), shows that this simple expression ceases to be useful as threshold energies are approached unless the number of oscillators is arbitrarily reduced, but for the present discussion we shall use the simplified expression which indicates a quasi-exponential dependence on the activation energy when $\epsilon_0/E \ll 1$. For this condition Eq. (1) reduces (Cooks *et al.*, 1973a) to

$$k = \nu e^{-(s-1)\epsilon_0/E}. \tag{2}$$

The rate constant k is defined for a particular value of E_x as the reciprocal of the mean time of fragmentation. Because fragmentation is statistically controlled, depending as it does upon the necessary energy becoming concentrated in the appropriate oscillators, it is not possible to obtain information on just a single rate constant even when the time of observation is set between narrow limits, and ions with only a single value of E_x are considered. The situation becomes still more complicated, of course, when measurements have necessarily to be made over a wider range of times and using ions that themselves contain an appreciable spread of internal energies.

Let us consider as a single example the case of a sector mass spectrometer in which ions are produced by electron bombardment. Three types of ions may be distinguished in the mass spectrum, molecular ions, fragment ions, and metastable ions, and we can illustrate (Chupka, 1959) with the aid of Fig. 2 the range of rate constants that contribute to the observation of these three species of ions. Molecular ions are defined as those ions that are formed with a sufficiently low internal energy that they reach the collector electrode, typically some 10^{-5} sec after formation, without fragmentation. Most of these ions will have internal energies corresponding to rate constants of 10^5 sec^{-1} or less but they will include a smaller fraction of those ions with larger internal energies, and thus larger rate constants which have taken more than the average time to decompose. It is important to remember that the molecular ions may include those that have isomerized to new structures without fragmentation. Such ions will not be distinguishable, in a normal mass spectrum, from other molecular ions. Fragment ions are those ions formed from molecular ions within the ion source. The residence time of ions within the source (Wallington, 1971) is of the order of 10^{-6} sec. The fragment ions must themselves contain a sufficiently small amount of

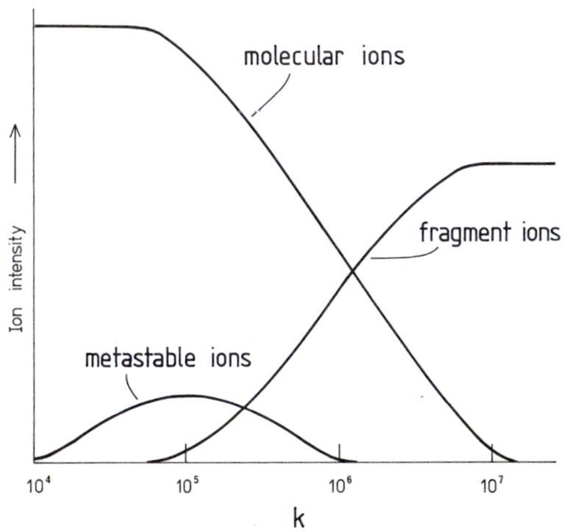

FIG. 2 Ion intensity as a function of k.

excess energy over the minimum necessary for their own fragmentation that they reach the collector unchanged. The number of fragment ions observed will thus increase, corresponding to molecular ions with rate constants greater than 10^6 sec^{-1}, although some fragment ions corresponding to rate constants less than 10^6 sec^{-1} will also be formed within the ionization chamber. Metastable ions correspond to molecular ions that fragment in the field-free regions of the mass spectrometer. Few molecular ions will decompose in this region if their rate constants for fragmentation are much greater or much smaller than the rate constant corresponding to the reciprocal of the mean arrival time of ions in this region of the mass spectrometer. This can be seen from Fig. 2.

In order to understand the fragmentation process in detail, it is necessary to be able to study the dynamics of the fragmentation process so as to measure, for example, the energy released as translational energy at the moment of ion decomposition. This can be done most easily by studying only those fragmentations (of metastable ions) that occur in the field-free regions (Cooks *et al.*, 1973a) of the instrument and thus after the ion has been accelerated. This is the basis of the methods used in ion kinetic energy (IKE) spectrometry (Beynon *et al.*, 1969) and will be discussed in greater detail below. As explained above, such studies are necessarily restricted, in sector-type instruments, to ion lifetimes of the order 10^{-5} sec.

To study fragmentations occurring in much shorter times, a mass spectrometer fitted with a field ionization source is used. This method (Beckey, 1977) relies upon that fact that in such a source, molecular ions are formed in a range of a few angstroms very near the tip and thus within a small range of potentials, ΔV below that of the tip itself, as shown in Fig. 3. The field in this region is extremely high and of the order of several volts per angstrom. Thus, as soon as it is formed, the molecular ion is accelerated away from the tip toward the entrance slit of the mass spectrometer. The profile of potential gradient between the tip and the earthed entrance slit is known, and the velocity profile of an ion falling through this field can therefore be calculated. In this way, the time taken for an ion to reach any particular potential can be worked out (Derrick, 1975). A mass spectrum of an organic molecule taken with a field ionization source usually shows fragment ions as well as the molecular ion. The peak due to the molecular ion is sharp, unless very high tip voltages are employed, while the peaks from fragment ions are sharp at the high-mass side, but are broadened in the direction of lower mass numbers. Typical peak shapes are shown in Fig. 4. The sharp portion of the fragment ion peak corresponds to formation of these ions in times of the order of 10^{-13} sec or less after formation of the molecular ions. Thus, these fragment ions fall through the voltage drop $(V - \Delta V)$ (see Fig. 3). On the other hand, ions at any position within the tail of the peak correspond to ions of lower momentum that have been formed at a potential $(V_1 + \Delta V)$ below that of the molecular ions and thus after a corresponding known time t_1. As ions from the various regions of the tail are collected successively and their abundances recorded, a plot can be made showing the rate of fragment ion formation as a function of time. Hence a time scale of 10^{-12}–$\sim 10^{-9}$ sec for fragmentations can be investigated. Unfortunately, it is not possible, using this method, to obtain information concerning translational energy release at the moment of fragmentation. The spread of fragment ion energies resulting from this energy release merely adds a small uncertainty to the measured time of fragmentation.

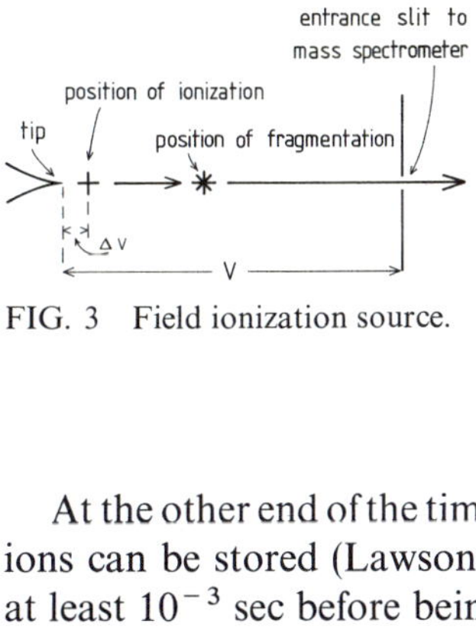

FIG. 3 Field ionization source.

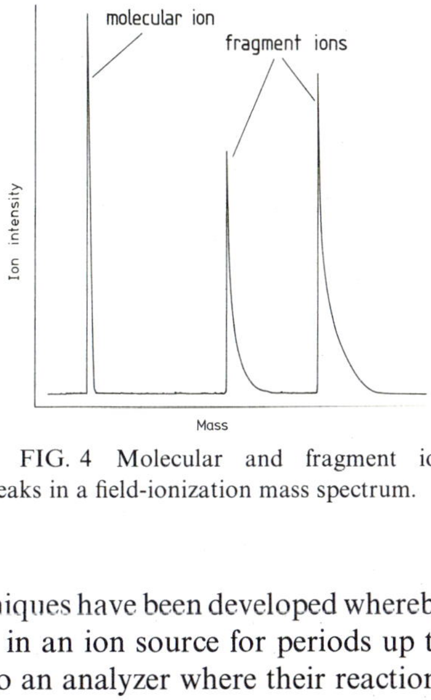

FIG. 4 Molecular and fragment ion peaks in a field-ionization mass spectrum.

At the other end of the time scale, techniques have been developed whereby ions can be stored (Lawson *et al.*, 1973) in an ion source for periods up to at least 10^{-3} sec before being pulsed into an analyzer where their reactions can be monitored. By varying the storage time, a picture can be built giving the rate of fragmentation as a function of time. A typical source used for studies at extended lifetimes in this way is the three-dimensional quadrupole source, the QUISTOR (Lawson *et al.*, 1973). Such a source may be used with any type of mass spectrometer, including a quadrupole analyzer, a sector analyzer, or an ion cyclotron resonance (ICR) instrument. This latter instrument (Lehman and Bursey, 1976) is itself inherently suitable for studying ion fragmentation reactions at times of a millisecond or longer after ion formation, this being the order of the flight time of an ion through the instrument. The particular advantage of an ICR instrument is that the identity of the parent fragmenting ion can be established. When the ion beam is irradiated at the resonant frequency of such a parent ion, the parent ion absorbs energy and the radius of motion is increased, causing it to be lost from the cell. Simultaneous observation of the daughter ion will show a reduction of its abundance due to removal of the parent.

When a molecule is ionized, for example, by bombardment with energetic electrons or photons, ions can be produced with a range of internal energies up to a maximum equal to the difference between the quantum energy of the bombarding radiation and the ionization energy. Fragment ions are formed by a series of competitive and consecutive processes, and one of the weaknesses of a conventional mass spectrum is that the details of the fragmentation pathway along which a fragment ion is formed are not accessible from it. Uncoupling the accelerating and electric sector voltages in a sector type of double-focusing instrument enables these pathways to be explored using the methods of IKE spectrometry, as already mentioned. The method (Cooks *et al.*, 1973a) depends on the fact that when an ion m_1^+ that has been accelerated and is traveling through the mass spectrometer with a velocity v undergoes fragmentation, the fragment ions m_2^+ will acquire a spread of energies but will continue through the instrument with a mean velocity v. Thus, the ratio of the translational energies of the parent and daughter ions is equal to the ratio of the masses. An electric sector can be used to explore the ratio of translational energy to charge of an ion beam; the energy-to-charge ratio is simply proportional to the voltage between the sector plates. The method consists of taking a monoenergetic beam of ions containing several different species $m_a^+, m_b^+, \cdots, m_i^+$, allowing it to enter an electric sector and scanning the voltage across this sector downwards, starting at the value at which all these initial ions are transmitted. These ions are said to constitute the main beam of stable ions. Any metastable ions that fragment in the field-free region preceding the electric sector will lose translational energy as described above. The electric sector voltage at which any daughter ion formed by fragmentation of a metastable ion is transmitted enables the mass ratio m_2/m_1 for this particular fragmentation to be determined. As the electric sector is scanned, a spectrum showing all the kinetic energies of the daughter ions present in the beam, and thus the various values of m_2/m_1, is obtained. A second measurement is necessary to establish the mass-to-charge ratio of the daughter ion. This is accomplished by allowing the transmitted m_2^+ ions to bypass the detector located after the electric sector and to traverse a magnetic sector whereby their momentum-to-charge ratio, and hence their mass-to-charge ratio can be measured.

There are two important features of the IKE spectra obtained as described above. Because no prior mass separation of the ion beam has been performed, daughter ions of more than one mass may contribute to a peak in the IKE spectrum. Also, because there is a release of translational energy at the moment of fragmentation, the peaks in the IKE spectrum are broadened and may overlap. For most chemical work, and to reduce these difficulties, it is preferable to reverse the order in which the ion beam traverses the electric and magnetic sectors and allow it to pass through the magnetic sector first. This reversed geometry offers the advantage that the magnetic sector separates the beam of stable and metastable ions passing through it into its individual constitutents. A single ion species is selected by the magnet. The IKE spectrum of the daughter ions arising from this single species can then be obtained by scanning the voltage applied to the electric sector plates. The entire fragmentation pattern of a complex molecule can then be explored systematically. First the mass selected IKE (or MIKE) (Beynon *et al.*, 1973) spectrum of the molecular ion is obtained. Each of the daughter ion species from this spectrum is then in turn selected by means of the magnet from the ions issuing from the source. A further series of MIKE spectra is obtained. By continuing this process, a detailed picture of the various fragmentation paths along which ions are formed may be built. The chance of overlapping of peaks is now greatly reduced and the chance of observing the detailed shape of a MIKE peak correspondingly increased, although it is still possible for adjacent peaks to interfere with one another due to the fact that they are broadened compared to the main beam.

The broadening of MIKE peaks is entirely due to the fact that, at the instant of fragmentation, part of the energy of the transition state is converted into translational energy which produces a spread of energies and of directions of motion of the fragment ions (Beynon *et al.*, 1965). By analyzing the shape of the resulting MIKE peak, it is possible to deduce the shape of the curve showing the probability of release of various amounts of energy. Figure 5 shows that as reaction proceeds the sum of the excess energy of the transition state $\epsilon^{\ddagger}$ and the reverse activation energy is partitioned between the translational energy T released and the internal energy of the charged and neutral fragments m_2^+ and m_3. Since we are dealing with a unimolecular process, if no energy is lost by radiation, the probability curve showing the values of T released will also give information about the internal energy retained in the products.

In order to obtain detailed information concerning the energy partitioning, it is necessary to be able to distinguish differences in translational energy release as the products are formed in various vibrational states. This requires careful design of the instrument used to give the necessary energy and angular resolution of the product ion beam (Schopman *et al.*, 1970). The energy resolution is determined by the radius of the electric sector, the sizes of the resolving slits on either side of this sector, and the spread of energies in the beam of reactant ions. It is worth noting that the requirement of a narrow slit between the electric and magnetic sectors in order to achieve

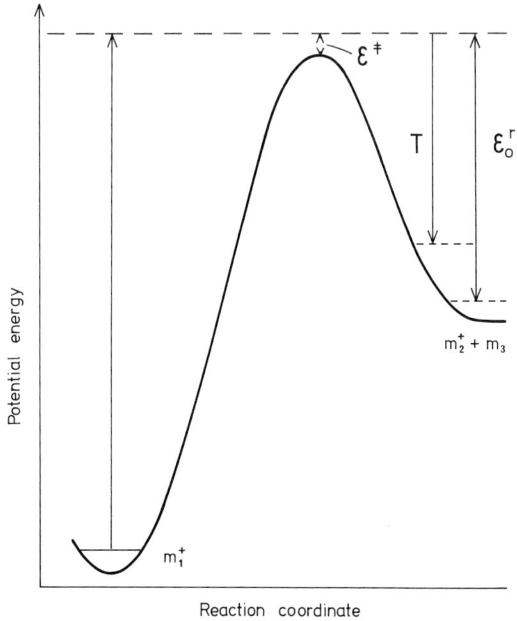

FIG. 5 Schematic potential energy diagram.

good energy resolution conflicts with the conditions for good sensitivity at high mass resolution, which requires that this slit be wide in order to give high sensitivity. Commercial high resolution mass spectrometers are thus unsuited for work at high energy resolution unless the intermediate slit width can be controlled and unless the stability and energy spread in the primary ion beam is improved to a sufficient degree.

A recently introduced technique that allows the excess energy in the transition state to be measured directly is photoelectron–photoion coincidence spectroscopy (Baer, 1979; Brehm and von Puttkamer, 1968). This is accomplished by measuring the energy with which the photoelectron is emitted at the moment of ionization by an energetic photon. This procedure establishes the internal energy of the ion formed. The arrival of the products of fragmentation of this ion at a detector are correlated with the production of the photoelectron by the coincidence method. Two different modifications of this experimental method have been described (Tsai *et al.*, 1975; Eland, 1974).

An interesting new development, useful in identifying all unimolecular fragmentations of an ion, is the technique of linked scanning in which two of the three fields of the mass spectrometer, accelerating (V), electric sector (E), or magnetic sector (B), are scanned simultaneously (Boyd and Beynon,

1977a). Perhaps the most important of these "linked scans" is that in which E and B are scanned downward so as to keep the ratio of E/B constant (Millington and Smith, 1977). The initial values of E and B are chosen so as to transmit a particular parent ion. As the scan proceeds, all daughters formed from this parent in the field-free region preceding both fields are transmitted in succession. The peak shape is, however, extremely narrow; any ions that have received an extra component of velocity in the xy plane during fragmentation are discriminated against. The technique offers the possibility of detecting all transitions from a particular ion without fear of overlapping peaks. It is thus particularly useful for analytical studies or other cases where energy release information is not required.

To fully elucidate the mechanism and energetics of unimolecular fragmentation of ions we need the following information concerning each elementary step: (1) the structural and electronic identity of reactant, transition state, and products; (2) heat of formation of reactant and products, and the activation energy; and (3) how the excess energy in the products is distributed amongst the vibrational, rotational, and translational degrees of freedom. In practice, there are many cases where the only information available is the elemental composition of the reactant and products. It is certainly true that for polyatomic ions little or no information is available that would enable us, with any degree of confidence, to determine the structure of an ion. We do not yet have the kind of spectroscopic data that is available for neutral species. Most structures are inferred from the structure of the neutral molecule before ionization or from a consideration of the reaction channels open to the ion. Both these approaches can be dangerous. Extensive rearrangements may occur on ionization, and the open reaction channels of an ion well reflect the properties of the transition state of the reacting ion; the structure of the nonreacting ion may well be different. However, in many cases it is now possible, with a fair degree of certainty, to say whether or not ions produced from different precursors, but with the same elemental composition, have similar structures and fragment by similar reaction mechanisms.

II. ENERGETICS AND MECHANISM

The first step in any investigation into the mechanism of an ion fragmentation reaction involves determining the elemental composition of the reactant and product ions. With a double-focusing mass spectrometer this determination presents few problems, as the exact empirical formula of the ions involved can usually be determined (Beynon, 1960; Kerr and Duckworth, 1958). However, with a single-focusing instrument, different techniques have to be resorted to. A distinction between different product ions of the same nominal mass but different elemental composition can sometimes be made by using isotopic labeling techniques. Recent work on the benzylamine ion (Bruins *et al.*, 1972; Gilbert and Al-Mashhadani, 1978) has shown that the $(M - 1)^+$ ion undergoes a rearrangement reaction, giving a metastable ion peak at 58.9 involving the loss of a neutral species with a mass of 27. The neutral species could have an empirical formula of either HCN or C_2H_3. The mass spectrum of fully deuterated benzylamine shows that the $(M - 2)^+$ ion gives a metastable ion peak only at 64.9, which corresponds to the loss of 28 mass units, thus indicating that the neutral species lost from the original $(M - 1)^+$ ion was HCN and not C_2H_3.

Four principal methods of approach are at present popular in the investigation of mechanisms and energetics of ion fragmentation reactions. They are isotopic labeling, shapes and abundances of metastable ion peaks, collision-induced decompositions, and ionization and appearance energy measurements.

A. *Isotopic Labeling Techniques*

The purpose of isotopic labeling experiments is to locate both the center of reaction and the atom or atoms involved at the center. The data obtained, which may also give evidence of primary isotope effects, can be used to gain information concerning the transition state. The majority of studies to date have used deuterium labeling but there is an increasing trend to use ^{13}C, ^{18}O, and ^{15}N labeled derivatives (Benoit and Holmes, 1972). The simultaneous scrambling reactions in isotopically labeled derivatives may cause complications. For details of the applications of this technique to mass spectroscopy the reader is referred to a review by Holmes (1975).

With the development of MIKE spectrometers it is now possible to perform labeling experiments using the natural isotopic abundances in unlabeled derivatives. Beynon *et al.* (1974) have recently studied methane loss from *n*-butane ions by this method.

B. *Metastable Ion Abundance Ratios*

When an ion undergoes two or more competing reactions to give metastable ions, the ratios of the metastable ion abundances may be used to characterize the ion structure and the potential energy surface over which the reaction occurs. This approach originates with the work of Rosenstock *et al.* (1964). These workers studied the metastable ion abundance ratio for the competing reactions of *n*-hexyl ions, viz,

$$C_6H_{13}^+ \rightarrow C_4H_9^+ + C_2H_4, \quad (3)$$

$$C_6H_{13}^+ \rightarrow C_3H_7^+ + C_3H_6. \quad (4)$$

The *n*-hexyl ions were produced from a variety of precursors such as normal alkanes, *n*-hexyl bromide, and di-*n*-hexyl ether. The measured abundance ratio was constant, indicating that the ions are decomposing from the same structure or mixture of structures. This constant ratio also indicates that the ion structure is independent of the method of preparation of the ion. Later work with secondary and tertiary hexyl bromides (Cole and Williams, 1968) showed a similar ratio to that obtained for the *n*-hexyl ions. Hence, the normal, secondary, and tertiary hexyl ions must isomerize to a common configuration before fragmenting. Presumably, the activation energy for isomerization is lower than that for either decomposition channel. Also, as ions produced from different precursors are formed with different internal energy distributions, the internal energy range (metastable window) giving metastable transitions must be so narrow as to make the shape of the energy distribution unimportant. However, later work using both quasi-equilibrium theory (QET) calculations and experimental investigations (Yeo and Williams, 1971) has shown that the mestastable ion abundance ratios for some competing reactions are dependent on electron beam energy and that the metastable window varies from 0.2 V for reactions with high frequency factors ($\sim 10^{13}$ sec^{-1}) up to several volts for reactions with low frequency factors ($\sim 10^8$ sec^{-1}). The authors (Yeo and Williams, 1971) suggest that small changes in the abundance ratio, up to a factor of ~ 5, is not reliable evidence for different structures, but that a similar ratio is good evidence that the process is occurring from a common structure or mixture of structures.

C. *Metastable Ion Peak Shapes and Kinetic Energy Release*

Metastable ion peaks are usually wider than normal peaks. This width is entirely due to the release of kinetic energy at the moment of fragmentation, and may therefore be used as a transition state probe. Figure 6 illustrates some typical peak shapes. A fragmentation reaction occurring, with release of kinetic energy, in front of a magnetic analyzer leads to an exploding sphere of ions. Those ions exploding with a velocity component parallel to the magnetic field are not focused by the analyzer; thus, as slits are of finite length, these ions will be detected with lower efficiency than those exploded in other directions. However, if the energy release is small then this discrimination against ions exploded parallel to the magnetic field will be very

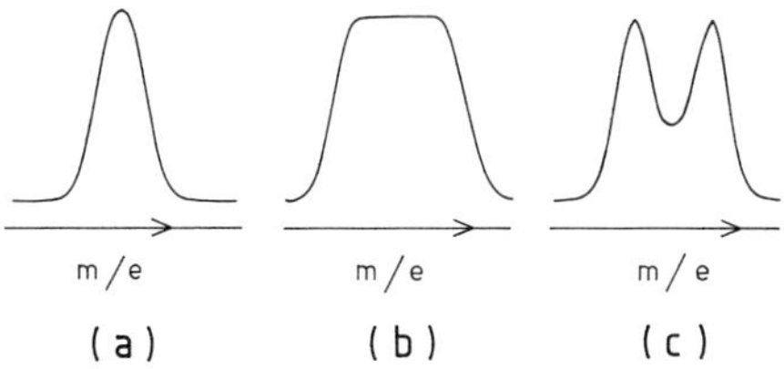

FIG. 6. Typical metastable ion peak shapes.

small and may be neglected. Assuming that this discrimination effect is negligible, it has been shown that for a single kinetic energy release, the metastable ion peak will be rectangular in shape, the width of which will depend on the amount of energy released. (Terwilliger *et al.*, 1975; Holmes *et al.*, 1976; Holmes and Osborne, 1977). In practice there will be a distribution of kinetic energy releases and thus the observed peak will be formed by a combination of rectangular peaks of different widths giving a typical "Gaussian" peak shown in Fig. 6a.

The measured kinetic energy release T is composed of two contributions, $T^{\ddagger}$, representing the fraction of excess energy $\epsilon^{\ddagger}$ in the transition state, and T^{r}, the fraction of the reverse activation energy $\epsilon_0{}^{r}$ that appears as translational energy (see Fig. 5). If $\epsilon_0{}^{r}$ is small (e.g., in simple bond fission reactions), then the value of $T^{\ddagger}$ depends on the magnitude of $\epsilon^{\ddagger}$. If $k(E)$ rises sharply at threshold then $\epsilon^{\ddagger}$ will be small, but if $k(E)$ rises slowly $\epsilon^{\ddagger}$ and $T^{\ddagger}$ may have large values. The measured value of $T^{\ddagger}$ will be insensitive to the internal energy distribution of the reactant ion if the metastable window is narrow, but may be strongly dependent on this distribution if the metastable window is wide. The internal energy distribution of fragment ions will be determined by the energy distribution of the precursor ion and the fraction of the excess energy above threshold that fluctuates into the fragment ion (Yin and Rabinovitch, 1970). Thus, fragment ions produced from different precursors may show different kinetic energy releases for a particular reaction even if they have the same structure and are reacting through the same transition state. Loudon (1978) has recently discussed the loss of HCN from aniline ions formed by loss of ketene from acetanilide ions (Uccela *et al.*, 1972; Gambol *et al.*, 1970),

$$C_6H_5NHCOCH_3{}^{+} \rightarrow C_6H_5NH_2{}^{+\cdot} + CH_2CO, \quad (5)$$

$$C_6H_5NH_2{}^{+\cdot} \rightarrow C_5H_6{}^{+\cdot} + HCN, \quad (6)$$

compared with HCN loss from directly ionized aniline. The kinetic energy releases are 82 and 103 meV for directly ionized aniline and aniline ions produced by prior fragmentation, respectively. This type of effect has been attributed to the shorter lifetimes, and thus higher internal energy, of the ions produced by prior fragmentation—the "kinetic delay effect." However, it is doubtful if the delay. will be large enough to account for the observed differences in kinetic energy release (McMaster, 1974). Moreover, a recent investigation by Hoffmann and Friesen (1976) into CO loss from $C_6H_6O^{+\cdot}$ ions has shown a reasonably constant kinetic energy release for $C_6H_6O^{+\cdot}$ produced by direct ionization of phenol and a variety of other precursors. The authors claim that these similar values of kinetic energy release are indicative of transition state structure and that the kinetic delay is not operating. However, the values of T for this system are much higher ($\sim$540 meV) than for the aniline system, which suggests a large reverse activation energy (see below).

Fragmentations involving a reverse activation energy are likely to convert a discrete fraction of $\epsilon_0{}^r$ into kinetic energy, and in many cases $T^\ddagger \ll T^r$. The metastable ion peaks observed are usually flat topped (see Fig. 6b) and have large values of T, in some cases up to 1 eV. For the large values of T, the peaks are often dish topped, as shown in Fig. 6c. This dishing is mainly due to discrimination against ions which are formed with a large velocity component parallel to the magnetic field of the analyzer. However, it must be pointed out that if the fraction of $\epsilon_0{}^r$ converted into kinetic energy is small then the excess energy term $T^\ddagger$ may predominate, giving Gaussian-type peak shapes as in Fig. 6(a).

Beynon (see Cooks *et al.*, 1973a; Cooks and Beynon, 1975) has suggested, on the basis of a survey of existing data, that the fraction of the reverse activation energy released as kinetic energy $T^r/\epsilon_0{}^r$ for reactions involving a cyclic transition state may be related to the ring size. The following generalization has been suggested (Boyd and Beynon, 1977b).

three-membered ring	$T^r_{1/2}/\epsilon_0{}^r$,	0.6–1.0,
four-membered ring	$T^r_{1/2}/\epsilon_0{}^r$,	0.1–0.75,
five- or six-membered ring	$T^r_{1/2}/\epsilon_0{}^r$,	0–0.1,

where $T^r_{1/2}$ is the kinetic energy release calculated from the half-peak width of the metastable ion peak.

To summarize, if large kinetic energy releases, due to reverse activation energy, are observed, then this probably reflects the properties of the transition state. However, if Gaussian-shaped metastable ion peaks are observed, then the width of the peak, i.e., kinetic energy release, could be dependent on the energy distribution as well as the properties of the transition state.

D. *Collision-Induced Decompositions (CID)*

Most of the early development of the use of collision-induced decompositions in mass spectrometry were carried out by Beynon (see Ast *et al.*, 1972;

Cooks and Beynon, 1971) and McLafferty *et al.* (1973a,b). The experiments are usually carried out by introducing a target gas into one of the field-free regions of the mass spectrometer (Cooks *et al.*, 1973a). With incident ion beam energies typically of 4–10 keV the transfers of relatively large amounts of energy, up to 12 eV in one particular case (Jennings, 1968), occurs from translational to internal degrees of freedom of the incident ion.

Normal metastable ions usually have a small range of internal energies just above threshold, whereas many of the collisionally activated ions can possess considerable amounts of internal energy. Hence, the frequency factor rather than the activation energy may become the dominating feature in collision-induced decompositions. This is probably the reason why simple bond cleavages are observed to be more probable in CID than in the unimolecular fragmentation of metastable ions. Collision-induced decompositions provide a useful complementary tool to normal metastable ion spectra in mechanistic studies because they are more likely to reflect the properties of the non reacting ions, thus giving a further point in the path from nonreacting ions to products via reacting ions and the transition state.

E. Ionization and Appearance Energies and Thermochemistry

In principle, the measurement of ionization and appearance energies enables us to calculate the heats of formation of reactant and products, and the activation energy of the ion fragmentations under investigation. This information may then be used to speculate, in a general way, about the potential energy surface over which the reaction proceeds. The thermochemistry involved in this approach for reactions (7) and (8) is illustrated in Fig. 7.

$$M + e \rightarrow M^+ + 2e \tag{7}$$

$$M^+ \rightarrow A^+ + N, \tag{8}$$

where M^+ is an ion fragmenting to give a daughter ion A^+ and neutral N; IP (ad) and IP (vert) are the adiabatic and vertical ionization energies, respectively; AP(expt) and AP(ad) are the experimental and adiabatic threshold energies for the appearance of fragment A^+.

The experimental difficulties associated with the measurement of ionization and appearance energies have been discussed by several authors (Kiser, 1965; Field and Franklin, 1957; Rosenstock, 1976). These problems are mainly due to the problems of detecting small ion currents, of calibrating the voltage scale, and the energy spread in the electron beam. However, there are some more fundamental problems associated with these measurements. The electron impact process is a "vertical" or Franck–Condon type of process, and hence the ions may not be formed in their vibrational ground

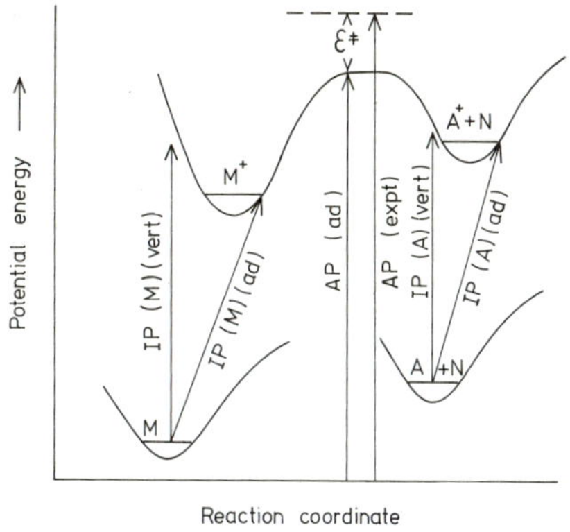

FIG. 7 Illustration of the various energy terms used in ion thermochemistry.

state. This difference, IP(vert) – IP(ad), can lead to errors in ionization energies of the order of 0.1–0.7 eV (Lossing, 1971).

In order to observe daughter ions A^+ as normal ions in a mass spectrum, these ions must be formed in the ionization chamber. Before acceleration an ion typically spends 10^{-6} sec in the ion source and thus an appreciable number of such ions must be formed in times less than 10^{-6} sec to ensure their detection. Therefore, the measured threshold corresponds to an energy at which a large number of ions are decomposing with a rate constant that has a value greater than 10^6 sec^{-1}. The excess energy $\epsilon^{\ddagger}$ necessary to achieve this rate of fragmentation is called the kinetic shift (Knewstubb and Reed, 1970). The magnitude of the kinetic shift depends on the slope of the $k(E)$ versus E curve and the residence time of ions in the source, which is a function of instrument design. McLafferty *et al.* (1970) have estimated kinetic shifts in a series of 1:2 diphenylethanes for the fragmentations (9),

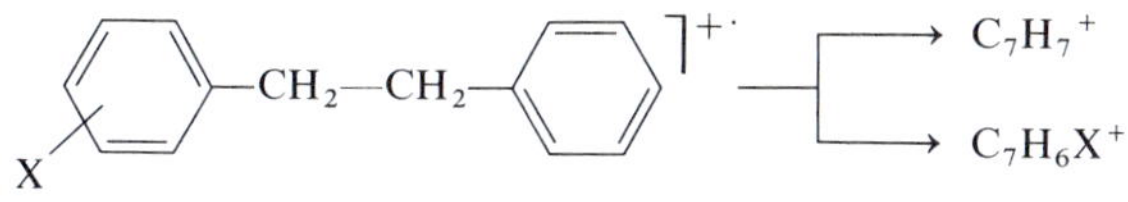

(9)

and values as high as 1.2 eV were calculated. The error introduced into appearance energy measurements by the kinetic shift can be reduced by measuring the appearance energy of metastable ions. This reduction is simply due to the longer lifetime, and hence lower excess energy, of metastable ions. However, the diffuse nature of metastable ion peaks gives rise to sensitivity problems and the possibility of collision-induced reactions may cause complications (Beynon *et al.*, 1975).

If competing reactions are occurring at the threshold of the reaction under study, the rise in the number of fragment ions A^+ with energy may be very slow. This situation can cause AP(expt) to be up to 2 eV above AP(ad) (Gilbert and Stace, 1974). This is the so-called "competitive shift" (Lifshitz and Long, 1964).

The common feature concerned with the three sources of error discussed above is that a sufficiently large number of ions be produced and hence detected. The Fanck–Condon restrictions do not ensure a zero probability of forming ions in their ground state, but a small one. As fragmentation is a statistical process, there will be a small, but nevertheless finite, probability of forming fragment ions in the source with internal energies corresponding to threshold. Similarly the competitive shift is due to the low probability of forming particular fragment ions when a faster competing process is occurring. Therefore, if a sufficiently high detection sensitivity could be achieved these errors could be considerably reduced (Bentley *et al.*, 1973).

The heats of formation and activation energy for the reactions (7) and (8) are given by the following equations:

$$\Delta H_f^{\,0}(\mathrm{A}^+) = \Delta H_f^{\,0}(\mathrm{M}) + \mathrm{AP(expt)} - \Delta H_f^{\,0}(\mathrm{N}) - (\epsilon_0^{\,\mathrm{r}} + \epsilon^{\ddagger}), \tag{10}$$

$$\Delta H_f^{\,0}(\mathrm{M}^+) = \Delta H_f^{\,0}(\mathrm{M}) + \mathrm{IP(ad)}, \tag{11}$$

$$\epsilon_0 = \mathrm{AP(ad)} - \mathrm{IP(A)(ad)}. \tag{12}$$

In practice it is often necessary to use AP(expt) and IP(vert), assuming that the errors discussed above are negligible or cancel. The major difficulties in using Eq. (10) to calculate heats of formation of fragment ions are the problems associated with estimating the value of the excess energy term $(\epsilon_0^{\,\mathrm{r}} + \epsilon^{\ddagger})$. Williams and Bowen (1976) have shown how three simple assumptions may be used to place limits on the value of this term. The assumptions may be listed as follows (Boyd and Beynon, 1977b).

(1) Absence of an abundant metastable ion peak indicates a low excess energy.

(2) Entropies of activation are relatively unimportant; one reaction will normally be observed in preference to a possible competing reaction provided that the activation energy is lower.

(3) For fragmentations where N of Eq. (8) is a radical the reverse activation energy $\epsilon_0^{\,\mathrm{r}}$ is negligible. On the other hand, for reactions when N is a molecule there may be appreciable reverse activation energy and this must be taken into account in any calculations of $\Delta H_f^{\,0}(\mathrm{A}^+)$. This approach (Williams and Bowen, 1976) was used to examine the fragmentation reactions of a series of monosubstituted benzenes; in all cases the thermochemically preferred reaction was observed to occur.

For reactions involving large reverse activation energies, the excess energy term may be dominated by the reverse activation energy. In these cases, clearly, if $T^{\mathrm{r}}/\epsilon_0^{\,\mathrm{r}}$ is known, or can be estimated, then the measurement of T^{r} from metastable ion peaks would enable $\epsilon_0^{\,\mathrm{r}}$ to be estimated.

III. SPECIFIC EXAMPLES

The use of these approaches will now be further illustrated by considering some specific examples.

A. $C_4H_9^+$ *Ions*

The isomeric *n*-, *sec*-, *iso*-, and *tert*-butyl cations have been shown to undergo the metastable transitions given in Eqs. (13) and (14) (Davis *et al.*, 1970; Yeo and Williams, 1970).

$$n\text{-}, iso\text{-}, sec\text{-}, tert\text{-}C_4H_9^+ \rightarrow C_3H_5^+ + CH_4 \quad (13)$$

$$n\text{-}, iso\text{-}, sec\text{-}, tert\text{-}C_4H_9^+ \rightarrow C_2H_5^+ + C_2H_4 \quad (14)$$

Irrespective of which ionic isomer is generated at threshold, they undergo reactions (13) and (14) with a metastable ion abundance ratio of approximately 24:1 with allyl-ion formation being the dominant process. This result indicates that the various isomers interconvert to a common structure or mixture of structures before undergoing fragmentations (13) and (14). This fast interconversion, which requires hydrogen and methyl shifts, is supported by the statistical loss of CH_4, CH_3D, and CH_2D_2 in metastable transitions, from *d*-butyl ions produced by ionization of and fragmentation of $CH_3CH_2CH_2CD_2Br$ (Dymerski and McLafferty, 1976). Moreover, the measured appearance energy of the alkyl ion is the same for all isomeric precursor ions (Yeo and Williams, 1970). Figure 8 shows the type of potential energy surface involved in this type of mechanism.

The collision-induced decomposition spectra of both the isomeric butyl cations and of derivatives produced by an ion–molecule reaction with acetyl chloride [Eq. (15)]

$$C_4H_9^+ + CH_3COCl \rightarrow C_6H_{11}O^+ + HCl \quad (15)$$

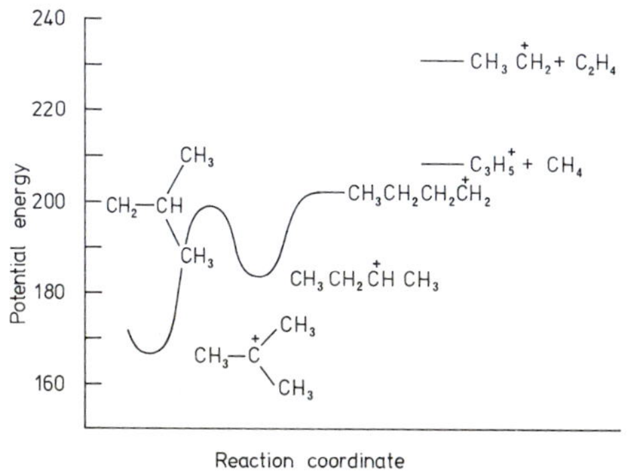

FIG. 8 Schematic potential energy diagram for the butyl cation.

show considerable differences depending on the structures of the isomers used as a precursor to $C_4H_9^+$ (Dymerski and McLafferty, 1976; Hass and Jennings, 1974). These results are in agreement with the potential-energy surface shown in Fig. 8, i.e., the ions are formed in potential wells and do not collapse to more stable structures without activation energy.

B. $C_3H_7O^+$ *Ions*

Methyl radical loss from the molecular ions of diethyl ether and methyl isopropyl ether produces the isomeric oxonium ions (a) and (b), respectively (Tsang and Harrison, 1973; Yeo and Williams, 1971).

$$\underset{\text{(a)}}{CH_3{-}CH_2{-}\overset{+}{O}{=}CH_2} \qquad \underset{\text{(b)}}{CH_3{-}CH{=}\overset{+}{O}{-}CH_3}$$

Tsang and Harrison (1970) determined the metastable ion abundance ratio for the unimolecular fragmentations

$$C_3H_7O^+ \left[\begin{array}{ll} \rightarrow C_3H_5^+ + H_2O & (16) \\ \rightarrow CH_2{=}\overset{+}{O}H_2 + C_2H_4 & (17) \\ \rightarrow C_2H_5^+ + CH_2O & (18) \end{array}\right.$$

and found different ratios for the two isomers; in particular, loss of CH_2O [Eq. (18)] does not occur from isomer (a). On the basis of this evidence, Tsang and Harrison tentatively suggested that isomers (a) and (b) do not fragment through a common intermediate.

Hvistendahl and Williams (1975) have recently made a detailed study of reactions (16)–(18) from isomers (a) and (b) and conclude that both decompose via (a), with a high rate-determining barrier to isomerization, (b) → (a). The findings of these authors may be summarised as follows:

(1) The metastable ion abundance ratios were measured in both the first and second field-free regions of an AEIMS9 mass spectrometer and are shown in Table I.

The fact that 95% of ions reacting from (a) produce $C_3H_5^+$ after 10 μsec (second field-free region) and 70% produce $C_3H_5^+$ after 5 μsec, whereas the competing reaction [$M^*(CH_3O^+)$] shows an abundance decrease from 30% to 5% with increase in the ion lifetime indicates that not only is reaction (16) the lowest energy process but also that the rate of increase of $k(E)$ with E is slower for reaction (16) that for reaction (17). This is to be expected in view of the extensive rearrangement, probably through a "tight" transition-

TABLE I

Relative Abundance of Metastable Ions

Ion	Field-free region	[M* ($C_3H_5^+$)]	[M* (CH_3O^+)]	[M* ($C_2H_5^+$)]
a	1st	70	30	0
a	2nd	95	5	0
b	1st	1.4	77	22
b	2nd	1.7	86	12

state complex, that is necessary before reaction (16) can occur, whereas reaction (17) may proceed after a simple H-atom transfer.

(2) Deuterium labeling experiments with derivatives (c) and (d)

$$\underset{\text{(c)}}{CH_3{-}CH_2{-}\overset{+}{O}{=}CD_2} \qquad \underset{\text{(d)}}{CH_3CH{=}\overset{+}{O}{-}CD_3}$$

show that from (c) ethylene is lost [Eq. (17)] as C_2H_4, C_2H_3D, and $C_2H_2D_2$ in the ratios 16:56:28, which indicates that (a) undergoes intramolecular rearrangement prior to ethylene loss, whereas (d) loses C_2H_3D with 80% specificity and CD_2O with greater than 90% specificity. These results are understandable if, as seems likely, the activation energy and frequency factor for intramolecular rearrangement are lower than those for fragmentation. Then if isomer (b) reacts via a high energy state of (a), the fragmentation reaction would proceed before intramolecular rearrangement had time to occur.

(3) The appearance potentials were determined from the metastable transitions in the first field-free region and coupled with tabulated heat of formation gave activation energies for reactions (16) and (17) from (a) of 23 and 28 kcal mole^{-1}, respectively. The activation energy for reaction (18) was calculated as 40 kcal mole^{-1} assuming zero activation energy ϵ_0^r for the reverse reaction. The measured activation energies for reactions (16)–(18) from isomer (b) were found to be 58 ± 3, 58 ± 3, and 62 ± 3 kcal mole^{-1}, respectively.

(4) The measured average kinetic energy releases, in the first field-free region, were found to be 1.0 and 0.4 kcal mole^{-1} for reactions (16) and (17) from isomer (a), and 1.6 and 1.3 kcal mole^{-1} from isomer (b).

These results suggest that the reactions of both isomers occur via structure (a) over a potential energy surface of the type shown in Fig. 9. These results are confirmed by the collision-induced decompositions (Jennings, 1968; Howe and McLafferty, 1970; Beynon and Boyd, 1978) (which weight some high energy processes) of isomer (a) where the observed fragment ion abundance ratios were 30:60:10 for reactions (16)–(18).

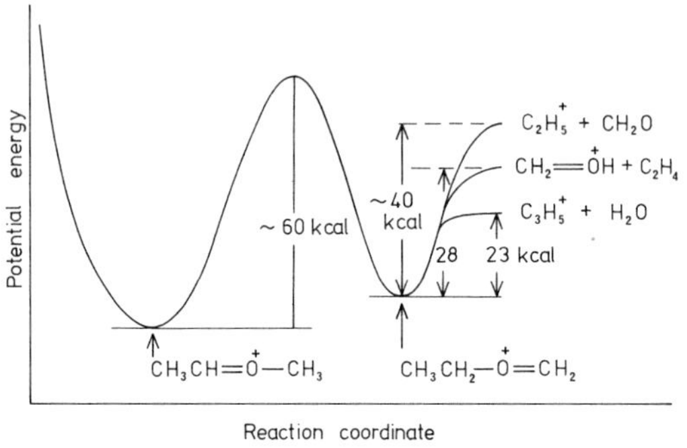

FIG. 9 Schematic potential energy diagram for $C_3H_7O^+$ oxonium ions.

C. *Substituted Anisole Ions*, $XC_6H_4OCH_3^{+\cdot}$

Studies of CH_2O loss from anisole ions (Barnes and Occolowitz, 1963; McLafferty and Bursey, 1968) have led to suggestions that the reaction occurs through a four-membered cyclic transition state [Eq. (19)].

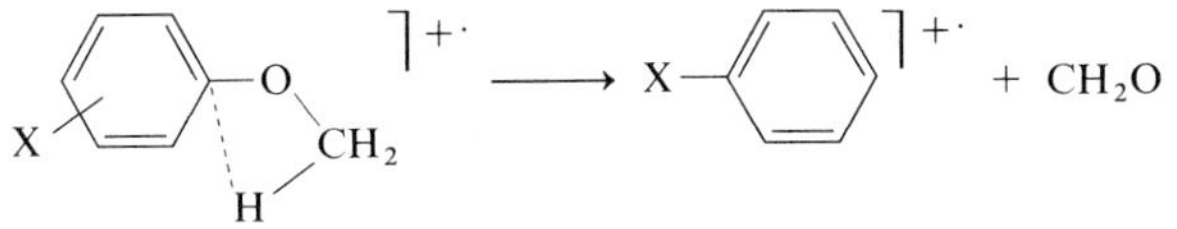

(19)

There is also evidence from labeled derivatives that supports the five-membered cyclic transition state (Lacey *et al.*, 1971) [Eq. (20)].

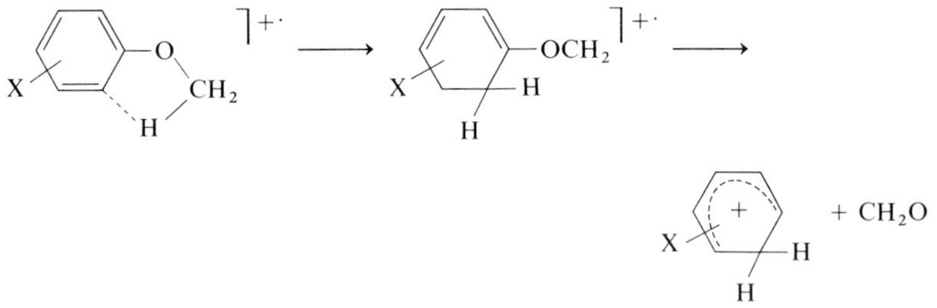

(20)

These mechanisms are not mutually exclusive and both reactions could be occurring together. Cooks *et al.* (1973a,b) found the metastable ion peak for loss of CH_2O from the anisole molecular ion to be composite, as illustrated in Fig. 10. This indicates that two separate processes contribute to the CH_2O elimination reaction. The average kinetic energy release for both components of the metastable ion peak was measured by deconvolution of the composite peak. The two values of kinetic energy release corresponding to the narrow and broad components of the metastable ion peak, designated T_s and T_l, respectively, were found to be relatively insensitive to a change in substituent X. Changing the electron energy had no effect on the shape

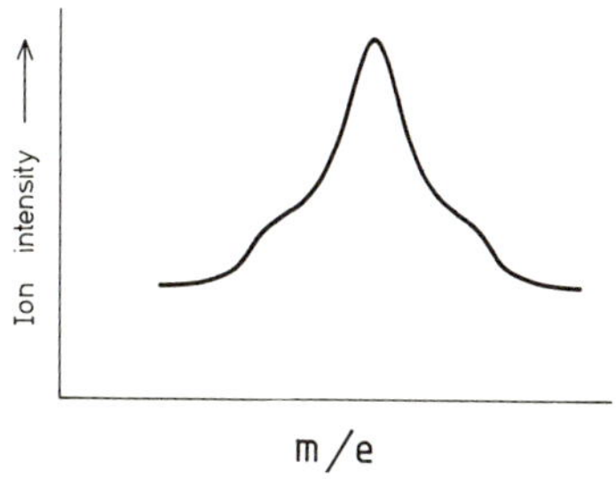

FIG. 10 Illustration of the composite metastable ion peak for the loss of CH_2O from a substituted anisole molecular ion.

of the peaks and generating the anisole ions from O-methyl oxime ether gave the same results as those obtained from directly ionized anisole. Finally, the peak shapes were unchanged by introduction of a collision gas into the field-free region, thus proving that both processes were unimolecular and not collision induced.

The composite peak could be due to two forms of the reactant ion, ionic product, or neutral product. The first excited state of formaldehyde, the neutral product, is 3.3 eV above the ground state, which is too high to permit the likelihood that formation of the ground and first excited states of the neutral are responsible for the composite peak. Formation of either ionic or neutral product in different vibrational states can be discounted because the observation of only two vibrational states is unlikely. Moreover, it would imply that composite metastable ion peaks should be the rule and not the exception. The fact that the relative contributions of the two processes do not change with electron energy indicates that two different reactant ions are unlikely to be involved. Also, the results of the experiments using alternative sources of reactant ion support the suggestion that two different reactant ions are not involved.

Therefore, these results suggest that the molecular ions of anisoles undergo CH_2O elimination by two reaction channels from the same reactant ion to produce two different isomers of the product ion. It seems likely, especially in view of the previously discussed evidence (Barnes and Occolowitz, 1963; McLafferty and Bursey, 1968; Lacey *et al.*, 1971), that the four-membered cyclic transition state which gives the stable benzene ion can be associated with the T_l value, and the five-membered cyclic transition state to give the unstable carbene associated with T_s. The difference between T_l and T_s is approximately 300 meV, which means that the differences in $(\epsilon^{\ddagger} + \epsilon_0{}^{r})$ for the two channels must be in excess of 300 meV. This is consistent with the suggestion that ionized carbene is the less stable product. The measured appearance energies for the $(M—CH_2O)^+$ ion can be used to derive the potential energy surface over which the reactions proceed. Such a surface is illustrated in Fig. 11.

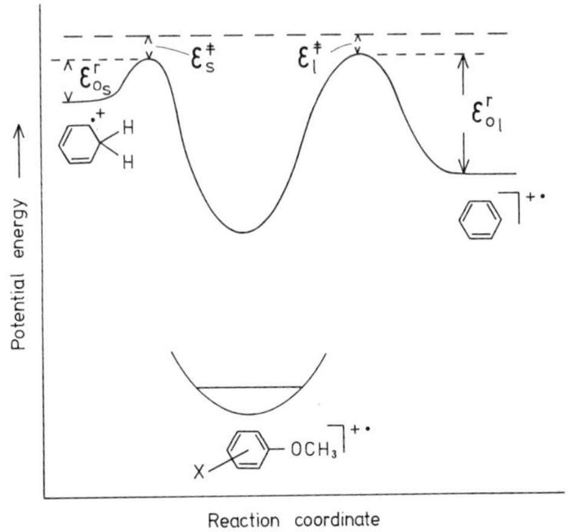

FIG. 11 Schematic potential energy diagram showing the thermochemistry relevant to CH_2O loss from anisole by four- and five-centered hydrogen rearrangement reactions.

The relative proportions of each channel were found to vary with substituent X. Electron withdrawing substituents decrease the contribution by the larger kinetic energy release, electron donating substituents increase it. Whereas the substituent has little effect on the value of T_s or T_l, these results are consistent with both reactions proceeding by hydrogen transfers to radical sites over a surface similar to Fig. 11.

D. *Benzyl Ions*, $C_7H_7^+$

The high abundance of the $C_7H_7^+$ ion in the mass spectra of toluene and its isomers, and many benzylic compounds, was originally explained (Rylander *et al.*, 1957) by proposing that it exists as the symmetrical tropylium cation rather than the benzyl cation.

Deuterium labeling studies by Meyerson *et al.* (1959, 1968) showed that for the fragmentation

$$C_7H_7^+ \rightarrow C_5H_5^+ + C_2H_2 \tag{21}$$

all hydrogen atoms had lost their positional identity when the precursor molecules for $C_7H_7^+$ were dibenzyl, benzyl alcohol, and ethyl benzene. Using ^{13}C labeled derivatives Rinehart *et al.* (1968) and Siegel (1970) showed that all the carbon atoms in $C_7H_7^+$ also lost their positional identity. These labeling results were consistent with the formation of a symmetric tropylium cation by insertion of the CH_2 group into the ring followed by hydrogen-atom scrambling. They were supported by later work of Howe and McLafferty

(1971) and Beynon *et al.* (1970), who obtained results suggesting that even at low internal energies scrambling was complete prior to fragmentation.

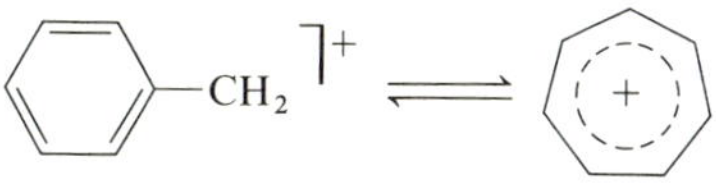

(22)

More recent work, however, has cast some doubt on the complete randomization of C and H in $C_7H_7^+$ and on the tropylium cation structure. Levsen *et al.* (1973) have performed a similar study to that of Howe and McLafferty (1971) and obtained evidence that scrambling is not complete before reaction. A study of the collision-induced decompositions of the $C_7H_7^+$ ion by McLafferty and Winkler (1974) has shown that for low energy ions having lifetimes greater than 10^{-5} sec 90% of these ions have a benzyl ion structure. Thus, $C_6H_5CD_2^+$ ions produced from $C_6H_5CD_2CD_2C_6H_5$ showed a very high specificity for loss of CD_2 on being activated by collision.

The results discussed above do not provide information as to the possible mechanism of formation of the tropylium cation. To shed some light upon this problem the decomposition of *p*-methylethylbenzene, labeled with deuterium in either of the two methyl groups, was investigated (Meyer and Harrison, 1964). It was found that either methyl could be lost without prior hydrogen rearrangement and that the ratio of the side chain methyl to ring methyl loss was 5:1. This ratio remained constant from the appearance energy up to 50 eV. To account for these observations it was proposed that one of the methylene hydrogens was transferred to the ring followed by insertion of the $\diagdown$CH—CH_3 groups between any two carbons of the ring

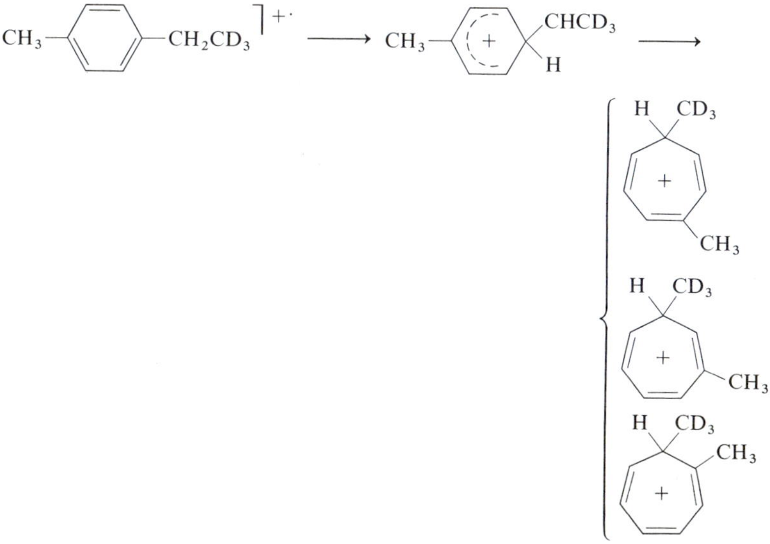

(23)

The hydrogen scrambling can now occur via a 1:2 hydrogen shift.

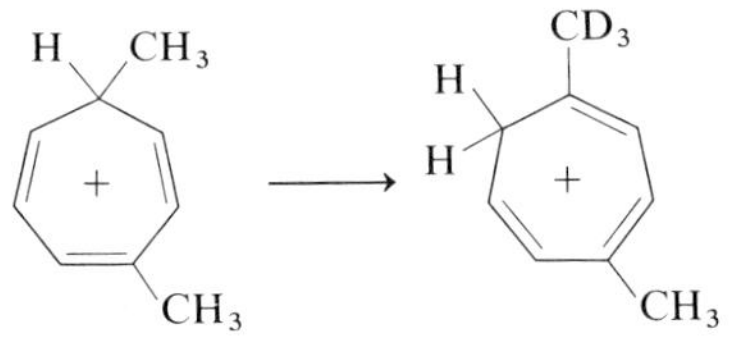

(24)

If methyl elimination occurs only from methylene positions, e.g.,

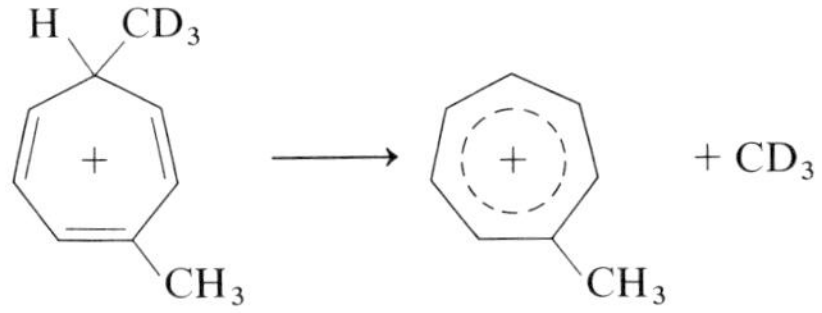

(25)

and a fast 1:2 hydrogen shift occurs in the 1,2-dimethyl tropylium ion,

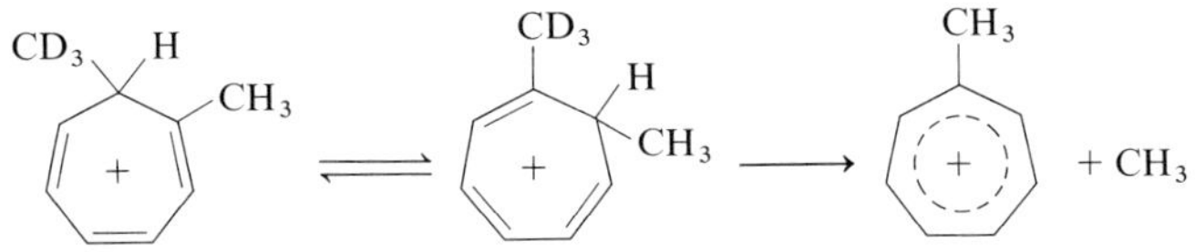

(26)

then this mechanism will account for the observed ratio.

It is of interest to compare the mechanisms of formation of tropylium cations from precursors produced by electron impact, as described above. with the results obtained using a field-ionization source. The field-ionization mass spectrometer gives data concerned with fragmentations occurring at times very close (10^{-10} sec) to ionization and reactions involving skeletal rearrangements are often too slow to be observed as normal peaks (Brown *et al.*, 1970; Beckey and Levsen, 1971). Levsen and Beckey (1972) have studied the formation of the tropylium cation, in a field-ionization mass spectrometer, from the benzyl alcohol and ethyl benzene molecular ions. Their observations demonstrate that this rearrangement starts with a field-induced benzyl ion formation (in contrast to the electron impact results)

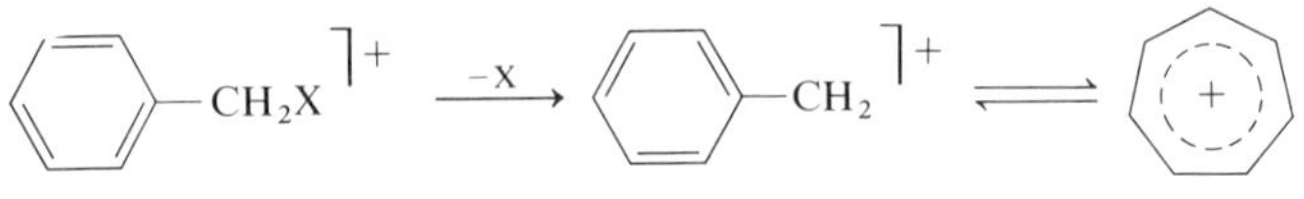

(27)

REFERENCES

Ast, T., Beynon, J. H., and Cooks, R. G. (1972). *J. Am. Chem. Soc.* **94**, 6611.

Baer, T. (1979). *In* "Gas Phase Ion Chemistry." (M. T. Bowers, ed.), Vol. 1, Ch. 5. Academic Press, New York.

Barnes, C. S., and Occolowitz, J. L. (1963). *Aust. J. Chem.* **16**, 219.
Beckey, H. D. (1977). "Principles of Field Ionisation and Field Desorption Mass Spectrometry." Pergamon, Oxford.
Beckey, H. D., and Levsen, K. (1971). *In* "Recent Topics in Mass Spectrometry" (R. I. Reed, ed.), p. 169. Gordon & Breach, New York.
Benoit, F., and Holmes, J. L. (1972). *Org. Mass Spectrom.* **6**, 549.
Bentley, T. W., Johnstone, R. A. W., and McMaster, B. N. (1973). *J. Chem. Soc. D* 510.
Beynon, J. H. (1960). "Mass Spectrometry and its Applications to Organic Chemistry." Elsevier, Amsterdam.
Beynon, J. H., and Boyd, R. K. (1978). *Adv. Mass Spectrom.* **7**, 1115.
Beynon, J. H., Saunders, R. A., and Williams, A. E. (1965). *Z. Naturforsch, Teil A* **20**. 180.
Beynon, J. H., Amy, J. W., and Baitinger, W. E. (1969). *Chem. Commun.* p. 723.
Beynon, J. H., Corn, J. E., Baitinger, W. E., Caprioli, R. M., and Benkeser, R. A. (1970). *Org. Mass Spectrom.* **3**, 1371.
Beynon, J. H., Cooks, R. G., Amy, J. W., Baitinger, W. E., and Ridley, T. Y. (1973). *Anal. Chem.* **45**, 1023A.
Beynon, J. H., Brothers, D. F., and Cooks, R. G. (1974). *Anal. Chem.* **46**, 1299.
Beynon, J. H., Cooks, R. G., Jennings, K. R., and Ferrer-Correra, A. J. (1975). *Int. J. Mass Spectrom. Ion Phys.* **18**, 87.
Boyd, R. K., and Beynon, J. H. (1977a). *Org. Mass Spectrom.* **12**, 163.
Boyd, R. K., and Beynon, J. H. (1977b). *Int. J. Mass Spectrom. Ion Phys.* **23**, 163.
Brehm, B., and von Puttkamer, E. (1968). *Adv. Mass Spectrom.* **4**, 591.
Brown, P., Pettit, G. R., and Robins, R. K. (1970). *Org. Mass Spectrom.* **3**, 67.
Bruins, A. P., Nibbering, N. M. M., and de Boer, T. J. (1972). *Tetrahedron Lett.* 1109.
Chupka, W. A. (1959). *J. Chem. Phys.* **30**, 191.
Cole, W. G., and Williams, D. H. (1968). *Chem. Commun.* p. 956.
Cooks, R. G., and Beynon, J. H. (1971). *Chem. Commun.* p. 1282.
Cooks, R. G., and Beynon, J. H. (1975). *In* "Mass Spectrometry" (A. Maccoll, ed.), MTP International Review of Science, Physical Chemistry, Series II, Vol. 5, p. 159. Butterworth, London.
Cooks, R. G., Beynon, J. H., Caprioli, R. M., and Lester, G. R. (1973a). "Metastable Ions." Elsevier, Amsterdam.
Cooks, R. G., Bertrand, M., Beynon, J. H., Rennekamp, M. E., and Setser, D. W. (1973b). *J. Am. Chem. Soc.* **95**, 1732.
Davis, B., Williams, D. H., and Yeo, A. N. H. (1970). *J. Chem. Soc. B* 81.
Derrick, P. J. (1975). "Mass Spectrometry" (A. Maccoll, ed.), MTP International Review of Science, Physical Chemistry, Series II, Vol. 5, p. 1. Butterworth, London.
Dymerski, P. P., and McLafferty, F. W. (1976). *J. Am. Chem. Soc.* **98**, 6070.
Eland, J. H. D. (1974). *Adv. Mass Spectrom.* **6**, 917.
Field, F. H., and Franklin, J. L. (1957). "Electron Impact Phenomena." Academic Press, New York.
Forst, W. (1973). "Theory of Unimolecular Reactions." Academic Press, New York.
Gambol, A. A., Gilbert, J. R., and Zillett, J. G. (1970). *Org. Mass Spectrom.* **3**, 1223.
Gilbert, J. R., and Al-Mashhadani, A. (1978). Unpublished observations.
Gilbert, J. R., and Stace, A. J. (1974). *Int. J. Mass Spectrom. Ion Phys.* **15**, 311.
Hass, J. R., and Jennings, K. R. (1974). *Adv. Mass Spectrom.* **6**, 711.
Hoffmann, M. K., and Friesen, M. D. (1976). *Org. Mass Spectrom.* **11**, 1101.
Holmes, J. L. (1975). *In* "Mass Spectrometry" (A. Maccoll, ed.), MTP International Review of Science, Physical Chemistry, Series II, Vol. 5, p. 207. Butterworth, London.
Holmes, J. L., and Osborne, A. D. (1977). *Int. J. Mass Spectrom. Ion Phys.* **23**, 189.

Holmes, J. L., Osborne, A. D., and Weese, G. M. (1976). *Int. J. Mass Spectrom. Ion Phys.* **19**, 207.
Howe, I., and McLafferty, F. W. (1970). *J. Am. Chem. Soc.* **92**, 3797.
Howe, I., and McLafferty, F. W. (1971). *J. Am. Chem. Soc.* **93**, 99.
Hvistendahl, G., and Williams, D. H. (1975). *J. Am. Chem. Soc.* **97**, 3097.
Jennings, K. R. (1968). *Int. J. Mass Spectrom. Ion Phys.* **1**, 227.
Kerr, J. T., and Duckworth, H. E. (1958). *Can. J. Phys.* **36**, 986.
Kiser, R. W. (1965). "Introduction to Mass Spectrometry and its Applications." Prentice-Hall, Englewood Cliffs, New Jersey.
Knewstubb, P. F., and Reed, N. W. (1970). *Int. J. Mass Spectrom. Ion Phys.* **5**, 361.
Lacey, M. J., MacDonald, C. G., and Shannon, J. S. (1971). *Org. Mass Spectrom.* **5**, 1391.
Lawson, G., Bonner, R. F., and Todd, J. F. J. (1973). *J. Phys. E* **6**, 357.
Lehman, T. A., and Bursey, M. M. (1976). "Ion Cyclotron Resonance Spectrometry." Wiley, New York.
Levsen, K., and Beckey, H. D. (1972). *Int. J. Mass Spectrom. Ion Phys.* **9**, 63.
Levsen, K., McLafferty, F. W., and Jerina, D. M. (1973). *J. Am. Chem. Soc.* **95**, 6332.
Lifshitz, C., and Long, F. A. (1964). *J. Chem. Phys.* **41**, 2468.
Lin, Y. N., and Rabinovitch, B. S. (1970). *J. Phys. Chem.* **74**, 1769.
Lossing, F. P. (1971). *Can. J. Chem.* **49**, 357.
Loudon, A. G. (1978). Personal communication.
McLafferty, F. W., and Bursey, M. M. (1968). *J. Org. Chem.* **33**, 124.
McLafferty, F. W., and Winkler, J. (1974). *J. Am. Chem. Soc.* **96**, 5182.
McLafferty, F. W., Wachs, T., Lifshitz, C., Innorta, G., and Irving, P. (1970). *J. Am. Chem. Soc.* **92**, 6867.
McLafferty, F. W., Bente, P. F., Kornfeld, R., Tsai, S. C., and Howe, I. (1973a). *J. Am. Chem. Soc.* **95**, 2120.
McLafferty, F. W., Kornfeld, R., Hadden, W. F., Levsen, K., Sakai, I., Bente, P. F., Tsai, S. C., and Schuddemage, H. D. R. (1973b). *J. Am. Chem. Soc.* **95**, 388.
McMaster, B. N. (1974). *In* "Mass Spectrometry" (R. A. W. Johnstone, senior reporter), Specialist Period Rep.," Vol. 3, p. 1. Chem. Soc., London.
Meyer, F., and Harrison, A. G. (1964). *J. Am. Chem. Soc.* **86**, 4757.
Meyerson, S., Rylander, P. N., Eliel, F. L., and McCollum, J. D. (1959). *J. Am. Chem. Soc.* **81**, 2606.
Meyerson, S., Hart, H., and Leitch, L. (1968). *J. Am. Chem. Soc.* **90**, 3419.
Millington, D. S., and Smith, J. (1977). *Org. Mass Spectrom.* **12**, 264.
Rinehart, K. L., Bucholz, A. C., Van Lear, G. E., and Cantrill, H. L. (1968). *J. Am. Chem. Soc.* **90**, 2983.
Rosenstock, H. M. (1976). *Int. J. Mass Spectrom. Ion Phys.* **20**, 139.
Rosenstock, H. M., Wallenstein, M. B., Wahrhaftig, A. L., and Eyring, H. (1952). *Proc. Natl. Acad. Sci. U.S.A.* **38**, 667.
Rosenstock, H. M., Dibeler, V. H., and Harlee, F. N. (1964). *J. Chem. Phys.* **40**, 591.
Rylander, P. N., Meyerson, S., and Grubb, H. M. (1957). *J. Am. Chem. Soc.* **79**, 842.
Schopman, J., Barua, A. K., and Los, J. (1970). "Recent Developments in Mass Spectrometry" (K. Ogata and T. Hayakawa, eds.) p. 888 Univ. Park Press, Baltimore, Maryland.
Siegel, A. S. (1970). *J. Am. Chem. Soc.* **92**, 5277.
Terwilliger, D. T., Elder, J. F., Beynon, J. H., and Cooks, R. G. (1975). *Int. J. Mass Spectrom. Ion Phys.* **16**, 225.
Tsai, B. T., Baer, T., Werner, A. S., and Lin, S. F. (1975). *J. Phys. Chem.* **79**, 570.
Tsang, C. W., and Harrison, A. G. (1970). *Org. Mass Spectrom.* **3**, 647.

Tsang, C. W., and Harrison, A. G. (1973). *Org. Mass Spectrom.* **7**, 1377.
Uccella, N. A., Howe, I., and Williams, D. H. (1972). *Org. Mass Spectrom.* **6**, 229.
Wallington, M. (1971). *J. Phys. E* **4**, 1.
Williams, D. H., and Bowen, R. D. (1976). *Org. Mass Spectrom.* **11**, 223.
Yeo, A. N. H., and Williams, D. H. (1970). *Chem. Commun.* p. 737.
Yeo, A. N. H., and Williams, D. H. (1971). *J. Am. Chem. Soc.* **93**, 395.

Chapter 14

Ion photodissociation

ROBERT C. DUNBAR

DEPARTMENT OF CHEMISTRY
CASE WESTERN RESERVE UNIVERSITY
CLEVELAND, OHIO

"From whence, fragment?"
(Shakespeare, Troilus and Cressida V, 1)

ISBN 0-12-120802-8

I. INTRODUCTION

Ion photodissociation is the process

$$AB^+ \xrightarrow{h\nu} A^+ + B \qquad (1)$$

and quite properly belongs in a book about ion–molecule reactions by virtue of being the inverse of the radiative association reaction. A very wide range of spectroscopic, thermodynamic, kinetic, and dynamic information about ions and their reactions can be drawn from photodissociation data, and the last few years have seen both spreading recognition of these possibilities and a rapid growth in the number of laboratories pursuing them. Five years ago a comprehensive review would have been brief indeed, while five years from now the literature seems bound to be formidably large for such a task. This chapter includes a comprehensive bibliography through 1977 of the published and some of the to-be-published primary literature concerned with photodissociation of gas-phase ions.

II. A SMALL HISTORY

Three distinct periods can be distinguished in the study of ion photodissociation. It is not surprising that the first work in the 1960s was done by physicists on very simple species: Dunn's laboratory (Dunn, 1964, 1968; von Busch and Dunn, 1972) made some cross section measurements mostly on H_2^+, while Dehmelt's laboratory (Dehmelt and Jefferts, 1962; Dehmelt, 1967, 1969a,b, 1976; Jefferts, 1968, 1969; Richardson *et al.*, 1968; Menasian and Dehmelt, 1973) used selective photodissociation as a convenient means of preparing spatially oriented H_2^+ populations, on which noteworthy hyperfine-level rf experiments have been carried out.

Beginning in 1970, our laboratory and subsequently several others began using the ion cyclotron resonance (ICR) technique to study a wide variety of ions, with a bias toward more complex species of chemical interest. Photodissociation has become a major area of interest to ICR spectroscopy, and there are at least six ICR laboratories currently active in such studies.

The third historical phase has been the renewed application of mass-filter type mass spectrometers to these problems, greatly aided by the availability of useful laser light sources. This area is still showing rapid growth following an initial surge of interest around 1973, and among the laboratories now active may be included those at Houston (moved from Utah), Stanford Research Institute, Hanscomb Air Force Base, Orsay (France), Amsterdam (Holland), Southampton (United Kingdom), Aberdeen Proving Ground, and Latrobe (Australia). The bias of most of this work has been toward very simple ions studied at a high level of detail, and often toward ions of atmospheric importance.

A review of the field appeared several years ago (Dunbar, 1975b), but progress since then has been rapid.

III. EXPERIMENTAL TECHNIQUES

Ion photodissociation experiments group immediately into two types, the ion trap methods and the ion beam methods, having differing sets of problems, opportunities, and practitioners. Only the briefest summary of techniques will be given, and the reader should consult the literature for fuller descriptions and elegant diagrams of the various instruments and techniques.

A. Ion Traps

A very few studies have been done in rf quadrupole traps, (e.g., Ensberg and Jefferts, 1975) but ion trap work is nearly synonymous with ICR instrumentation (Dunbar, 1974). Appreciation of the excellence of the ICR cell as an ion trap of the Penning variety came slowly and largely accidentally into the consciousness of the ICR community, but a number of ICR groups, at CWRU, Stanford (Richardson *et al.*, 1974a,b), Cal Tech (Freiser and Beauchamp, 1974), Florida (Eyler, 1974), and Purdue (Freiser *et al.*, 1978) are actively using it in this way for ion photodissociation.

The original technique (Dunbar, 1971), not yet entirely abandoned, was the continuous mode of operation, in which ions are continuously produced at the electron beam, stored in the trap, and dissipated by various processes. The steady-state ion abundance at a chosen mass is continuously monitored by the detector. When the light is turned on, an additional ion-loss mechanism is introduced, and the photodissociation rate is determined by analyzing the resulting perturbation of the steady-state kinetics. The details of these kinetics have been well worked out, and this approach has served to carry out a large amount of exploratory work (most of it correct).

A pulsed mode of ICR operation, first seriously applied to photodissociation by the Cal Tech group (Freiser and Beauchamp, 1974), is in most ways superior. The trap is filled by a brief pulse of the electron beam. The trap is then irradiated with the light for a reaction period, at the end of which the cell contents are rapidly sampled by the detector to ascertain the extent of photodissociation (and other processes) during the reaction period. Then the cell is swept clean preparatory to a new cycle.

All ICR approaches share the double resonance capability, which allows rapid removal from the cell of a chosen ion mass. This often serves to verify the parent–daughter lineage of a particular photodissociation process, by showing that the daughter-ion signal disappears when the parent ion is ejected. In addition, one of the serious problems with ion-trap techniques is the occurrence of various kinds of ion–molecule chemistry involving the ions (particularly the photodissociation daughter ions) of interest. Double resonance ejection serves to map out such photochemistry, and often to suppress ion–molecule reaction processes which interfere with the measurements.

The "time-resolved" ICR technique (Fu *et al.*, 1976) (which can be more accurately given the ponderous title "exhaustive photodissociation") is an additional tool of importance. In many photodissociation studies of chemical interest, there is uncertainty about the ion structure, and about whether the population of ions in the trap all possess the same structure. In the time-resolved approach, ion production is terminated, the ions in the trap are strongly irradiated at a photodissociating wavelength, and the disappearance of the photodissociating ions is continously monitored, usually for 10 sec or more. A homogeneous ion population will be observed to decay exponentially to zero under such conditions; whereas if a fraction of the ions possess a different, nonphotodissociating structure, the ion signal will decay to a new, but not zero, level. This method was first used to show the existence of a mixed ion population in benzyl chloride parent ion (Fu *et al.*, 1976). Similar methods can be used in a pulsed ICR mode of operation for the same purpose.

B. Ion Beams

There are seven active laboratories looking at the photodissociation of ions in a collimated ion beam [Orsay (Ozenne *et al.*, 1972), Amersterdam (Van Asselt *et al.*, 1974a), Houston (Vestal and Futrell, 1974), Hanscomb AFB (Thomas *et al.*, 1977) Southampton (Carrington *et al.*, 1976), La Trobe (McGilvery and Morrison, 1978), and SRI (Huber *et al.*, 1977a,b), in rough chronological order]. There are many variations, but they have in common the formation and mass selection of a beam of primary ions, an interaction region where the beam encounters the light, and a final product mass selection. The mass selection methods for primary ions and product ions encompass a number of combinations of magnetic deflection, electrostatic deflection, and quadrupole filter arrangements. Both pulsed and continuous laser sources are used, the advantage of a pulsed laser being the potential (realized at Hanscomb) for wavelength doubling into the uv. The instruments at Houston, Hanscomb, Southampton, and SRI are designed for high-source-pressure operation, so that cluster ions and other species requiring high pressure for their formation can be studied.

Velocity analysis of the product ions is of interest for determining the amount of energy partitioned into product ion translation (in fact at Orsay and Amsterdam it is the feature of primary concern), and all of the instruments except the quadrupoles at Latrobe and Houston are capable of product-ion energy measurement (the quadrupole detector at Hanscomb being supplemented by time-or-flight analysis).

All of the instruments have been used in crossed-beam configurations, in which the light beam crosses perpendicular to the ion beam. This can give cross section measurements, velocity distribution of product ions, and, by rotating the plane of polarization of the light, angular dependence information about the product ion velocities. A recent development which has aroused great interest is the merged, or coaxial, beam configuration (Carrington *et al.*, 1976), in which the ion and laser beams travel in parallel along the interaction region. As shown at Southampton, the merged beam configuration allows the laser to be operated on a fixed-wavelength narrow line, while the ions are "Doppler tuned" through a small wavelength range by varying their velocity. Because the ion velocities are so high compared to the ion thermal velocities, the Doppler broadening of spectral lines due to ion thermal motion becomes insignificant, and optical spectra can be obtained at an optical resolution of ~ 100 MHz.

The extensive study of atmospheric ions at SRI has been done using a unique drift-tube instrument (Moseley *et al.*, 1974): ions are produced in a high-pressure source and drift down a drift tube, undergoing, if desired, ion–molecule reactions. Just at the end of the drift tube the ions pass through the intracavity laser beam, and are then extracted and mass analyzed in a quadrupole mass filter. This device has given large quantities of accurate photodissociation cross section information. Vanderhoff (1977) and co-workers at Aberdeen Proving Ground have used a similar instrument for work on some cluster ions of atmospheric interest.

C. *Photon Sources*

"What I have to do will want true colour . . ."

(Shakespeare, Hamlet iii, 4)

Many light sources have been used in photodissociation work. Much ICR work has used arc sources, convenient for wide-wavelength-range spectroscopy. Almost all groups have done at least some work with tunable dye lasers: continuous dye lasers are well matched to ICR experiments, although pulsed lasers can be used, and a pulsed-laser pulsed-ICR arrangement (Eyler, 1974) has potential advantages. Most ion beam studies have taken advantage of the high overlap which can be achieved between a tight ion beam and a focused laser, and have been restricted to laser light sources.

IV. ION SPECTROSCOPY

"Bothe were they faire and bright of hewe."

(Chaucer, Romaunt de la Rose 855)

It is as a window into the optical spectroscopy of gas-phase ions that photodissociation has aroused the most widespread interest. To suggest why the spectroscopist has welcomed this concept so eagerly, one need only observe that the maximum concentration of mass-selected ions which can be assembled in a spectrometer cell with current methods is of the order of 10^{-14} *M*, which is discouraging to attempts to obtain direct optical absorption spectra. Since photodissociation methods deal usefully with ion populations of tens or even ones of ions/cm^3, their potential spectroscopic importance is clear.

A. Small Ions

It is useful to discuss small and large ions separately, because the interpretive points of view are so different. With small ions it is feasible to consider a complete specification of all the quantum numbers of each state; high quality potential-surface calculations are available or at least thinkable, and a great deal is independently known about optical transitions in many small ions from emission spectroscopy. With large ions, exact quantum descriptions and potential surface calculations are out of the question, and the language is that of molecular orbital theory; emission data are only rarely available for comparison.

The smallest molecular ion H_2^+ is very well understood and not very interesting spectroscopically (von Busch and Dunn, 1972). Photodissociation is observed with gradually increasing cross section from the infrared to below 250 nm: from theory one expects a maximum near 200 nm. The transition is ${}^2\Sigma_u{}^+ \rightarrow {}^2\Sigma_g{}^+$, and in line with its being a bound-free transition, no vibrational structure has been observed. The shape of the curve reflects the distribution of vibrational states of the electron-impact produced H_2^+ population and the Franck–Condon factors for transitions from each to the repulsive excited state. The chief unresolved question, requiring data of extreme precision, is the extent to which a Franck–Condon model of H_2 ionization correctly gives the vibrational population distribution of the ground-state ions.

Ground-state O_2^+ is inaccessible to most photodissociation instruments, having a dissociation threshold of 183 nm. The recent observation of photodissociation thus occasioned considerable interest (Vestal *et al.*, 1975; Tabché-Fouhaillé *et al.*, 1976). The metastable excited state which gives

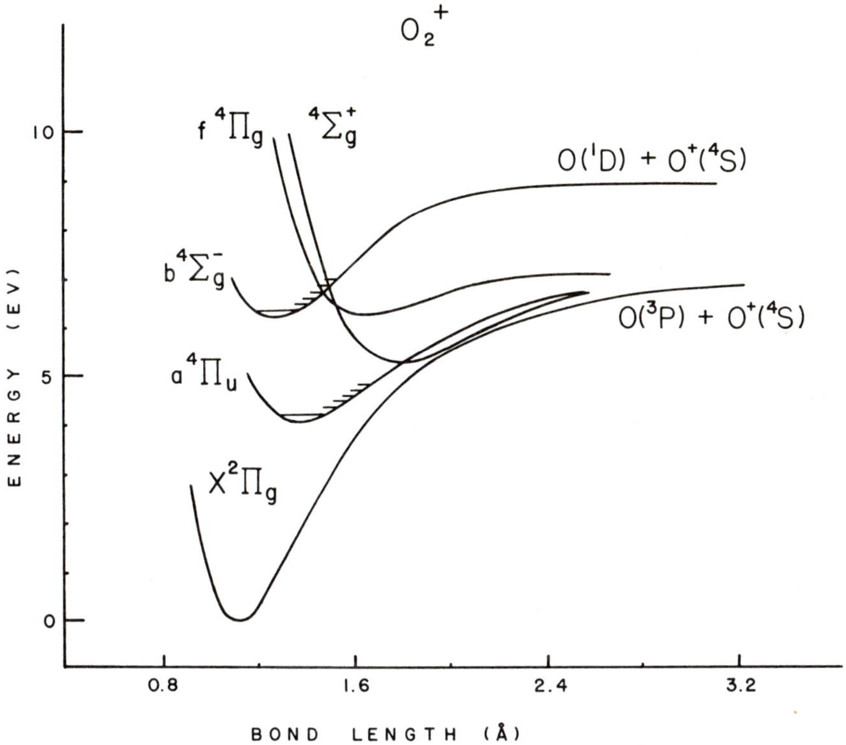

FIG. 1 Potential curves for O_2^+ states which may be involved in photodissociation processes originating in the $\tilde{a}$ state. [From Carrington *et al.* (1978).]

rise to this is produced in the electron-impact source, and survives long enough to reach the irradiation region in ion beam type instruments: it has been assigned (Tabché-Fouhaillé *et al.*, 1976) as $\tilde{a}^4\Pi_u$, comprising 30% of the ions. There are several sets of experimental results on the spectroscopy of this state. Figure 1 shows the relevant known electronic states of the O_2^+ quartet manifold, including the $\tilde{f}^4\Pi_g$ state recently postulated by Tabché-Fouhaillé *et al.* (1976), to account for their photofragment results and also indicates the spectroscopic transitions thought to have been observed in these photodissociation experiments.

The SRI group (Tadjeddine *et al.*, 1978), continuing work begun by the Orsay group (Moseley *et al.*, 1976b), has used a tunable dye laser in a threshold photofragment arrangement to examine transitions from rotational and fine-structure states of the $V = 4$ level of $\tilde{a}^4\Pi_u$ up to the dissociation threshold, with a resolution of the order of 2 meV in photofragment energy and 0.02 nm in laser wavelength. It was shown that the transitions observed are predissociative: a peak is observed when the laser is on resonance with a transition between some level of $\tilde{a}^4\Pi_u$ and a predissociation level of the upper state lying not more than 1 meV above the dissociation threshold. Based on photofragmentation work discussed below, Tabché-Fouhaillé *et al.* (1976) assigned the upper state as a $^4\Pi_g$ state which they named $\tilde{f}^4\Pi_g$, but Tadjeddine *et al.* have concluded that not all of the observed transitions can be ascribed to this upper state, and propose that the threshold photofragment experiment is in fact displaying a superposition of $\tilde{f}^4\Pi_g \leftarrow \tilde{a}^4\Pi_u(V = 4)$, $\tilde{b}^4\Sigma_g^-(V = 4) \leftarrow \tilde{a}^4\Pi_u(V = 4)$ and $\tilde{b}^4\Sigma_g^-(V = 5) \leftarrow \tilde{a}^4\Pi_u(V = 5)$ transitions. As something like 50 peaks are discernible between 580.8 and 583.5 nm in the spectrum of Tadjeddine *et al.* (1978) at 0.02 nm resolution, with every likelihood of more being resolved at higher resolutions, the sorting out of this spectrum will be a formidable task.

The Southampton group (Carrington *et al.*, 1977) has used the Doppler-shift coaxial beam technique to do spectroscopy of O_2^+ at extreme resolution, limited by the natural linewidth. The transitions observed in this region are rotational lines of the $\tilde{b}^4\Sigma_g^-(V = 4) \leftarrow \tilde{a}^4\Pi_u(V = 1)$ transition, with a constant background dissociation arising from structureless bound-free transitions to the continuum of $\tilde{f}^4\Pi_g$. Carrington *et al.* (1977) report observation of 22 lines, having widths varying from 240 to 1200 MHz, which have been analyzed with reasonable success. As these authors point out, the results provide a striking illustration of the potential power of this experimental technique: the optical resolution is unlimited by Doppler broadening and is far better than the Doppler width of 1–2 GHz which would limit a conventional spectrographic study. The widths of the rotational lines observed here are much greater than experimental resolution, and are a pure measure of the lifetime of the $\tilde{b}^4\Sigma_g^-(V = 4)$ rotational components against predissociation. It is believed that predissociation occurs through the $\tilde{f}^4\Pi_g$ state.

The photodissociation spectrum obtained by McGilvery *et al.* (1978a) at 3 cm^{-1} resolution across the entire visible is also in accord with O_2^+ photodissociation arising from $\tilde{b}^4\Sigma_g^- \leftarrow \tilde{a}^4\Pi_u$ excitations followed by predissociation through $\tilde{f}^4\Pi_g$. The vibrational structure in their spectrum is in good agreement with calculation.

Photodissociation of Cl_2^- has been observed by several groups. Depending on the means of forming the ion, some vibrationally excited species may be present, but Sullivan *et al.* (1977) believe they have observed the spectrum of the ground-state ion. A symmetrical, unstructured peak centered at 350 nm is observed, which is attributed to the bound free process $^2\Sigma_g^+ \leftarrow {}^2\Sigma_u^+$, and which is in reasonable agreement with the calculated and observed (in solid state) absorption spectrum of Cl_2^-.

N_2O^+ has been studied across the visible/uv in a low-resolution ICR experiment (Orth and Dunbar, 1977), and at high resolution near 300 nm in a beam experiment (Thomas *et al.*, 1977). The ICR work found peaks corresponding to the $\tilde{A}^2\Sigma^+$ state at 3.65 eV (peak cross section at 14 nm resolution of 4.9×10^{-19} cm^2), and to the $\tilde{B}^2\Pi$ state at 5.16 eV (6.5×10^{-18} cm^2). From photoionization work (Eland, 1973), the $\tilde{A}^2\Sigma^+$ (000) vibrational state is known not to dissociate, while the (100) state has a quantum yield for dissociation of 0.4, which agrees with the photodissociation peak position. Since fluorescence for the $\tilde{B}$ state has not been observed, it is reasonable to assume its photodissociation quantum yield to be unity, which leads to an estimate of 0.14 for the oscillator strength of $\tilde{B}^2\Pi \leftarrow \tilde{X}^2\Pi$.

Thomas *et al.*'s (1977) experiment using the crossed laser/ion beam apparatus at Hanscom yielded a multitude of sharp lines at ~0.2 nm resolution for the $\tilde{A}^2\Sigma^+ \leftarrow \tilde{X}^2\Pi$ transition in the range from 300 to 343 nm. A detailed analysis is given, in terms of transitions originating in $^2\Pi_{1/2}$ and $^2\Pi_{3/2}$ components of the ground state, with the possibility of one or more quanta in the bending vibration (making necessary a careful consideration of Renner effect splittings) and terminating in various vibrationally excited states of $^2\Sigma^+$. The spectrum fits the analysis well, and the peak positions are in good agreement with the emission and photoelectron (PES) spectra. Careful cross section measurements were made, and Thomas *et al.* address the question of the absolute values of optical absorption cross sections and estimate some absolute numbers.

The photodissociation spectrum of CO_3^- (Moseley *et al.*, 1974, 1976a) shows rich structure between ~1.8 and 2.4 eV. Polarization measurements indicate y polarization and support an assignment of $1^2A_1 \leftarrow 1^2B_2$ for the electronic transition involved. Vibrational assignments were made using modes of energies 990, 1470, and 880 cm^{-1} and sufficient confidence was placed in the vibrational assignment to place the O–O transition energy at 1.520 eV for this transition, although the observed photodissociation threshold is near 1.8 eV.

Cosby *et al.* (1978a,b) find the O_3^- photodissociation spectrum in the 640–508 nm region to have a number of broad peaks, which they assign to vibrational progressions of an excited electronic state, and they assign a bending and a stretching mode in both ground and excited states. They suggest that the broadness of the vibrational features indicates that the excited electronic state is directly dissociative along the asymmetric stretch coordinate, but bound along symmetric stretch and bending coordinates. If correct, this interesting interpretation of the O_3^- spectrum means that one should not assume that vibrational structure will be absent for direct dissociations in polyatomic molecules (which would be assumed following Herzberg, 1966).

Several groups have looked at CH_4^+ photodissociation (Riggin and Dunbar, 1975; Ensberg and Jefferts, 1975; McGilvery *et al.*, 1978b). The broad, weak, and structureless feature rising across the visible and near uv to (perhaps) a peak of cross section 3×10^{-19} cm^2 near 340 nm is attributed to transitions among the three Jahn–Teller split components of the threefold degenerate ion ground state (a $^2E \leftarrow {}^2B_2$ electronic transition.) The hoped-for peak near 443 nm, sought as an explanation of the diffuse interstellar line at that wavelength, and other hoped-for visible peaks are not observed.

Some of the most beautiful laser photodissociation spectra were obtained by McGilvery and Morrison (1977) on CH_3I^+ and CD_3I^+. Long series of vibrational progressions are observed corresponding to optical excitations $\tilde{X}^2E_{1/2} \rightarrow \tilde{A}^2E_{1/2}$ and $\tilde{X}^2E_{3/2} \rightarrow \tilde{A}^2E_{1/2}$. Assymmetry of the peaks suggests almost-resolved rotational structure.

B. *Large Ions*

With large ions it begins to make sense to think in terms of chromophoric groups, and the general spectral properties of classes of molecules. In fact, our point of view has quite recently changed from regarding the photodissociation spectroscopy of radical ions on a case-by-case basis to a conceptual grouping of classes of ions sharing a common chromophore, and to some understanding of the properties of several chromophores. So it is natural, if not according to the original chronology, to survey this field according to several important chromophores.

1. *Alkanes and Saturated Hydrocarbons*

Accustomed as one is to thinking of alkanes as colorless, it is a surprise on first sight to find the alkane radical cations to be strongly absorbing in the visible region (Dunbar, 1976a). In fact, the alkane ions from butane to octane all show very broad and more or less intense photodissociations (Benz and Dunbar, 1978). In butane, the cross section has its maximum in the uv, in pentane it has shifted to 420 nm and shifts progressively further to the red, with octane cation absorbing strongly across the whole visible and peaking at 640 nm. The bands are somewhat irregular and in appearance suggest a superposition of many absorbing transitions.

In part, the broadness and irregularity of the bands arises from the many configurations available to the flexible ion. The numerous low-energy transitions observed in the ions are understood readily: the neutral alkanes have a number of occupied MOs lying near the highest occupied MO, with a gap of 5 or 6 eV to the lowest empty MO (as in the left-hand side of Diagram

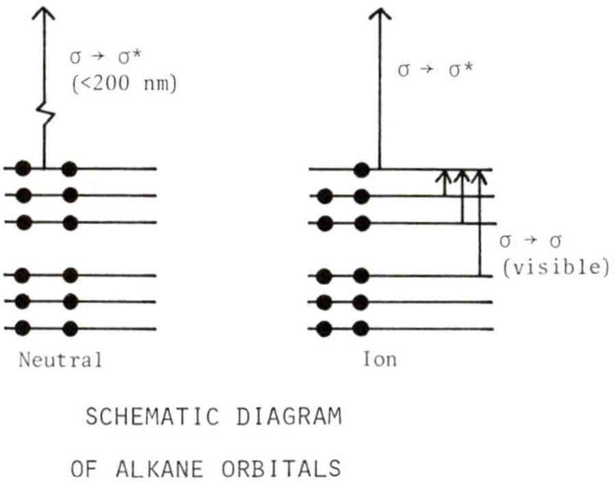

Diagram 1

1). The spectrum consists of high-energy $\sigma \rightarrow \sigma^*$ transitions. The radical ion, however, has an electron removed so that a number of low-lying $\sigma \rightarrow \sigma$ transitions can occur, many of them optically allowed. Alkane PES spectra confirm this picture, as do semiempirical (MINDO/3) calculations. In fact, by crudely choosing some representative geometrical configurations, and assuming that all the $\sigma \rightarrow \sigma$ transitions are equally allowed, the MINDO/3 calculations reproduce the general features of the observed ion photodissociation spectra in a reasonably quantitative way. A few spectra of cyclic alkanes have been taken, and fit into the patterns established for the linear alkanes.

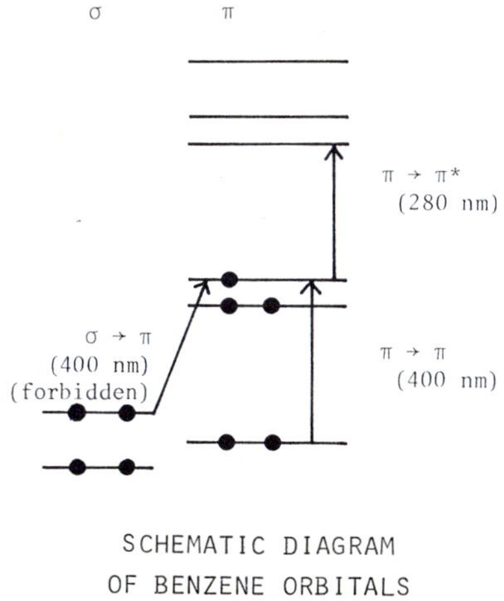

Diagram 2

2. Benzene Derivatives

Benzene ion and its simple derivatives have two spectral absorptions, which are readily understood as π electronic transitions within a Hückel picture (Digram 2). The $\pi \rightarrow \pi^*$ transition is analogous to the uv spectral transitions of the neutral and occurs near 270 nm. The $\pi \rightarrow \pi$ transition,

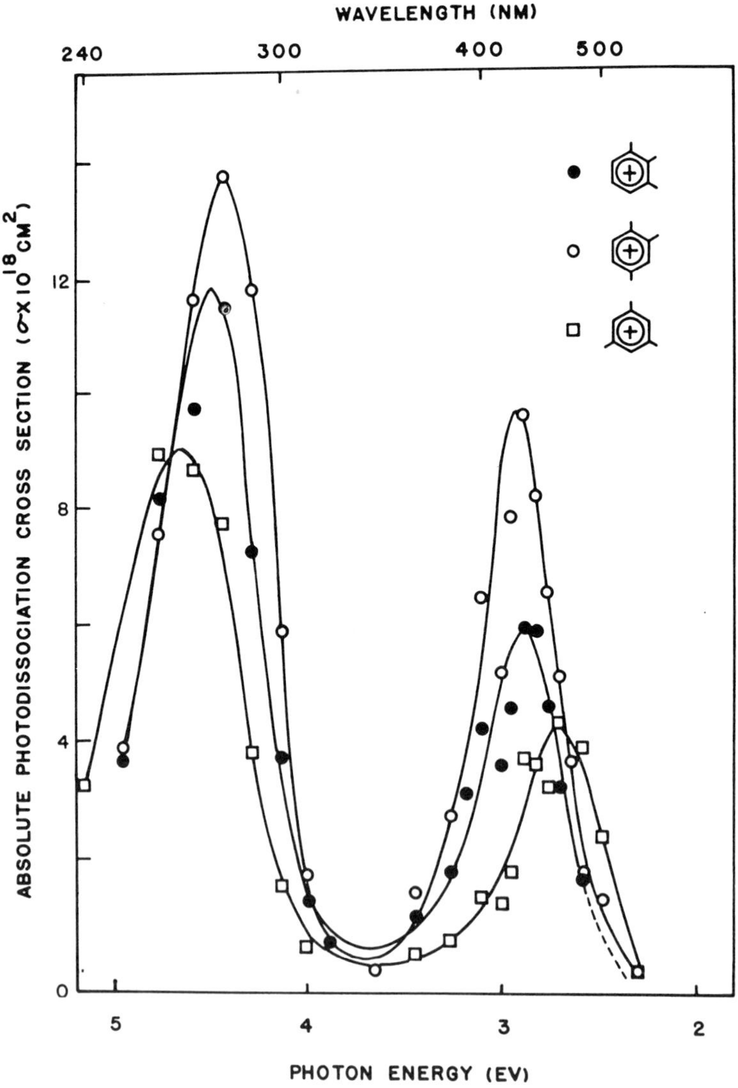

FIG. 2 Photodissociation spectra of the three isomeric trimethylbenzene ions.

which is characteristically ionic, occurs near 400 nm. The peak cross sections vary with substitution, but are both of the order of 5×10^{-18} cm^2. (The $\pi \leftarrow \pi$ transition in benzene ion itself is a two-photon photodissociation and has not been exactly located, but it is well understood by comparison with various substituted ions for which it is a one-photon process.) The benzene cation and the anion, which is predicted by the pairing theorem to have the same spectrum, have been the subjects of much calculation (e.g., Gardner, 1966), and the spectrum is well understood.

An elegant accomplishment of molecular spectroscopic theory has been the understanding, within a perturbation approach involving a small number of parameters, of the shifts in position and intensity of the $^1L_b(^1B_{2u})$ transition in neutral benzene derivatives as a function of the number and arrangement of substituents (Petruska, 1961). The peaks in the spectra of the methylbenzene cations show variations in position and intensity, suggesting that a similar perturbation treatment might be successful. Because both the $\pi \rightarrow \pi$ and $\pi \rightarrow \pi^*$ transitions are degenerate the perturbation treatment is rather involved, but has been carried out (Dunbar, 1978), and a comparison made between theory and experiment (Teng and Dunbar, 1978). Simple first-order perturbations by the methyl groups appear to give a satisfactory account of the observed substituent shifts in the $\pi \rightarrow \pi$ visible transition. It was surprising to find that this simple model is entirely unable to account for the $\pi \rightarrow \pi^*$ peak shifts, and a rather elaborate second-order perturbation model was needed to rationalize the experimental results within the perturbation-theory framework. Because both optical transitions in the cation are allowed, in contrast to the forbidden ^1L_b transition in the neutrals, the intensity perturbations are not expected to be as significant, but some useful correlations with theory were found nevertheless (Teng and Dunbar, 1978). Figure 2 shows representative spectra for three of the ions in this series. It has become clear in the course of this work that substituent effects in the benzene ion derivatives constitute an exceedingly complex question, and a good deal of experimental and theoretical work is clearly needed. The question of Jahn–Teller effects, for instance, has not been analyzed at all.

When substituents like halogen having their own spectroscopically signicant orbitals are put on benzene cation, the spectroscopy is more drastically altered. The halotoluene cations (Fu *et al.*, 1976) all have a strong photodissociation peak in the visible region, progressing smoothly from 410 nm in fluorotoluene to 580 nm in iodotoluene. Comparison with the PES spectra of the neutral molecules is particularly useful for this set of $\pi \rightarrow \pi$ peaks (Dymerski *et al.*, 1974). Figure 3 shows the excellent correspondence between the visible photodissociation peak and the higher of the two sharp PES peaks between 2 and 3 eV binding energy. These PES peaks

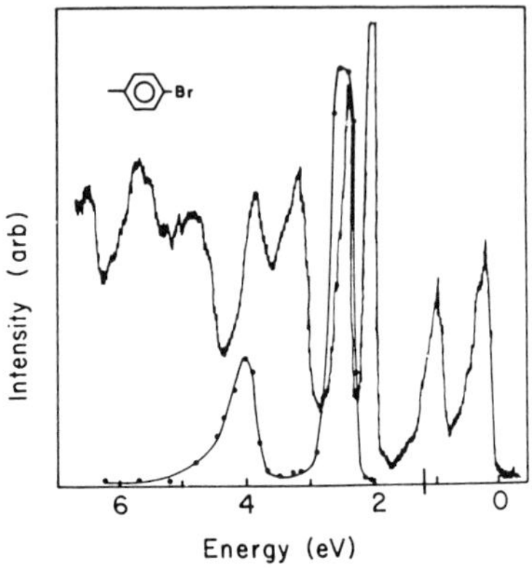

FIG. 3 The photodissociation spectrum (—●—) of *p*-bromotoluene ion and the photoelectron spectrum (———) of the neutral compound.

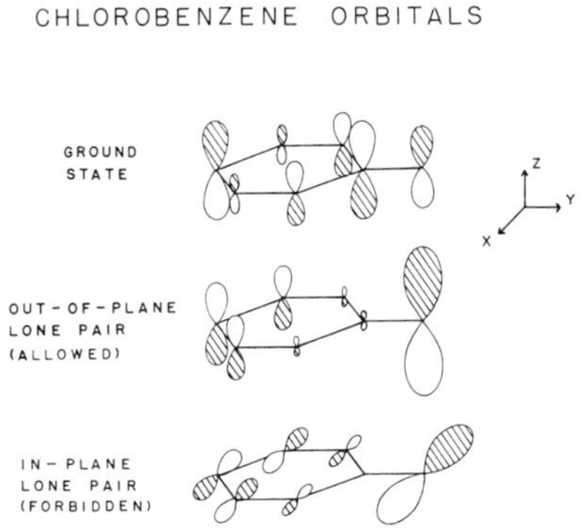

Diagram 3

are in the in-plane and out-of-plane lone-pair bromine orbitals, the out-of-plane orbitals being shifted to higher binding energy (lower orbital energy) by interaction with the benzene $E_{2g}\pi$ orbital. Diagram 3 shows the singly occupied orbital for the three relevant states, the ground state ($^2B_1$), the in-plane n state ($^2B_2$), and the out-of-plane n state ($^2B_1$) taking a MINDO/3 calculation on chlorobenzene as a typical model for this class of compounds. It is obvious by inspection, as well as from symmetry considerations, that the in-plane n state $^2B_2$ is optically forbidden from the ground state, while the out-of-plane n state is allowed in z polarization. The visible peak in bromotoluene thus has the character of charge transfer, with charge moving from the bromine to the ring. The situation is similar for chlorotoluene and iodotoluene, but not so clear for fluorotoluene ions. Similar considerations

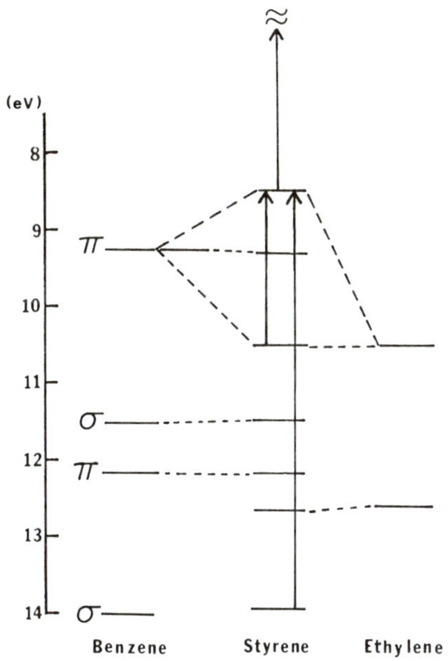

Diagram 4

also apply to the halobenzene ions, but work on them has been slower because the visible peaks are two-photon transitions.

The styrene ion (Fu and Dunbar, 1978) illustrates the interaction of a π system with the benzene ring. As Diagram 4 suggests, the two highest occupied ethylene orbitals form a three-orbital interaction problem. The resulting picture predicts $\pi \to \pi$ transitions at 2 and 3 eV, the first having a good deal of ethylene-to-ring charge transfer character, the second being more a ring-to-ring excitation. The effect of the ethylene substituent on the benzene ion $\pi \to \pi^*$ peak is less well understood, but it believed to result in small red shift so that the 3 eV $\pi \to \pi$ transition and the $\pi \to \pi^*$ transition near 4 eV partially overlap. The photodissociation spectrum of several substituted styrene ions is in excellent accord with this picture, with a visible peak near 600 nm and an unsymmetrical or split uV peak in the 300–400 nm region.

3. *Alkenes and Polyenes*

The alkene cations share a consistent, if unspectactular, spectroscopy: they have in common an intense, very broad $\pi \to \pi^*$ transition which, if it peaks at all in the useful range of the Hg/Xe arc source, does so near 38,000–40,000 cm^{-1} (Dunbar, 1976a). Some spectra, particularly for the butene ions (Kramer and Dunbar, 1973), appear as well to have a weak visible absorption which, if it in fact does not represent extreme broadening of the $\pi \rightarrow \pi^*$ peak, must be a $\pi \rightarrow \sigma$ transition and is in any case a bit surprising.

The polyene ions are more interesting. Not only are there strong, convenient spectral peaks (Dunbar, 1976a), but also the hexatriene ion was found to have prominant resolved vibrational structure in its laser visible absorption (Dunbar, 1976b), and other conjugated polyene ions will be interesting to examine at high resolution. Hexatriene ions have two prominent vibrational progressions in the visible peak, corresponding to a skeletal bend at 350 cm^{-1} and a skeletal stretch at 1200 cm^{-1}. The progression of peaks for the 350 cm^{-1} mode shows a hot band to the red of the O–O peak, and this gives the interesting possibility of using the hot band intensity as a measure of the ion vibrational temperature. In this case, comparison of the intensities of the first few peaks in the spectrum leads to an estimate of 500 ± 200 K for the temperature of $C_6H_8^{+\cdot}$ (Dunbar and Teng, 1978), corresponding to an excess energy of 6 ± 6 kcal in the ion. This approach seems likely to give useful new insights into the difficult question of ion internal excitation.

There is a satisfying regularity to conjugated diene ion spectra. The butadiene ion has peaks at 33,000 and 19,000 cm^{-1} (both about 4×10^{-18} cm^2) (Dunbar, 1975d). With substitution, a trend is seen toward a red shift and intensity increase in the uv peak, and a blue shift in the visible peak, so that the 1,1,4,4-tetramethyl butadiene ion has peaks at 28,000 cm^{-1} (5×10^{-17} cm^2) and 21,000 cm^{-1} (5×10^{-18} cm^2) (Dunbar, 1976a). This pair of peaks is well understood from MO calculations, the visible peak being $\pi \rightarrow \pi$, and uv peak $\pi \rightarrow \pi^*$ in character.

Of trienes, only the hexatriene ion has been studied (Dunbar and Teng, 1978). The visible transition is to the red of the diene peak, having its band maximum near 16,000 cm^{-1} (5×10^{-18} cm^2). The first $\pi \rightarrow \pi^*$ transition is also red shifted relative to dienes, at 28,000 cm^{-1} (5×10^{-17} cm^2), and there is also a suggestion of an expected $\pi \rightarrow \pi^*$ peak near 40,000 cm^{-1} ($\gtrsim 10^{-17}$ cm^2). Again, MO calculations on the system account well for this spectrum, although for triene and higher polyene cations the substantial number of close-lying excited states makes a substantial configuration–interaction calculation mandatory (Zahradnik and Carsky, 1970).

Spectroscopy of several other unsaturated hydrocarbon parent ions has received some attention, including cyclooctatetraene (Fu and Dunbar, 1978), barrelene (Dunbar *et al.*, 1977), cycloheptatriene (Dunbar and Fu, 1973), and norbornadiene (Dunbar and Fu, 1973). With these thermodynamically unstable molecules, the structure of the parent ion is the first concern, with the photodissociation spectrum serving largely to fingerprint various structures. Consideration of the significance of the spectra must at this point be rather speculative.

4. *Organometallics*

Undoubtedly ions of transition-metal complexes will prove to be an area of exceptional interest for photodissociation spectroscopy, although the unwieldy physical properties of most such compounds have so far deterred most investigators. Photodisappearance spectra (it was not reliably established for most ions whether dissociation or electron photodetachment was the initial process) were reported for a series of transition-metal carbonyl anions $M(CO)_n^-$ (Dunbar and Hutchinson, 1974). One of the ions, $Ni(CO)_3^-$, was suggested to be anomalous because the highest metal electron is 4s rather than 3d, but for the other nine examples, two trends were clear in the position of the photodisappearance onset wavelength: the photodisappearance onset energy moved higher (1) with increasing atomic number of the transition metal, and (2) with increasing number of carbonyls n. These trends were suggested to reflect the increasing stabilization of the d- orbital occupied by the relevant electron in its initial state by either increasing nuclear charge or increasing complexation. The results suggested that one additional nuclear proton, or one additional CO ligand, stabilizes the highest occupied metal d level by perhaps half an electron volt.

Freiser has reported studies of photodissociation of lithiated ferrocene (Freiser and Beauchamp, 1978) and $(C_5H_5)NiNO^+$ (Burnier and Freiser, 1978), and increasing interest in the transition-metal cations and anions can be expected.

5. *Protonated Molecules*

Freiser (and to a small extent our laboratory) has looked extensively at the photodissociation spectroscopy of protonated aromatics (and the very similar Li^+-bound aromatics) mostly from the point of view of the excited-state thermochemistry discussed below (Freiser and Beauchamp, 1976a, 1976b, 1977; Freiser *et al.*, 1976; Dunbar *et al.*, 1977). The spectra can usually be directly compared with the solution spectrum and, aside from small (potentially interesting) solvent shifts, match very closely. The most significant point about the spectroscopy is the strong similarity to the spectrum of the unprotonated neutral molecule—this is not at all surprising for the substituent-protonated molecules where the proton (or Li^+) functions effectively as a charge label, but for those molecules like mesitylene which ring protonate, the similarity in spectral pattern (aside from the substantial red shifts discussed below) is somewhat unexpected.

V. ION STRUCTURES

Information about ion structures constitutes one of the most fruitful and important aspects of photodissociation studies. The preceding section on spectroscopy has implicitly assumed many structures in discussing the spectra—we will be concerned here with the opposite face of the coin, considering the various interesting cases where the ion structure is the least certain variable and the chief goal of the investigation.

An early and particularly graphic example of ion structure determination was a comparison of the parent $C_7H_8^+$ ions obtained from ionizing toluene, cycloheptatriene, and norbornadiene (Dunbar and Fu, 1973). The parent ions had been thought to share a common structure, based on their similar fragmentation behavior, but the three photodissociation spectra were clearly different, and proved that little or no interconversion of structures occurred on a time scale of seconds. As we have seen above, the spectrum of the parent ion from toluene shows that the neutral structure is definitely retained in this case; the ion from cycloheptatriene gave a single peak near 500 nm, in excellent accord with the first $\pi \rightarrow \pi$ transition predicted on the basis of the PES spectrum of the neutral, and it seems likely that this structure also is retained upon ionization. By contrast, the ion from norbornadiene gave a spectrum quite unlike that suggested by the PES spectrum, and the presumption is that rearrangement occurs to an as yet unassigned structure.

More recently, an interesting addition to the $C_7H_8^+$ family of structures was investigated (Dunbar and Klein, 1977). It is accepted that the parent ions of *n*-butylbenzene (**I**) and 2-phenylethanol (**II**) can undergo the McLafferty-rearrangement process (1).

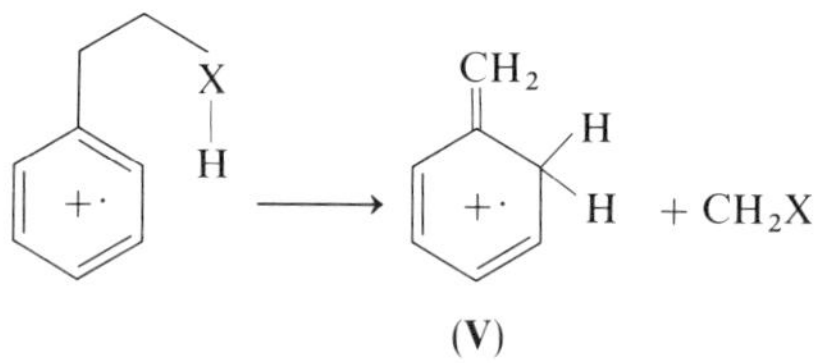

(1)

(**V**)

$X = CH - CH_3$ (**I**)
$X = O$ (**II**)

Whether the methylenecyclohexadiene cation (**V**) has more than transitory stability was, however, quite uncertain. Collisional activation results (Levsen *et al.*, 1973) suggested but did not prove conversion within microseconds to cycloheptatriene (**III**) ions, while ion–molecule reaction studies (Bursey *et al.*, 1971) had shown only that conversion to toluene ions (**IV**) did not take place within 10^{-2} sec. The photodissociation spectrum of the product ion (**V**) was quite definitive: the spectrum of the ion was very different from those of (**III**) or (**IV**) ions, and it was possible to rule out conversion to either of these alternative structures to an extent of more than 25% on a time scale of seconds. This experiment involved a particularly successful application of the time-resolved technique: because of the low photodissociation cross section of (**III**) ions, the ordinary photodissociation spectrum of (**V**) did not do an effective job of ruling out a substantial presence of (**III**) ions in the (**V**) population. However, time-resolved exhaustive photodissociation of the (**V**) population at 405 nm showed that more than 75% of the ions were uniformly dissociated at this wavelength, and provided the most definitive proof that the rearrangement (**V**) → (**III**) was not important.

The $C_7H_8^+$ studies described show that at least four different structures of this ion are stable for seconds, even though the electron impact ionization used is not particularly gentle, and is likely to deposit several kilocalories per mole of extra energy in the ions. It may seem that this conclusion is at variance with the mass spectrometric evidence that the fragmentation pattern of $C_7H_8^+$ is virtually independent of its source. There is no real contradiction here: the mass spectrometric studies necessarily select out of the total $C_7H_8^+$ population *only* those ions having sufficient internal energy to fragment, and show that for ions this energetic the potential surface is widely sampled. The photodissociation experiment, however, selects *only* those ions which cannot dissociate spontaneously, and demonstrates that these ions are not structurally labile. This important point is illustrated with a hypothetical construction of the $C_7H_8^+$ potential surface in Diagram 5. (The heats of formation of known species are shown on the diagram, but the barrier

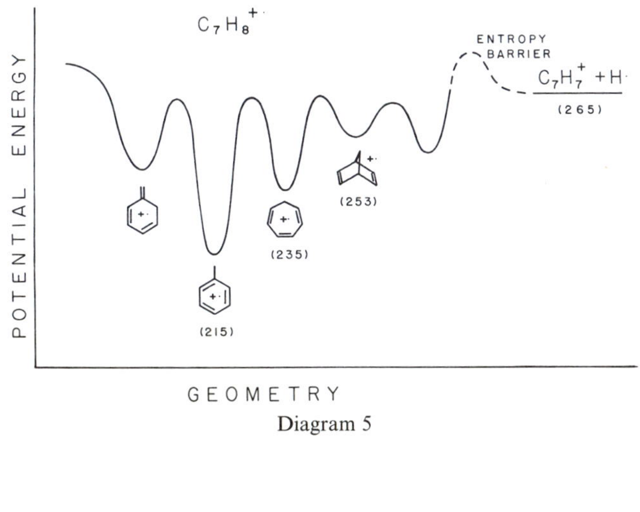

Diagram 5

heights are unknown, and it can only be inferred that they are rather high from the lack of rearrangements of isomeric $C_7H_8^+$ ions.)

It is clear that generation of a nondissociating ion on this surface will not usually give facile rearrangement, but an ion generated with an energy slightly above E_{diss} will be likely to traverse the potential surface widely before overcoming the entropy barrier to dissociation. This is true whether the ion is put into the dissociative energy state by ionization of the neutral, or by photon impact on the nondissociative ion, so that the extensive isotope scrambling prior to dissociation in both electron impact (Howe and McLafferty, 1972) and photodissociation studies (Dunbar, 1973b) is reasonable. So also is the reduced scrambling observed in very highly energetic ions (Howe and McLafferty, 1972), where the rate of dissociation becomes comparable to the rate of rearrangement in the excited ions. It is worth noting that the nondissociating norbornadiene ions can, from the heats of formation in the diagram, lie not more than 12 kcal below their dissociation threshold. Presuming the barriers to rearrangement to be lower than the dissociation energy, rearrangement can occur with $\lesssim$ 12 kcal activation energy, and we would expect this ion to rearrange rapidly, as is suggested by its photodissociation spectrum.

Quite a lot of recent work has been directed at unsaturated hydrocarbons. The reliable relationships between the chromophoric groups in the ion and its spectrum (Dunbar, 1976a) are beginning to be useful in a structure–analysis sense. For instance, we are certain that the 1,7-octadiene ion does not rearrange into conjugation, because its spectrum (showing a very broad peak above 40,000 cm^{-1}) is entirely characteristic of isolated double bonds, but entirely uncharacteristic of a conjugated diene ion. In the same way, 1,4-cyclohexadiene certainly does rearrange to the conjugated 1,3-cyclohexadiene structure (Hays and Dunbar, 1978), because the parent ions from both of these neutrals give spectra which are identical in peak positions and heights, and which are entirely characteristic of the conjugated diene chromophore, showing a peak of moderate intensity at 500 nm, and an intense peak at 360 nm.

Apparently a ring-opening rearrangement such as

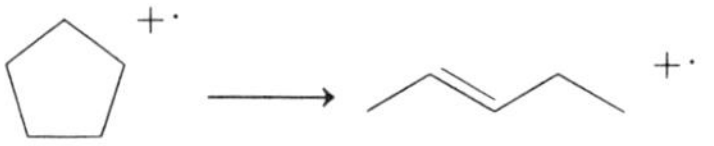

(VI) (VII) (2)

occurs in cyclopentane upon electron impact ionization, as the spectrum of the ion obtained from cyclopentane bears no resemblance to those of other alkane and cycloalkane ions, but instead looks characteristic of a pentene structure [(VII) or one of its isomers] (Benz and Dunbar, 1978). However, another likely ring opening, the rearrangement of 1,3-cyclohexadiene, does

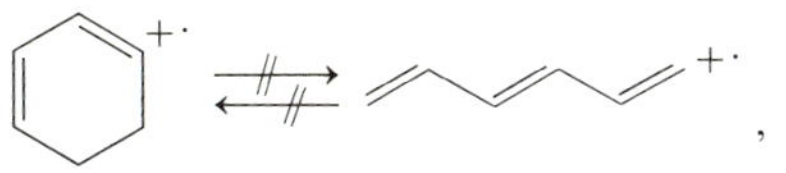

(3)

not occur. The spectra of cyclohexadiene and hexatriene ions are entirely distinct, and rearrangement in either direction can be ruled out to a very high level of precision (Hays and Dunbar, 1978). The rearrangements of reaction (3) are thermally unfavorable on orbital symmetry conservation grounds, so the negligible occurrence of either of these reactions is reasonable. However, reaction (3) is photochemically allowed. In solution, reaction (3) is driven to the right-hand side photochemically by irradiation of 1,3-cyclohexadiene cation in its absorption band near 500 nm. This process was searched for in the gas phase (Hays and Dunbar, 1978), using simultaneous laser irradiation at 514.5 and 600 nm, but no evidence for photochemical formation of stable hexatriene was found: apparently under the collisionless conditions of the gas-phase experiment, any hexatriene ions which are formed photochemically are still sufficiently energetic to dissociate, and are not observed.

While photodissociation spectra of quite a few even-electron cations are now known, not much has been attempted in exploring their gas-phase rearrangements. However, two studies of protonated isomeric hydrocarbons may be mentioned:

Hexamethyldewarbenzene (**VIII**) does not yield a measurable

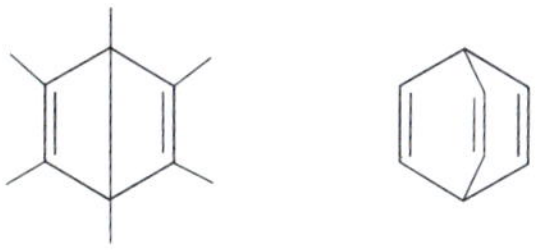

(**VIII**) (**IX**)

parent ion signal under our trapped-ion conditions, but it gives abundant $(M + 1)^+$ ions. (Dunbar *et al.*, 1977). The photodissociation spectrum of this ion was found to be identical with that of protonated hexamethylbenzene, and apparently the rearrangement of (**VIII**) to the more stable benzene structure is rapid and complete. Similarly, the spectra of protonated cyclooctatetraene and protonated barrelene (**IX**) (protonated by H_3O^+) are identical with the spectrum of protonated styrene (Dunbar *et al.*, 1979), indicating rapid rearrangement to the latter structure, which is undoubtedly the most stable $C_8H_9{}^+$ ion.

On the other hand, lest one begin to think that such rearrangements are universal among even-electron cations, the $C_7H_7{}^+$ ion has at least two distinct structures. This was clearly shown to be the case for $C_7H_7{}^+$ formed by photodissociation of toluene ion (Dunbar, 1975c):

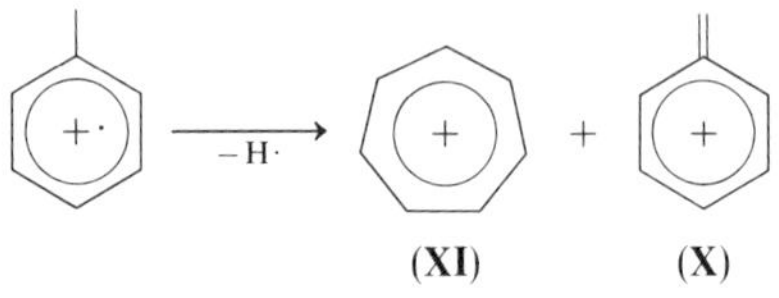

(4)

Part of the ions formed react rapidly with toluene and are assigned as benzyl ions (**X**), while the remainder are not at all reactive and are probably tropylium (**XI**). The abundance ratio of the two structures was found to be a function of the energy deposited in the precursor toluene ions, and the wavelength variation of this ratio was found to be in quantitative accord with a quasi-equilibrium calculation. The presence of both benzyl and tropylium cations among the fragmentation products of toluene ions is supported by collisional activation results (Winkler and McLafferty, 1973) and has received a further convincing confirmation in the recent report by McCrery and Freiser (1978) of the photodissociation spectrum of $C_7H_7^+$ formed by electron impact on toluene: the observed peak at 290 nm is both broader than and red shifted from that of tropylium ion in solution, and is probably attributable in part at least to benzyl ion photodissociation. Still more recently, Freiser *et al.* (1978) have observed two-photon dissociation of $C_7H_7^+$ ions at visible wavelengths. Since the tropylium ion surely has no optical absorption in the visible, this certainly reflects the presence of benzyl ions. In summary, the existence of the stable gas-phase isomers (**X**) and (**XI**) is beyond doubt, the relative abundances depending on the precursor neutral and on its internal energy.

There was some early hope that photodissociation spectroscopy would be a reliable means of distinguishing isomeric disubstituted benzene ions. The spectra of three isomeric xylene cations (Dunbar, 1973a), and the spectra of the various isomeric chloro-, bromo-, and iodotoluene ions (Fu *et al.*, 1976) do show small but discernible differences. It would be possible to identify reliably a pure sample of one isomer of these ions, but probably not to deal with a mixture of isomers. Higher spectral resolution seems unlikely to improve this situation, since the high-resolution laser photodissociation spectra which have been reported for toluene (Eyler, 1976) and iodotoluene (Fu *et al.*, 1976) show no resolved spectral structure which could help in the structure-determination problem.

The driving force for movement of double bonds into conjugation can be expected to be large when the conjugated system is extensive, as in aromatics. The styrene aromatic system provides a convenient model system for such rearrangements, since the spectrum of styrene cations is quite different from that of alkylbenzene cations. The spectrum of (**XII**) was found to be identical with **(XIII)**, which is in turn similar to that of the styrene ion **(XIV)**, indicating that, as expected, the 1,2 shift of the double bond into conjugation is complete (Fu and Dunbar, 1978).

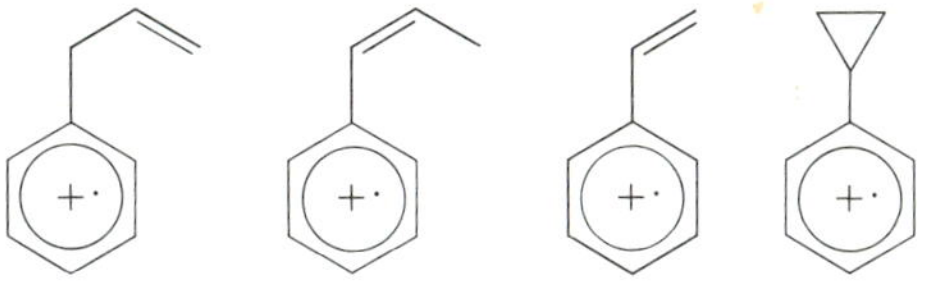

(XII) (XIII) (XIV) (XV)

However, **(XV)** provided a small surprise, showing a spectrum entirely unlike that of **(XIII)** and indicating that in spite of the driving force of ~10 kcal/mole, no rearrangement of cyclopropyl benzene cation takes place.

The series **(XVI)**–**(XVIII)** (Fu and Dunbar, 1978) provides a further example of double-bond shifting propensity:

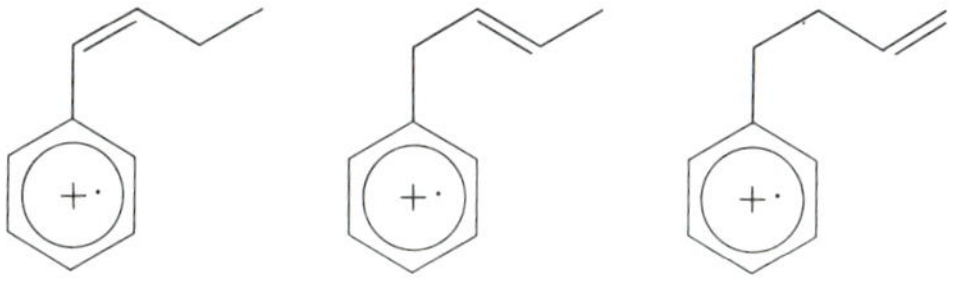

(XVI) (XVII) (XVIII)

The spectrum of **(XVII)** appeared to be a superposition of spectra corresponding to a mixed ion population: the characteristic peaks of the styrenelike structure were present, and also another peak attributable to unrearranged **(XVII)**, suggesting partial but not complete double-bond migration. The spectrum of **(XVIII)** is entirely distinct from **(XVI)**, looking quite similar to the butylbenzene ion spectrum, and apparently the 1,3 shift of the double bond does not occur.

In the course of these studies of substituted styrene ions, we also examined two other $C_8H_8^+$ isomers, the parent ions of cyclooctatetraene and barrelene (Fu and Dunbar, 1978). Each gave an entirely unique spectrum and suggested that on the $C_8H_8^+$ potential surface, just as on the $C_7H_8^+$ surface, the various local minima corresponding to stable neutral structures are separated by substantial energy barriers.

The photodissociation of the benzyl chloride parent ion was a source of some confusion when it was first obtained (Fu *et al.*, 1976), since it corresponded poorly to the PES spectrum of the neutral, but it was soon realized that it represented a superposition of two spectra, and it was to clarify this situation that the time-resolved technique was devised. Time-resolved curves clearly showed that about 40% of the population had a structure, presumably unrearranged parent ion, giving a peak near 600 nm, while the remaining 60%

of the population had rearranged to another structure spectroscopically similar to chlorotoluene ions. It appeared that the benzyl fluoride ion undergoes a similar partial rearrangement.

For several of the atmospheric ions it has been thought that two structures may exist, a normal form and a peroxy form, as, for example, the three cases in Diagram 6:

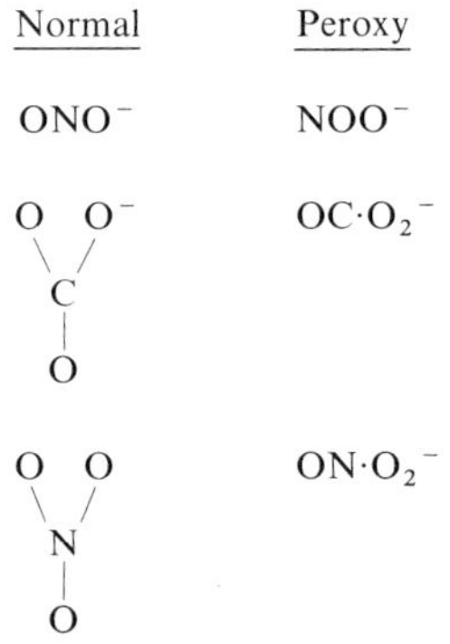

It had been thought that photodetachment threshold measurements provided evidence for both forms of NO_2^- (Herbst *et al.*, 1974; Richardson *et al.*, 1974a,b; Pearson *et al.*, 1974), but a recent reexamination of this ion by the SRI group (Huber *et al.*, 1977a,b) suggests that a single ion structure with varying degrees of vibrational excitation is more likely. If this is true, the SRI data indicate cross sections for vibrational quenching of NO_2^- by O_2 and CO_2 of the order of 10^{-12} cm^3/sec, so that the vibrational excitation can survive hundreds or thousands of ion–neutral collisions. An earlier suggestion (Burt, 1973) of the peroxy CO_3^- structure has also not stood up, and the photodissociation spectrum is consistent with a single, normal structure (Moseley *et al.*, 1974, 1976a).

In the case of NO_3^-, the existence of the peroxy form is known from ion–molecule reaction studies. The NO_3^- ion produced in an O_2/N_2O mixture photodissociates to $O_2^- + NO$ in the visible region from 600 nm to at least 510 nm, and seems likely to have the peroxy structure, since it dissociates at wavelengths below 2 eV, while normal NO_3^- should have a threshold ≥ 4.3 eV (Cosby *et al.*, 1977).

Laser photodissociation spectra have by now resolved vibrational structure in the spectra of quite a few ions, and it has been clear that such vibrational structure can form the basis for highly secure, unambiguous structural assignments. In one instance, at least, this promise has been fulfilled (Orth and Dunbar, 1978b): the laser photodissociation spectra of four chloropropene cation isomers are markedly different, (although of these only allyl chloride ion would be readily distinguishable on the basis of low-resolution spectroscopy) and all (except allyl chloride) show resolved vibrational progressions consisting of several peaks. One point of interest is the clear difference in the spectra of the *cis*- and *trans*-isomers (**XIX**) and (**XX**)

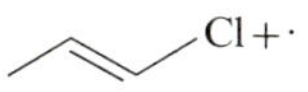

(**XIX**)

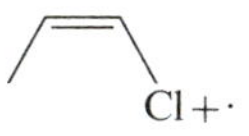

(**XX**)

In general the question of the cis–trans barrier in olefin radical cations is open; spectra of *cis*- and *trans*-2-butene ions are identical, albeit not very distinctive (Kramer and Dunbar, 1973), suggesting (weakly) that isomerization occurs in that case. However, the chloropropene evidence for retention of a cis–trans configuration is unambiguous.

Ions of the chloropropene composition are formed in the ion molecule reactions (Orth and Dunbar, 1978b):

(**XXI**) (5)

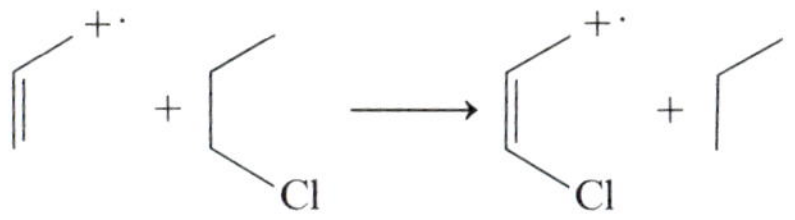

(**XXII**) (6)

The photodissociation spectrum of (**XXI**), the product of reaction of 2-chloropropane, was identical to the spectrum of the 2-chloropropene ion, having the same vibrational structure. The spectrum of (**XXII**), the product of reaction with 1-chloropropane, was not the same as any of the authentic 1-chloropropene isomers; however, the reaction might be expected to yield a mixture of *cis* and *trans*-isomers. This is evidently the case, and a 1:1 superposition of the spectra of authentic (**XXI**) and (**XXII**) gave a composite spectrum identical with the observed spectrum of product ion (**XXII**). These results clearly show that the skeleton of the reactant chloropropane is passed to the product ion in reactions (5) and (6), with the geometry in reaction (6) being trapped in approximately equal cis and trans configurations. There is precedent for olefin ions reacting by formal H_2^- abstraction (Henis, 1970; Beauchamp *et al.*, 1972): the results here, however, show not only that this is the mechanism, but also that the movement of hydrogen proceeds without disturbance of the heavy-atom skeleton of the neutral, and illustrate the potential utility of spectroscopic methods for analyzing products of ion–molecule reactions.

VI. UNIMOLECULAR DECOMPOSITION CHEMISTRY

To break the perilous alliance . . ."

(Chaucer, Complaint to Pity 83).

The kinetics, products, and other features of decomposition following deposition of energy in an ion have been a chief concern of mass spectrometrists since the beginning. Energy deposition via electron impact ionization, field ionization, or photoionization of the neutral, and via impact of a nonreactive projectile (neutral molecule or electron) on the ion, are processes that have been explored extensively from many points of view. Photodissociation provides an approach complementary to all of these in interesting ways, giving as it does the capability for taking a mass selected ion, suitably prepared with regard to neutral precursor, state of internal excitation, orientation, etc., depositing in it a photon of precisely known energy at a known time, and observing the details of the ensuing unimolecular decay as a function of time. This ideal experiment is far from full realization, but some useful insights into unimolecular decay have already been possible.

In a polyatomic ion with numerous electronic potential surfaces and vibrations, a number of different types of photodissociation can be anticipated, as Fig. 4 attempts to illustrate. The simplest type (d) is direct photoexcitation to a repulsive surface, which will lead to dissociation in $\leq 10^{-13}$ sec

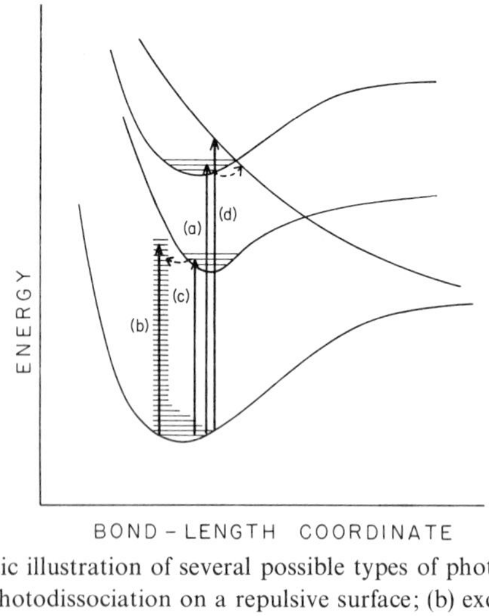

FIG. 4 A schematic illustration of several possible types of photodissociation of a polyatomic ion; (a) direct photodissociation on a repulsive surface; (b) excitation to a dissociative vibrational level of the ground electronic state; (c) vibrational predissociation; and (d) electronic predissociation. The attempt to represent a polyatomic, with its many vibrational modes, on a one-dimensional potential surface leads to the odd-looking necessity of drawing ground-state vibrational levels above the dissociation limit, which would of course not make sense for a diatomic.

along the reaction coordinate with respect to which the potential surface is repulsive. This is the situation for H_2^+ (von Busch and Dunn, 1972), for Ar_2^+, Kr_2^+, Xe_2^+ (Abouaf *et al.*, 1978; Moseley, 1977), Cl_2^- (Sullivan *et al.*, 1978; Asubiojo *et al.*, 1978), and probably for CH_3Cl^+ (Orth and Dunbar, 1978a).

CH_3Cl^+ is interesting because the exclusively observed dissociation process (7) is energetically less favorable than dissociation to $CH_2Cl^+ + H$,

$$CH_3Cl^+ \rightarrow CH_3^+ + Cl. \tag{7}$$

However, as suggested by the angular dependence of the photodissociation, the optical excitation is probably to the $\tilde{B}^2E_{3/2}$ electronic surface which is C—Cl antibonding but C—H bonding. Thus the direct dissociation (7) with large kinetic energy release predominates over the less endothermic loss of H which would require a predissociative mechanism.

The process denoted (b) in Fig. 4 is the direct excitation to a high predissociating vibrational level of the ground state, and is in most cases unfavorable because of the extreme violation of vibrational selection rules. However, these same levels can be reached as in process (c) in two steps by photoexcitation to a bound excited level, followed by internal conversion (dashed arrow) to the vibrationally hot ground state. It appears as if this internal conversion or vibrational predissociation mechanism dominates the dissociation of the bulk of the larger polyatomic ions so far studied. The dissociation of vibrationally hot ion ground states is a subject which has received overwhelming attention from mass spectroscopists; it is thought that this situation is usually well described by the energy randomization hypothesis forming the basis of the quasiequilibrium (QET) theory of unimolecular decay (see Robinson and Holbrook, 1972, for discussion). To the extent that QET theory describes a given photodissociation, one expects the following characteristics: the least endothermic sets of dissociation products will be strongly favored; the dissociation will be much slower than the 10^{-14} sec characteristic time of vibrational motion, will have its rate strongly dependent on photon energy, and will show little kinetic energy release and extensive rotational averaging of its angular dependence. The breakdown patterns for this type of dissociation should be similar to those for electron-impact or photoionization mass spectra at appropriate ionizing energies, and should be calculable by RRKM (Rice-Ramsperger-Kassel-Marcus) theory.

The most careful study of this type of photodissociation was a study in our laboratory (Riggin *et al.*, 1976; Kramer and Dunbar, 1973) of the dissociation of $C_4H_8^+$ ions by the competing pathways.

$$C_4H_8^+ \xrightarrow{h\nu} \begin{cases} C_4H_7^+ + H \\ C_3H_5^+ + CH_3^{\cdot} \end{cases} \tag{8}$$

The two paths differ by only 0.05 eV in endothermicity and compete over the entire range of wavelengths used, from 500 to 300 nm. This energy range corresponds to a range from 0.5 to 2.1 eV above the dissociation energy, a range sufficiently broad to have a significant effect on the competition between the pathways. The wavelength variation of the ratio of the two possible products, Fig. 5, was found to be fully in accordance with QET predictions, with the more energetically favorable $C_4H_7^+$ product being increasingly abundant at the longest wavelengths, and the ratio leveling off at shorter wavelengths. Acceptable quantitative agreement was found with RRKM calculations, but in view of the highly adjustable nature of the theory this was not considered very meaningful. What does seem important is the fact that the competition (8) is apparently quantitatively consistent with the QET statistical hypothesis, which supports the presumption that this dissociation does proceed through a vibrationally hot ground state. Further support for a QET mechanism for this dissociation comes from the quantitative agreement with the RRKM prediction of the extent of the further decomposition of the $C_3H_5^+$ ion,

$$C_4H_8^+ \xrightarrow[-CH_3]{h\nu} C_3H_5^{+*} \rightarrow C_3H_3^+ + H_2, \tag{9}$$

at high photon energies.

There is somewhat less direct evidence for dissociation via the hot ground electronic state for some other species: Toluene ion labeled with deuterium on either ring or methyl positions shows complete scrambling of ring and methyl hydrogens prior to dissociation, with H/D loss ratios being statistical

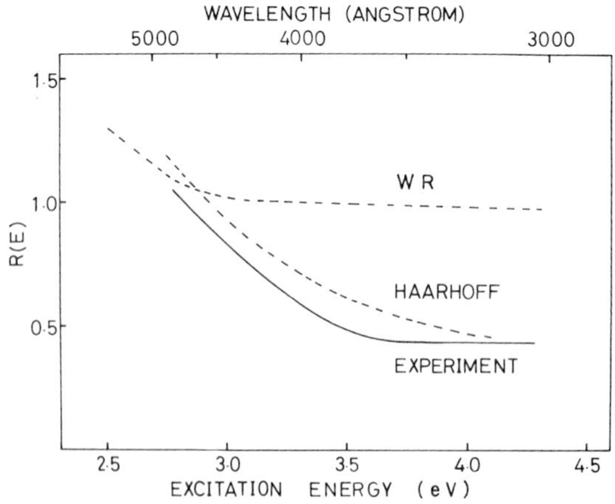

FIG. 5 The experimental ratio of the cross sections for production of $C_4H_7^+$ and $C_3H_5^+$ from 2-$C_4H_8^+$ (solid curve) is shown as a function of excitation energy and compared with two different models (dashed curves) of the dissociative process.

(aside from the isotope effect) (Dunbar, 1973b). The extensive molecular rearrangement implied by this scrambling is clearly inconsistent with a direct dissociation mechanism. The H/D isotope effect favors loss of H, and this isotope effect shows a strong wavelength dependence which is in quantitative agreement with RRKM predictions for the ground electronic state. Moreover, the initial electronic excitation process in the toluene ion near 400 nm is certainly to a π excited electronic state which involves little weakening of σ bonding, so that internal conversion to the ground state (or conceivably to the lowest σ excited state near 3 eV) is undoubtedly a prerequisite to dissociation.

There are a number of other known ions which photodissociate through more or less extensive rearrangement processes clearly ruling out direct dissociation on a repulsive potential surface: among these may be mentioned loss of CH_3 from $C_4H_6{}^+$ (1,3-butadiene ion) (Dunbar, 1975d).

The final, and least direct, suggestion that QET dissociation on the ground-state surface is very common is the general observation in a large number of known cases that the energetically most favored dissociation product is the only one observed. [Freiser and Beauchamp (1978) have stated this differently, to the effect that the product is the same as the most abundant metastable product in the mass spectrum; but in most instances these two rules give the same prediction.] This is in accord with the QET expectation; while if direct photodissociation, or dissociation from excited electronic states, were the rule, one would anticipate the common occurence of energetically less favored products.

The observation of a product ion in much more than RRKM abundance probably indicates dissociation involving an excited electronic state. (CH_3Cl^+ is a striking illustration.) In addition to the direct process (d) in Fig. 4, this can occur as in process (a) with excitation into a bound excited state which is predissociated by a nearby dissociative state. (This probably occurs in $O_2{}^+$ (Moseley *et al.*, 1976b).) While the various processes in Fig. 4 (and others not discussed) are hard to distinguish experimentally, it can probably be assumed that process (d) will not give resolved vibrational structure in the spectrum (although Moseley *et al.* (1977a,b) have raised doubt about this as well).

In a recent study having implications for the photochemistry of large polyatomic ions (Gooden and Brauman, 1977), it was reported that the butyrophenone ion photodissociates to two different sets of products, depending on the wavelength, and thus implying dissociation from two different electronic states of the ion. However, subsequent work in our laboratory (Kim and Dunbar, 1978) indicates a more complex photochemical situation than was realized, and suggests no reason to question the applicability of QET on the ground electronic surface to this fragmentation.

VII. PHOTOFRAGMENTATION DYNAMICS

The observation of the product energies and angular distributions in a neutral-molecule photodissociation is a fairly heroic undertaking. By contrast, the ease of detecting single ions makes the same experiment relatively straightforward in ion photodissociation.

Velocity and angular distribution data for dissociation products have somewhat different significance, depending on whether the dissociation is slow (vibrational or electronic predissociation) or fast (direct dissociation, or possibly very rapid electronic predissociation).

Direct dissociation is fast compared to ion rotation, so that the product ion emerges from the photoexcited primary ion with a direction determined by the ion's orientation at the moment of photon impact. However, in cases where the photodissociation takes $\sim 10^{-12}$ sec or longer, rotation of the ion begins to smear out the angular distribution, and for very slow dissociations, such as those for which QET treatment is valid, extensive rotational averaging occurs (Zare, 1972; Busch and Wilson, 1972; Yang and Bersohn, 1974). [It is not always appreciated that (because of angular-momentum conservation) even in the limit of complete rotational averaging an initially anisotropic product ion angular distribution does not become completely randomized, but always retains some "memory" of the initial ion orientation.] The shape of the product ion angular distribution serves to determine the orientation of the transition dipole moment in the ion (in diatomics, it indicates whether the optical absorption is polarized parallel or perpendicular to the molecular axis), while the extent of rotational averaging can serve as a probe of the rate of dissociation relative to the ion rotational frequency.

The product ion velocity distribution will ultimately serve to probe many details of the fragmentation dynamics, but for the present its first use will be to differentiate between direct and predissociative dissociations: a direct dissociation from an excited electronic state is expected to partition a substantial amount of the available excess energy into product ion translation, giving typically hundreds of meV of product ion energy. By contrast, if vibrational predissociation occurs, giving a QET type of dissociation, the product ion energies will be very small, as amply confirmed by observation of metastable ion decompositions (Cooks *et al.*, 1973). The observation of product ion energy in O_2^+ photodissociation has similarly served to elucidate the electronic predissociation occurring there (Moseley *et al.*, 1976b).

Product velocities and angular distributions for methyl chloride ion dissociation,

$$CH_3Cl^+ \xrightarrow{h\nu} CH_3^+ + Cl, \qquad (10)$$

were studied by the ICR ion trap approach (Dunbar and Kramer, 1973; Orth and Dunbar, 1978a). This is an interesting ion because there are two excited electronic states which could correspond to the optical transition peaking at 300 nm which leads to photodissociation. These are the $^2A_1$ state ($\tilde{A}$ state), optically allowed in perpendicular polarization, having its PES maximum 3 eV (400 mm) above the ground state, and the ^2E state, ($\tilde{B}$ state), allowed in both perpendicular and parallel polarization, with PES peak 4.1 eV (300 nm) above the ground state. The ICR anisotropy experiment showed that the observed transition at 366 nm has parallel polarization to high precision: for a diatomic or pseudodiatomic, the dissociation anisotropy is characterized by the parameter β, ranging from -1 for dissociation perpendicular to the transition moment, through 0 for isotropic dissociation, to $+2$ for dissociation parallel to the transition moment. For CH_3Cl^+, β was found to be greater than 1.5, showing parallel polarization, and further setting an upper limit of 20° on the average angle through which the ion may rotate prior to dissociation, showing dissociation to be faster than $\sim 10^{-13}$ sec^{-1}. Thus (barring some unlikely possibilities introduced by spin–orbit and vibronic coupling) the dissociation must initially involve optical excitation to the $\tilde{B}(^2E)$ state. On the other hand, adiabatic correlation arguments, and evidence from photoelectron–photoion coincidence measurements (Eland *et al.*, 1976), suggest that the dissociation itself is likely to be from the $\tilde{A}(^2A_1)$ state, which would therefore imply a predissociative crossing from $\tilde{B}(^2E)$ to $\tilde{A}(^2A_1)$ before dissociation. The photodissociation results are consistent with this mechanism.

In these same experiments, the average energy of the product $CH_3{}^+$ ion was measured as 0.58 eV (Orth and Dunbar, 1978a). This is substantially less than the maximum of 0.85 eV which could be released if all the available energy went into translation, but of course is much larger than the very small kinetic energy predicted for a QET type of dissociation, so that some kind of direct dissociation is clearly involved. Orth and Dunbar (1978a) discussed in detail several varieties of the "impulsive" model of CH_3Cl^+ dissociation, and concluded that the observed partitioning between kinetic and internal energy of the products is well within the range of possibilities allowed by such models.

The polarization of the CH_3Cl^+ photodissociation is still parallel out to 514 nm (Carrington *et al.*, 1976), and a kinetic energy release of the order of 0.5 eV is observed even at these low excitation energies. The nature of the electronic states involved at visible wavelengths is not yet clear.

Moseley *et al.* (1977a,b) have made a detailed study using the beam technique of the product energies and angular anisotropies in the $Ar_2{}^+$ photodissociation. The ground state of $Ar_2{}^+$ is $^2\Sigma_u{}^+$, which is bound by

1.33 eV, and peaks in the photofragment energy distribution spectra are observed corresponding to transitions to the purely repulsive ${}^2\Sigma_g{}^+$ and ${}^2\Pi_g$ potential curves. In the absence of spin-orbit coupling, the ${}^2\Pi_g \leftarrow {}^2\Sigma_u{}^+$ transition is perpendicular polarized, so the observation of a parallel component of this transition had to be explained in terms of spin–orbit coupling in the ${}^2\Pi_g$ state. Using their data along with theoretical curves, they were able to construct a complete set of potential curves for the six spin–orbit curves corresponding to ${}^2\Sigma_u{}^+$, ${}^2\Sigma_g{}^+$, ${}^2\Pi_u$, and ${}^2\Pi_g$.

VIII. TWO-PHOTON DISSOCIATION

"By this light he changes more and more."

(Shakespeare, Much Ado About Nothing V, 1)

While irradiating benzene ion well below its thermochemical photodissociation threshold, searching for photon-induced chemical reactivity, Freiser and Beauchamp (1975) unexpectedly observed rapid ion dissociation. They realized quickly that the dissociation was a two-photon process, and proposed a mechanism, which has been widely accepted, involving sequential absorption of two photons,

$$C_6H_6{}^+ \xrightarrow{h\nu} C_6H_6^{+*} \longrightarrow C_6H_6{}^{+v} \xrightarrow{h\nu} C_6H_5{}^+ + H, \tag{11}$$

M

where the initial photon excites to the excited electronic state $C_6H_6{}^{+*}$, which undergoes internal conversion to a vibrationally hot ground state $C_6H_6{}^{+v}$ which in turn is photodissociated by the second photon. The vibrationally excited ion $C_6H_6{}^{+v}$ has a lifetime against radiative decay of the order of seconds (Dunbar, 1975a), which means it has ample time for absorption of the second photon at moderate light intensities, and makes this process feasible with normal continuous light sources (hundreds of mW/cm^2); it also means that the vibrationally hot ion can be intercepted and deactivated by collision with a neutral molecule. Both our group (for bromobenzene ions, Dunbar and Fu, 1977), and the Cal Tech group (for cyanobenzene ions, Orlowski *et al.*, 1976) have carried out detailed studies of the pressure and light-intensity dependence of the two-photon photodissociation rate, and the quantitative agreement obtained with the postulated mechanism (11) gives strong confirmation of its correctness.

This description suggests the two features of two-photon dissociation which make it interesting: one is the capability it opens up for doing photodissociation spectroscopy at wavelengths below the thermochemical threshold. This capability was exploited in looking at the low- and high-resolution laser photodissociation spectra of methylnaphthalene cations in the 600 nm region (Dunbar and Klein, 1976). The other useful feature is accessibility of the rate of collisional quenching of the vibrationally hot ion [$C_6H_6^{+v}$ in (11), for instance]. Analysis of the pressure dependence of two photon dissociation yields a rate of collisional deactivation, and provides a convenient and reliable method for measuring absolute collisional quenching rates. The studies reported thus far have shown that deactivation of the excited ion by its own parent neutral compound is very rapid, presumably because the rapid resonant charge transfer,

$$M^{+v} + M' \rightarrow M^{v} + M'^{+}, \quad (12)$$

is highly efficient. Results on quenching by other neutrals will undoubtedly be reported soon.

IX. ION THERMOCHEMISTRY

A. *Ground States*

It would seem at first glance that a great deal of thermochemical information should be available from measurements of threshold wavelengths for dissociation of a parent ion of known heat of formation to give a product ion of uncertain heat of formation. However, as must be clear from much of the emphasis of this chapter, photodissociation spectra and thresholds generally reflect the spectroscopic, not the energetic, properties of the ions. Only when the thermochemical photodissociation threshold chances to lie within an optical absorption band of the parent ion can thermochemical information be anticipated. Adding to this limitation is the problem that photodissociation can occur *below* the thermochemical threshold, either through photodissociation of vibrationally excited parent ions, or through two-photon dissociation. Therefore, a photodissociation threshold can be interpreted as reflecting at least an upper limit to the thermochemical dissociation threshold, on condition that the parent ions are carefully thermalized, and that the process is characterized as being one photon. One case where these conditions were met was CO_3^- (Moseley *et al.*, 1976a), where the vibrationally resolved photodissociation spectrum was analyzed to give a reliable limit of 1.85 eV for $D(CO_2—O^-)$. This was combined with equilibrium-constant data to give a value of 1.80 $\pm$ 0.1 eV for this bond energy, and an electron affinity of CO_3 of 2.9 $\pm$ 0.3 eV.

In cases where the thermochemical threshold is suspected to lie slightly below a photodissociation band, an isotope effect can be a useful tool, as was illustrated in the reaction $C_7H_8 + \xrightarrow{h\nu} C_7H_7^+ + H\cdot$ (Dunbar, 1973b). It was shown that this dissociation occurs with extensive hydrogen scrambling

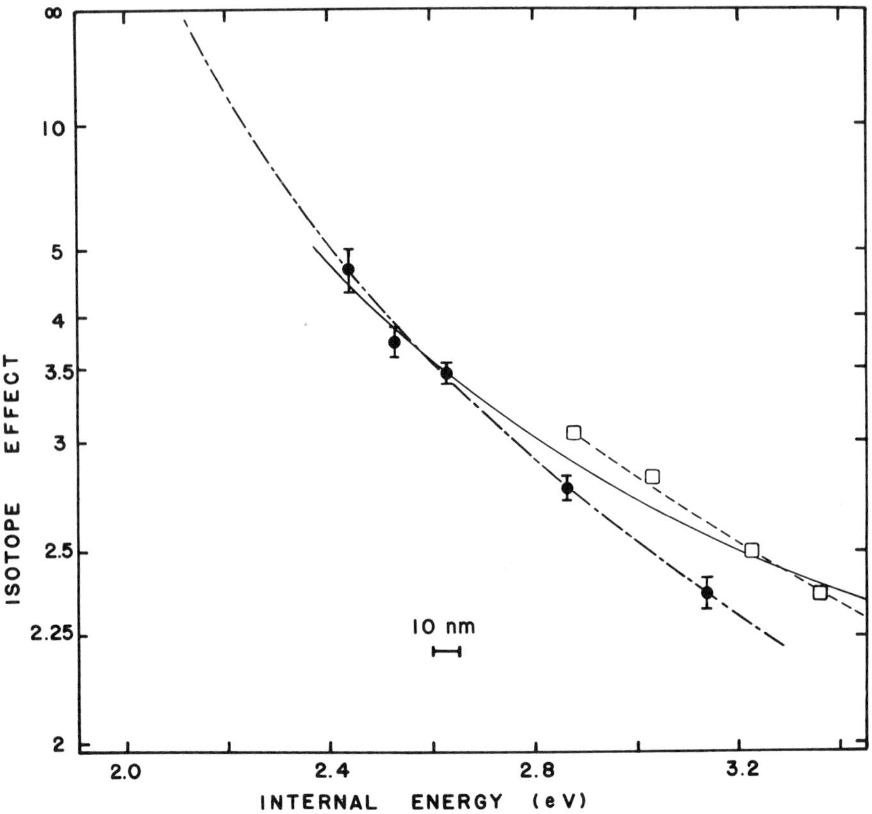

FIG. 6. Extrapolation of the H/D isotope effect for $C_6D_5CH_3$ decomposition. (–●–) ion photodissociation results, (—□—) mass spectrometric fragmentation, (———) QET calculation.

throughout the $C_7H_8^+$ parent, indicating a QET type of dissociation. The H/D isotope effect was measured as a function of wavelength in the 400–540 nm range for both $CD_3C_6H_5^+$ and $CH_3C_6D_5^+$ dissociations, giving a plot as shown in Fig. 6. The isotope effect is predicted to approach infinity at the thermochemical threshold for dissociation. Using RRKM calculations to guide the extrapolation of the isotope effect beyond the photodissociation band, the thermochemical threshold was estimated to lie at 590 nm, which gave a value of 212 kcal for the heat of formation of $C_7H_7^+$ (tropylium). This is uncertain to 2 or 3 kcal, because of the uncertainty in the internal energy of the thermalized $C_7H_8^+$ ions, but within these limits compares well with the most careful appearance potential measurements, and illustrates the good accuracy that such extrapolated isotope-effect calculations may be expected to give for QET-type dissociations.

B. *Excited States*

Freiser and Beauchamp (1976a, 1976b, 1978) have described an elegant approach to finding the proton affinity of excited states of aromatic molecules. The thermochemical cycle shown in Diagram 7 forms the basis for this: two of the energies in the diagram, PA(B), (the proton affinity of neutral B) and E_1, are known accurately from ion-equilibrium measurements and uv spectra, respectively. The quantity of interest, PA(B*), can thus be calculated if the excitation energy E_2 can be measured. E_2 is determined from the photodissociation band corresponding to the optical transition $BH^+ \xrightarrow{h\nu} (BH^+)^*$.

Diagram 7

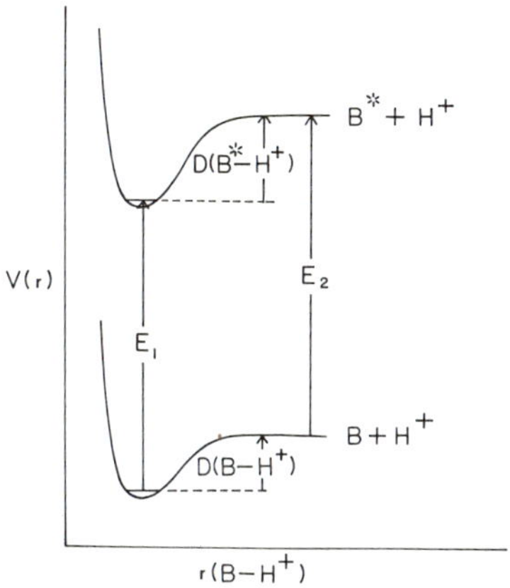

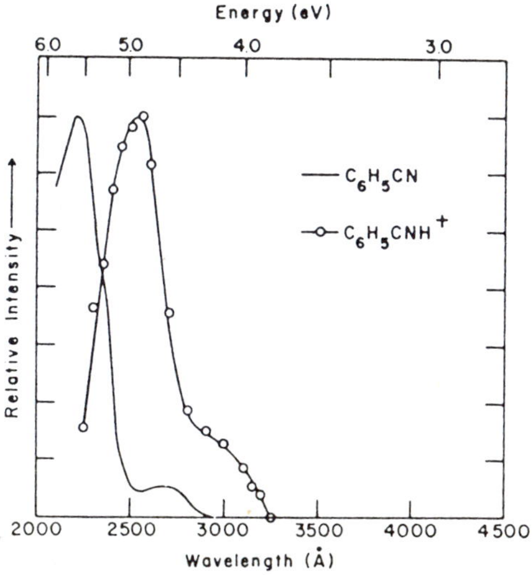

FIG. 7 Photodissociation spectrum of protonated nitrobenzene and the uV spectrum of the neutral compound. [From Freiser and Beauchamp (1978).]

The assignment of the E_2 optical transition is easy, since as illustrated in Fig. 7 the optical spectra of many neutral aromatics are strikingly similar in form to the photodissociation spectra of the corresponding protonated molecules, except for an energy shift. (In principle. the energy balance of Diagram 7 should be carried out for ground vibrational states of the species using O—O optical transitions, but in practice the peak positions are used, a procedure which should give little error where the neutral and ion spectra have similar shapes and widths.)

The protonated-neutral photodissociation peaks are usually red shifted from the neutral spectral peaks by amounts varying from zero to nearly 2 eV, indicating that the excited states usually have higher proton affinity than the ground state. Peaks are often seen and proton affinities assigned for two excited states, which in most cases are probably the ^1L_b and ^1L_a $\pi \rightarrow \pi^*$ excited states, and the proton affinities of the two excited states are usually similar. The excited-state proton-affinity enhancements correlate qualitatively with the dipole-moment enhancement of the neutral molecules, and are interpreted as reflecting a straightforward increase in electrostatic proton binding energy resulting from charge separation in the excited state.

Li^+ binding to aromatics was investigated in the same way, and with very similar results (Freiser *et al.*, 1976). There was some evidence that covalent binding is more important for H^+ than for Li^+ binding energies. It seems that Li^+ is on the whole a better probe of excited-state charge redistribution effects, since its lesser covalent binding makes it more purely a test of electrostatic binding changes.

X. CATALYSIS

An intriguing sidelight of ion photochemistry, first pointed out by Freiser and Beauchamp (1974), is the ion-catalyzed photolysis of neutral molecules. Two photochemical sequences from literature illustrate this idea:

$$C_3F_5^+ \xrightarrow{h\nu} CF_3^+ + C_2F_2 \xrightarrow{C_3F_6} C_3F_5^+ + CF_4 + C_2F_2 \tag{13}$$

(Freiser and Beauchamp, 1974), and

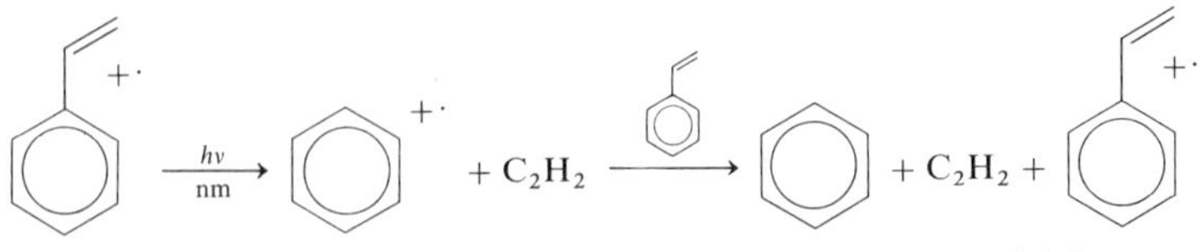

(14)

(Fu and Dunbar, 1978). Both of these reaction sequences, which occur with high efficiency, are seen to have as the net chemical result the decomposition of a molecule of neutral compound: in the first case the conversion of C_3F_6

into CF_4 and C_2F_2, and in the second case the cleavage of styrene to benzene and acetylene. The initial ionic species is regenerated by an ion–molecule reaction at the end of the cycle, and thus functions as a true catalyst, with the function of making the photolysis highly efficient at a convenient wavelength at which the neutral compound itself is transparent. Of course the photolyses (13) and (14) are of little practical interest, but one might envision the future discovery of useful chemical transformations catalyzed in a similar fashion, in which case the fact that the energy-expensive ion is not consumed makes such reactions synthetically interesting.

ACKNOWLEDGMENTS

Work in the author's laboratory described in this chapter has been supported by the National Science Foundation, the donors of the Petroleum Research Fund, administered by the American Chemical Society, and by the U.S. Air Force Geophysical Laboratory.

REFERENCES

The bibliography for ion photodissociation is fairly comprehensive through 1977. References preceded by an asterisk are not ion photodissociation papers.

Abouaf, R., Huber, B. A., Cosby, P. C., Saxon, R. P., and Moseley, J. T. (1978). *J. Chem. Phys.* **68**, 2406.

Asubiojo, O. I., McPeters, H. L., Olmstead, W. N., and Brauman, J. I. (1978). *Chem. Phys. Lett.* **48**, 127.

*Beauchamp, J. L., Holtz, D., Woodgate, S. D., and Patt, S. L. (1972). *J. Am. Chem. Soc.* **94**, 2798.

Benz, R., and Dunbar, R. C. (1978). To be published.

Beyer, R. A. and Vanderhoff, J. A. (1976). *J. Chem. Phys.* **65**, 2313.

Burnier, R. C., and Freiser, B. S. (1978). *J. Am. Chem. Soc.* (submitted).

*Bursey, M. M., Hoffman, M. K., and Benezra, S. A. (1971). *J. Chem. Soc. D* p. 1417.

*Burt, J. A. (1973). *J. Chem. Phys.* **59**, 1567.

*Busch, G. E., and Wilson, K. R. (1972). *J. Chem. Phys.* **56**, 3638.

Carrington, A, Roberts, P. G., and Sarre, P. J. (1978). *Mol. Phys.* **35**, 1523.

Carrington, A., Milverton, D. R. J., and Sarre, P. J. (1976). *Mol. Phys.* **32**, 297.

Carrington, A., Roberts, P. G., and Sarre, P. J. (1977). *Mol. Phys.* **34**, 291.

Cooks, R. G., Beynon, J. H., Caprioli, R. M., and Lester, G. R. (1973). "Metastable Ions," Elsevier, New York.

Cosby, P. C., Bennett, R. A., Peterson, J. R., and Moseley, J. T. (1975). *J. Chem. Phys.* **63**, 1612.

Cosby, P. C., Ling, J. H., Peterson, J. R., and Moseley, J. T. (1976). *J. Chem. Phys.* **65**, 5267.

Cosby, P. C., Smith, G. P., and Moseley, J. T. (1978a). *J. Chem. Phys.* **69**, 2779.

Cosby, P. C., Moseley, J. T., Peterson, J. R., and Ling, J. H. (1978b). *J. Chem. Phys.* **69**, 2771.

Dehmelt, H. G. (1967). *Adv. At. Mol. Phys.* **3**, 53.

Dehmelt, H. G. (1969a). *Adv. At. Mol. Phys.* **5**, 109.

Dehmelt, H. G. (1969b). "Atomic Physics," p. 475. Plenum, New York.

Dehmelt, H. G. (1976). *In* "The Physics of Electronic and Atomic Collision" (J. S. Risley and R. Geballe, eds.), Univ. of Washington Press, Seattle.

Dehmelt, H. G., and Jefferts, K. B. (1962). *Phys. Rev.* **125**, 1318.

Dunbar, R. C. (1971). *J. Am. Chem. Soc.* **93**, 4354.

Dunbar, R. C. (1973a). *J. Am. Chem. Soc.* **95**, 6191.
Dunbar, R. C. (1973b). *J. Am. Chem. Soc.* **95**, 472.
Dunbar, R. C. (1974). *In* "Chemical Reactivity and Reaction Paths" (G. Klopman, ed.), p. 339 Wiley (Interscience), New York.
Dunbar, R. C. (1975a). *Spectrochim. Acta, Part A* **31**, 797.
Dunbar, R. C. (1975b). *In* "Interactions Between Ions and Molecules" (P. Ausloos, ed.), p. 579. Plenum, New York.
Dunbar, R. C. (1975c). *J. Am. Chem. Soc.* **97**, 1382.
Dunbar, R. C. (1975d). *Chem. Phys. Lett.* **32**, 508.
Dunbar, R. C. (1976a). *Anal. Chem.* **48**, 723.
Dunbar, R. C. (1976b). *J. Am. Chem. Soc.* **98**, 4671.
Dunbar, R. C. (1978). *J. Chem. Phys.* **68**, 3125.
Dunbar, R. C., and Armentrout, P. (1977). *Int. J. Mass Spectrom. Ion Phys.* **24**, 465.
Dunbar, R. C., and Fu, E. W. (1973). *J. Am. Chem. Soc.* **95**, 2716.
Dunbar, R. C., and Fu, E. W. (1977). *J. Phys. Chem.* **81**, 1531.
Dunbar, R. C., and Hutchinson, B. B. (1974). *J. Am. Chem. Soc.* **96**, 3816.
Dunbar, R. C., and Klein, R. (1976). *J. Am. Chem. Soc.* **98**, 7994.
Dunbar, R. C., and Klein, R. (1977). *J. Am. Chem. Soc.* **99**, 3744.
Dunbar, R. C., and Kramer, J. M. (1973). *J. Chem. Phys.* **58**, 1266.
Dunbar, R. C., and Teng, H. H. (1978). *J. Am. Chem. Soc.* **100**, 2279.
Dunbar, R. C., Fu, E. W., and Olah, G. A. (1977). *J. Am. Chem. Soc.* **99**, 7502.
Dunbar, R. C., Kim, M. S., and Olah, G. A. (1979). *J. Am. Chem. Soc.* (in press).
Dunn, G. H. (1964). *In* "Atomic Collision Processes" (M. R. C. McDowell, ed.), p. 997. North-Holland Publ., Amsterdam.
Dunn, G. H. (1968). *Phys. Rev.* **172**, 1.
Durup, J. (1978). *Colloq. Int. CNRS Aussois* (to be published).
Dymerski, P. P., Fu, E. W., and Dunbar, R. C. (1974). *J. Am. Chem. Soc.* **96**, 4109.
*Eland, J. H. D. (1973). *Int. J. Mass Spectrom. Ion Phys.* **12**, 389.
*Eland, J. H. D., Frey, R., Kuestler, A., Schulte, H., and Brehm, B. (1976). *Int. J. Mass Spectrom. Ion Phys.* **22**, 155.
Ensberg, E. S., and Jefferts, K. B. (1975). *Astrophys. J.* **195**, L89.
Eyler, J. R. (1974). *Rev. Sci. Instrum.* **45**, 1154.
Eyler, J. R. (1976). *J. Am. Chem. Soc.* **98**, 6831.
Freiser, B. S., and Beauchamp, J. L. (1974). *J. Am. Chem. Soc.* **96**, 6260.
Freiser, B. S., and Beauchamp, J. L. (1975). *Chem. Phys. Lett.* **35**, 35.
Freiser, B. S., and Beauchamp, J. L. (1976a). *J. Am. Chem. Soc.* **98**, 265.
Freiser, B. S., and Beauchamp, J. L. (1976b). *J. Am. Chem. Soc.* **98**, 3136.
Freiser, B. S., and Beauchamp, J. L. (1977). *J. Am. Chem. Soc.* **99**, 3214.
Freiser, B. S., and Beauchamp, J. L. (1978). To be published.
Freiser, B. S., Staley, R. H., and Beauchamp, J. L. (1976). *Chem. Phys. Lett.* **39**, 49.
Freiser, B. S., McCrery, D. A., Burnier, R. C., Cody, R. B., and Reents, W. D., Jr. (1978). To be published.
Fu, E. W. (1976). Ph.D. Thesis, Case Western Reserve Univ., Cleveland, Ohio.
Fu, F. W., and Dunbar, R. C. (1978). *J. Am. Chem. Soc.* **100**, 2283.
Fu, E. W., Dymerski, P. P., and Dunbar, R. C. (1976). *J. Am. Chem. Soc.* **98**, 337.
*Gardner, C. L. (1966). *J. Chem. Phys.* **45**, 572.
Gooden, R., and Brauman, J. I. (1977). *J. Am. Chem. Soc.* **99**, 1977.
Hays, J., and Dunbar, R. C. (1978). To be published.
*Henis, J. M. S. (1970). *J. Chem. Phys.* **52**, 282.
*Herbst, E., Patterson, T. A., and Lineberger, W. C., Jr. (1974). *J. Chem. Phys.* **61**, 1300.

*Herzberg, G. (1966). "Molecular Spectra and Molecular Structure," Vol. 3. Van Nostrand-Reinhold, Princeton, New Jersey.
*Howe, I., and McLafferty, F. W. (1972). *J. Am. Chem. Soc.* **93**, 99.
Huber, B. A., Miller, T. M., Cosby, P. C., Zeman, H. D., Leon, R. L., Moseley, J. T., and Peterson, J. R. (1977a). *Rev. Sci. Instrum.* **48**, 1306.
Huber, B. A., Cosby, P. C., Peterson, J. R., and Moseley, J. T. (1977b). *J. Chem. Phys.* **66**, 4520.
Jefferts, K. B. (1968). *Phys. Rev. Lett.* **20**, 39.
Jefferts, K. B. (1969). *Phys. Rev. Lett.* **23**, 1476.
Kim, M. S., and Dunbar, R. C. (1978). *J. Am. Chem. Soc.* (in press).
Kramer, J. M., and Dunbar, R. C. (1972). *J. Am. Chem. Soc.* **94**, 4346.
Kramer, J. M., and Dunbar, R. C. (1973). *J. Chem. Phys.* **59**, 3092.
Kramer, J. M., and Dunbar, R. C. (1974). *J. Chem. Phys.* **60**, 5122.
*Levsen, K., McLafferty, F. W., and Jerina, D. M. (1973). *J. Am. Chem. Soc.* **95**, 6332.
McCrery, D. A., and Freiser, B. S. (1978). *J. Am. Chem. Soc.* **100**, 2902.
McGilvery, D. C., and Morrison, J. D. (1977). *J. Chem. Phys.* **67**, 368.
McGilvery, D. C., and Morrison, J. D. (1978). *Int. J. Mass Spectrom. Ion Phys.* (submitted).
McGilvery, D. C., Morrison, J. D., and Smith, D. L. (1978a). *J. Chem. Phys.* **68**, 4759.
McGilvery, D. C., Morrison, J. D., and Smith, D. L. (1978b). *J. Chem. Phys.* **68**, 3949.
Menasian, S. C., and Dehmelt, H. G. (1973). *Am. Phys. Soc. Meet. San Diego, Calif.*
Miller, T. M., Ling, J. H., Saxon, R. P., and Moseley, J. T. (1976). *Phys. Rev. A* **13**, 2171.
Moseley, J. T. (1977). *Abstr. Pap. Int. Conf. Phys. Electron. At. Collisions, 10th, Commis. Energie At., Paris* p. 681.
Moseley, J. J., Bennett, R. A., and Peterson, J. R., (1974). *Chem. Phys. Lett.* **26**, 288.
Moseley, J. T., Cosby, P. C., Bennett, R. H., and Peterson, J. R. (1975). *J. Chem. Phys.* **62**, 4826.
Moseley, J. T., Cosby, P. C., and Peterson, J. R. (1976a). *J. Chem. Phys.* **65**, 2512.
Moseley, J. T., Tadjeddine, M., Durup, J., Ozenne, J. B., Pernot, C., and Tabché-Fouhaillé, A. (1976b). *Phys. Rev. Lett.* **37**, 891.
Moseley, J. T., Cosby, P. C., and Peterson, J. R. (1977a). Final Rep. U.S. Army Res. Office.
Moseley, J. T., Saxon, R. P., Huber, B. A., Cosby, P. C., Abouaf, R., and Tadjeddine, M. (1977b). *J. Chem. Phys.* **67**, 1659.
Orlowski, T. E., Frieser, B. S., and Beauchamp, J. L. (1976). *Chem. Phys.* **18**, 439.
Orth, R. G., and Dunbar, R. C. (1977). *J. Chem. Phys.* **66**, 1616.
Orth, R. G., and Dunbar, R. C. (1978a). *J. Chem. Phys.* **68**, 3254.
Orth, R. G., and Dunbar, R. C. (1978b). *J. Am. Chem. Soc.* **100**, 5949.
Orth, R. G., Dunbar, R. C., and Riggin, M. T. (1977). *Chem. Phys.* **19**, 279.
Ozenne, J. B., Pham, D., and Durup, J. (1972). *Chem. Phys. Lett.* **17**, 422.
Ozenne, J. B., Durup, J., Odom, R. W., Pernot, C., Tabché-Fouhaillé, A., and Tadjeddine, M. (1976). *Chem. Phys.* **16**, 75.
*Pearson, P. K., Schaefer, H. F., III, Richardson, J. H., Stephenson, L. M., and Brauman, J. I. (1974). *J. Am. Chem. Soc.* **96**, 6778.
*Petruska, J. (1961). *J. Chem. Phys.* **34**, 1111, 1121.
Richardson, C. B., Jefferts, K. B., and Dehmelt, H. G. (1968). *Phys. Rev.* **65**, 80.
Richardson, J. H. (1976). *Appl. Spectrosc. Rev.* **12**, 159.
*Richardson, J. H., Stephenson, L. M., and Brauman, J. I. (1974a). *Chem. Phys. Lett.* **25**, 318.
*Richardson, J. H., Stephenson, L. M., and Brauman, J. I. (1974b). *J. Am. Chem. Soc.* **96**, 3671.
Riggin, M. T., and Dunbar, R. C. (1975). *Chem. Phys. Lett.* **31**, 539.
Riggin, M. T, Orth, R. G., and Dunbar, R. C. (1976). *J. Chem. Phys.* **65**, 3365.
*Robinson, P. J., and Holbrook, K. A. (1972). "Unimolecular Reactions," Wiley, New York.
*Shida, T., Kato, T., and Nosaka, Y. (1977). *J. Phys. Chem.* **81**, 1095.
Smith, G. P., Cosby, P. C., and Moseley, J. T. (1977). *J. Chem. Phys.* **67**, 3818.

Smith, G. P., Lee, L. C., Cosby, P. C., Peterson, J. R., and Moseley, J. T. (1978). *J. Chem. Phys.* **68**, 3818.

Sullivan, S. A., Freiser, B. S., and Beauchamp, J. L. (1977). *Chem. Phys. Lett.* **48**, 294.

Tabché-Fouhaillé, A., Durup, J., Moseley, J. T., Ozenne, J. B., Pernot, C., and Tadjeddine, M. (1976). *Chem. Phys.* **17**, 81.

Tadjeddine, M., Huber, B. A., Abouaf, R., Cosby, P. C., and Moseley, J. T. (1978). *J. Chem. Phys.* **69**, 710.

Teng, H. H., and Dunbar, R. C. (1978). *J. Chem. Phys.* **68**, 3133.

Thomas, T. F., Dale, F., and Paulson, J. F. (1977). *J. Chem. Phys.* **67**, 793.

Van Asselt, N. P. F. B., Maas, J. G., and Los, J. (1974a). *Chem. Phys. Lett.* **24**, 555.

Van Asselt, N. P. F. B., Maas, J. G., and Los, J. (1974b). *Chem. Phys.* **5**, 429.

Van Asselt, N. P. F. B., Maas, J. G., and Los, J. (1975). *Chem. Phys.* **11**, 253.

Vanderhoff, J. A. (1977). *J. Chem. Phys.* **67**, 2332.

Vanderhoff, J. A. and Beyer, R. A. (1976). *Chem. Phys. Lett.* **38**, 532.

Vestal, M. L., and Futrell, J. H. (1974). *Chem. Phys. Lett.* **28**, 559.

Vestal, M. L., and Mauclaire, G. H. (1977a). *J. Chem. Phys.* **67**, 3767.

Vestal, M. L., and Mauclaire, G. H. (1977b). *J. Chem. Phys.* **67**, 3758.

Vestal, M. L., Mauclaire, G. H., and Futrell, J. H. (1975). *Annu. Conf. Mass Spectrom., 23rd Houston, Tex.*

von Busch, F., and Dunn, G. H. (1972). *Phys. Rev. A* **5**, 1726.

*Winkler, J., and McLafferty, F. W. (1973). *J. Am. Chem. Soc.* **95**, 7533.

*Yang, S., and Bersohn, R. (1974). *J. Chem. Phys.* **61**, 4400.

*Zahradnik, R., and Carsky, P. (1970). *J. Phys. Chem.* **74**, 1240.

*Zare, R. N. (1972). *Mol. Photochem.* **4**, 1.

Chapter 15

Molecular beam techniques: applications to the study of ion–molecule collisions

W. RONALD GENTRY

CHEMICAL DYNAMICS LABORATORY
DEPARTMENT OF CHEMISTRY
UNIVERSITY OF MINNESOTA
MINNEAPOLIS, MINNESOTA

I. INTRODUCTION

The goal of this review is to summarize the state of the art in the application of molecular beam techniques to the study of ion–molecule collisions, in

ISBN 0-12-120802-8

the light of advances which have been made within the last five years. Even though earlier work is, in some cases, still representative of the present state of the art, the work prior to 1973 has been discussed in several previous general reviews (Ausloos, 1975; Ferguson, 1975; Franklin, 1972; McDaniel *et al.*, 1970) and in other reviews of more specialized aspects of ion–molecule collision phenomena (Mahan, 1976; Toennies, 1976; Koski, 1975; Doering, 1973; Henglein, 1970; Neynaber, 1969), and therefore will usually be omitted from mention here. The primary focus of attention will be on molecular rearrangement collisions (chemical reactions), but some very important progress in the study of charge transfer and energy transfer processes will also be discussed.

It is gratifying to see the extent to which the technologies of molecular beam experiments with ion beams and with neutral beams are now becoming integrated. Not only do we find many ion–molecule crossed beam experiments which utilize the same supersonic beam technology usually employed in crossed neutral beam experiments, but we also find examples in which ion beams are used to generate (by charge exchange) neutral beams of highly labile or energetic species which are difficult to obtain by any other means. One of the principal advantages that ion beam experiments have had historically for the study of microscopic reaction dynamics is the large range of relative kinetic energy which is accessible by simply varying the ion acceleration potential. In order to obtain a complete picture of the dynamical process it is important to be able to probe the collision dynamics in different energy regimes in which the influences of various features of the process in question may change in relative importance. For example, the magnitudes of the cross sections for many ion–molecule reactions at thermal energies are controlled principally by the long-range attractive part of the potential energy surface. At kinetic energies comparable to the bond energies, however, the shape of the potential energy surface in the vicinity of the well may be the most important dynamical influence. At even higher kinetic energies, the nature of the repulsive forces at very small separations or the opening of channels leading to different electronic states of the products may determine the outcome of the collision. We will limit our discussion here to experiments in the "chemical" energy range, i.e., from thermal energies up to the range of ionization energies for simple molecules.

As a final boundary condition on the coverage of this chapter, it seems advisable to define what is meant by the term "molecular beam technique." There is the possibility of some confusion in applying this term to ion beam experiments because, after all, every sector-field mass spectrometer contains an ion beam. We will use the term in the sense understood by investigators in this field to mean a method in which a beam of at least one of the reactants is prepared prior to the collision in a state of well-defined energy and angle, in which only single collisions occur between the two selected reactant species, and in which some provision has been made for measuring dynamical properties of the scattered products. Thus, we do not include single-source mass spectrometer experiments, various "swarm" methods, or conventional tandem mass spectrometer methods.

A. *The "Ideal" Beam Experiment*

The principal motivation for doing molecular beam experiments is to obtain directly *microscopic* information on the details of molecular interactions. The object is to remove as many as possible of the averaging steps which stand between a single elementary molecular collision event and the actual experimental result which is observed. The ideal molecular beam experiment, therefore, would be one in which the two intersecting primary beams were selected with respect to all the molecular internal quantum numbers (Bromberg *et al*., 1975), were collimated and speed selected to within the uncertainty-principle limit, and in which the product angle distribution were measured within the uncertainty-principle limit for each fully-resolved internal quantum state of both collision products. The ideal molecular beam apparatus would permit experiments of this type to be performed with any chemical species in any selected internal quantum state and in any range of kinetic energy. There are, of course, no such apparatuses available, and no such experiments yet performed. The fundamental conflict which arises in the design and construction of every new instrument is always between resolution and intensity. In molecular beam experiments (as in many others), it has often happened that too much greed for the former has resulted in an unmeasurably small amount of the latter. In every successful molecular beam apparatus, both the resolution in some degrees of freedom, and also some degree of versatility with respect to chemical systems, has been sacrificed in order to achieve a more-or-less optimum configuration for performing a limited range of experiments. Thus, the state of the art is represented not by any single instrument, but by a large collection of different instruments and techniques, each having capabilities which complement the others.

B. *Approximations to the Ideal*

As an introduction to the more detailed discussion of particular instruments which is given in Section II, we offer here a brief summary of the salient features of each general type of apparatus now in use, and a comparison of the advantages and disadvantages of different techniques.

1. Ion Beam–Collision Cell Instruments

In this type of apparatus an ion beam, which may be mass and energy selected, is focused into a scattering cell containing the target gas at some steady-state density. In order to avoid multiple collisions, the target gas pressure is typically chosen to be sufficiently small that the attenuation of the ion beam by the scattering gas is $\lesssim 1\%$. Since the average velocity vector of the target gas in the laboratory frame is zero, the nominal value of the initial center of mass (cm) kinetic energy of collision is

$$E_i = \tfrac{1}{2}\mu v_{\mathrm{I}}^2 = \frac{m_{\mathrm{N}}}{m_{\mathrm{I}} + m_{\mathrm{N}}} \mathscr{E}_{\mathrm{I}}, \tag{1}$$

where m_{I} and m_{N} are the ion and neutral masses, respectively, μ is the reactant reduced mass, and v_{I} and $\mathscr{E}_{\mathrm{I}}$ are, respectively, the laboratory speed and kinetic energy of the ion. Thus, the cm and laboratory kinetic energies differ by a simple mass factor. The lower limit to the ion kinetic energy is determined by the defocusing effects of the ion beam space charge and by the surface charge variations or "patch effects" which exist on all (except possibly superconducting) metal surfaces. There is a great deal of art and superstition involved in the preparation of lens surfaces for low-energy ion beam experiments, but so far no apparatus of this type has performed well and reliably at ion energies $\mathscr{E}_{\mathrm{I}}$ very much below 1 eV, and many instruments are limited to much higher energies. Nevertheless, very small cm kinetic energies can be obtained if the mass ratio in Eq. (1) is sufficiently small. Since the ion energy is usually very much larger than the thermal average kinetic energy of the neutral molecules, it might be naively presumed that the thermal energy distribution of the target gas makes a negligible contribution to the cm kinetic energy resolution. However, this is often not the case. It is the spread in relative velocity vectors of the ion beam and the target gas molecules which is important, and in fact the average speed of a target gas molecule at 300 K is about the same as the spread in speeds of a 10 eV ion beam of the same mass which has an energy spread of 1 eV. In some instruments the collision cell may be cooled in order to reduce the contribution of the target gas to the spread in cm kinetic energy.

An ion beam–collision cell apparatus can be set up to measure the product ion angle distributions, either by having a wide slot in the collision cell wall for the product ions to emerge, or by having a collision cell exit aperture which rotates with the detector. Usually some provision is also made for energy analyzing the product ions. One of the advantages of using a collision cell is that the integral of the target gas density along the primary ion beam path is nearly constant and can be determined accurately by simply measuring the gas pressure in the cell. The pressure is usually in the 10^{-4}–10^{-2}

torr range, where modern capacitance manometers are capable of high accuracy. Thus this technique has the potential capability of measuring *absolute* differential scattering cross sections. However, the additional problems of determining the absolute detector efficiency and acceptance function have in most cases discouraged investigators from attempting absolute measurements.

2. *Crossed Beam Instruments*

Crossed beam apparatuses are similar to beam–collision cell instruments except that the target gas scattering cell is replaced by a neutral molecular beam which crosses the ion beam, usually at a 90° intersection angle. In favorable cases, the neutral beam source should preferably be of the "supersonic nozzle" type, since the high-pressure hydrodynamic expansion in this type of source results in both an increase in beam intensity and a decrease in the velocity spread compared to an effusive source. However, care must be exercised because of the tendency of some species to form polymers or clusters in supersonic jets. With only moderate collimation of the neutral beam, the contribution of the target beam velocity spread to the total spread in relative velocity of collision can be made very much smaller than is the case with the collision cell arrangement. The crossed beam geometry is therefore to be preferred for experiments in which very high energy and angle resolution is required. It is, however, necessary to sacrifice some sensitivity for the detection of scattered products. The usable intensity of the neutral beam at the crossing point is usually limited by the requirement that the background pressure of the scattering gas be kept low, and therefore is determined by the available differential pumping capacity for the neutral beam source. Despite the many ingenious designs which have appeared for arranging several large vacuum pumps around the beam path, the attenuation of the ion beam by the target gas is typically only 10^{-5}–10^{-4} in the crossed beam arrangement, rather than the attenuation of about 10^{-2} which one normally would like to have in order to obtain the greatest possible scattered intensity without a significant probability of multiple collisions. Using a molecular beam for the neutral gas target does offer the advantage that control over the internal state populations (and sometimes the population of highly labile chemical species) is often easier to achieve under collision-free molecular beam conditions than in a thermally equilibrated collision cell. Absolute cross sections have been measured in crossed beam instruments, but absolute measurements are more difficult than in the collision cell arrangement because the absolute neutral beam speed and angle distributions must be determined, rather than simply the target gas pressure and temperature.

3. *Merged Beam Instruments*

In merged beam experiments the neutral beam is formed by neutralization (usually by charge transfer) of a high-energy ion beam. A second ion beam is merged with the neutral beam by deflection in a magnetic or electric field, so that the two beams become coaxial in the interaction zone. If the beams are well collimated, so that their transverse components of velocity are negligible, then the nominal cm kinetic energy of collision is

$$E_i = \tfrac{1}{2}\mu(v_{\mathrm{I}} - v_{\mathrm{N}})^2 = \tfrac{1}{2}\mu g_i^2, \qquad (2)$$

which can be very much smaller than the laboratory beam energies. In fact, there is a considerable advantage in making the laboratory energies of both beams as large as possible, consistent with other experimental requirements. In the electrostatic acceleration of an ion beam, each ion experiences the same potential energy difference and therefore gains the same increment in laboratory kinetic energy. Except for possible chromatic effects in the beam focusing optics, the energy spread $\Delta\mathscr{E}$ of the beam remains constant. The velocity spread

$$\Delta v = \Delta\mathscr{E}/mv \qquad (3)$$

therefore shrinks as v increases. At a given fixed value of relative speed $(v_{\mathrm{I}} - v_{\mathrm{N}})$, the spread in initial relative kinetic energy is approximately proportional to $\mathscr{E}^{-1/2}$ for either beam. For sufficiently high laboratory energies, the increase in kinetic energy resolution in the cm system can be very great. For example, with beams of equal mass and equal laboratory energy spreads $\Delta\mathscr{E}$, the spread in cm kinetic energy is approximately

$$\Delta E \simeq \Delta\mathscr{E}(E/\bar{\mathscr{E}})^{1/2}, \qquad \Delta E \ll E, \qquad (4)$$

where $\bar{\mathscr{E}}$ is the average laboratory kinetic energy of the two beams. Thus two beams, each with $\Delta\mathscr{E} = 1$ eV, may be merged at $\bar{\mathscr{E}} = 4000$ eV to obtain a relative kinetic energy $E = 1$ eV, and the resulting cm energy resolution will be $\Delta E = 0.05$ eV. The absolute resolution ΔE improves with decreasing E; therefore, the merged beam technique extends to collision energies much lower than are available with other ion beam methods. Under conditions where the nominal ion and neutral velocities are made equal, the actual average kinetic energy of collision has in some cases been 0.002 eV or less.

Because of the small ratio of the relative translational speed g_i to the center of mass laboratory speed v_c, the products in a merged beam experiment are confined to a small range of laboratory speed and scattering angle. It is therefore relatively easy to collect all of the ionic products and thus to determine the total reaction cross section. The product speed and angle distributions in the cm coordinate system cannot easily be measured separately, but a high-resolution measurement of the product laboratory kinetic energy distribution yields the distribution in axial components of the product cm velocity, which still gives a great deal of information on the reaction dynamics. Even though the beam attenuation in merged beam experiments is extremely small ($\sim 10^{-10}$–10^{-14}) the beam intensities and product collection efficiency can be sufficiently large to enable the study of processes with small cross sections. Determination of the absolute cross sections requires an accurate measurement of the three-dimensional spatial intensity distributions of both beams in the collision volume. However, high-energy ion and neutral beams can both be detected with high efficiency and low noise, so absolute measurements are fairly routine.

Another important feature of the merged beam technique is the range of chemical systems which can be studied. Most ion–molecule crossed beam experiments have been performed with neutral beam species which are available at the turn of a tap, and rather extraordinary means are required to obtain pure beams of highly labile neutral species. By contrast, it is relatively easy to produce the corresponding ion of most simple neutral molecules, then to remove interfering species in a mass spectrometer and to neutralize the resulting ion beam by charge transfer. Because of the ion beam space charge limitations, this method would not produce sufficient intensity for crossed beam experiments at low kinetic energies. At the high laboratory energies of merged beam experiments, however, usable beams of many exotic species can easily be obtained. A problem associated with this method of forming the neutral reactant is that it is relatively difficult to characterize the internal state distribution in the beam. The consecutive processes of ionization and charge transfer may each populate vibrationally and electronically excited states. It is possible in special cases to deduce the state distribution from other data, or to control the populations by selection of the ion source operating conditions and the charge transfer gas, but in general this is a difficult problem which requires the careful selection of systems for study.

4. *Guided Beam Instruments*

Guided beam experiments are similar to beam–collision cell experiments, except that high frequency electric multipole fields along the primary and secondary ion paths are used to provide a channel-like effective potential for the ion motion, thus guiding (rather than focusing) the beam from the source to the collision zone, and from the collision zone to the detector. Energy and mass filtering also can be achieved by the appropriate selection of dc and rf fields. The relatively strong rf guiding field dominates the small surface charge effects which afflict traditional electrostatic ion optics, and permits lower ion beam energies to be achieved. The ability to trap ions for long residence times in the source makes it possible to relax the primary ion internal modes by many collisions with a bath gas. The sensitivity of the guided beam method for small cross section processes can be very high, since the product collection and detection efficiencies are nearly unity. The use of rf guide fields in the collision zone precludes the measurement of angular distributions; however, absolute total cross sections can be measured very accurately.

5. *Beam Fluorescence Instruments*

In an ion beam–fluorescence experiment, a mass and energy selected ion beam collides with a neutral target gas and the fluorescence of the collision products is spectrally resolved and detected. Even a relatively crude optical detector provides much better resolution for the product state distribution than can be obtained with translational energy selectors. For example, at 5000 Å a wavelength resolution of 1 Å is equivalent to an energy resolution of 0.0005 eV. Of course, fluorescence detection can only be used for products in excited states having transitions within the acceptable spectral range, and with radiative lifetimes which are shorter than about 10^{-4} sec. Under normal conditions, the radiative distribution is expected to be spatially isotropic, and there will be no information available about the product speed and angle distributions. It is possible to determine the detector response function for such an experiment so that absolute total cross sections for production of specific excited states can be measured. In practice, however, absolute calibration is difficult because of the large number of geometrical and optical factors involved.

6. *Summary of Comparisons*

Table I presents an assessment of the relative strengths and weaknesses of these five general methods, on a qualitative scale of 0 to 10 for each of several features. The estimated merits of each technique are based on a personal judgment by the author of the potential capabilities of that method, even if for some reason that capability has not yet been completely realized. Naturally, individual instruments of a particular type may vary considerably in performance and it is always possible for new developments to occur which render previous assessments obsolete. For any particular experimental study, of course, the importance of the various individual performance features may be weighted very unevenly. It does not make sense to say that one general method is better than another except for some specific limited purpose.

TABLE I

Estimated Relative Merits of Different Types of Ion Beam Instruments, on a Qualitative Scale from 0 (Lowest) to 10 (Highest)

	Beam-gas	Crossed beams	Merged beams	Guided beam	Beam fluorescence
A. Resolution					
(1) Reactant KE	6	8	10	2	4
(2) Product KE or state	3	5	3	2	10
(3) Scattering angle	8	10	3	1	0
B. Sensitivity	5	2	6	10	5
C. Accuracy (absolute cross sections)	5	2	9	10	6
D. Versatility					
(1) Range of systems	5	8	10	5	5
(2) Energy range	3	3	10	8	3
(3) Control of internal states	6	10	2	8	6

II. RECENT ADVANCES

In this section we survey recent developments in instrumentation, giving brief descriptions of the newest instruments and techniques and illustrations of the type of chemical information being obtained. Within each general category of experimental method, the discussion is organized in terms of the type of collision process being studied.

A. *Crossed Beam and Beam–Collision Cell Experiments*

1. *Rearrangement Reactions*

Most of the emphasis in crossed beam and beam–collision cell experiments has been on investigating the dynamics of atom-rearrangement reactions. The last five years have witnessed not only the continued development of experimental techniques, but also an increasingly broader perspective on the choice of systems for study. Incisive choices of experiments during this period have resulted in the exploration of many new chemical phenomena.

Figure 1 shows an apparatus recently placed in operation at the University of Kansas by Hierl and co-workers (Hierl *et al.*, 1973; Wyatt *et al.*, 1975a). The electron impact ion source in this apparatus operates over the range of laboratory energy from about 0.5 to 100 eV, and can be rotated with

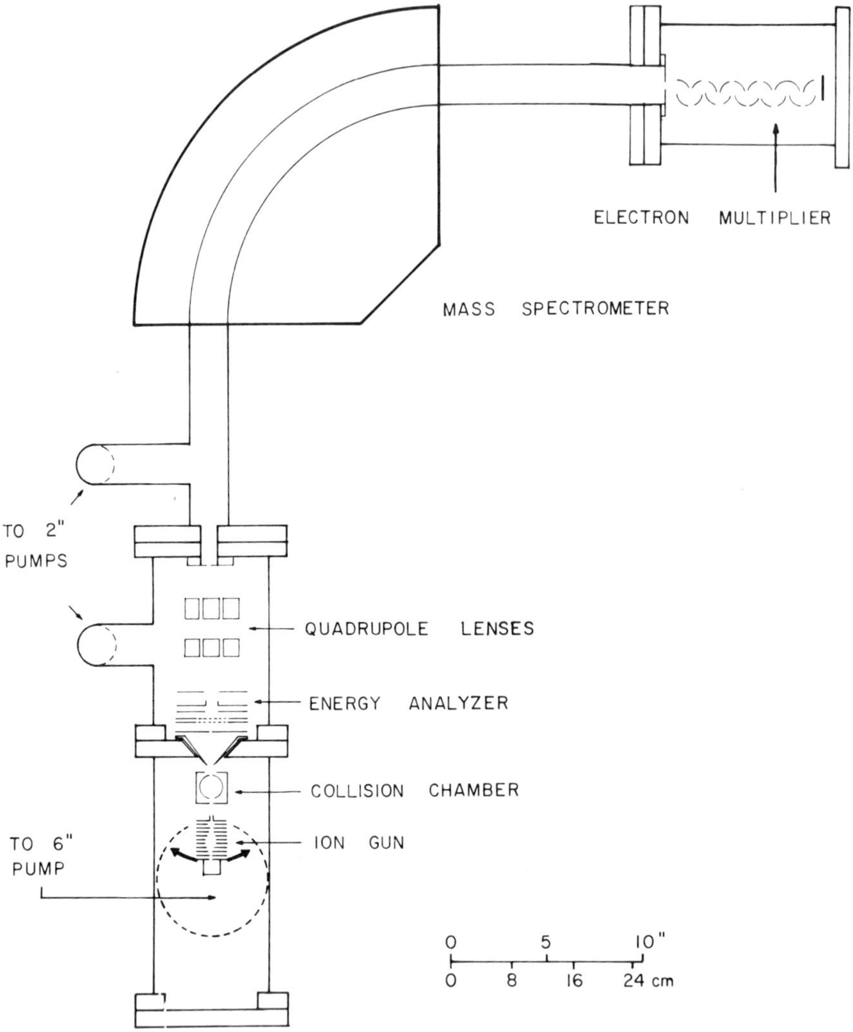

FIG. 1 Schematic diagram of the ion beam–collision cell apparatus at the University of Kansas. [From Wyatt *et al.* (1975a).]

respect to the fixed detector in order to measure product ion angular distributions. A modulated retarding potential analyzer is used to measure product energy distributions. The collision cell pressure is monitored with a capacitance manometer to permit determination of absolute cross sections. The collision cell temperature is controlled by heating. With an Ar^+ primary beam, the ion beam laboratory kinetic energy spread is typically $\Delta\mathscr{E}_I = 0.013\mathscr{E}_I + 0.15$ eV, and the overall laboratory angle resolution is a few degrees. The lack of primary ion mass selection in the present instrument limits the range of ion species which can be used without contamination of the beam, but results in relatively simple ion optics and a small source-to-target distance, thus making it possible to obtain sufficient ion beam intensity over a broader range of source operating conditions.

One of the most interesting systems investigated by this group is the reaction

$$Ar^+ + CH_4 \rightarrow ArH^+ + CH_3, \qquad \Delta H_0 = -1.7 \text{ eV}, \tag{5}$$

which was studied in various isotopic combinations over the range of cm kinetic energy from 0.4 to 30 eV (Wyatt *et al.*, 1974, 1975a,b). Figure 2 shows the total cross section as a function of kinetic energy for the reactions with CH_4 and CD_4. Even though this abstraction reaction is very exothermic, it displays a distinct kinetic energy threshold, which is attributed to competition with the (unobserved) dissociative charge transfer reaction. This accounts for the very small magnitude of the total cross section (for which the absolute uncertainty is estimated to be $\pm 40\%$). The reaction

$$N_2{}^+ + CH_4 \rightarrow N_2H^+ + CH_3$$

is similar to reaction (5) in the dependence of the cross section on kinetic energy (Wyatt *et al.*, 1976a). The isotope effect favoring the transfer of an H atom over a D atom is seen in reactions of the mixed isotope CH_2D_2 as well as in the comparison between CH_4 and CD_4. Very recently, this group has been investigating isotope effects in the reactions

$$Kr^+ + HD \rightarrow KrH^+, KrD^+ + D, H$$

(Chivalak and Hierl, 1977a,b; Wyatt *et al.*, 1976b).

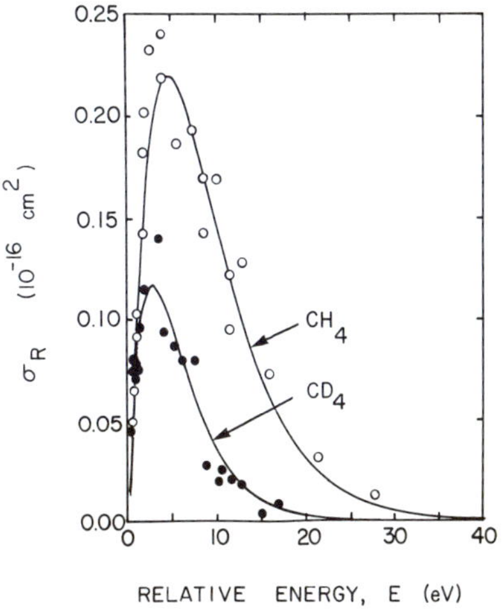

FIG. 2 Total reaction cross section σ_R as a function of cm kinetic energy for the reaction $Ar^+ + CH_4, CD_4 \rightarrow ArH^+, ArD^+ + CH_3, CD_3$. The solid lines are empirical fits to the data. [From Wyatt *et al.* (1975a).]

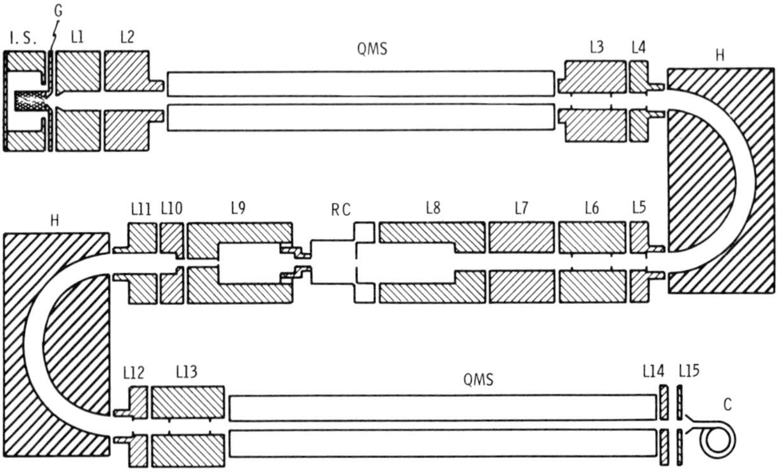

FIG. 3 Schematic diagram of the ion beam–collision cell apparatus at Johns Hopkins University [Wendell *et al.* (1975)]. IS is the ion source, L_1–L_{15} the ion lens elements, *H* the hemispherical plate energy analyzer, *QMS* the quadrupole mass filter, *RC* the reaction cell, and *C* the channeltron electron multiplier.

A very different type of ion beam–collision cell apparatus constructed by Koski and co-workers at Johns Hopkins University is shown in Fig. 3 (Wendell *et al.*, 1975). Ions formed by electron impact are first mass selected in a quadrupole mass filter, and energy selected in a hemispherical electrostatic deflector. The beam is then focused into the collision cell, which permits rotation through a laboratory scattering angle range of $\pm 45°$. A similar hemispherical sector and quadrupole mass filter are used to analyze the energy and mass of the product ions. The primary ion laboratory kinetic energy range extends to slightly below 1 eV, with a spread reported to be typically 0.070 eV.

This apparatus has recently been used to study the reactions of F^+ and Cl^+ with H_2 and D_2 (Jones *et al.*, 1977a,b; Wendell *et al.*, 1975), the reactions of ground-state and excited B^+ with D_2 and N_2O (Lin *et al.*, 1974a,b) and the reaction $H_2O^+ + D_2 \rightarrow H_2DO^+ + H$ (Jones *et al.*, 1975).

An especially impressive illustration of the capability of this instrument comes from a recently published report (Jones *et al.*, 1977c) on the reaction

$$C^+ + H_2 \rightarrow CH^+, CH^{+*} + H, \qquad \Delta H_0 = 0.4 \text{ eV}. \tag{6}$$

Figure 4 shows product ion energy distributions taken at 0° scattering angle for the H_2 and D_2 isotopic reactions. Despite the fact that a collision cell

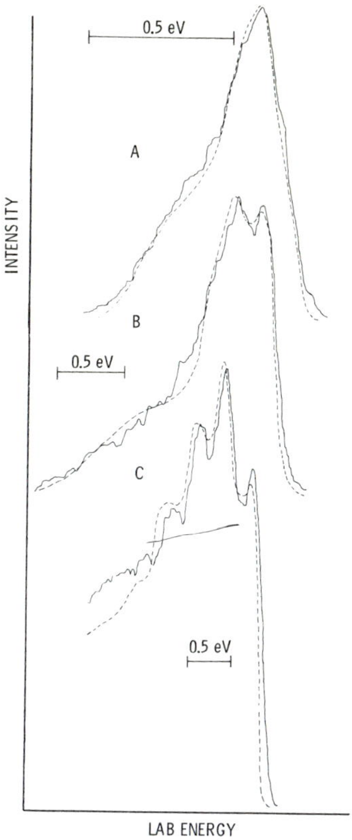

FIG. 4 Product ion laboratory energy spectra at 0° laboratory scattering angle (Jones *et al.*, 1977c.) (a) $C^+ + D_2 \rightarrow CD^+ + D$, $E_i = 0.30$ eV; (b) $C^+ + D_2 \rightarrow CD^+ + D$, $E_i = 0.85$ eV; (c) $C^+ + H_2 \rightarrow CH^+ + H$, $E_i = 1.15$ eV. The dashed curves are theoretical predictions.

rather than a crossed beam is used, sufficient resolution in the product kinetic energy is achieved by working near the threshold for this endothermic reaction to permit the CH^+ product vibrational structure to be resolved in the kinetic energy spectra. The vibrational state distribution appears to be consistent with the results of a phase space calculation by Truhlar (1969). By using beam attenuation methods to characterize the electronic-state populations of the ion beam, the same group (Jones *et al.*, 1976) was also able to study the exothermic excited-state reaction

$$C^+(^4P) + H_2 \rightarrow CH^+ + H.$$

A somewhat more complex hydrocarbon ion reaction has recently been studied by Henglein and co-workers, using a modified version of an earlier ion beam–collision cell apparatus (Bosse *et al.*, 1971). Eisele *et al.* (1974) measured product ion energy and angle distributions for the reactions

$$CH_2^+ + H_2, D_2 \rightarrow CH_3^+, CH_2D^+, CHD_2^+, CHD^+ + \text{other products} \qquad (7)$$

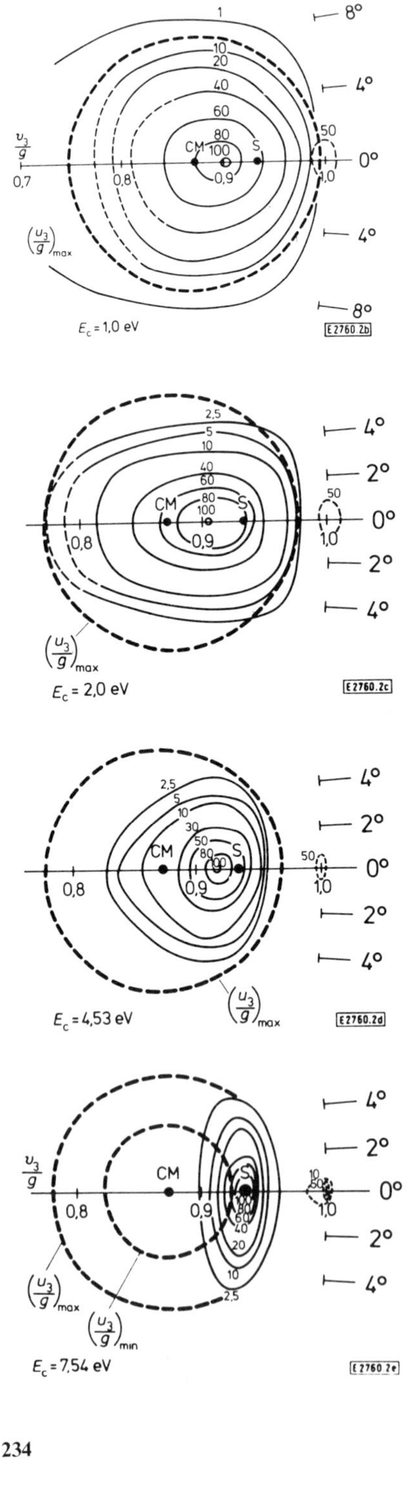

FIG. 5 Product ion intensity contour maps for the reaction $CH_2^+ + H_2 \rightarrow CH_3^+ + H$ (Eisele *et al.*, 1974). The center of mass velocity and the product velocity expected from the spectator stripping mechanism are marked CM and S, respectively. The primary beam intensity contours are shown at the right of each distribution. The heavy dashed circles are energetic limits.

over the cm kinetic energy range from 0.54 to 13.3 eV. Figure 5 shows contour maps in product ion velocity vector space for the reaction

$$CH_2{}^+ + H_2 \rightarrow CH_3{}^+ + H \qquad (8)$$

at several different collision energies. The product ion velocity distributions show distinct forward scattering (relative to $CH_2{}^+$) at high energies, and a tendency to become spherically symmetric in the cm system at the lowest collision energies accessible. This tendency was interpreted as a trend toward formation of a long-lived intermediate complex at sufficiently low energy. Evidence in support of this conclusion is provided by the product branching ratio for the analogous $CH_2{}^+ + D_2$ reaction. The abstraction product CH_2D^+ and the substitution product $CHD_2{}^+$ were found to be formed with equal probability at low kinetic energies. Eisele and Henglein (1974) have also studied hydrogen atom abstraction reactions in the series $NH_3{}^+$, $NH_2{}^+, NH^+ + H_2 \rightarrow NH_4{}^+, NH_3{}^+, NH_2{}^+ + H$, finding that these reactions each proceed by a direct pickup mechanism.

Several interesting ion–molecule reactions have recently been investigated by Mahan and co-workers, using both an early-vintage ion beam–collision cell apparatus (Gentry *et al.*, 1968) and a newer beam–collision cell instrument which operates at lower ion kinetic energies (Fair and Mahan, 1975). Mahan and Schubart (1977) studied the reactions

$$CO_2{}^+ + D_2 \rightarrow DCO_2^+, DCO^+ + D, OD \qquad (9)$$

over the cm kinetic energy range from 1 to 15 eV. The abstraction product $DCO_2{}^+$ is formed predominately in a direct process, while the exchange product DCO^+ appears to be formed only from the decay of a persistent intermediate complex, except at very high collision energies. Interestingly, the other energetically accessible products CO^+, O^+, D_2O^+, and OD^+ are not produced with significant probability. This group has also performed studies of the reactions $O^+ + H_2 \rightarrow OH^+ + H$ (Gillen *et al.*, 1973a,b), $C^+ + H_2 \rightarrow CH^+ + H$ (Mahan and Sloane, 1973), and $N^+ + H_2 \rightarrow NH^+ + H$ (Mahan and Ruska, 1976; Farrar *et al.*, 1976; Fair and Mahan, 1975). The very interesting $C^+ + H_2$ reaction shows a tendency toward symmetric cm velocity vector distributions as the collision energy is lowered, but does not reach this condition in the experimentally accessible energy range (down to $E_i = 2.9$ eV). Subsequent measurements by Herbst *et al.* (1975) extend this same conclusion to energies as low as $E_i = 0.9$ eV. In the case of the $N^+ + H_2$ reaction, markedly different dynamics for the reaction of ground state $N^+(^3P)$ and metastable $N^+(^1D)$ were discovered. Figure 6 shows the product ion cm velocity vector distributions for two different N^+ beam electronic-state populations: (a) N^+ formed from N_2 by 160 eV electron impact, producing both ground and metastable excited states, and (b) N^+ formed from N_2 in

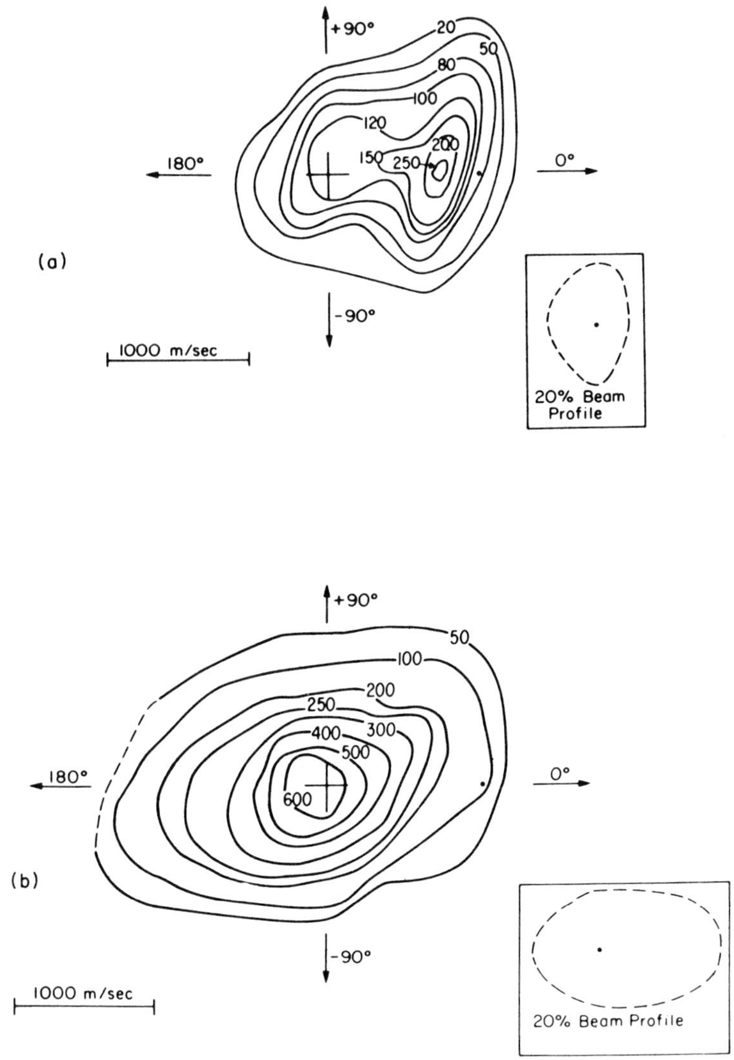

FIG. 6 Product-ion intensity contour maps for the reaction $N^+ + H_2 \rightarrow NH^+ + H$, drawn in the cm velocity vector coordinate system (Farrar *et al.*, 1976). Angles are specified relative to the initial N^+ velocity vector. (a) N^+ formed by 160 eV electron impact on N_2 (b) N^+ formed in a microwave discharge in N_2

a microwave discharge, which produces only ground-state N^+ in significant amounts. Part (b) shows that the ground-state ion produces a product distribution symmetric about $\pm 90°$ in the cm system, which is consistent with the formation of a long-lived NH_2^+ intermediate complex. Part (a), by contrast, shows a bimodal character, with one component symmetric about the cm origin and the other peaked strongly at 0°. The latter process is attributed to a direct reaction mechanism involving $N^+(^1D)$. The observed reaction dynamics are consistent with the adiabatic state correlations for this system, which show that the deep $NH_2^+(^3B_1 - {}^3\Sigma)$ ground-state potential well is adiabatically accessible to the $N^+(^3P) + H_2(^1\Sigma)$ reactants, but not to $N^+(^1D) + H_2(^1\Sigma)$.

A new crossed beam apparatus has recently been constructed by Futrell and colleagues at the University of Utah (Vestal *et al.*, 1976a). A schematic diagram is shown in Fig. 7. A special feature of this apparatus is a high-pressure chemical ionization source, in which the average collision number of ions can be made large enough to achieve some degree of internal relaxation. Space-charge-limited ion beam intensities can be produced at laboratory kinetic energies less than 1 eV, with a full width at half maximum (FWHM) energy spread $\Delta\mathscr{E}_I = 0.3$ eV $+ 0.084\mathscr{E}_I$. The neutral gas target is a differentially pumped supersonic beam having a 4° divergence angle and a typical effective density at the crossing point corresponding to about $(1\text{–}2) \times 10^{-5}$ torr. The rotatable detector uses a quadrupole mass spectrometer and an electron multiplier–based ion counting system. In the original instrument, product ion energy analysis was accomplished with a retarding potential device. The present apparatus (Blakley *et al.*, 1977) incorporates a 90° cylindrical electrostatic deflector.

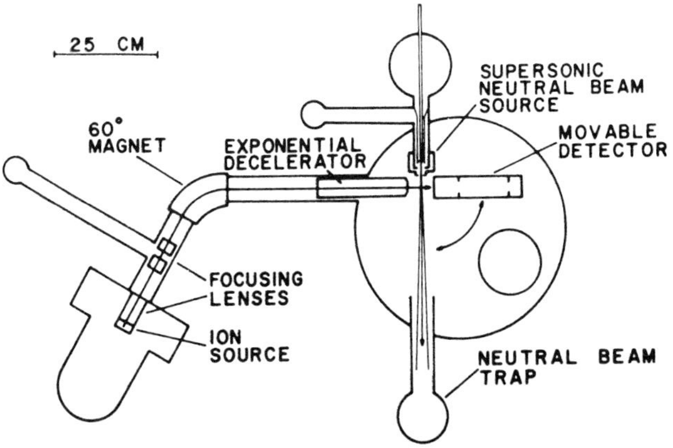

FIG. 7 Schematic diagram of the crossed beam apparatus at the University of Utah. [From Blakley *et al.* (1977).]

This group has studied several proton-transfer reactions of H_3^+, including the reactions $H_3^+(D_3^+) + D_2(H_2) \rightarrow D_2H^+(H_2D^+) + H_2(D_2)$ (Vestal *et al.*, 1974, 1976b), and $H_3^+ + N_2 \rightarrow N_2H^+ + H_2$ (Vestal *et al.*, 1976b). The reaction

$$H_3^+ + Ar \rightleftarrows ArH^+ + H_2, \qquad \Delta H_0 \sim 0.6 \text{ eV} \tag{10}$$

(Vestal *et al.*, 1976b; Blakley *et al.*, 1977) was studied in both the forward and reverse directions. Both reactions are found to proceed by direct mechanisms over the entire experimental cm energy range from 0.87 to 9.7 eV and from 0.18 to 6.7 eV, respectively. The product ion kinetic energy distribution in the forward reaction is found to be too broad to be consistent with the estimated reaction endothermicity and the assumption of fully relaxed H_3^+ reactants. This result is surprising in view of the large estimated collision number in the ion source and the expectation that excited H_3^+ should be able to relax rapidly by proton exchange with thermal H_2. Either the vibrational relaxation rate is much smaller than expected or some mechanism exists for repopulating the H_3^+ excited states, such as, for example, collisions between H_3^+ and H_2 in the ion source extraction zone. The most probable energy of the forward-scattered product ions in both reactions agrees with the spectator stripping model prediction within the experimental error. The cm kinetic energy distribution is, however, narrower for the reverse (exothermic) process than for the forward (endothermic) process, perhaps indicating that ArH^+ is more effectively vibrationally relaxed in the source than is H_3^+.

A substantial increase in the complexity of the *chemical system* being studied has been tackled by Safron and co-workers at Florida State University. This group has investigated the dynamics of the reactions of Cs^+ with some very large (by molecular beam standards) halogenated hydrocarbons. Abbreviating the benzyl radical $C_6H_5CH_2\cdot$ as Bz and the tolyl radical $CH_3C_6H_4\cdot$ as Tol, the prototype reactions are

$$Cs^+ + BzCl \rightarrow CsCl + C_7H_7^+ \tag{11a}$$

$$\rightarrow CsCl + C_2H_2 + C_5H_5^+ \tag{11b}$$

(Safron *et al.*, 1976) and

$$Cs^+ + \text{o-, m-, p-TolCl} \rightarrow CsCl + C_7H_7^+ \tag{12a}$$

$$\rightarrow CsCl + C_2H_2 + C_5H_5^+ \tag{12b}$$

(Miller and Safron, 1976). Schematic diagrams of the crossed beam apparatus are shown in Fig. 8. The Cs^+ beam originates in a porous tungsten surface ionization source which, without subsequent energy selection other than a

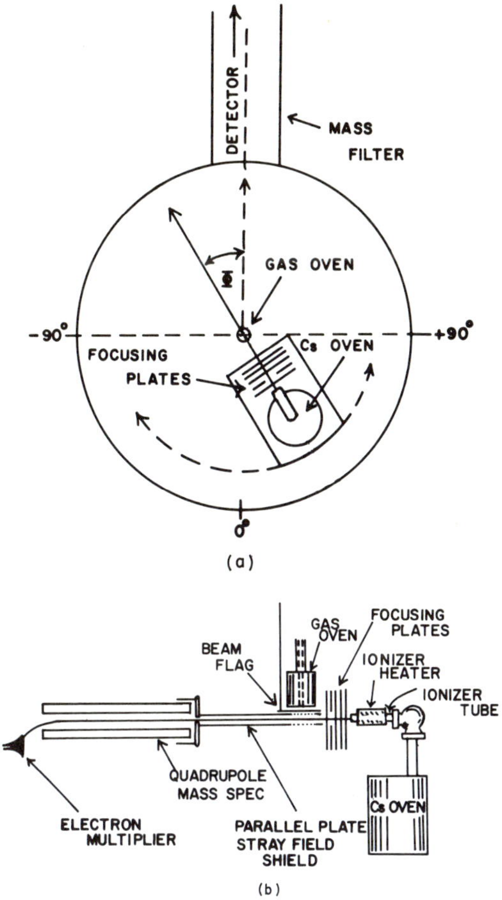

FIG. 8 Schematic diagram of the ion beam collision cell apparatus at Florida State University [Safron *et al.* (1976)]. (a) Top view. (b) Side view, with the detector energy analyzer not shown.

simple focusing lens, furnishes a beam of about 5% energy spread and 4°–8° FWHM angle spread. A heated effusive source is used for the hydrocarbon, with the target beam out of plane with respect to the ion beam and detector. The Cs^+ source rotates relative to the stationary detector in order to measure angle distributions. Energy analysis of the product ions is accomplished with a retarding-potential analyzer, which is followed by a quadrupole mass filter and an electron multiplier pulse-counting system.

Figure 9 shows a product ion intensity contour map measured for reaction (11a) at an ion laboratory energy $\mathscr{E}_{Cs^+} = 16$ eV. The two dominant features of the distribution are both peaked along the relative velocity vector. There

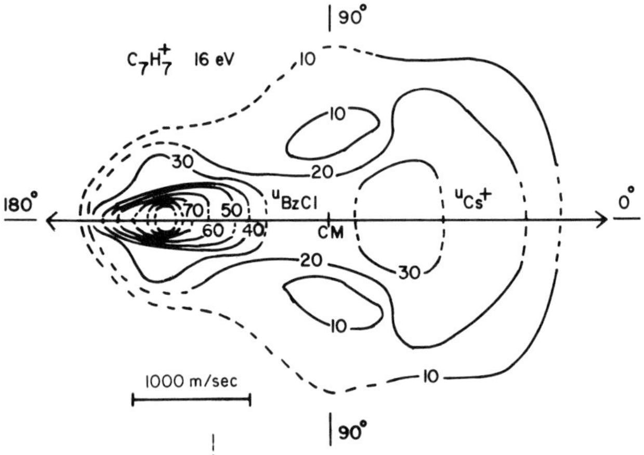

FIG. 9 Product ion intensity contour map for the reaction $Cs^+ + C_6H_5CH_2Cl \rightarrow C_7H_7^+ + CsCl$ for a Cs^+ laboratory kinetic energy of 16 eV ($E_i = 7.8$ eV). u_{Cs^+} and u_{BzCl} are the respective reactant cm velocity vectors and CM is the origin of the cm velocity vector coordinate system. [From Safron *et al.* (1976).]

is a very sharp forward-scattered peak (180° in the cm coordinate system as defined) which is qualitatively attributable to a stripping process, and there is also a backward-scattering peak which is much broader in angle and speed than the forward peak and which corresponds to greater product internal excitation energy. At $\mathscr{E}_{Cs^+} = 20$ eV the relative magnitudes of the two peaks have switched, so that backward scattering is the dominant process. One may reasonably infer that there are two distinct mechanisms for the reaction, which vary in relative importance with the collision energy. Safron *et al.* conclude that the backward-scattered component arises from collisions of Cs^+ with the Bz end of the molecule, resulting in large translational–internal energy transfer in the rearrangement process. Further dissociation represented by Eq. (11b) is thought to occur then by unimolecular decomposition of the excited Bz^+. Unlike most ion–molecule reactions which have been studied, the reactants in this case are both in singlet electronic states. It is interesting that several energetically accessible products which have triplet ground states are not observed experimentally, implying that spin conservation holds for this relatively complicated (16-atom, 114-electron) system, while spin is found not to be conserved in some relatively "simple" three-atom systems. The results of the chlorotoluene reactions (12a) and (12b) are qualitatively similar to those just discussed, in that two simultaneous mechanisms appear to compete. The difference in the reaction dynamics for *o*-TolCl compared to *m*-TolCl and *p*-TolCl appears to be direct confirmation of the expected "steric effect" of the adjacent methyl group in partially blocking the Cl^- stripping reaction.

The last five years have also witnessed the application of beam methods to the study of negative ion–molecule reaction dynamics. An apparatus designed specifically for the study of halide ion reactions has recently been placed in operation by Kashihira *et al.* (1976) at Kernforschungsanlage Jülich, GmbH. The ion source (Kashihira *et al.*, 1977) operates on a supply of alkali halide vapor and produces halide ions by surface ionization on a LaB_6 plug at about 1300°C. A small magnetic field in the ion beam path removes the thermally emitted electrons. The source furnishes a Br^- beam of about 1×10^{-9} to 2×10^{-8} A over the energy range 10–100 eV and with an energy spread of about 0.6 eV FWHM. The halide beam is focused into a scattering cell without additional mass or energy selection, and the product ions emerging near the ion beam axis are extracted and focused into a quadrupole mass spectrometer detector. The energy dependence of the cross section measured for the endothermic reaction

$$Br^- + CH_4 \rightarrow CH_2Br^- + H_2 \tag{13}$$

is shown in Fig. 10 (Kashihira *et al.*, 1976). Although the reaction cross section rises steeply above the threshold energy, it reaches a maximum value

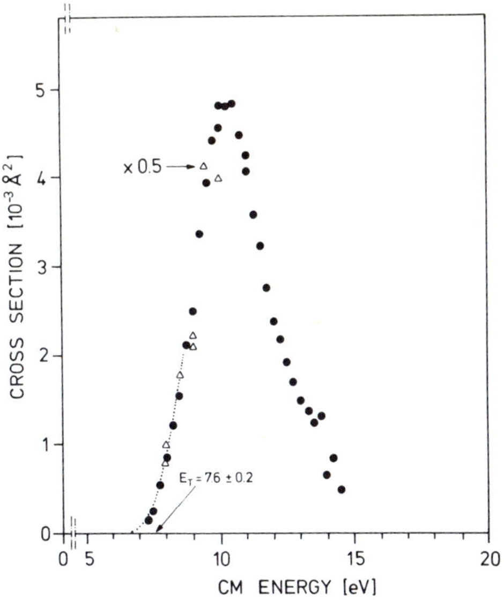

FIG. 10 Energy dependence of the total cross section for the reactions

$$Br^- + CH_4 \rightarrow CH_2Br^- + \cdots \quad (\bullet)$$
$$\rightarrow H^- + \cdots \quad (\triangle)$$

[Kashihira *et al.* (1976)]. The arrow shows the estimated energy threshold for this endothermic reaction.

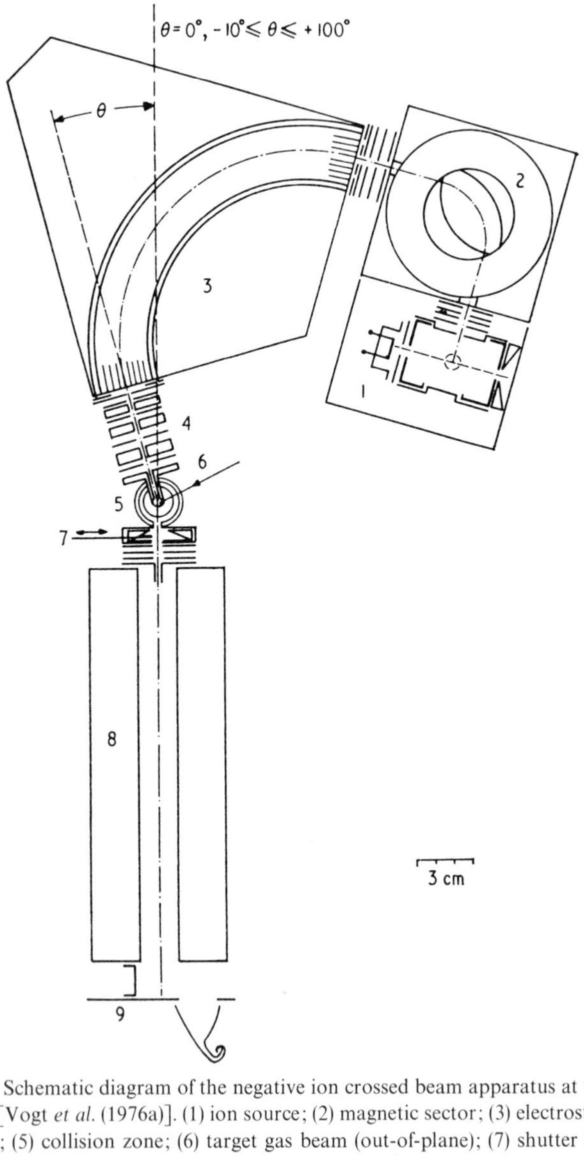

FIG. 11 Schematic diagram of the negative ion crossed beam apparatus at the University of Hamburg [Vogt *et al.* (1976a)]. (1) ion source; (2) magnetic sector; (3) electrostatic analyzer; (4) ion optics; (5) collision zone; (6) target gas beam (out-of-plane); (7) shutter for measuring primary ion beam current; (8) quadrupole mass filter; (9) channeltron detector.

of only about 4×10^{-3} Å^2 and then falls rapidly again. One of the reasons for this very small cross section is competition with the alternative reaction channel $Br^- + CH_4 \rightarrow CH_3Br + H^-$, the cross section for which cannot be determined accurately with this apparatus because of the kinematics, which cause the H^- product to be distributed through all laboratory scattering angles, while the detector looks only at angles within about 4° of the primary Br^- beam. However, the authors estimate that the cross section for this reaction is about 25 times larger than that for reaction (13). Other energetically accessible products, such as C^-, CH^-, CH_2^-, CH_3^-, and $CHBr^-$, are not observed.

Vogt and co-workers at the University of Hamburg have recently described a new negative ion crossed beam apparatus capable of differential cross section measurements (Vogt *et al.*, 1976a). The apparatus is shown in Fig. 11. Negative ions are formed in an electron bombardment source, usually by an ion pair formation mechanism. The ion beam is subjected to both momentum and energy analysis in tandem magnetic and electrostatic sectors, and is then focused using zoom optics into a shielded zone where the target gas beam intersects the ion beam at right angles. The ion source assembly rotates with respect to the detector in the out-of-plane geometry. Laboratory ion energies in the range from 0.2 to 100 eV are accessible, with an energy spread of about 0.2 eV FWHM and an intensity of about 10^{-11} A at energies above 2 eV. Product ions are mass selected with a quadrupole mass filter and counted digitally. The present instrument does not incorporate provision for energy analyzing the scattered products. The laboratory angular resolution is about 3°.

An example of the data is shown in Fig. 12, which gives the differential cross section for 0° laboratory scattering angle as a function of primary beam laboratory kinetic energy, for the exothermic reaction

$$H^- + H_2O \rightarrow OH^- + H_2. \tag{14}$$

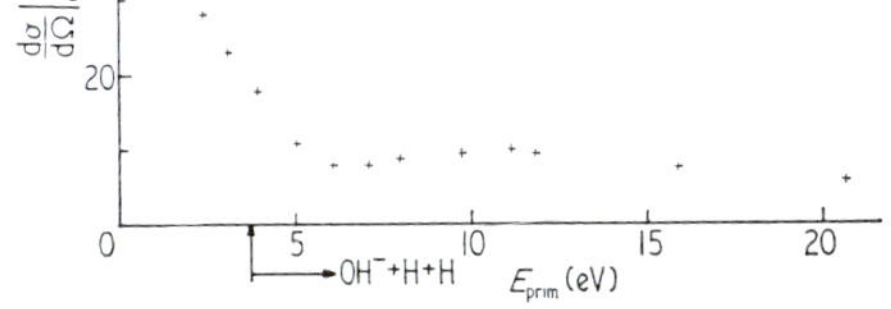

FIG. 12 Differential cross section at 0° scattering angle as a function of primary beam energy, for the reaction $H^- + H_2O \rightarrow OH^- + H_2$. The arrow marks the calculated threshold energy for the dissociative channel which produces $OH^- + H + H$. [From Vogt *et al.* (1976a).]

The rapid drop in cross section with energy at low energies is similar to the typical result for exothermic positive ion reactions. This group has also measured differential cross sections for the endothermic reactions Cl^-, Br^-, $I^- + O_2 \rightarrow ClO^-$, BrO^-, $IO^- + O$ (Vogt *et al.*, 1977b), as well as similar charge transfer and atom transfer reactions in the $I^- + CO$ system (Vogt and Mischke, 1977) and the O^-, $O_2^- + O_2$ systems (Vogt *et al.*, 1976b, 1977a).

The first molecular beam study of a negative ion–molecule reaction was an investigation of the reaction $O^- + D_2 \rightarrow OD^- + D$ by Doverspike *et al.* (1973). Very recently a further study of the $O^- + D_2$ reaction has been completed by Cross and co-workers at Yale University (Johnson *et al.*, 1978). The apparatus used was the early crossed beam instrument EVA (Herman *et al.*, 1969), which was modified to incorporate a negative ion source and detector. O^- is generated by dissociative electron attachment to N_2O in a small magnetic field which helps to confine the electrons. Contour maps of the product ion energy and angle distributions were measured over the range of initial cm kinetic energy from 1.2 to 4.7 eV. At the low end of the energy range the product ion distribution is centered at the cm velocity vector, and at higher energies the product distribution shifts to a distinct peak at 0° in the cm system relative to the initial velocity vector of O^-. The authors conclude that a long-lived D_2O^- complex is formed at cm energies less than about 2.5 eV. The alternative product channel $D^- + OD$ is energetically accessible, but is not observed. Other recent work carried out with this apparatus includes studies of the "simplest" reaction $H^+ + H_2 \rightarrow H_2^+ + H$ in several isotopic combinations (Krenos *et al.*, 1974). In this reaction transitions can occur between the two lowest-lying singlet electronic potential energy surfaces of H_3^+. Classical trajectory surface-hopping calculations were carried out to interpret the scattering data. In an investigation of the reaction

$$O^+ + N_2 \rightarrow NO^+ + N \tag{15}$$

(Smith and Cross, 1974), it was concluded that this controversial and much-studied reaction occurs by a single direct mechanism over the cm energy range from 1.5 to 15.8 eV, and yields ground-state products. Other processes studied recently by the Yale group are the reactions $C_2H^+ + D_2 \rightarrow C_2D^+$, C_2HD^+, $C_2D_2^+$ + other products (Smith *et al.*, 1975), and H_2^+, $D_2^+ + D_2$, $H_2 \rightarrow D_2H^+$, $H_2D^+ + H$, D (Krenos *et al.*, 1976).

The latter reactions are additional examples of direct reactions in which two electronic surfaces (corresponding asymptotically to the two nearly resonant charge transfer states) participate. Calculations of the potential-energy surfaces and the nonadiabatic coupling matrix elements indicate that electronic transitions (charge transfer) occur readily in the entrance channel.

At the Czechoslovak Academy of Sciences, Herman and colleagues have constructed a new version of the Yale crossed-beam apparatus, which they call EVAII (Hierl *et al.*, 1977). Ions produced in an electron impact source are momentum selected in a 90° deflection magnetic sector and are then decelerated to the desired laboratory kinetic energy, which may be as low as 0.5 eV. Typical primary beam energy and angle spreads are 0.2 eV FWHM and 1° FWHM, respectively. A chopped neutral beam intersects the ion beam at right angles, and the sources rotate with respect to the stationary detector in the in-plane geometry. Retarding potential measurements are used to energy analyze the product ions, which then pass into a magnetic sector mass spectrometer and ion counting system.

This apparatus and another similar ion beam–collision cell instrument (Herman and Birkinshaw, 1973) have been used recently for the study of both charge transfer reactions and rearrangement reactions. Among the rearrangement processes investigated is the endothermic reaction

$$H_2^+ + He \rightarrow HeH^+ + H, \qquad \Delta H_0 = 0.8 \text{ eV} \tag{16}$$

(Pacak *et al.*, 1977; Herman and Pacak, 1977; Schneider *et al.*, 1976, 1977). These workers were able to study the influence of H_2^+ reactant vibrational energy on the reaction dynamics by producing the H_2^+ in either a simple electron impact process (which gives a broad, approximately Franck–Condon, distribution of H_2^+ vibrational states) or in a source containing a relatively high pressure of Ne, which reacts rapidly with vibrationally excited H_2^+ for all the exothermic channels,

$$H_2^+(v \geq 2) + Ne \rightarrow NeH^+ + H. \tag{17}$$

Thus, most of the H_2^+ in excited states $v \geq 2$ are removed either by reaction or by collisional relaxation. The comparison of these two cases shows that the reaction is direct with either vibrationally hot or vibrationally cold H_2^+, but that small-angle scattering (qualitatively stripping) is much more prominent in the reaction with vibrationally excited H_2^+. The difference in the reaction dynamics is associated with the potential energy requirement for an endoergic process, which requires a small impact parameter (or small centrifugal contribution to the effective potential) in order for the translational energy to be effective in surmounting the potential energy barrier to reaction. With a vibrationally excited reactant molecule this constraint is relaxed somewhat. This group has also investigated the more complicated reactions

$$\begin{aligned} C_2H_2^+ + C_2H_4 &\rightarrow C_3H_3^+ + CH_3 \\ &\rightarrow C_4H_5^+ + H \end{aligned} \tag{18}$$

and

$$I_2^+ + C_2H_4 \rightarrow C_2H_3I^+ + HI \tag{19}$$

(Herman and Birkinshaw, 1973). Reaction (19) represents the relatively rarely studied class of "four-center" reactions, in which both reactants lose one atom or group and gain another. The results indicate a complicated reaction mechanism possibly involving both complex- and direct-mode processes.

Finally, we should mention in this section the studies of rearrangement reactions carried out recently by Rutherford and Vroom at IRT Corporation, using a crossed beam, total cross section apparatus (Rutherford *et al.*, 1971). These investigators were able to study the influence of H_2^+ vibrational energy on the total cross section for reaction (16) by varying the electron energy in the H_2^+ source in the range near threshold for excited state production (Rutherford and Vroom, 1973a). The phenomenological reaction cross section increases, of course, with increasing vibrational excitation at low kinetic energies (below $E_i = 9$ eV), but the vibrational excitation appears to make little difference in the reaction cross section at very large collision energies. The cross sections for the reaction $HeH^+ + H_2 \rightarrow H_3^+ + He$ (Rutherford and Vroom, 1973b) were found to be fairly typical for exothermic proton transfer reactions. Using an iridium furnace to dissociate O_2, Rutherford and Vroom (1974) measured the energy dependence of the cross section for the reaction $N_2^+ + O \rightarrow NO^+ + N$, which is interesting because it explores the same set of N_2O^+ intermediate potential energy surfaces as the important $O^+ + N_2$ reaction (15), but from a different initial electronic state. It is also one of a very small number of reactions which have been observed experimentally between an ion and an atom which possesses a permanent quadrupole moment (see Section II.B).

Other reactions recently investigated by this group include $C^+ + CO_2 \rightarrow CO^+ + CO$ (Rutherford and Vroom, 1976a), $O^+ + NO \rightarrow O_2^+ + N$ (Rutherford and Vroom, 1976b), and a series of Al^+ reactions, including $Al^+ + O_2 \rightarrow AlO^+ + O$ and $Al^+ + N_2O \rightarrow AlO^+ + N_2$ (Rutherford and Vroom, 1976c). The Al^+ beam was generated by surface ionization of $AlCl_3$ on a tungsten filament. The thermodynamics of the Al^+ reactions are not known quantitatively. However, the cross sections for both of these reactions exhibit threshold behavior which is qualitatively consistent with the expected endothermicity. The cross sections for the reaction $Al^+ + N_2 \rightarrow AlN^+ + N$ were found to be unmeasurably small. Total cross sections for a variety of charge transfer processes have also been measured in this laboratory, including negative ion charge transfer to O atoms and to O_3 (Rutherford and Vroom, 1974, 1975; Rutherford *et al.*, 1973).

2. *Charge Transfer Reactions*

Processes in which electrons are transferred between atoms or molecules can sometimes be as complex dynamically as those in which atoms are transferred. Beam techniques have proven to be very useful in understanding

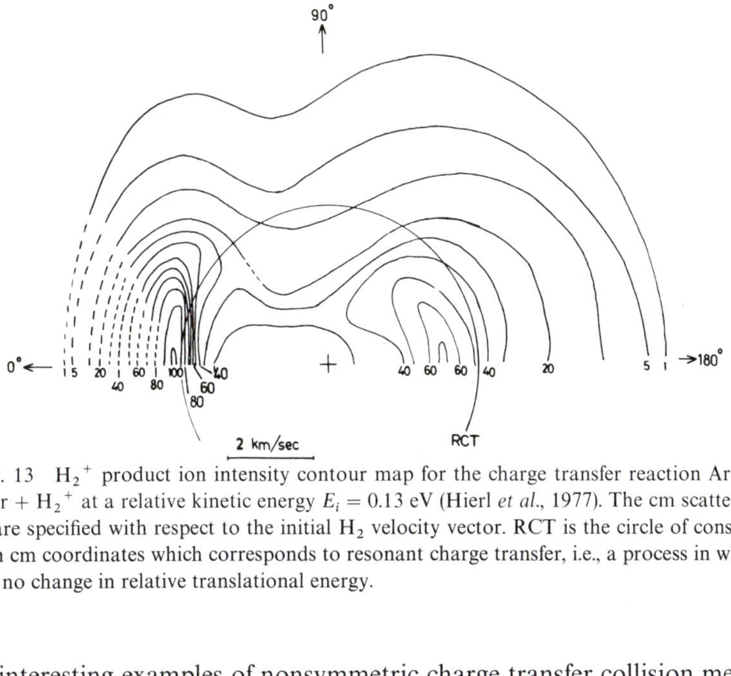

FIG. 13 H_2^+ product ion intensity contour map for the charge transfer reaction $Ar^+ + H_2 \rightarrow Ar + H_2^+$ at a relative kinetic energy $E_i = 0.13$ eV (Hierl *et al.*, 1977). The cm scattering angles are specified with respect to the initial H_2 velocity vector. RCT is the circle of constant speed in cm coordinates which corresponds to resonant charge transfer, i.e., a process in which there is no change in relative translational energy.

some interesting examples of nonsymmetric charge transfer collision mechanisms.

Herman and co-workers have recently used the crossed beam apparatus described earlier to study the dynamics of the charge transfer reactions

$$Ar^+ + H_2 \rightarrow Ar + H_2^+ \tag{20}$$

(Hierl *et al.*, 1977) and

$$Ar^+ + NO \rightarrow Ar + NO^+ \tag{21}$$

(Herman *et al.*, 1976). The results of their lowest-energy experiment on reaction (20) are shown as an H_2^+ product ion intensity contour map in Fig. 13. The narrow and intense peak at 0° corresponds to a large impact parameter process which occurs with small translational–internal energy transfer. These H_2^+ ions therefore have small laboratory kinetic energies, and the dashed contour lines can only be estimated. It is very interesting, however, that H_2^+ products are formed in comparable intensity at all other scattering angles as well, despite the fact that the charge transfer channel competes with the atom transfer channel $Ar^+ + H_2 \rightarrow ArH^+ + H$, which has a very large cross section at low collision energies. Although this group did not determine the relative integral probabilities for the charge transfer and atom transfer processes, it is apparent that the rearrangement channel does not consume all the reactants which collide with small impact parameters. The charge transfer reaction (20) produces H_2^+ in more highly excited states at large scattering angles than at small scattering angles. It is to be expected for intimate, small impact parameter collisions on a reactive potential energy surface that the excursion of the molecular vibrational coordinate during the collision will be large, thus enhancing the probability of translational-to-vibrational energy transfer at large scattering angles (Giese and Gentry, 1974). With an increase in the collision energy to 3.44 eV, the probability of the large-angle charge transfer process relative to that for the small-angle process decreases by a factor of about 3500, for reasons which are not entirely clear. The data on the $Ar^+ + NO$ charge transfer reaction (21) indicate both a long-range electron transfer and a long-lived intermediate complex mechanism. At low energies both mechanisms are thought to yield excited NO^+ in the $a^3\Sigma^+$ state, for which the charge transfer process is nearly resonant. Higher excited NO^+ electronic states are populated as the collision energy increases above 1.5 eV.

The dynamics of several dissociative charge transfer reactions have been investigated by Udseth, Giese, and Gentry at the University of Minnesota. The general-purpose ion beam–collision cell apparatus used in this work is shown in Fig. 14 (Udseth *et al.*, 1973). Ions are formed in a miniature coaxial

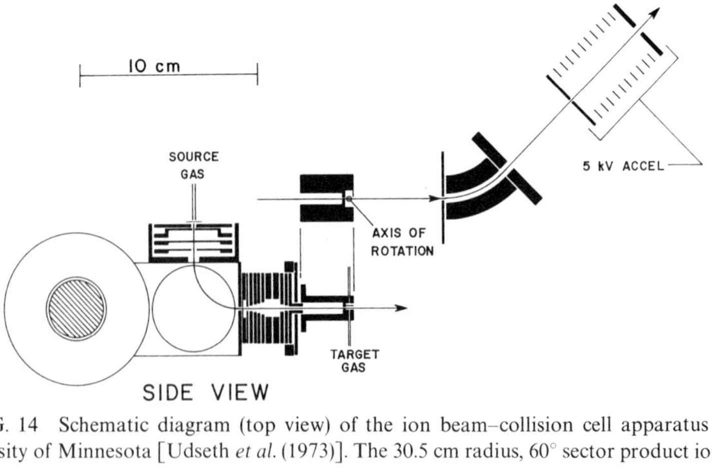

FIG. 14 Schematic diagram (top view) of the ion beam–collision cell apparatus at the University of Minnesota [Udseth *et al.* (1973)]. The 30.5 cm radius, 60° sector product ion mass spectrometer and ion counting system are not shown.

electron impact source and are extracted through the tungsten mesh cathode. The beam is then momentum selected in a 90° magnetic deflector and refocused into the collision cell. Ions scattered into the detector are energy analyzed by a 45°-deflection cylindrical electrostatic sector and are then accelerated to 5 keV and mass analyzed with a 30.5 cm radius, 60° sector mass spectrometer. A scintillation detector is used for ion counting. The collision cell can be cooled to 77 K with liquid nitrogen in order to reduce the spread in target molecule velocities. Typical effective FWHM instrumental resolutions (taking into account both the ion beam and the detector contributions) are roughly 0.1 eV + $0.01\mathscr{E}$ in energy and 1.5° in angle. Useful ion beam intensity can be obtained at energies down to about 0.5 eV. A unique feature of this apparatus is that both the laboratory scattering angle and the energy analyzer settings are controlled digitally. Data scans over a large number of points are acquired automatically and accumulated in a multiscaler memory. By using an off-line computer to generate the necessary laboratory to cm transformations prior to the experiment, scans at either constant cm angle or constant cm speed may be acquired by simultaneously adjusting both the laboratory angle and laboratory energy at each point in such a way as to keep the appropriate cm variable fixed.

One of the charge transfer reactions which was studied,

$$He^+ + O_2 \rightarrow He + O + O^+ \tag{22}$$

(Udseth *et al.*, 1979), occurs through an interesting mechanism which is revealed in the energy and angle distributions of the O^+ product. Figure 15 shows laboratory O^+ energy spectra measured at a cm collision energy of 1.36 eV and at several values of laboratory scattering angle. The large peak at 1.7 eV, in the 0° scan only, is an artifact due to the primary He^+ beam which passes through the detector energy analyzer at 1.7 eV and 0° and then reacts with O_2 background gas to produce O^+ in the acceleration lens of the product mass spectrometer. The four peaks seen at other scattering angles each correspond to different electronic states of the $O + O^+$ products. Table II lists the product states which are energetically accessible, and the exothermicity of each channel relative to the 24.58 eV ionization energy of He. The observed peak energies are, within experimental error, independent of laboratory scattering angle, implying that the O^+ product energy distribution is approximately symmetric about the *laboratory*, not the cm, origin. Furthermore, the O^+ peak energies are also virtually independent of He^+ kinetic energy below about 3 eV. Taken together, these data imply that the charge transfer reaction occurs with very little concomitant momentum transfer to the oxygen species from the incoming He^+. To a first approximation, the O_2 molecule transfers its electron to He^+ at a sufficiently large separation that the interaction potential is small at this point. The resulting

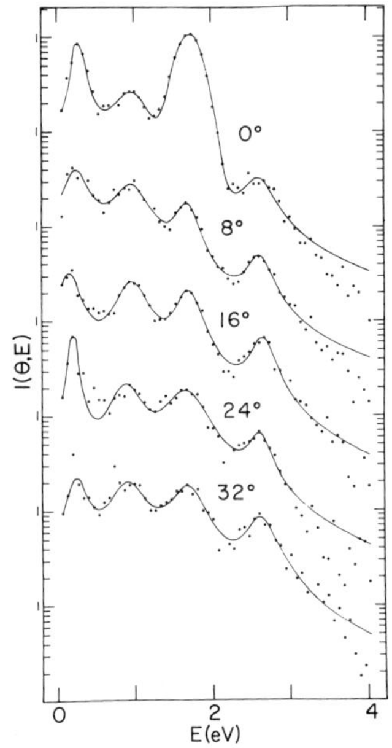

FIG. 15 Laboratory kinetic energy spectra for O^+ produced in the reaction $He^+ + O_2 \rightarrow He + O + O^+$ at $E_i = 1.36$ eV [Udseth *et al.* (1979)]. The spectra for the various indicated laboratory scattering angles are arbitrarily displaced from each other by about one order of magnitude in the logarithmic intensity scale. The intensity plotted is the relative differential cross section for scattering into the element $d^3\mathrm{v}$ in laboratory velocity vector space.

TABLE II

Energetics of Various Product Channels in the Reaction $He^+ + O_2 \rightarrow He^+ + O + O^+$.[a]

Product state	Translational exoergicity (eV)	Predicted O^+ energy (eV)	Observed energy (eV)
(a) $O(^3P) + O^+(^4S)$	5.86	2.93	2.67
(b) $O(^1D) + O^+(^4S)$	3.89	1.95	1.79
(c) $O(^3P) + O^+(^2D)$	2.53	1.27	0.96
(d) $O(^1S) + O^+(^4S)$	1.67	0.84	absent
(e) $O(^3P) + O^+(^2P)$	0.84	0.42	} 0.3
(f) $O(^1D) + O^+(^2D)$	0.57	0.29	

[a] For each exothermic channel is given the translational exoergicity at a total energy of 24.58 eV above the O_2 ground state, the O^+ laboratory energy predicted from the simplest model of the dynamics to be $\frac{1}{2}$ the translational exoergicity, and the observed O^+ energy for the experimental peaks assigned to the various product states.

O_2^+ ion is left in a dissociative state about 24.58 eV above the O_2 ground state. As the O_2^+ explodes about its own center of mass (which is nearly stationary in laboratory coordinates), the O atom and O^+ ion each receive half of whatever reaction exoergicity is available to translation. This simple prediction is also listed in Table II for each product channel. The O^+ energies which are actually observed correlate well with the predicted energies, but are slightly lower in each case. This slight discrepancy can be attributed to the long-range attractive force between He^+ and O_2, which imparts a small backward laboratory velocity to O_2 prior to the moment of charge transfer. The two product states (e) and (f) in Table II cannot be resolved in the O^+ energy spectra and appear together in unknown relative intensity in the lowest energy experimental peak. The three lowest internal energy states of $O + O^+$ are seen clearly in the experimental data, with probabilites which increase monotonically with increasing product internal energy. The small variation of the peak intensities with scattering angle is not experimentally significant. The product channel (d) $O(^1S) + O^+(^4S)$ is conspicuous by its absence in all the experimental energy spectra.

One can imagine either of two charge transfer mechanisms to be responsible for the observed reaction dynamics. At the moment of charge transfer the O_2 molecule could make a transition directly to one of several repulsive $O_2{}^+$ potential energy functions which cross the resonance energy (24.58 eV above the O_2 ground state) at separations near the O_2 potential minimum. Alternatively, the transition could occur to a bound excited state of $O_2{}^+$ which has a short predissociation lifetime, or which radiates to a repulsive or predissociated state. Evidence on this point is provided by comparison of the $He^+ + O_2$ charge transfer results with those of O_2 photoionization and electron impact ionization experiments in which the O^+ fragment energies were measured. The charge transfer ionization mechanism requires that the O_2 transition energy be very close to 24.58 eV and that the $O + O^+$ products carry away all of the translational energy available to each product channel. In electron impact ionization or photoionization experiments, the free electron or electrons may receive an arbitrary fraction of the available energy, and there is therefore no reason to expect the same discrete O^+ kinetic energy peaks seen in the dissociative charge transfer experiments, *unless the dissociation proceeds by predissociation of the same* $O_2{}^+$ *bound state in all cases*. In fact, the data from photoionization experiments (Gardner and Samson, 1975) and electron impact experiments (Schopman and Locht, 1974) have peaks in the O^+ energy spectra which look remarkably similar to those shown in Fig. 15. The most obvious candidate for the intermediate $O_2{}^+$ state is $O_2{}^+(c^4\Sigma_u{}^-)$, which in its ground vibrational state is almost exactly resonant with the He ionization energy. Spectroscopic evidence, however, indicates that the low vibrational states of $O_2{}^+(c^4\Sigma_u{}^-)$ are not completely predissociated, and may possibly radiate to stable bound $O_2{}^+$ states. Despite a very careful search, no $O_2{}^+$ products were detected in our low energy $He^+ + O_2$ charge transfer experiments. This result is consistent with recent ion cyclotron resonance (ICR) data (Anicich *et al.*, 1977). While other simultaneous processes are not completely ruled out, the preponderance of the evidence indicates that the dominant reaction mechanism is charge transfer to a $^2\Pi_u$ state, which predissociates to the observed products in a time comparable to or less than the vibrational period. The absence of the product state (d) $O(^1S) + O^+(^4S)$ can be attributed to the failure of the corresponding $^4\Sigma_u^-$ state potential to cross the $^2\Pi_u$ state potential in the appropriate separation range, or to the lack of sufficient spin–orbit coupling to induce the transition.

Some interesting dissociative charge transfer reactions have recently been studied by Maier (1975) at the Los Alamos Scientific Laboratory, using a beam–collision cell apparatus for measuring total cross sections. Charge transfer in the rare-gas systems

$$He_2^+ + Ne, Ar, Kr, Xe \rightarrow Ne^+, Ar^+, Kr^+, Xe^+ + 2He \tag{23}$$

and

$$Ar_2^+ + Xe \rightarrow Xe^+ + 2Ar \tag{24}$$

occurs by dissociation of the unbound or very weakly bound neutral molecule rather than of the ion. The cross sections for these processes show a rapid rise with decreasing collision energy in the range 1–40 eV. Maier was also able to observe the rearrangement channels $He_2^+ + Ne \rightarrow HeNe^+ + He$ and $Ar_2^+ + Ne \rightarrow ArNe^+ + Ar$.

3. *Energy Transfer Processes*

Ion beam experiments during the last several years have made major contributions to the understanding of inelastic nonreactive molecular collisions. In the most successful of these experiments, the inelastic energy defect of the collision is determined from the cm translational energy change, $Q = E_f - E_i$, which is much easier to measure experimentally for an ion than for a neutral species. With sufficient instrumental resolution, differential cross sections for discrete vibrational and even rotational transitions can be measured in favorable cases. Although there were some previous examples in which vibrational structure was resolved in integral cross section measurements (e.g., Moore and Doering, 1969) and in differential cross sections for a mixture of molecular ion states (e.g., Cosby and Moran, 1970), the first measurements of differential cross sections for distinct resolved vibrational state-to-state transitions were reported by Udseth *et al.* (1971), for the system

$$H^+ + H_2(v = 0) \rightarrow H^+ + H_2(v = 0, 1, 2, 3). \tag{25}$$

The apparatus used was the same one shown in Fig. 14. Subsequent experimental work (Udseth *et al.*, 1973) extended these measurements over the cm kinetic energy range from 4 to 16 eV and to the additional isotopic target molecules HD and D_2. Theoretical analysis using an accurate fit to a large number of calculated points on the potential energy surface (Giese and Gentry, 1974; Gentry and Giese, 1975a) revealed the mechanism of vibrational excitation in these small scattering angle experiments to be a stretching force between the molecular nuclei, caused by dilution of the molecular bond as the passing proton withdraws electron density from the molecule. Thus, the excitation process in this case is very different from the traditional concept of vibrational excitation, in which the molecular vibrational potential remains undistorted, and vibrational excitation is presumed to take place by compression of the molecular bond in small impact-parameter repulsive collisions. Another interesting excitation mechanism was discovered for $H^+ + CH_4$, CD_4 collisions (Gentry *et al.*, 1975a). Although discrete vibrational states were not resolved in this case, the inelastic energy loss spectra showed a bimodal structure with one sharp peak corresponding to very small average energy transfer, and another broad peak corresponding to an average excitation of many vibrational quanta. It was conjectured that a potential energy surface-hopping mechanism may be involved, and that the low-excitation peak is due to collisions in which the system remains on the diabatic $H^+ + CH_4$ surface, while the high-excitation peak is caused by a temporary transition to a surface which could be described approximately as $H + CH_4{}^+$ in character. The $CH_4{}^+$ ion is Jahn–Teller distorted from the tetrahedral geometry, and the $\sim$2 eV distortion energy implies that the vibrational excitation probabilities would be very large for an abrupt $CH_4 \rightarrow CH_4{}^+ \rightarrow CH_4$ electronic transition.

Udseth *et al.* (1974) also studied the inelastic collisions of protons with the polar molecules HF, HCl, and CO. Figure 16 shows an energy spectrum for $H^+ + HF$ collisions at constant cm scattering angle $\chi = 5°$ for $E_i = 10$ eV. The overall experimental energy resolution in this example is 0.23 eV FWHM; therefore, pure vibrational transitions ($\Delta E = 0.49$ eV) of the type seen in the $H^+ + H_2$ scattering would be clearly resolved. However, deconvolution of the experimental data yields a simple monotonically decreasing energy-loss distribution. Although the inelastic energy transfer is large, no vibrational structure is seen in the data, indicating that (unresolved) rotational excitation is the dominant inelastic process. This result is to be expected because of the very strong and long-range torque on the molecule arising from the charge–dipole interaction. The average excitation energy for small angle $H^+ + HF$ scattering is in semiquantitative agreement with a classical perturbation theory calculation (Gentry, 1974) in which the interaction potential is simplified to include only the charge–dipole and charge-induced dipole terms.

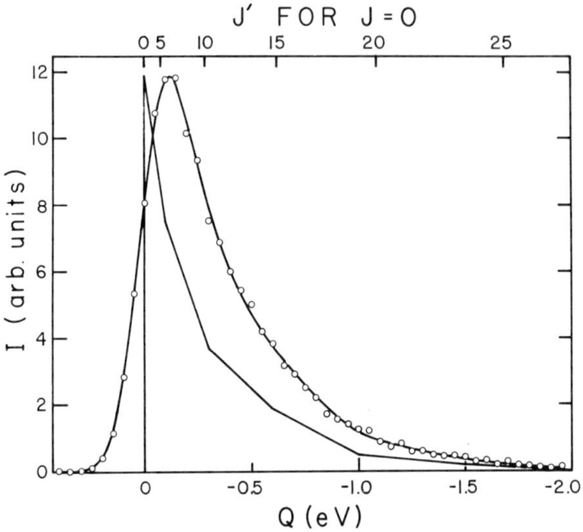

FIG. 16 Experimental energy distribution for $H^+ + HF$ collisions at $E_i = 10$ eV and $\chi = 5°$. Q is the translational energy defect in the cm system, $Q = E_f - E_i$. The data are compared with the results of a deconvolution procedure which gives the best piecewise linear approximation to the excitation energy distribution. The segmented curve is the deconvoluted function, and the solid line through the data points is the result of the least-squares fit of the convolution to the data.

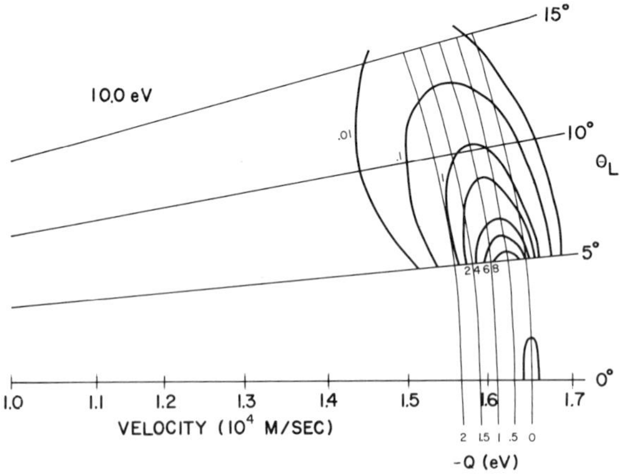

FIG. 17 Intensity contour map for nonreactive scattering of Li^+ from H_2O at a Li^+ initial laboratory kinetic energy $\mathscr{E}_I = 10.0$ eV (Siskind *et al.*, 1976). The angles and speeds indicated are in laboratory coordinates. The circular arcs shown indicate constant cm values of $Q = E_f - E_i$ between 0 and -2 eV. The contour line terminating on the 0° axis is the 50%-of-peak contour for the primary $^6Li^+$ beam.

Another example of ion scattering from polar molecules has been provided recently by Coplan and colleagues at the University of Maryland (Siskind *et al.*, 1976). Alkali metal ions are generated by surface ionization on a heated porous tungsten plug and are accelerated and focused into the collision cell without additional mass or energy selection. The collision cell is sealed with a brass bellows and the assembly is articulated to permit rotation of the source through $\pm 45°$. The scattered ions are focused into a hemispherical electrostatic energy analyzer and are counted by means of a channeltron detector. Typical laboratory resolutions are 2% FWHM in energy and 2° FWHM in angle.

Figure 17 shows an interesting contour map for $^6Li^+$ scattered from H_2O at $\mathscr{E}_I = 10.0$ eV ($E_i = 7.4$ eV). It can be seen that an energy-loss profile at constant cm. scattering angle would look qualitatively very much like that shown in Fig. 16 for $H^+ + HF$ collisions. No fine structure is resolved in the experimental distributions, but a broad inelastic feature is observed which shows that the most probable excitation energy increases with increasing scattering angle. Because of the closely spaced vibrational levels of H_2O, it is not expected that even pure vibrational excitation would be well resolved in this experiment; however, it seems likely that the translational energy loss is mostly into rotational degrees of freedom. The small angle differential cross section is proportional to χ^{-3}, as expected for a potential dominated by the R^{-2} charge–dipole term.

Toennies and co-workers at the Max-Planck-Institut für Strömungsforschung have been studying Li^+ collisions with various molecules using a very different type of apparatus, shown in Fig. 18 (Eastes *et al.*, 1977). $^7Li^+$ ions are produced by surface ionization and are then energy selected in a 127° cylindrical electrostatic deflector. The energy selected beam is then electrically chopped to create 0.25 μsec duration pulses of ions with a pulse interval of typically 20 μsec. The chopping pulse starts a time-to-amplitude converter which is stopped by the arrival of an ion pulse at the detector, located at a distance of 90–115 cm from the collision zone. A 400-channel pulse-height analyzer is used to accumulate the resulting time-of-flight spectrum. A supersonic beam of the neutral target molecules is formed by a nozzle–skimmer arrangement, and the detector flight tube rotates about the beam crossing point in the out-of-plane geometry to measure angular distributions. Because the chopped ion beam intensity is very low ($\sim 10^4$ ions/sec) long data accumulation times are required. A typical spectrum might require 36 hr to obtain. However, very good kinetic energy resolution is achieved because of the long flight path.

The Göttingen group has carried out pioneering work on vibrationally state-resolved inelastic scattering in $Li^+ + H_2$ collisions at large scattering angles (David *et al.*, 1973) and on rotationally state-resolved $Li^+ + H_2$ and

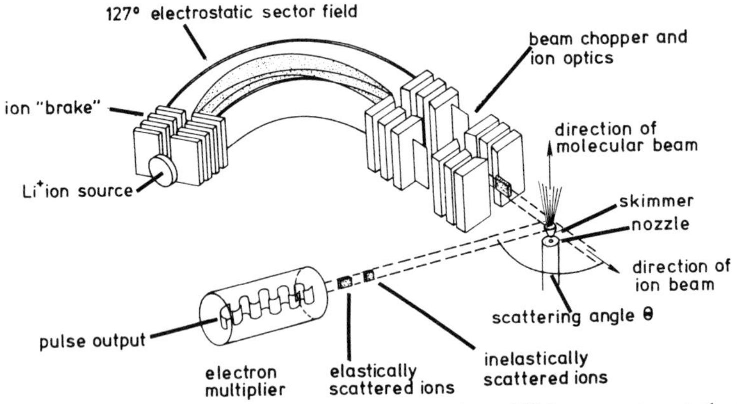

FIG. 18 Schematic diagram of the crossed beam time-of-flight apparatus at the Max-Planck-Institut für Strömungsforschung. [From Eastes *et al.* (1977).]

$H^+ + H_2$ collisions at small scattering angles (van den Bergh *et al.*, 1973; McGuire *et al.*, 1976). The vibrational excitation mechanism in the $Li^+ + H_2$ system is very different from that in the $H^+ + H_2$ system. Vibrational excitation occurs with high probability only for small impact parameter, large scattering angle collisions in which the dominant force is repulsive rather than attractive (Gentry and Giese, 1975b; Barg *et al.*, 1976). Classical trajectory calculations show very strong rotational–vibrational coupling in slightly noncollinear collisions, which produces simultaneous vibrational and multiquantum rotational excitation. Experiments have also been carried out to study the dynamics of collision-induced dissociation in the $Li^+ + H_2$ and $Na^+ + H_2$ systems (Schöttler and Toennies, 1973, 1974). [Inelastic scattering of Na^+ by D_2 and HD has also been investigated by Dimfl and Mahan (1974)]. More recently, vibrational excitation has been observed in the $Li^+ + N_2$, CO systems (Böttner *et al.*, 1976) and in the $Li^+ + CO_2$, N_2O systems (Eastes *et al.*, 1977). The N_2 and CO target molecules yielded spectra showing unresolved rotational excitation superimposed on the discrete quantum vibrational structure. Relatively little rotational excitation was seen in the CO_2 and N_2O systems, making the assignment of peaks in the measured spectra to particular vibrational transitions relatively easy. A typical spectrum for $Li^+ + CO_2$ collisions is presented in Fig. 19. The laboratory scattering angle $\theta = 30°$ in this experiment is much larger than the rainbow angle ($\theta_r = 13°$); therefore, one expects the scattering to be dominated by the repulsive part of the potential energy surface. The effective

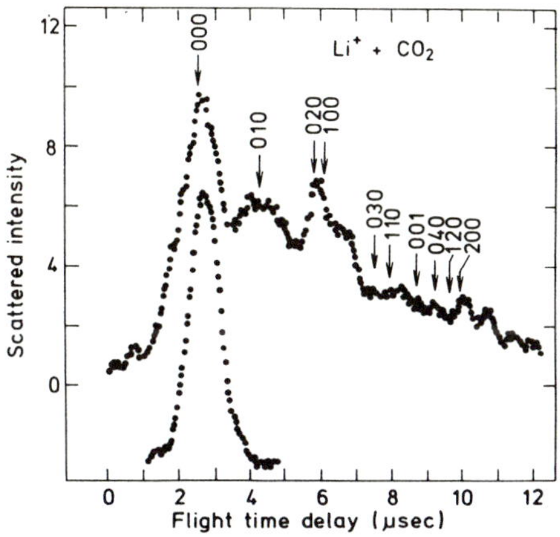

FIG. 19 Time-of-flight spectrum of $^7Li^+$ ions scattered from CO_2 at an ion energy $\mathscr{E}_I = 3.25$ eV ($E_i = 2.80$ eV) and at a laboratory scattering angle of 30° ($\chi \sim 34.6°$). Times calculated for various final CO_2 vibrational states are indicated with arrows. The lower single peak shows the elastic scattering spectrum for $Li^+ + Ar$ collisions at the same energy and scattering angle. [From Eastes *et al.* (1977).]

cm energy resolution may be judged from the spacing of the (000) and (010) peaks, for which $\Delta E = 0.083$ eV. Virtually all observed transitions originate in the (000) ground vibrational state and from rotational states which are low because of the cooling which accompanies the neutral beam adiabatic expansion. The excitation probability appears to be largest for the bending-mode vibrational state ($0n0$) for which the energy spacing is relatively small. The width of the vibrationally elastic (000) peak is about the same as that seen for $Li^+ + Ar$ scattering, indicating that there is little contribution from pure (unresolved) rotational excitation, despite what must be considerable anisotropy in the repulsive part of the potential. Qualitatively, it appears that the strong excitation of the bending mode and the very weak excitation of the asymmetric stretch mode may be correlated with the nature of the long-range interaction, even though the scattering angle is determined mostly by the short-range forces. The quadrupole moment of CO_2 arises from the (symmetric) polarity of the two C—O bonds, in which the C atom and the two O atoms may be thought of as possessing partial positive and negative charges, respectively. At most orientations, an interaction with the Li^+ ion will therefore tend to bend the CO_2 molecule: the positive charge repels the C atom and attracts the two O atoms. Detailed dynamical calculations will undoubtedly be required to provide a definitive interpretation of the data. The N_2O data show much weaker vibrational excitation than that for CO_2 and evidence of greater rotational excitation, which might be expected

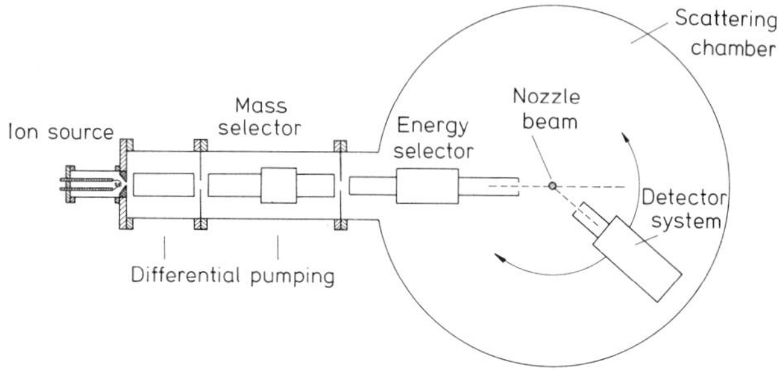

FIG. 20 Schematic diagram of the crossed beam apparatus at the University of Kaiserslautern [Hermann *et al.* (1978)]. General layout.

because of the (small) dipole moment of N_2O. The $Li^+ + CO_2$ system has recently been studied at higher energies by Kobayashi *et al.* (1977).

Linder and co-workers at the University of Kaiserslautern have very recently placed in operation an apparatus which probably, as of this writing, represents the highest state of the art for inelastic ion–molecule scattering experiments. Diagrams of the apparatus (Hermann *et al.*, 1978) are given in Figs. 20 and 21. An ion beam from a conventional dc discharge source is mass selected at an energy of about 200 eV with a Wien filter, then decelerated to about 3 eV and energy selected in a tandem arrangement of two 127° cylindrical electrostatic sectors. Zoom optics are then used to accelerate the beam to the desired final kinetic energy. A symmetric configuration of ion optics and tandem electrostatic sector fields is used to energy analyze the scattered product ions. The primary ion beam typically has an energy spread of about 0.030 eV FWHM, an angular divergence of about 1° FWHM, and an intensity of about 2×10^{-10} A. The overall experimental energy resolution is typically about 0.050 eV. The neutral gas target is a supersonic beam of about 3° FWHM angular width, which crosses the ion beam at a 90° intersection angle in the out-of-plane geometry. The vacuum system is unusual in that extensive differential pumping is provided for both the ion beam source and the neutral beam source. The scattering chamber is pumped with a turbomolecular pump, and the ion optics can be baked to 450°C. Presumably, the extreme surface cleanliness of the ion optics in this apparatus is largely responsible for the high ion beam intensity and high energy resolution which is achieved.

So far, this group has investigated inelastic collisions of H^+ with H_2 (Schmidt *et al.*, 1976; McGuire *et al.*, 1977; Schinke *et al.*, 1977; Hermann *et al.*, 1978), with CO (Krutein and Linder, 1977a), with CO_2 (Krutein and Linder, 1977b), and with CH_4 (Krutein and Linder, 1977c). An example of the $H^+ + H_2$ data is given in Fig. 22 (Schmidt *et al.*, 1978). The upper part of the figure shows a high-resolution spectrum at a relatively small scattering

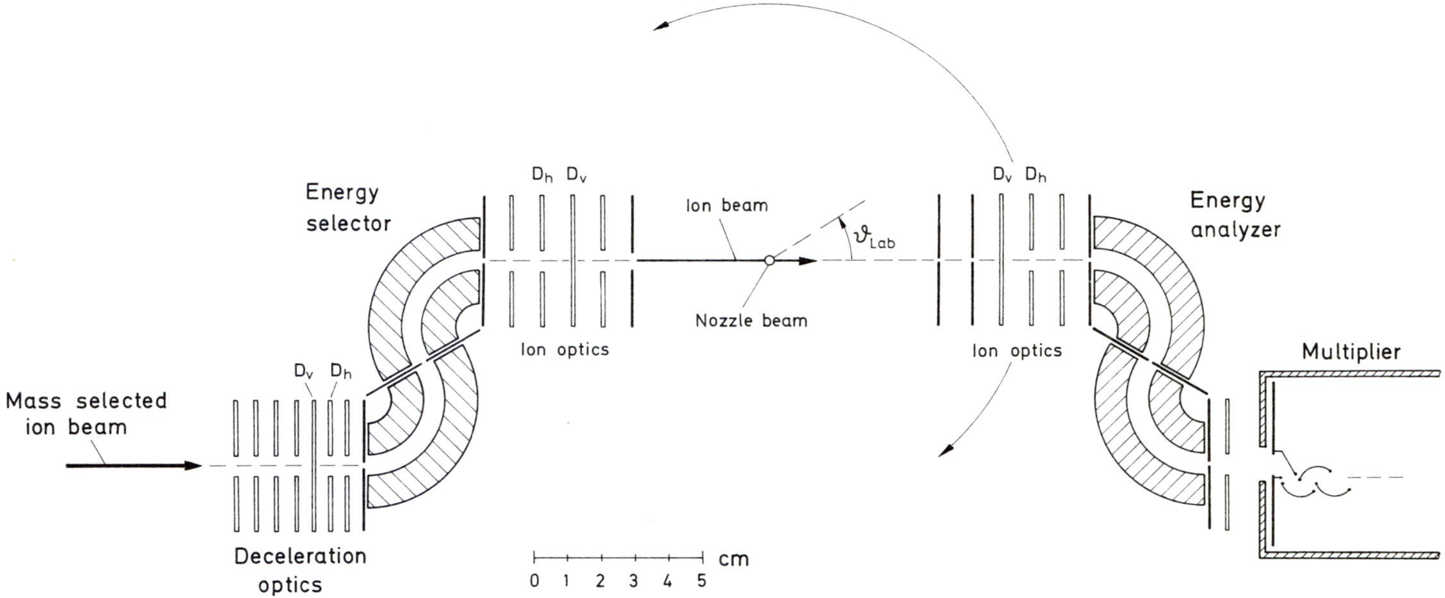

FIG. 21 Schematic diagram of the crossed beam apparatus at the University of Kaiserslautern [Hermann *et al.* (1978)]. Ion optics.

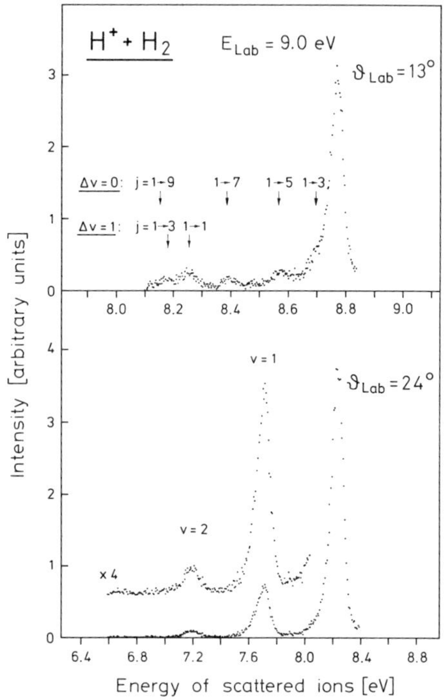

FIG. 22 Energy spectra for $H^+ + H_2$ scattering at a H^+ laboratory energy of 9.0 eV ($E_i = 6.0$ eV) and at laboratory scattering angles of 13° and 24°. The arrows indicate the expected energies for the indicated vibrational and rotational transitions. [From Schmidt *et al.* (1978).]

angle. The individual rotational transitions are resolved well enough to permit the discrete multiquantum rotational state-to-state transition probabilities to be extracted by deconvolution. The lower part of the figure shows an energy distribution at a larger scattering angle, which was measured with lower energy resolution. Although rotational states are not resolved in this case, the baseline resolution of the vibrational states makes possible a very accurate determination of the vibrational transition probabilities. Some of the data on vibrational transitions at $E_i = 10$ eV (Schmidt *et al.*, 1976) is summarized in Fig. 23, which compares the data of the Kaiserslautern group with the lower-resolution data of the Minnesota group. Even though the two instruments and techniques are quite dissimilar, the data are in excellent agreement. This provides a very useful confirmation of the experimental results. In a field such as molecular beams, in which every apparatus is to some extent idiosyncratic, verification of results using different instruments is an important and too-often-neglected duty.

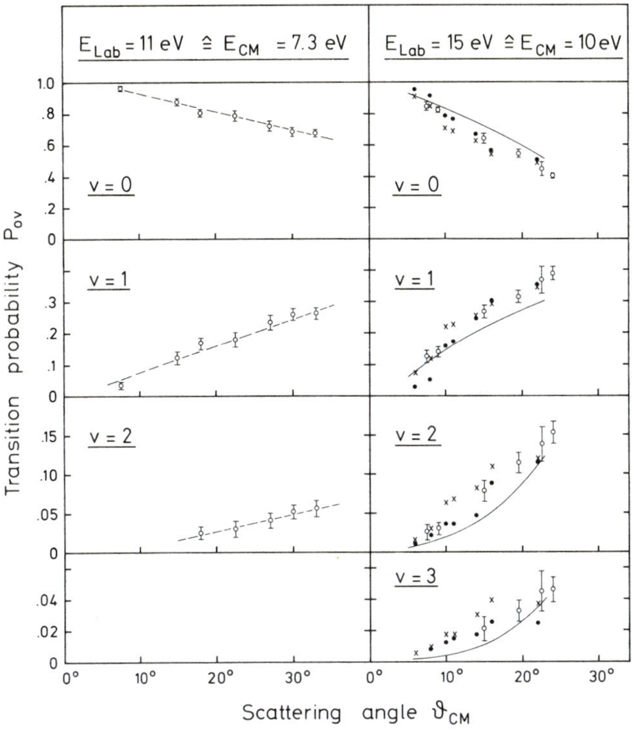

FIG. 23 Vibrational transition probabilities for $H^+ + H_2$ ($v = 0$) collisions at $E_i = 7.3$ eV and 10 eV, as a function of cm scattering angle. The open circles with error bars are the results of Schmidt *et al.* 1976). The crosses and the closed circles are the results of Udseth *et al.* (1973) uncorrected (crosses) and corrected (closed circles) for rotational contributions on the basis of the theoretical calculations of Giese and Gentry (1974). The solid lines are the DECENT model theoretical predictions. [From Giese and Gentry (1974).]

B. *Merged Beam Experiments*

1. *Rearrangement Reactions*

The advantages of the merged beam technique for achieving a very broad range of collision energies with high kinetic energy resolution were first pointed out by Trujillo *et al.* (1966). This feature, together with the suitability of the method for studying processes involving exotic neutral species, gives the merged beam method unique capabilities for contributing to the understanding of ion–molecule reactions.

A merged beam apparatus currently being used at the University of Minnesota by the author and colleagues (Gentry *et al.*, 1975b) is shown

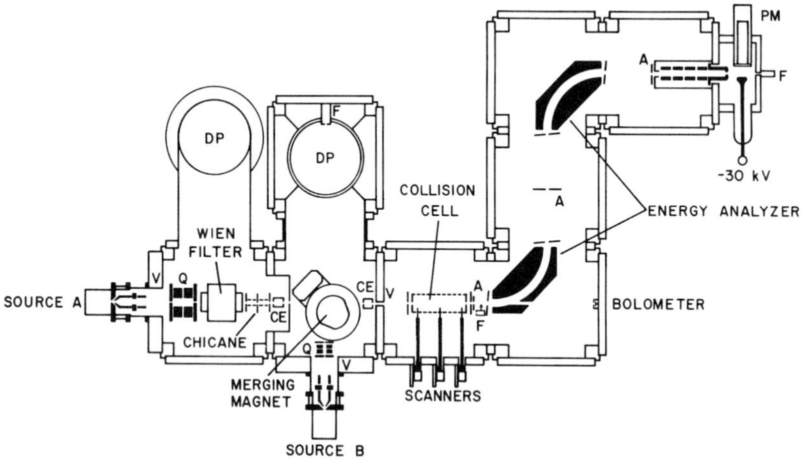

FIG. 24 Schematic diagram of the merged beam apparatus at the University of Minnesota [Gentry *et al.* (1975b)]. Apertures and other small features are not drawn to scale. *A* is an adjustable aperture, *CE* a charge exchange cell, *DP* a diffusion pump, *F* a Faraday cup detector, *Q* a quadrupole doublet lens, and *V* an isolation valve. The edge length of each cubical vacuum chamber module is 30.5 cm.

in Fig. 24. The two ion sources may be either electron bombardment or microwave discharge sources. Both ion beams are extracted at the final desired laboratory kinetic energy in the range 1–7 keV, and are focused by means of einzel lenses followed by electrostatic quadrupole doublets. The beam from source A is mass selected with a Wien filter, passed through a chicane to remove any high-energy neutrals which may be present, and then is neutralized by charge exchange with the selected gas. The neutral beam then passes between the poles of the merging magnet, proceeds through the collision cell, and finally is detected with a thermistor bolometer. The ion beam from source B is mass selected and deflected by 90° in the merging magnet, and thus becomes collinear with the neutral beam from source A. A second charge exchange cell is provided so that the B beam may also be neutralized, if desired, for studies of chemi-ionization reactions. The relative kinetic energy of the A and B beams is normally chosen so that the reaction products to be observed are not formed in the merging region, where the beams are not sufficiently well collimated. For endothermic processes this may be done by merging the beams at a very small value of E_i. For exothermic reactions, the usual practice is to make the nominal value of E_i in the merging region so large that any incipient product ions which may be formed in this region will be unstable with respect to dissociation. A potential applied to the center element of the collision cell accelerates or decelerates the ions relative to the neutrals and establishes the desired cm kinetic energy in a region of well-defined length where the beam diameters are typically about 1 mm. The intensity profiles of the two beams in planes perpendicular to the beam axis are measured with 0.050 mm resolution at three axial locations in the collision cell. Thus, the three-dimensional overlap integral for the two beam densities may be determined directly, for the purpose of establishing the absolute magnitude of the reaction cross section. In a merged beam experiment, the two reactant species and the reaction products all have nearly the same laboratory speed. The laboratory kinetic energies of the various species present are therefore approximately proportional to their respective masses. The energy analyzer used to separate the products from the reactants consists of two 90°-deflection spherical electrostatic sectors, in tandem. The tandem arrangement provides the very high discrimination necessary to remove noise caused by various indirect processes (e.g., surface scattering) involving the primary beams. Reaction cross sections have been measured in some cases in which the ratio of product signal to primary beam intensity has been less than 10^{-14}. In addition to simply separating the product ions from the reactants, high-resolution product ion energy analysis also gives information on the product energy and angle distribution in cm coordinates. Adjustable apertures in the energy analyzer system permit the energy resolution to be varied between 0.04 and 0.8% FWHM. For measurement of the total reaction cross section in cases where the product ion energy range exceeds the unit-transmission range of the energy analyzer, a rapid linear energy sweep is used to integrate electronically over all product energies. The product ions are counted with a scintillation detector having nearly unit efficiency and a noise level of about 5 min^{-1}.

Since the merged beam method is unique in its ability to achieve extremely small relative kinetic energies, we expend considerable effort to exploit this capability by minimizing the transverse components of the primary beam laboratory velocity, which modify the nominal value of E_i from Eq. (2) by adding another term (Gentry *et al.*, 1975b). By using the beam scanners in the collision cell to tune each beam for minimum angular divergence and to spatially match the peak beam intensities, relative kinetic energies less than 0.002 eV can be achieved. Figure 25 shows estimates of the actual average cm kinetic energy and of the rms spread in cm kinetic energy for typical experiments on the reaction

$$D_2^+ + C \rightarrow CD^+ + D \tag{26}$$

(Schuette and Gentry, 1978). The calculation was done by Monte Carlo averaging over the distribution of relative velocity vectors, using as data the measured laboratory energy and angular distributions of both beams. The actual mean kinetic energy of collision is very nearly equal to the nominal

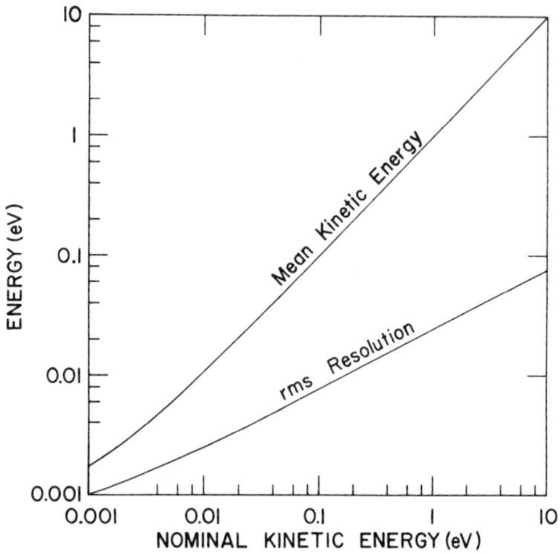

FIG. 25 Average relative kinetic energy $\bar{E}_i$ and the rms energy spread ΔE_i(rms), as functions of the nominal cm kinetic energy from Eq. (2). These results were calculated from the measured primary beam energy and angle spreads for an experiment on the reaction $D_2^+ + C \rightarrow CD^+ + D$. [From Schuette and Gentry (1979).]

value at all energies higher than about 0.01 eV, but at lower energies a correction must be added to the nominal value. Significant fractional resolution $\Delta E_i/E_i$ exists above 0.01 eV. At energies lower than 0.01 eV, the fractional energy resolution is degraded somewhat, but the effective mean kinetic energy can clearly be extended as low as ~0.001 eV.

At the lowest energies accessible to the merged beam technique, the close-collision cross section for a typical ion–molecule system is determined principally by the long-range intermolecular potential rather than by the short-range valence forces. Provided that the reaction is exothermic and that the reduced mass of the products is not much less than that of the reactants, the reaction probability for those collisions which penetrate the centrifugal barrier should be approximately constant. Under these conditions, the reaction cross section σ_R will be simply

$$\sigma_R = P\sigma_0, \tag{27}$$

where σ_0 is the capture cross section, to be calculated from the long-range potential function, and P the reaction probability per close collision. Thus, provided that the long-range potential is known sufficiently accurately, P can be determined from a calculation of σ_0 and a measurement of σ_R (McClure *et al.*, 1977a).

An example of the usefulness of this approach can be seen in the data for reaction (26). Figure 26 shows the energy dependence of the total reaction

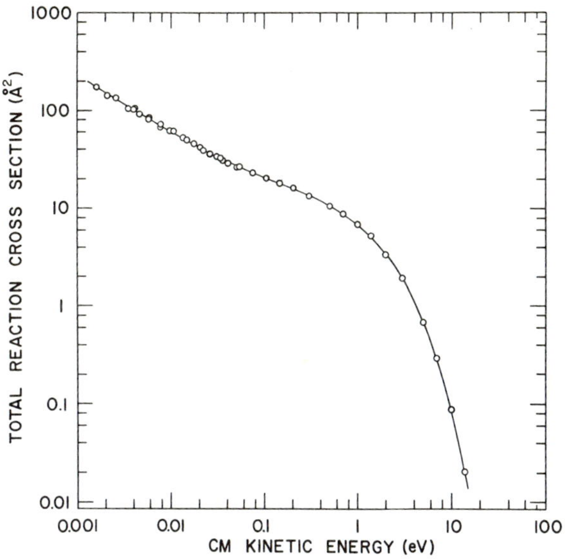

FIG. 26 Energy dependence of the total cross section for the reaction $D_2^+ + C \rightarrow CD^+ + D$. [From Schuette and Gentry (1979).]

cross section over more than four orders of magnitude in cm kinetic energy (Schuette and Gentry, 1978). The reaction cross section declines relatively slowly with increasing energy up to about 1 eV, then falls rapidly due to the onset of the three-body dissociation channel. The C atom beam for this experiment was produced by charge exchange of ground state C^+ with Xe, which is expected on the basis of energy defect considerations to give principally the ground state $C(^3P)$ neutral product. The merged beam data on (1) the energy dependence of the relative total reaction cross section, and (2) the product ion energy distributions, measured with several different charge exchange gases, indicate that only $C(^3P)$ reacts with significant probability to give CD^+. Now in order to interpret the low-energy cross section data in terms of Eq. (27) we need to know the long-range potential function for $D_2^+ + C(^3P)$ interactions. For a spherically symmetric atom such as $N(^4S)$, the long-range potential is the familiar charge-induced dipole potential

$$V(R) = -q^2\alpha/2R^4, \tag{28}$$

where q is the ion charge, α the dipole polarizability, and R the ion–atom separation. The predicted capture cross section is proportional to $E_i^{-1/2}$, and indeed this limiting energy dependence was found at low energies for the reaction $D_2^+ + N(^4S) \rightarrow ND^+ + D$ (McClure *et al.*, 1977a). This provides direct experimental confirmation of a constant low-energy reaction probability. However, atoms in P states such as $C(^3P)$ and $O(^3P)$ will in general possess permanent electrostatic quadrupole moments Q_M, where M denotes a dependence of Q on an angular momentum projection quantum number. With the charge-quadrupole term included, the long-range potential becomes

$$V_M(R) = \frac{qQ_M}{2R^3} - \frac{q^2\alpha_M}{2R^4}. \tag{29}$$

The longer-range charge-quadrupole term (which may be either attractive or repulsive, depending on the sign of qQ_M) will always dominate the potential at sufficiently large R. Because an atom has no internal symmetry axis of its own, the potential in Eq. (29) is a function only of R, in contrast to the case of the interaction of an ion with a molecular quadrupole, where the potential also depends on the molecular orientation angle. However, Q_M and α_M may themselves depend on R through the spin–orbit coupling of electronic angular momenta in the atom. Analytic expressions for the long-range ion–atom potential functions which include the spin–orbit interaction have been derived recently by Gentry and Giese (1977a). At large separations, where the spin–orbit coupling energy is large compared to the interaction potential, each atomic state of total electronic angular momentum J and projection $|M_J|$ on the ion–atom axis leads to a different potential energy function. Thus there are three different long-range potential functions for ^2P atoms (e.g., B and F) and six different long-range potential functions for ^3P atoms (e.g., C and O).

As the ion–atom separation decreases and the anisotropy of the interaction potential becomes larger than the spin–orbit energy, the interaction potentials go over smoothly to ones which correspond to the separate quantization of the orbital angular momentum projection $|M_L|$. These can be described as Σ in character ($M_L = 0$) or Π in character ($|M_L| = 1$) if the ion at large separations is considered to be a point charge. The results for positive and negative ion interactions with $C(^3P)$, $O(^3P)$, and $F(^2P)$ are shown in Fig. 27.

In the case of positive ion interactions with $C(^3P)$, the Σ potentials are strongly repulsive at moderate separations. The Π potentials are very attractive, since in this case both potential terms in Eq. (29) are attractive at moderate separations. Statistically, $\frac{2}{3}$ of the reactants will approach on Π surfaces and $\frac{1}{3}$ on Σ surfaces. For interactions of $C(^3P)$ with $H_2^+(^2\Sigma_g^+)$, in the valence interaction region the Σ- and Π-type long-range potentials lead to Σ and Π states of the three-atom system in collinear approaches, and to A_2 and (B_1, B_2) states, respectively, in C_{2v} approaches. Although the shorter-range valence forces may lower the repulsive barrier for the Σ surfaces somewhat, it is clear that at kinetic energies much less than 0.1 eV

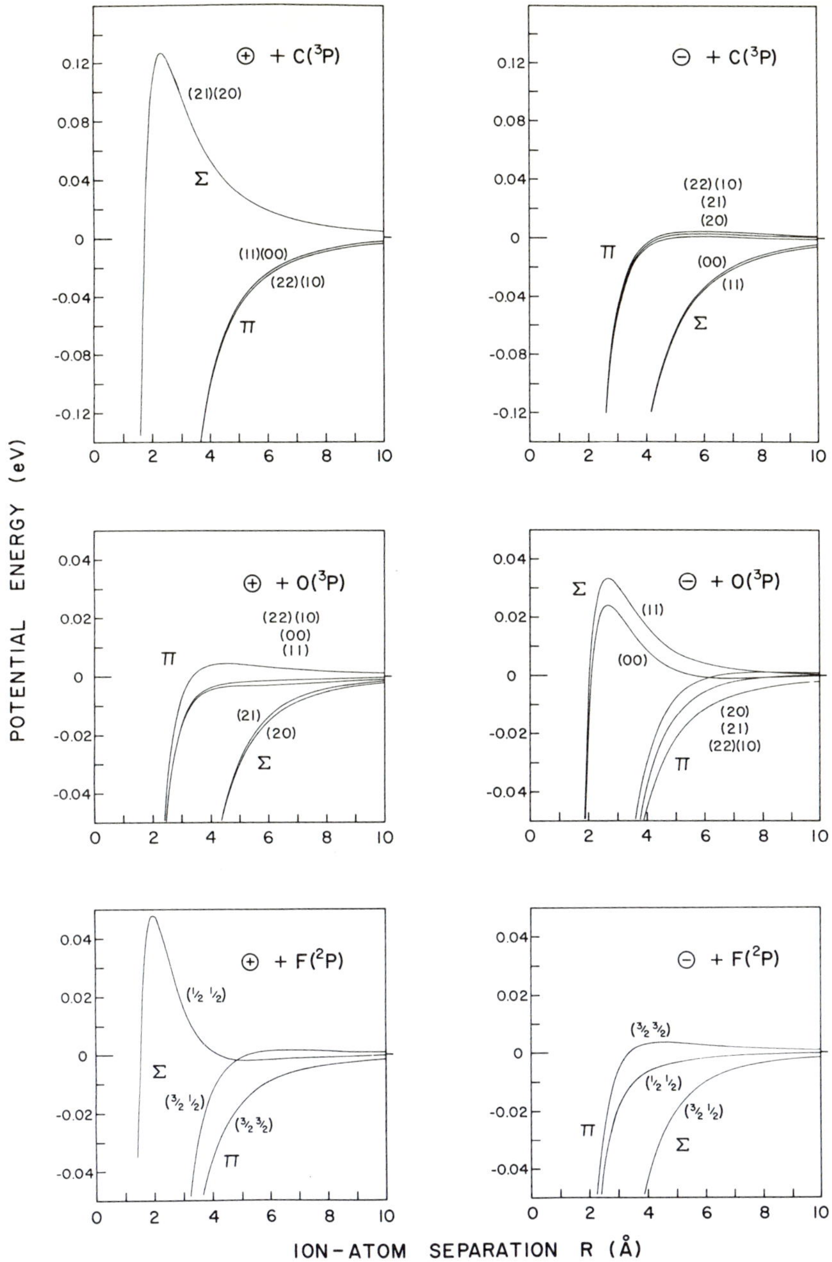

FIG. 27 Potential energy functions for interactions of positive and negative ions with $C(^3P)$, $O(^3P)$, and $F(^2P)$, where the ions are taken to be structureless point charges. The individual curves are labeled with the asymptotic values of the quantum numbers (JM_J), and groups of curves are labeled Σ or Π according to the M_L correlations for strong interactions. Where two curves are not resolvable on the scale of the figure the two labels are written on the same horizontal line. In order to make small differences in the potentials more apparent, the asymptotic spin–orbit energy has been subtracted from each curve so that every potential function approaches zero at large R. The potentials with the spin–orbit energies included do not cross. [From Gentry and Giese (1977a).]

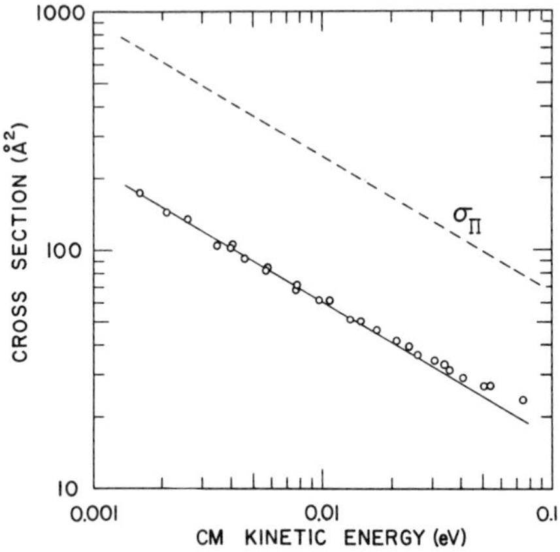

FIG. 28 Energy dependence of the total cross section for the reaction $D_2^+ + C \rightarrow CD^+ + D$ in the very low energy regime. σ_Π is $\frac{2}{3}$ of the cross section for capture collisions on Π-type potential surfaces (see Fig. 27), and the solid line through the low-energy data points is drawn at $\sigma_R = 0.25\sigma_\Pi$. [From Schuette and Gentry (1979).]

for this system only collisions on Π surfaces can contribute to the observed reaction cross section. It is possible that the positive curvature seen in $\sigma_R(E_i)$ at about 0.1 eV is due to the onset of contributions from Σ-surface collisions which penetrate the potential barrier. Figure 28 shows the experimental data in the very low energy regime, and a comparison with the cross section σ_Π which would be expected if the reaction probability per close collision were unity for Π surface collisions and zero for Σ surface collisions. The solid line which fits the data points in the limit of low collision energy is drawn at $0.25\sigma_\Pi$, indicating a constant reaction probability for Π surface approaches of $P_\Pi = 0.25$. If a small fraction of the C atom beam in these experiments were in nonreactive metastable excited states, the calculated reaction probability would be slightly higher. Of the $H_2^+(^2\Sigma_g^+) + C(^3P)$ approaches which occur on Π surfaces, $\frac{1}{3}$ will be $^2\Pi$ and $\frac{2}{3}$ will be $^4\Pi$ in the valence region. The electronic-state correlation diagram for CH_2^+ (Schuette and Gentry, 1979) shows that the $^2\Pi$ surface correlates with the $CH^+ + H$ products which are observed. Therefore, a reaction probability of 0.25 is consistent with an electronically adiabatic mechanism for this reaction. The reaction

$$D_2^+ + O(^3P) \rightarrow OD^+ + D, \tag{30}$$

on the other hand, occurs by a nonadiabatic mechanism (McClure *et al.*, 1977b).

Those ion–molecule reactions chosen for study by beam methods have tended to be atom-abstraction processes with large cross sections. Most of these, as one would expect, have been found to occur via direct mechanisms which give predominately forward scattering of the abstraction product with respect to the abstracting reactant. For example, the abstraction process

$$H_3^+ + D_2 \rightarrow D_2H^+ + H_2 \tag{31}$$

(Douglass and Gentry, 1979a) is direct, with the D_2H^+ product scattered preferentially in the direction of the initial D_2 velocity vector in cm coordinates. The reaction probability per close collision is 0.70 at low collision energy, and would probably be even larger except for the large vibrational energies of the reactants (Douglass *et al.*, 1977). Complete isotopic scrambling would, by a simple atom count, give twice as much H_2D^+ as D_2H^+, and therefore the expected probability for reaction (31) would be much smaller. We may therefore conclude that the proton abstraction reaction occurs in peripheral collisions, on a time scale short compared to that required for atom exchange or isotopic scrambling. Similar conclusions were reached for the abstraction reactions

$$\begin{array}{lcl} HD^+ + D_2 & \longrightarrow & D_2H^+ + D \\ & \times & \\ D_2^+ + HD & \longrightarrow & D_3^+ + H \end{array} \tag{32}$$

(Douglass *et al.*, 1977; Gentry and Ringer, 1977).

Consider now the reaction

$$HD_2^+ + HD \rightarrow D_3^+ + H_2 \tag{33}$$

(Douglass and Gentry, 1979b). This reaction provides an interesting contrast to the previous example of a reaction in a H_5^+ system. In this case, the D_3^+

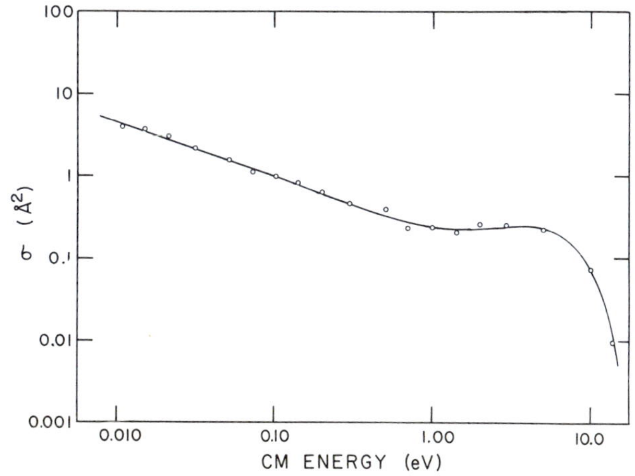

FIG. 29 Energy dependence of the total cross section for the reaction $HD_2^+ + HD \rightarrow D_3^+ + H_2$. [From Douglass and Gentry (1979b).]

product can be formed only by exchange of the D atom in HD with the H atom in HD_2^+. The abstraction products, on the other hand, will be either H_2D^+ or HD_2^+. One feels intuitively that an exchange reaction will necessarily require a more intimate encounter than does an abstraction reaction, and that perhaps an intermediate complex of fairly long lifetime might be required. Also, one feels that the steric requirements for the reaction will be more stringent, leading to a smaller reaction probability.

Figure 29, which shows the energy dependence of the total reaction cross section for D_3^+ formation from reaction (33), certainly shows the latter expectation to be correct. While the abstraction cross section at low energies is a large fraction of the capture cross section, the probability of the exchange

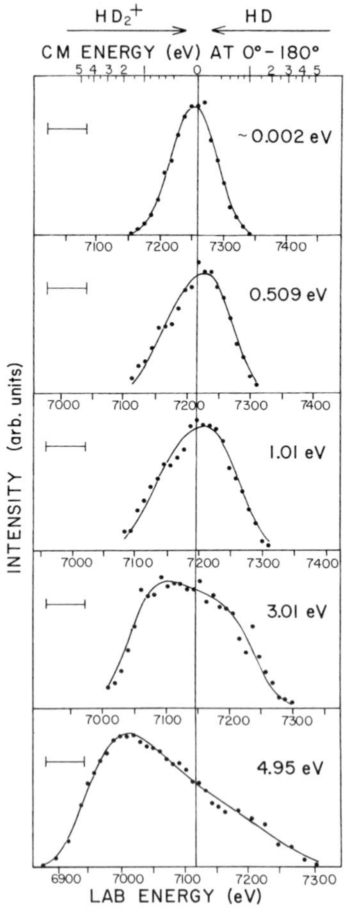

FIG. 30 Product ion energy distributions for the reaction $HD_2^+ + HD \rightarrow D_3^+ + H_2$ at the indicated initial cm kinetic energies [Douglass and Gentry (1979b)]. The central vertical line is the "center energy," the energy the product ion would have if it had the same laboratory speed as the system center of mass. The horizontal bars give the FWHM energy resolution.

reaction per capture collision at low kinetic energy is only 1.7%. Furthermore the energy dependence of the reaction cross section displays an unusual feature. The reaction cross section is proportional to the capture cross section at low energies and it falls off as is usual at sufficiently high energies. In the range from 1 to 5 eV, however, it actually levels off and perhaps even increases slightly with an increase in kinetic energy. This suggests that perhaps a change in the reaction mechanism occurs in this energy range.

The form of the product energy distributions, shown in Fig. 30, supports this conclusion. The distributions are symmetric in the cm coordinate system at very low kinetic energies, then shift slightly toward the initial direction of the HD_2^+ reactant. This behavior persists up to an energy of ~1 eV, then the product distributions shift toward the initial direction of the HD reactant. Symmetry about the center energy for the product ion distribution is consistent with reaction occuring through the formation of a complex which lives for at least several rotational periods. There is, however, no reason why a direct reaction cannot also give a symmetric product distribution in some cases. Deconvolution of the product ion energy distributions (McClure *et al.*, 1977a), for reaction (33) shows that, even at lowest initial kinetic energies, only a very small fraction (about 0.3 eV) of the ~3 eV total energy available from vibrational excitation of the reactants appears as translational energy of the products. The translational and internal degrees of freedom, therefore, do not appear to be strongly coupled, contrary to what one would expect for a long-lived complex. This reaction would be a very interesting case for classical trajectory calculations.

Another interesting type of reaction mechanism is seen in the substitution reaction

$$H_3^+ + D \rightarrow H_2D^+ + H \tag{34}$$

(Douglass and Gentry, 1979c).

In this reaction a competing abstraction process is not possible at low kinetic energies because of energetic constraints, therefore, the substitution product is the only one possible. The ground-state reaction exothermicity of 0.051 eV is due only to the zero-point energy difference. The energy dependence of the total reaction cross section, shown in Fig. 31, indicates that the reaction probability for low-energy capture collisions is 0.75, consistent with a zeroth-order statistical distribution of isotopes in the products ($n_H = 3n_D$). Apparently a low-energy substitution reaction can occur as *efficiently* as an abstraction reaction, even though reactions of the latter type are usually characterized by strong forward peaking of the reaction products, implying atom transfer at relatively large distances in peripheral collisions. However, the *dynamics* of the substitution reaction are very

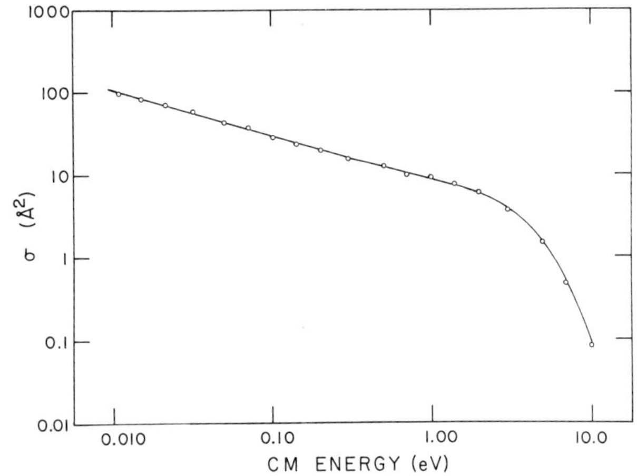

FIG. 31 Energy dependence of the total cross section for the reaction $H_3^+ + D \rightarrow H_2D^+ + H$. [From Douglass and Gentry (1979c).]

different, as indicated by the energy distributions shown in Fig. 32. Unlike the exchange reaction discussed previously, which displayed small deviations from symmetry about the center energy below 1 eV, the substitution reaction (34) is characterized by laboratory energy distributions which are perfectly symmetric about the center energy, within the experimental uncertainty, up to a kinetic energy of 1 eV. Also note that there is very little change in the width of the product energy distribution as the initial kinetic energy is increased at low energies. The deconvolution results show that the data for low collision energies is consistent with angle distributions which are either isotropic or slightly peaked at the 0° and 180° poles of the cm coordinate system. The average product translational energy at $E_i \sim 0.002$ eV is 0.5 eV, which is roughly half of the total available energy. This may be contrasted with the result for the exchange reaction, in which the average product translational energy was 0.3 eV—less than 10% of the total energy available. On this basis it appears that the substitution reaction (34) really proceeds at low kinetic energies through a strongly coupled complex, while the exchange reaction (33) does not. Whether this complex is one which actually lives for several rotational periods is another question. Potential energy surface calculations (Poshusta and Zetik, 1973) indicate that H_4^+ is bound by only 0.045 eV, which seems much too small a well depth to support a long-lived complex with large internal energy and angular momentum. Again, this would be an interesting system for classical trajectory calculations.

Some merged beam studies of rearrangement reactions in hydrogen systems have also been carried out by Rol and co-workers at Wayne State

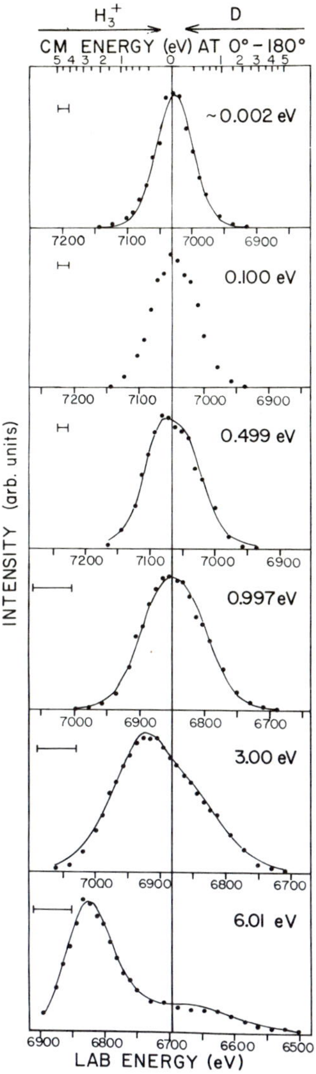

FIG. 32 Product ion energy distributions for the reaction $H_3^+ + D \rightarrow H_2D^+ + H$. [From Douglass and Gentry (1979c).]

University. This group measured product-ion energy distributions for reaction (31) (Lees and Rol, 1975b), and also for the abstraction reactions $H_2^+ + H_2 \rightarrow H_3^+ + H$ and $H_2^+ + D_2 \rightarrow D_2H^+ + H$ (Lees and Rol, 1974); $D^+ + H_2 \rightarrow HD^+ + H$ (Lees and Rol, 1975a); and $H_2^+ + D \rightarrow HD^+ + H$ (Wendell and Rol, 1974).

The latter two reactions are particularly interesting because the same electronic state of the products is reached from two different electronic states of the reactants. Thus, these reactions are examples of a "surface-hopping" reaction mechanism (Krenos *et al.*, 1974). The merged beam technique permits the dynamics to be studied from both of the two entrance channels. This group has also studied the reactions of the atmospheric ions O_2^+, N_2^+, CO^+, and NO^+ with Na, K, and Mg (Rol *et al.*, 1974).

R. H. Neynaber and G. D. Magnuson at IRT Corporation have performed merged beam studies of the reactions $O^+ + N_2 \rightarrow NO^+ + N$ (Neynaber and Magnuson, 1973a) and $H_2^+ + He \rightarrow HeH^+ + H$ (Neynaber and Magnuson, 1973b). In the first reaction, evidence was obtained for the influence of both electronic and vibrational excitation of the reactants on the reaction cross section and dynamics.

2. *Chemi-Ionization Reactions*

The merged beam technique is nicely suited to the study of chemi-ionization phenomena, since the ionic product is easily separated from the neutral reactants, and since many of the processes of interest involve metastable excited states of rare-gas atoms, which can conveniently be produced by the approximately resonant charge exchange of a high-energy rare-gas ion beam with an alkali-metal-atom vapor. A large number of such processes have been studied recently by Neynaber and Magnuson, including associative ionization (AI) and Penning ionization (PI) reactions in He* + H, D, N, O collisions (Magnuson and Neynaber, 1974; Neynaber and Magnuson, 1975c, 1976a), in Ne* + Ar, Kr collisions (Neynaber and Magnuson, 1975a, 1976b), in Ar* + Na collisions (Neynaber and Magnuson, 1977). Other studies focused on the reactions

$$He^* + Ne^* \rightarrow HeNe^+ + e^- \quad (35a)$$

$$\rightarrow He + Ne^+ + e^- \quad (35b)$$

(Neynaber and Magnuson, 1975b; Neynaber and Tang, 1977), and

$$He^* + H_2 \rightarrow HeH^+ + H + e^- \quad (36a)$$

$$\rightarrow HeH_2^+ + e^- \quad (36b)$$

(Neynaber and Magnuson, 1974a). In general, these exothermic reactions have cross sections which decrease monotonically as the collision energy increases. The PI reactions have larger cross sections than the AI reactions as expected from general phase space considerations (three free product particles rather than two), and from angular-momentum constraints on the magnitude of the AI cross section due to the extremely small product reduced mass. The AI reaction (35a), which involves two electronically excited reactants, is unusual in that the energy dependence of the reaction cross section shows an abrupt increase at $E_i \sim 0.7$ eV and a maximum at about $E_i = 1.5$ eV. This might be due to the onset of a second reaction mechanism, perhaps the threshold for the system to reach a second potential energy function crossing point. Another exceptional system is the AI reaction (36b) between He* and H_2. In this case the reaction cross section is approximately constant at energies $E_i \sim 0.01$ eV, rises slightly to a broad maximum at $E_i \sim 0.1$ eV, then declines rapidly at high energies. The three-atom AI process is, of course, likely to be much more complicated than the two-atom AI reactions. The interpretation of the data is further complicated in this case by the large average vibrational energy of the H_2 from the ionization and subsequent charge transfer neutralization processes used to form the high-energy neutral beam.

Ringer and Gentry (1977, 1979) have recently completed a study of the endothermic AI reaction

$$N(^2D) + O(^3P) \rightarrow NO^+ + e^-, \qquad \Delta H_0 = 0.38 \text{ eV} \tag{37}$$

using the same merged beam apparatus shown in Fig. 24. A beam of virtually pure ground state $O(^3P)$ was produced by forming $O^+(^4S)$ in a microwave discharge source and neutralizing the O^+ beam by charge exchange with Kr (McClure *et al.*, 1977b). The fractional population of metastable $N(^2D)$ in the N atom beam could be varied between 18 and 60% by using different charge exchange gases. It was possible to determine the absolute excited and ground-state populations by comparing the phenomenological cross sections obtained with various N^+ charge exchange gases for *both* reaction (37) and the ground-state reaction $D_2^+ + N(^4S) \rightarrow ND^+ + D$. Figure 33 shows the energy dependence of the AI reaction cross section over the range of E_i between 0 and 21 eV. The details of the cross section near threshold, measured with an experimental sensitivity of about 10^{-3} Å^2, show a threshold energy for the AI reaction which agrees within experimental error with the calculated endothermicity of reaction (37). The ground-state AI reaction

$$N(^4S) + O(^3P) \rightarrow NO^+ + e^- \tag{38}$$

has a thermodynamic threshold energy of 2.76 eV; however, no evidence of this reaction was found. The data place an upper limit on the contribution of reaction (38) to the measured cross section (for $E_i > 2.76$ eV) of about 15%. There is also no evidence for participation of the $N(^2P)$ metastable state, for which the AI reaction would be exothermic by 0.82 eV.

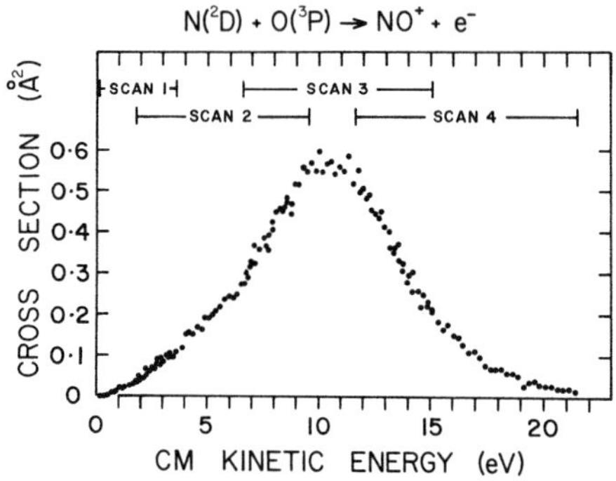

FIG. 33 Energy dependence of the total cross section for the reaction $N(^2D) + O(^3P) \rightarrow NO^+ + e^-$. The data was acquired in four separate scans of cm kinetic energy, which were normalized to each other in the overlap regions. The uncertainty in the absolute value of the cross sections is $\pm 30\%$. [From Ringer and Gentry (1979).]

Even though only two atoms are involved, the interpretation of the reaction dynamics is complicated by the large number of NO molecular potential functions which correlate with the $N(^2D) + O(^3P)$ reactants, some of which may intersect the $NO^+(X^1\Sigma^+)$ potential at points other than near the potential minimum. The break in the slope of the cross section function at about $E_i = 6.5$ eV was reproduced in several experiments and is thought to be significant. This feature might possibly be due to the opening up of a new channel for formation of $NO^+(X^1\Sigma^+)$, to a threshold process for formation of an NO^+ electronically excited state, or to some other dynamical feature. The maximum in the AI cross section at about $E_i = 10.5$ eV correlates with the opening of the competitive channels

$$N(^2D) + O(^3P) \rightarrow N^+(^3P) + O^-(^2P), \qquad \Delta H_0 = 10.68 \text{ eV} \tag{39}$$

and

$$N(^2D) + O(^3P) \rightarrow N(^4S) + O^+(^4S) + e^-, \qquad \Delta H_0 = 11.23 \text{ eV}. \tag{40}$$

A theoretical treatment of AI and PI in the N + O system has recently been completed by Dahler and Nielsen (1979). The theory is cast in the semiclassical framework proposed by Miller (1970), and uses the potential functions calculated by Michels (1974, 1975). The agreement with the experimental data is particularly good in the threshold region, where there is relatively little uncertainty about the electronic states of NO and NO^+ which are involved. The AI reaction (37) is related by microscopic reversibility to the reverse process—dissociative recombination (DR) of $e^- + NO^+$, a reaction of considerable aeronomic importance. Direct comparison of the AI and DR data is difficult because the AI process can produce NO^+ in internally excited states (except very close to threshold) while the DR process is usually studied with near-thermal NO^+. There is experimental evidence, however, that the near-thermal DR reaction yields predominantly $N(^2D)$ + $O(^3P)$ products (Kley *et al.*, 1977). This result is at least qualitatively consistent with the AI data, in which NO^+ formation is observed only with $N(^2D) + O(^3P)$ reactants.

3. *Charge Transfer Reactions*

Neynaber and Magnuson (1974b) used their merged beam instrument to investigate the dynamics of the charge transfer reaction

$$N_2{}^+ + O_2 \rightarrow N_2 + O_2{}^+, \qquad \Delta H_0 = -3.5 \text{ eV} \tag{41}$$

over the cm energy range from 0.1 to 5 eV. The data showed that this highly exothermic process occurs with very slight momentum transfer, and therefore gives mostly 0° scattering in the cm system. For the reaction to be translationally thermoneutral, the reaction exoergicity must be absorbed in internal excitation of one or both of the products. This electronic-to-internal energy transfer occurs very efficiently in this system.

A special-purpose merged beam instrument for investigating collision processes which produce two ionized products has recently been constructed by Brouillard and colleagues at the University of Louvain (Burniaux *et al.*, 1977). The initial experiments have focused on the charge transfer reactions

$$^3He^{2+} + H^*(n) \rightarrow {}^3He^+(n') + H^+, \tag{42}$$

in which the H* reactant is in a Rydberg state with principal quantum number $8 \leq n \leq 24$. Figure 34 shows a diagram of this unique apparatus. Ion beams in the laboratory energy range 5–15 keV are mass selected in the crossed-field analyzer (3). A proton beam (1) is neutralized by charge exchange with Ar in cell (4), producing a broad distribution of neutral-atom excited states. An electric field at (5) which is variable up to 60 kV/cm is used to selectively ionize the highest H-atom excited states in order to remove these states from the beam. A $^3He^{2+}$ beam (2) was chosen rather than $^4He^{2+}$ to avoid the problem of separating $^4He^{2+}$ from $H_2{}^+$ in the magnetic product ion analyzer (9). (However, it nevertheless was possible to do some experiments with $^4He^{2+}$.) The $^3He^{2+}$ beam is merged with the H atom beam by means of a series of electrostatic deflectors (6). The aperture plates 17–21 define the spatial extent of the beams. A voltage applied to the collision cell (7) serves to establish the desired relative kinetic energy and to provide a well-defined collision volume for creation of the observed products. This region is kept at a pressure of about 5×10^{-10} torr. The collision energy is scanned

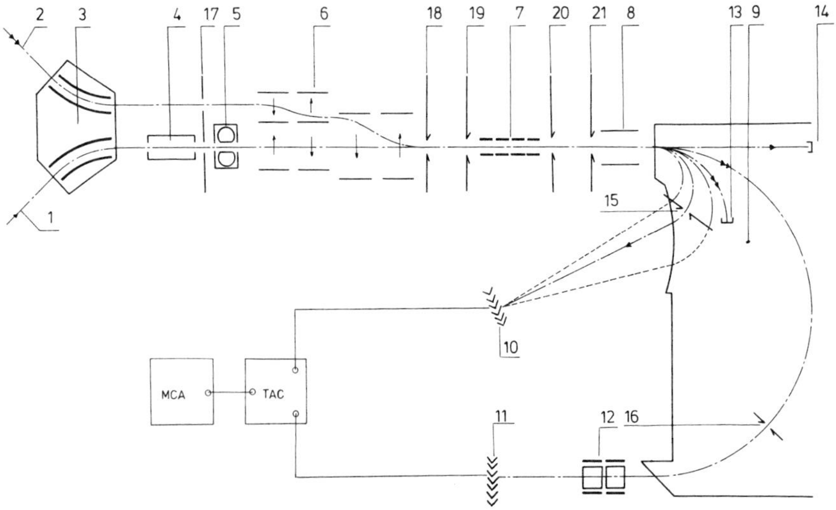

FIG. 34 Schematic diagram of the merged beam apparatus at the University of Louvain [Burniaux *et al.* (1977)]. The numbered components are described in the text.

by varying the acceleration potential for the H^+ beam. An einzel lens (8) provides the final beam focusing as the product ions enter the magnetic analyzer (9). The primary ion and neutral beams are monitored with detectors at (13) and (14), respectively. The populations of excited states in the H* beam were established approximately by using an axial ionizing field at (7) as a probe of the Rydberg states remaining in the beam after passage through the first ionizer at (5). The most unusual feature of the instrument is that it takes advantage of coincidence counting of the two ionic products in order to vastly improve the signal-to-noise ratio. The $^3He^+$ product ions emerge through the slit (16), are focused by a quadrupole doublet lens (12), and are counted using a particle multiplier (11). The H^+ products are deflected with a smaller radius in the magnetic analyzer, pass through the adjustable slit (15), and are counted using a second particle multiplier (10). The pulse pair coincidence signal is typically about a factor of four larger than the background coincidence level.

By measuring the phenomenological charge transfer cross sections as a function of the ionization field strength at (5), this group was able to determine the average charge transfer cross sections for different groups of H-atom Rydberg states over the cm kinetic energy range from 0.25 to 478 eV. The results are shown in Fig. 35 (Burniaux *et al.*, 1977). The charge transfer cross sections for the higher Rydberg states are very large ($\sim 10^4$ Å^2), and appear to become approximately constant at collision energies less than

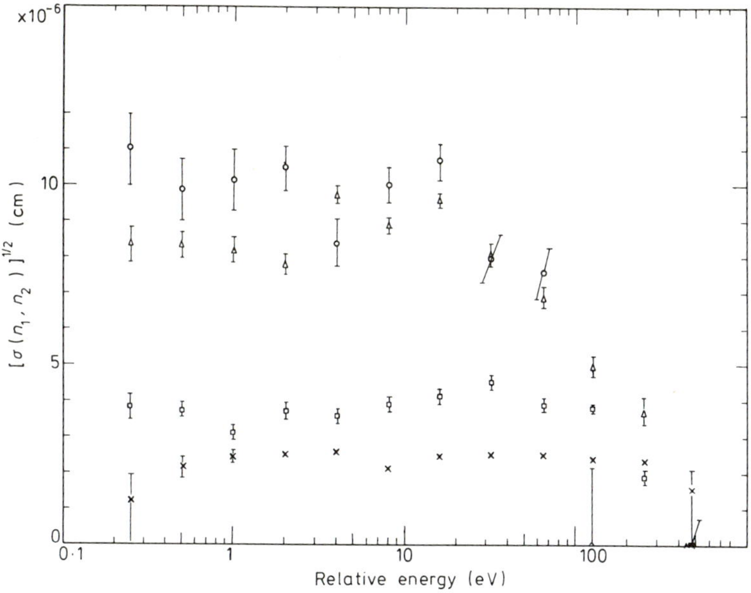

FIG. 35 Energy dependence of the square root of the average total cross sections for the charge transfer processes $^3He^{2+} + H^*(n) \rightarrow He^+ + H^+$ [Burniaux *et al.* (1977)]. × :$8 \leq n \leq 9$; □ : $10 \leq n \leq 14$; △: $15 \leq n \leq 19$; ○ : $20 \leq n \leq 24$.

10 eV. Because of the asymmetric nature of this one-electron system, the results are of great theoretical interest. Theoretical interpretation is complicated, however, by the lack of a precisely defined H* reactant state, and by uncertainty about the final state of He^+. Except for a very small difference due to the electron–nuclear reduced mass change, reaction (42) will be energetically resonant whenever $n' = 2n$. One might therefore expect these nearly resonant processes to dominate the charge transfer cross sections at sufficiently low collision energies. However, the detection system discriminates against very highly excited He^+ ($n' \gtrsim 40$), because these products are ionized to He^{2+} in the 2250 V/cm Lorentz field of the magnetic analyzer. Future experiments are planned to better define the reactant state and to explore the effect of energy resonance on the charge transfer cross sections. One type of experiment which is planned will use laser excitation (Doppler shifted by the H-atom velocity) to populate specific single Rydberg states of H* (Brouillard, 1978).

Various ion–ion charge transfer processes have also recently been studied by Peart *et al.* (1976a,b, 1977a,b), using an "inclined beams" apparatus having an $8\frac{1}{2}°$ beam intersection angle.

C. *Special Methods*

The experimental techniques described in this section are the relatively new methods of guided beams and ion beam experiments utilizing fluorescence detection of the products. Both of these methods have distinct advantages for certain types of studies and will undoubtedly find much more extensive application in the near future.

1. *Guided Beam Experiments*

The guided beam technique was originated by Teloy and Gerlich (1974) at the University of Freiburg. Most investigators in the field of ion–molecule reactions are familiar with quadrupole mass spectrometers, which achieve high mass resolution by precise control of dc and rf voltages applied to the poles. At a specific ratio of dc to rf voltage, which depends on the field geometry, stable trajectories along the axis of the pole assembly will be found for only one value of the mass-to-charge ratio at a time. In the absence of a dc field, however, stable trajectories will occur for any mass-to-charge ratio above some critical value m_c, where m_c becomes smaller with an increase in the rf frequency. In addition, the geometry of the rf field is then no longer critical and can be adapted to the other special requirements of a particular experiment. Provided that the rf frequency is sufficiently high, the motion of an ion near the axis of the multipole field can be considered to occur in an "effective potential" which is nearly zero at the axis, is flat for a distance off axis which depends on the geometry, and which becomes steeply repulsive near the boundaries of the pole assembly. In the apparatus of Teloy and co-workers this channel-like effective potential is used to guide the primary ion beam to the collision zone and then to collect and guide the product ions to the detector.

A simple example of the use of an rf guiding field can be found in the design of the ion source, which is shown in Fig. 36. The source is constructed of 10 molybdenum plates. The eight plates in the middle of the assembly have *U*-shaped channels cut in them and the middle three plates have an opening at one end of the *U* for ions to escape. The two end plates have slots on one side of the *U* only for an electron beam pass through. A positive dc potential of about 1 V applied to the end plates prevents ions formed by electron impact from escaping transversely, and an rf potential on the central plates ($\sim$10 MHz, 30–200 V, with opposite instantaneous polarities on the odd- and even-numbered plates in the stack) provides an effective potential which guides ions around the *U* and out of the exit hole. Typical ion residence times in the source are from 10^{-4} to 10^{-3} sec, which is sufficient under the proper gas mixture and pressure conditions to obtain fully internally relaxed primary ions. Typical primary ion kinetic energy spreads are

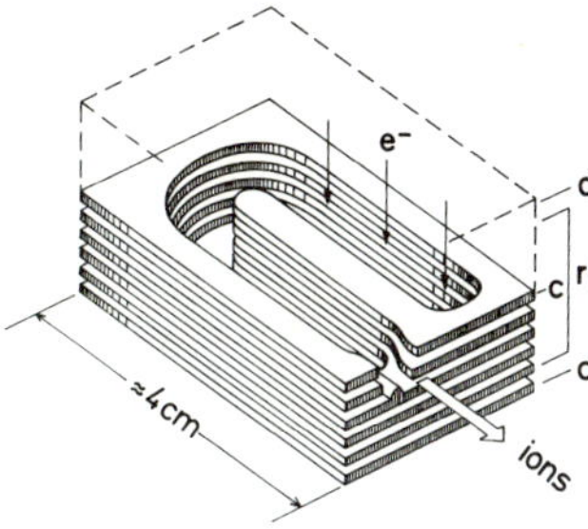

FIG. 36 Geometry of the storage ion source for the guided beam apparatus of Teloy and Gerlich (1974). The source consists of ten plates, the top four of which are not shown in the diagram.

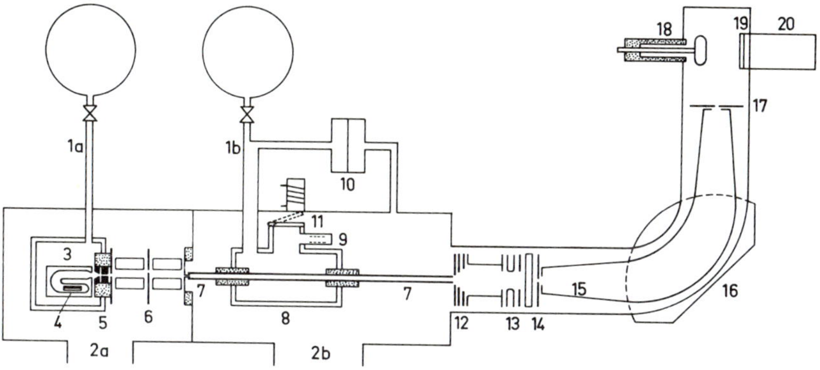

FIG. 37 Simplified schematic diagram of the guided beam apparatus at the University of Freiburg [Teloy and Gerlich (1974)]. Components are not drawn to scale: (1a, 1b) gas inlets for ion source and scattering chamber; (2a,2b) to cold traps and mercury diffusion pumps; (3) storage ion source; (4) cathode and electron beam; (5) electrodes for shaping and pulsing of the ion beam; (6) rf mass and velocity filter (8 cm long); (7) rf octopole (38.7 cm long, free diameter 0.52 cm); (8) scattering chamber (effective length 9.1 cm); (9) ionization gauge; (10) capacitance manometer; (11) discharge valve for the scattering chamber, operated by an electromagnet; (12) acceleration; (13) lenses for the shaping of the ion beam; (14) entrance slit of the mass spectrometer (0.15 × 1.4 cm); (15) metal box, defining the electrical potential in the mass spectrometer (−3 kV); (16) poles of the 90° magnet (radius 12.5 cm); (17) exit slit (0.3 × 1.4 cm); (18) voltage feedthrough and converter (−30 kV); (19) scintillator and quartz window; (20) photomultiplier.

0.1–0.2 eV FWHM. Figure 37 shows the remainder of the apparatus. The ion beam, which may be extracted dc or in a pulse from the ion source, is focused at an energy of 0.5–1 eV into a pair of tandem rf quadrupole fields. The mass-dependent focusing of ions between apertures at the entrance, the middle, and the exit of the quadrupole assembly provides the ion beam mass selection as well as some velocity filtering. Some degree of ion beam collimation is obtained at the exit aperture, which is a tube 1 mm in diameter and 6 mm long. Ions which pass through this tube are accelerated to the desired kinetic energy and enter an rf octupole guiding field which encloses the scattering region. The octupole field generates an effective potential which is proportional to r^6, where r is the distance from the central axis. The effective potential is therefore relatively flat in the central valley and steeply repulsive at the edges. The rf field is not completely benign, however, since it does impart an additional transverse velocity distribution to the primary ion beam. This transverse velocity distribution for a given trajectory is sharply peaked at the nominal value determined by the angle at which the ion enters the field, but it also has broad tails at both lower and higher speeds. The octupole field collects all product ions (except any which might have very large transverse velocity components) and guides them to the product mass spectrometer, where they are accelerated to 3 keV, mass analyzed with a 90° magnetic sector, and counted with scintillation counter. In order to collect any product ions which might fly backward toward the ion source, the ion source–mass filter assembly may be pulsed so as to inject a burst of ions into the scattering chamber, after which the dc potential on the source is raised to reflect back toward the detector any ions which approach the entrance end of the collision zone. In the pulsed mode of operation, the time-of-flight distribution of the product ions can be analyzed to give some information about the distribution of axial components of product ion velocity.

Early work by this group includes studies of the reactions $Ar^+ + D_2 \rightarrow ArD^+ + D$ and $Ne^+ + CO \rightarrow Ne + C^+ + O$ (Teloy and Gerlich, 1974) and all three product ion channels of the important prototype systems

$$H^+ + D_2 \rightarrow D^+ + HD, \qquad \Delta H_0 = +0.04 \text{ eV} \tag{43a}$$
$$\rightarrow D^+ + H + D, \qquad \Delta H_0 = +4.55 \text{ eV}$$
$$\rightarrow HD^+ + D, \qquad \Delta H_0 = +1.87 \text{ eV} \tag{43b}$$
$$\rightarrow D_2^+ + H, \qquad \Delta H_0 = +1.85 \text{ eV} \tag{43c}$$

(Ochs and Teloy, 1974). The data for reactions (43) are shown in Fig. 38. By repetitive scanning, an excellent signal-to-noise ratio in the cross section data is obtained for a very large number of points, even though the primary ion beam intensity is only about 10^6 sec^{-1} (10^{-13} A). The error limits for the absolute cross section values are estimated to be $\pm 20\%$. At low kinetic energies the rearrangement channel (43a) dominates, with a total cross section which falls rapidly with increasing energy. The HD^+ and $D_2{}^+$ product channels show the expected threshold behavior and some additional structure which is not yet understood. At three values of collision energy the data are in excellent agreement with the trajectory-surface-hopping calculations of Krenos *et al.* (1974), which take into account motion on and transitions between the two low-lying singlet potential energy surfaces for this system.

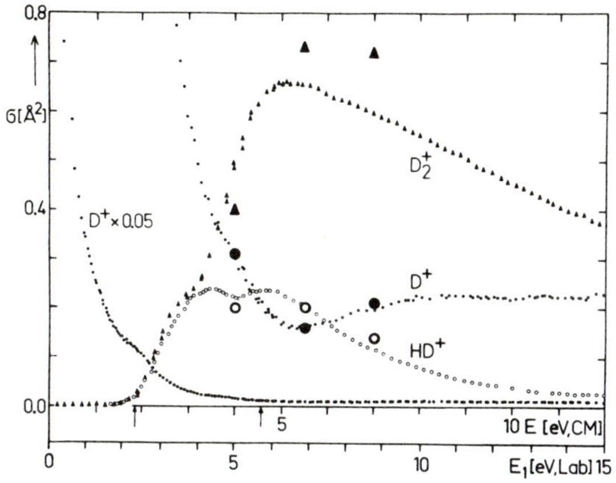

FIG. 38 Energy dependences of the total cross sections for the reactions

$$\begin{aligned} H^+ + D_2 &\rightarrow D^+ + HD \text{ or } (H + D) && (\bullet\bullet\bullet) \\ &\rightarrow HD^+ + D && (\circ\circ\circ) \\ &\rightarrow {D_2}^+ + H && (\blacktriangle\blacktriangle\blacktriangle) \end{aligned}$$

[Ochs and Teloy (1974)]. The small symbols are the experimental data and the large symbols are the results of trajectory-surface-hopping calculations. [From Krenos *et al.* (1974).]

Very recently this group has studied reactions of N^+ with CO in which they observed all five possible product channels:

$$N^+ + CO \rightarrow CO^+ + N, \qquad \Delta H_0 = -0.52 \text{ eV} \tag{44}$$

$$\rightarrow NO^+ + C, \qquad \Delta H_0 = -0.65 \text{ eV} \tag{45}$$

$$\rightarrow C^+ + NO, \qquad \Delta H_0 = +1.35 \text{ eV} \tag{46}$$

$$\rightarrow O^+ + CN, \qquad \Delta H_0 = +2.61 \text{ eV} \tag{47}$$

$$\rightarrow CN^+ + O, \qquad \Delta H_0 = +2.90 \text{ eV} \tag{48}$$

(Frobin *et al.*, 1977). The cross sections for the exothermic channels (44) and (45) show the expected decline with increasing kinetic energy for small E_i, but then rise again to broad maxima as the endothermic channels open up. At $E_i = 12$ eV, the endothermic channel (46) dominates over all other processes except the charge transfer channel (44). The electronic structures of the NCO^+ intermediates for this system are extremely complex. It is clear, however, that the simplest adiabatic electronic-state correlations are of little help in predicting the relative cross sections for the different product channels. Both adiabatic and diabatic reactions apparently occur with comparable facility when energetically allowed. Structure in the energy dependence of the total cross sections suggests that there may even be multiple dynamical mechanisms responsible for a single product channel. Hopefully, data of the extremely high quality produced by the guided beam technique will prove to be an irresistible challenge to theoreticians.

2. *Ion Beam Experiments with Fluorescence Detection*

Ion–molecule collision experiments which employ spectrally resolved detection of product fluorescence have achieved direct determinations, generally not possible by any other experimental method, of excited-state populations. This highly detailed information is obtained at a sacrifice of knowledge about the nonfluorescent channels and about the product angular distribution.

One apparatus of this type has been constructed by Ottinger and coworkers at the Max-Plank-Institut fur Strömungsforchung (Brandt *et al.*, 1973). A schematic diagram is given in Fig. 39. Ions produced in a dc discharge source are mass selected in a 60° sector magnet at a laboratory kinetic energy of about 1 keV. Deceleration optics are then used to retard the ions to kinetic energies as low as 2 eV, with an energy spread of about 1.5 eV. After passing through the target gas collision cell, the primary beam is reaccelerated and energy analyzed in a 90° electrostatic sector. A second magnetic sector is provided for eventual use in detecting ionic reaction products as in a conventional tandem mass spectrometer. Fluorescence radiation emitted from the collision cell normal to the ion beam axis is focused into a monochro-

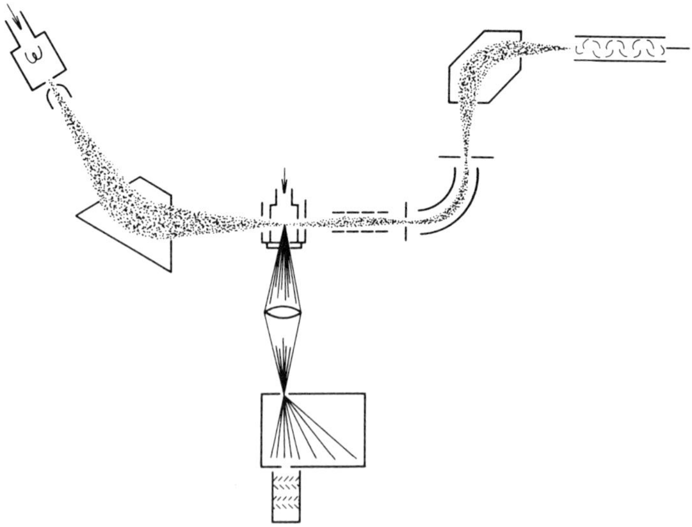

FIG. 39 Schematic diagram of the ion beam–fluorescence apparatus at the Max-Planck-Institut für Strömungsforschung. [From Brandt *et al.* (1973).]

mator with a photon counting detector. The electronics permit multichannel signal averaging over many hours, for recording weak emissions.

Early work by this group (Brandt *et al.*, 1973) included studies of a variety of different collision processes: direct product excitation,

$$C^+, N^+, O^+ + NO \rightarrow C^+, N^+, O^+ + NO^*, \tag{49}$$

rearrangement reaction chemiluminescence,

$$N^+ + NO \rightarrow N_2{}^{+*} + O, \tag{50}$$

$$\begin{aligned} &C^+ + NO \rightarrow C + NO^{+*}, \\ &C + NO \rightarrow CN^* + O, \end{aligned} \tag{51}$$

charge transfer excitation,

$$Ar^+, Ne^+ + CO \rightarrow CO^{+*} + Ar, Ne, \tag{52}$$

$$Ar^+, Ne^+ + N_2 \rightarrow N_2{}^{+*} + Ar, Ne, \tag{53}$$

and dissociative charge transfer to excited product states,

$$Ne^+ + CO \rightarrow Ne + C^{+*} + O, \tag{54}$$

$$Ne^+ + N_2 \rightarrow Ne + N^{+*} + N. \tag{55}$$

The direct excitation and charge transfer excitation processes were generally found to produce excited-state vibrational distributions markedly different from what one would predict on the basis of the Franck–Condon factors alone, thus demonstrating the "chemical" nature of the interactions. The chemiluminescent process (50) and the second reaction in the two-step sequence (51) are isoelectronic; however, the kinetic energy dependences of the chemiluminescence cross sections for these two cases are quite different. The N_2^{+*} emission decreases rapidly as the kinetic energy E_i is lowered from 6.8 to 3.4 eV while the CN* emission does not (Brandt and Ottinger, 1973). In more recent work Appell *et al.* (1975) studied the reaction

$$C^+(^2P) + H_2 \rightarrow CH^+(A^1\Pi) + H(^2S) \tag{56}$$

over the cm kinetic energy range from 2 to 20 eV. This electronically nonadiabatic reaction occurs with a cross section which peaks sharply at $E_i = 5$ eV. A very rough estimate of the absolute emission cross section at the peak yielded a value of 0.04 Å^2, showing that the observed excited-state product accounts for only a minor fraction ($\sim 1\%$) of the total CH^+ formation cross section. Rotationally resolved data on the reaction

$$C^+ + O_2 \rightarrow CO^+(A^2\Pi) + O \tag{57}$$

(Ottinger and Simonis, 1975) showed very high rotational excitation of the CO^+ product, corresponding to an average rotational quantum number of

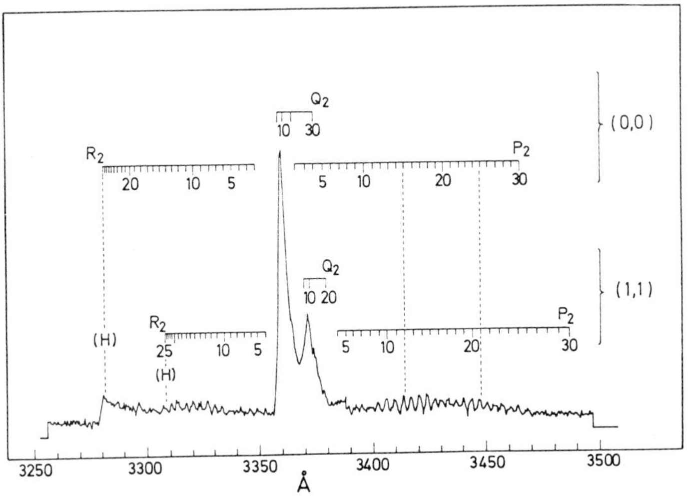

FIG. 40 Rotationally resolved $NH(A^3\Pi \rightarrow X^3\Sigma^-)$ spectrum (2 Å *FWHM*) from N^+ + C_2H_6 (20 eV_{Lab}) collisions [Kusunoki *et al.* (1976)]. Line positions are indicated for the P_2, Q_2, R_2 branches of the (0,0) and (1,1) bands. Lines of the P_1, Q_1, R_1, P_3, Q_3, R_3 branches nearly coincide with the ones shown. Lines (H) mark the *R* heads. The dashed lines in the P branches mark near coincidences between lines of the two bands. These lines differ by three and four units, respectively, of their rotational quantum number.

about 78 at $E_i = 3.6$ eV. The average product rotational energy exceeded the average product vibrational energy in this case. High rotational energies were also found in the particularly interesting series of reactions

$$N^+ + H_2, CH_4, C_2H_4, C_2H_6, C_3H_6 \rightarrow NH(A^3\Pi) + \text{other products} \tag{58}$$

(Kusunoki *et al.*, 1976). Figure 40 shows the NH emission spectrum obtained from the $N^+ + C_2H_6$ reaction at $\mathscr{E}_{N^+} = 20$ eV ($E_i = 13.6$ eV), measured with a wavelength resolution of 2 Å. The resolved rotational structure in the P and R branch transitions corresponds to an approximately Boltzmann distribution at a rotational temperature of 2×10^4 K. The authors suggest that a sequential hard-sphere collision model may be applicable to these reactions. It is interesting that the observed $NH(A^3\Pi)$ product is not adiabatically correlated with the reactants in any of the five cases given in Eq. (58), while other product states which are correlated adiabatically with the reactants were not seen experimentally even at low collision energies.

Harris, Leventhal, and co-workers at the University of Missouri at St. Louis have investigated a variety of ion–molecule collision processes by means of fluorescence detection, using the apparatus diagrammed in Fig. 41 (Harris and Leventhal, 1976). Ions from an electron impact source are extracted, focused, and mass selected at relatively low kinetic energy in a

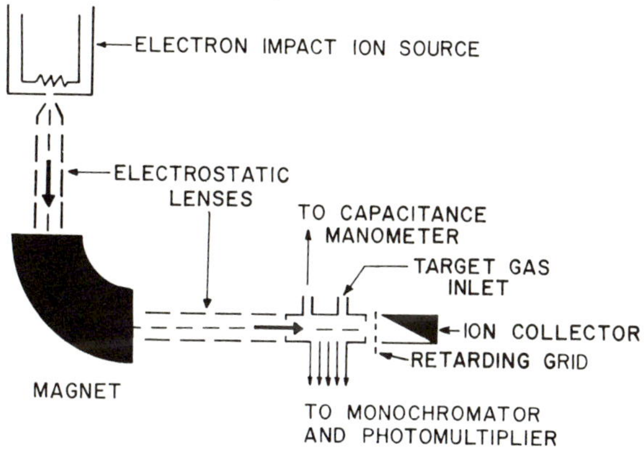

FIG. 41 Schematic diagram of the ion beam–fluorescence apparatus at the University of Missouri at St. Louis. [From Harris and Leventhal (1976).]

4.2 cm radius magnetic sector mass spectrometer. The average ion beam laboratory energy in the collision cell may be as low as ~1 eV, and the energy spread is typically several electron volts, as determined by retarding potential analysis. Photons emerging from a slot in the collision cell enter a 0.25-m, f/3.7 scanning monochromator which is digitally controlled. A multiscaler is used to accumulate the dispersed signal over many repetitive scans.

This group has studied the direct excitation of molecular Rydberg states in the system

$$H^+ + NO \rightarrow H^+ + NO^* \qquad (59)$$

(Bearman *et al.*, 1976a), which occurs at cm kinetic energies as low as 10 eV while similar excitations of the molecules N_2, CO, and O_2 are not observed at all. A large number of charge transfer excitation processes were also studied (Bearman *et al.*, 1976b, 1977; Ranjbar *et al.*, 1977; Leventhal *et al.*, 1975), including the reactions

$$He^+ + O_2 \rightarrow O_2^{+*}, O^{+*}, O^*, He^{+*}, He^* + \text{other products} \qquad (60)$$

(Harris *et al.*, 1974). In the $He^+ + O_2$ system, the nearly resonant charge transfer to the $O_2^+(c^4\Sigma_u^-)$ state is *not* observed.

Investigations of the chemiluminescent reactions

$$\begin{aligned} O^+ + H_2, D_2 &\rightarrow OH^{+*}, OD^{+*} + H, D \\ &\rightarrow OH^*, OD^* + H^+, D^+ \end{aligned} \qquad (61)$$

(Harris and Leventhal, 1973, 1976), indicate that at low cm kinetic energies $OH^+(A^3\Pi)$ and $OH(A^2\Sigma^+)$ are formed only from metastable excited O^+

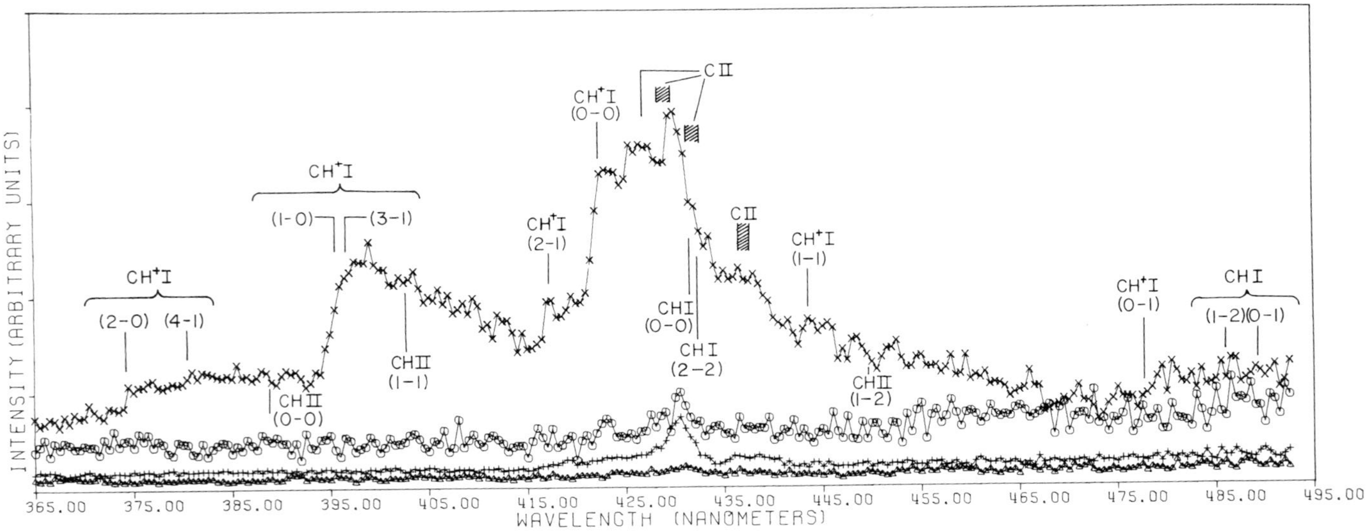

FIG. 42 Emission spectra from $C^+ + H_2$, ×, $E_i = 5$ eV; +, $E_i = 18.3$ eV; ○ and △, ion beam background signal [Harris *et al.* (1975)]. The band systems for CH^+ are identified with roman numerals. The CH transitions indicated [I($A^2\Delta \rightarrow X^2\Pi$); II($B^2\Sigma \rightarrow X^2\Pi$)] are not energetically accessible in the low-energy spectrum, $E_i = 5$ eV.

reactants. Figure 42 shows one portion of the emission spectrum observed for the reaction $C^+ + H_2 \rightarrow CH^{+*} + H$ (Harris *et al.*, 1975). Together with other spectra at smaller wavelengths, the data indicate that the $A(^1\Pi)$, $b(^3\Sigma)$, and $B(^1\Delta)$ states of CH^+ are formed in relative populations which depend on the collision energy. No emission is seen which can definitely be attributed to CH neutrals. Chemiluminescence data of this sort provides much more direct information on product internal states than is obtained by kinematic analysis.

Chemiluminescence in relatively high-energy He^+, $H_2{}^+ + CO_2$ charge transfer collisions was studied by Bregman-Reisler and Doering (1974). Similar experiments on the $Ar^+ + CO_2$, CH_4, C_2H_4, C_2H_6 systems were recently reported by Hedrick and Moran (1976).

III. FUTURE PROSPECTS

Upon surveying the field covered by this review one cannot help but be impressed with the ingenuity and vigor with which new experimental problems have been formulated, attacked, and solved during the last five years. During this period instrumental developments and improvements have permitted many microscopic details of ion–molecule collision phenomena to be studied for the first time, such as in fully state-resolved experiments on vibrational and rotational excitation of simple molecules, and measurements of the electronic, vibrational, and rotational populations of reaction products. At the same time, the range of chemical systems studied by beam techniques has expanded enormously, as a glance at the set of reactions cited in this review will quickly show.

It seems natural at this point to indulge in a bit of speculation about what directions the field will take in the near future. Central to the issue of future progress is the question of what technological developments either on the horizon or just beyond it seem likely to be useful for ion–molecule collision studies.

A. Auspicious Developments

1. Sources

The molecular beam experimentalist is constantly searching for more intense beam sources and the means for controlling the chemical and internal state populations of the beams. The conventional dc, rf, and microwave discharge sources are capable of furnishing an enormous variety of ionic species, but new types of sources will continue to be useful for special cases.

An example of a recent development is the laser-blowoff ion source of Friichtenicht *et al.* (1976), which produces extremely high-intensity pulsed ion beams of such metals as Al, Fe, and Ti. Another example is the source developed by Henkes *et al.* (1977) specifically for producing large cluster ions, in which there is growing interest (e.g., Nestler *et al.*, 1977). New sources of negative ions are very much in demand because of the relative difficulty of producing most negatively charged species. The best candidate for a source of internally state–controlled positive ions appears to be photoionization. The conventional broadband vacuum–uv sources have not been used so far in beam experiments because of intensity limitations. After the radiation is passed through a monochromator of sufficient resolution to obtain significant internal-state selection, the photon flux from these sources is just too small to provide the ion beam intensity needed for most experiments. However, it seems likely that such a source could be used for guided beam experiments, in which the extremely high intrinsic sensitivity of the method makes the required ion intensity much smaller than in other types of experiments. It is possible that the new high-intensity synchrotron radiation sources presently under construction may be usable for ion beam experiments for those who can obtain access to these facilities. A much less versatile technique, but one which might be applicable in some special cases is the use of rare-gas resonance lamps to furnish narrow-band radiation of the necessary energy without requiring wavelength selection. A general-purpose high-intensity photoionization source will probably have to await further development and applications of laser technology. Ideally one wants a tunable vuv laser in the photon energy range from, say, 8 to 30 eV. The few vuv lasers which do exist at present have very small duty factors, and pretty much require the full attention of the experimenter just to make the laser work. This makes it depressing to consider an experiment which combines the considerable experimental difficulties of both vuv laser and low-energy ion beam technologies. An alternative is to use multiphoton ionization at longer wavelengths, where the available lasers are more tractable and where the commercial development is more advanced. At present it appears that high molecular photoionization efficiency can be achieved only by using a frequency which matches at least one absorption resonance of the molecule. If a two-laser two-photon ionization were used, only one of the lasers would have to be tunable, since the second absorption would be in the ionization continuum. However, wavelengths short enough for tunable two-photon ionization of most molecules are also not available commerically. This picture may change considerably in the next few years. In the meantime, there may be some special cases in which three- or four-photon ionization processes occur with sufficient probability to permit beam experiments with the ionized products. The first step in this direction is to establish a data base on gas-phase multiphoton ionization spectroscopy, and considerable effort is currently being expended on studies of this kind.

In ion–molecule crossed beam experiments, of course, one is concerned not only with the ion beam but also the neutral beam. The basic technology of supersonic molecular beams has not changed very much in the last decade, although many variations have been developed which extend the kinetic energy range or permit exotic species to be produced. The maximum neutral beam intensity which can be used in a crossed beam experiment has generally been limited by vacuum pumping speed and background signal considerations. Very recently, Gentry and Giese (1978) developed a *pulsed* molecular beam source which is capable of producing instantaneous beam intensities at least two orders of magnitude greater than those which are normally feasible for continuous crossed beam experiments. In high-resolution neutral–neutral scattering experiments (Gentry and Giese, 1977b,c) the signal-to-background ratio was improved by at least a factor of 10^3 over that expected from conventional continuous-beam methods. This new technology may also be useful for some ion-molecule beam experiments, particularly those in which it is advantageous for some reason to use a pulsed ion beam.

2. Detectors

Most modern ion beam collision experiments employ some sort of particle-multiplier detector and ion counting technique. The efficiency of these devices is usually close to unity, and the residual background signal is often determined mostly by the cosmic radiation background, so there is little left to do in the way of improving the sensitivity. However, for differential cross section measurements the scattered signal is dispersed in angle or energy or both, while the usual sort of detector is capable of measuring the signal for only one product ion energy and/or angle at a time. The overall experimental sensitivity would therefore be vastly improved if one could measure the entire distribution in energy or angle simultaneously, as is done, for example, in Fourier transform optical spectroscopy. One method for doing this is to use time-of-flight (TOF) spectroscopy to measure product ion speed distributions. Although the TOF method has many advantages, it does have the disadvantage of a small experimental duty factor. Another method, which in principle achieves the same effect without a loss of duty factor, has been used by Wijnaendts van Resant *et al.* (1977a,b) to measure angular distributions in alkali ion–atom collisions. The detector in this case is a "channelplate" which provides true simultaneous multichannel amplification of a spatially dispersed signal. It would be a straightforward matter to adapt the same sort of idea to measurement of product ion energy distributions, using an electrostatic field to transform the energy distribution into a spatial distribution. With the further improvement of multichannel image intensifiers, methods similar to this will find increased application to collision experiments.

Further advances are also likely in the area of product fluorescence detection. Very recently Ding and Redpath (1977) succeeded in resolving infrared emission from ion–molecule collisions. Their high-sensitivity infrared spectrometer covers the spectral range from 1.7 to 5.5 μm, with a resolution of 1 cm^{-1} at 2500 cm^{-1}. The resolution is sufficient to separate single vibration–rotation lines of simple molecules. The initial experiments show high vibrational and rotational excitation of CO^* arising from $CO^+ + CO$ collisions in a bulk discharge. In favorable cases it may also be possible to apply similar methods to beam experiments. Another real possibility for the future is the application of the laser-induced fluorescence method (Kinsey, 1977) to ion–molecule reactions, which would permit measurements of the product state populations for channels which are not spontaneously fluorescent. So far this method has been applied only to neutral beam collision experiments, but in principle the sensitivity should be great enough in favorable cases to enable the lower product densities in ion–molecule reactions to be detected.

3. *State Selection*

We have already mentioned the possibility of populating selected ion internal states by high-resolution photoionization. Another possibility is to radiatively excite ions to the desired state in the beam. For example, laser excitation of vibrational states in ions should be approximately as easy (or as difficult) to accomplish as similar excitation of neutrals. Because the speed of an ion beam is greater, the typical irradiation time will be shorter than for a neutral beam. On the other hand, ion beams can be more easily Doppler tuned into resonance with a laser line. Vibrational spectroscopy of HD^+ beams has already been carried out (Wing *et al.*, 1976) using a fixed-frequency CO laser and Doppler tuning.

B. *Putting It All Together*

It is a good rule of thumb that every new component added to an apparatus which improves the resolution in some experimental parameter or which enhances the versatility of the apparatus exacts a price in decreased sensitivity or resolution in some other experimental parameter. It is not foreseeable at this point that a new instrument can be developed which combines the kinetic energy range and resolution of the merged beam technique, the angular resolution of the crossed beam technique, the product state resolution of the fluorescence detection technique, and the sensitivity of the guided beam technique. However, it is foreseeable that the art (or science) of compromise will continue to become more highly refined in the development of new instruments. At the present time it appears that the most fruitful direction for future instrumental improvements will be the synthesis of kinematic and optical methods to permit the product internal quantum state distributions to be correlated with the product angular distributions, or to make possible correlation of the product energy and angular distributions with the internal states of the reactants. Whatever directions future research in this area actually takes, one prediction can be made with confidence—that the next five years will be even more exciting than the last.

ACKNOWLEDGMENTS

Different aspects of the author's research in the field covered by this review have been supported by the Department of Energy (Office of Basic Energy Sciences) and by the National Science Foundation. A Research Fellowship from the Alfred P. Sloan Foundation is also gratefully acknowledged. The author wishes to express his appreciation to the many investigators who contributed to this review by providing preprints and copies of figures, and to Ms. Carol Buckley for her invaluable aid in the preparation of the manuscript.

REFERENCES

Anicich, V. G., Laudenslager, J. F., Huntress, W. T., and Futrell, J. H. (1977). *J. Chem. Phys.* **67**, 4340.

Appell, J., Brandt, D., and Ottinger, C. (1975). *Chem. Phys. Lett.* **33**, 131–135.

Ausloos, P., ed. (1975). "Interactions between Ions and Molecules." Plenum, New York.

Barg, G. D., Kendall, G. M., and Toennies, J. P. (1976). *Chem. Phys.* **16**, 243–268.

Bearman, G. H., Earl, J. D., Harris, H. H., James, P. B., and Leventhal, J. J. (1976a). *Chem. Phys. Lett.* **44**, 471–474.

Bearman, G. H., Earl, J. D., Pieper, R. J., Harris, H. H., and Leventhal, J. J. (1976b). *Phys. Rev. A* **13**, 1734–1742.

Bearman, G. H., Harris, H. H., and Leventhal, J. J. (1977). *J. Chem. Phys.* **66**, 4111–4115.

Blakley, C. R., Vestal, M. L., and Futrell, J. H. (1977). *J. Chem. Phys.* **66**, 2392–2398.

Böttner, R., Ross, U., and Toennies, J. P. (1976). *J. Chem. Phys.* **65**, 733–746.

Bosse, G., Ding, A., and Henglein, A. (1971). *Ber Bunsenges. Phys. Chem.* **75**, 413–420.

Brandt, D., and Ottinger, C. (1973). *Chem. Phys. Lett.* **23**, 257–261.

Brandt, D., Ottinger, C. and Simonis, J. (1973). *Ber. Bunsenges. Phys. Chem.* **77**, 648–655.

Bregman-Reisler, H., and Doering, J. P. (1974). *Chem. Phys. Lett.* **27**, 199–202.

Bromberg, E. E. A., Proctor, A. E., and Bernstein, R. B. (1975). *J. Chem. Phys.* **63**, 3287–3294.

Brouillard, F. (1978). Personal communication.

Burniaux, M., Brouillard, F., Jognaux, A., Govers, T. R., and Szucs, S. (1977). *J. Phys. B* **10**, 2421–2430.

Chivalak, S., and Hierl, P. M. (1977a). *J. Chem. Phys.* **67**, 4654–4664.

Chivalak, S., and Hierl, P. M. (1977b). *Chem. Phys. Lett.* **45**, 99–102.

Cosby, P. C., and Moran, T. F. (1970). *J. Chem. Phys.* **52**, 6157–6165.

Dahler, J. S., and Nielsen, S. E. (1979). *J. Chem. Phys.* (to be published.)

David, R., Faubel, M., and Toennies, J. P. (1973). *Chem. Phys. Lett.* **18**, 87–90.

Dimpfl, W. L., and Mahan, B. H. (1974). *J. Chem. Phys.* **60**, 3238–3250.

Ding, A., and Redpath, A. (1977). *Abstr. Pap. Int. Conf. Phys. Electron. At. Collisions 10th, Commis. Energie At., Paris* p. 759.
Doering, J. P. (1973). *Ber. Bunsenges. Phys. Chem.* **77**, 593–605.
Douglass, C. H., and Gentry, W. R. (1979a). To be published.
Douglass, C. H., and Gentry, W. R. (1979b). To be published.
Douglass, C. H., and Gentry. W. R. (1979c). To be published.
Douglass, C. H., McClure, D. J., and Gentry, W. R. (1977). *J. Chem. Phys.* **67**, 4931–4940.
Doverspike, L. D., Champion, R. L., and Lam, S. K. (1973). *J. Chem. Phys.* **58**, 1248–1250.
Eastes, W., Ross, U., and Toennies, J. P. (1977). *J. Chem. Phys.* **66**, 1919–1928.
Eisele, G., and Henglein, A. (1974). *Ber. Bunsenges. Phys. Chem.* **78**, 1090–1097.
Eisele, G., Henglein, A., and Bosse, G. (1974). *Ber. Bunsenges. Phys. Chem.* **78**, 140–145.
Fair, J. A., and Mahan, B. H. (1975). *J. Chem. Phys.* **62**, 515–519.
Farrar, J. M., Hansen, S. G., and Mahan, B. H. (1976). *J. Chem. Phys.* **65**, 2908–2910.
Ferguson, E. E. (1975). *Annu. Rev. Phys. Chem.* **26**, 17–38.
Franklin, J. L., ed. (1972). "Ion–Molecule Reactions," Vols. 1 and 2. Plenum, New York.
Friichtenicht, J. F., Utterback, N. G., and Valles, J. R. (1976). *Rev. Sci. Instrum.* **47**, 1489–1495.
Frobin, W., Schlier, C., Strein, K., and Teloy, E. (1977). *J. Chem. Phys.* **67**, 5505–5516.
Gardner, J. L., and Samson, J. A. R. (1975). *J. Chem. Phys.* **62**, 4460–4463.
Gentry, W. R. (1974). *J. Chem. Phys.* **60**, 2547–2553.
Gentry, W. R., and Giese, C. F. (1975a). *Phys. Rev. A* **11**, 90–96.
Gentry, W. R., and Giese, C. F. (1975b). *J. Chem. Phys.* **62**, 1364–1371.
Gentry, W. R., and Giese, C. F. (1977a). *J. Chem. Phys.* **67**, 2355–2361.
Gentry, W. R., and Giese, C. F. (1977b). *J. Chem. Phys.* **67**, 5389–5391.
Gentry, W. R., and Giese, C. F. (1977c). *Phys. Rev. Lett.* **39**, 1259–1262.
Gentry, W. R., and Giese, C. F. (1978). *Rev. Sci. Instrum.* **49**, 595–600.
Gentry, W. R., and Ringer, G. (1977). *J. Chem. Phys.* **67**, 5398–5399.
Gentry, W. R., Gislason, E. A., Mahan, B. H., and Tsao, C. W. (1968). *J. Chem. Phys.* **49**, 3058–3070.
Gentry, W. R., Udseth, H., and Giese, C. F. (1975a). *Chem. Phys. Lett.* **36**, 671–673.
Gentry, W. R., McClure, D. J., and Douglass, C. H. (1975b). *Rev. Sci. Instrum.* **46**, 367–375.
Giese, C. F., and Gentry, W. R. (1974). *Phys. Rev. A* **10**, 2156–2173.
Gillen, K. T., Mahan, B. H., and Winn, J. S. (1973a). *J. Chem. Phys.* **58**, 5373–5384.
Gillen, K. T., Mahan, B. H., and Winn, J. S. (1973b). *J. Chem. Phys.* **59**, 6380–6396.
Harris, H. H., and Leventhal, J. J. (1973). *J. Chem. Phys.* **58**, 2333–2335.
Harris, H. H., and Leventhal, J. J. (1976). *J. Chem. Phys.* **64**, 3185–3191.
Harris, H. H., Crowley, M. G., and Leventhal, J. J. (1974). *Chem. Phys. Lett.* **29**, 540–546.
Harris, H. H., Crowley, M. G., and Leventhal, J. J. (1975). *Phys. Rev. Lett.* **34**, 67–70.
Hedrick, A. F., and Moran, T. F. (1976). *J. Chem. Phys.* **64**, 1858–1860.
Henglein, A. (1970). *In* "Molecular Beams and Reaction Kinetics: Proceedings of the International School of Physics Enrico Fermi" (C. Schlier, ed.), pp. 139–183. Academic Press, New York.
Henkes, W., Hoffman, V., and Mikosch, F. (1977). *Rev. Sci. Instrum.* **48**, 675–681.
Herbst, E., Champion, R. L., and Doverspike, L. D. (1975). *J. Chem. Phys.* **63**, 3677–3678.
Herman, Z., and Birkinshaw, K. (1973). *Ber. Bunsenges. Phys. Chem.* **77**, 566–575.
Herman, Z., and Pacak, V. (1977). *Int. J. Mass Spectrom. Ion Phys.* **24**, 355–358.
Herman, Z., Kerstetter, J. D., Rose, T. L., and Wolfgang, R. (1969). *Rev. Sci. Instrum.* **40**, 538–544.
Herman, Z., Pacak, V., Yencha, A. J., and Futrell, J. (1976). *Chem. Phys. Lett.* **37**, 329–334.
Hermann, V., Schmidt, H., and Linder, F. (1978). *J. Phys. B* **11**, 493–505.
Hierl, P. M., Strattan, L. W., and Wyatt, J. R. (1973). *Int. J. Mass Spectrom. Ion Phys.* **10**, 385–403.

Hierl, P. M., Pacak, V., and Herman, Z. (1977). *J. Chem. Phys.* **67**, 2678–2686.

Johnson, S. G., Kremer, L. N., Metral, C. J., and Cross, R. J. (1978). *J. Chem. Phys.* **68**, 144–1447.

Jones, C. A., Wendell, K. L., and Koski, W. S. (1975). *J. Chem. Phys.* **63**, 2254–2255.

Jones, C. A., Wendell, K. L., Kaufman, J. J., and Koski, W. S. (1976). *J. Chem. Phys.* **65**, 2345–2352.

Jones, C. A., Sauers, I., Kaufman, J. J., and Koski, W. S. (1977a). *J. Chem. Phys.* **67**, 3599–3607.

Jones, C. A., Wendell, K. L., and Koski, W. S. (1977b). *J. Chem. Phys.* **67**, 4917–4922.

Jones, C. A., Wendell, K. L., and Koski, W. S. (1977c). *J. Chem. Phys.* **66**, 5325–5329.

Kashihira, N., Vietzke, E., and Zellermann, G. (1976). *Chem. Phys. Lett.* **39**, 316–319.

Kashihira, N., Vietzke, E., and Zellermann, G. (1977). *Rev. Sci. Instrum.* **48**, 171–172.

Kinsey, J. L. (1977). *Annu. Rev. Phys. Chem.* **28**, 349–372.

Kley, D., Lawrence, G. M., and Stone, E. J. (1977). *J. Chem. Phys.* **66**, 4157–4165.

Kobayashi, N., Itoh, Y., and Kaneko, Y. (1977). *Abstr. Pap., Int. Conf. Phys. Electron. At. Collisions, 10th, Commis. Energie At., Paris* pp. 782–783.

Koski, W. S. (1975). *Adv. Chem. Phys.* **30**, 185–246.

Krenos, J. R., Preston, R. K., Wolfgang, R., and Tully, J. C. (1974). *J. Chem. Phys.* **60**, 1634–1659.

Krenos, J. R., Lehmann, K. K., Tully, J. C., Hierl, P. M., and Smith, G. P. (1976). *Chem. Phys.* **16**, 109–116.

Krutein, J., and Linder, F. (1977a). *Chem. Phys. Lett.* **51**, 597–599.

Krutein, J., and Linder, F. (1977b). *J. Phys. B* **10**, 1363–1375.

Krutein, J., and Linder, F. (1977c). *Abstr. Pap. Int. Conf. Phys. Electron. At. Collisions, 10th, Commis. Energie At., Paris* pp. 784–785.

Kusunoki, I., Ottinger, C., and Simonis, J. (1976). *Chem. Phys. Lett.* **41**, 601–605.

Lees, A. B., and Rol, P. K. (1974). *J. Chem. Phys.* **61**, 4444–4449.

Lees, A. B., and Rol, P. K. (1975a). *J. Chem. Phys.* **63**, 2461–2465.

Lees, A. B., and Rol, P. K. (1975b). *J. Chem. Phys.* **63**, 5077–5080.

Leventhal, J. J., Earl, J. D., and Harris, H. H. (1975). *Phys. Rev. Lett.* **35**, 719–722.

Lin, K., Cotter, R. J., and Koski, W. S. (1974a). *J. Chem. Phys.* **60**, 3412–3416.

Lin, K., Watkins, H. P., Cotter, R. J., and Koski, W. S. (1974b). *J. Chem. Phys.* **60**, 5134–5135.

McClure, D. J., Douglass, C. H., and Gentry, W. R. (1977a). *J. Chem. Phys.* **66**, 2079–2093.

McClure, D. J., Douglass, C. H., and Gentry, W. R. (1977b). *J. Chem. Phys.* **67**, 2362–2370.

McDaniel, E. W., Cermak, V., Dalgarno, A., Ferguson, E. E., and Friedman, L. (1970). "Ion–Molecule Reactions." Wiley (Interscience), New York.

McGuire, P., Rudolph, K., and Toennies, J. P. (1976). *J. Chem. Phys.* **65**, 5522–5523.

McGuire, P., Schmidt, H., Hermann, V., and Linder, F. (1977). *J. Chem. Phys.* **66**, 4243–4244.

Magnuson, G. D., and Neynaber, R. H. (1974). *J. Chem. Phys.* **60**, 3385–3387.

Mahan, B. H. (1976). *In* "International Review of Science, Physical Chemistry, Series Two" (A. D. Buckingham, D. R. Herschbach, eds.), Vol. 9, pp. 25–66. Butterworth. London.

Mahan, B. H., and Ruska, W. E. W. (1976). *J. Chem. Phys.* **65**, 5044–5051.

Mahan, B. H., and Schubart, P. J. (1977). *J. Chem. Phys.* **66**, 3155–3162.

Mahan, B. H., and Sloane, T. M. (1973). *J. Chem. Phys.* **59**, 5661–5675.

Maier, W. B. (1975). *J. Chem. Phys.* **62**, 4615–4622.

Michels, H. H. (1974). Rep. AFWL-TR-73-288. Air Force Cambridge Res. Lab., Cambridge, Massachusetts.

Michels, H. H. (1975). Rep. AFCRL-TR-75-0509. Air Force Cambridge Res. Lab., Cambridge, Massachusetts.

Miller, G. D., and Safron, S. A. (1976). *J. Chem. Phys.* **64**, 5065–5072.

Miller, W. H. (1970). *J. Chem. Phys.* **52**, 3563–3572.

Moore, J. H., and Doering, J. P. (1969). *Phys. Rev. Lett.* **23**, 564–567.

Nestler, V., Betz, B., and Warneck, P. (1977). *Ber. Bunsenges. Phys. Chem.* **81**, 13–22.
Neynaber, R. H. (1969). *Adv. At. Mol. Phys.* **5**, 57–108.
Neynaber, R. H., and Magnuson, G. D. (1973a). *J. Chem. Phys.* **58**, 4586–4598.
Neynaber, R. H., and Magnuson, G. D. (1973b). *J. Chem. Phys.* **59**, 825–831.
Neynaber, R. H., and Magnuson, G. D. (1974a). *J. Chem. Phys.* **61**, 749–750.
Neynaber, R. H., and Magnuson, G. D. (1974b). *J. Chem. Phys.* **61**, 3490–3491.
Neynaber, R. H., and Magnuson, G. D. (1975a). *Phys. Rev. A* **11**, 865–871.
Neynaber, R. H., and Magnuson, G. D. (1975b). *Phys. Rev. A* **12**, 891–894.
Neynaber, R. H., and Magnuson, G. D. (1975c). *J. Chem. Phys.* **62**, 4953–4954.
Neynaber, R. H., and Magnuson, G. D. (1976a). *J. Chem. Phys.* **64**, 2840–2844.
Neynaber, R. H., and Magnuson, G. D. (1976b). *Phys. Rev. A* **14**, 961–964.
Neynaber, R. H., and Magnuson, G. D. (1977). *J. Chem. Phys.* **67**, 430–433.
Neynaber, R. H., and Tang, S. Y. (1977). *J. Chem. Phys.* **67**, 5619–5623.
Ochs, G., and Teloy, E. (1974). *J. Chem. Phys.* **61**, 4930–4931.
Ottinger, C., and Simonis, J. (1975). *Phys. Rev. Lett.* **35**, 924–927.
Pacak, V., Havemann, U., Herman, Z., Schneider, F., and Zülicke, L. (1977). *Chem. Phys. Lett.* **49**, 273–276.
Peart, B., Grey, R., and Dolder, K. T. (1976a). *J. Phys. B* **9**, L373–L375.
Peart, B., Grey, R., and Dolder, K. T. (1976b). *J. Phys. B* **9**, L369–L372.
Peart, B., Grey, R., and Dolder, K. T. (1977a). *J. Phys. B* **10**, 2675–2682.
Peart, B., Gee, D. M., and Dolder, K. T. (1977b). *J. Phys. B* **10**, 2683–2687.
Poshusta, R. D., and Zetik, D. F. (1973). *J. Chem. Phys.* **58**, 118–125.
Ranjbar, F., Harris, H. H., and Leventhal, J. J. (1977). *Appl. Phys. Lett.* **31**, 385–387.
Ringer, G., and Gentry, W. R. (1977). *Abstr. Pap. Int. Conf. Phys. Electron. At. Collisions, 10th, Commis. Energie At., Paris* pp. 1146–1147.
Ringer, G., and Gentry, W. R. (1979). *J. Chem. Phys.* (to be published).
Rol. P. K., Entemann, E. A., and Wendell, K. L. (1974). *J. Chem. Phys.* **61**, 2050–2058.
Rutherford, J. A., and Vroom, D. A. (1973a). *J. Chem. Pbys.*ı76–4079.
Rutherford, J. A., and Vroom, D. A. (1973b). *J. Chem. Phys.* **59**, 4561–4562.
Rutherford, J. A., and Vroom, D. A. (1974). *J. Chem. Phys.* **61**, 2514–2519.
Rutherford, J. A., and Vroom, D. A. (1975). *J. Chem. Phys.* **62**, 1460–1462.
Rutherford, J. A., and Vroom, D. A. (1976a). *J. Chem. Phys.* **64**, 1251–1252.
Rutherford, J. A., and Vroom, D. A. (1976b). *J. Chem. Phys.* **64**, 3057–3059.
Rutherford, J. A., and Vroom, D. A. (1976c). *J. Chem. Phys.* **65**, 4445–4446.
Rutherford, J. A., Mathis, R. F., Turner, B. R., and Vroom, D. A. (1971). *J. Chem. Phys.* **55**, 3785–3793.
Rutherford, J. A., Turner, B. R., and Vroom, D. A. (1973). *J. Chem. Phys.* **58**, 5267–5271.
Safron, S. A., Miller, G. D., Rideout, F. A., and Horvat, R. C. (1976). *J. Chem. Phys.* **64**, 5051–5064.
Schinke, R., Krüger, H., Hermann, V., Schmidt, H., and Linder, F. (1977). *J. Chem. Phys.* **67**, 1187–1190.
Schmidt, H., Hermann, V., and Linder, F. (1976). *Chem. Phys. Lett.* **41**, 365–369.
Schmidt, H., Hermann, V., and Linder, F. (1978). Unpublished observations.
Schneider, F., Havemann, U., Zülicke, L., and Herman, Z. (1977). *Chem. Phys. Lett.* **48**, 439–442.
Schöttler, J., and Toennies, J. P. (1973). *Chem. Phys.* **2**, 137–157.
Schöttler, J., and Toennies, J. P. (1974). *Chem. Phys.* **4**, 24–35.
Schopman, J., and Locht, R. (1974). *Chem. Phys. Lett.* **26**, 596–600.
Schuette, G. F., and Gentry, W. R. (1979). To be published.
Siskind, B., Alexander, M. H., and Coplan, M. A. (1976). *J. Chem. Phys.* **65**, 1063–1070.
Smith, G. P. K., and Cross, R. J. (1974). *J. Chem. Phys.* **60**, 2125–2131.

Smith, G. P. K., Carlson, R. D., and Cross, R. J. (1975). *J. Chem. Phys.* **62**, 1999–2000.

Teloy, E., and Gerlich, D. (1974). *Chem. Phys.* **4**, 417–427.

Toennies, J. P. (1976). *Annu. Rev. Phys. Chem.* **27**, 225–260.

Truhlar, D. G. (1969). *J. Chem. Phys.* **51**, 4617–4623.

Trujillo, S. M., Neynaber, R. H., and Rothe, E. W. (1966). *Rev. Sci. Instrum.* **37**, 1655–1661.

Udseth, H., Giese, C. F., and Gentry, W. R. (1971). *J. Chem. Phys.* **54**, 3642–3643.

Udseth, H., Giese, C. F., and Gentry, W. R. (1973). *Phys. Rev. A* **8**, 2483–2493.

Udseth, H., Giese, C. F., and Gentry, W. R. (1974). *J. Chem. Phys.* **60**, 3051–3056.

Udseth, H., Giese, C. F., and Gentry, W. R. (1979). To be published.

van den Bergh, H. E., Faubel, M., and Toennies, J. P. (1973). *Faraday Discuss. Chem. Soc.* **55**, 203–210.

Vestal, M. L., Blakley, C. R., Ryan, P. W., and Futrell, J. H. (1974). *Chem. Phys. Lett.* **27**, 490–492.

Vestal, M. L., Blakley, C. R., Ryan, P. W., and Futrell, J. H. (1976a). *Rev. Sci. Instrum.* **47**, 15–26.

Vestal, M. L., Blakley, C. R., Ryan, P. W., and Futrell, J. H. (1976b). *J. Chem. Phys.* **64**, 2094–2111.

Vogt, D., and Mischke, J. (1977). *Phys. Lett. A* **60**, 19–20.

Vogt, D., Mischke, J., and Dreves, W. (1976a). *J. Phys. E* **9**, 38–40.

Vogt, D., Mischke, J., and Dreves, W. (1976b). *Phys. Lett. A* **59**, 113–114.

Vogt, D., Dreves, W., and Mischke, J. (1977a). *Z. Naturforsch., Teil A* **32**, 13–16.

Vogt, D., Dreves, W., and Mischke, J. (1977b). *Int. J. Mass Spectrom. Ion Phys.* **24**, 285–296.

Wendell, K. L., and Rol, P. K. (1974). *J. Chem. Phys.* **61**, 2059–2061.

Wendell, K. L., Jones, C. A., Kaufman, J. J., and Koski, W. S. (1975). *J. Chem. Phys.* **63**, 750–756 (erratum **64**, 913).

Wijnaendts van Resandt, R. W., de Vreugd, C., Champion, R. L., and Los, J. (1977a). *Chem. Phys.* **26**, 223–235.

Wijnaendts van Resant, R. W., Champion, R. L., and Los, J. (1977b). *Chem. Phys.* **20**, 107–119.

Wing, W. H., Ruff, G. A., Lamb, W. E., and Speseski, J. J. (1976). *Phys. Rev. Lett.* **36**, 1488–1491.

Wyatt, J. R., Strattan, L. W., Snyder, S. C., and Hierl, P. M. (1974). *J. Chem. Phys.* **60**, 3702–3703.

Wyatt, J. R., Strattan, L. W., Snyder, S. C., and Hierl, P. M. (1975a). *J. Chem. Phys.* **62**, 2555–2565.

Wyatt, J. R., Strattan, L. W., Chivalak, S., and Hierl, P. M. (1975b). *J. Chem. Phys.* **63**, 4582–4591.

Wyatt, J. R., Strattan, L. W., Snyder, S. C., and Hierl, P. M. (1976a). *J. Chem. Phys.* **64**, 3757–3762.

Wyatt, J. R., Strattan, L. W., and Hierl, P. M. (1976b). *J. Chem. Phys.* **65**, 1593–1594.

Chapter 16

Variational treatment of charge transfer reactions*

ARTHUR B. WEGLEIN[†] *AND DONALD RAPP*

THE UNIVERSITY OF TEXAS AT DALLAS
RICHARDSON, TEXAS

* Support under Grant No. AT-464 by The Robert A. Welch Foundation is gratefully acknowledged.

[†] Present address: Cities Service Oil Company, Exploration and Production Research, Tulsa, Oklahoma.

ISBN 0-12-120802-8

I. REVIEW OF CHARGE EXCHANGE THEORY

Charge exchange is the inelastic collision process in which an electron is captured from a target atom or molecule by an incident atomic projectile. Most theoretical work on charge exchange has been concentrated on incident ions capturing an electron from target atoms, and this chapter will be limited to that scope. Charge exchange is described as symmetric when the atomic target is the atomic precursor of the projectile ion.

The proper treatment of charge exchange collisions involves a time-independent scattered wave description with the evaluation of phase shifts used to determine the cross sections. It is found that at sufficiently high collision velocities, the number of phase shifts that contribute substantially to the cross section becomes large (see, e.g., Rapp, 1971). Under such conditions, the summations used in evaluating cross sections can be closely approximated by integrals, and this can be shown to be mathematically equivalent to a "semiclassical" procedure (Rapp, 1971). In the semiclassical procedure, the relative motion of the nuclei are treated as classical, the motion producing an interaction between projectile and target that is time dependent. The time-dependent Schrödinger equation is then used to describe the electrons in the atoms, with the time-dependent perturbation due to nuclear motion acting to produce quantum transitions in the states of the collision partners. A comparison of the scattered wave and semiclassical approaches for specific examples leads to the conclusion that the semiclassical approach should be a good approximation at velocities in excess of about 10^5–10^6 cm/sec. Since the scattered wave approach is usually intractable, pragmatism demands the use of the semiclassical calculation. The semiclassical approach only makes sense if the incident collisional kinetic energy is large compared to the energy gaps between states undergoing transition. Therefore, it is common to use the impact parameter approximation within the semiclassical approach, in which the paths of the colliding partners are treated as straight lines with constant velocity. All calculations described herein make use of the impact parameter semiclassical approach. The impact parameter approach is illustrated in Figs. 1 and 2. Representing the charge exchange process as

$$A^+ + (B^+ + e^-) \rightarrow (A^+ + e^-) + B^+, \tag{1}$$

a flux of A^+ ions approaches targets B at a variety of impact parameters b. For a fixed collision velocity v, one must calculate the transition probability $P(b, v)$ for any transition as a function of b. Then the effective cross section at velocity v for that transition is given by

$$\sigma(v) = \int_0^\infty P(b, v) 2\pi b \, db. \tag{2}$$

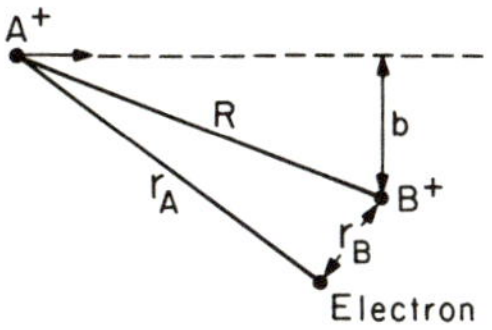

FIG. 1 Coordinate system for impact-parameter treatment of charge transfer.

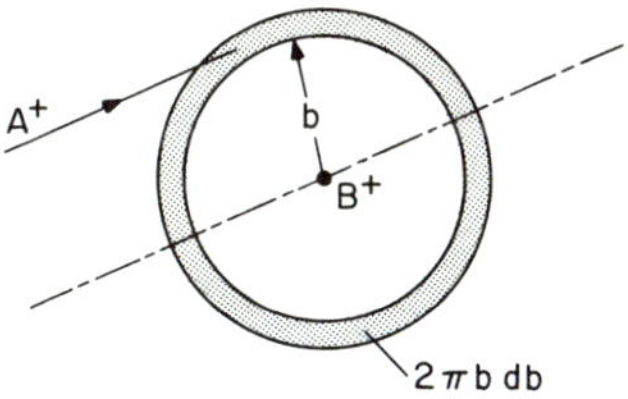

FIG. 2 Illustration of contribution of an annular element $2\pi b\,db$ to the cross section in the impact-parameter treatment.

The calculation of $P(b, v)$ is made by solution of the appropriate time-dependent Schrödinger equation.

For many purposes it is convenient to utilize a one-electron model in which only a single "active" electron is considered. The net effective potential (due to nucleus plus inner electrons) attracting the active electron to A^+ or B^+ is $V_A(r_A)$ or $V_B(r_B)$. For a nucleus without inner electrons, such as H^+ or He^{2+}, $V_A(r_A)$ is rigorously given by a Coulomb term $-z/r_A$, whereas for nuclei with inner electrons, a pseudopotential is called for. Then the Schrödinger equation for the active electron is

$$H_{el}\psi = [-\tfrac{1}{2}\nabla^2 + V_A(r_A) + V_B(r_B)]\psi = i\frac{\partial\psi}{\partial t}. \tag{3}$$

This equation must be solved subject to the appropriate boundary conditions. The time dependence of V_A and V_B is due to the implicit dependence of r_A and r_B on $R(t)$.

In order to provide a framework for approximate solution of Eq. (3), one would usually want to express ψ, the wave function of the active electron, as an expansion in some set of suitable functions:

$$\psi = \sum_j \alpha_j(t)\psi_j(r, t). \tag{4}$$

By choosing one of the ψ_j to correspond to the initial state ψ_0 as $t \to -\infty$, the initial condition is then given as

$$\begin{aligned} \alpha_0(-\infty) &= 1, \\ \alpha_j(-\infty) &= 0, \qquad j \neq 0. \end{aligned} \tag{5}$$

The problem then becomes to calculate $\alpha_j(t)$ as $t \to +\infty$, with the transition probability then given by $P(b, v) = |\alpha_j(\infty)|^2$ for transition to state j.

In choosing the expansion functions $\psi_j(r, t)$, one might be tempted to use *atomic* wave functions centered on each nucleus. Thus, in a two-state approximation, one might use

$$\psi(r, t) = c_A(t)\phi_A(r_A)e^{-i\epsilon_A t} + c_B(t)\phi_B(r_B)e^{-i\epsilon_B t} \tag{6}$$

If the electron is initially on center B in state ϕ_B, the initial condition is $c_B(-\infty) = 1$, $c_A(-\infty) = 0$, and the transition probability is $|c_A(\infty)|^2$. On the other hand, one might wish to expand in *molecular* wave functions $\Phi_1(r_A, r_B)$ and $\Phi_2(r_A, r_B)$ such that the Φ_j go to the ϕ_j (or linear combinations of ϕ_k) as $t \to \pm\infty$. There are a variety of other functions ("pseudostates") that might be used as expansion functions. There is a problem, however, in attempting to use expansion functions as given by Eq. (6), or its counterpart in molecular functions. This problem is fundamentally related to the proper asymptotic boundary conditions at large separations. Trial functions of the form of Eq. (6) violate fundamental requirements of the boundary conditions, and are therefore improper. This violation is reflected by several observations. One is that the semiclassical model requires that the transition probabilities should be independent of whether the nuclear potential $z_A z_B/R^2$ is included in the Hamiltonian H_{el}. Interactions between the nuclei can only affect the nuclear trajectories, not the transition probabilities, along straight line paths in the impact parameter treatment. Another unsavory effect of Eq. (6) is that such functions lead to transition probabilities that depend on the origin of coordinates. When the transition probabilities calculated with improper expansion functions are compared with those calculated with proper functions, it is found that the differences tend to decrease at low collision velocities. Thus, trial functions such as Eq. (6) may be regarded as approximations for low collision velocities.

The proper treatment of the asymptotic conditions requires that we first derive the wave function for an electron relative to a moving center. Consider a hydrogen atom translating to the right at constant velocity v. An origin is fixed in space, and the coordinates of the electron relative to the fixed origin are $\mathbf{r}$, and relative to the proton are $\mathbf{r}'(t)$. The position of the proton is $\mathbf{R}(t)$, as shown in Fig. 3. The Schrödinger equation is

$$\left(-\tfrac{1}{2}\nabla^2 - \frac{1}{r'(t)}\right)\psi(r, t) = i\,\frac{\partial\psi(r, t)}{\partial t}, \tag{7}$$

where $\partial/\partial t$ is carried out at fixed r. If the nuclear motion is along the z axis, it can be shown (Rapp, 1971) that the solution of Eq. (7) can be expressed in terms of the solution of a stationary hydrogen atom. For a stationary hydrogen atom, $\mathbf{r}' = \mathbf{r}$, and $\mathbf{R} = 0$. Thus the solution may be written

$$\psi(\mathbf{r}, t) = \phi(\mathbf{r})e^{-i\epsilon t} \tag{8}$$

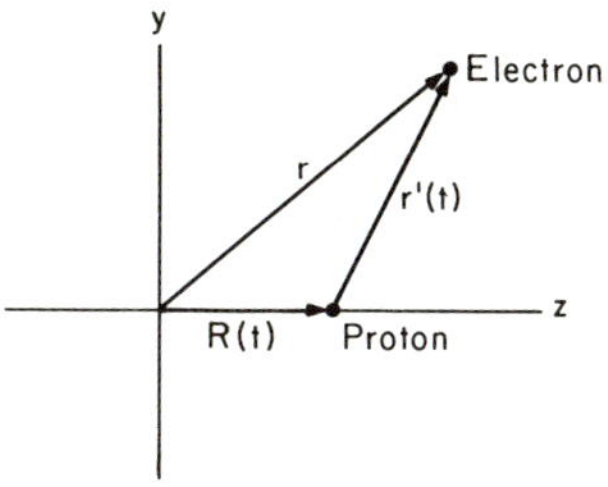

FIG. 3 Coordinates for describing an electron on a moving proton in both fixed and moving frames.

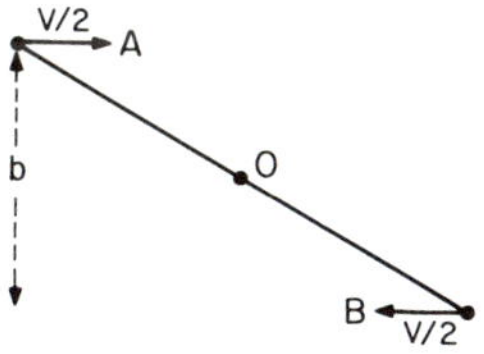

FIG. 4 Illustration of the kinematics in the charge transfer calculation with fixed center of mass.

for any arbitrary state. For the moving atom, it can be shown that

$$\psi(\mathbf{r}, t) = \phi(\mathbf{r}')e^{-i\epsilon t}e^{ivz}e^{-iv^2t/2}, \tag{9}$$

where $\phi(\mathbf{r}')e^{-i\epsilon t}$ is an ordinary atomic wave function relative to a moving origin, and $e^{ivz}e^{-iv^2t/2}$ is a translational factor to account for motion of the origin.

The application of traveling atomic orbitals, as expressed in Eq. (9), to charge exchange is obtained by considering a fixed center of coordinates at the midpoint between the two nuclear centers. The nuclei move in opposite directions at velocity $v/2$, resulting in a relative velocity v, as shown in Fig. 4. The distance between A and B is $R(t)$, and the distances of the electron from the moving centers are r_A and r_B. Wave functions centered on A and B then have the form

$$\begin{aligned} \chi_A &= \phi(r_A)e^{-i\epsilon_A t}e^{ivz/2}e^{-iv^2t/8}, \\ \chi_B &= \phi(r_B)e^{-i\epsilon_B t}e^{-ivz/2}e^{-iv^2t/8}, \end{aligned} \tag{10}$$

and these should be used in a trial function of the form

$$\psi(r, t) = c_A(t)\chi_A + c_B(t)\chi_B \tag{11}$$

instead of Eq. (6). In multistate calculations, one should use these functions as the ψ_j in Eq. (4). If one examines Eq. (7), it will be found that the right-hand side contains terms due to explicit functions of t such as $e^{-i\epsilon t}$, as well as implicit functions of t such as

$$\frac{\partial\psi}{\partial\mathbf{r}'}\cdot\frac{\partial\mathbf{r}'}{\partial t}. \tag{12}$$

The only way that Eq. (6) can be made to approximately satisfy Eq. (7) at large internuclear separations is by neglecting terms such as (12). This becomes a good approximation at low velocities where e^{ivz} is nearly constant across the dimensions of an atom. However, Eq. (11) satisfies Eq. (7) for all velocities at large internuclear separations. When using molecular functions as expansion functions, it is very difficult to choose the molecular functions so they have the proper asymptotic form as expressed in Eqs. (10). Therefore, most molecular functions have been chosen to go simply to Eq. (8), and the calculations are only applicable to low velocity collisions. This is not necessarily very limiting because the velocity range where the traveling factors begin to have significant effect tends to be about 3×10^7 cm/sec and above.

For moderate collision velocities, the molecular expansion functions without traveling wave factors appear to offer the best general approach if the molecular functions are available in a form for allowing their utilization. Such functions are not generally available for many systems, and are difficult to approximate by simple functions. For the special case of symmetric charge exchange between A^+ and A, where only a single asymptotic atomic state is considered, there exist gerade and ungerade molecular states which go asymptotically to $\phi_A + \phi_B$ and $\phi_A - \phi_B$ at large internuclear distances. It can be shown (Rapp, 1971) that use of an expansion in these molecular states with neglect of the traveling wave factors leads to the transition probability

$$P(b, v) = \sin^2 \int_{-\infty}^{\infty} (E_u - E_m) \frac{dx}{v}, \tag{13}$$

where dx is an element of path along the nuclear motion, and $E_u - E_g$ is the difference between energies of ungerade and gerade molecular states at any point along the trajectory. In systems for which $E_u - E_g$ increases monotonically as R is reduced, it follows that $P(b, v)$ oscillates with b ever more rapidly as b is reduced, and the cross section will have a form similar to

$$\sigma(v) = \alpha - \beta v, \tag{14}$$

where α and β are constants. If $E_u - E_g$ passes through a maximum as R is reduced, the oscillations in the function $P(b, v)$ with b will be nonuniform, and although σ will tend to increase with decreasing velocity, there will be characteristic oscillations of σ with v about the basic trend line for $\sigma(v)$ (Rapp, 1971). When excited states are included in molecular calculations, the situation becomes more complicated. For simple asymmetric exchange between the ground state of B and the ground state of A, a formula similar to Eq. (13) is obtained except that a factor

$$\exp[i(E_B - E_A)x/v] \tag{15}$$

occurs in the integral, E_B and E_A replace E_u and E_g, and E_B and E_A are energies of molecular states that go to atomic states on B and A at large internuclear separations. The effect of the term in Eq. (15) is to reduce $\sigma(v)$ below that given in Eq. (14) as the velocity is reduced below the point where Eq. (15) makes more than one oscillation across the range of interaction.

In general, the expansion in atomic orbitals affords the simplest approach to deal with a wide range of systems over a wide range of velocities. One major drawback of the atomic approach is that the atomic expansion functions do not generally give a good description of the compound system formed during the collision. For example, if one uses a two-state atomic expansion for symmetric charge exchange, the transition probability is found to have a form similar to that in Eq. (13) except that $E_u - E_g$ is replaced by the LCAO (linear combination of atomic orbitals) approximation to $E_u - E_g$. For H_2^+ at separations of (1–5) a_0, this can lead to errors of almost a factor of two in $E_u - E_g$ (Rapp, 1971). One way of bridging the gap between adequacy of description of the compound system in the molecular and atomic expansions is to employ parameters in the atomic functions that vary with internuclear separation. At large separations they have atomic properties, whereas at small separations they have molecular character. By using this technique, one can satisfy the proper boundary conditions using traveling orbitals centered on atoms, but not be limited to asymptotic atomic functions during the middle of the collision. One can set these parameters by using "chemical intuition" based on knowledge of variation of structure of molecular ions, as in pseudostate calculations. However, it appears more desirable to set up procedures for adjusting these parameters during a collision in order to "optimize" their values according to a consistent concept of quality of a calculation. The present paper is a review of procedures that have been used to optimize parameters in charge exchange calculations employing traveling atomic orbitals with variable parameters. We are concerned with evaluating the relative quality of various procedures, finding the optimum procedure, and determining bounds on the exact transition probabilities imposed by inexact calculations.

II. INTRODUCTION AND DEFINITION

A. *Differential Equations and Stationary Integrals*

Throughout physics it is found that basic principles are expressed in terms of both differential equations and stationary integrals. The best known example of this dualism is in classical mechanics where Newton's second law is the differential equation and Hamilton's principle is the stationary integral.

In quantum mechanics, this dualism can be expressed in terms of the time-dependent Schrödinger equation

$$G\psi = \left(H - i\frac{\partial}{\partial t}\right)\psi = 0. \tag{16}$$

The exact solution ψ of this equation also satisfies the equivalent stationary integral equation

$$\delta \int \langle \psi | G\psi \rangle \, dt = 0. \tag{17}$$

In most applications one seeks an approximation to ψ in terms of a trial solution, ψ_{tr}, which depends on time-dependent parameters $\alpha_1, \alpha_2, \ldots$. The problem is to select a procedure for evaluating the parameters so that ψ_{tr} will approximate ψ as closely as possible within the limited subspace spanned by ψ_{tr}.

This article is concerned with various procedures that have been developed for "optimizing" the approximate trial solutions of the Schrödinger equation. There is often a wide gap between the quality of solution that is desired and the quality of a solution that is calculable. As a result, there is a range of possible levels of effort that can be expended in refining approximate solutions. In some cases, the procedures do not necessarily bring the approximate solution closer to the exact solution unless the approximate solution is already close to the exact solution. Indeed, some "optimization" procedures may actually move the trial solution further from the exact solution when the trial function starts far from the exact solution. Thus the word "optimization" is a rather broad term covering many possibilities. We have defined several "levels of optimization" which describe the various approaches as a regular progression of procedures leading to better results as more effort is expended.

B. *Levels of Optimization*

For any arbitrary trial function, the pinnacle of optimization procedures is the requirement that $|G\psi_{tr}|^2$ (where ψ_{tr} is a trial solution) be a minimum at every positional coordinate and time. Since $G\psi = 0$ for the exact solution, the smaller $|G\psi_{tr}|^2$ is made, the closer ψ_{tr} is to ψ in this sense. If ψ_{tr} is optimized by this requirement we refer to it as "level A optimized."

Level A optimization can not be achieved with the usual forms of trial solutions. In an attempt to define a less demanding (but perhaps achievable) criterion for optimization, one may ask for the integral of $|G\psi_{tr}|^2$ over all space to be minimized at any t. We define the deviation $D(t)$ as

$$D(t) = \int |G\psi_{tr}(\mathbf{r}, t)|^2 \, d\mathbf{r}$$

and define the trial solution which minimized $D(t)$ at every instant as "level B optimized." Unfortunately, the forms of trial solutions used in typical applications are not sufficiently flexible to achieve even level B optimization. One may then lower this demand and ask that only the time *and* space integral of $|G\psi_{\text{tr}}|^2$ be minimized. That is, one may require that the integral

$$J(\psi_{\text{tr}}) = \iint |G\psi_{\text{tr}}|^2 \, d\mathbf{r} \, dt = \int_{-\infty}^{+\infty} D(t) \, dt$$

be minimized. Typical forms of trial solutions are adequate so that one can attempt to achieve this goal. We refer to the solution which minimizes $J(\psi_{\text{tr}})$ as "level C optimized."

We will include within level C optimization the minimization of the space and time integral of any power of $D(t)$. In Section III.E it will be shown how $J(\psi_{\text{tr}})$, the space, and time-averaged error, can be minimized. Optimization at levels A, B, and C is not based upon the assumption that the trial solution is close to the exact solution.

The next lower category of approximations will be referred to as "level D optimization." Rather than seeking to minimize some measure of the error involved in the approximate solution, this category of approximation assumes that the form of the trial solution is sufficiently comprehensive to span the space of the exact solution. One may then impose a property of the exact solution upon the approximate solution, hoping that this will lead to optimized parameters in the approximate solution. Alternatively, a principle governing the exact solution may be applied to the approximate solution by analogy.

For example, several authors (Sil, 1960; Mittleman, 1961; Cheshire, 1968; McCaroll *et al.*, 1970) have assumed that by forcing the integral analogous to Eq. (17) to be stationary,

$$\delta I = \delta \int \langle \psi_{\text{tr}} | G\psi_{\text{tr}} \rangle \, dt = 0, \tag{18}$$

"optimized" values of the parameters will be obtained. The success of this method critically depends on the extent to which the space spanned by the trial solution includes the exact solution. In actual cases, the difficulty is that one rarely knows the adequacy of the space spanned by ψ_{tr}. Equation (18) does not guarantee the "best" choice of parameters in ψ_{tr} in any consistent sense when the trial solution is not a close approximation to the exact solution. Only if this space is sufficiently comprehensive will this method "optimize" the solution.

Optimization at level D is the most easily used "optimization" criterion. Its application has the great advantage of being both numerically achievable and straightforward and therefore of practical significance. Before discussing the details of level C and D optimization, it is useful to present a brief survey of the works on bounds on physically significant quantities. These bounds have a relationship to the optimization criteria.

III. DISCUSSION OF OPTIMIZATION THEORY

A. *Bounds on Properties of the Exact Solution*

One major problem involved in the use of approximate solutions of the Schrödinger equation is that it is usually difficult to assess the accuracy of an approximate calculation. If a procedure can be found that utilizes the approximate solution to place bounds on properties of the exact solution, the approximate solution can be more valuable. For some cases, bounds have been established on properties of, and on the error made in some estimate of a property of, the exact solution.

Consider a system described by a Hamiltonian H_0 with eigenfunctions $\phi_1, \phi_2, \ldots$ and eigenvalues $\epsilon_1, \epsilon_2, \ldots$. Let a perturbation $V(\mathbf{r}, t)$ with the property $V \to 0$ as $t \to \pm\infty$ act on the system, where $\mathbf{r}$ represents the positional coordinates.

The exact solution of the equation

$$G\psi = \left(H_0 + V - i\frac{\partial}{\partial t}\right)\psi = 0 \tag{19}$$

can be expanded in terms of the ϕ_i as

$$\psi = \sum_{j=1}^{\infty} a_j(t)\phi_j(\mathbf{r})e^{-i\epsilon_j t}, \tag{20}$$

where

$$\dot{a}_k = \frac{1}{i}\sum_{j=1}^{\infty} \langle k|V|j\rangle a_j = \frac{1}{i}\sum_{j=1}^{\infty} V_{kj}e^{i\omega_{kj}t}a_j \tag{21}$$

and

$$|j\rangle = \phi_j e^{-i\epsilon_j t}, \qquad \omega_{ij} = \epsilon_i - \epsilon_j,$$

$$V_{kj}(t) \equiv \int d\mathbf{r}\ \phi_k^*(\mathbf{r})V(\mathbf{r}, t)\phi_j(\mathbf{r}). \tag{22}$$

B. *Systems Possessing Only Two States*

Consider a hypothetical system that has only two possible states ϕ_1 and ϕ_2. In addition to being of purely academic interest, it is also representative of a certain class of problems having practical significance. For example, this model is used for an atom in an external magnetic field or as an approximation to systems where the coupling due to the perturbation $\langle i|V|j\rangle$ is particularly strong between the two states of interest.

For a system with only two states, $a_j(t) \equiv 0$ for $j > 2$, and Eq. (21) reduces to

$$\dot{a}_k = \frac{1}{i} \sum_{j=1}^{2} \langle k|V|j\rangle a_j. \tag{23}$$

It is assumed that the system is definitely known to be in state 1 initially ($t = -\infty$) and therefore the initial condition is

$$\begin{aligned} |a_1(-\infty)|^2 &= 1, \\ a_2(-\infty) &= 0. \end{aligned} \tag{24}$$

The probability of a transition to state 2 caused by V is

$$P_{1\to 2} = |a_2(+\infty)|^2.$$

For the case where ϕ_1 and ϕ_2 have the same energy (exact resonance) and $V_{11} = V_{22}$, $V_{12} = V_{21}$, the exact solution is

$$P_{1\to 2} = |a_2(+\infty)|^2 = \sin^2 I, \tag{25}$$

where

$$I \equiv \int_{-\infty}^{+\infty} V_{12}(t)\, dt.$$

It is convenient to rearrange Eqs. (21) so as not to involve V_{ii}, the diagonal matrix elements. To achieve this, we change the representation of the wave function by defining

$$b_j(t) \equiv a_j(t) e^{+i \int_{-\infty}^{t} V_{jj}(t')\, dt'}. \tag{26}$$

Since

$$|b_j(t)|^2 = |a_j(t)|^2, \qquad j = 1, 2,$$

it follows that transition probabilities can be calculated with either representation. Equations (21) then take the form (when $V_{12} = V_{21}$)

$$\begin{aligned} \dot{b}_1 &= (1/i) V_{12} b_2 e^{-i\omega_{21} t}, \\ \dot{b}_2 &= (1/i) V_{12} b_1 e^{i\omega_{21} t}. \end{aligned} \tag{27}$$

Rosen and Zener (1932) show that for the special case

$$V_{12}(t) = V_{12}(0)\, \mathrm{sech}(\pi t/\tau),$$

where τ determines the width of the potential, Eqs. (27) can be solved exactly. The transition probability in this case is

$$P_{1\to 2} = \sin^2 I \cdot \operatorname{sech}^2(\pi\tau\omega_{12}). \tag{28}$$

Since $\operatorname{sech}^2(\pi\tau\omega_{12})$ is always less than or equal to unity, $\sin^2 I$ is a bound on $P_{1\to 2}$ for this potential.

For small transition probabilities, in the first-order perturbation approximation to Eqs. (27) for any potential

$$P_{1\to 2} \cong \left| \int_{-\infty}^{+\infty} V_{12} e^{i\omega_{21}t}\, dt \right|^2 \tag{29}$$

and if in addition $\omega_{12} \simeq 0$,

$$P_{1\to 2} \cong \left| \int_{-\infty}^{+\infty} V_{12}\, dt \right|^2. \tag{30}$$

When $V_{12}(t)$ is either symmetric in time or has a constant sign for all time, Eq. (30) is an upper bound on the $P_{1\to 2}$ of Eq. (29).

Aspinall and Percival (1967) derived an upper bound on the transition probability out of the initial state of a two-state system for which $\epsilon_1 \neq \epsilon_2$. To avoid the presence of the diagonal matrix elements in the equations for the probability amplitudes, they used the representation given in Eqs. (27). The second of these equations,

$$(1/i)V_{12} b_1 e^{i\omega_{21}t} = \dot{b}_2, \tag{31}$$

can be used to derive the desired bound. It is convenient to write the various complex factors in this equation in terms of their phases and moduli. Five complex numbers q_n $(n = 1, 2, \ldots, 5)$ of modulus unity are defined as follows:

$$\begin{aligned} q_1 &= e^{i\omega_{21}t}, \\ q_2 &= b_1/i|b_1|, \\ q_3 &= b_2/|b_2|, \\ q_4 &= \langle 1|V|2\rangle/|\langle 1|V|2\rangle|, \\ q_5 &= q_4 q_2 q_3{}^* = q_4 q_2/q_3. \end{aligned} \tag{32}$$

In terms of these factors, Eq. (31) can be written

$$q_2 V_{12}(1 - |b_2|^2)^{1/2} = \frac{d}{dt}(|b_2| q_3), \tag{33}$$

where the relation $|b_1|^2 + |b_2|^2 = 1$ was used to substitute for $|b_1|$.

By expanding the right-hand side of (3.15) and multiplying by $q_3^* = 1/q_3$, it is found that

$$q_5|V_{12}|(1 - |b_2|^2)^{1/2} = q_3^*|b_2| \frac{dq_3}{dt} + \frac{d}{dt}|b_2|. \tag{34}$$

Since q_3 is of modulus unity

$$q_3 q_3^* = 1$$

and

$$\frac{d}{dt}(q_3 q_3^*) = 0, \qquad \mathrm{Re}\left(q_3^* \frac{dq_3}{dt}\right) = 0.$$

Therefore, taking the real part of Eq. (33) we find

$$\mathrm{Re}[q_5|V_{12}|(1 - |b_2|^2)^{1/2}] = \frac{d}{dt}|b_2|. \tag{35}$$

From Eq. (35) it follows that

$$\left|\frac{d}{dt}|b_2|\right| \leq |V_{12}|(1 - |b_2|^2)^{1/2}. \tag{36}$$

Aspinall and Percival (1967) used Eq. (36) to derive the desired bound. The angle between the wave function ψ and the first state ϕ_1 of this two-state system is defined as α, as shown in Fig. 5. Assuming that the two-state approximation is normalized, $\alpha = \sin^{-1}|b_2|$. Aspinall and Percival (1967) defined the *total* integrated angular change of state of the system at time t as a continuously increasing function $\omega(t)$ of time such that

$$\omega(t) \equiv \int_{-\infty}^{t} dt' \left|\frac{d}{dt}\sin^{-1}|b_2(t')|\right|. \tag{37}$$

From Eqs. (3.18) and (3.19) it follows that

$$\frac{d\omega}{dt} \leq |V_{12}|,$$

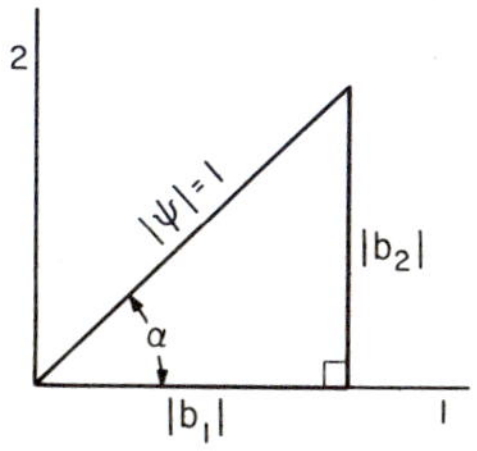

FIG. 5 Illustration of two-dimensional Hilbert space and the reference angle α.

or in integrated form,

$$\sin^{-1}|b_2(t)| \leq \omega(t) \leq \int_{-\infty}^{t} dt'|V_{12}| \equiv I(t).$$

The bound on the transition probability $p_2(t)$ ($p_2(t) = |b_2(t)|^2$) is

$$p_2(t) \leq \sin^2 I(t), \qquad I(t) < \pi/2.$$

If $I(t) \geq \pi/2$, the only bound on $p_2(t)$ is the trivial bound of unity. When $t = +\infty$, the bound on the transition probability is

$$P_{1\to 2} \leq \sin^2 I,$$

where

$$I \equiv I(+\infty) \equiv \int_{-\infty}^{+\infty} |V_{12}|\, dt.$$

This is satisfying if not unexpected result. Usually (though not always) the closer two states are in energy the greater is the transition probability between them. Therefore, in this sense, the case of exact resonance represents a "natural bound" for the transition probability in a two-state system.

C. *Systems With More Than Two States*

Aspinall and Percival (1968) generalized the work on bounds for a two-state system to systems with any finite or infinite number of states.

Consider a system described by the time-dependent Schrödinger equation

$$H(t)|\psi\rangle \equiv [H_0 + V(t)]|\psi\rangle = i\frac{\partial}{\partial t}|\psi\rangle, \tag{38}$$

where H_0 is the unperturbed Hamiltonian, $V(t)$ the interaction potential that induces transitions, and $|\psi\rangle$ a solution. It is now convenient to define an arbitrary complete orthonormal set of functions $|j\rangle$ which are not necessarily solutions of Eq. (38). In particular, one of these functions, $|o\rangle$ is specified as a reference function. The projection of $|\psi\rangle$ on the set of functions $|j\rangle$ is defined in terms of the projection on $|o\rangle$, and upon the remainder of the set $|j\rangle$ excluding $|o\rangle$. Thus, the projection operators $P(t)$ and $Q(t)$ are defined by

$$\begin{aligned} P(t)|\psi\rangle &= |o \times o|\psi\rangle, \\ Q(t)|\psi\rangle &= \sum_{j\neq 0} |j \times j|\psi\rangle, \end{aligned} \tag{39}$$

with

$$\begin{aligned} P + Q &= 1, \\ P^2 &= P, \\ Q^2 &= Q. \end{aligned} \tag{40}$$

The object of the approach by Aspinall and Percival was to derive an upper bound for $\|Q(t)|\psi\rangle\|^2$. Aspinall and Percival were able to derive such a bound from the Cauchy–Schwarz inequality. The form of their result is

$$p(t) \equiv \|Q(t)|\psi\rangle\|^2 < \sin^2 \min \begin{cases} \pi/2 \\ \int_{-\infty}^{t} d\tau \|P(\tau)G(\tau)Q(\tau)\| \end{cases}, \tag{41}$$

where

$$G(t) = H(t) - i\frac{d}{dt}.$$

The utility of the result of Aspinal and Percival lies in the generality of the arbitrary basis set of functions $|j\rangle$, and in particular, the specific function $|o\rangle$. One can choose these functions for any purpose in defining limits to transition probabilities. For example, suppose that $|o\rangle$ is chosen to approach the eigenstate of the unperturbed Hamiltonian corresponding to the initial state at large negative and positive t, when $V(t) = 0$. Then at large positive times, $\|P(t)|\psi\rangle\|^2$ represents the probability of finding the system described by Eq. (38) in state $|o\rangle$, and $p(t)$ is the probability of finding the system in an eigenstate of the unperturbed Hamiltonian other than $|o\rangle$. Thus, with this choice of $|o\rangle$, Eq. (41) bounds the total probability out of the initial state at large t, for the system described by Eq. (38). To make Eq. (41) more useful for calculational purposes, we note that

$$\|PGQ\| = \left(\sum_{j \neq 0} |\langle o|G|j\rangle|^2 \right)^{1/2},$$

and using the completeness of $|j\rangle$ it follows that

$$\|PGQ\| = (\|G|o\rangle\|^2 - \langle o|G|o\rangle^2)^{1/2}. \tag{42}$$

Thus, Eq. (41) can be written

$$p(t) \leqq \sin^2 \min \begin{cases} \pi/2 \\ \int_{-\infty}^{t} (\|G|o\rangle\|^2 - \langle o|G|o\rangle^2)^{1/2}\, d\tau \end{cases} \tag{43}$$

By integrating Eq. (43), one can obtain an upper bound to the flow out of the initial state when $V(t)$ acts on a system, without ever solving Eq. (38).

In those cases where the integral in Eq. (43) is considerably less than $\pi/2$, this might be useful in practice. On the other hand, for cases where the integral is not small compared to $\pi/2$, Eq. (43) is of no practical value. There does not appear to be any way to determine in advance how large the integral is. One must simply calculate the integral and observe whether it is small enough to be useful.

If $|o\rangle$ is chosen to be the ground state of the unperturbed Hamiltonian for all time, then

$$p(\infty) \leq \sin^2 \min \begin{cases} \pi/2 \\ \int_{-\infty}^{+\infty} dt(\langle o|V^2|o\rangle - |\langle o|V|o\rangle|^2)^{1/2} \end{cases}. \tag{44}$$

The variance of the potential in the ground state appears in Eq. (44). This variance expression (or its generalization) will reappear throughout our optimization procedures whenever this type of trial solution is considered. Thus, whenever a linear combination of unperturbed eigenfunctions is chosen as the trial solution the variance will play a crucial role in determining the adequacy of this approximation.

Another useful choice for $|o\rangle$ is based on the fact that one might attempt an approximate solution of Eq. (38) which can be denoted $|\psi_{tr}\rangle$. This approximate solution is based upon a form of the trial function which is arbitrarily chosen. One is never certain how adequate the approximate solution is. Therefore, it would be useful to choose $|o\rangle$ as $|\psi_{tr}\rangle$ for all t. Then the integral in Eq. (43) represents a bound on the total probability in that part of Hilbert space orthogonal to $|\psi_{tr}\rangle$. This is a measure of the degree to which ψ_{tr} does not span the Hilbert space spanned by the exact solution ψ. Thus, it is a measure of the adequacy of $|\psi_{tr}\rangle$ as an approximation to ψ.

In summary, Aspinall and Percival (1967, 1968) have found: (1) a bound on $P_{1\to 2}$ in a two-state system and (2) a bound (in a general quantum mechanical system) for the *total* transition probability out of the initial state.

In an attempt to gain more detailed information about *specific* transition probabilities for general quantum mechanical systems, Spruch (1968) and later Storm (1973) found bounds on the errors involved in estimates of the corresponding transition amplitudes.

D. *Nonstationary and Stationary Bounds for the Error in Estimates of Transition Amplitudes*

The first part of this section is a review of the pioneering work of Spruch (1968) on bounding the exact transition amplitude in terms of factors that depend on approximate solutions of the Schrödinger equation. The latter part of this section deals with Storm's alternate treatment of the same subject.

Spruch's derivation of this bound is based on the time evolution operator formalism of wave mechanics. The time evolution operator $U(t, \tau)$ is defined as the operator which operates on the exact wave function at time τ, $\psi(\tau)$, to produce the exact wave function at time t, $\psi(t)$. Thus

$$\psi(t) = U(t, \tau)\psi(\tau). \tag{45}$$

An equivalent definition of $U(t, \tau)$ is given by the pair of equations

$$G_t U(t, \tau) = 0, \tag{46}$$

$$U(\tau, \tau) = 1, \tag{47}$$

where

$$G_t = H - i\frac{\partial}{\partial t}.$$

It can be shown that

$$U^\dagger(t, \tau) = U(\tau, t) \tag{48}$$

and

$$U^\dagger(t, \tau)U(t, \tau) = 1, \tag{49}$$

where † denotes the adjoint of a given operator.

It is assumed that the Hamiltonian can be written in the form of either of two arrangement channels H_α and H_β in the asymptotic region where the interaction potential has vanished. Initial and final states are defined which are eigenfunctions of these asymptotic Hamiltonians.

$$\begin{aligned} H_\alpha|\alpha\rangle &= E_\alpha|\alpha\rangle, \\ H_\beta|\beta\rangle &= E_\beta|\beta\rangle, \end{aligned} \tag{50}$$

where $|\alpha\rangle$ is the initial state.

Time-dependent functions are defined by

$$\begin{aligned} |\alpha, t\rangle &= \exp(-iH_\alpha t)|\alpha\rangle, \\ |\beta, t\rangle &= \exp(-iH_\beta t)|\beta\rangle. \end{aligned} \tag{51}$$

At time t_1, the amplitude for transition from the initial state $|\alpha, t_0\rangle$ to $|\beta, t_1\rangle$ is defined as

$$A(t_1, t_0) = \langle\beta, t_1|\psi(t_1; \alpha, t_0)\rangle, \tag{52}$$

where $|\psi(t_1; \alpha, t_0)\rangle$ is the state of the system at time t_1 which evolved from the initial state $|\alpha\rangle$ at time t_0. From Eq. (45) it follows that

$$A(t_1, t_0) = \langle\beta, t_1|U(t_1, t_0)|\alpha, t_0\rangle. \tag{53}$$

In actual practice, one cannot evaluate this exact transition amplitude. Instead, one performs some sort of approximate integration of the Schrödinger equation to obtain an approximate solution. This trial solution can be conceptually described by a time evolution operator $\tilde{U}(t_1, t_0)$ which when applied to $|\alpha, t_0\rangle$ generates a trial solution at time t_1. $\tilde{U}(t_1, t_0)$ represents either $\bar{U}$ or $\hat{U}$, two approximate trial estimates of the exact U which may or may not be the same and which can contain parameters. The use of two approximate trial solutions allows an added flexibility which is exploited after the bound is found in terms of these two approximations. The only requirement imposed upon $\tilde{U}$ is that

$$\tilde{U}(t, t) = 1. \tag{54}$$

Our goal is to bound the difference between the exact and approximate probability amplitude. Therefore, we derive a relationship between the exact and approximate time evolution operators. From Eq. (47) it follows that

$$G_{t_1}\bar{U}(t_1, t_0) = U(t_1, t_1)G_{t_1}\bar{U}(t_1, t_0). \tag{55}$$

Equation (55) can be rewritten in the form

$$G_{t_1}\bar{U}(t_1, t_0) = iG_{t_1}\int_{t_0}^{t_1} dt\, U(t_1, t)G_t\bar{U}(t, t_0) \tag{56}$$

by using Eq. (46) and the well-known theorem for differentiating integrals

$$\frac{\partial}{\partial x}\int_a^x F(x, z)\, dz = F(x, x) + \int_a^x \frac{\partial F(x, z)}{\partial x}\, dz.$$

Upon bringing all of the terms in Eq. (56) to the left-hand side of the equation, the result is

$$G_{t_1}\left(\bar{U}(t_1, t_0) - i\int_{t_0}^{t_1} dt\, U(t_1, t)G_t\bar{U}(t_1, t_0)\right) = G_{t_1}(V(t_1, t_0)) = 0, \tag{57}$$

where

$$V(t_1, t_0) \equiv \bar{U}(t_1, t_0) - i\int_{t_0}^{t_1} dt\, U(t_1, t)G_t\bar{U}(t, t_0). \tag{58}$$

Equations (54) and (58) imply that

$$V(t_0, t_0) = 1. \tag{59}$$

Equations (57), (59), and the uniqueness of the time evolution operator as defined by (46) and (47), imply that

$$V(t_1, t_0) = U(t_1, t_0)$$

and consequently from (58)

$$U(t_1, t_0) = \bar{U}(t_1, t_0) - i\int_{t_0}^{t_1} dt\, U(t_1, t)G_t\bar{U}(t, t_0). \tag{60}$$

Equation (60) will be used in the derivation of both stationary and nonstationary bounds. The nonstationary bound will follow immediately from Eq. (60). However, to derive the stationary bound we first rewrite (60) in terms of the other approximation $\hat{U}$, replace t_1 by t, t_0 by t_1, t by τ, take the adjoint, and interchange the limits of τ, and we find

$$U(t_1, t) = \hat{U}^\dagger(t, t_1) - i \int_t^{t_1} d\tau [G_\tau \hat{U}(\tau, t_1)]^\dagger U(\tau, t). \qquad (61)$$

We then perform an iteration on Eq. (60) by substituting Eq. (61) for U in the integrand of Eq. (62) and find

$$U(t_1, t_0) = \bar{U}(t_1, t_0) - i \int_{t_0}^{t} dt\, \hat{U}^\dagger(t, t_1) G_t \bar{U}(t, t_0)$$
$$- \int_{t_0}^{t_1} dt \int_t^{t_1} dt [G_\tau \hat{U}(\tau, t_1)]^\dagger U(\tau, t) G_t \bar{U}(t, t_0). \qquad (62)$$

Spruch used Eq. (60) to find a nonstationary bound and Eq. (62) to find a stationary bound on the error involved in an approximate form of the transition amplitude.

1. *Nonstationary Bound on the Error in Approximate Transition Amplitudes*

The exact solution satisfies the property

$$|\psi(t; \delta, \tau)\rangle = U(t, \tau)|\delta, \tau\rangle,$$

where $\delta = \alpha$ or β (i.e., δ indicates explicitly the state of the system at time τ). One then introduces $\tilde{\psi}$, an approximation to ψ, which is defined by

$$\tilde{\psi}(t; \delta, \tau) = \tilde{U}(t, \tau)|\delta, \tau\rangle.$$

The only demand placed on $\tilde{U}$ is Eq. (54). Substituting (55) into (54) (with $\tilde{U}$ replacing $\hat{U}$), we obtain

$$A(\gamma) = A_0(\gamma) + A_1{}^{\mathrm{F}}(\gamma),$$

where

$$A_0(\gamma) \equiv \langle \beta, t_1 | \tilde{U}^\dagger(t_0, t_1) | \alpha, t_0 \rangle = \langle \tilde{\psi}(t_0; \beta, t_1) | \alpha, t_0 \rangle$$

is a zeroth-order estimate of $A(\gamma)$, and $A_1{}^{\mathrm{F}}(\gamma)$ is defined by

$$A_1{}^{\mathrm{F}}(\gamma) \equiv -i \int_{t_0}^{t_1} dt \langle G_t \tilde{\psi}(t; \beta, t_1) | \psi(t; \alpha, t_0) \rangle. \qquad (63)$$

For convenience we have let γ represent the set (t_1, t_0). The superscript F indicates the formal character of the result due to the presence of the exact state ψ. The subscript 1 indicates "first order." The power of $G_t \tilde{\psi}$ in any expression defines its "order." This definition is motivated by the fact that

$$G_t \tilde{\psi} = G_t(\tilde{\psi} - \psi) = G_t \delta\psi$$

is "first order" in the error $\delta\psi$. This terminology suggests that terms of higher order will tend to be smaller. Of course this will only be true for $\tilde{\psi}$ "sufficiently close" to ψ. Equation (60) will now be used to find a nonstationary bound on the error involved in estimating $|A(\gamma)|$. Utilizing the Cauchy–Schwarz inequality we have

$$|\langle G_t\tilde{\psi}|\psi\rangle| \leq |\langle G_t\tilde{\psi}|G_t\tilde{\psi}\rangle|^{1/2}|\langle\psi|\psi\rangle|^{1/2}. \tag{64}$$

From Eq. (63) it follows that

$$|A_1{}^{\mathrm{F}}(\gamma)| = \left|\int\langle G_t\tilde{\psi}|\psi\rangle\,dt\right| \leq \int|\langle G_t\tilde{\psi}|\psi\rangle|\,dt. \tag{65}$$

When Eq. (64) is used in Eq. (65), the result is

$$|A_1{}^{\mathrm{F}}(\gamma)| \leq \int\langle G_t\tilde{\psi}|G_t\tilde{\psi}\rangle^{1/2}\,dt \tag{66}$$

in which it has been assumed that ψ is normalized. The approximate solution $\tilde{\psi}$ is not assumed to be normalized. Thus, the exact solution conserves probability (as it must), but the trial solution does not necessarily satisfy this property. We now adopt the notation convention for the norm of a vector $|\phi\rangle$ as

$$\langle\phi|\phi\rangle = \||\phi\rangle\|^2$$

and write Eq. (66) in the form

$$|A_1{}^{\mathrm{F}}(\gamma)| \leq \int\||G_t\tilde{\psi}(t;\beta,t_1)\rangle\|\,dt \equiv A_1{}^{\mathrm{F}}(\gamma)_{\mathrm{up}}. \tag{67}$$

From the definition of $A(\gamma)$ as the sum of $A_0(\gamma)$ and $A_1{}^{\mathrm{F}}(\gamma)$ together with Eq. (67), one can show that

$$|A_0(\gamma)| - A_1{}^{\mathrm{F}}(\gamma)_{\mathrm{up}} \leq |A(\gamma)| \leq |A_0(\gamma)| + A_1{}^{\mathrm{F}}(\gamma)_{\mathrm{up}}. \tag{68}$$

Clearly, these bounds on the transition amplitude will be useful only if they lie between zero and one. We have used the bound on the error in an approximate transition amplitude to bound the exact transition amplitude. The fact that these bounds only depend on the approximate wave function is the reason that they are of potential practical significance. In general, one approximates an exact solution ψ by a trial solution $\tilde{\psi}$, which has an error $\delta\psi = \tilde{\psi} - \psi$. The trial solution may be used to calculate a bound on the error in the approximate transition amplitude which may be of the same order or higher order than $\delta\psi$. When $\delta\psi$, the error in $\tilde{\psi}$ is small, the higher the order of the bound, the more accurate the bound will be. The nonstationary bound $A_1{}^{\mathrm{F}}(\gamma)$ is of first order in $\delta\psi$, and is not as desirable (when $\delta\psi$ is small) as a stationary bound which is second order in $\delta\psi$. When $\delta\psi$ is not small, there is no reason to give preference to a stationary over a nonstationary bound. We procede to derive a stationary bound understanding both the potential benefits and possible pitfalls of such a procedure.

2. *Stationary Bound on the Error in Approximate Transition Amplitude*

To derive a stationary bound we substitute Eq. (62) into Eq. (53) and the result is

$$A(\gamma) = \bar{A}_0'(\gamma) + A_1(\gamma) + A_2^{\mathrm{F}}(\gamma) \equiv A_v(\gamma) + A_2^{\mathrm{F}}(\gamma),$$

where

$$\bar{A}_0'(\gamma) = \langle \beta, t_1 | \bar{\psi}(t; \alpha, t_0) \rangle$$

and

$$A_1(\gamma) \equiv -i\, dt \langle \hat{\psi}(t; \beta, t_1) | G_t \bar{\psi}(t; \alpha, t_0).$$

The term $A_v(\gamma)$ is a stationary estimate of $A(\gamma)$, and the second-order term is given formally by

$$A_2^{\mathrm{F}}(\gamma) = -\int dt \int d\tau \langle G_\tau \hat{\psi}(\tau; \beta, t_1) | U(\tau, t) | G_t \bar{\psi}(t; \alpha, t_0) \rangle.$$

Use of the Cauchy–Schwarz inequality in this expression leads to

$$|A_2^{\mathrm{F}}(\gamma)| < \int dt \bar{N}(t; \alpha, t_0) \int d\tau \hat{N}(\tau; \beta, t_1) \equiv A_2^{\mathrm{F}}(\gamma)_{\mathrm{up}}, \tag{69}$$

where the norm $\tilde{N}$ is defined by

$$\tilde{N}(\tau'; \delta, \tau'') = \| | G_{\tau'} \tilde{\psi}(\tau'; \delta, \tau'') \rangle \|.$$

This bound, $A_2^{\mathrm{F}}(\gamma)_{\mathrm{up}}$, is second order in G operating on the error in the trial solution.

Utilizing this bound, we find that

$$|A_v(\gamma)| - A_2^{\mathrm{F}}(\gamma)_{\mathrm{up}} \leq |A(\gamma)| \leq |A_v(\gamma)| + A_2^{\mathrm{F}}(\gamma)_{\mathrm{up}}. \tag{70}$$

In principle both (67) and (70) are true for any trial solution. However, as in the case of nonstationary bounds, this is a useful statement only when $A_2^{\mathrm{F}}(\gamma)_{\mathrm{up}}$ lies between zero and one. Thus, if the trial function is sufficiently close to the exact solution that $\delta\psi$ is small, the second-order procedure actually improves upon the trial function in the accuracy of bounding the transition amplitude. However, if $\delta\psi$ is moderately large, the second-order property of $A_2^{\mathrm{F}}(\gamma)$ is of no value in reducing the error in estimating the transition amplitude. The second-order term (and its bound) produced by this particular iterative procedure is a function of the particular iteration one chooses (Spruch, 1968). Equations (68) and (69) are the most convenient for the purposes of calculation. One could also carry out further iterations on Eq. (60) obtaining $A = A_0 + A_1 + A_2 + A_3{}^{\mathrm{F}}$ and bound the term $A_3{}^{\mathrm{F}}$ of $\mathrm{O}(\delta\psi^3)$, where $A_0 + A_1 + A_2$ is the approximate $A(\gamma)$. However, the higher the order of the term being bound the more difficult will be the calculational problem of evaluating its bound. Therefore, we will restrict ourselves to the estimates discussed by Spruch (1968) and consequently only treat $O(\delta\psi^2)$ bounds for the error.

The evaluation of $A_2{}^{\mathrm{F}}(\gamma)_{\mathrm{up}}$ simplifies considerably for certain cases. Assume, for example, that $|\alpha\rangle = |\alpha\rangle^{\mathrm{T}}$, the time reversed state of $|\alpha\rangle$, and that $H^*(t) = H(-t)$. For $|\beta\rangle = |\alpha\rangle$ and $t_0 = -t_1$ the exact solution satisfies

$$\psi(t;\beta,t_1) = \psi(t;\alpha,t_1) = \psi^*(-t,\alpha,-t_1).$$

We now exploit the freedom involved in our use of two approximate trial solutions to arrange for a single "time-reversal invariant appearing approximation." If $\bar{\psi}$ satisfied the same time-reversal property as ψ we could write

$$\bar{\psi}^*(-t) = \bar{\psi}_{\mathrm{T}}(t).$$

There is no reason to suspect that in general this will be the case. However, we can choose the other trial solution $\hat{\psi}$ such that

$$\hat{\psi}(\tau;\beta,t_1) = \bar{\psi}^*(-\tau;\alpha,-t_1) \tag{71}$$

and use the time-reversal properties to simplify the second-order bound $A_2{}^{\mathrm{F}}(\gamma)_{\mathrm{up}}$. Utilizing the fact that the norm is invariant under complex conjugation (and substituting $\tau = -\tau'$), the integrand of the double integral in Eq. (69) can be written in a form symmetric in τ' and τ:

$$A_2{}^{\mathrm{F}}(\alpha,t_1;\alpha,-t_1)_{\mathrm{up}} = \int_{-t_1}^{t_1} dt \int_{-t_1}^{-t} d\tau' \bar{N}(t;\alpha,-t_1)\bar{N}(\tau';\alpha,-t_1). \tag{72}$$

In order to rewrite Eq. (72) in terms of a one-dimensional integral, we use the fact that for any function $F(t)$

$$\int_{-1}^{+1} dt \int_{-1}^{t} dt' F(t)F(t') = \frac{1}{2}\int_{-1}^{+1}\int_{-1}^{+1} dt\, dt' F(t)F(t'),$$

and consequently Eq. (72) can be written as

$$\begin{aligned} A_2{}^{\mathrm{F}}(\alpha,t_1;\alpha,-t_1)_{\mathrm{up}} &= \frac{1}{2}\int_{-t_1}^{t_1}\int_{-t_1}^{t_1} dt\, d\tau' \bar{N}(t;\alpha,-t_1)\bar{N}(\tau';\alpha,-t_1) \\ &= \frac{1}{2}\left(\int_{-t_1}^{t_1} \bar{N}(t;\alpha,-t_1)\, dt\right)^2. \end{aligned} \tag{73}$$

Spruch, at this point, made the additional assumption that $\bar{N}(t;\alpha,-t_1) = \bar{N}(-t;\alpha,-t_1)$. It will be seen in Section III.E (even when $H(t) = H^*(-t)$

that this assumption is not in general valid. However, if this assumption is made, i.e., $\bar{N}(t) = \bar{N}(-t)$, one can write Eq. (72) as

$$A_2{}^{\mathrm{F}}(\alpha, t_1; \alpha, -t_1)_{\mathrm{up}} = 2\left(\int_0^{t_1} \bar{N}(t; \alpha, -t_1)\, dt\right)^2. \tag{74}$$

Therefore, for this particular case it is no more difficult to calculationally obtain stationary than nonstationary bounds.

Storm used a different approach to finding bounds on the error involved in approximate transition amplitudes. Storm (1973) dealt directly with the state vectors rather than the time evolution operator formalism.

Following Storm (1973) we write the Schrödinger equation

$$H\psi = [H_0 + V(t)]\psi = i\frac{\partial\psi}{\partial t}, \tag{75}$$

where H_0 is the unperturbed Hamiltonian with eigenfunctions

$$\Phi_n = \phi_n e^{-i\epsilon_n t}$$

Let ψ_n be the exact solution of (75) with boundary conditions

$$\begin{aligned} \psi_n &\to \Phi_n \qquad \text{as} \quad t \to -\infty \\ &\to \Phi_n + \sum_k A_{kn}\Phi_k \qquad \text{as} \quad t \to +\infty \end{aligned}$$

and let ψ_{nR} be the exact time-reversed solution subject to the boundary conditions

$$\begin{aligned} \psi_{mR} &\to \Phi_{mR} \qquad \text{as} \quad t \to +\infty \\ &\to \Phi_{mR} + \sum_k A_{kmR}\Phi_{kR} \qquad \text{as} \quad t \to -\infty. \end{aligned}$$

A_{kn} and A_{kmR} are exact transition amplitudes. H_0 and V are assumed to be real and invariant under time reversal, i.e.,

$$\begin{aligned} H(t) &= H(-t) = H^*(-t), \\ \Phi_{mR} &= \phi_m{}^* e^{-i\epsilon_m t}, \\ \Phi_{mR}(t) &= \Phi_m{}^*(-t). \end{aligned}$$

In order to arrive at a bound, Storm defined the functional R as

$$R(\chi_{mR}, \chi_n) \equiv \int_{-\infty}^{+\infty} dt \langle \chi_{mR} | G | \chi_n \rangle,$$

where χ_n and χ_{mR} are any functions that satisfy the form of the boundary conditions with A_{kn} and A_{kmR} replaced by the approximate C_{kn} and C_{kmR}, respectively.

The variation in R brought about by keeping ψ_n fixed and varying ψ_{mR} is defined as $\hat{\delta}R(\psi_{mR}, \psi_n)$:

$$\hat{\delta}R = \int_{-\infty}^{+\infty} \langle \psi_{mR} + \delta\psi_{mR} | G\psi_n \rangle \, dt.$$

Since $G\psi_n = 0$, $\delta R(\psi_{mR}, \psi_N) = 0$ for the exact solutions ψ_{mR} and ψ_n.

In general, consider variations about the exact solution

$$\chi_n = \psi_n + \delta\psi_n, \qquad \delta\psi_n \to 0 \quad \text{as} \quad t \to -\infty,$$
$$\chi_{mR} = \psi_{mR} + \delta\psi_{mR}, \qquad \delta\psi_{mR} \to 0 \quad \text{as} \quad t \to +\infty.$$

One can write R in terms of ψ_n, ψ_{mR}, $\delta\psi_n$, and $\delta\psi_{mR}$ as

$$R = \int dt \langle \delta\psi_n | G_t | \psi_{mR} \rangle^* - i \langle \delta\psi_{mR} | \delta\psi_n \rangle \Big|_{-\infty}^{+\infty} + \mathrm{O}(\delta\psi^2),$$

where $\mathrm{O}(\delta\psi^2)$ is a "second-order" term.

To evaluate the second term we use the asymtotic conditions

$$\psi_{mR} \to \Phi_{mR},$$
$$\delta\psi_n \to \sum_k (C_{kn} - A_{kn})\Phi_k \qquad \text{as} \quad t \to \infty$$

and it is found that

$$\langle \psi_{mR} | \delta\psi_n \rangle \to \sum_k (C_{kn} - A_{kn}) \langle \Phi_{mR} | \Phi_k \rangle \qquad \text{as} \quad t \to +\infty.$$

The time-reversed state is written as

$$\Phi_{mR} = \phi_m{}^* e^{-i\epsilon_m t}.$$

Storm chose the ϕ_m to be real. This assumption reduces the sum over k to a single term. With this assumption and using the facts that $G\psi_{mR} = 0$ and that as $t \to -\infty$, $\delta\psi_n \to 0$ we have

$$R = -i(C_{mn} - A_{mn}) + \int dt \langle \delta\psi_{mR} | G | \delta\psi_n \rangle.$$

Solving for the approximate transition amplitude A_{mn}, we have

$$A_{mn} = C_{mn} - iR(\chi_{mR}, \chi_n) + i \int dt \langle \delta\psi_{mR} | G | \delta\psi_n \rangle. \tag{76}$$

We define Δ as

$$\Delta \equiv i \int dt \langle \delta\psi_{mR} | G | \, \delta\psi_n \rangle,$$

a "second-order" term. Note that Spruch counted powers of $G\delta\psi$ to find "the order" of a term, whereas Storm considered $\delta\psi_{mR}$ and $G\delta\psi_n$ as equivalent in determining or defining "the order" of a given expression.

We now follow Storm's procedure to derive a bound on Δ. Let

$$\epsilon_n = G\chi_n = G(\delta\psi_n - \psi_n) = G\delta\psi_n,$$

$$|\Delta| = \left|\int dt \langle \delta\psi_{mR} | \epsilon_n \rangle\right|,$$

and it follows that

$$|\Delta| \leq \int dt |\langle \delta\psi_{mR} | \epsilon_n \rangle|.$$

Using the Cauchy–Schwarz inequality we have

$$|\Delta| \leq \int dt \|\delta\psi_{mR}\| \, \|\epsilon_n\|. \tag{77}$$

At this point, Storm (1973) assumed that the trial solution conserves probability. This assumption is in marked contrast with the work of Spruch (1968) who assumed that the *exact* solution conserves probability (which it must) but *never* assumed the trial solution is normalized. This assumption turns out to be critical in the analysis and comparison of the techniques of these two authors.

Following Storm, we write

$$\begin{aligned} \|\delta\psi_{mR}\| &\equiv \|\chi_{mR} - \psi_{mR}\| \leq \|\chi_{mR}\| + \|\psi_{mR}\|, \\ \|\chi_{mR}\| &= \|\psi_{mR}\| = 1 \end{aligned} \tag{78}$$

and find from (78) that

$$\|\delta\psi_{mR}\| \leq 2,$$

whence

$$|\Delta| \leq \int \|\epsilon_n\| 2 \, dt \equiv \Delta_{\max},$$

where

$$\Delta_{\max} \equiv 2 \int_{-\infty}^{+\infty} \langle G\chi_n | G\chi_n \rangle^{1/2} \, dt$$

and therefore we find from Eq. (76) that

$$|C_{mn} - R| - \Delta_{\max} \leq |A_{mn}| \leq |C_{mn} - R| + \Delta_{\max}. \tag{79}$$

Storm assumed that $\chi_n{}^0$ and χ_{mR}^0 exist for which

$$\begin{aligned} \hat{\delta} R(\chi_{mR}^0, \chi_n{}^0) &= 0, \\ \hat{\delta} R^*(\chi_n{}^0, \chi_{mR}^0) &= 0. \end{aligned} \tag{80}$$

in the limited portion of Hilbert space spanned by the trial solutions. This statement is a level D optimization procedure; i.e., Storm assumed that by having the trial solutions satisfy a property in a limited portion of Hilbert space that the exact solution satisfies throughout Hilbert space, an optimization of some sort is reached. Storm then showed that in order to satisfy (80) for trial solutions of the form

$$\chi_n = \sum_{k=1}^{N} [\delta_{kn} + C_{kn}(t)]\Phi_k,$$

$$\chi_n{}^0 = \sum_{k=1}^{N} [\delta_{kn} + C_{kn}^0(t)]\Phi_k$$

and where the variations are generated by the expansion coefficients

$$C_{kn}(t) = C_{kn}^0(t) + \delta C_{kn}$$

it is necessary to have

$$\langle \Phi_k | G | \chi_n{}^0 \rangle = 0 \tag{81}$$

for each k value.

These are the "usual coupled equations" and will be discussed in detail in Section III.E.1. Storm then showed that by imposing condition (81), it follows that R will vanish. Notice that Eq. (81) can be written

$$\langle \Phi_k | G | \chi_n{}^0 \rangle = \langle \Phi_k | G | \chi_n{}^0 - \psi_n \rangle = \langle \Phi_k | G | \delta\chi_n{}^0 \rangle, \tag{82}$$

where $\delta\chi_n{}^0 = \chi_n{}^0 - \psi_n$, a "first-order" variation about the exact solution.

Storm (1974) in a later paper found a relationship between the bound on his second-order expression and the bound on Spruch's second-order term. Starting with the expressions

$$\Delta = \int_{-\infty}^{+\infty} \langle \delta\psi_{mR} | G | \delta\psi_n \rangle \, dt$$

and

$$|\Delta| < \int dt \|\delta\psi_{mR}\| \, \|\varepsilon_n\|$$

Storm (1974) no longer assumed that the trial solution is normalized. Instead he wrote

$$\|\delta\psi_{mR}\| = -\int_t^{\infty} \frac{d}{dt'} \|\delta\psi_{mR}(t')\| \, dt', \tag{83}$$

which follows from

$$\delta\psi_{mR}(+\infty) = \delta\psi_m{}^*(-\infty) = 0.$$

The integrand in Eq. (83) can be written

$$\frac{d}{dt}\langle\delta\psi_{mR}|\delta\psi_{mR}\rangle^{1/2} = \operatorname{Im}\frac{\langle\delta\dot{\psi}_{mR}|G|\chi_{mR}\rangle}{\|\delta\psi_{mR}\|}$$
$$\leq \|G(\chi_{mR})\|,$$

where a Cauchy–Schwarz inequality was used in the last step.

Therefore, we find the result

$$\Delta \leq \Delta_2 = \int_{-\infty}^{+\infty} dt\|G(\chi_n)\| \int_t^{\infty} dt'\|G(\chi_{mR})\|,$$

which (under the restrictions imposed by time-reversal invariance) is the second-order bound found by Spruch. When, in addition, the final state $|\beta\rangle$ is the same as the initial state $|\alpha\rangle$, Spruch showed that

$$\Delta_2 = \tfrac{1}{2}\Delta_1{}^2,$$

where

$$\Delta_1 = \int_{-\infty}^{+\infty} \|G\chi_n\|\, dt.$$

Now if $0 < \Delta_1 < 1$ it follows that

$$0 < \Delta_1{}^2 < \Delta_1$$

and

$$0 < \tfrac{1}{2}\Delta_1{}^2 < \tfrac{1}{2}\Delta_1 < 2\Delta_1.$$

Note that $2\Delta_1$ is the expression Storm originally derived (Storm, 1973) to bound his second-order term (when he assumed that the trial solution was normalizable). For $\Delta_1 < 1$ (and with the time-reversal and final-state restrictions) Spruch's technique achieves a lower upper bound $\Delta_2 = \frac{1}{2}\Delta_1{}^2$ on the second-order term.

3. *Summary of Bounds on Transition Amplitudes*

Bounds on transition amplitudes are only useful when they are considerably less than unity. If the trial function is sufficiently comprehensive, it is possible that the calculated bound will be small enough to closely bracket the exact amplitude. For example, if the calculated amplitude is, say, 0.5 and the bound is 0.05, then the exact amplitude is bracketed to within the range 0.5 ± 0.05. In real cases, the theoretician may not be so fortunate, and the calculated bound may be as large as, or larger than, the calculated amplitude. In such cases the bound does not function directly to bracket the exact amplitude to within useful limits. One may nevertheless conjecture that even in such cases the magnitude of the bound is a measure of the "quality" of an approximate calculation. Thus, in comparing several approximate trial solutions, one may be tempted to prefer the result with the smallest calculated bound. However, the bound is calculated from a summation of *absolute* errors in the trial solution over time, whereas the error in the actual calculated amplitude is a result of summing errors in the trial solution, including the signs of the errors. Thus it is possible that some solutions can have large bounds but accurate amplitudes due to a fortunate cancellation of positive and negative errors. Other solutions with smaller bounds might have less accurate amplitudes. Nevertheless, in some semi-quantitative way, one can attempt to associate calculations with smaller bounds with a higher probability of more accurate calculated amplitudes. In doing this, one must remain aware of the possibilities for unusual functions. For example, a trial solution might remain close to the exact solution for most of the range of time involved in a collision, but wander from it as time approaches the asymptotic region (large t). Yet, another trial solution might deviate considerably from the exact solution for most of time, but approach the exact solution as time approaches the asymptotic region. The latter solution will have a larger bound but be more asymptotically accurate. This argument is illustrated in Fig. 6. Another example is a trial solution that oscillates about the exact solution with a small amplitude and large frequency. Such a solution will have a large bound but will be quite accurate.

Despite these possible pitfalls, the most natural procedure for optimizing parameters in a trial solution appears to be the adoption of the criterion that $\int_{-\infty}^{\infty} D^{1/2}\,dt$ be minimized. This criterion will lead to a minimized

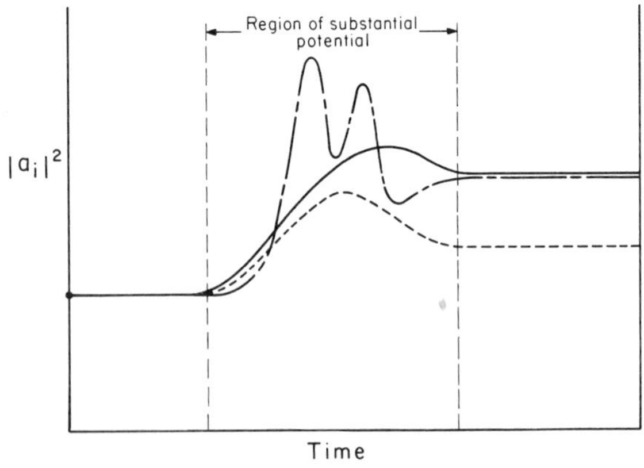

FIG. 6 Illustration of (———) the exact $|a_i|^2$, (— — —) trial $|a_i|^2$, which deviates from the exact $|a_i|^2$ most of the time but approaches the exact solution in the asymptotic region, and (-----) trial $|a_i|^2$, which remains close to the exact $|a_i|^2$ but wanders from it as time approaches the asymptotic region.

nonstationary bound, and a minimized stationary bound for the cases discussed previously. When the bounds are small compared to the amplitudes, they are directly useful in bracketing the exact solution. When the bounds are large, one must hope that the trial solution does not have unusual pathological properties to vitiate a reasonable relation between the bound and the accuracy of the calculation.

E. *Level C Optimization*

It has been shown that the expression $\int D^{1/2}\,dt$ is always a nonstationary bound, and under certain conditions is also a stationary bound on the error associated with an approximate transition amplitude. Therefore, it would appear that a natural candidate for a level C optimization criterion is the requirement that this integral be minimized. Unfortunately, the equations resulting from an analytical attempt to minimize this integral are very cumbersome and are not sufficiently tractable for solution. However, there is another integral, namely, $\int D\,dt$, which can be minimized analytically, and which is closely related to $\int D^{1/2}\,dt$. Chang and Rapp (1973) suggested that minimization of $D\,dt$ might be a meaningful criterion for optimization of parameters in a trial solution. On purely heuristic grounds, it is believed that the parameters resulting from minimization of $D\,dt$ will not be very different from those resulting from minimization of $\int D^{1/2}\,dt$. Chang and Rapp defined

$$J(\psi_{\rm tr}) = \int_{-\infty}^{\infty} D\,dt \tag{84}$$

and suggested two methods for seeking a minimum in J. In a simple approximate method, one quite arbitrarily sets

$$\frac{\partial D}{\partial \dot{\alpha}} = 0 \tag{85}$$

for every real or complex parameter. This leads to the equation

$$\operatorname{Im}\left\langle \frac{\partial \psi_{\rm tr}}{\partial \alpha} \middle/ G\psi_{\rm tr} \right\rangle = 0 \tag{86}$$

for a real parameter α, and $\langle \partial\psi_{\rm tr}/\partial\alpha \,|\, G\psi_{\rm tr}\rangle = 0$. Chang and Rapp (1973) also described a more proper method, in which the Euler–Lagrange procedure was applied to $J(\psi_{\rm tr})$. The result for a real parameter was

$$\operatorname{Re}\left\langle \frac{\partial \psi_{\rm tr}}{\partial \alpha} \middle| G^2\psi_{\rm tr} \right\rangle = 0 \tag{87}$$

Equation (87) gives the exact analytic function $\alpha(t)$ which minimizes J.

In a recent article Weglein (1978) developed an alternative approach based on Hamilton's equations for determining parameters to exactly minimize $J(\psi_{tr})$. This method leads to equations that are more easily integrated than Eq. (87) (or its complex counterpart) in actual applications. Weglein (1978) has also discovered new insights in the relation between ψ and ψ_{tr} and has explored the relationship between level C and D optimization. The method of Hamilton's equations was applied by Weglein to a specific simple example, namely, a hypothetical perturbed hydrogen atom with a nuclear charge that varies with time. A similar problem was originally treated by Chang (1973). The comparison of the various methods when applied to this problem leads to interesting insights into the useful regions of applicability of the methods.

Although level D optimization has been applied and discussed in the literature, it is felt that certain aspects of this method have not been clarified, and the following discussion is therefore presented.

1. Review of Level D Optimization

The exact solution ψ of the time-dependent Schrödinger equation

$$G\psi = \left(H - i\frac{\partial}{\partial t}\right)\psi = 0 \tag{88}$$

also satisfies the equivalent stationary integral equation

$$\delta \int \langle \psi | G\psi \rangle \, dt = 0. \tag{89}$$

In most applications one seeks an approximation to ψ in terms of a trial solution ψ_{tr}, which depends on time-dependent parameters $\alpha_1, \alpha_2, \ldots$. The problem is to select a procedure for evaluating the parameters so that ψ_{tr} will approximate ψ as closely as possible within the limited subspace spanned by ψ_{tr}. Level D optimization is the approach in which one assumes that the trial function either satisfies a principle analogous to the exact equation, or a property satisfied by the exact solution. For example, several authors have assumed that by forcing the integral analogous to Eq. (89) to be stationary,

$$\delta I = \delta \int \langle \psi_{tr} | G\psi_{tr} \rangle \, dt = 0, \tag{90}$$

"optimized" values of the parameters will be obtained. As we mentioned in Section II.B the success of this method critically depends on the extent to which the space spanned by the trial solution includes the exact solution. In actual cases, the difficulty is that one never knows the adequacy of the space spanned by ψ_{tr}. Equation (90) does not guarantee the "best choice" of parameters in ψ_{tr} in any consistent sense.

2. *Comment on the Method of Equation (90)*

We examine the use of Eq. (90) for real and complex parameters. Consider arbitrary complex variations which either (a) conserve normalization (i.e., $N = \langle\psi_{tr}|\psi_{tr}\rangle$, $\delta N = 0$) or (b) vanish at the end points in time. For such arbitrary complex variations, one can imply that for any allowed variation $\delta\psi_{tr}$, $i\delta\psi_{tr}$ is also an allowable variation and consequently Eq. (90) can be written

$$\mathrm{Re}\int\langle\delta\psi_{tr}|G\psi_{tr}\rangle\,dt = 0. \tag{91}$$

Substituting $i\delta\psi_{tr}$ for the variation in Eq. (91) and adding and subtracting this form Eq. (91):

$$\int\langle\delta\psi_{tr}|G\psi_{tr}\rangle\,dt = 0 \tag{92}$$

and

$$\int\langle G\psi_{tr}|\delta\psi_{tr}\rangle\,dt = 0. \tag{93}$$

That Eqs. (92) and (93) can be derived from Eq. (91) demonstrates that even though the variations $\delta\psi_{tr}$ and $\delta\psi_{tr}^*$ are not independent (Messiah, 1966), they can be treated as if they were. When arguments analogous to those leading to Eqs. (92) and (93) are used for the exact solution ψ and $\delta\psi$, Eq. (89) can be equivalent to Eq. (88).

For a variation of ψ_{tr} due to a complex parameter α,

$$\delta\psi_{tr} = \frac{\partial\psi_{tr}}{\partial\alpha}\,\partial\alpha.$$

For any arbitrarily chosen ψ_{tr}, $\partial\psi_{tr}/\partial\alpha$ is a specific complex function. However, since $\delta\alpha$ is arbitrary and complex, $\delta\psi_{tr}$ is an arbitrary complex variation. Then a new variation $i\delta\psi_{tr}$ can be considered which is generated by $i\delta\alpha$. Therefore, for variations in ψ_{tr} produced by variations in a complex parameter, both $\delta\psi_{tr}$ and $i\delta\psi_{tr}$ may be taken as variations. For variations in ψ_{tr} due to variations in a real parameter,

$$\delta\psi_{tr} = \frac{\partial\psi_{tr}}{\partial z}\,\delta z,$$

where for an arbitrarily chosen ψ_{tr}, $\partial\psi_{tr}/\partial z$ is a specific complex function. Now, however, δz is an arbitrary real variation and consequently $\delta\psi_{tr}$ is not a general complex variation. In this case $\delta\psi_{tr}$ is a general real multiple of $\partial\psi_{tr}/\partial z$. Therefore, if the variation δz produces a variation $\delta\psi_{tr}$, this does not imply that $i\delta\psi_{tr}$ is also a variation. Arbitrary complex variations of ψ_{tr} cannot be generated by variations in a real parameter. The use of the analog of Eq. (89) for a ψ_{tr} depending on a real parameter necessitates the imposition of a property of the exact solution on the trial solution (since $G\psi$ vanishing implies $\mathrm{Re}\int\langle\delta\psi\,|\,G\psi\rangle\,dt$ vanishes). Therefore, for a real parameter satisfying condition (b), Eq. (91) is appropriate, whereas Eqs. (92) and (93) are not. For the case of a real parameter z, Eq. (91) takes the form

$$\mathrm{Re}\left\langle\frac{\partial\psi_{tr}}{\partial z}\middle|\left[H - i\left(\frac{\partial}{\partial t}\right)_{r,z} - i\dot{z}\frac{\partial}{\partial z}\right]\psi_{tr}\right\rangle = 0. \tag{94}$$

Since z and $\langle\partial\psi_{tr}/\partial z\,|\,\partial\psi_{tr}/\partial z\rangle$ are real, Eq. (94) becomes

$$\mathrm{Re}\left\langle\frac{\partial\psi_{tr}}{\partial z}\middle|\left[H - i\left(\frac{\partial}{\partial t}\right)_{r,z}\right]\psi_{tr}\right\rangle = 0. \tag{95}$$

This is an algebraic equation for $z(t)$.

A shortcoming of the use of Eq. (90) for variations due to a real parameter is that the result Eq. (91) is the analog of a necessary but not sufficient condition that ψ be an exact solution of Eq. (88). It always results in an algebraic equation rather than a differential equation. One should expect a differential equation to result from the analog of the variational principle which is equivalent to the Schrödinger partial differential equation. To approximate a differential equation by an algebraic one is a much more serious alteration than replacing it with a simpler differential equation (as occurs for complex parameters).

3. *The Method of Hamilton Equations for Minimizing $J(\psi_{tr})$*

There are two mathematically equivalent methods which will make $J(\psi_{tr})$ stationary with respect to variations in the parameters. One is the method in which the Euler–Lagrange equation

$$\frac{d}{dt}\frac{\partial D}{\partial\dot{\alpha}_i} - \frac{\partial D}{\partial\alpha_i} = 0 \tag{96}$$

is applied to independent variations of α_i. A problem with Eq. (96) is the presence of the G^2 operator, which leads to a nonlinear $\dot{\alpha}^2$ term in the differential equation.

A mathematically equivalent approach is derived using Hamilton's equations, in which a "momentum" is defined conjugate to each parameter. For a ψ_{tr} that depends on N complex parameters α_j and M real parameters λ_i one defines

$$P_{\alpha_j} \equiv \frac{\partial D}{\partial\dot{\alpha}_j}\ \text{(complex)}, \qquad j = 1, \ldots, N, \tag{97}$$

$$P_{\lambda_i} \equiv \frac{\partial D}{\partial\dot{\lambda}_i}\ \text{(real)}, \qquad i = 1, \ldots, M, \tag{98}$$

and a psuedo-Hamiltonian K,

$$K = \sum_{j=1}^{N} (\dot{\alpha}_j P_{\alpha_j} + \dot{\alpha}_j^* P_{\alpha_j}^*) + \sum_{i=1}^{M} \dot{\lambda}_i P_{\lambda_i} - D.$$

Hamilton's equations are then

$$\dot{\alpha}_j = \frac{\partial K}{\partial P_{\alpha_j}} \qquad (j = 1, 2, \ldots, N), \tag{99a}$$

$$\dot{P}_{\alpha_j} = \frac{-\partial K}{\partial \alpha_j} = \frac{\partial D}{\partial \alpha_j} \qquad (j = 1, 2, \ldots, N), \tag{99b}$$

$$\dot{\lambda}_i = \frac{\partial K}{\partial P_{\lambda_i}} \qquad (i = 1, 2, \ldots, M), \tag{99c}$$

$$\dot{P}_{\lambda_i} = -\frac{\partial K}{\partial \lambda_i} = \frac{\partial D}{\partial \lambda_i} \qquad (i = 1, 2, \ldots, M). \tag{99d}$$

The simultaneous solution of the first-order differential equations (99) determines the ψ_{tr} that makes $J(\psi_{\text{tr}})$ stationary for parameters α_j and λ_i that have no variation at the end points.

It turns out that more is achieved than merely the replacement of a number of second-order Euler–Lagrange differential equations by twice as many first-order Hamilton differential equations. The momenta conjugate to the parameters can be interpreted as a measure of how well the subspace of the trial solution covers the space of the exact solution.

4. *Hamilton's Equations for the Expansion Coefficients*

Consider a system described by a Hamiltonian H_0 with known eigenfunctions $\phi_1, \phi_2, \ldots$. Let a perturbation $V(\mathbf{r}, t)$, with the property $V = 0$ as $t \to \pm\infty$, act on the system, where $\mathbf{r}$ represents the positional coordinates.

The exact solution of the equation

$$G\psi = \left(H_0 + V - i\frac{\partial}{\partial t}\right)\psi \tag{100}$$

can be expanded in terms of the ϕ_i as

$$\psi = \sum_{j=1}^{\infty} a_j(t)\phi_j(\mathbf{r}, t),$$

where

$$\dot{a}_k = \frac{1}{i}\sum_{j=1}^{\infty} \langle k|V|j\rangle \alpha_j. \tag{101}$$

Now suppose an approximate solution is sought in terms of a trial function that is a truncated expansion in the ϕ_j:

$$\psi_{\text{tr}} = \alpha_1\phi_1 + \alpha_2\phi_2. \tag{102}$$

One common method for estimating $\alpha_1(t)$ and $\alpha_2(t)$ is to use truncated forms of Eq. (101):

$$\dot{\alpha}_k = -i \sum_{j=1}^{2} \langle k|V|j\rangle\alpha_j. \tag{103}$$

This method is based on the concept that as the space spanned by ψ_{tr} is increased by including more of the ϕ_j, the α_j will approach the α_j. However, for any finite truncation, there is no guarantee that the α_j produced by Eq. (103) are the "best" functions of time. Equation (103) may give very poor values of $\alpha_j(t)$ [in the sense that $\alpha_j(t)$ is not close to $a_j(t)$].

The Hamilton's equations method will now be used to obtain optimized $\alpha_j(t)$ in the sense that for the choice of Eq. (102) as ψ_{tr}, $J(\psi_{\text{tr}})$ is minimized.

Let the set of expansion coefficients that minimize $J(\psi_{\text{tr}})$ be denoted as $\alpha_j(t)$. A necessary condition for a minimum $J(\psi_{\text{tr}})$ is

$$\left.\frac{dJ}{d\epsilon}\right|_{\epsilon=0} = \sum_{j=1}^{2} \int_{t_i}^{t_f} \left(\frac{d}{dt}\frac{\partial D}{\partial \alpha_j} - \frac{\partial D}{\partial \alpha_j}\right)\eta_j(t)\,dt + \sum_{j=1}^{2} \eta_j(t)\left.\frac{\partial D}{\partial \alpha_j}\right|_{t_i}^{t_f} = 0, \tag{104}$$

where $\delta\alpha_j$ is characterized by the arbitrary function $\eta_j(t)$ and the parameter ε (Goldstein, 1965).

Assuming that $\psi \to \phi_1$ at $t = t_i$, the trial function becomes exact at this time and one can set $\eta_j(t_i) = 0$ for each value of j. At time t_f, the values of $a_j(t_f)$ are not usually known, and therefore $\eta_j(t_f)$ cannot be set equal to zero. In order to apply the Euler–Lagrange or Hamilton's equations methods to the determination of $\alpha_j(t)$, it is necessary to force

$$\left.\frac{\partial D}{\partial \dot{\alpha}_j}\eta_j(t)\right|_{t=t_f} = 0 \qquad (j = 1, 2). \tag{105}$$

Equation (105) will be satisfied if the momenta conjugate to α_j are made to vanish at $t = t_f$. Boundary conditions given by Eq. (105) are well known (Gelfand and Fomin, 1963) in calculus of variations problems with fixed and free end points. For the trial function in Eq. (102) with G defined in Eq. (100),

$$D(t) = \sum_{k=1}^{2} |\dot{\alpha}_k|^2 + \sum_{k,m=1}^{2} \dot{\alpha}_k{}^*\alpha_m\langle k|V^2|m\rangle$$
$$+ i \sum_{k,m=1}^{2} \dot{\alpha}_k{}^*\alpha_m\langle k|V|m\rangle - i \sum_{k,m=1}^{2} \dot{\alpha}_k\alpha_m{}^*\langle m|V|k\rangle. \tag{106}$$

The momenta conjugate to $\alpha_k{}^*$ $(k = 1, 2)$ are

$$P_{\alpha_k{}^*} = \frac{\partial D}{\partial \dot{\alpha}_k} = \dot{\alpha}_k + i \sum_{j=1}^{2} \langle k|V|j\rangle \alpha_j = (P_{\alpha_k})^*. \tag{107}$$

The first pair of Hamilton's equations can be found from inverting Eq. (107), obtaining

$$\dot{\alpha}_k = P_{\alpha_k{}^*} + \frac{1}{i} \sum_{j=1}^{2} \langle k|V|j\rangle. \tag{108}$$

The second pair of Hamilton's equations are

$$\begin{aligned}\dot{P}_{\alpha_k} = \frac{\partial D}{\partial \alpha_k} &= i(P_k\langle k|V|k\rangle + P_m\langle m|V|k\rangle) \\ &+ \alpha_k{}^*(\langle k|V^2|k\rangle - |\langle k|V|k\rangle|^2 - |\langle k|V|m\rangle|^2) \\ &+ \alpha_m{}^*(\langle m|V^2|m\rangle - \langle m|V|k\rangle\langle k|V|k\rangle - \langle m|V|m\rangle\langle m|V|k\rangle),\end{aligned} \tag{109}$$

in which $m = 2$ if $k = 1$, and $m = 1$ if $k = 2$. By inserting the operator $\sum_{n=1}^{\infty} |nxn|$ in the V^2 terms in Eq. (109), one finds

$$\begin{aligned}\dot{P}_{\alpha_k} &= i(P_k\langle k|V|k\rangle + P_m\langle m|V|k\rangle) \\ &+ \alpha_k{}^* \sum_{n=3}^{\infty} \langle k|V|n \times n|V|k\rangle + \alpha_m{}^* \sum_{n=3}^{\infty} \langle m|V|n \times n|V|k\rangle,\end{aligned} \tag{110}$$

where again $m = 2$ if $k = 1$ and $m = 1$ if $k = 2$.

This equation shows that in some way the values of P_{α_1} and P_{α_2} partly compensate for the effect of the missing states $\phi_3, \phi_4, \ldots$, which were not included in ψ_{tr}. Comparing Eqs. (108) and (101) for $\dot{\alpha}_k$ and $\dot{a}_k$, respectively, one can make the identification of α_k as the best approximation to a_k in the sense of minimizing $J(\psi_{\text{tr}})$, and in the same sense identify the $\dot{P}_{\alpha_k}$ with the best single-term approximation to $(1/i) \sum_{j=3}^{\infty} \langle k|V|j\rangle a_j$. Further support for this interpretation is found from using Eq. (108) to obtain

$$\frac{d}{dt}(|\alpha_1|^2 + |\alpha_2|^2) = \alpha_1 P_{\alpha_1} + \alpha_1{}^* P_{\alpha_1}^* + \alpha_2 P_{\alpha_2} + \alpha_2{}^* P_{\alpha_2}^*, \tag{111}$$

which shows that probability is not conserved in this model due to the implicit flow of probability between states (1, 2) and (3, 4, 5, . . .). This demand for an average contribution from all the higher states results in our inability to have a normalized two-state approximation which minimizes $J(\psi_{\text{tr}})$. However, it is necessary that

$$\frac{d}{dt}|\alpha_1|^2 = \frac{d}{dt}|\alpha_2|^2 = 0$$

as $V(r,t) \to 0$. At $t = t_i$, $|\alpha_1|^2 = 1$ and $|\alpha_2|^2 = 0$. Setting $\alpha_1 = e^{i\theta}$ and $\alpha_2 = 0$ at t_i, one can show from (108) that at $t = t_i$ it is required that

$$\mathrm{Re}(P_{\alpha_1}) = \mathrm{Im}(P_{\alpha_1}) \tan\theta$$

to insure

$$\frac{d}{dt}|\alpha_1|^2 = \frac{d}{dt}|\alpha_2|^2 = 0$$

at t_i.

Now one can choose θ, $\mathrm{Im}(P_{\alpha_1})$, $\mathrm{Re}(P_{\alpha_2})$, and $\mathrm{Im}(P_{\alpha_2})$ at $t = t_i$, so that $\mathrm{Re}(P_{\alpha_1})$, $\mathrm{Im}(P_{\alpha_1})$, $\mathrm{Re}(P_{\alpha_2})$, and $\mathrm{Im}(P_{\alpha_2})$ all vanish at $t = t_f$. Then one can choose $\mathrm{Re}(P_{\alpha_1})$ at $t = t_i$ equal to $\mathrm{Im}(P_{\alpha_1})\tan\theta$. The arrangement of P_{α_k} such that they vanish at $t = t_f$ will not only lead from Eq. (104) to the Euler–Lagrange or Hamilton equations minimizing $J(\psi_{\mathrm{tr}})$ but also to

$$\frac{d}{dt}|\alpha_1|^2 = \frac{d}{dt}|\alpha_2|^2 = 0$$

at $t = t_f$.

Contrary to what has been claimed in the literature (Fraenkel, 1934; McLachlan, 1964), not only will Eqs. (103) not minimize $D(t)$, but in addition they will not minimize $\int_{-\infty}^{+\infty} D\,dt = J(\psi_{\mathrm{tr}})$. The usual coupled equations (103) can be derived from Eq. (108) if the P_{α_k} are made to vanish identically. This would imply from Eq. (97) that D does not depend on the $\dot{\alpha}_k$. If P_{α_k} vanishes identically it implies that $\dot{P}_{\alpha_k}$ and $\partial D/\partial\alpha$ vanish. Therefore, the usual coupled equations could only minimize $J(\psi_{\mathrm{tr}})$ if D did not depend on either α_k or $\dot{\alpha}_k$. Since any meaningful D must depend on either the α_k and/or the $\dot{\alpha}_k$ the usual coupled equations can never minimize $J(\psi_{\mathrm{tr}})$. Since D is always greater than or equal to zero, the fact that Eqs. (103) cannot minimize $J(\psi_{\mathrm{tr}})$ implies that Eqs. (103) cannot minimize $D(t)$. The trouble with all the reported proofs (Fraenkel, 1934) which claim to show that $D(t)$ is minimized by Eqs. (103) is their attributing to every time t is a property, $\delta\psi_{\mathrm{tr}} = 0$, which is only justified for the initial time.

While the Euler–Lagrange equation for the α_k would certainly be mathematically equivalent to (108) and (109), the relationship between (103) and these second-order differential equations would not be nearly as apparent as between (103) and the combined Hamilton's equations (108) and (109).

In summary, Eqs. (108) and (109) constitute a set of four coupled linear differential equations, which when solved subject to the aforementioned boundary conditions yield a set of parameters $\alpha_1(t)$, $\alpha_2(t)$, $P_{\alpha_1}(t)$, and $P_{\alpha_2}(t)$, which minimize $J(\psi_{\mathrm{tr}})$. These equations (or their generalization for more than two states) would be particularly well suited for problems in which either (a) a limited set $\phi_1, \phi_2, \ldots, \phi_m$ are known, whereas the higher states are important but not well known, or (b) the important states ϕ_k are known, but are of such great number or complexity that to avoid an impractical calculational problem a truncated wave function is chosen as ψ_{tr}. In this second case, only some of the important wave functions ϕ_k are treated explicitly and an average effect of the other states is treated through the conjugate momenta. This technique necessitates the inclusion in ψ_{tr} of all the states whose $\alpha_k(t)$ are to be optimized.

5. *Application of Hamilton's Equations to Internal Parameters of the Expansion Functions*

In some applications the form of the approximate solution is conveniently chosen as

$$\psi_{\text{tr}} = \sum_j \alpha_j(t)\chi_j(r, t, z_{j_1}, z_{j_2}, \ldots),$$

in which the $z_{j_1}, z_{j_2}, \ldots$ are parameters internal to the expansion functions, and for which $\chi_j \to \phi_j$ as $t \to t_i$ or t_f. The Hamilton's equations technique to minimize $J(\psi_{\text{tr}})$ can be used to optimize the parameters $z_{j_1}, z_{j_2}, \ldots$ as well as the $\alpha_j(t)$. Such internal parameters are important in the theoretical treatment of charge exchange (Sil, 1960; Mittleman, 1961; Cheshire, 1968; McCarroll *et al.*, 1970), and it is desired to demonstrate the relative effectiveness of different choices of procedures for "optimizing" them.

The method will be illustrated in terms of a simple example, consisting of a hypothetical perturbed hydrogen atom with a time-dependent nuclear charge $\beta(t)$ that varies with time, with the end points satisfying $\beta(\pm\infty) = 1$. The Hamiltonian for this problem is

$$H = -\nabla^2 - 2\beta(t)/r, \tag{112}$$

where $\beta(t)$ can be selected arbitrarily, provided it goes to 1 as $t \to \pm\infty$. A trial function was selected in the form of a hydrogenlike function

$$\psi_{\text{tr}} = [z^3(t)/\pi]^{1/2} e^{-z(t)r} e^{+iz^2(t)t}, \tag{113}$$

where $z(t)$ is a parameter representing a scale factor which is to be optimized in the sense that $J(\psi_{\text{tr}})$ is minimized for any $\beta(t) \to 1$ for large positive and negative times. It follows that $z(t)$ must also go to unity and that $\psi_{\text{tr}} = \psi$ at such times. Since we are dealing with a problem of two fixed end points, the calculus of the variations problem is straightforward. Although simple, this example has certain important characteristics found in the standard H^+–H charge exchange model. One could imagine the perturbed nuclear charge $\beta(t)$ being caused by the passing proton, and the form of the trial function is the archetype of that used in H^+–H charge exchange.

It is desired to utilize each of the previously described methods of evaluating $z(t)$ and $J(\psi_{tr})$ to compare the adequacy of various approaches. For any arbitrary function $z(t)$, the deviation is determined from Eqs. (112) and (113) as

$$D(t) = 8z^2[(z - \beta)^2 + (z - \beta)\dot{z}t] + \dot{z}^2(4t^2z^2 + \tfrac{3}{4}z^2). \tag{114}$$

This is evaluated and integrated over time to obtain $J(\psi_{tr})$ for each method.

First consider the method used by Cheshire (1968) and others as given in Eq. (90). A simple calculation shows that this method yields the result

$$z_1(t) = \beta(t)$$

for all time. The deviation D based on this method is then

$$D_1(t) = \dot{\beta}^2(4t^2\beta^2) + \tfrac{3}{4}\beta^2 \tag{115}$$

The integral over time of $D_1(t)$ is referred to as J_1.

The approach to minimization of J by means of Hamilton's equations is described next. Hamilton's equations are

$$\dot{z} = (8t^2z^2 + \tfrac{3}{2}z^2)^{-1}[P_z - 8z^2(z - \beta)t], \tag{116}$$

$$\dot{P}_z = \frac{\partial D}{\partial z} = 16z[(z - \beta)^2 + (z - \beta)\dot{z}t] + 8z^2[2(z - \beta) + \dot{z}t] + \dot{z}^2(8t^2z - \tfrac{3}{2}z^3). \tag{117}$$

The simultaneous solution of Eqs. (116) and (117) with P_z and $\dot{P}_z$ set equal to zero is $z_1(t)$, the solution derived from Eq. (90). When Eqs. (116) and (117) are simultaneously integrated, and the resulting $z(t)$, $z_2(t)$ is used to calculate the deviation, it is designated as $D_2(t)$, and the time integral is J_2. The initial value of z is unity, but the initial value of P_z is unknown. The actual numerical integration is carried out from a large negative time as a starting point, and the initial value of P_z is adjusted until the resulting integration produces a $z(t)$ at large positive time that approaches unity.

In the method suggested by Chang and Rapp (1973), as given in Eq. (75), the basis is the $\partial D/\partial \dot{z}$ set equal to zero at all time, and one obtains the differential equation

$$\dot{z} = \frac{-4z^2(z - \beta)t}{4t^2z^2 + \tfrac{3}{4}z^2}. \tag{118}$$

This was numerically integrated and the resulting $z(t)$, $z_3(t)$ used to obtain $D_3(t)$ and its time integral J_3.

A very simple approximation was tried in which $z(t)$ was set equal to unity for all time. The resulting values of the deviation are denoted $D_4(t)$,

TABLE I

Comparison of the Deviation for Various Techniques in Section II.E. D_1, D_2, D_3, and D_4 are the $D(t)$ Corresponding to Solutions Found from Eq. (90), Hamilton's Equations, Eq. (118) and $z(t) = 1$, Respectively ($A = 0.04$, $B = 0.1$, Wide Potential.)

Time	D_1	D_2	D_3	D_4
−15.0	0.790(−6)	0.848(−3)	0.0	0.0
−12.0	0.211(−3)	0.160(−2)	0.421(−6)	0.733(−6)
−9.0	0.104(−1)	0.316(−2)	0.786(−4)	0.122(−3)
−6.0	0.784(−1)	0.439(−2)	0.363(−2)	0.449(−2)
−3.0	0.469(−1)	0.430(−2)	0.590(−1)	0.389(−1)
−1.0	0.128(−2)	0.253(−2)	0.411	0.738(−1)
0.0	0.0	0.336(−3)	0.728	0.780(−1)
1.0	0.128(−2)	0.236(−2)	0.411	0.738(−1)
3.0	0.469(−1)	0.391(−2)	0.590(−1)	0.389(−1)
6.0	0.784(−1)	0.413(−2)	0.363(−2)	0.449(−2)
9.0	0.104(−1)	0.306(−2)	0.786(−3)	0.122(−3)
12.0	0.211(−3)	0.152(−2)	0.421(−6)	0.733(−6)
15.0	0.790(−6)	0.757(−3)	0.101(−13)	0.0
Time integral J	0.814[a]	0.957(−1)[a]	0.205(1)	0.501

[a] These J values were calculated with $t_i = -30.0$ and $t_f = 30.0$ as in Fig. 1.

and the time integral is J_4. Thus four different calculations of $D(t)$ and $J(\psi_{tr})$ were performed and compared.

The function selected for $\beta(t)$ was

$$\beta(t) = 1 + Be^{-At^2}.$$

Values of B equal to 0.1 and 0.2 were found to be sufficient to test the principles involved. Three values of A were chosen to represent slow, moderate, and fast variations of B with time ($A = 0.04$, 1.0, and 100.0). For these values of A and B, calculations of $D(t)$ and $J(\psi_{tr})$ were made for each of the four $z(t)$ functions described above. Differential equations were integrated on a digital computer using a predictor–corrector method. It was found that there was no substantial qualitative difference between the results for $B = 0.1$ and $B = 0.2$. To save space, only the results for $B = 0.1$ are presented here.

The resulting values of $D(t)$ at selected values of t (atomic units) are given in Table I for $A = 0.04$ and $B = 0.1$. A comparison of the deviations D_1 and D_2, and the effective nuclear charges z_1 and z_2 are shown for $A = 1.0$ and $A = 100.0$, ($B = 0.1$) in Figs. 7 and 8. The J values for the three A values and $B = 0.1$ are given in Table II. In each case, as expected, the lowest value of $J(\psi_{tr})$ results from the use of Hamilton's equations to determine

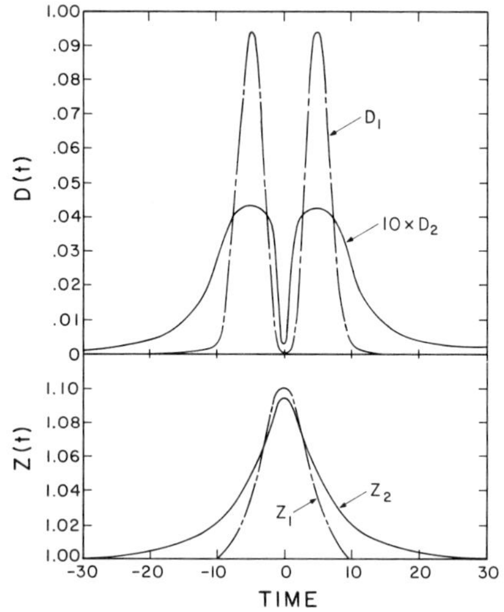

FIG. 7 Comparison of the deviations $D_1(t)$ and $D_2(t)$ and the effective nuclear charges $z_1(t)$ and $z_2(t)$ ($A = 1.0$, $B = 0.1$, intermediate width).

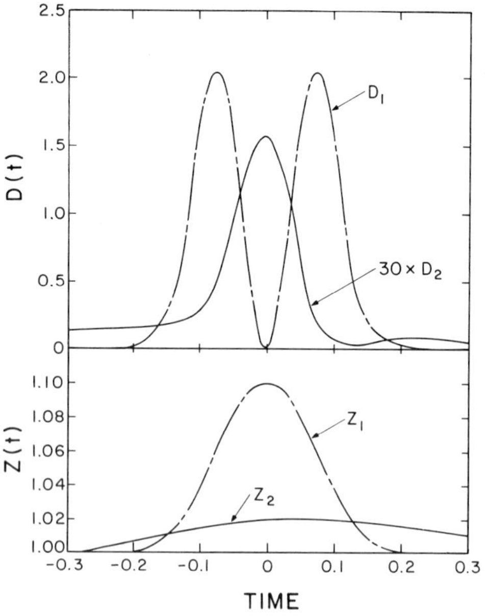

FIG. 8 Comparison of the deviations $D_1(t)$ and $D_2(t)$ and the effective nuclear charges $z_1(t)$ and $z_2(t)$ ($A = 100.0$, $B = 0.1$, narrow potential).

TABLE II

Comparison of the Measure of the Integrated Error J for the Intermediate and Narrow Potential Cases in Section II.E. J_1, J_2, J_3, and J_4 Correspond to the J Values Found with Eq. (90). Hamilton's Equations, Eq. (118), and Simply Setting $z(t) = 1$, Respectively.

	J_1	J_2	J_3	J_4
$A = 0.04$, $B = 0.04$ (wide width)	0.814	0.957(−1)	0.205(1)	0.501
$A = 1.0$, $B = 0.1$ (intermediate width)	0.195	0.228(−1)	0.243	0.100
$A = 100.0$, $B = 0.1$ (narrow width)	0.355	0.796(−2)	0.104(−1)	0.100(−1)

$z(t)$ (method 2). $z_2(t)$ is always wider in time, but not as high as $z_1(t)$. It appears that as the width of $\beta(t)$ becomes large, $z_2(t)$ will approach $\beta(t)$, whereas when $\beta(t)$ becomes exceedingly narrow, $z_2(t)$ will approach unity for all t. Thus, when the perturbation varies rapidly the wave function does not have time to readjust to the perturbation, and when the perturbation varies slowly the wave function can adjust almost adiabatically. Since $\beta(t) \to 0$ as $t \to 0$, $D_1(t)$ vanishes and $z_1(t)$ must be exact as $t = 0$. Nevertheless, $J_1(\psi_{tr})$ exceeds $J_2(\psi_{tr})$ even though $D_1(t) < D_2(t)$ for some periods of time. On a relative basis, the method of Hamilton's equations is the greatest improvement over the method of Eq. (90) when the perturbation varies most rapidly. Method 3, utilizing Eq. (118) gives a time-varying $z(t)$ that decreases below unity instead of rising above one. This appears to be a fundamental imperfection in this method, although it leads to better results than method 1 when the perturbation varies rapidly. Method 3 has the virtue that it leads to $z(t) \to 1$ for all t as the perturbation varies very rapidly, whereas method 1 is qualitatively incorrect in this region. On the other hand, method 3 is qualitatively incorrect for slowly varying $\beta(t)$, while method 1 is a better approximation.

The $z(t)$ and $D(t)$ produced by methods 1, 3, and 4 are symmetric about $t = 0$. However, z_2 and D_2 are not symmetric about $t = 0$, and thus P_z can be interpreted as the factor responsible for the time delay or adjustment time necessary to minimize $J(\psi_{tr})$.

To further demonstrate that $J_2(\psi_{tr})$ is a relative minimum, a calculation was performed in which the arbitrary term $0.002e^{-AT^2}$ was added or subtracted to $z_2(t)$ for all t, and $D(t)$ calculated in each case. The results are given in Table III. When small positive or negative changes are made about $z_2(t)$, $J(\psi_{tr})$ increases.

It may be concluded that (i) the method of Hamilton's equations produces a minimum integrated deviation; (ii) the method of Eq. (90) best approximates the Hamilton's equations method for slowly varying perturbations, where it eventually goes to the adiabatic limit at very slow rates of change; (iii) the

TABLE III

Dependence of the Deviation on Small Variations ($\pm 0.002\ exp(-At^2)$ about the Function $a_2(t)$ Found from Hamilton's Equations. D_2^+ and D_2^- are the $D(t)$ Corresponding to z_2^+ and z_2^-, Respectively ($A = 100.0$, $B = 0.1$).

Time	D_2^-	D_2	D_2^+
−0.300	0.467(−2)	0.467(−2)	0.467(−2)
−0.200	0.500(−2)	0.519(−2)	0.539(−2)
−0.100	0.718(−2)	0.849(−2)	0.101(−1)
−0.050	0.327(−1)	0.323(−1)	0.324(−1)
−0.025	0.500(−1)	0.483(−1)	0.467(−1)
0.000	0.570(−1)	0.545(−1)	0.521(−1)
0.025	0.481(−1)	0.459(−1)	0.439(−1)
0.050	0.294(−1)	0.282(−1)	0.274(−1)
0.100	0.322(−2)	0.368(−2)	0.446(−2)
0.200	0.210(−2)	0.218(−2)	0.228(−2)
0.300	0.128(−2)	0.128(−2)	0.128(−2)
Time integral J	0.80019(−2)[a]	0.79636(−2)[a]	0.80030(−2)[a]

[a] These J values were calculated with $t_i = -0.3$ and $t_f = 0.95$.

method of Eq. (85) best approximates the Hamilton's equations method for rapidly varying perturbations, where it eventually leads to constant parameters for very rapid rates of change; and (iv) using $z(t) = 1$ for rapidly changing perturbations is much better than the use of Eq. (90).

Note that the "best" solution is "best" in a time-averaged sense and that at any particular time the other techniques can lead to solutions having a lower $D(t)$. None of these techniques provide a minimum of $D(t)$ at every instant of time, i.e., none of these techniques can provide a level B optimum. In fact, no purely time-dependent parameter could ever lead to $D(t)$ being minimized at every instant t. If the parameter is a function of the positional coordinates and t, then this goal might be attainable. From this, one might infer for the problem H^+–H charge exchange that as the velocity of the incident proton increases, the method of Eq. (90) will lead to progressively less accurate results. Since the Hamilton's equations method is always the most accurate, and in particular when the potential is rapidly varying is far superior to the technique of Eq. (90), it is natural for application to this problem in the region of high velocities.

IV. SUMMARY AND OVERVIEW

As we have seen, the work on bounds by Aspinall and Percival (1967, 1968), Spruch (1968), and Storm (1973, 1974) leads to a reasonable choice of level C optimization criteria, i.e., that $\int D^{1/2}\,dt$ be minimized for trial solutions of the Schrödinger equation. The work of Chang and Rapp (1973) and Weglein (1978) develops the mathematical tools to achieve a different form of level C optimization, i.e., $\int D\,dt$ is minimized for standard forms of trial solutions. The level of optimization that is sought in any particular problem is as much a practical as it is a conceptual problem. We feel that the application of level C optimization to a proton–hydrogen charge transfer calculation would provide a natural, interesting, and useful extension of the effective nuclear charge problem of Section III.E. The realization of level C optimization criteria could make a significant contribution to our understanding of physical systems.

REFERENCES.

Aspinall, L., and Percival, I. C. (1967). *Proc. Phys. Soc.* **90**, 315.
Aspinall, L., and Percival, I. C. (1968). *J. Phys. B* **1**, 589.
Chang, C. (1973). Ph.D. Thesis, Univ. of Texas at Dallas, Richardson, Texas.
Chang, C., and Rapp, D. (1973). *J. Chem. Phys.* **59**, 972.
Cheshire, I. M. (1968). *J. Phys. B* **1**, 428.
Fraenkel, J. (1934). "Wave Mechanics, Advanced General Theory." Oxford Univ. Press (Clarendon), London.
Gelfand, I. M., and Fomin, S. V. (1963). "Calculus of Variations." Prentice-Hall, Englewood Cliffs, New Jersey.
Goldstein, H. (1965). "Classical Mechanics." Addison-Wesley, Reading, Massachusetts.
McCarroll, R., Piacentini, R. D., and Salin, A. (1970). *J. Phys. B* **3**, 137.
McLachlan, A. D. (1964). *Mol. Phys.* **8**, 39.
Messiah, A. (1966). "Quantum Mechanics." Vol. 2, p. 764. Wiley, New York.
Mittleman, M. M. (1961). *Phys. Rev.* **122**, 1930.
Rapp, D. (1971). "Quantum Mechanics." Holt, New York.
Rosen, N., and Zener, C. (1932). *Phys. Rev.* **40**, 502.
Sil, N. C. (1960). *Proc. R. Soc. London* **75**, 194.
Spruch, L. (1968). In "Lectures in Theoretical Physics—Atomic Collisions" (S. Geltman, K. T. Mahanthappa, and W. E. Britten, eds.), Vol. XIC, pp. 77–87. Gordon & Breach, New York.
Storm, D. (1973). *Phys. Rev. A* **8**, 1789.
Storm, D. (1974). *Phys. Rev. A* **10**, 1008.
Weglein, A. (1978). *Phys. Rev. A* **17**, 1810.

Index